This series aims to report new developments in physical research and teaching – quickly, informally, and at a high level. The type of material considered for publication includes:

1. Preliminary drafts of original papers and monographs
2. Lectures on a new field, or presenting a new angle on a classical field
3. collections of seminar papers
4. Reports of meetings

Texts which are out of print but still in demand may also be considered if they fall within these categories.

The timeliness of a manuscript is more important than its form, which may be unfinished or tentative. Thus, in some instances, proofs may be merely outlined and results presented which have been or will later be published elsewhere.

Publication of *Lecture Notes* is intended as a service to the international physical community, in that a commercial publisher, Springer-Verlag, can offer a wider distribution to documents which would otherwise have a restricted readership. Once published and copyrighted, they can be documented in the scientific libraries.

Manuscripts

Manuscripts are reproduced by a photographic process; they must therefore be typed with extreme care. Symbols not on the typewriter should be inserted by hand in indelible black ink. Corrections to the typescript should be made by sticking the amended text over the old one, or by obliterating errors with white correcting fluid. The figures (in the original size) ready for reproduction should be inserted into the text. Should the text, or any part of it, have to be retyped, the author will be reimbursed upon publication of the volume. Authors receive 50 free copies.

The typescript is reduced slightly in size during reproduction, therefore a large size of type should be used; best results will not be obtained unless the text on any one page is kept within the overall limit of 18 x 26.5 cm (7 x 10½ inches). **The publishers will be pleased to supply on request special stationery with the typing area outlined.**

Manuscripts in English, German or French should be sent to Springer-Verlag, 6900 Heidelberg, Postfach 1780.

Die „*Lecture Notes*" sollen rasch und informell, aber auf hohem Niveau, über neue Entwicklungen in der Physik berichten. Zur Veröffentlichung kommen:

1. Vorläufige Fassungen von Originalarbeiten und Monographien.
2. Spezielle Vorlesungen über ein neues Gebiet oder ein klassisches Gebiet in neuer Betrachtungsweise.
3. Seminarausarbeitungen.
4. Vorträge von Tagungen.

Ferner kommen auch ältere vergriffene spezielle Vorlesungen, Seminare und Berichte in Frage, wenn nach ihnen eine anhaltende Nachfrage besteht.

Die Beiträge dürfen im Interesse einer größeren Aktualität durchaus den Charakter des Unfertigen und Vorläufigen haben. Sie brauchen Beweise unter Umständen nur zu skizzieren und dürfen auch Ergebnisse enthalten, die in ähnlicher Form schon erschienen sind oder später erscheinen sollen.

Die Herausgabe der „*Lecture Notes*" Serie durch den Springer-Verlag stellt eine Dienstleistung an die physikalischen Institute dar, indem der Springer-Verlag für ausreichende Lagerhaltung sorgt und einen großen internationalen Kreis von Interessenten erfassen kann. Durch Anzeigen in Fachzeitschriften, Aufnahme in Kataloge und durch Anmeldung zum Copyright sowie durch die Versendung von Besprechungsexemplaren wird eine lückenlose Dokumentation in den wissenschaftlichen Bibliotheken ermöglicht.

Lecture Notes in Physics

Edited by J. Ehlers, Austin, K. Hepp, Zürich and
H. A. Weidenmüller, Heidelberg
Managing Editor: W. Beiglböck, Heidelberg

8

Proceedings of the Second International Conference on Numerical Methods in Fluid Dynamics

September 15–19, 1970
University of California, Berkeley
Edited by Maurice Holt

Springer-Verlag Berlin Heidelberg GmbH 1971

ISBN 978-3-540-05407-8   ISBN 978-3-540-36533-4 (eBook)
DOI 10.1007/978-3-540-36533-4

 Library of Congress Catalog Card Number 71-159649.

Originally published by Springer-Verlag Berlin Heidelberg New York in 1971.

Editor's Preface

This volume of Lecture Notes in Physics is devoted to the Proceedings of the Second International Conference on Numerical Methods in Fluid Dynamics held at the University of California, Berkeley, September 15-19, 1970. A total of 65 papers were presented and they are published here in full, with a minimum of editorial changes. The Conference was divided into seven sessions with two devoted to new fundamental numerical techniques, two to viscous flow problems, two to high speed compressible flow and one to incompressible flow. Contributions from many countries were made, including important papers from the USA, the USSR, France, Germany, England, Holland, Canada and Australia.

The Conference was organized within a one year period, following the First Conference held in Novosibirsk, USSR, in August 1969. Valuable financial support was provided through a grant from the U.S. Office of Naval Research and the U.S. Air Force Office of Scientific Research. Many services of the Berkeley campus of the University were made available to us, including the use of the Physical Sciences Lecture Hall for all Conference sessions, and housing in the Halls of Residence. The Northrop Corporation kindly provided refreshments at the opening reception of the Conference.

I wish to thank the many students and colleagues who worked to make the Conference a success. Special mention should be made of Mrs. Arlene Martin, who did all the secretarial work, Mr. William F. Ballhaus, Jr., a graduate student, who supervised most of the arrangements, and Drs. Mark Wilkins and Robert L. Street who served on the program committee.

Finally I am indebted to Dr. W. Beiglböck, Editor of Lecture Notes in Physics, and to Dr. Klaus Peters of Springer-Verlag, for arranging the early publication of these Proceedings in this series.

January 29, 1971

Maurice Holt
Chairman, Second Conference
Berkeley

Contents

Session I
Fundamental Numerical Techniques
A.J. CHORIN, Chairman

1. N.N. YANENKO and Y.I. SHOKIN: On the Group Classification of Difference Schemes for Systems of Equations in Gas Dynamics....3

2. K. ROESNER: Numerical Integration of the Euler-Equations for Three-Dimensional Unsteady Flows....18

3. P. KUTLER and H. LOMAX: The Computation of Supersonic Flow Fields about Wing-Body Combinations by "Shock-Capturing" Finite Difference Techniques....24

4. M.L. WILKINS, S.J. FRENCH and M. SOREM: Finite Difference Schemes for Calculating Problems in Three Space Dimensions and Time....30

5. A. SAKURAI: Foundation of Approximate Solutions....34

6. W.P. CROWLEY: FLAG: A Free-Lagrange Method for Numerically Simulating Hydrodynamic Flows in Two Dimensions....37

7. S.Z. BURSTEIN and A.A. MIRIN: Difference Methods for Hyperbolic Equations Using Space and Time Split Difference Operators of Third Order Accuracy....44

8. P.J. ROACHE: A New Direct Method for the Discretized Poisson Equation....48

9. G. MORETTI: Initial Conditions and Imbedded Shocks in the Numerical Analysis of Transonic Flows....54

Session II
Numerical Techniques and Applications
V.V. RUSANOV, Chairman

1. J. BOUJOT, J.L. SOULÉ and R. TEMAM: Traitement Numérique d'un Problème de Magnétohydrodynamique....61

2. A.D. GOSMAN and D.B. SPALDING: Computation of Laminar Recirculating Flow Between Shrouded Rotating Discs....67

3. B.W. THOMPSON: Some Semi-Analytical Methods in Numerical Fluid Dynamics....73

4. H.J. LUGT and H.J. HAUSSLING: Laminar Flows Past a Flat Plate at Various Angles of Attack....78

5. B.J. DALY and F.H. HARLOW: Inclusion of Turbulence Effects in Numerical Fluid Dynamics....84

6. H.B. KELLER and T. CEBECI: Accurate Numerical Methods for Boundary Layer Flows. I. Two Dimensional Laminar Flows........92

7. R. DE VOGELAERE: The Reduction of the Stefan Problem to the Solution of an Ordinary Differential Equation............101

8. C.P. KENTZER: Discretization of Boundary Conditions on Moving Discontinuities.....................................108

9. P.J. ZANDBERGEN: The Viscous Flow Around a Circular Cylinder...114

10. J.E. FROMM: A Numerical Study of Buoyancy Driven Flows in Room Enclosures..120

Session III
Boundary Layers
R. TIMMAN, Chairman

1. O.R. BURGGRAF and K. STEWARTSON: The Structure of the Laminar Boundary Layer Under a Potential Vortex...............129

2. E. KRAUSE and E.H. HIRSCHEL: Exact Numerical Solutions for Three-Dimensional Boundary Layers.........................132

3. R.A. WAGSTAFF and S.S. LEE: Higher Order Effects in Laminar Boundary Layer Theory for Curved Surfaces 138

4. F.G. BLOTTNER: Finite-Difference Methods for Solving the Boundary Layer Equations with Second-Order Accuracy..........144

5. R.W. MACCORMACK: Numerical Solution of the Interaction of a Shock Wave with a Laminar Boundary Layer....................151

6. B.B. ROSS and S.I. CHENG: A Numerical Solution of the Planar Supersonic Near-Wake with its Error Analysis..........164

7. R.C. ACKERBERG: Boundary-Layer Separation at a Free Streamline-Finite Difference Calculations.....................170

8. G.O. ROBERTS: Computational Meshes for Boundary Layer Problems...171

9. K.L.E. NICKEL: Error Bounds in Boundary Layer Theory.........178

Session IV
Flow Field Calculations
O.M. BELOTSERKOVSKII, Chairman

1. P. LAVAL: Time-Dependent Calculation Method for Transonic Nozzle Flows...187

2. J.L. STEGER and H. LOMAX: Generalized Relaxation Methods Applied to Problems in Transonic Flow.........................193

3. E.M. MURMAN and J.A. KRUPP: Solution of the Transonic Potential Equation Using a Mixed Finite Difference Scheme....199

4. M. HOLT and B.S. MASSON: The Calculation of High Subsonic Flow Past Bodies by the Method of Integral Relations.........207

5. B.D. MOISEENKO and B.L. ROZHDESTVENSKII: The Calculation of Hydrodynamic Forces with Tangential Discontinuities.......215

6. A.P. BAZZHIN: Some Results of Calculations of Flows Around Conical Bodies at Large Incidence Angles.....................223

7. N.E. HOSKIN and B.D. LAMBOURN: The Computation of General Problems in One Dimensional Unsteady Flow by the Method of Characteristics...230

8. V.H. RANSOM, H.D. THOMPSON and J.D. HOFFMAN: Stability and Accuracy Studies on a Second-Order Method of Characteristics Scheme for Three-Dimensional, Steady, Supersonic Flow........236

9. R.E. MELNIK and D.C. IVES: Subcritical Flows Over Two Dimensional Airfoils by a Multistrip Method of Integral Relations...243

Session V
1. Shock Waves, 2. Turbulence
H. CABANNES, Chairman

1. O.M. BELOTSERKOVSKII: On the Calculation of Gas Flows with Secondary Floating Shocks...................................255

2. R. COLLINS and HSIANG-TEH CHEN: Motion of a Shock Wave Through a Nonuniform Fluid.................................264

3. V.V. RUSANOV: Non-linear Analysis of the Shock Profile in Difference Schemes...270

4. B.M. SEGAL and J.H. FERZIGER: Shock Wave Structure by Several New Modeled Boltzmann Equations......................279

5. A.J. CHORIN: Computational Aspects of the Turbulence Problem...285

6. C.E. LEITH: Two-Dimensional Turbulence and Atmospheric Predictability..290

7. R. VAGLIO-LAURIN and G. MILLER: A Heuristic Approach to Three-Dimensional Boundary Layers.............................296

8. T.H. GAWAIN and G.D. O'BRIEN, JR.: Numerical Simulation of Transition and Turbulence in Plane Poiseuille Flow........308

9. L.D. TYLER: Heuristic Analysis of Convective Finite Difference Techniques..314

10. A.L. GONOR, V.I. LAPYGIN and N.A. OSTAPENKO: The Conical Wing in Hypersonic Flow.....................................320

Session VI
Navier-Stokes Equations.
Fully Viscous Flows
K. STEWARTSON, Chairman

1. M. FORTIN, R. PEYRET and R. TEMAM: Calcul des Ecoulements d'un Fluide Visqueux Incompressible....337

2. S.C.R. DENNIS and A.N. STANIFORTH: A Numerical Method for Calculating the Initial Flow Past a Cylinder in a Viscous Fluid....343

3. C.W. HIRT: An Arbitrary Lagrangian-Eulerian Computing Technique....350

4. T.D. TAYLOR and E. NDEFO: Computation of Viscous Flow in a Channel by the Method of Splitting....356

5. H.H. BOSSEL: Study of Vortex Flows at High Swirl by an Integral Method Using Exponentials....365

6. B.D. NICHOLS: Recent Extensions to the Marker-and-Cell Method for Incompressible Fluid Flows....371

7. W.R. BRILEY and H.A. WALLS: A Numerical Study of Time-Dependent Rotating Flow in a Cylindrical Container at Low and Moderate Reynolds Numbers....377

8. R. GLOWINSKI: Méthodes Numériques pour l'Ecoulement Stationnaire d'un Fluide Rigide Visco-Plastique Incompressible....385

9. B.B. NOVACK and HSIEN KEI CHENG: Numerical Analysis and Modeling of Slip Flows at Very High Mach Numbers....395

Session VII
Incompressible Flow Problems
K. ROESNER, Chairman

1. C. BRENNEN: Some Numerical Solutions of Unsteady Free Surface Wave Problems Using the Lagrangian Description of the Flow....403

2. O.F. VASILIEV: Numerical Solution of the Non-Linear Problems of Unsteady Flows in Open Channels....410

3. J.W. PRITCHETT: Incompressible Calculations of Underwater Explosion Phenomena....422

4. R.K.-C. CHAN, R.L. STREET and J.E. FROMM: The Digital Simulation of Water Waves - An Evaluation of SUMMAC....429

5. T.D. BUTLER: Linc Method Extensions....435

6. T. STRELKOFF: An Exact Numerical Solution of the Solitary Wave....441

7. F.H. HARLOW, A.A. AMSDEN and C.W. HIRT: Numerical Calculation of Fluid Flows at Arbitrary Mach Number..........447

8. W.E. PRACHT: Implicit Solution of Creeping Flows, with Application to Continental Drift..............................452

9. A.K. WHITNEY: The Numerical Solution of Unsteady Free Surface Flows by Conformal Mapping..........................458

Session I

Fundamental Numerical Techniques

A.J. Chorin, Chairman

ON THE GROUP CLASSIFICATION OF DIFFERENCE SCHEMES FOR SYSTEMS OF EQUATIONS IN GAS DYNAMICS

N. N. Yanenko and Y. I. Shokin

Computing Center, Academy of Sciences, USSR
Siberian Branch, Novosibirsk

INTRODUCTION

The subject of this paper is the group classification of difference schemes approximating the equations of gas dynamics. It is known that the equations of gas dynamics are invariant with respect to a certain group of point transformations in the space of independent and dependent variables. This invariance follows from the invariance of the conservation laws, which are the basis for the equations of gas dynamics. Any given difference scheme is utilized in connection with some particular grid which in itself upsets the invariance of a computational algorithm. This lack of invariance can be demonstrated, for example, in the calculation of critical features of the flow (shock waves, contact surfaces, weak discontinuities), which move with various inclinations to the grid lines. The introduction of the difference scheme makes a group analysis difficult because difference operators possess group properties different from differential operators. Consequently, it appeared desirable to carry out a group classification of difference schemes on the basis of their first differential approximation. An explanation of the first differential approximation was given in References [1] - [3] and proved fruitful in examining properties of stability and especially dissipation of difference schemes. Inasmuch as the first differential approximation is in fact a differential equation with coefficients containing the parameters of the scheme, it occupies an intermediate position between the basic equations of gas dynamics and the difference scheme approximating them: In its hyperbolic part it preserves information concerning the basic equations while the difference scheme is reflected in the parabolic part. In consequence the obvious question is, to what extent does the first differential approximation preserve the group characteristics of the equations of gas dynamics. In this connection, all schemes can be divided into two classes: those which preserve the group characteristics and those which do not. In this paper we formulate those conditions under which a parabolic system of equations of the first differential approximation admits all groups of transformations [4], allowed by the basie system of gas dynamic equations. Systems of equations in one space variable are studied in both Eulerian and Lagrangian coordinates; in the case of two space dimensions only the Eulerian point of view is taken. Furthermore, it is noted in which systems of Lagrangian equations of the first differential approximation the law of conservation of mass is observed and where it is violated. Stability of the classes of schemes constructed is tested by the method of the first differential approximation.

1. THE CASE OF ONE SPACE DIMENSION

1. Let us consider the system of equations of gas dynamics in Eulerian coordinates in the case of one space variable.

$$w_t = f_x \quad , \tag{1.1}$$

in which

$$w = \begin{pmatrix} \rho u \\ \rho \\ \rho E \end{pmatrix} , \quad f = \begin{pmatrix} - p - \rho u^2 \\ - \rho u \\ - \rho u E - up \end{pmatrix} , \quad E = \varepsilon + \frac{1}{2} u^2 ,$$

u is the gas velocity, p is the pressure, ρ is density, ε is the specific internal energy. The equation of state of the gas has the form

$$\varepsilon = \varepsilon(p,\rho) \quad .$$

The system of equations (1) is approximated by the following difference scheme:

$$\frac{\Delta_o w^n(x)}{\tau} = \frac{\Delta_1 + \Delta_{-1}}{2h} f^n(x) + \frac{\Omega(x + \frac{h}{2}) \frac{\Delta_1}{h} - \Omega(x - \frac{h}{2}) \frac{\Delta_{-1}}{h}}{h} w^n(x) \quad , \tag{1.2}$$

in which $t = n\tau$, τ is the step in time t, h is the length step along the x axis, $w^n(x) = w(x,n\tau)$, $\Omega = ||\Omega_{ij}||_1^3$ is a matrix as yet unknown, for which $||\Omega|| = O(\tau)$,

$$\Delta_o = T_o - E \quad , \quad \Delta_1 = T_1 - E \quad , \quad \Delta_{-1} = E - T_{-1} \quad ,$$

T_o is the displacement operator in t, T_1 is the displacement operator in x, E is the identity operator, $T_{-1} = T_1^{-1}$. The difference scheme (2) has at least first order accuracy.

The hyperbolic and parabolic forms of the first differential approximation of the difference scheme (1.2) have, respectively, the forms:

$$\frac{\tau}{2} w_{tt} + w_t = f_x + (\Omega w_x)_x \quad , \tag{1.3}$$

$$w_t = f_x + (C w_x)_x \quad , \tag{1.4}$$

in which

$$C = \Omega - \frac{\tau}{2} A^2 = ||\mu_{ij}||_1^3 \quad , \quad A = \frac{df}{dw} \quad ,$$

$$\mu_{11} = \Omega_{11} - \frac{\tau}{2}(3u^2 + \theta + z\eta - 3u^2 z) \quad ,$$

$$\mu_{12} = \Omega_{12} + \frac{\tau}{2} u[2u^2 - 2\theta - 3u^2 z + 3Ez + z\eta] \quad ,$$

$$\mu_{13} = \Omega_{13} - \frac{3}{2}\tau u z \quad , \quad \mu_{21} = \Omega_{21} - \frac{\tau}{2} u(2-z) \quad ,$$

$$\mu_{22} = \Omega_{22} + \frac{\tau}{2}(u^2 - \theta - u^2 z + Ez) \quad ,$$

$$\mu_{23} = \Omega_{23} - \frac{\tau}{2} z \quad ,$$

$$\mu_{31} = \Omega_{31} - \tau u(E + \eta - u^2 z) - \frac{\tau}{2} u(\theta - Ez) \quad ,$$

$$\mu_{32} = \Omega_{32} + \frac{\tau}{2}[(2u^2 + Ez)(E + \eta) - (u^2 + E)\theta - \eta\theta - u^2(2u^2 - E)z] \quad ,$$

$$\mu_{33} = \Omega_{33} - \frac{\tau}{2} u^2(1 + 2z) - \frac{\tau}{2} z(E + \eta) \quad ,$$

$$z = \frac{p_\varepsilon}{\rho} \quad , \quad \eta = \frac{p}{\rho} \quad , \quad \theta = p_\rho \quad ,$$

$$\Omega_{ij} = \Omega_{ij}(t, x, w, w_t, w_x, \ldots) \quad .$$

The system of equations (1.4) can be written in the form

$$w_t = f_x + N_x \quad , \tag{1.5}$$

in which

$$N = \begin{pmatrix} N_1 \\ N_2 \\ N_3 \end{pmatrix} = \bar{C}\,\bar{w}_x = C\,w_x \quad , \quad \bar{w} = \begin{pmatrix} u \\ \rho \\ p \end{pmatrix} \quad , \quad \bar{C} = \| \delta_{ij} \|_1^3 \quad ,$$

$$N_k = \delta_{k1}\, u_x + \delta_{k2}\, \rho_x + \delta_{k3}\, p_x \quad ,$$

$$\delta_{k1} = \rho(\mu_{k1} + u\,\mu_{k3}) \quad ,$$

$$\delta_{k2} = u\,\mu_{k1} + \mu_{k2} + (E + \rho\,\varepsilon_\rho)\mu_{k3} \quad ,$$

$$\delta_{k3} = \rho\,\varepsilon_p\,\mu_{k3} \quad .$$

Expression of the functions w, f in terms of the function $\bar{w}$ leads to the result that the system of equations (1.5) is equivalent to the system

$$F_1 = u_t + u\,u_x + \frac{1}{\rho}\,p_x - \frac{1}{\rho}\,N_{1x} + \frac{u}{\rho}\,N_{2x} = 0 \quad ,$$

$$F_2 = \rho_t + u\,\rho_x + \rho\,u_x - N_{2x} = 0 \quad , \tag{1.6}$$

$$F_3 = p_t + u\,p_x + a^2\,u_x + b\,N_{2x} + \frac{u}{\rho\,\varepsilon_p}\,N_{1x} - \frac{1}{\rho\,\varepsilon_p}\,N_{3x} = 0 \quad ,$$

in which

$$a^2 = \frac{p - \rho^2\,\varepsilon_\rho}{\rho\,\varepsilon_p} \quad , \qquad b = \frac{\varepsilon + \rho\,\varepsilon_\rho - \frac{1}{2}u^2}{\rho\,\varepsilon_p} \quad .$$

2. The system of equations (1.1) admits a space of operators with the basis [4]:

$$L_1 = \frac{\partial}{\partial t} \quad , \quad L_2 = \frac{\partial}{\partial x} \quad , \quad L_3 = t\,\frac{\partial}{\partial x} + \frac{\partial}{\partial u} \quad , \quad L_4 = t\,\frac{\partial}{\partial t} + x\,\frac{\partial}{\partial x} \quad , \tag{1.7}$$

which represent, respectively, the following finite transformations preserving the system of equations (1.1):

1) translation in time,
2) translation in a space coordinate,
3) Galilean transformation,
4) Similarity transformation.

The requirement that the system of equations (1.4) or the equivalent system (1.6) admit a space of operators with the basis (1.7) leads to certain restrictions on the choice of matrix Ω.

The system of equations (1.6) admits a space of operators with the basis (1.7) if, and only if, the following equations are satisfied (see [4]):

$$\tilde{L}_\alpha\, F_k \Big|_{F_1=0,\,F_2=0,\,F_3=0} = 0 \quad ,$$

$$\alpha = 1, 2, 3, 4 \quad ; \quad k = 1, 2, 3 \quad ,$$

in which $\tilde{L}_\alpha$ is the extended operator obtained by extending the operator L_α.

In the given case

$$\tilde{L}_1 = L_1 = \frac{\partial}{\partial t} \quad , \quad \tilde{L}_2 = L_2 = \frac{\partial}{\partial x} \quad ,$$

$$\tilde{L}_3 = t\frac{\partial}{\partial x} + \frac{\partial}{\partial u} - u_x \frac{\partial}{\partial u_t} - \rho_x \frac{\partial}{\partial \rho_t} - p_x \frac{\partial}{\partial p_t} \quad ,$$

$$\tilde{L}_4 = t\frac{\partial}{\partial t} + x\frac{\partial}{\partial x} - u_t \frac{\partial}{\partial u_t} - \rho_t \frac{\partial}{\partial \rho_t} - p_t \frac{\partial}{\partial p_t} -$$

$$- u_x \frac{\partial}{\partial u_x} - \rho_x \frac{\partial}{\partial \rho_x} - p_x \frac{\partial}{\partial p_x} \quad .$$

<u>Lemma</u>. If in the difference scheme (1.2) the elements of the matrix Ω are independent of t and x, the system of equations (1.6) is invariant with respect to transformations consisting of translation in either the time or space coordinates.

Indeed, in satisfying the conditions of the lemma, it follows that

$$\tilde{L}_1F_1 = -\frac{1}{\rho}\frac{\partial}{\partial t}N_{1x} + \frac{u}{\rho}\frac{\partial}{\partial t}N_{2x} = 0 \quad ,$$

$$\tilde{L}_2F_1 = -\frac{1}{\rho}\frac{\partial}{\partial x}N_{1x} + \frac{u}{\rho}\frac{\partial}{\partial x}N_{2x} = 0 \quad ,$$

$$\tilde{L}_1F_2 = -\frac{\partial}{\partial t}N_{2x} = 0 \quad , \quad \tilde{L}_2F_2 = -\frac{\partial}{\partial x}N_{2x} = 0 \quad ,$$

$$L_1F_3 = b\frac{\partial}{\partial t}N_{2x} + \frac{u}{\rho\,\varepsilon_p}\frac{\partial}{\partial t}N_{1x} - \frac{1}{\rho\,\varepsilon_p}\frac{\partial}{\partial t}N_{3x} = 0 \quad ,$$

$$\tilde{L}_2F_3 = b\frac{\partial}{\partial x}N_{2x} + \frac{u}{\rho\,\varepsilon_p}\frac{\partial}{\partial x}N_{1x} - \frac{1}{\rho\,\varepsilon_p}\frac{\partial}{\partial x}N_{3x} = 0 \quad ,$$

inasmuch as the independence of the elements of the matrix Ω on x and t implies also independence of δ_{ij} on x,t, thus proving the lemma.

<u>Theorem 1.1</u>. If the conditions of the lemma are satisfied, and if in addition,

$$\frac{\partial}{\partial u_t}\Omega_{ij} = 0 \quad , \quad \frac{\partial}{\partial \rho_t}\Omega_{ij} = 0 \quad , \quad \frac{\partial}{\partial p_t}\Omega_{ij} = 0 \quad (i,j = 1,2,3) \quad , \qquad (1.8)$$

$$\frac{\partial}{\partial u}N_{1x} = N_{2x} \quad , \quad \frac{\partial}{\partial u}N_{2x} = 0 \quad , \quad \frac{\partial}{\partial u}N_{3x} = N_{1x} \quad , \qquad (1.9)$$

$$(u_x\frac{\partial}{\partial u_x} + \rho_x\frac{\partial}{\partial \rho_x} + p_x\frac{\partial}{\partial p_x})N_{\alpha x} = N_{\alpha x} \quad (\alpha = 1,2,3) \quad , \qquad (1.10)$$

the system of equations (1.6) of the first differential approximation of the difference scheme (1.2) admits a space of operators with the basis (1.7).

<u>Proof</u>. Under our assumptions it follows from the lemma that

$$\tilde{L}_1F_k = 0 \quad , \quad \tilde{L}_2F_k = 0 \quad (k = 1,2,3) \quad .$$

A further examination of the operator $\tilde{L}_3$ shows that

$$\tilde{L}_3F_1 = -u_x + u_x - \frac{1}{\rho}\tilde{L}_3N_{1x} + \frac{u}{\rho}\tilde{L}_3N_{2x} + \frac{1}{\rho}N_{2x} = -\frac{1}{\rho}(\tilde{L}_3N_{1x} - N_{2x}) +$$

$$+ \frac{u}{\rho}\tilde{L}_3N_{2x} = -\frac{1}{\rho}\left(\frac{\partial}{\partial u}N_{1x} - N_{2x}\right) + \frac{u}{\rho}\frac{\partial}{\partial u}N_{2x} = 0 \quad ,$$

$$\tilde{L}_3F_2 = -\rho_x + \rho_x - \tilde{L}_3N_{2x} = -\frac{\partial}{\partial u}N_{2x} = 0 \quad ,$$

$$\tilde{L}_3F_3 = -p_x + p_x + b\,\tilde{L}_3N_{2x} - \frac{u}{\rho\varepsilon_p}N_{2x} + \frac{u}{\rho\varepsilon_p}\tilde{L}_3N_{1x} + \frac{1}{\rho\varepsilon_p}N_{1x} -$$

$$-\frac{1}{\rho\varepsilon_p}\tilde{L}_3N_{3x} = \frac{u}{\rho\varepsilon_p}\left(\frac{\partial}{\partial u}N_{1x} - N_{2x}\right) + \frac{1}{\rho\varepsilon_p}\left(N_{1x} - \frac{\partial}{\partial u}N_{3x}\right) = 0$$

Let

$$L_4 = u_x\frac{\partial}{\partial u_x} + \rho_x\frac{\partial}{\partial \rho_x} + p_x\frac{\partial}{\partial p_x} \quad ,$$

Then

$$\tilde{L}_4 = t\frac{\partial}{\partial t} + x\frac{\partial}{\partial x} - u_t\frac{\partial}{\partial u_t} - \rho_t\frac{\partial}{\partial \rho_t} - p_t\frac{\partial}{\partial p_t} - L_4 \quad ,$$

$$\tilde{L}_4F_2 = -\rho_t - u\rho_x - \rho u_x - \tilde{L}_4N_{2x} = -N_{2x} + L_4N_{2x} = 0 \quad ,$$

$$\tilde{L}_4F_1 = -u_t - uu_x - \frac{1}{\rho}p_x - \frac{1}{\rho}\tilde{L}_4N_{1x} + \frac{u}{\rho}\tilde{L}_4N_{2x} =$$

$$= -\frac{1}{\rho}N_{1x} + \frac{u}{\rho}N_{2x} + \frac{1}{\rho}L_4N_{1x} - \frac{u}{\rho}L_4N_{2x} = 0 \quad ,$$

$$\tilde{L}_4F_3 = b(n_{2x} - L_4N_{2x}) - \frac{u}{\rho\varepsilon_p}(L_4N_{1x} - N_{1x}) + \frac{1}{\rho\varepsilon_p}(L_4N_{3x} - N_{3x}) = 0$$

and the theorem is proved.

The conditions $(\partial/\partial u)N_{2x} = 0$ imply that N_{2x} does not depend explicitly on the function u. Then Eqs. (1.9) can be written in the form:

$$\frac{\partial}{\partial u}N_{2z} = 0 \quad , \quad N_{1x} = u\,N_{2x} + R_1 \quad , \quad N_{3x} = \frac{1}{2}u^2N_{2x} + u\,R_1 + R_2 \quad ,$$

$$\frac{\partial}{\partial u}R_1 = 0 \quad , \quad \frac{\partial}{\partial u}R_2 = 0 \quad . \tag{1.11}$$

If

$$\Omega_{21} = \overline{\Omega}_{21} + \frac{\tau}{2}u(2-z) - u(\overline{\Omega}_{23} - \frac{\tau}{2}z) \quad ,$$

$$\Omega_{22} = \overline{\Omega}_{22} - u[\overline{\Omega}_{21} - u(\overline{\Omega}_{23} - \frac{\tau}{2}z)] - \frac{\tau}{4}u^2(2-z) - \frac{1}{2}u^2(\overline{\Omega}_{23} - \frac{\tau}{2}z) \quad , \tag{1.12}$$

$$\Omega_{23} = \overline{\Omega}_{23} \quad ,$$

in which

$$\frac{\partial}{\partial u}\overline{\Omega}_{2k} = 0 \qquad (k = 1,2,3) \quad , \tag{1.13}$$

then

$$\frac{\partial}{\partial u} N_{2x} = 0.$$

Thus Theorem 1.1 is still valid if, in its formulation, conditions (1.9) are replaced by the conditions (1.11) to (1.13).

3. Let it be required that, in the first differential approximation of the difference scheme (1.2), the law of conservation of mass is satisfied, that is, that equation

$$N_2 = 0$$

holds, and consequently,

$$\Omega_{21} = \frac{\tau}{2} u(2-z) \quad ,$$
$$\Omega_{22} = \frac{\tau}{2} (\theta - \varepsilon z - u^2 + u^2 z) \quad , \tag{1.14}$$
$$\Omega_{23} = \frac{\tau}{2} z \quad .$$

Then the system of equations (1.6) assumes the form:

$$F_1 = u_t + uu_x + \frac{1}{\rho} p_x - \frac{1}{\rho} N_{1x} = 0 \quad ,$$
$$F_2 = \rho_t + u\rho_x + \rho u_x = 0 \quad , \tag{1.15}$$
$$F_3 = p_t + up_x + a^2 u_x + \frac{u}{\rho\varepsilon_p} N_{1x} - \frac{1}{\rho\varepsilon_p} N_{3x} = 0 \quad .$$

Theorem 1.2. If the conditions of the lemma, Eqs. (1.8), (1.10), (1.14), are satisfied, and if in addition,

$$\frac{\partial}{\partial u} N_{1x} = 0 \quad , \quad N_{3x} = (u N_1)_x + R - u_x N_1 \quad , \quad \frac{\partial}{\partial u} R = 0 \quad , \tag{1.16}$$

then the system of equations (1.15) of the first differential approximation to difference scheme (1.2) admits a space of operators with the basis (1.7) and, in this approximation, the law of conservation of mass is satisfied.

Indeed, it follows from (1.14) that

$$\delta_{2k} = 0 \qquad (k = 1,2,3) \quad ,$$

That is, $N_2 = 0$, and hence, in the first differential approximation the law of conservation of mass is satisfied. In addition the validity of Theorem 1.2 follows on the basis of Theorem 1.1, since in this case the condition (1.9) assumes the form

$$\frac{\partial}{\partial u} N_{1x} = 0 \quad , \quad \frac{\partial}{\partial u} N_{3x} = N_{1x} \quad ,$$

that is,

$$N_{3x} = u N_{1x} + R = (u N_1)_x + R - u_x N_1 \quad , \quad \frac{\partial}{\partial u} R = 0 \quad . \tag{1.17}$$

The fact that N_{1x} is independent of the function u implies that its coefficients δ_{1k} (k=1,2,3) do not depend on the function u. Setting

$$\Omega_{11} = \bar{\Omega}_{11} - u\,\bar{\Omega}_{13} + \frac{3}{2}\tau\, u(1-z) \quad ,$$

$$\Omega_{12} = \bar{\Omega}_{12} - u\,\bar{\Omega}_{11} + \frac{1}{2}u^2\,\bar{\Omega}_{13} - $$

$$- \frac{\tau}{2} u[2u^2 - 2\theta - 3u^2 z + 3Ez + \eta z] \quad , \qquad (1.18)$$

$$\Omega_{13} = \bar{\Omega}_{13} + \frac{3}{2}\tau\, u\, z \quad ,$$

$$\frac{\partial}{\partial u}\bar{\Omega}_{1k} = 0 \qquad (k = 1,2,3) \quad ,$$

we find that

$$\delta_{11} = \rho\,\bar{\Omega}_{11} - \frac{\tau}{2}\rho(\theta + z\eta) \quad ,$$

$$\delta_{12} = \bar{\Omega}_{12} + (\varepsilon + \rho\varepsilon_\rho)\bar{\Omega}_{13} \quad ,$$

$$\delta_{13} = \rho\varepsilon_p\,\bar{\Omega}_{13}$$

are also independent of u. Thus it is seen that in the formulation of Theorem 1.2 the condition (1.16) can be replaced by the conditions (1.17) and (1.18).

It follows from the system of equations (1.11) that

$$S_t + u\,S_x = 0 \qquad (1.19)$$

expressing the conservation of entropy S along the characteristic with slope $(dx/dt) = u$ (the trajectory). Formally, this equation is obtained by multiplying Equation (1.1) by the vector $X = (1/S)(-u,\ u^2 - E - \eta,\ 1)$, representing the left eigenvector of the matrix A corresponding to the eigenvalue $-u$. Here the second law of thermodynamics is used.

As is known, if the physical fluid viscosity enters in addition in p, then the entropy is increased, so that instead of Equation (1.19), the following result is valid

$$S_t + u\,S_x = \mu\, u_x^2 = q \quad , \qquad q > 0 \quad .$$

If in the difference scheme (1.2) the elements of the matrix Ω do not depend upon t,x and, furthermore, Equations (1.8), (1.10), (1.14), (1.17), (1.18) and $R = u_x N_1$, are satisfied, then the approximating viscosity enters additively into p in the system of equations of the first differential approximation and, under the condition $u_x N_1 > 0$, this viscosity leads to an increase in entropy.

Indeed, in this case the system of equations (1.15) can be written in the form:

$$(\rho u)_t + (\bar{p} + \rho u^2)_x = 0 \quad ,$$

$$\rho_t + (\rho u)_x = 0 \quad , \qquad (1.20)$$

$$(\rho E)_t + (\rho u\, E + u\,\bar{p})_x = 0 \quad .$$

in which $\bar{p} = p + N_1$. Multiplication of this system of equations by the vector X leads to the result

$$S_t + u\,S_x = \frac{1}{\rho T}\,u_x\,N_1 > 0 \quad .$$

In view of Theorem 1.2 the system (1.20) also admits a space of operators with the basis (1.7).

It is noteworthy that if the condition $R = u_x N_1$ is not satisfied, then the approximating viscosity no longer enters additively into p, analogous to physical viscosity, and it follows then that

$$T(S_t + u\,S_x) = \frac{1}{\rho}\,R \quad .$$

From the conditions (1.9) or the equivalent (1.16), it follows in particular that if the approximating viscosity enters into the equation of motion $(N_1 \neq 0)$, then it must necessarily enter into the energy equation as well $(N_3 \neq 0)$. Conversely, $(\partial/\partial x)N_{3x} = 0$, then $N_{1x} = 0$, but $N_{3x} \neq 0$, that is, there exists an approximating viscosity entering only into the energy equation $(N_3 \neq 0)$, but not entering into the equation of motion $(N_1 = 0)$. Difference schemes with an analogous viscosity are used in the numerical solution of a variety of problems in gas dynamics [5].

5. In references [2] - [3], [6] the property K of difference schemes approximating hyperbolic systems of equations with constant coefficients is determined. Let us introduce an analogous property into the given case as well.

Definition. The difference scheme (1.2) has a property K if the equation $XC = 0$ is satisfied.

If $N_2 = 0$, then

$$C = \begin{pmatrix} \mu_{11} & \mu_{12} & \mu_{13} \\ 0 & 0 & 0 \\ \mu_{31} & \mu_{32} & \mu_{33} \end{pmatrix} \quad .$$

The equality $XC = 0$ implies that

$$-u\,\mu_{1k} + \mu_{3k} = 0 \qquad (k = 1,2,3)$$

and consequently,

$$\Omega_{31} = u\,\Omega_{11} + \tau u(\varepsilon + \eta - u^2 + \frac{1}{2}\,z\eta + \frac{1}{2}\,u^2 z - \frac{1}{2}\,Ez) \quad ,$$

$$\Omega_{32} = u\,\Omega_{12} + \frac{\tau}{2}\,u[2u^2 - 2\theta - 3u^2 z + 3Ez + z\eta] -$$

$$- \frac{\tau}{2}\,[(2u^2 + Ez)(E + \eta) - (u^2 + E)\theta - \eta\theta - u^2(2u^2 - E)z] \quad , \qquad (1.20a)$$

$$\Omega_{33} = u\,\Omega_{13} + \frac{\tau}{2}\,z(E + \eta) + \frac{\tau}{2}\,u^2(1 - z) \quad .$$

Thus, if $N_2 = 0$ and Equations (1.20a) are satisfied, the difference scheme possesses the property K.

Let $N_2 = 0$ and $XC = 0$. Then

$$XC\,w_x = XN = \frac{1}{\rho}\,(-u\,N_1 + N_3) = 0 \quad ,$$

that is,

$$N_3 = u\, N_1 \qquad (1.21)$$

and consequently,

$$(\delta_{31} - u\,\delta_{11})u_x + (\delta_{32} - u\,\delta_{12})\rho_x + (\delta_{33} - u\,\delta_{13})p_x = 0 \quad .$$

The last equation will apply provided

$$\mu_{3k} = u\,\mu_{1k} \qquad (k = 1,2,3)$$

and, consequently, as previously shown, provided Equations (1.20a) are satisfied.

We have thus justified

Theorem 1.3. If in the difference scheme (1.2) the elements of the matrix Ω satisfy the conditions of the lemma and Equations (1.8), (1.10), (1.14), (1.17), (1.18), (1.21), are satisfied, then the difference scheme possesses the property K, the law of conservation of mass is satisfied in its first differential approximation, and the system of equations of the first differential approximation admits a space of operators with the basis (1.7).

6. Stability of the given family of difference schemes was studied by the method of first differential approximation, that is, conditions were found under which the system of equations of the first differential approximation were partially parabolic $(C \geq 0)$, and the domain of dependence of the hyperbolic system of equations of the first differential approximation

$$\frac{\tau}{2} w_{tt} + w_t = \Omega\, w_{xx} + (A + \Omega_x)w_x$$

does not overlap the domain of dependence of the difference scheme. These conditions lead to the inequalities

$$\frac{\tau}{2} A^2 \leq \Omega \leq \frac{h^2}{2\tau} I \quad ,$$

which comprise additional limitations on the choice of the matrix Ω and grid steps τ,h. In particular, it follows that for stability it is necessary that

$$\frac{\tau^2 u^2}{h^2} \leq 1 \quad , \qquad \frac{\tau^2(u \pm a)^2}{h^2} \leq 1 \quad .$$

Group classification of difference schemes for systems of equations of gas dynamics in Lagrangian coordinates, and also implicit difference schemes, is carried out in analogous fashion.

2. A CASE OF TWO SPACE VARIABLES

1. The system of equations of gas dynamics in the case of two space dimensions has the form

$$w_t + f_x + g_y \quad , \qquad (2.1)$$

in which

$$w = \begin{pmatrix} \rho u \\ \rho v \\ \rho \\ \rho E \end{pmatrix} , \quad f = \begin{pmatrix} \rho u^2 + p \\ \rho uv \\ \rho u \\ \rho uE + up \end{pmatrix} , \quad g = \begin{pmatrix} \rho uv \\ \rho v^2 + p \\ \rho v \\ \rho vE + vp \end{pmatrix} ,$$

Here u and w are the components of vector velocity in the x and y directions, respectively, p is the gas pressure, ρ is the density, $E = \varepsilon + (1/2)(u^2 + v^2)$, ε is the specific internal energy. The equation of state of the gas has the form:

$$p = p(\varepsilon,\rho) \quad .$$

The system of equations (2.1) is approximated by the following difference scheme:

$$\begin{aligned} \frac{\Delta_o w^n(x,y)}{\tau} + \frac{\Delta_1 + \Delta_{-1}}{2h_1} f^n(x,y) + \frac{\Delta_2 + \Delta_{-2}}{2h_2} g^n(x,y) &= \\ = \{ \frac{1}{h_1^2} [T_1^{1/2} \Omega_{11}(x,y)\bar{\Delta}_1 - T_{-1}^{1/2} \Omega_{11}(x,y)\bar{\Delta}_1] &+ \\ + \frac{1}{h_2^2} [T_2^{1/2} \Omega_{22}(x,y)\bar{\Delta}_2 - T_{-2}^{1/2} \Omega_{22}(x,y)\bar{\Delta}_2] &+ \\ + \frac{1}{h_1h_2} [T_1^{1/2} \Omega_{12}(x,y)\bar{\Delta}_2 - T_{-1}^{1/2} \Omega_{12}(x,y)\bar{\Delta}_2] &+ \\ + \frac{1}{h_1h_2} [T_2^{1/2} \Omega_{21}(x,y)\bar{\Delta}_1 - T_{-2}^{1/2} \Omega_{21}(x,y)\bar{\Delta}_1]\} w^n(x,y) &\quad , \end{aligned} \tag{2.2}$$

in which h_1 and h_2 are the step lengths on the difference grid in x and y, respectively, T_1 is the displacement operator on the x axis, T_2 the displacement operator in the y direction,

$$\Delta_i = T_i - E \quad , \qquad \Delta_{-i} = E - T_i^{-1} \quad ,$$

$$\bar{\Delta}_i = T_i^{1/2} - T_i^{-1/2} \quad , \qquad T_{-i} = T_i^{-1} \qquad (i = 1,2) \quad ,$$

Ω_{ij} is a 4 x 4 matrix, as yet undetermined, for which $||\Omega_{ij}|| = O(\tau)$.

The hyperbolic and parabolic forms of the first differential approximation to the difference scheme (2.2) have, respectively, the forms

$$\begin{aligned} \frac{\tau}{2} w_{tt} + w_t + f_x + g_y &= \\ = \frac{\partial}{\partial x} (\Omega_{11}w_x + \Omega_{12}w_y) + \frac{\partial}{\partial y} (\Omega_{21}w_x + \Omega_{22}w_y) &\quad , \end{aligned} \tag{2.3}$$

$$\begin{aligned} w_t + f_x + g_y &= \\ = \frac{\partial}{\partial x} (C_{11}w_x + C_{12}w_y) + \frac{\partial}{\partial y} (C_{21}w_x + C_{22}w_y) &\quad , \end{aligned} \tag{2.4}$$

in which

$$C_{ij} = \Omega_{ij} - \frac{\tau}{2} A_iA_j \qquad (i,j = 1,2) \quad ,$$

$$A_1 = \frac{df}{dw} \quad , \qquad A_2 = \frac{dg}{dw} \quad .$$

By means of algebraic transformations the system of equations (2.4) can be brought to the form

$$
\begin{aligned}
F_1 &= u_t + uu_x + vu_y + \frac{1}{\rho} p_x - \frac{1}{\rho} n_1 + \frac{u}{\rho} n_3 = 0 , \\
F_2 &= v_t + uv_x + vv_y + \frac{1}{\rho} p_y - \frac{1}{\rho} n_2 + \frac{v}{\rho} n_3 = 0 , \\
F_3 &= \rho_t + u\rho_x + v\rho_y + \rho u_x + \rho v_y - n_3 = 0 , \\
F_4 &= p_t + up_x + vp_y + a^2 u_x + a^2 u_y + bn_3 + \\
&\quad + \frac{u}{\rho\varepsilon_p} n_1 + \frac{v}{\rho\varepsilon_p} n_2 - \frac{1}{\rho\varepsilon_p} n_4 = 0 .
\end{aligned}
\tag{2.5}
$$

Here

$$
a^2 = \frac{p - \rho^2 \varepsilon_\rho}{\rho\varepsilon_p} , \qquad b = \frac{E + \rho\varepsilon_\rho - u^2 - v^2}{\rho\varepsilon_p} ,
$$

$$
n = \begin{pmatrix} n_1 \\ n_2 \\ n_3 \\ n_4 \end{pmatrix} = \begin{pmatrix} N^{(1)}_{1x} + N^{(2)}_{1y} \\ N^{(1)}_{2x} + N^{(2)}_{2y} \\ N^{(1)}_{3x} + N^{(2)}_{3y} \\ N^{(1)}_{4x} + N^{(2)}_{4y} \end{pmatrix} = N^{(1)}_x + N^{(2)}_y ,
$$

$$
N^{(1)} = C_{11} w_x + C_{12} w_y = ||N_k^{(1)}||_1^4 ,
$$

$$
N^{(2)} = C_{21} w_x + C_{22} w_y = ||N_k^{(2)}||_1^4 ,
$$

$$
\begin{aligned}
N_k^{(i)} &= \delta_{k1}^{i1} u_x + \delta_{k2}^{i1} v_x + \delta_{k3}^{i1} \rho_x + \delta_{k4}^{i1} p_x + \\
&\quad + \delta_{k1}^{i2} u_y + \delta_{k2}^{i2} v_y + \delta_{k3}^{i2} \rho_y + \delta_{k4}^{i2} p_y ,
\end{aligned}
$$

$$
\delta_{k1}^{ij} = \rho(\omega_{k1}^{ij} + u\, \omega_{k4}^{ij}) ,
$$

$$
\delta_{k2}^{ij} = \rho(\omega_{k2}^{ij} + v\, \omega_{k4}^{ij}) ,
$$

$$
\delta_{k3}^{ij} = u\, \omega_{k1}^{ij} + v\, \omega_{k2}^{ij} + \omega_{k3}^{ij} + E\, \omega_{k4}^{ij} + \rho\varepsilon_\rho\, \omega_{k4}^{ij} ,
$$

$$
\delta_{k4}^{ij} = \rho\varepsilon_p\, \omega_{k4}^{ij} ,
$$

$$
C_{ij} = ||\omega_{k\ell}^{ij}||_1^4 \qquad (i,j = 1,2;\ k,\ell = 1,2,3,4) .
$$

2. The greatest possible Lie group of point transformations allowed by the system of equations (2.1) is of the seventh order. The basis of the corresponding Lie algebra consists of the operators [4]:

$$L_1 = \frac{\partial}{\partial t} \quad , \quad L_2 = \frac{\partial}{\partial x} \quad , \quad L_3 = \frac{\partial}{\partial y} \quad ,$$

$$L_4 = t\frac{\partial}{\partial x} + \frac{\partial}{\partial u} \quad , \quad L_5 = t\frac{\partial}{\partial y} + \frac{\partial}{\partial v} \quad ,$$

$$L_6 = y\frac{\partial}{\partial x} - x\frac{\partial}{\partial y} + v\frac{\partial}{\partial u} - u\frac{\partial}{\partial v} \quad , \tag{2.6}$$

$$L_7 = t\frac{\partial}{\partial t} + x\frac{\partial}{\partial x} + y\frac{\partial}{\partial y} \quad ,$$

which reflect, respectively, the following finite transformations, preserving the system of Equations (2.1):

1) a shift in time,
2) a shift along the x-coordinate,
3) a shift along the y-coordinate,
4) Galilean transformation along the x-axis,
5) Galilean transformation along the y-axis,
6) Rotation transformation,
7) Similarity transformation.

3. Next we shall find the conditions under which the system of equations (2.4) or, what is equivalent. (2.5) permits transformations given by the operator (2.6). As is shown in Reference [4], this will be true if and only if

$$\tilde{L}_\alpha F_k \Big|_{F_1=0,F_2=0,F_3=0,F_4=0} = 0 \quad , \quad \alpha = 1,2,3,4,5,6,7 \; ; \; k = 1,2,3,4,$$

in which $\tilde{L}_\alpha$ is the extended operator obtained by extending the operator L_α.

In the given case

$$\tilde{L}_1 = L_1 \quad , \quad \tilde{L}_2 = L_2 \quad , \quad \tilde{L}_3 = L_3 \quad ,$$

$$\tilde{L}_4 = L_4 - u_x\frac{\partial}{\partial u_t} - v_x\frac{\partial}{\partial v_t} - \rho_x\frac{\partial}{\partial \rho_t} - p_x\frac{\partial}{\partial p_t} \quad ,$$

$$\tilde{L}_5 = L_5 - u_y\frac{\partial}{\partial u_t} - v_y\frac{\partial}{\partial v_t} - \rho_y\frac{\partial}{\partial \rho_t} - p_y\frac{\partial}{\partial p_t} \quad ,$$

$$\tilde{L}_6 = L_6 + v_t\frac{\partial}{\partial u_t} - u_t\frac{\partial}{\partial v_t} + (v_x + u_y)\left(\frac{\partial}{\partial u_x} - \frac{\partial}{\partial v_y}\right) + (v_y - u_x)\left(\frac{\partial}{\partial u_y} + \frac{\partial}{\partial v_x}\right) + \rho_y\frac{\partial}{\partial \rho_x} - \rho_x\frac{\partial}{\partial \rho_y} + p_y\frac{\partial}{\partial p_x} - p_x\frac{\partial}{\partial p_y} \quad ,$$

$$\tilde{L}_7 = L_7 - u_t\frac{\partial}{\partial u_t} - u_x\frac{\partial}{\partial u_x} - u_y\frac{\partial}{\partial u_y} - \rho_t\frac{\partial}{\partial \rho_t} - \rho_x\frac{\partial}{\partial \rho_x} - \rho_y\frac{\partial}{\partial \rho_y} - v_x\frac{\partial}{\partial v_t} - v_t\frac{\partial}{\partial v_t} - v_y\frac{\partial}{\partial v_y} - p_t\frac{\partial}{\partial p_t} - p_x\frac{\partial}{\partial p_x} - p_y\frac{\partial}{\partial p_y} \quad .$$

Theorem 2.1. If in the difference scheme (2.2) the elements of the matrix Ω_{ij} are independent of t,x,y, and if in addition

$$\frac{\partial}{\partial u_t}\Omega^{ij}_{k\ell} = 0 \quad , \quad \frac{\partial}{\partial \rho_t}\Omega^{ij}_{k\ell} = 0 \quad ,$$

$$\frac{\partial}{\partial v_t}\Omega^{ij}_{k\ell} = 0 \quad , \quad \frac{\partial}{\partial p_t}\Omega^{ij}_{k\ell} = 0 \quad (i,j = 1,2 \; ; \; k,\ell = 1,2,3,4) \quad , \tag{2.7}$$

$$n_1 = un_3 + R_1 \quad , \quad n_2 = vn_3 + R_2 \quad ,$$

$$n_4 = \frac{1}{2}u^2 n_3 + \frac{1}{2}v^2 n_3 + uR_1 + vR_2 + R \quad ,$$

$$\frac{\partial}{\partial u} n_3 = 0 \quad , \quad \frac{\partial}{\partial u} R_1 = 0 \quad , \quad \frac{\partial}{\partial u} R_2 = 0 \quad , \quad \frac{\partial}{\partial u} R = 0 \quad ,$$

$$\frac{\partial}{\partial v} n_3 = 0 \quad , \quad \frac{\partial}{\partial v} R_1 = 0 \quad , \quad \frac{\partial}{\partial v} R_2 = 0 \quad , \quad \frac{\partial}{\partial v} R = 0 \quad ,$$

$$\bar{L}_6 n_1 = n_2 - vn_3 \quad , \quad \bar{L}_6 n_2 = un_3 - n_1 \quad , \quad \bar{L}_6 n_3 = 0 \quad ,$$

$$\bar{L}_6 n_4 = - vn_1 + un_2 \quad ,$$

$$\bar{L}_7 n_k = n_k \qquad (k = 1,2,3,4) \quad ,$$

in which

$$\bar{L}_6 = (v_x + u_y)\left(\frac{\partial}{\partial u_x} - \frac{\partial}{\partial v_y} \right) + (v_y - u_x)\left(\frac{\partial}{\partial u_y} + \frac{\partial}{\partial v_x} \right) + $$
$$+ \rho_y \frac{\partial}{\partial \rho_x} - \rho_x \frac{\partial}{\partial \rho_y} + p_y \frac{\partial}{\partial p_x} - p_x \frac{\partial}{\partial p_y} \quad ,$$

$$\bar{L}_7 = u_x \frac{\partial}{\partial u_x} + u_y \frac{\partial}{\partial u_y} + v_x \frac{\partial}{\partial v_x} + v_y \frac{\partial}{\partial v_y} +$$
$$+ \rho_x \frac{\partial}{\partial \rho_x} + \rho_y \frac{\partial}{\partial \rho_y} + p_x \frac{\partial}{\partial p_x} + p_y \frac{\partial}{\partial p_y} \quad ,$$

then the system of equations (2.5) admits a space of operators with a base of (2.6).

Proof. In satisfying the conditions of the theorem it is evident that

$$\tilde{L}_k F_1 = - \frac{1}{\rho} L_k n_1 + \frac{u}{\rho} L_k n_3 = 0 \quad ,$$

$$\tilde{L}_k F_2 = - \frac{1}{\rho} L_k n_2 + \frac{v}{\rho} L_k n_3 = 0 \quad ,$$

$$\tilde{L}_k F_3 = - L_k F_3 = 0 \quad ,$$

$$\tilde{L}_k F_4 = bL_k n_3 + \frac{u}{\rho \varepsilon_p} L_k n_1 + \frac{v}{\rho \varepsilon_p} L_k n_2 - \frac{1}{\rho \varepsilon_p} L_k n_4 = 0$$

for $k = 1,2,3$. Furthermore,

$$\tilde{L}_4 F_1 = u_x - u_x - \frac{1}{\rho} \tilde{L}_4 n_1 + \frac{1}{\rho} \tilde{L}_4 (un_3) = \frac{v}{\rho} n_3 - \frac{1}{\rho} \tilde{L}_4 n_1 = \frac{1}{\rho}(vn_3 - L_4 n_1) =$$
$$= \frac{v}{\rho} n_3 - \left(v \frac{\partial}{\partial u} - u \frac{\partial}{\partial v}\right) n_1 = \frac{v}{\rho} n_3 - \left(v \frac{\partial}{\partial u} - u \frac{\partial}{\partial v}\right)(un_3 + R_1) = 0 \quad ,$$

$$\tilde{L}_4 F_2 = v_x - v_x - \frac{1}{\rho} \tilde{L}_4 n_2 + \frac{1}{\rho} \tilde{L}_4 (vn_3) = - \frac{u}{\rho} n_3 - \frac{1}{\rho} L_4 n_2 =$$
$$= - \frac{u}{\rho} n_3 - \frac{1}{\rho} \left(v \frac{\partial}{\partial u} - u \frac{\partial}{\partial v}\right) n_2 = 0 \quad ,$$

$$\tilde{L}_4 F_3 = \rho_x - \rho_x - \tilde{L}_4 n_3 = - L_4 n_3 = - \left(v \frac{\partial}{\partial u} - u \frac{\partial}{\partial v}\right) n_3 = 0 \quad ,$$

$$\tilde{L}_4F_4 = p_x - p_x + b\tilde{L}_4n_3 - \frac{u}{\rho\varepsilon_p} n_3 + \frac{u}{\rho\varepsilon_p}\tilde{L}_4n_1 + $$
$$+ \frac{1}{\rho\varepsilon_p} n_1 + \frac{v}{\rho\varepsilon_p}\tilde{L}_4n_2 - \frac{1}{\rho\varepsilon_p}\tilde{L}_4n_4 = 0 \quad .$$

In an analogous way it can be shown that in satisfying the conditions of the theorem:

$$\tilde{L}_\alpha F_k\Big|_{F_1=0,F_2=0,F_3=0,F_4=0} = 0 \qquad (\alpha = 5,6,7 \; ; \; k = 1,2,3,4) \quad .$$

and the theorem is proved.

4. We shall state that the law of conservation of mass will be satisfied in the system of equations of the first differential equation to the difference scheme (2.2) provided

$$n_3 = 0 \quad . \tag{2.8}$$

It should be noted that if $\Omega_{12} = \Omega_{21} = 0$, the first law of conservation of mass is not satisfied in the system of equations of first differential approximation to the difference scheme.

Indeed, in this case it can be shown that $n_3 \neq 0$ for any choice of matrices Ω_{11}, Ω_{22}.

It is easy to show that if

$$\begin{aligned}
&\Omega^{11}_{31} = \frac{\tau}{2} u(2-z) \;, \quad \Omega^{11}_{32} = -\frac{\tau}{2} vz \;, \\
&\Omega^{11}_{33} = \frac{\tau}{2}[-u^2 + \theta + (v^2 + u^2 - E)z] \;, \quad \Omega^{11}_{34} = \frac{\tau}{2} z \;, \\
&\Omega^{12}_{31} = \frac{\tau}{2} v \;, \quad \Omega^{12}_{32} = \frac{\tau}{2} u \;, \quad \Omega^{12}_{33} = -\frac{\tau}{2} uv \;, \quad \Omega^{13}_{34} = 0 \;, \\
&\Omega^{21}_{31} = \frac{\tau}{2} v \;, \quad \Omega^{21}_{32} = \frac{\tau}{2} u \;, \quad \Omega^{21}_{33} = -\frac{\tau}{2} uv \;, \quad \Omega^{12}_{34} = 0 \;, \\
&\Omega^{22}_{31} = -\frac{\tau}{2} uz \;, \quad \Omega^{22}_{32} = \frac{\tau}{2} v(2-z) \;, \\
&\Omega^{22}_{33} = \frac{\tau}{2}[-v^2 + \theta + (v^2 + u^2 - E)z] \;, \quad \Omega^{22}_{34} = \frac{\tau}{2} z \;,
\end{aligned} \tag{2.9}$$

in which

$$z = \frac{p_\varepsilon}{\rho} \;, \quad \theta = \frac{p}{\rho} \;, \quad \Omega_{ij} = ||\Omega^{ij}_{k\ell}||^4_1 \quad (i,j = 1,2) \;,$$

then Equation (2.8) follows.

<u>Theorem 2.2.</u> If in the difference scheme (2.2) the elements of the matrix Ω_{ij} $(i,j = 1,2)$, are independent of t,x,y, and if, in addition, Equations (2.7), (2.9) and

$$n_4 = un_1 + vn_2 + R \;,$$

$$\frac{\partial}{\partial u} n_k = 0 \;, \quad \frac{\partial}{\partial v} n_k = 0 \qquad (k = 1,2) \;,$$

$$\frac{\partial}{\partial u} R = 0 \;, \quad \frac{\partial}{\partial v} R = 0 \;,$$

$$\overline{L}_6 n_1 = n_2 \ , \quad \overline{L}_6 n_2 = - n_1 \ , \quad \overline{L}_6 n_4 = - v n_1 + u n_2 \ ,$$

$$\overline{L}_7 n_j = n_j \qquad (j = 1,2,3,4) \ ,$$

are satisfied, then the system of equations of first differential approximation to the difference scheme admit a space of operators with the basis (2.6) and thereby the law of conservation of mass is satisfied.

The proof of the theorem is achieved in a manner analogous to the proof of Theorem 2.1.

REFERENCES

1. N. N. Yanenko, Y. I. Shokin. "On the correctness of first differential approximations of difference schemes," Doklady AN SSSF 182, 4 (1968), 776-778.

2. N. N. Yanenko and Y. I. Shokin. "On the first differential approximation of difference schemes for hyperbolic systems of equations," Siberian Mathematical Journal 10, 4 (1969), 1174-1188.

3. N. N. Yanenko and Y. I. Shokin. "First differential approximation method and approximate viscosity of difference schemes." High-Speed Computing in Fluid Dynamics. The Physics of Fluids, Supplement II, New York, 1969.

4. L. V. Ovsyannikov. "Group properties of differential equations," Novosibirsk, 1962.

5. N. N. Yanenko, N. N. Anuchina, V. E. Petrenko and Y. I. Shokin. "On methods of calculating problems of gas dynamics with large deformations," Informational Bulletin, Numerical Methods of Mechanics of a Continuous Medium 1, 1 (1970).

6. N. N. Yanenko and Y. I. Shokin. "On the approximating viscosity of difference schemes," Doklady AN SSSR, 182, 2 (1968), 280-281.

NUMERICAL INTEGRATION OF THE EULER-EQUATIONS FOR THREE-DIMENSIONAL UNSTEADY FLOWS

K. Roesner

Max-Planck-Institut fuer Stroemungsforschung

Goettingen

An extending of the well known method of characteristics is given for the treatment of three-dimensional time-dependent gasdynamical flows without discontinuities. This method is compared with a difference scheme based on the method of fractional steps. The unsteady flow of an ideal gas through a nozzle is computed numerically.

I. INTRODUCTION

The method of characteristics (m.c.) is an often applied tool for the computation of time-dependent gas flows. In the absence of any kind of discontinuities the calculation of the flow field variables is straightforward if the problem depends only on a single space variable. Difficulties result from an increasing number of spatial coordinates.

There are many papers on this subject but most of them do not take into account the whole number of characteristic manifolds. Only a few methods are known which introduce into the numerical procedure all the available characteristic manifolds. This conception is realized in the computational methods given in the papers of G. Groß [1], K. Roesner [3], V. V. Rusanov [5], H. Sauerwein [6], and M. Schäfer [7].

Contrary to the m.c. there are finite difference methods based on the method of fractional steps (m.f.s.) which are used, too, for the numerical treatment of various problems in fluid dynamics. Reference is given here only to the books of N. N. Janenko [2] and B. L. Rozhdestvenskii [4]. Concerning the applicability of the m.f.s. to nonlinear problems in gasdynamics we refer to the paper of N. N. Yanenko, V. P. Frolov, and V. E. Neuvazhaev [8].

In the following we present an iterative procedure for the numerical treatment of Cauchy's initial value problem and mixed problems for multidimensional hyperbolic systems of quasilinear partial differential equations. This m.c. is compared with the m.f.s. with respect to their usage on computing machines. It is a matter of fact that the m.f.s. is applicable to a much wider class of differential equations than the m.c. But the latter can easily be applied to compute field points on fixed or moving boundaries.

In the example given below we have tested the algorithm based on the m.c. by applying it to the calculation of the flow quantities of the steady rarefaction flow inside a nozzle induced by the accelerated motion of a piston.

The results show that the m.c. can successfully be applied to gasdynamical problems not including discontinuities.

II. BASIC EQUATIONS

We are investigating the isentropic motion of an ideal gas with constant ratio of specific heats. The influence of viscosity, heat conduction and exterior forces is neglected. Hence the basic equations of gasdynamics can be written in the form:

$$\frac{\partial \vec{f}(\vec{x},t)}{\partial t} = \Omega(\vec{x},t;\vec{f})\,\vec{f}(\vec{x},t) \, , \qquad (1)$$

$\vec{f} = \{u(\vec{x},t), v(\vec{x},t), w(\vec{x},t), \pi(\vec{x},t)\}$ is a vector function; u,v,w the cartesian components of the velocity; $\pi = \ln(p/p_o)^{1/\kappa}$, $\kappa = c_p/c_v$ the ratio of the specific heats, and p/p_o the pressure ratio related to the pressure of the gas at rest.

$$\Omega(\vec{x},t;\vec{f}) = -\begin{pmatrix} u\partial_x + v\partial_y + w\partial_z & 0 & 0 & e^{0.4\pi}\partial_x \\ 0 & u\partial_x + v\partial_y + w\partial_z & 0 & e^{0.4\pi}\partial_y \\ 0 & 0 & u\partial_x + v\partial_y + w\partial_z & e^{0.4\pi}\partial_z \\ \partial_x & \partial_y & \partial_z & u\partial_x + v\partial_y + w\partial_z \end{pmatrix};$$

$\partial_x, \partial_y, \partial_z$ denote partial derivatives with respect to x,y,and z.

As boundary conditions we demand that the normal component of the gas velocity at the wall and the normal component of the wall velocity coincide.

III. ITERATIVE PROCEDURE

Transformation of (1) from cartesian coordinates to arbitrary curvilinear coordinates makes it possible to introduce characteristic hypersurfaces. They are deduced by the demand that derivatives of the flow field variables across the characteristic hypersurfaces are canceled. Introducing finite differences the coordinates of a new field point $\vec{x}_N=(x_N,y_N,z_N,t_N)$ are given by the so called 'edge' equations, which are l i n e a r with respect to $\vec{x}_N$:

$$X\cdot\vec{x}_N = \vec{r}\,. \qquad (2)$$

This equation connects the unknown coordinates $\vec{x}_N$ with the coordinates of the four edges of a tetrahedron in the initial value distribution. X is a 4×4 matrix and $\vec{r}$ a vector, both depending on the initial values and the unknown solution $\vec{f}_N$ in the new field point.

The flow field variables are given by the q u a s i l i n e a r system:

$$V\cdot\vec{f}_N = \vec{R}\,, \qquad (3)$$

where V and $\vec{R}$ are depending on $\vec{f}_N$ itself and the chosen initial values in the basic field points P_i, i=1,2,3,4.

We start the iteration procedure by choosing arbitrary values of $\vec{f}_N^{\,0}$ getting a first approximation of $\vec{x}_N^{\,1}$ and $\vec{f}_N^{\,1}$ by (2) and (3). This process is continued till a prescribed accuracy is reached for the difference of two succeeding values of $\vec{x}_N^{\,k}$ and $\vec{f}_N^{\,k}$.

Points on the boundary of the flow field can be situated either on a surface of arbitrary shape or on the intersection line of two such surfaces or in the intersection point of three surfaces. In each of these cases one, two or three of the edge equations are replaced by the equations of the boundary surfaces. This leads generally to non-linear equations for x_N^*. To find f_N^* on the boundary we have to take into account the boundary conditions on each of the surfaces.

IV. NUMERICAL RESULTS

The m.c. mentioned above is applied to the calculation of the flow field of a gas through a three-dimensional nozzle without any rotational symmetry. The small and the large semiaxis of the elliptical cross section of the nozzle vary according to a circular arc. The difference between A(x) and B(x) is constant along the x-axis. A fourth of the whole channel is shown in the figure below. We have confined our calculation to this part of the flow field. Figure 1 gives a picture of the nozzle shape. The variation of the cross section along the x-axis is demonstrated in figure 2. The gas at rest inside the nozzle is forced to follow the motion of an accelerated piston moving along the channel with constant elliptical cross section. The head of the first rarefaction wave is assumed to be a filar evolvent in the curved part of the nozzle.

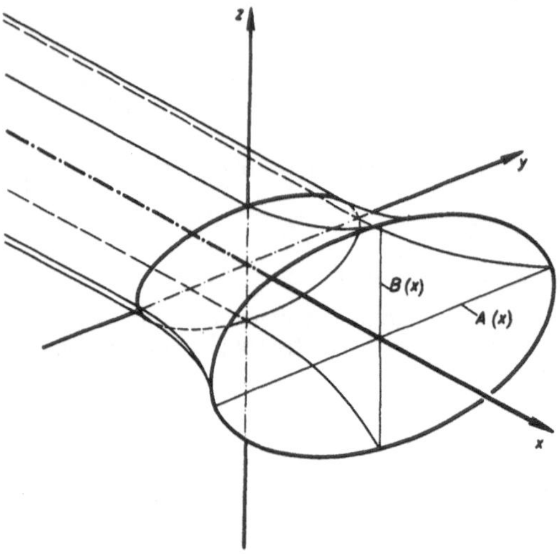

Fig. 1. Nozzle shape

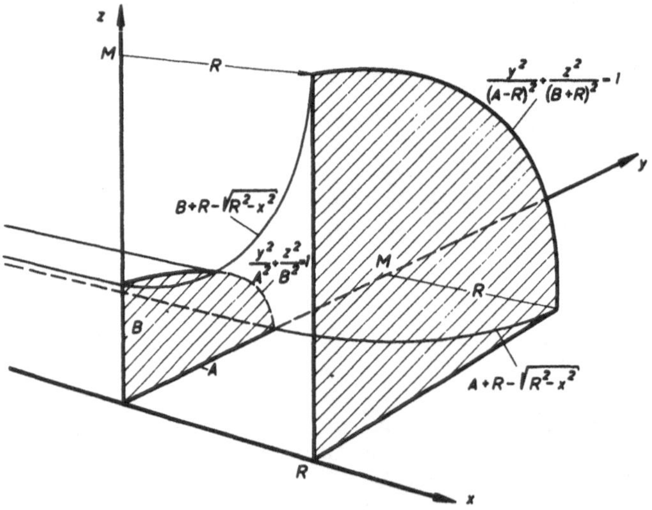

Fig. 2. Nozzle cross sections

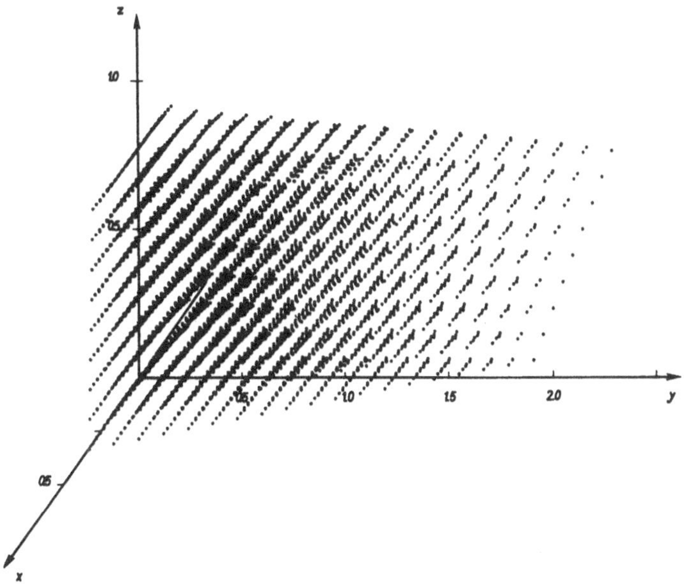

Fig. 3. Calculated field points inside the nozzle

A plot of all computed field points inside a small part of the nozzle appears in figure 3. The time varies between 0. and 0.5. This perspective view of the flow field suggests the conjecture that the computed points are arranged in a regular manner. As it is shown in figure 4 all the points are situated on planes which are parallel to the x-axis and inclined to the y-axis. The interpolation can be carried out within each of the planes. In figure 5 the chronological evolution of some lines of constant Mach-number is demonstrated for the plane given in figure 4.

V. CONCLUSIONS

A comparison of the m.c. and the m.f.s. leads to the following statements:

(1) The formalism of the m.c. is more complicated than that of the m.f.s..

(2) The applicability of the m.c. is confined to hyperbolic problems without any discontinuity whereas the m.f.s. can be applied to all differential equations regardless whether discontinuities occur or not.

(3) The calculation of boundary points based on the m.c. is nearly straightforward. Difficulties can arise from the application of the m.f.s. to the calculation of boundary points unless the boundary surfaces can be embedded into a specially chosen coordinate system.

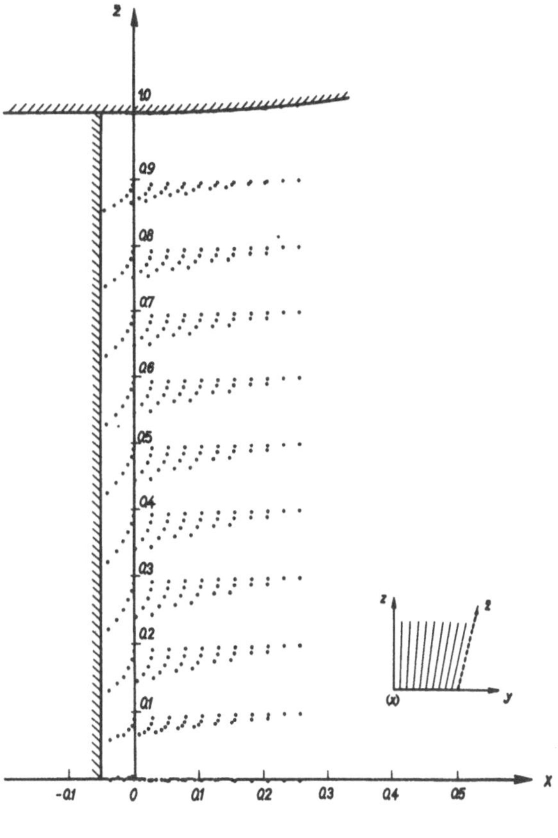

Fig. 4.
Points on an inclined plane

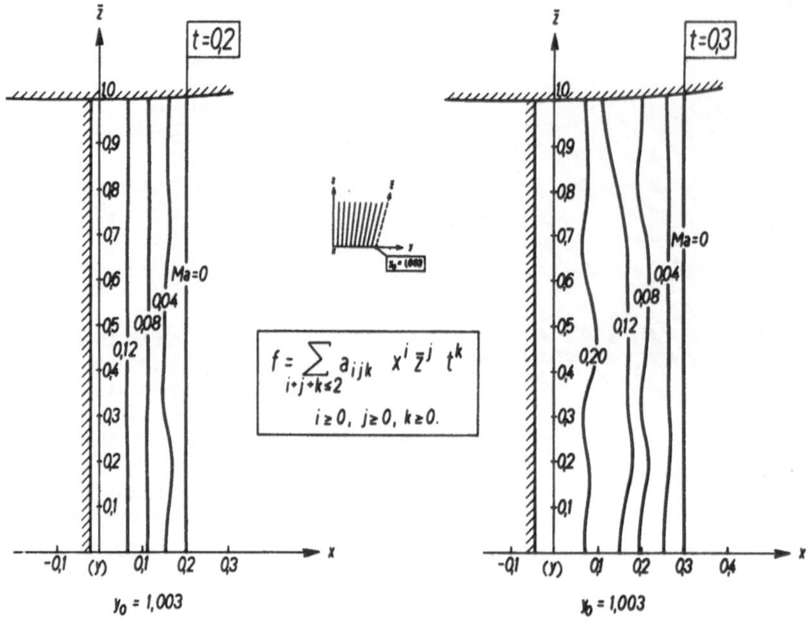

Fig. 5. Lines of constant Mach-number

(4) As we are interested mainly in the distribution of flow field variables for an instant value of time the m.c. produces an unnecessary large number of field points compared with the m.f.s. which introduces a lattice of points with constant distances. So we get the flow quantities immediately in all the net points simultaneously.

(5) With respect to the application of both methods to linear or nonlinear problems there is no difference. Nonlinear problems lead necessarily to iterative procedures in both cases.

As a result we might say that the m.f.s. is better than the m.c. if we want to solve problems with a simple shape of the boundaries for large values of the time. This is due to the stability and accuracy of the m.f.s.

REFERENCES

[1] Groß, G. — Charakteristikenverfahren für zweidimensionale instationäre Strömungen eines kompressiblen Gases

Diplomarbeit, Göttingen (1958)

[2] Janenko, N. N. — Die Zwischenschrittmethode zur Lösung mehrdimensionaler Probleme der mathematischen Physik

Lecture Notes in Mathematics, vol.91,1969 Springer-Verlag Berlin-Heidelberg-New York

[3] Roesner, K. — Die Berechnung dreidimensionaler, instationärer Strömungsfelder kompressibler Medien

Dissertation, Göttingen (1967)

[4] Rozhdestvenskii, B. L. and Yanenko, N. N. — Системы квазилинейных уравнений и их приложения к газовой динамике

Изд. «Наука», Москва 1968

[5] Rusanov, V. V. — Characteristics of the general equations of gas dynamics

Zh. vych. mat., 3, No. 3 (1963), p. 508-527

[6] Sauerwein, H. — Numerical Calculations of Multidimensional and Unsteady Flows by the Method of Characteristics

J. Comp. Phys. 1, 406-432 (1967)

[7] Schäfer, M. — Zur Anwendung der mehrdimensionalen Charakteristikentheorie in der Gasdynamik

Miszellaneen der Angewandten Mechanik (Festschrift W. Tollmien zum 60. Geburtstag)

Akademie-Verlag, Berlin (1962)

[8] Yanenko, N. N., Frolov, V. P., and Neuvazhaev, V. E. — О применении метода расщепления для численного движений теплопроводного газа в криволинейных координатах.

(The application of the splitting method for the numerical calculation of gas flows with heat conduction in curvilinear coordinates)

Изв. сиб. отд. А.Н. СССР, серия технических наук вып. 2, 1967, No.8

THE COMPUTATION OF SUPERSONIC FLOW FIELDS ABOUT WING–BODY COMBINATIONS BY "SHOCK–CAPTURING" FINITE DIFFERENCE TECHNIQUES

Paul Kutler and Harvard Lomax

Ames Research Center, NASA
Moffett Field, Calif., 94035

INTRODUCTION

Numerous numerical techniques for the solution of flows with embedded shocks have been developed in recent years. These techniques are based on both the Eulerian and Lagrangian systems and their combination. In this paper we consider those numerical methods that advance the solution of hyperbolic partial differential equations by use of differencing schemes constructed on a fixed Eulerian mesh. We also consider physical problems in which the shocks form as the computations proceed, and limit our attention to those methods that have the advantage of predicting the shock strength and location without special recourse to additional balances, such as the Rankine-Hugoniot relations. Such "shock-capturing" techniques have received extensive study and application during the past decade, resulting in the construction of many differencing schemes with second-order accuracy. The acceptance of such shock capturing numerical methods is far from universal, however, principal objections being their inability to resolve the shock location with absolute precision, and their tendency to give spurious fluctuations for the magnitudes of the dependent variables in the shock's vicinity. The purpose of this paper is twofold: first to indicate how the objections just mentioned can be lessened, and second to show by examples what can be expected for the computation of the supersonic flow about complicated three-dimensional aerodynamic objects using second-order, shock-capturing methods presently available.

IMPROVEMENT OF SHOCK–CAPTURING TECHNIQUES

There are three fundamental ways in which the resolution of the shock position can be improved: (a) increase the order of the differencing scheme; (b) increase the number of node points in the differencing mesh; (c) adjust some parameter, or even the mesh itself, so that in the shock vicinity the mesh is in some sense compatible with the local characteristics. The latter condition is purposely vague for the general, nonlinear case, but can be made precise for linear computations by restating it as the requirement that the local Courant number be equal to one. Using conditions (a) and (b) to increase the "sharpness" of the shock generally causes the fluctuations in the vicinity of the shock to be more severe. One can compensate for this effect by adroitly applying (c) or by deliberately adding dissipation to higher order terms. Let us inspect these possibilities in more detail.

The methods we will consider are all outgrowths of Runge-Kutta techniques. A necessary condition (Lomax) for the existence of a Runge-Kutta method is that it produce only one characteristic root when applied to the equation

$$u' \equiv u_t = \sigma u + e^{\mu t} \tag{1}$$

and that this root be identical to the value of the Taylor series expansion of $e^{\sigma\Delta t}$ truncated at the term corresponding to the particular order chosen for the method. Second-order forms of interest are given by

$$\left.\begin{aligned} \tilde{u} \equiv \tilde{u}_{n+\alpha} &= u_n + \Delta t(\alpha u_n') \\ u_{n+1} &= \beta_1 u_n + \beta_2 \tilde{u} + \Delta t(\beta_3 u_n' + \beta_4 \tilde{u}') \end{aligned}\right\} \tag{2}$$

where the α and β are given constants restrained only by the requirement that the integration of Eq. (1) has second-order accuracy in both the complementary and particular solution. Predictor-corrector formulations of the Lax-Wendroff method are based on these equations. For example: $\alpha = 1/2$, $\beta_1 = \beta_4 = 1$, $\beta_2 = \beta_3 = 0$ is the basis for the method given on page 303 in Richtmyer and Morton, $\alpha = \beta_1 = 1$, $\beta_2 = 0$,

$\beta_3 = \beta_4 = 1/2$ is the basis for the method proposed by Burstein; and $\alpha = 1$, $\beta_1 = \beta_2 = \beta_4 = 1/2$, $\beta_3 = 0$ is the basis for the method proposed by MacCormack. When applied to the simple linear wave equation

$$u_t + cu_x = 0 \tag{3}$$

all of these give (with appropriate differencing in the space coordinate)

$$\begin{aligned} u_j^{n+1} = & \left(u_j^n \right) \\ & - \nu \left(\frac{1}{2} u_{j+1}^n - \frac{1}{2} u_{j-1}^n \right) \\ & + \nu^2 \left(\frac{1}{2} u_{j+1}^n - u_j^n + \frac{1}{2} u_{j-1}^n \right) \end{aligned} \tag{4}$$

where $\nu \equiv c\Delta t/\Delta x$ is the Courant number and n and j are the time and space indices, respectively.

Next, consider the following procedure regarding Eq. (4). First, expand each term in a Taylor series about the point (n,j). This results in a partial differential equation with an infinite number of terms. Next, eliminate all t derivatives, except u_t, by algebraic reduction. Let us refer to the result as the modified partial differential equation, in contrast to Eq. (3) which is the basic partial differential equation. Within the limits of the convergence of the Taylor series expansion, Eq. (4) gives the exact solution of the modified equation. This concept can be quite helpful in gaining an insight to the effects on the physical problem of the numerical approximations. It has been used by Hirt; and it is well known (see Richtmyer and Morton, p. 332) that the result for Eq. (4) is

$$u_t + cu_x = \frac{-c\Delta x^2}{6}(1 - \nu^2)u_{xxx} - \frac{c\Delta x^3}{8}(\nu - \nu^3)u_{xxxx} + \ldots \tag{5}$$

All the well-known properties of the Lax-Wendroff method with regard to order, dispersion, and dissipation appear in this result.

Consider Eq. (5) for the special cases $\nu = \pm 1$. In these cases the terms on the right-hand side all become zero. This corresponds to the fact that Eq. (4) satisfies the "shift condition"; that is for $\nu = \pm 1$ the data at one time interval are simply shifted without modification one space index to give the solution for the next time interval, and this, of course, is the exact solution of Eq. (3). If a method satisfies the shift condition when applied to Eq. (3), it will satisfy condition (c) above for the two-dimensional linear case. Numerical experiments indicate that, if properly used, it can also be a good shock capturing method for nonlinear cases.

Let us examine the contents of the preceding paragraph in another light. Consider the second-order Runge-Kutta method that follows from: using Eq. (2) with $\alpha = 1/2$, $\beta_1 = \beta_4 = 1$, $\beta_2 = \beta_3 = 0$; and approximating all space differentials with central differences. To this result let us add a term proportional to a fourth derivative in x. The result can be written as

$$\begin{aligned} u_j^{n+1} = & \left(u_j^n \right) \\ & + \delta \left(-\frac{1}{8} u_{j+2}^n + \frac{1}{2} u_{j+1}^n - \frac{3}{4} u_j^n + \frac{1}{2} u_{j-1}^n - \frac{1}{8} u_{j-2}^n \right) \\ & - \nu \left(\frac{1}{2} u_{j+1}^n - \frac{1}{2} u_{j-1}^n \right) \\ & + \nu^2 \left(\frac{1}{8} u_{j+2}^n - \frac{1}{4} u_j^n + \frac{1}{8} u_{j-2}^n \right) \end{aligned} \tag{6}$$

If $\delta = 0$, it is well known that this method is unstable for all ν and it is also apparent that the shift condition is violated. On the other hand, if $\delta = 1$, the shift condition is satisfied and the method is stable for all $|\nu| \leqslant 1$ as indicated by the modified partial differential equation

$$u_t + cu_x = -\frac{c\Delta x^2}{6}(1 - \nu^2)u_{xxx} - \frac{c\Delta x^3}{8}\left(\frac{\delta}{\nu} - \nu^3\right)u_{xxxx} + \ldots \tag{7}$$

Observe also that to the lowest order the error in dispersion is identical to that given by Eq. (5).

Current research in the development of shock-capturing methods involves the construction of higher-order methods. Predictor-corrector methods that are third order in both the time and space variables have been proposed by Rusanov, and Burstein & Mirin. When applied to Eq. (3) these reduce to the following difference scheme:

$$\begin{aligned} u_j^{n+1} = {} & \left(u_j^n \right) \\ & + \delta\left(-\frac{1}{8}u_{j+2}^n + \frac{3}{6}u_{j+1}^n - \frac{3}{4}u_j^n + \frac{3}{6}u_{j-1}^n - \frac{1}{8}u_{j-2}^n\right) \\ & - \nu\left(-\frac{1}{12}u_{j+2}^n + \frac{4}{6}u_{j+1}^n - \frac{4}{6}u_{j-1}^n + \frac{1}{12}u_{j-2}^n\right) \\ & + \nu^2\left(\frac{1}{8}u_{j+2}^n - \frac{1}{4}u_j^n + \frac{1}{8}u_{j-2}^n\right) \\ & - \nu^3\left(\frac{1}{12}u_{j+2}^n - \frac{1}{6}u_{j+1}^n + \frac{1}{6}u_{j-1}^n - \frac{1}{12}u_{j-2}^n\right) \end{aligned} \tag{8}$$

having the modified partial differential equation

$$\begin{aligned} u_t + cu_x = {} & -\frac{c\Delta x^3}{24}\left(\frac{3\delta}{\nu} - 4\nu + \nu^3\right)u_{xxxx} \\ & - \frac{c\Delta x^4}{120}[15\delta - 4 - 15\nu^2 + 4\nu^4]u_{xxxxx} + \ldots \end{aligned} \tag{9}$$

It is important to notice that for the particular cases $\nu = \pm 1$, the improvement in shock resolution found by adding the δ term to Eqs. (6) and (8) (with $\delta = 1$) has nothing to do with adding dissipation to the systems. Rather its value in such cases is that the numerical technique has become identical to the method of characteristics for Eq. (3). Furthermore, for "off design" cases when $|\nu| < 1$, our hypothesis is that the best choice of δ occurs when the shift condition is satisfied as closely as possible, again regardless of the added dissipation. This hypothesis is extended to include in Eq. (4) the same δ term appearing in Eqs. (6) and (8). Such a hypothesis is supported by inspecting the C_k in

$$u_j^{n+m} = \sum_{k=-2m}^{2m} C_k u_{j+k}^n \tag{10}$$

which represents the summed values of Eqs. (4), (6), or (8) after m time steps. For example, after 10 steps the values of C_k representing the exact solution for $\nu = 0.5$ are all zero except for C_{-5} which is 1. The actual values of C_k for the three methods are shown in Fig. 1 where in each case δ was chosen to minimize the expression

$$|C_{-5} - 1| + \sum_{k \neq -5} |C_k| \tag{11}$$

EXAMPLE RESULTS USING SECOND–ORDER SHOCK–CAPTURING TECHNIQUES

To illustrate the power of shock-capturing methods over a wide range of applications, the results for four sets of computations are included. In every case the numerical shock-capturing method used was the second-order variant of the Lax-Wendroff technique developed by MacCormack, and the model equations were the conservation-law form of the inviscid gasdynamic equations.

Figure 2 shows the pressure distribution on the surface of a cone at various angles of attack (α), at a Mach number (M) equal to 5 and a cone half-angle (σ) equal to 10°. The embedded crossflow shock is clearly evident at the highest angles of attack, and it coincides with an appropriate entropy increase. The results are in excellent agreement with those published by Jones, and Holt & Ndefo (using Telenin's method) although no attempt was made in our calculations to resolve the entropy layer at the body surface.

Figure 3 shows the pressure distribution on the compression and expansion side of a delta wing having supersonic leading edges and a wing sweep (Λ) equal to 50° and 45°. Comparisons are made with calculations made by South and Klunker, Babaev, and Beeman and Powers. South and Klunker, in turn, compared their results with a variety of others not shown in Fig. 3 and found all of these to agree well. The evidence indicates that the dashed curves in Fig. 3 are inaccurate in the region where they do not agree with the symbols.

Figure 4 presents the computed results for the pressure on the surface of, and the shock surrounding a conical, wing-body combination at angle of attack. The sonic-edged wing has 5° of dihedral. Only the compression side is shown, although the expansion side was also computed.

The real power of a shock-capturing method is illustrated in Fig. 5. This displays the variation of pressure along vertical lines in the wake region behind a delta wing at 5° angle of attack. Figure 5(a) shows the upper expansion region and lower compression region just behind the trailing edge. A conical coordinate system was used so that the outer shock and expansion fan remain in the original grid without intermediate interpolation. Figures 5(b) and 5(c) are 0.3 and 4.3 chord lengths behind the trailing edge, respectively, and show the formation and continuation of the shocks in the inviscid wake.

BIBLIOGRAPHY

Babaev, D. A. *J. Comput. Math. Math. Phys.*, 2, 278-289 (1962).

Babaev, D. A. *AIAA J.* 1 (1963).

Beeman, E. R., and Powers, S. A. *AIAA Paper* 69-646 (1969).

Burstein, S. Z. *J. Comput. Phys.* 2, 178-196 (1967).

Burstein, S. Z., and Mirin, A. A. *J. Comput. Phys.* 5, 547-571 (1970).

Hirt, C. W. *J. Comput. Phys.* 2, 339-355 (1968).

Holt, M., and Ndefo, D. E. *J. Comput. Phys.* 5, 463-486 (1970).

Jones, D. J. *AGARDograph* 137 (1969).

Lomax, H. *NASA TR R–262* (1967).

MacCormack, R. W. *AIAA Paper* 69-354 (1969).

Richtmyer, R. D., and Morton, K. W. *Difference Methods for Initial-Value Problems*, Second ed., Interscience Publishers, Inc. (1967).

Rusanov, V. V. *J. Comput. Phys.* 5, 507-516 (1970).

South, J. C., and Klunker, E. B. *NASA SP–228*, 131-158 (1969).

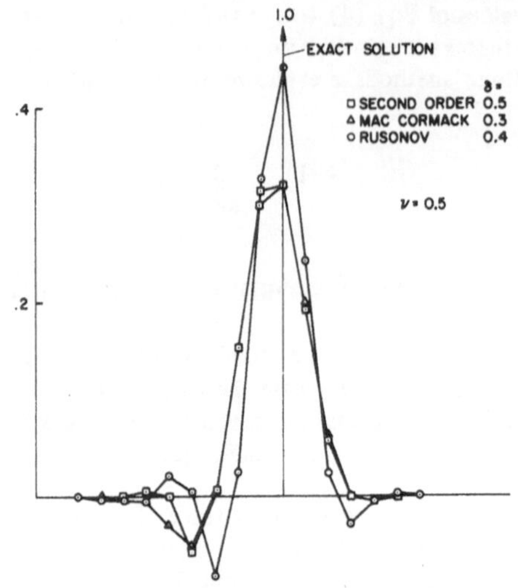

Fig. 1.- Solution for an initial delta function after m = 10 steps.

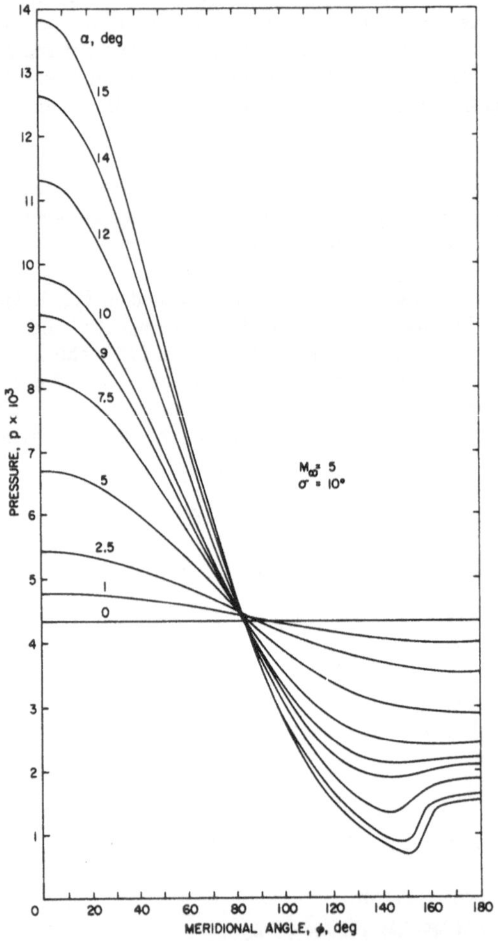

Fig. 2.- Surface pressure distribution on cone at angle of attack.

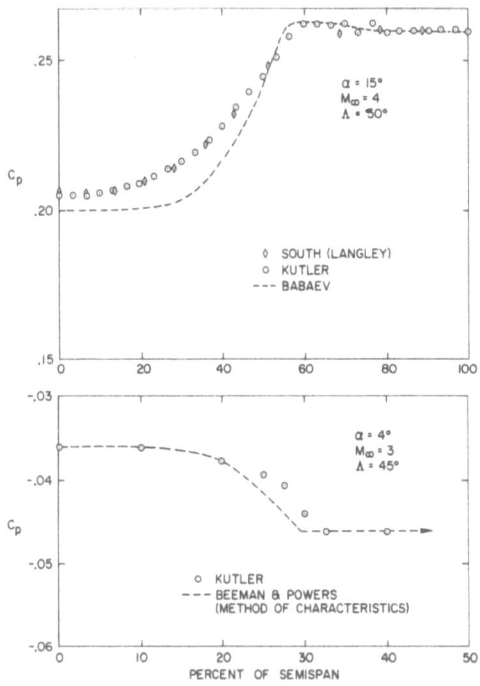

Fig. 3.- Spanwise pressure distribution on the compression (top plot) and expansion side of a planar delta wing.

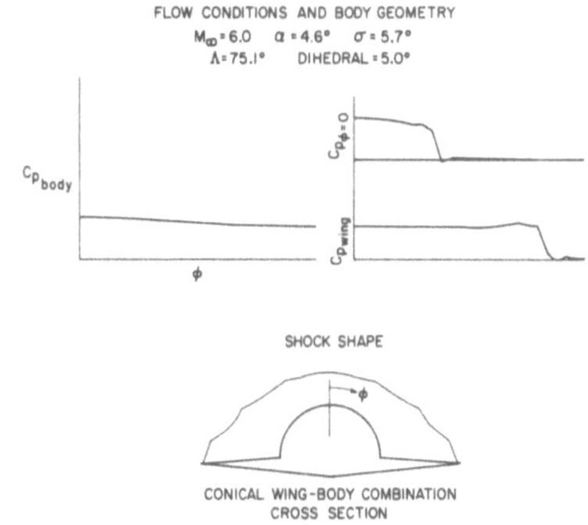

Fig. 4.- Pressure distribution and shock shape for a conical wing-body combination.

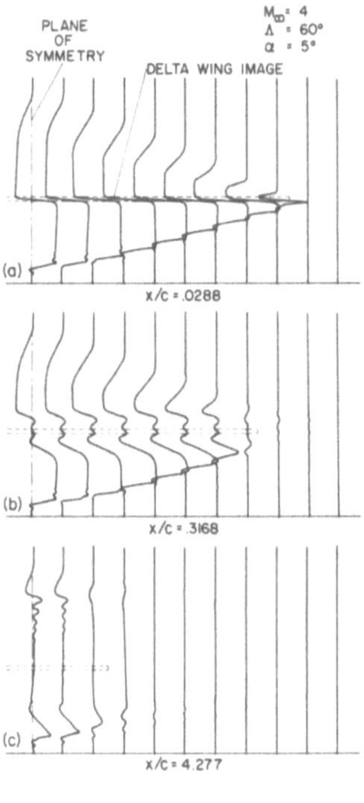

Fig. 5.- Vertical pressure profiles in the wake of a planar delta wing.

FINITE DIFFERENCE SCHEME FOR CALCULATING PROBLEMS IN THREE SPACE DIMENSIONS AND TIME*

M. L. Wilkins, S. J. French, M. Sorem

Lawrence Radiation Laboratory, University of California
Livermore, California 94550

INTRODUCTION

In previously published papers (Wilkins, 1969; Wilkins, 1970) details are given for a second-order difference scheme that can be used to solve the partial differential equations of continuum mechanics formulated in two space dimensions and time. The scheme is an extension into two dimensions of the Von Neumann and Richtmyer artificial viscosity idea originally used to solve problems in one-dimensional hydrodynamics (Von Neumann and Richtmyer, 1950). In this paper the two-dimensional scheme is in turn extended into a three-dimensional difference scheme. A computer program using the difference scheme has been operating for several years at the Lawrence Radiation Laboratory on the CDC 6600 high-speed computer. Recently it has been programmed for the CDC 7600 high-speed computer. The three-dimensional program or 3D code follows the same form as the 2D HEMP code (Wilkins, 1969; Wilkins, 1970) so that problems in elasticity, elastic-plastic flow, and gas dynamics can be solved when the initial conditions and boundary conditions are given.

In practice the large number of Lagrange zones and hence machine time required for 3D calculations, as compared to similar 2D calculations, limit the class of problems that are reasonable to solve on even the largest present-day computer. However, problems of current interest that do not require a large number of zones include studies of elastic-plastic material behavior for materials subjected to combined stresses and problems involving anisotropic yield surfaces.

3D DIFFERENCE OPERATOR

The 3D difference operator is the analogue of the 2D difference operator described for the HEMP code (Wilkins, 1969; Wilkins, 1970). The space derivatives are defined as the summation of the normal component of a flux around an enclosed volume. Thus, conservation form is implicit in the difference equations. A two-time step and two-grid systems are used, analogous to the original Von Neumann-Richtmyer scheme. One grid is used to calculate gradients of pressure stresses and the other to calculate the divergence of the velocity vector. The pressure and stress gradients are centered at grid nodes at integral time intervals. The divergence of the velocity is centered at midpoints of the three-dimensional grid at one-half time intervals.

Figure 1 shows the Lagrange grid used to define a mass zone. The following integral definitions (Reddick and Miller, 1955) of partial derivatives are used to develop the difference operators:

$$\frac{\partial P}{\partial X} = \lim_{V \to 0} \frac{\iint_S P(\hat{N} \cdot \hat{i})\, ds}{V} \qquad \text{(Component of } \nabla P\text{)} \qquad (1)$$

P = pressure or stress

similar for

$$\frac{\partial P}{\partial Y} \text{ and } \frac{\partial P}{\partial Z}$$

*This work was performed under the auspices of the U. S. Atomic Energy Commission.

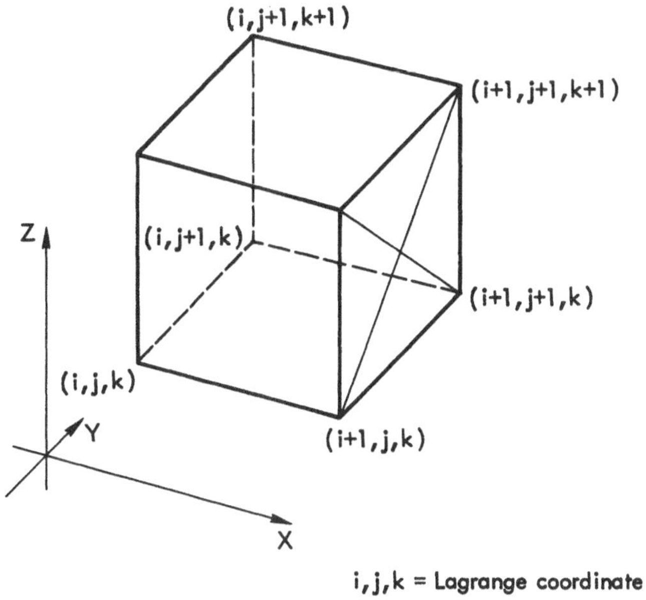

Fig. 1. Grid formed from Lagrange coordinates used to define a mass zone. The grid is used to calculate $\nabla \cdot \vec{W}$, where $\vec{W}$ is the velocity vector and $\nabla \cdot \vec{W}$ is considered to be at the center of the mass zone $(i + 1/2,\ j + 1/2,\ k + 1/2)$.

$$\frac{\partial \dot{X}}{\partial X} = \lim_{V \to 0} \frac{\iint_S \dot{X}(\hat{N} \cdot \hat{i})\, ds}{V} \qquad \text{(Component of } \nabla \cdot \vec{W}\text{)} \tag{2}$$

$\vec{W}$ = velocity vector

similar for

$\frac{\partial \dot{Y}}{\partial Y}$ and $\frac{\partial \dot{Z}}{\partial Z}$ $\qquad \vec{W} = \dot{X}\hat{i} + \dot{Y}\hat{j} + \dot{Z}\hat{k}$

Here $\hat{N}$ is the outward normal to the surface element ds where the surface s encloses a volume V.

Pressure and stress are defined at the center of a mass zone, and space derivatives of these quantities are defined at zone nodes. Velocity is defined at zone nodes, and the space derivatives of velocities are defined at zone centers.

Figure 2 shows the eight sided figure that represents the surface s of Eq. (1). The surface integral of Eq. (1) is approximated by summing the eight products formed by multiplying the pressure or stress variables by the projection of the corresponding triangular area onto the coordinate plane that is normal to the space coordinate of the derivative considered. The procedure maps two-thirds of the volume of a given zone twice. It is necessary that the entire volume be mapped. This is achieved by scaling so that all of the volume of a given zone is included in

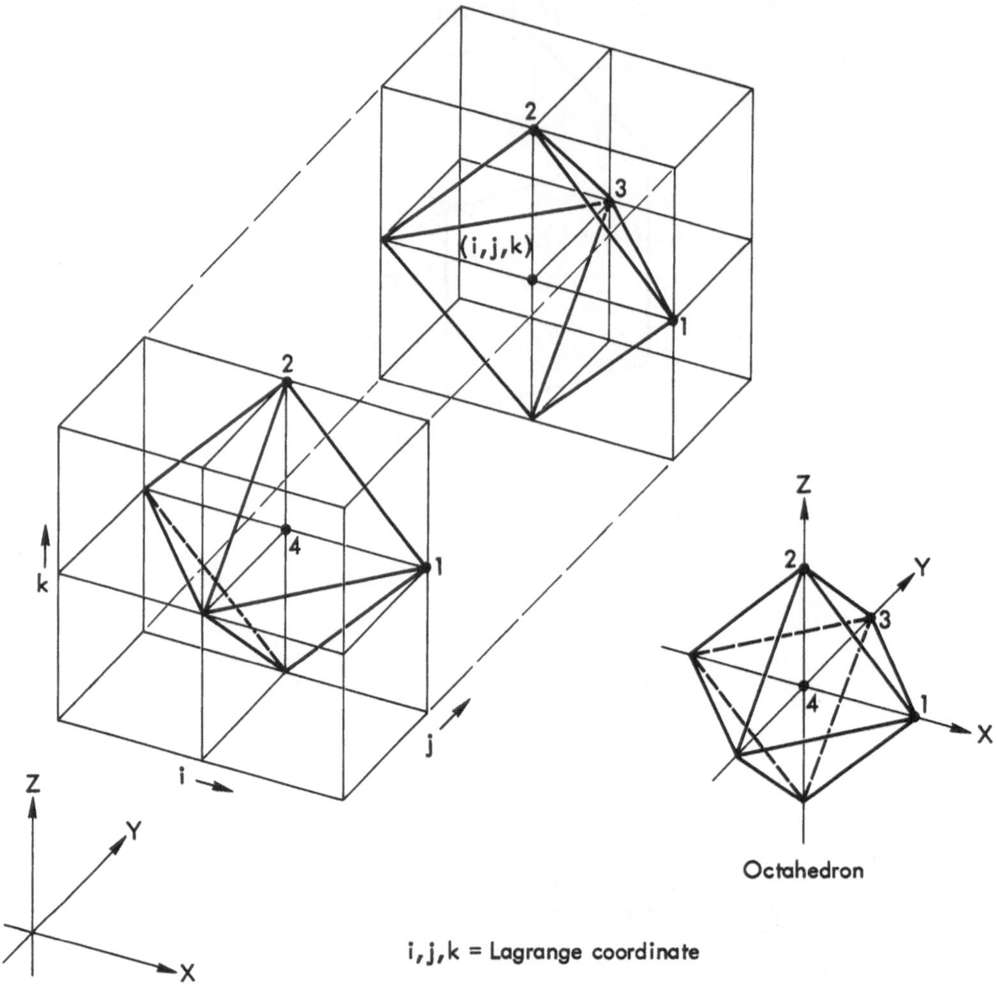

Fig. 2. Octahedron grid formed from the Lagrange coordinates. The grid is used to calculate gradients of stresses. The stresses are centered at zone mid-points, i + 1/2, j + 1/2, k + 1/2, and the gradient is centered at zone nodes, i, j, k (point 4).

the surface integral. The derivatives of pressure and stress in the equations of motion appear with the density ρ.

$$\frac{1}{\rho}\frac{\partial P}{\partial X}, \text{ etc.}$$

Taking this into account we have:

$$\frac{1}{\rho}\frac{\partial P}{\partial X} = \frac{\sum_{i}^{8} (P \times A)_i}{2\overline{M}},$$

where A is the area of triangles projected on the YZ plane (Fig. 2), and $\overline{M}$ is the average of the mass of the eight zones surrounding point 4 in Fig. 2.

The components of $\nabla \cdot \vec{W}$ are approximated using the grid shown in Fig. 1. The surface integral of Eq. (2) is obtained by summing the product of the velocity times the projection of the corresponding surface onto the coordinate axis normal to the direction of the derivative. The surfaces are projected as triangles, and the velocity is the average of the three velocities at each corner. This procedure maps the entire zone and the volume in Eq. (2) is the actual zone volume.

In the limit of two-dimensional motion the above difference scheme reduces to the difference scheme in the 2D HEMP program.

The artificial viscosity q used to calculate shocks and to stabilize the grid has the usual form:

$$q = C_0^2 \rho L^2 \left(\frac{\dot{V}}{V}\right)^2$$

where

C_0^2 = constant

ρ = zone density

V = zone volume

L = characteristic zone length = $\sqrt[3]{V}$

The characteristic zone L used here does not in general reduce to the characteristic length used in the 2D HEMP Code. For a one-to-one comparison with a 2D code, the length L can be made a constant for a given problem.

For quasi-static problems in applied mechanics a linear q is required to damp nonphysical spurious oscillations. The angle q described in Wilkins (1970), generalized to three dimensions, has proved to be very effective.

As a check of the accuracy of the code, a calculation was done of a rod vibrating in three dimensions so as to activate all six components of the stress tensor. The calculation reproduced the fundamental frequency and verified that the numerical method did not introduce nonphysical damping into the result. A calculation of the impact of an elastic-plastic cylinder on a rigid boundary was done using three orthogonal coordinates. The results were identical to a similar calculation using a two-dimensional axial symmetric computer program.

REFERENCES

Reddick, H. W., and Miller, F. H., Advanced Mathematics for Engineers (John Wiley, New York, 1955), 3rd ed.

Von Neumann, J. and Richtmyer, R. D., J. Appl. Phys. 21, 232 (1950).

Wilkins, M. L., Calculation of Elastic-Plastic Flow, Lawrence Radiation Laboratory, Livermore, Rept. UCRL-7322, Rev. I (1969).

Wilkins, M. L., "Finite Difference Scheme for Calculating Problems in Two Space Dimensions and Time," J. Comput. Phys. (May 1970).

FOUNDATION OF APPROXIMATE SOLUTIONS

Akira Sakurai

Waterways Experiment Station, Vicksburg, Mississippi

INTRODUCTION

The large majority of the problems in applied mathematics are concerned with solution of the equations representing the various basic laws of sciences. Because of the general difficulty in o-taining an exact analytical solution of these equations, it is common practice to utilize approximate solutions derived from varieties of approximation methods or techniques. These are especially the case in mathematical studies in fluid dynamics, whose nonlinear nature inevitably requires the use of various approximation methods. These methods are usually considered less rigorous from the viewpoint of ordinary mathematics as demonstrated by the question, "How is it possible to find the solution approximately without even knowing its existence?" The fact is, however, that these methods have been developed regardless of the criticism and have been used in spite of their seeming imperfections to furnish results that are satisfactory for practical purposes.

The object of this paper is to investigate the situation for a better understanding of the true meaning of these approximate solutions and to utilize it for improvements of an approximation method.

APPROPRIATE SOLUTION

Underlying principles for various approximation methods are thus explored in relation with the reality of the problem without regard for its mathematical idealization.

The most important starting point for this task is probably to realize the fact that what is required of the solution u of a given equation $L(u) = 0$ in reality is not necessarily the solution satisfying the equation exactly, but is the one satisfying $L(u) =$ "small", since the equation is, of course, usually only an idealized approximation to the real nature, which can be written down more adequately by the condition above, where the various negligibly small effects are represented as "small" (Sakurai).

Then, naturally, the next question arises: How small should the "small" term be? The answer to this can be seen in the general reason for discarding a small amount in practice, on the ground that its contribution to the final result is insignificant. Adapting this to the present problem, the smallness of the above "small" term can be determined in the sense that the term causes no significant difference to the resulting solution. The concept of this smallness is, in fact, related to the actual procedure in most of the approximation methods. It is also related to the stability of the equation, which is indispensable to the system as a model of a real problem (Courant, Isaacson-Keller).

Now, an ordering is needed for solutions to determine whether they differ significantly from each other or not. Semiordering appears to be suited for the purpose. It is used primarily in the field of mathematical psychology to cope with such judgments of indifference concerning some attribute of stimuli (Suppes-Zinnes), and is defined by a binary system $\langle A,P\rangle$ in which the following three axioms are satisfied for all a,b,c,d in A ; i) Not aPa : ii) if aPb and cPd then either aPd or cPb: iii) if aPb and bPc then either aPd or dPc. Then, Indifferent relation, aIb is defined by not aPb and not bPa.

Furthermore, a solution u obtained in the usual practice of an approximation method is regarded as one from a certain set D (Approximants, Davis), such as a class of functions, which are quite often the solutions of an auxiliary equation.

It is noted that the semiordering is introduced easily to the set D which appears in the usual practice of approximation methods. It is also noted that the procedure is similar, in some aspects, to that for the Pattern Recognition (Nagy, Levine).

These aspects above are now combined to provide a new concept of the solution (Appropriate solution), which is more suitable to represent reality than by the solution in the classical sense; a u_0 of $u_0 \in D$ is regarded as a solution of $L(u) = 0$ even if $L(u_0) \neq 0$, if there is a u_1 of $u_1 \in D$, $u_0 \ I \ u_1$, and $L(u_0) \ P \ L(u_1)$, $L(u_1) \ I \ 0$, expressed symbolically as $u_0 \sim u_1$ and $L(u_0) >> L(u_1)$. In other words, it says that $L(u_0)$ is small since the ~~difference between~~ $L(u_0)$ and 0 is practically the same as the difference between $L(u_0)$ and $L(u_1)$, and this difference causes no significant difference between u_0 and u_1.

Using this concept of the solution, various conventional means of judging the valicity of an approximate solution can be unified under a simple principal and the validity of a solution can be judged by a definite criterion, in contrast to the fact that the validity under the classical solution concept often remained obscure.

APPROXIMATION METHODS AND APPROPRIATE SOLUTION

There are varieties of approximation methods being used for the problems of applied mathematics. They include such means as perturbation methods, iterations, Rayleigh-Ritz and Galerkin methods, and numerical methods. The real nature of the approximate solutions as given by these approximation methods is better understood as that of the solution defined above.

First, it is readily seen that the typical procedure in the perturbation and iteration methods is actually constructing the above defined solution systematically, in fact, consider a typical case of the equation including a small parameter ε as $L(u;\varepsilon) = 0$ and its solution obtained as $u = u^{(o)} + \varepsilon u^{(1)} + ---$, and put $u_0 = u^{(o)}$, and $u_1 = u^{(o)} + \varepsilon u^{(1)}$, then $u_0 - u_1 = O(\varepsilon)$ while $L(u_0) = O(\varepsilon)$ and $L(u_1) = O(\varepsilon^2)$ so that $u_0 \sim u_1$ and $L(u_0) >> L(u_1)$ for small ε values.

More generally, the following concept of Indifferent equation is useful to the type of equations as given above. Given two equations, $L(u) = 0$ and $L_1(u) = 0$, then the equation $L_1(u) = 0$ is said to be indifferent from $L(u) = 0$ at u_0 if $L(u) - L_1(u) << L(u_0)$ for a u of $u \in Su_o$, where Su_o is a significant interval around u_0, defined as $Sa = (x \mid xPc$ and bPx or xIb or xIc for some b,c satisfying bPa and aPc).

Furthermore, if $L_1(u) = 0$ has a solution u_1 and $u_0 \sim u_1$, it is obvious that u_0 is an appropriate solution of $L(u) = 0$.

The equation $L(u; \ ) = 0$ above is often in practice related with $L(u;0) = 0$ in the manner as described above. Another example, the finite difference approximation $y_{i+1} - y_{i-1} = 2h \ f(x_i, y_i)$ to the ordinary differential equation $dy/dx = f(x,y)$ is obtained from the approximation $dy/dx = f(x_i, y_i)$, $x_{i-1} \leq x \leq x_{i+1}$, $i = 1,2,---,n$, which is indifferent from the original equation for a smooth f if h is small enough.

Auxiliary equations to determine the approximate solutions in some methods such as Rayleigh-Ritz, Galerkin and the finite difference approximation to a partial difference equation are generally not so closely related to their original equations as is the indifferent equation above. Nevertheless, the above concept of the solution makes it possible to judge the validity of these solutions without any knowledge about their exact solutions. This is also true for solutions obtained by any ad-hoc approximations. This is illustrated in the following simple examples.

Take the solution of $L(u) \equiv d^2u/dx^2 + (1+x^2)u + 1 = 0$, $u(\pm 1) = 0$ determined from

$$\int_{-1}^{1} (du/dx)^2 - (1 + x^2)u^2 - 2u)\, dx = \min.$$

by assuming

$$u = u_n = \sum_{i=1}^{n} a_i (1 - x^{2i}) \quad .$$

The validity of the solution u_1 is determined from u_2 by examining $u_1 \sim u_2$ and $L(u_1) \gg L(u_2)$. It turns out that $u_1 = 0.921\ (1-x^2)$, $u_2 = 0.934 - 0.988\ x^2 + 0.054\ x^4$, $L(u_1) = 0.08 - 0.921\ x^4$, $L(u_2 = -0.04 + 0.59\ x^2 - 0.934\ x^4 + 0.05\ x^6$ so that u_1, u_2 are close enough but $L(u_2)$ is not close enough to 0 compared with $L(u_1)$. The next approximation u_3 gives $(L(u_3))_{x=0} = 0.0008$, which is satisfactory.

Consider a solution of the finite difference approximation to $\Delta u = 0$:

$$4u_{i,j} = u_{i+1,j} + u_{i-1,j} + u_{i,j+1} + u_{i,j-1}$$

and assume, in consistence with the origin of the approximation, that

$$u = u(h) \equiv \frac{1}{2} h^{-2} (u_{i+1,j} + u_{i-1,j} - 2u_{i,j})(x^2 - y^2)$$

$$+ \frac{1}{2} h^{-1} (u_{i+1,j} - u_{i-1,j})x + \frac{1}{2} h^{-1} (u_{i,j+1} - u_{i,j-1})y + u_{i,j}$$

for each of the cells at (i,j). Then, the validity of the solution $u(h)$ can be judged from another solution $u(h')$ of different mesh size h'. This case, $u(h)$ satisfies $\Delta u = 0$ exactly (Boundary Method, Collatz), and the only error arises as $\delta u(h)$ from a boundary condition and the incongruity between the overlapping solutions. Take, for example, a simple case of a unit square boundary on which $u = 1$ at its one side otherwise $u = 0$, and put $h = 1/2$ and $h' = 1/3$. It is readily found that $\text{Max.}|u(1/2) - u(1/3)|$ (at mesh points) = 0.19 and the average errors at the boundary are $|\delta u(1/2)| = 0.17$, $|\delta u(1/3)| = 0.13$. They are, of course, hardly acceptable as $u(h) \sim u(h')$ and $\delta u(h) \gg \delta u(h')$.

Sometimes a direct use of the definition above can lead to a required solution more directly and more efficiently than by use of the conventional method. For example, consider a periodic solution of $L(u) \equiv d^2u/dt^2 + u + \varepsilon u^3 = 0$ for small ε. Put $u_0 = a \cos t$, then $L(u_0) = \varepsilon a^3 \cos^3 t = O(\varepsilon)$, now assume $u_1 = u_0 + \varepsilon b \cos 3\omega t$ and determine b, ω to make $L(u_1) = O(\varepsilon^2)$ so that $L(u_1)$ is much closer to 0. This results directly in the non-secular solution.

REFERENCES

Collatz, L. The Numerical Treatment of Differential Equations. Springer-Verlag, 28 (1960).

Courant, R. "Hyperbolic Partial Differential Equations and Applications", in Modern Mathematics for Engineers, McGraw-Hill, 92 (1956).

Davis, P. J. Interpolation and Approximation, Blaisdell Publishing Co. 4 (1965).

Isaacson, E. and Keller, H. B. Analysis of Numerical Methods, John Wiley & Sons, 22 (1966).

Levine, M. D. "Feature Extraction: A Survey," Proc. of The IEEE 57, 1392 (1969).

Nagy, G. "State of the Art in Pattern Recognition," Proc. of The IEEE 56, 837 (1968).

Sakurai, A. "Appropriate Solution and its Application to Problems in Fluid Dynamics," Fluid Dynamics Transactions, Warsaw, 4, 127-133 (1969).

Suppes, P. and Zinnes, J. L. "Basic Measurement Theory," in Handbook of Mathematical Psychology, Wiley and Co. Vol. 1, 30 (1963).

FLAG: A FREE-LAGRANGE METHOD FOR NUMERICALLY SIMULATING HYDRODYNAMIC FLOWS IN TWO DIMENSIONS*

W. P. Crowley

Lawrence Radiation Laboratory, University of California
Livermore, California

INTRODUCTION

Traditionally, time-dependent hydrodynamic motions are numerically simulated in two ways. The Lagrangian formulation is used for relatively smooth flows, while the Eulerian description must be used for flows involving violent motions. We consider here a quasi-Lagrangian approach which eliminates the mesh constraints traditionally associated with the Lagrangian method.

Mesh scrambling in traditional Lagrangian calculations happens because mesh connections are permanent entities. If logically connected fluid particles move substantial distances apart during a calculation, the resulting mesh distortions reduce the integration time step and decrease the accuracy of the solution.

The basic advantage in the Free-Lagrange (FLAG) method is that mesh points are not tied together for the duration of the calculation. As each cycle begins the mesh is optimized by an algorithm which links "nearest" neighbors together. Thus points are free to drift apart without the usual attendant mesh scrambling.

The FLAG code consists of three major parts: (1) Mesh-maker, (2) mesh optimization, (3) equations of motion. Each part will be described separately.

The FLAG concept applies to both compressible and incompressible fluid motions. We consider an incompressible, inviscid form of the equations of motion with streamfunction and vorticity as dependent variables. Incompressible, inviscid calculations give qualitative verification of a vorticity cascade toward higher wavenumbers and an energy cascade into lower wavenumbers currently postulated for two-dimensional hydrodynamic motions.

MESH-MAKER

The problem faced by the mesh-maker is interconnecting a random two-dimensional distribution of points to form a single-valued mesh. Single-valued means: if point i is connected to point j, then j is connected to i. A simple way to do this is to construct a mesh with triangular mesh elements.

The mesh-maker proceeds in two stages: (1) the points are connected to form concentric convex rings (this essentially orders the points); (2) the points on adjacent rings are interconnected to form triangular mesh elements. After step (2), each point is connected to several others and all non-boundary points are surrounded by triangles.

Convex rings are constructed as follows: let S be the set of all points, and let $\hat{X}$: $(\hat{x},\hat{y})$ be the point of minimum y, i.e., for all $y_i \in S$, $y_i \geq \hat{y}$. Take $X_1^1 = \hat{X}$ as the first member of the first (outer) ring R_1. (The nth point on the ith ring is denoted X_i^n.) To find the second point on R_1, construct the horizontal vector V^{01} originating at X_1^1 and terminating at $X_1^1 + (1,0)$. Next consider all other vectors V^{i1} originating at X_1^1 and terminating at $X_i \in S$. Choose for the second member of R_1 (viz. X_1^2) the one minimizing the angle between V^{01} and V^{i1}. Remove point 2 from S and place it in R_1; define the vector $V^{21} = X_1^2 - X_1^1$. Choose the third member of R_1 by

*Work performed under the auspices of the U.S. Atomic Energy Commission.

minimizing the angle between V^{21} and the vectors originating at X_1^2 and terminating at all remaining $X_i \in S$. Continue until the point chosen is $\hat{X}$ then remove X_1^1 from S. The outer ring is now closed. The points along this first ring are the boundary points for the region. The members of the second ring R_2 are chosen similarly from the set $S - R_1$; those of the third ring are chosen from $S - R_1 - R_2$ and so on. Finally all points originally in S belong to some ring R_m; they are ordered since we know both which ring they are on and the spatial relation of one ring to another. Further, the ordering is counterclockwise on each ring, and each point has two neighbors.

The second stage of the mesh-maker interconnects the rings. We specifically consider connecting the two outer rings; the others are connected similarly. To start, connect X_1^1 to X_1^2; this gives three logical sides of a quadrilateral (Fig. 1). To form a triangle choose one of the diagonals $X_2^1 X_1^2$ or $X_1^1 X_2^2$. If both diagonals are interior to the quadrilateral, choose the shorter; otherwise choose the interior one.

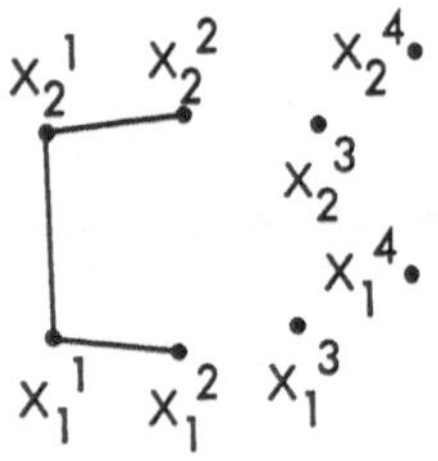

Figure 1

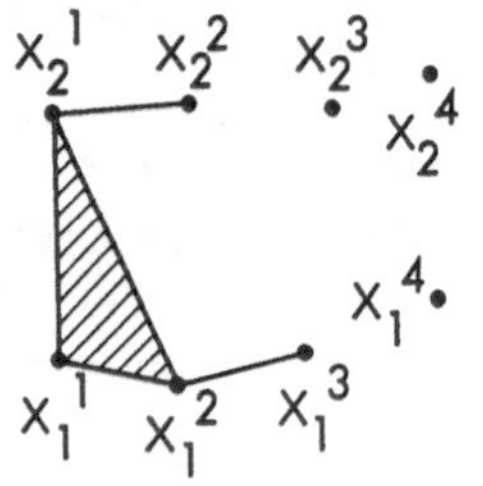

Figure 2

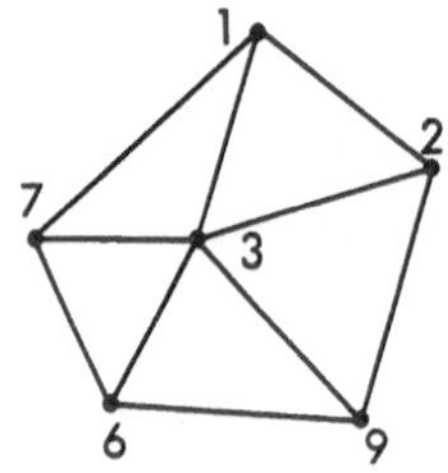

Figure 3

Suppose it is $X_1^2 X_2^1$. Move along R_1 and R_2 counterclockwise (ignoring X_1^1 for the moment) and consider the quadrilateral with vertices X_1^3, X_1^2, X_2^1, X_2^2 (Fig. 2). Again choose the shortest interior diagonal, and calculate through R_1 and R_2. Finally the end points are connected; the region between R_1 and R_2 is covered by a mesh of triangles. Use the same procedure to connect the members of R_2 and R_3 starting with $X_2^1 X_3^1$. Finally the region enclosed by R_1 is covered by triangles and the mesh is complete.

Each point now has a set of points as neighbors and the only remaining job is logically ordering the points in a manner consistent with their physical ordering.

At this point we abandon all information about the ring structure and go from global to local knowledge, retaining at each point only information about the neighbors of that point. In Fig. 3 for example, we see that point 3 has as neighbors only points 1, 2, 6, 7, and 9. The "order" routine orders these points counterclockwise. It is used both in the final stage of the mesh-maker and in mesh-optimization (described below).

The order routine proceeds as follows. Let a subscript i label a point and a subscript ℓ label its neighbors in some given order. The components of the vector W_ℓ are given by W_ℓ: $(x_\ell - x_i, y_\ell - y_i)$ and we associate an angle θ_ℓ with each vector $\cos\theta_\ell = W_\ell \cdot W_1 / |W_\ell||W_1|$, $\ell > 1$, namely the angle between W_ℓ and W_1. To eliminate using a square root, we define two functions

$$F(\theta_\ell) = (\cos\theta_\ell)^2 \left(\frac{\cos\theta_\ell}{|\cos\theta_\ell|}\right) = \frac{(W_\ell \cdot W_1)^2}{W_\ell^2 W_1^2} \left(\frac{W_\ell \cdot W_1}{|W_\ell \cdot W_1|}\right)$$

and

$$G(\theta_\ell) = \begin{cases} F(\theta_\ell) & \sin\theta_\ell \geq 0 \\ -2-F(\theta_\ell) & \sin\theta_\ell < 0 \end{cases}$$

The function $G(\theta_\ell)$ monotonically decreases as θ goes from 0 to 2π. Ordering the neighbors of point i with respect to $G(\theta_\ell)$ puts then in the desired counterclockwise arrangement.

MESH OPTIMIZATION

There are many ways to interconnect a set of points to form a single-valued triangular mesh. The mesh-maker calculation gives simply one such construction, not necessarily the most desirable one.

The mesh optimization in FLAG is not based on global decisions, though the iterative nature of the calculation gives it global status. The local decisions result from viewing each mesh connection as the diagonal of a quadrilateral. Mesh optimization then consists of selecting one of two diagonals which is the more desirable by certain criteria. In Fig. 4, we define B to be a normal quadrilateral (it has two interior diagonals); A is not-normal (it has only one interior diagonal, viz. 1-3). For normal quadrilaterals the diagonal is chosen that satisfies some criterion, two of which are described below. For non-normal quadrilaterals no further action is taken.

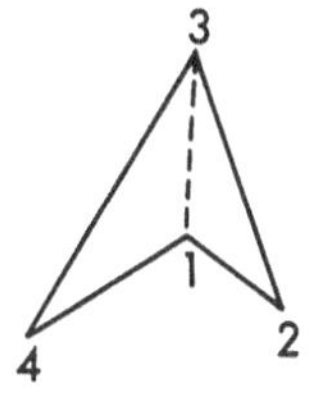

Figure 4A

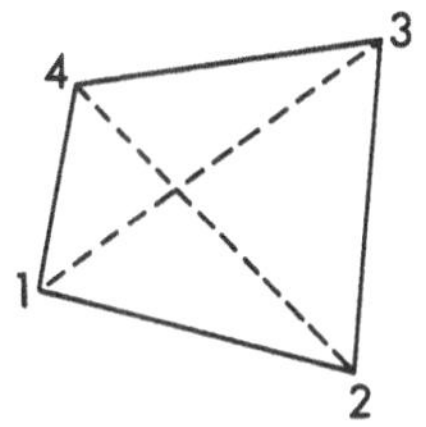

Figure 4B

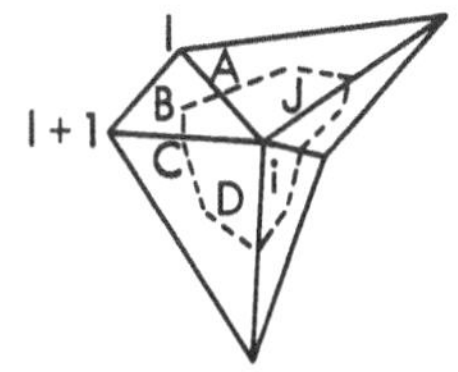

Figure 5

A mesh is a foundation for derivative approximations. The most accurate approximations are obtained on meshes in which a point's neighbors are taken to be those points closest to it. This suggests the "nearest-neighbor" algorithm in which (for a normal quadrilateral) the shorter diagonal is always chosen. (R. N. Byrne pointed out that the iteration based on this algorithm would probably converge since the algorithm would correspond to the minimization of potential energy in the case of a mesh of springs.)

The "smooth" mesh concept provides another algorithm. In this case, the diagonal chosen divides the largest interior angle of each normal quadrilateral.

Iterations based on both algorithms converge in a few cycles. The second one involves more arithmetic operations but usually converges faster. We haven't enough experience to tell which one gives better results.

In FLAG, the mesh optimizer is used both after the mesh-maker has completed its calculation and after each time-step in the hydrodynamic calculation as well. Thus the mesh optimizer constructs the new mesh after the points have moved, putting more constraints on the optimizer and complicating somewhat the rather idyllic algorithms mentioned above. For example, since a normal quadrilateral may turn into a "boomerang" (Fig. 4A) in a time-step, selection of the proper diagonal must be assured. This possibility also indicates the need for a control on the integration time-step.

EQUATIONS OF MOTION AND FINITE DIFFERENCE EQUATIONS

In two dimensions the incompressible Lagrangian form of the equations of hydrodynamics is

$$\frac{d\zeta}{dt} = \nabla \cdot \nu \nabla \zeta \tag{1}$$

$$\nabla^2 \psi = \zeta \tag{2}$$

$$\frac{d}{dt} \underset{\sim}{X} = \underset{\sim}{U} = \underset{\sim}{k} \wedge \nabla \psi \tag{3}$$

where $\zeta = \underset{\sim}{k} \cdot (\nabla \wedge \underset{\sim}{U})$ is the vorticity, ψ is the streamfunction, $\underset{\sim}{k}$ is a unit vector out of the plane of the motion and ν is the kinematic viscosity.

The construction of finite difference equations is based largely on Green's theorem for converting divergence expressions into line integrals (see Noh[4] and Winslow[5]) and for evaluating derivatives. If R is some plane region enclosed by a curve C, then the mean value (somewhere in R) of the divergence of a vector $\underset{\sim}{G}$ is given by

$$\nabla \cdot \underset{\sim}{G} = \frac{\oint_C \underset{\sim}{G} \cdot \underset{\sim}{N} d\sigma}{\oint_C x dy} \tag{4}$$

where N is the unit outward normal to C and $d\sigma$ is a line element along C. From Eq. (4) we obtain mean values of the partial derivatives of a smooth function f on R:

$$\frac{\partial f}{\partial x} = \frac{\oint f dy}{\oint x dy} , \qquad \frac{\partial f}{\partial y} = - \frac{\oint f dx}{\oint x dy} \tag{5}$$

We use Eqs. (5) to obtain finite difference approximations to Eqs. (2) and (3). These calculations are being carried out for inviscid flows, so Eq. (1) is trivial. However, for viscous flows, its difference approximation will be similar to that for Eq. (2), and the inclusion of a nonlinear eddy viscosity (Leith's[3] for example is based on $\nabla \zeta$) would be straightforward.

For the incompressible case, the vorticity and streamfunction are initially specified at each mesh point, and each point lies within a rectilinear vortex. With time the points move according to Eq. (3), and the streamfunction changes at each mesh point due to mesh motion.

We have a primary mesh of triangles with vertices at mesh points, and to invert Eq. (2) we construct a secondary mesh based on the intersection of vertex bisectors (Fig. 5). If the values of ψ are given at mesh points, and if we use linear interpolation, then $\nabla\psi$ is constant over a triangle and we have for triangle $(i, \ell, \ell+1)$ $\int_{ABC} \nabla\psi \cdot \underset{\sim}{N} d\sigma = \frac{\partial \psi}{\partial x}(y_C - y_A) - \frac{\partial \psi}{\partial y}(x_C - x_A)$ where the integration path lies along $\overline{AB}$ and $\overline{BC}$. When the derivatives are approximated by Eqs. (5) and the line integrals are taken around the appropriate triangles, we have $\int_{ABC} \nabla\psi \cdot \underset{\sim}{N} d\sigma = \left[-\psi_i (W_{\ell+1} - W_\ell) \cdot (W_{\ell+1} - W_\ell) + \psi_\ell W_{\ell+1} \cdot (W_{\ell+1} - W_\ell) - \psi_{\ell+1} W_\ell \cdot (W_{\ell+1} - W_\ell) \right] / 4A_{\ell+1/2}$ where $2A_{\ell+1/2} = k \cdot (W_\ell \wedge W_{\ell+1})$. Using this expression to evaluate the flux crossing the surface ABC ... J and applying Eq. (4) to Eq. (2) we have $\oint \nabla\psi \cdot \underset{\sim}{N} d\sigma = -\psi_i \sum_\ell \beta_{i\ell} + \sum_\ell \psi_\ell \beta_{i\ell} = \int_R \zeta dxdy$ which becomes

$$\psi_i = \left[\sum_\ell \psi_\ell \beta_{i\ell} - \frac{1}{3} \zeta_i \sum_\ell A_{\ell+1/2}\right] \Big/ \sum_\ell \beta_{i\ell} \tag{6}$$

a set of simultaneous equations for ψ_i at all non-boundary mesh points. The ℓ sums are taken only over the neighbors of point i. Coupling coefficients $\beta_{i\ell}$ for the points i and ℓ are given by $4\beta_{i\ell} = W_{\ell+1} \cdot (W_{\ell+1} - W_\ell) / A_{\ell+1/2} - W_\ell \cdot (W_\ell - W_{\ell-1}) / A_{\ell-1/2}$ and it can be shown that $\sum_\ell \beta_{i\ell} > 0$. The simultaneous Eqs. (6) are solved by successive over-relaxation; this part of the calculation consumes about 90% of the computer time used per problem.

Velocities are obtained from the streamfunction with Eqs.(5); this locates them within the triangles, and we must interpolate to obtain velocities at mesh points. For a point with L neighbors, $V_i = \sum_\ell V_{\ell+1/2} S_{\ell+1/2}$ where $S_{\ell+1/2} = \frac{2}{L} - \left[h_{\ell+1/2}\right] / \sum_\ell h_{\ell+1/2}$ and $V_{\ell+1/2}$ is the velocity component at the triangle center. The weight $h_{\ell+1/2}$ is the distance from the point i to the center of $(i, \ell, \ell+1)$. In one dimension this corresponds to quadratic interpolation on ψ.

Having velocities at the mesh points lets us advance the calculation in time. We found that both centered and forward time differences gave undesirable results. An iterative scheme is now used, and it seems satisfactory, if expensive. Thus an inner iteration inverts Eq. (2) and an outer iteration advances the mesh in time. If μ is the iteration counter for the outer iteration, the equations are solved in the following order in going from times $(n-1)\Delta t$ to $n\Delta t$:

$${}^\mu \underset{\sim}{X}{}_i^n = \underset{\sim}{X}{}_i^{n-1} + \left({}^{\mu-1}\underset{\sim}{V}{}_i^n + \underset{\sim}{V}{}_i^{n-1}\right) \frac{\Delta t}{2} \tag{7}$$

$${}^\mu\psi_i^n = \left\{\sum_\ell \left({}^\mu\psi_\ell^n\right)\left({}^\mu\beta_{i\ell}^n\right) - \frac{1}{3} \zeta_i \sum_\ell {}^\mu A_{\ell+1/2}^n\right\} \Big/ \sum_\ell {}^\mu\beta_{i\ell}^n \tag{8}$$

$${}^\mu \underset{\sim}{V}{}^n = f\left({}^\mu\psi^n\right) \tag{9}$$

where ${}^0\underset{\sim}{V}{}^n = \underset{\sim}{V}{}^{n-1}$. Experience shows that one outer iteration suffices for most purposes.

NUMERICAL RESULTS

Two simple test problems are described in this section. Both involve inviscid, incompressible flow in a square (13 units on a side) with boundary conditions $\psi = \zeta = 0$ along the edges. The problems differ in their initial conditions.

For problem one, the initial condition is $\psi = \sin 2x' \sin y'$ where $x' = \pi x/13$, $y' = \pi y/13$. Since $J(\psi, \zeta) = 0$, the solution is $\frac{\partial\psi}{\partial t} = \frac{\partial\zeta}{\partial t} = 0$; i.e., the solution is stationary.

In Fig. 6A and B, we have the initial streamline pattern and the initial mesh configuration. In Fig. 7A, B and C, we have the streamlines, level lines of vorticity and the mesh at t = 46.7, corresponding to 150 calculation cycles. The small scales of motion that have shown up in the vorticity map indicate the need for an artificial viscosity, a not uncommon experience in non-linear hydrodynamic calculations.

In the second problem, the initial condition is $\psi = \sin 2x' (\sin y' + \sin 3y')$. According to Fjørtoft[1] if three or more non-zero modes are present, they interact to transfer energy primarily toward larger scales. Thus we may expect a non-stationary solution.

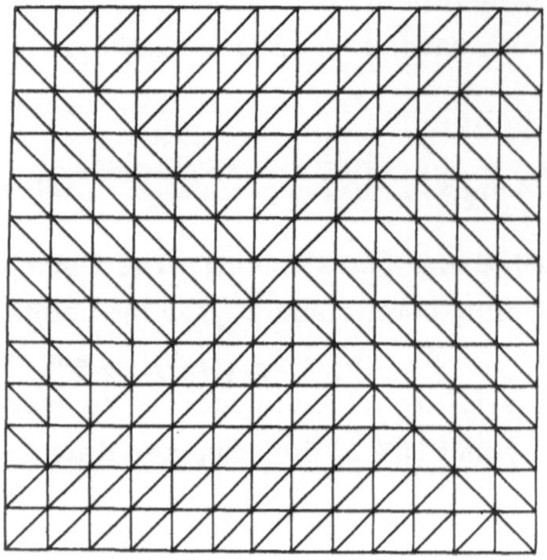

Figure 6A

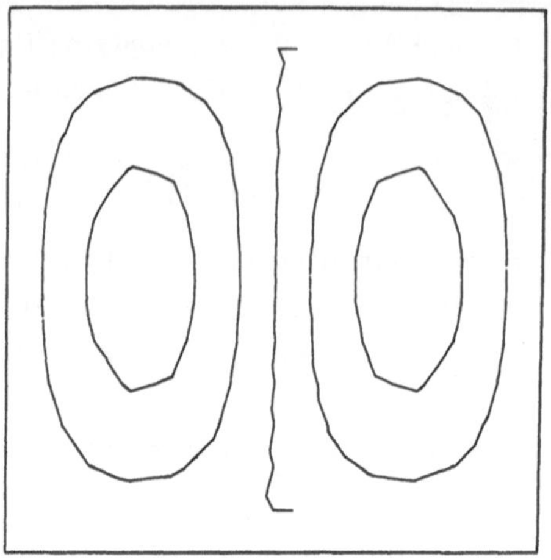

Figure 6B

Figure 7A

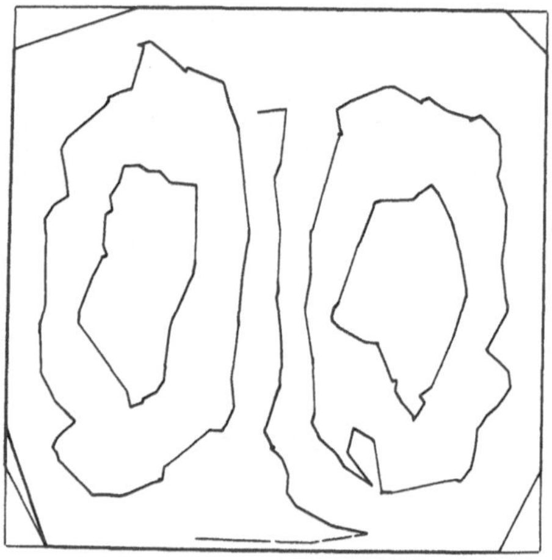

Figure 7B

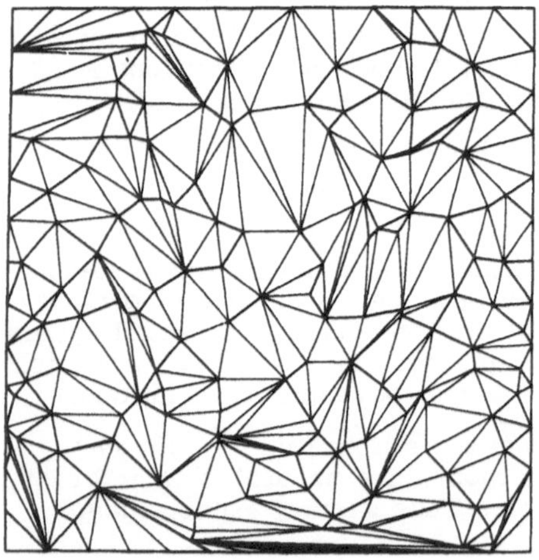

Figure 7C

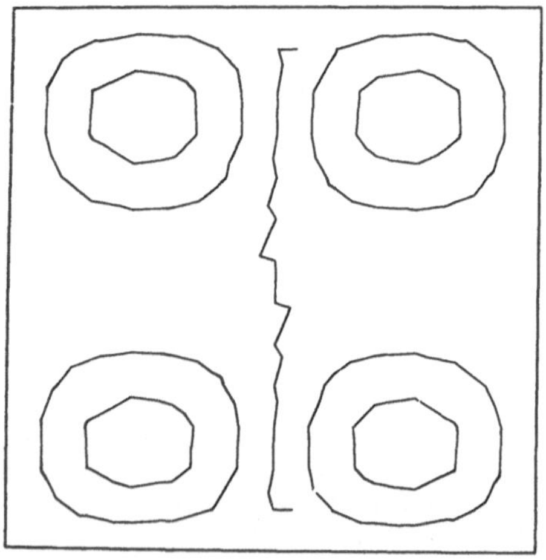

Figure 8

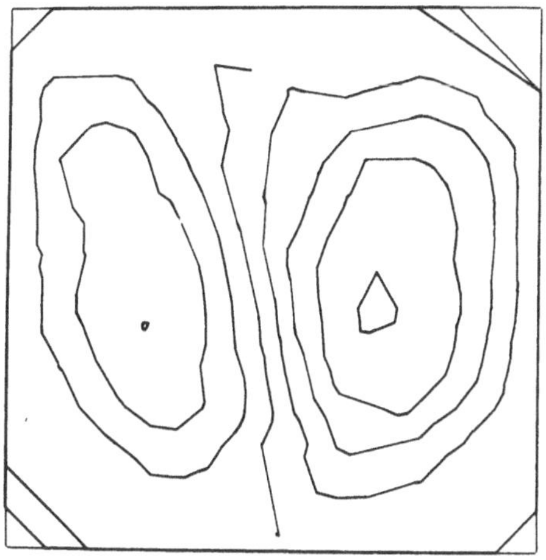

Figure 9A

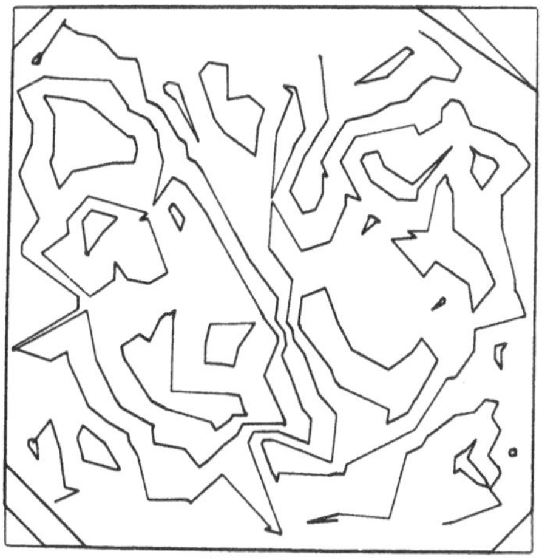

Figure 9B

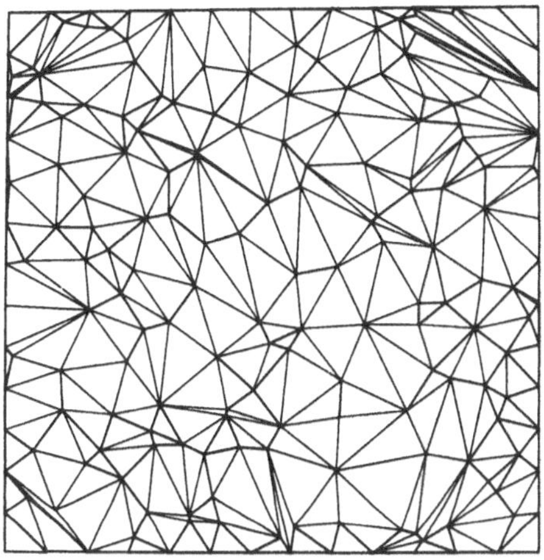

Figure 9C

Figure 8 shows the initial streamline pattern for problem two. In Figs. 9A, B and C we have the streamlines, level lines of vorticity and mesh configuration at time 24, corresponding to 330 calculation cycles. These results agree qualitatively with Kraichnan's[2] conjectures on the appearance of a dual energy spectrum with energy transfer toward low wavenumbers and vorticity transfer toward high wavenumbers.

ACKNOWLEDGEMENT

Many of the FLAG concepts originated with C. E. Leith and N. Hardy some years ago. Discussions with Leith, A. Winslow and D. Bailey played a significant part in the early stages of FLAG.

REFERENCES

1. Fjørtoft, R., Tellus, 5, 225-230 (1953).
2. Kraichnan, R. H., Physics of Fluids, 10, 1417-1423 (1967).
3. Leith, C. E., Properties of Matter Under Unusual Conditions, Interscience (1969).
4. Noh, W. F., Methods of Computational Physics, Vol. III, Academic Press (1964).
5. Winslow, A. M., J. Comp. Physics, 1, 149-172 (1966).

DIFFERENCE METHODS FOR HYPERBOLIC EQUATIONS USING SPACE AND TIME SPLIT DIFFERENCE OPERATORS OF THIRD ORDER ACCURACY*

Samuel Z. Burstein
Courant Institute of Mathematical Sciences, New York, New York

and

Arthur A. Mirin
University of California, Lawrence Radiation Laboratory, Livermore, California

INTRODUCTION

This note indicates some additional results obtained using third order difference methods which were constructed in Ref. 1. In that reference, we tested the method on a one-dimensional shock problem and a two-dimensional scalar problem. The linear stability theory gave good bounds on the range of allowable time steps and allowed for accurate determination of the artificial viscosity parameters. Here we indicate some results on a nonlinear two-dimensional problem requiring the integration of a system of equations--the conservation laws of fluid dynamics.

DIFFERENTIAL AND DIFFERENCE EQUATIONS

The system of equations we wish to solve is the set of conservation laws defining the motion of a compressible fluid through the system

$$u_t + f_x + g_y = 0, \tag{1}$$

subject to the initial conditions $u(x,y,0) = u_o$. The vector u has components of mass, momentum and energy while the vectors f and g are corresponding fluxes of these quantities.

Equation (1) can be rewritten, using the chain rule

$$u_t + Pu = 0, \tag{2}$$

where the partial differential operator P is given by

$$P = \sum_i A_i \frac{\partial}{\partial x_i}, \tag{3}$$

and where $x_1 = x$ and $x_2 = y$. The matrices A_i, which are functions of u, are non-commuting and have real eigenvalues.

If the differential operator P can be localized, Eq. (2) may be integrated to yield

$$u(t + \Delta t) = e^{-P\Delta t}u(t). \tag{4}$$

A difference analogue to Eq. (4) defined in terms of w, the difference approximation to u, can be written as

$$w(t + \Delta t) = Lw(t), \tag{5}$$

where the difference operator L can be expressed in the form

$$L = \sum_j c_j D_{x_1}^{\tau_{1_j}} \dots D_{x_m}^{\tau_{m_j}}. \tag{6}$$

*Work performed under the auspices of the U. S. Atomic Energy Commission.

Each of the matricies $D_{x_1}^{\tau_{kj}}$ represent difference operators, defined in a <u>single</u> coordinate direction, x_j, each of which advance the solution vector w a distance $\tau_{kj}\Delta t$.

In order that $|u - w| = O(\Delta t^4)$, the c_j and $D_{x_i}^{\tau_{kj}}$ must be chosen so that the expansion of L in powers of Δt will be identical to the noncommuting Taylor series for $e^{-P\Delta t}$ through terms $O(\Delta t^3)$. Carrying out this procedure (1) results in specifying the difference operator L to be

$$L = \frac{9}{8} D^{1/3}(A_1)D^{2/3}(A_2)D^{2/3}(A_1)D^{1/3}(A_2) - \frac{1}{8} D^1(A_1)D^1(A_2). \tag{7}$$

Here $D_{x_1}^{\tau} = D^{\tau}(A_1)$ and $D_{x_2}^{\tau} = D^{\tau}(A_2)$.

We see the advantage of splitting the difference operator into one-dimensional difference operators. In problems where there is a need for nonuniform mesh spacing, i.e., entropy layers, the programming is much simpler when only one-dimensional operators are to be considered. Hence if $\Delta x_1 \ll \Delta x_2$, in order to insure that the time steps in each direction will be equal, write

$$D^{\tau}(A_1) = [D^{\tau/M}(A_1)]^M, \tag{8}$$

where $M\Delta x_1 = \Delta x_2$.

There is no need to preserve the explicitness of the operator D. An alternative to Eq. (8) would be to use the implicit form

$$D^{\tau}(A_1) = D[A_1(t,t+\tau\Delta t)], \tag{9}$$

which uses data at two time levels and which would allow unconditional stability in marching in the x_1 direction. Since we use five points to approximate D, a quidiagonal matrix would have to be invented for each row of mesh points. In addition, one could construct operators which are essentially iterative in nature.

For the calculations presented below $\Delta x_1 = \Delta x_2$ and the operators $D^{\tau}(A_1)$ are split in time into three steps:

$$D^{\tau}[A_1(t+\tau\Delta t)] = D[A_1(t+\tfrac{2}{3}\tau\Delta t)]D[A_1(t+\tfrac{1}{3}\tau\Delta t)]D[A_1(t)]. \tag{10}$$

The arguments in Eq. (10) indicate the time level of the data which is operated upon. The product in Eq. (10) is constructed in such a way that the first operator is first order accurate, the second is second order accurate, while the third generates a third order accurate (see Ref. 1) solution.

NUMERICAL RESULTS

We have applied second and third order accurate difference operators in Eq. (5) to a test problem which has initial data w_o defining supersonic impulsive starting flow about a flat faced body in cartesian coordinates. The boundary condition of vanishing of the normal velocity on the body surface was obtained by using reflection rules on two rows of virtual interior points which are placed parallel to the body surface. The steady state (asymptotic, $t \to \infty$) solution was obtained with a free stream Mach number of 2.5 with a specific heat ratio of 1.4; the free stream pressure and density were 1.0.

In the table below, the third order accurate steady solution can be compared with the second order accurate and exact solution at the stagnation point on the body at the axis of symmetry. For the second order method Eq. (7) becomes

$$L = D^{\frac{1}{2}}(A_1)D^1(A_2)D^{\frac{1}{2}}(A_1), \tag{11}$$

and Eq. (10) takes the form

$$D[A_i(t+\tau\Delta t)] = D[A_i(t+\tfrac{1}{2}\tau\Delta t)]D[A_i(t)]. \tag{12}$$

The difference operator defined by Eq. (12) need be only second order accurate so that the coding of the first two steps of the third order method is used. In order to obtain stable solutions for this second order method an artificial viscosity was needed. The nonlinear viscosity used by Lapidus (2), which is defined in each separate coordinate direction, was used in Eq. (12).

The exact stagnation density and pressure are obtained by solving the jump conditions across the normal shock at the axis of symmetry and then connecting the post shock state to the stagnation point through the isentropic flow relations.

Conditions on Body Face

Scheme	Stagnation Density	Stagnation Pressure
Second Order	3.6947	8.3635
Third Order	3.9084	8.5223
Exact	3.7894	8.5268

The computational speed of the third order method was measured at 35,000 to 40,000 mesh points per minute on the CDC 7600 whereas approximately three times this number of mesh points per minute could be processed using the second order algorithm.

The second problem considered was a shock interaction problem in which a blunt body, moving supersonically through the air at a Mach number of 2.5 behind a detached shock wave, intercepts a plane incident shock moving at a Mach number of 2.0 with respect to the free stream. Since the incident shock is placed several mesh points to the left of the detached shock at the start of the computation, the interaction starts at t = 1.164. This allows the profile of the incident shock to be well formed as it propagates towards the subsonic region. The incident wave moves through the detached shock and intercepts the body face at t = 4.397. During this time interval the detached shock moves in rapidly toward the body face. This shock reaches a minimum value of the detachment parameter (ratio of shock standoff distance to body height) of 1.647; the steady state value is 2.0. At this point (t = 6.039) the body face experiences a very rapid rise in pressure to a peak level at the stagnation point of approximately five times the steady level (see Fig. 1). The pressure level drops as the incident shock reflects off the body face and moves out toward the detached shock which has remained stationary at a minimum value of detachment parameter. However it quickly moves out away from the body, when the reflected shock intercepts, to a standoff distance of 2.588, which exceeds the steady state standoff distance. The detached shock finally collapses in and reaches its steady standoff distance at t = 25.000. It was found that the third order accurate difference operator, in general, could only be used on flows with a lower strength shock wave than could be computed with second order methods. This is due to the more delicate question of computational stability of operators, Eq. (7), with nonpositive weights. Computational instability was observed using Eq. (7) for strong shock flows although Eq. (11) was seen to be stable for the same flows.

CONCLUSIONS

We have shown that third order accurate space and time split difference operators can be used in hydrodynamic flows that contain shock waves. Although third order methods give somewhat more accurate results compared with second order methods, their accuracy is not exhibited in flows with discontinuities. However if smooth flows are computed and if "weak" discontinuities arise, the third order difference operators shown here will be stable and the time dependent solution can be continued without the use of special procedures at the discontinuities.

REFERENCES

1. S. Z. Burstein and A. A. Mirin, "Third Order Difference Methods for Hyperbolic Equations," J. Comp. Phys. 5, No. 3 (June 1970).
2. R. D. Richtmyer and K. W. Morton, "Difference Methods for Initial-Value Problems," Wiley (Interscience), New York (1967).

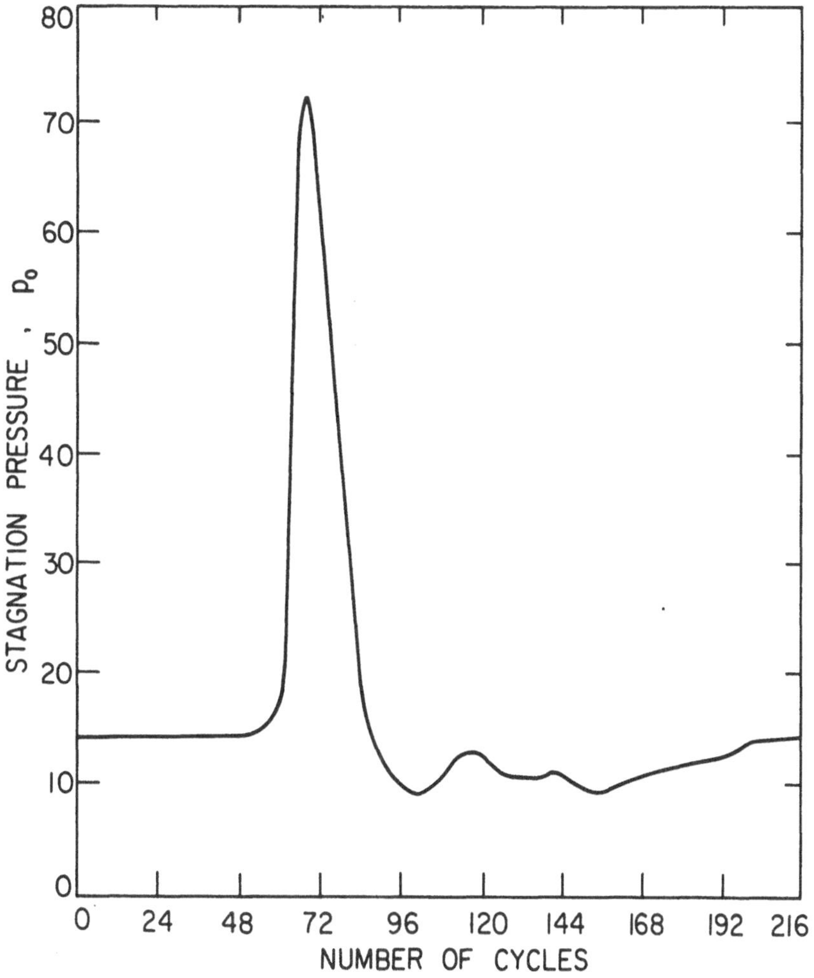

Fig. 1.

A NEW DIRECT METHOD FOR THE DISCRETIZED POISSON EQUATION

Patrick J. Roache, Numerical Fluid Dynamics Division
Sandia Laboratories, Albuquerque, New Mexico

ABSTRACT

An influence coefficient matrix method, called the error vector propagation or EVP method, for solving the discretized Poisson equation is described and its characteristics are investigated. This direct method is especially well suited to solving a family of problems, each of which has different non-homogenous terms and different boundary values, but which are all solved in the same mesh with the same class of linear boundary conditions. This is the case in incompressible fluid dynamics problems using the stream function-vorticity approach.

INTRODUCTION

We are concerned with obtaining numerical solutions of discretized versions of elliptic equations such as the Poisson equation,

$$\nabla^2\Psi = \zeta \tag{1}$$

In incompressible fluid dynamics problems in Cartesian coordinates, Ψ is the stream function and ζ is the vorticity. There are many other applications, of course, (1) being one of the most common equations of classical physics.

The most popular methods of solving the discrete Poisson equation are iterative methods, such as successive over-relaxation (SOR) and alternating direction-implicit (ADI) methods (see Forsythe and Wasow, 1960). For the large-mesh problems of interest, these methods are superior to the brute strength direct solution by the usual Gaussian elimination method. Other direct methods exist which can be considerably faster than iterative methods. As pointed out in the recent survey paper by Dorr (1970), most of these methods depend on a great degree of regularity in the block structure of the matrix equation, so that they are generally limited to simple rectilinear regions, such as rectangles and L- or T-shaped domains, and to Cartesian coordinates. (The work of Angel (1968 A and B) is the exception.) Some are limited in application to small field size due to adverse propagation of round-off error, some are limited to the two-dimensional problem, and most have obscure conceptual origins.

The present error vector propagation method makes use of only the most elementary principles of linear algebra and is, we believe, conceptually very simple. The method is similar to the methods of Ishizaki (1957) and of Lucey and Hansen (1964)-see also Froelich (1966)-but is algorithmically distinct from and less limited than these. It does have the disadvantage of being limited by the field size of the problem, but has none of the other disadvantages of other direct methods. We introduce the method via the one-dimensional problem.

ONE-DIMENSIONAL PROBLEM

We consider the one-dimensional Cartesian form of (1), and consider the problem with Dirichlet two-point boundary conditions.

$$d^2\Psi/dy^2 = \zeta \tag{2}$$

$$\Psi(0) = a, \quad \Psi(1) = b \tag{3}$$

Using the second-order accurate centered difference approximation gives

$$\Psi_{j+1} - 2\Psi_j + \Psi_{j-1} = \Delta y^2 \ \zeta_j \tag{4}$$

where j runs from 1 to J, and $\Delta y = 1/(J - 1)$. The boundary conditions are

$$\Psi_1 = a, \quad \Psi_J = b \tag{5}$$

We now pick an arbitrary value of Ψ_2', where the prime denotes a provisional value, say $\Psi_2' = \Psi_1 = a$. This Ψ_2' is in error from the true value Ψ_2 by the unit error e, that is,

$$\Psi_2 = \Psi_2' + e \tag{6}$$

The remaining provisional values up through J are now marched out in one sweep, starting at j = 3, by rearrangement of (4).

$$\Psi_{j+1}' = \Delta y^2 \zeta j + 2\Psi_j' - \Psi_{j-1}' \tag{7}$$

These provisional values Ψ_j' are in error by e_j, that is,

$$\Psi_j = \Psi_j' + e_j \tag{8}$$

Substituting (8) into (4) and using (7), we obtain the recursion relation for the error propagation as

$$e_{j+1} = 2e_j - e_{j-1} \tag{9}$$

which is seen to be independent of the non-homogeneous term ζ_j. For the presently considered boundary conditions, we have $e_1 = 0$ and $e_2 = e$, and (9) then gives

$$e_j = (j - 1)\cdot e \tag{10}$$

At the end of the first sweep, the unit error e is calculated from the known boundary value b and (8) and (10) as

$$e = \frac{b - \Psi_J'}{J - 1} \tag{11}$$

With e so determined, the provisional values are now corrected to the final values in a second sweep, using (10) and (8).

$$\Psi_j = \Psi_j' + (j - 1)\cdot e \tag{12}$$

The effect of different boundary conditions may easily be determined (Roache, 1970).

Since the recursion relation (9) for the error propagation is linear in j, there is no practical danger of generating excessively large $\Psi'(j)$ and thereby destroying accuracy due to machine round-off error. The dimensional extension of this method is not free from this shortcoming.

THE REFERENCE TWO-DIMENSIONAL PROBLEM

We now generalize the above technique to two dimensions, considering as a reference problem the Poisson equation in Cartesian coordinates

$$\frac{\partial^2\Psi}{\partial x^2} + \frac{\partial^2\Psi}{\partial y^2} = \zeta(x,y) \tag{13}$$

We consider a rectangular domain (Figure 1) of dimensions X and Y, with constant Δx and Δy, using the usual second-order accurate 5-point difference analogue of the Laplacian operator.

$$\frac{\Psi_{i+1,j} - 2\Psi_{ij} + \Psi_{i-1,j}}{\Delta x^2} + \frac{\Psi_{i,j+1} - 2\Psi_{ij} + \Psi_{i,j-1}}{\Delta y^2} = \zeta_{ij} \tag{14}$$

where $\Delta x = X/(I - 1)$ and $\Delta y = Y/(J - 1)$. For concreteness, we consider Dirichlet boundary conditions at all boundaries.

We now pick an arbitrary vector of provisional values $\Psi_{i,2}'$ just inside boundary B1, say $\Psi_{i,2}' = \Psi_{i,1}$. This $\Psi_{i,2}'$ is in error by the unit error vector $e_{i,2}$.

$$\Psi_{i,2} = \Psi'_{i,2} + e_{i,2} \tag{15}$$

With $\Psi'_{i,2}$ so chosen, the remaining provisional values for $2 \le i \le I - 1$ and j up to J (boundary B2) are calculated in one sweep, starting at (i,3), by rearrangement of (14).

$$\Psi'_{i,j+1} = \Delta y^2\ \zeta_{ij} + 2(1 + \alpha)\ \Psi'_{ij} - \alpha[\Psi'_{i+1,j} + \Psi'_{i-1,j}] - \Psi'_{i,j-1} \tag{16}$$

where $\alpha = (\Delta y/\Delta x)^2$. The correct boundary values $\Psi_{1,j}$ at B3 and $\Psi_{I,j}$ at B4 are used in (23) when needed. The error propagation equation equation is then

$$e_{i,j+1} = 2(1 + \alpha)e_{ij} - \alpha e_{i+1,j} - \alpha e_{i-1,j} - e_{i,j-1} \tag{17}$$

with boundary values along B1, B3 and B4 of

$$e_{i,1} = e_{1,j} = e_{I,j} = 0 \tag{18}$$

After the first sweep, the final errors are calculated from

$$e_{i,J} = \Psi_{i,J} - \Psi'_{i,J} \tag{19}$$

where $\Psi_{i,J}$ is the known boundary value vector along B2. From (17), a linear relation may be established, allowing the solution for $e_{i,2}$ from $e_{i,J}$. The correct values of $\Psi_{i,2}$ are then solved from (15), and a second sweep using the recursion relation (16) (with Ψ replacing Ψ') establishes the final solution.

To establish the linear relation allowing the solution for $e_{i,2}$ in terms of $e_{i,J}$, it is convenient to introduce two vectors, shown in Figure 1. The final error vector is defined as $F_\ell = e_{i,J}$, where $\ell = i - 1$ runs from 1 to I - 2. The initial error vector is defined as $E_m = e_{i,2}$, where $m = i - 1$ also runs from 1 to I - 2. The influence coefficient matrix $C = [C_{\ell m}]$ is defined by

$$F_\ell = C_{\ell m}\ E_m \tag{20}$$

Equation (20) is the two-dimensional counterpart of the one-dimensional equation (10), in which C degenerates to the number (J - 1).

Unlike the one-dimensional case, no convenient equation exists for the $C_{\ell m}$. The matrix is established prior to the solution of a particular problem by the following process. Taking a particular value m_1 of m, we set $E_{m_1} = e_{m_1+1,2} = 1$, all other $E_m = 0$. Then the propagation of the error vector E into e_{ij} is calculated by application of (17) and (18), and the resulting final error vector is $F_\ell = e_{i,J}$, where $\ell = i - 1$ runs from 1 to I - 2. The m_1-column of C is so determined as

$$C_{\ell,m_1} = F_\ell \tag{21}$$

Repeating the generation of $E_{m_1} = 1$, all other $E_m = 0$, for m_1 ranging from 1 to I - 2 fills in the influence coefficient matrix C. Finally, to solve for $e_{i,2}$ in terms of $e_{i,J}$, we invert (20) using direct Gaussian elimination and obtain

$$E_m = C^{-1}_{m\ell}\ F_\ell \tag{22}$$

and finally

$$e_{m+1,2} = C^{-1}_{m\ell}\ e_{\ell+1,J} \tag{23}$$

When applicable, this method is clearly more efficient than direct Gaussian elimination. The original problem of (I - 2) x (J - 2) simultaneous equations with a pentadiagonal matrix form has been reduced to that of solving (I - 2) equations to find C^{-1}, and additionally doing the equivalent work of three Richardson iterations: two sweeps of Ψ', Ψ in (16) and one sweep of e in (17). Further utility accrues to the method when the Poisson equation is to be solved many times in the same mesh with the same class of boundary conditions. Since the error propagation equation (17) is independent of the non-homogeneous term ζ_{ij}, and its boundary conditions (18) do not depend on the boundary values of Ψ but only on their specification as Dirichlet boundary conditions, the sweep of e via (17) and (18) and the inversion of C need be done only once for a family of solutions in the same mesh with the same type of boundary conditions, but with different boundary values of Ψ and different ζ. This is precisely the case in fluid dynamics problems.

Compared to iterative methods, we can easily make EVP look 10-100 times faster, depending on the mesh size and the convergence criteria used. But EVP does have the disadvantage of requiring considerably more core memory storage, and the method cannot compete with the simplicity of the SOR method. Most important, it is field size limited, due to its error propagation characteristics.

ERROR PROPAGATION CHARACTERISTICS FOR THE REFERENCE PROBLEM

There is apparently a tendency for C to become ill-conditioned as J becomes large, intuitively seen as a difficulty in discriminating an error at (i,2) from errors at (i $\pm$ 1,2) as the source of any error at J. But in practice, it is not this behavior which limits the method, but the following. Unlike the recursion relation (9) for the one-dimensional error propagation, which is linear in j, the two-dimensional version (17) gives a value F_{ic} (ic is at the I-center of the mesh) which increases exponentially in j. For J large, this means that the ability of the method to resolve the error at j = 2 is limited by machine round-off error.

This behavior puts an absolute ceiling on the resolution ability of the method, even if the inversion of C were to be accomplished with perfect accuracy. In applications of the method to date, we have used a Gaussian elimination routine* which performs a double-precision iteration to reduce this error to an entirely negligible level. Also, the details of the particular problem (i.e., Ψ and ζ values) have not significantly affected the error propagation, provided the boundary values of Ψ are reasonably scaled.** Consequently, an order of magnitude estimate of the practical limitation of the method can be found by numerically marching out (17) and (18) with unit errors at j = 2.

The error propagation has several fortunate aspects. The largest (and therefore most limiting) error occurs in the center of the mesh, so that we need only consider F_{ic}. Also, the effect of various conditions along the boundaries i = 1 and i = I adjacent to the march have negligible effect on the center value even for I as small as 7, so we may neglect the I dimension and the adjacent boundary conditions as parameters of the error propagation. Finally, there is a strong effect of mesh aspect ratio $\beta = \Delta x/\Delta y$, which may be used to advantage. Small β has an adverse effect on error propagation, while large β has a favorable effect. (The leading term in F_{ic} at J is $2(1 + \alpha)^{J-2}$, where $\alpha = \beta^{-2}$.) In the limit of large β, the error propagation approaches that of the one-dimensional problem, which is merely linear in J.

The dimensionless length of practical interest in determining the applicability of the method is $Y/\Delta x$ (that is, the number of x-increments that we can go in the y-direction). For a unit error $E_{ic} = 1$, the value $P = \log_{10}[F_{ic}]$ is plotted in Figure 2 with $\beta = \Delta x/\Delta y$ as the parameter. The resolution level (number of significant figures) S of several current U. S. computers is also shown. (The prefixes SP, DP and HDP respectively refer to single precision, double precision, and hardware double precision.) As an example of interpreting this figure, consider a CDC 6600 with

*"GERPIR", by C. B. Bailey, Math. Computing Services Division, Sandia Laboratories. See Bailey (1970).

**Errors from these other sources may be somewhat reduced by overall iteration of the entire EVP method. But in computations performed to date, this procedure improved the final error by less than the improvement obtained by reducing J by one, and began to diverge beyond the second iteration.

single precision, $S = 14.45$. For $\beta = 1$, $S \simeq P$ at $Y/\Delta x = 20$. At this condition ($J = 21$), we may expect resolution errors in $\Psi_{i,J}$ of order unity, which is ordinarily unacceptable. But at $Y/\Delta x = 10$ with $\beta = 2$ (still $J = 21$), we find $P \simeq 7.5$. The difference $S - P \simeq 14.5 - 7.5 = 7$ indicates that we may expect resolution errors in $\Psi_{i,J}$ of order 10^{-7}, which is generally acceptable. We have used the method in production runs on a CDC 6600 in a 67 x 31 mesh, with $\beta = 2\frac{1}{2}$; the final resolution error was of order 10^{-6}, and each solution was obtained in 0.27 seconds. Note that for $2 \lesssim \beta \lesssim 10$, the maximum attainable $Y/\Delta x$ becomes approximately independent of β and of the field size J. (But for β large enough, the error propagation is linear in J, and $y/\Delta x \rightarrow 0$.)

This resolution error of the EVP method differs from a convergence criteria used in iterative methods. The largest resolution errors in the EVP method appear on the single boundary at the end of the march, while the errors at interior points are much smaller. The largest magnitude convergence errors of iterative methods appear at internal points while the specified boundary values remain intact. Also, when SOR is used with the near-optimum over-relaxation parameter, a high frequency "hash" error of this magnitude appears at internal points. This is of some consequence in fluid dynamic calculations, since the dynamic use of Ψ is in determination of velocity components by numerical differentiation of Ψ. Thus, a resolution error in Ψ of 10^{-6} at the final boundary in EVP is not directly comparable to, but is better than, an iterative change in Ψ of 10^{-6} in ADI or SOR.

IRREGULAR GEOMETRIES

One of the advantages of the EVP method over other direct methods is its simple adaption to irregular boundary geometries and varied combinations of boundary conditions. The only explanation required for the adaption to irregular geometries is the definition of the initial and final error vectors E and F. Several examples are given in Figure 3. The indexing of E and F are not unique.

The EVP characteristics would have to be worked out for each case. But since the presence of boundaries more than 4 or 5 cells from an interior march path has a relatively slight favorable effect on e, it may be expected that the method will frequently be limited by the P of Figure 2, based on the longest march path of the problem.

Note also that partial cell treatment of irregular boundaries (Salvadori and Baron, 1961) may be easily accomodated in the march equation (16).

OTHER EXTENSIONS TO THE REFERENCE PROBLEM

Other extensions of the EVP method are possible. Neumann boundary conditions (which give zero gradient conditions on e) are easily treated. They have a negligible effect on the error characteristics when applied along B3 or B4, a slight favorable effect when applied along B1, and a seriously adverse effect when applied along B4. Robbin's conditions are also easily treated, as are irregular mesh systems. The 3- and n-dimensional problems are readily formulated, although the inversion of C may become difficult. Although the jump from 1 to 2 dimensions greatly deteriorates the error characteristics, higher dimensions have a rapidly diminishing adverse effect.

The diagonal unit square or the nine-point analogues of the Laplacian may be used in EVP. With these, an implicit march scheme worsens the error characteristics, while an explicit diagonal march (solving for $\Psi_{i+1,j+1}$) improves them, for I small. Another worthwhile adaption is the use of a fourth-order accurate five point analogue to $\partial^2\Psi/\partial x^2$, transverse to the march direction. This produces about a 12% increase in P for $\beta = 1$, but may allow the use of larger β with a consequent decrease in P. The EVP method is also applicable to other linear elliptic equations of fluid dynamics, and can be used in an iterative fashion to solve variable coefficient and nonlinear Poisson equations. For details, see Roache (1970).

CONCLUSIONS

The error vector propagation method, or EVP method, has been shown to be a viable method for solving the discretized Poisson equation. Although field size limited, it possesses demonstrated utility in large fields, provided that the mesh aspect ratio is favorable. It is conceptually simple, and is especially attractive for irregular boundary geometries and for irregular meshes. In some cases, it is practical for other than the usual 5-point analogue for the Laplacian operator. In a rectangular mesh, the computational time may be 10-100 times faster than the popular ADI methods.

ACKNOWLEDGEMENTS

The author gratefully acknowledges the contributions of C. Bailey, B. Hulme, F. Dorr, F. G. Blottner, and especially A. J. Russo.

REFERENCES

Angel, E., J. Math. Anal. Appl. 23, 471-484 (1968A)
Angel, E., J. Math. Anal. Appl. 23, 628-630 (1968B)
Aziz, K., and Hellums, J. D., The Phys. of Fluids 10, 314-324 (1967)
Bailey, C. B., Sandia Laboratories, Albuquerque, N.M., Rept. SC-M-69-337, Jan.1970.
Dorr, F. W., SIAM Review 12, 248-263 (1970)
Forsythe, G. E., and Wasow, W. R., Finite-Difference Methods for Partial Differential Equations, Wiley, N. Y., 1960.
Froelich, R., General Atomic, San Diego, Calif., Rept. GA-7164, Parts I and II(1966)
Ishizaki, H., Kyoto University, Bull. Disaster Prevention Research Institute, Bull. No. 18 (1957)
Lucey, J. W., and Hansen, K. F., Trans. Amer. Nuclear Soc. 7, 259 (1964)
Roache, P. J., Sandia Laboratories, Albuquerque, N.M., Rept. SC-RR-70-579, Sept.1970.
Salvadori, M. G., and Baron, M. L., Numerical Methods in Engineering, Prentice-Hall, Inc., Englewood Cliffs, N. J., Second Edition, 1961.

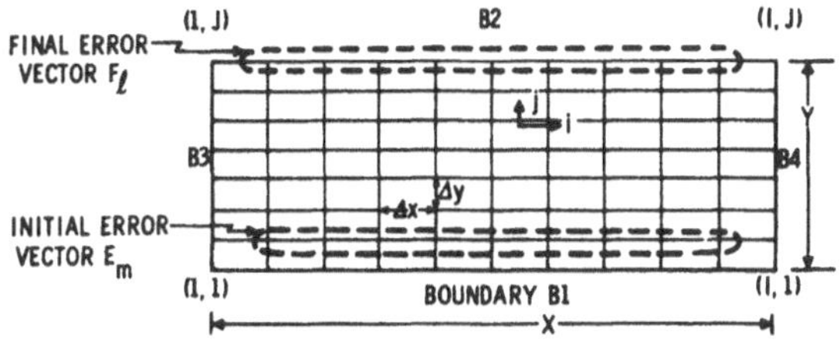

FIGURE 1. GEOMETRY OF THE REFERENCE PROBLEM

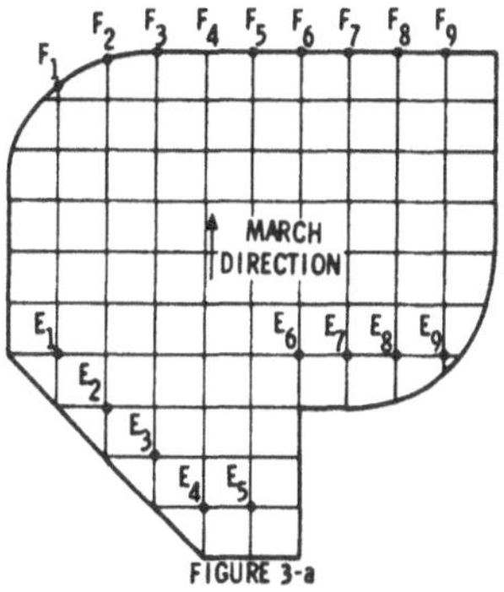

FIGURE 3-a

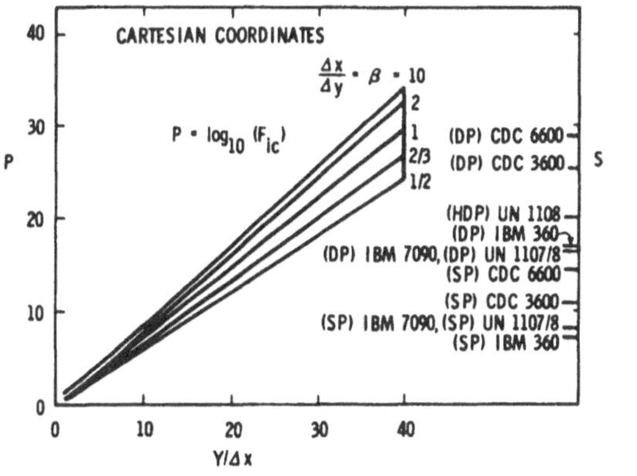

ERROR PROPAGATION CHARACTERISTICS FOR THE EVP METHOD APPLIED TO THE POISSON EQUATION IN CARTESIAN COORDINATES (x, y)

FIGURE 2.

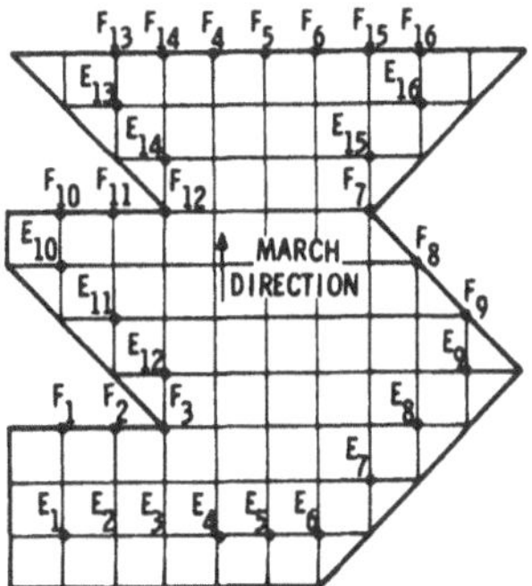

FIGURE 3-b

EVP METHOD FOR IRREGULAR GEOMETRIES. E_m = COMPONENTS OF THE INITIAL ERROR VECTOR, F_l = COMPONENTS OF THE FINAL ERROR VECTOR

INITIAL CONDITIONS AND IMBEDDED SHOCKS IN THE NUMERICAL ANALYSIS OF TRANSONIC FLOWS

Gino Moretti

Polytechnic Institute of Brooklyn
Graduate Center, Farmingdale, New York

Many practical problems in gas dynamics involve the determination of a steady flow. Numerical procedures to approximate the solution are generally iterative: one starts with an initial guess of the flow field and tries to correct the errors in successive steps, until the computed values vary from step to step by less than an accepted tolerance. Assuming that one succeeds in his intent, the question still remains open, whether the numerically steady flow approximates a physical steady flow, consistent with the assumed physical model and the boundary conditions. Existence and uniqueness theorems are generally not available for the original system of partial differential equations. Assuming, though, that a steady physical solution exists and is unique (a single point P in a functional space), there is no way of proving that the functional point defined at each step by the numerical procedure converges to P. I would like here to show why and how time-dependent techniques may help us to get a clearer idea of the correctness of the numerical solution.

Relaxation techniques applied to the steady flow equations are a powerful computational tool, which unfortunately may be difficult to use as long as it relies on the skill of the numerical analyst. Time-dependent techniques are based on a different concept: the unsteady flow equations are used to reach a steady flow asymptotically. The concept has been often misused or, at least, not used to exploit all its advantages. Consider a numerical procedure which, at each computational step, closely approximates the unsteady flow equations. In other words, there are no arbitrary elements which become inefficient only if and when the flow becomes steady.* We may expect, then, the computation to describe a physical evolution starting from a given set of initial conditions. This is what I like to call a time-dependent technique. In applying it, we expect all perturbations to propagate according to the laws of physics. We can easily pinpoint sources of errors. The local details of the physical evolution of a flow are well known and can be used to carry on a continuous check of the numerical procedure. It seems to me obvious that, if a steady state exists and has a unique description, and if it can be reached asymptotically in time by a natural evolution from an initial state, a consistent numerical computation, starting at the same initial state, should lead us to the same steady state or, at least, to a neighboring state, to within a prescribed tolerance.

An analysis of this kind shifts the emphasis from mathematics to physics. Without underestimating the role of mathematics in the analysis of a technique, we should acknowledge that a formal evaluation is, at least, so far, strongly limited by too many restrictive assumptions. Physical intuition (which seems to be the ability to connect separate pieces of knowledge to get an overall picture of a phenomenon) is often a more efficient tool for analyzing

*An example of a numerical procedure which is based on the unsteady Navier-Stokes equations, but does not describe a physical evolution of the flow in time because of the introduction of arbitrary elements can be found in Ref. 1.

complicated problems. In this connection, it is rather distressing to see clumsy numerical techniques and their poor results accepted by gas dynamicists as a consequence of a certain lack of communication between them and the numerical analysts. I believe that the computer can perform numerical manipulations which are as far-reaching and expressive as the formal and numerical manipulations which our Masters in gas dynamics used to perform before the electronic era. Only too often one who has no experience in numerical analysis surrenders to what seem to be unobjectionable conclusions of a mathematical study and accepts results which he does not like only because he thinks that nothing better can be done numerically. Or, if he does not accept them, he draws the conclusion that numerical techniques are just inefficient-a negative attitude which should not be assumed since numerical analysis of gas dynamics is yet in its infancy. I would like to see all gas dynamicists to ask for every little detail to be carried out to their complete satisfaction, even if this implies a slowdown in the production of reports.

In dealing with time-dependent techniques, these points should be kept in mind if the computation is required to describe a physical evolution:

1. An overall consistency of the numerical technique with the mathematical model of the flow,

2. A proper treatment of boundary conditions (which is a particular aspect of the preceding requirement), and

3. A choice of initial conditions which leads to a natural evolution of the flow to a steady state.

I have briefly commented on point 2 in a preceding paper (Ref. 2). With regard to point 1, I would like to say that, besides satisfying obvious requirements for numerical stability, the integration scheme should strictly observe the Courant-Friedrichs-Lewy rule (Ref. 3),and the existence of discontinuities in the flow field should be taken into account. An inviscid flow may have, and generally has, discontinuities particularly in the form of shock waves and contact surfaces. This is a consequence of the mathematical model of an inviscid flow and I see no reason for neglecting it in a numerical computation. In addition, there is a practical justification for the introduction of discontinuities in a numerical technique. Limitations in computational time and memory storage make the usage of coarse meshes imperative. In regions of steep gradients and high curvatures of the physical parameters the truncation errors may become gigantic. Simple examples may be given where the Taylor expansions which are used to define truncation errors do not even converge in a mesh interval (Ref. 4). Consequently, the computed functions depart more and more from the physical patterns, to such an extent sometimes to produce instability. Introduction of a weak shock, for example, within a mesh interval to partially replace untreatably steep gradients may allow the computation to proceed smoothly without refining the mesh. A weak shock, however, does not change the nature of the flow since the difference between a shock and an acoustic wave is of the third order in the shock strength.

Obviously, the treatment of discontinuities involves certain complications in the logic of the numerical technique. The payoff, however, is in greater accuracy and shorter computational time (the latter by orders of magnitude). Imbedded shocks are considered to be the major difficulty in the analysis of supercritical flows past flying vehicles. On the basis of my own experience, I can say that the logical difficulties can be kept within reasonable limits and that topological difficulties can be overcome by letting discontinuities travel within a prefixed mesh. An example of such a treatment of imbedded shock waves can be found in Ref. 5. Similar applications of the same basic idea will be reported in forthcoming papers.

Let us discuss now briefly the role of initial conditions. Does a computed steady state depend on the choice of initial conditions? or is it a unique solution? Authors who present results of numerical computations are often asked these questions, mostly to their great embarrassment. It is obvious that, as long as no attempt is made to approximate a physical evolution by numerical techniques, existence and uniqueness of the numerical solution are unrelated to existence and uniqueness of a physical pattern. Therefore, the questions above seem to be legitimate and the embarrassment is justified insofar as no theorems have been proven. However, if the technique is time-dependent in the strict sense defined above, the existence of a steady state solution is related to the physical possibility of reaching a steady state, starting from a given initial state. In this case, physical intuition will supply the answer. Nobody has ever challenged, for example, the assumption of a steady flow past a model in a wind tunnel, after a certain transient (regardless of the tunnel's driving technique) and the identity of such a flow with the steady flow past the same model in free flight.

For the sake of a well-running, easy-to-understand computation, thus, the initial state should make physical sense. The choice is not necessarily unique, though, but it may be narrowed down by reasons of convenience.

In a supersonic blunt-body problem, for example, an arbitrary choice of the bow shock and a reasonable and smooth choice of values in the shock layer will be sufficient to let the computation proceed smoothly. Even if the initial state is not completely self-consistent, all ensuing perturbations will propagate in a way which will soon become physical and eventually will be flushed away, together with any initial information, through the supersonic boundary of the region of interest.

In problems which involve a vehicle traveling in an infinite domain at a subsonic speed, the situation is more delicate. Computations are often started in an impulsive way: the obstacle is suddenly introduced into a uniform flow. By so doing, a strong perturbation is introduced in the vicinity of the body at the initial time and shock waves tend to form shortly after. Unless the computational program has a special capability for handling shock waves temporarily, the numerical values degenerate quickly. In supercritical cases, the computation tends to become unstable. In subcritical cases the errors may be kept confined for a while. However, they propate outwards from the body until they reach a region where the mesh resolution is too poor to describe them. There they are amplified and new perturbations travel inwards (the flow is subsonic) until they reach the body. At this stage the results rapidly become unstable. A steady state is never reached. A steady state seems to be reached near the body before the final degenerating process, but it is hard to accept it since it cannot be justified physically.

A different way of starting the computation consists of assuming that, at the initial time, the vehicle stands at rest in a gas at rest. Undoubtedly, this is the easiest and more exact set of initial conditions among all imaginable sets. Then, the vehicle starts moving with a prescribed law until it reaches the required speed. The initial perturbations move away from the body across a space whose physical characteristics are exactly defined. Compression waves may eventually coalesce into a single shock wave on the perturbation front. At any instant of time following a short transient, the flow picture is as follows:

1. An external, infinite region of uniform flow (as seen from the vehicle moving at a constant speed),

2. A perturbation front, moving outwards and propagating the signals from the initial accelerating process,

3. An intermediate region, also moving outwards and propagating a readjustment process, and

4. A steady state region, internally bounded by the body and continuously growing in size.

From a theoretical standpoint, it is interesting to note that the steady state does not appear as the solution of a boundary value problem in an infinite domain, but is a solution valid in a finite region only. We witness the curious phenomenon of a steady state which propagates from the body, a mathematical contradiction in terms but, nevertheless, a clear physical fact which is the closest to our daily experience. Simple numerical computations can track it (Ref. 6) with great accuracy in a minimum of computational time. The great practical advantage of this procedure resides in the fact that the onset of the steady state near the body is made clear on physical ground. Therefore, if the vicinity of the body is the only region of practical interest, one can interrupt the computation with confidence in the results as soon as the near field is stabilized, in this way avoiding waste of computational time and error feedback due to poor resolution in the far field.

References:

1. Crocco, L., AIAA J. 3, 1824-1832 (1965)
2. Moretti, G., Phys. of Fl., II, 13-20 (1969.)
3. Courant, R., Friedrichs, K.O., and Lewy, H., Math. Ann., 100, 32-74, (1928).
4. Moretti, G., PIBAL Reprt. 69-26 (1969).
5. Moretti, G., PIBAL Reprt. 70-50 (1970).
6. Moretti, G., PIBAL Reprt. 70-20 (1970).

Session II

Numerical Techniques and Applications

V.V. Rusanov, Chairman

TRAITEMENT NUMERIQUE D'UN PROBLEME DE MAGNETOHYDRODYNAMIQUE

J.P. BOUJOT*, J.L. SOULE*, R. TEMAM**

O. INTRODUCTION :

Une tentative est faite ici d'étudier numériquement et de décrire l'expansion d'un plasma confiné par un champ magnétique intense. Par l'action d'un faisceau laser, une poussière Li H en suspension dans le vide donne naissance à un plasma dense dont l'expansion libre est très rapide. Lorsqu'on soumet ce plasma à un champ magnétique longitudinal de forte intensité, l'expansion radiale (i.e. perpendiculaire aux lignes de champ) est freinée ; le bord du plasma est alors animé d'un mouvement oscillatoire et on y note une augmentation importante de la température. Notre étude porte plus spécialement sur l'étude de certains phénomènes correspondant à cette expérience.

Au §1. nous donnons les équations et conditions aux limites utilisées pour décrire le phénomène. Ces équations ont été obtenues dans [1] à partir d'une description hydrodynamique de l'évolution du plasma et tiennent compte des symétries et particularité du problème. Ce système est constitué par le couplage d'un système hyperbolique et d'un système parabolique analogue au système couplé son et chaleur considéré par Richtmyer [4] et étudié d'un autre point de vue dans le cas linéaire par Lions [2]. Sous cette forme, et comme dans [2], le système se prête bien à un traitement numérique par des méthodes de décomposition analogues aux pas fractionnaires (cf. [3] [6] [5]). L'un des schémas auxquels on est ainsi conduit est décrit au §2. Le §3. donne quelques résultats numériques et leur confrontation avec des résultats expérimentaux. Cette note est seulement une introduction au sujet et nous renvoyons à des travaux ultérieurs pour des développements plus importants.

.../...

---

* Commissariat à l'Energie Atomique, 91-Saclay, France.
** Université de Paris, 91-Orsay, France.

1. EQUATIONS DU PROBLEME :

L'espace est rapporté aux variables cylindriques r,θ,z ; toutes les fonctions inconnues sont indépendantes de θ, le problème admettant une symètrie de révolution ; plusieurs fonctions inconnues sont aussi indépendantes de z et cette variable n'apparaitra pas en fait dans les équations ; $r\in[o,r_{max}]$, t est le temps.

Dans la formulation lagrangienne du problème les fonctions inconnues sont R(r,t) le rayon eulérien, H(r,t) la jacobien, B(r,t) le champ magnétique longitudinal, T(r,t) la température, V(r,t) la vitesse radiale. Nous choisissons comme inconnues principales R,H,V, P=BRH, $Q=T^{3/2}RH(t+1)$ et nous notons $\mathcal{U}$ le vecteur de composantes H,V,P,Q.

Les équations (sans dimension) s'écrivent alors :

(1.1) $$NRH(t+1)=r\, n_o(r)$$

(1.2) $$\frac{\partial R}{\partial t} - V = 0.$$

(1.3) $$\frac{\partial \mathcal{U}}{\partial t} + \mathcal{R}_1 \mathcal{U} + \mathcal{R}_2\, \mathcal{U} = \mathcal{F}$$

où $\mathcal{R}_1$, $\mathcal{R}_2$ et $\mathcal{F}$ sont des opérateurs différentiels :

$$\mathcal{R}_1 = \begin{pmatrix} 0 & -\frac{\partial}{\partial r} & 0 & 0 \\ -A_1 \frac{I}{H^2}\frac{\partial}{\partial r} & 0 & 0 & 0 \\ 0 & 0 & 0 & 0 \\ 0 & 0 & 0 & 0 \end{pmatrix}$$

$$\mathcal{R}_2 = \begin{pmatrix} 0 & 0 & 0 & 0 \\ 0 & 0 & 0 & 0 \\ 0 & 0 & -A_5\frac{\partial}{\partial r}\frac{1}{H^2T^{3/2}}\frac{\partial}{\partial r} & 0 \\ 0 & 0 & 0 & -A_3\frac{T^{1/2}}{N}\frac{\partial}{\partial r}\frac{T^2}{H^2}\frac{\partial}{\partial r} \end{pmatrix}$$

.../...

$$\mathcal{F} = \begin{pmatrix} 0 \\ A_1 TN\frac{\partial}{\partial r}\left(\frac{1}{NH}\right) - A_1 \frac{\partial T}{\partial r} - \frac{A2}{NH}\frac{\partial B^2}{\partial r} \\ \frac{\partial}{\partial r}\left(\frac{B}{H^2 T^{3/2}} \frac{\partial(RH)}{\partial r}\right) \\ -A_3 \frac{T^{\frac{1}{2}}}{N} \frac{\partial}{\partial r} \frac{T^{7/2}}{H^2} \frac{\partial}{\partial r}(RH(t+1)) + A_4 \frac{T^3 R(t+1)}{NH} \frac{\partial B}{\partial r}^2 \end{pmatrix}$$

Les quantités $A_1,\ldots,A_5$ sont des nombres sans dimension.

Conditions initiales : elles s'écrivent en variables adimensionnelles

(1.4) $R(r,o)=r,\ H(r,o)=1,\ V(r,o)=r,\ T(r,o)=1,\ B(r,o)=1,$

d'où

(1.5) $P(r,o)=r,\ Q(r,o)=r.$

Conditions aux limites : On a $o \leq r \leq r_{max}$; au point $r=r_{max}$,

(1.6) $B(r_{max},t)=1,\ \frac{\partial T}{\partial r}(r_{max},t)=o,\ \frac{\partial H}{\partial r}(r_{max},t)=o$

d'où

(1.7) $P(r_{max},t)= R(r_{max},t/.H(r_{max},t)$

(1.8) $\frac{\partial Q}{\partial r}(r_{max},t) = T^{3/2}(r_{max},t).H^2(r_{max},t).(t+1).$

Au point r=o,

(1.9) $\frac{\partial B}{\partial r}(o,t) = \frac{\partial T}{\partial r}(o,t) = V(o,t)=o$

d'où

(1.10) $P(o,t) = Q(o,t) = o$

Nous renvoyons à [1] pour l'obtention des équations et une analyse de la validité des conditions initiales et aux limites précédentes. Le système (1.3) est déjà écrit sous une forme adaptée à l'étude numérique.

.../...

2. TECHNIQUE DE RESOLUTION :

L'intégration de (1.2) se fait à l'aide du schéma explicite

$$(2.1)\quad \frac{R_h^{n+1}-R_h^n}{\Delta t} + V_h^n = 0$$

h désignant la maille d'espace, l'indice supérieur n le temps ($t=n\Delta t$).

Les deux premières équations de (1.3) sont à caractère hyperbolique, les deux dernières à caractère parabolique. L'intégration par des schémas explicites de la partie hyperbolique du système conduisait à des instabilités rapidement incontrôlables, et nous avons ainsi été amené à rajouter un terme de pseudo viscosité. Cette pseudo viscosité est choisie de façon à améliorer l'ordre de troncature dans la discrétisation de $\frac{\partial V}{\partial t}$ (cf. [1]) et s'écrit :

$$\epsilon_v \mathcal{B} u \ , \ \epsilon_v = \frac{\Delta t}{2} A_1 \ \frac{T}{H^2}$$

$$\mathcal{B} = \begin{pmatrix} 0 & 0 & 0 & 0 \\ 0 & -\frac{\partial^2}{\partial r^2} & 0 & 0 \\ 0 & 0 & 0 & 0 \\ 0 & 0 & 0 & 0 \end{pmatrix}$$

Notons $\mathcal{F}=\mathcal{F}_1+\mathcal{F}_2$, $\mathcal{F}_1 = \begin{pmatrix} 1 & 0 & 0 & 0 \\ 0 & 1 & 0 & 0 \\ 0 & 0 & 0 & 0 \\ 0 & 0 & 0 & 0 \end{pmatrix} \mathcal{F}$;

L'intégration par un schéma explicite des parties hyperboliques et paraboliques conduisent à des conditions de stabilité très différentes, il est souhaitable alors d'utiliser des pas de temps différents, pour chacun de ces sous-systèmes. Cela est très pratique à l'aide des techniques de pas fractionnaire ; chaque intervalle Δt est découpé en M sous-intervalles pour l'intégration de la partie parabolique et l'on est conduit alors au schéma suivant :

$$\frac{1}{\Delta t}(u_h^{n+\frac{1}{2}} - u_h^n) + (\mathcal{R}_{1h} + \epsilon_v \mathcal{B}_h)\, u_h^n = \mathcal{F}_{1h}.$$

$$\frac{M}{\Delta t}(u_h^{n+\frac{1}{2}+\frac{1}{2}M} - u_h^{n+\frac{1}{2}}) + \mathcal{R}_{2h}, u_n^{n+\frac{1}{2}} = \mathcal{F}_{2h}.$$

$$\frac{M}{\Delta t}(u_h^{n+1} - u_h^{n+\frac{1}{2}+\frac{M-1}{2M}}) + \mathcal{R}_{2h}\, u_h^{n+\frac{1}{2}+\frac{M-1}{2M}} = 0$$

.../...

3. RESULTATS NUMERIQUES :

Le problème a été traité numériquement dans le cas physique suivant :

Sachant que les fonctions principales sont données sous forme réduite par

$G = \frac{g}{G_o}$,

$N_o = 10^{22}\ cm^{-3}$ avec $N(r,o) = e^{-r^2}$

$V_o = 1{,}1.10^7$ cm/s $T_o = 7{,}75.10^5\ {}^oK$

$B_o = 160000$ G $t_o = 1{,}43$ nanosec.

$r \in [o,3]$

- Les résultats sont regroupés sur les schémas suivants :

Fig.I : Les courbes d'évolution de

$V(t,rmax)=\frac{v}{V_o}$

$T_{max}=\frac{\theta max}{T_o}$

R(t,rmax)

Fig.I →

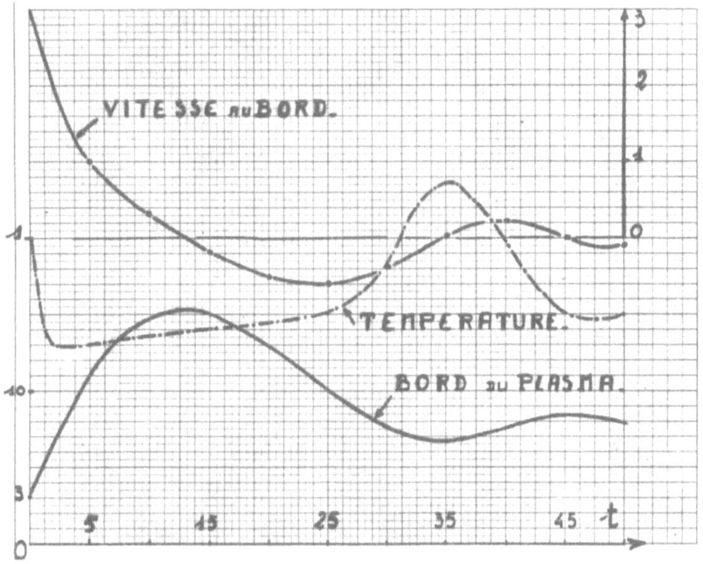

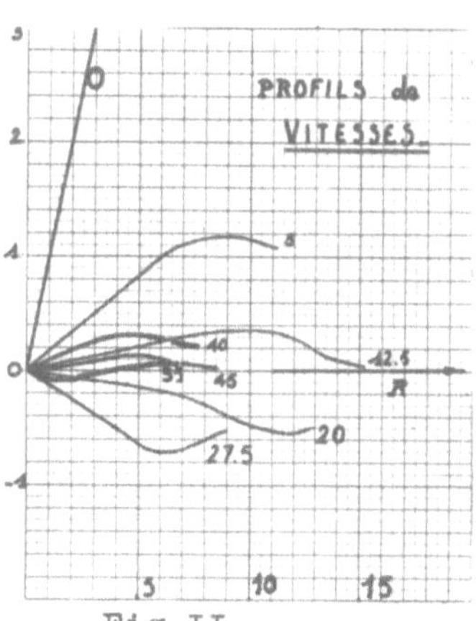

Fig.II

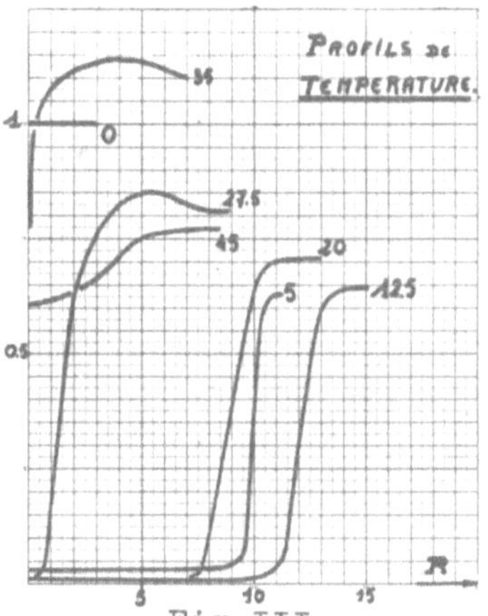

Fig.III

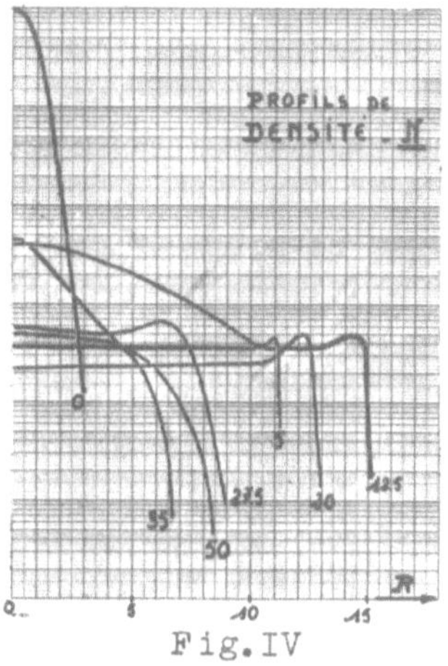

Fig.IV

Ces figures rapportent les profils de V,T,N.

Cette étude numérique montre que le sous découpage du système de nature parabolique entraine un gain de temps calcul appréciable.
Pour M = 4 ce gain est d'environ 15%.

BIBLIOGRAPHIE

[1] J.P. BAYARD, J.P. BOUJOT, J.L. SOULE.
Description hydrodynamique d'un plasma soumis à un champ magnétique.
RAPPORT. C.E.A. - SACLAY - (A paraitre).

[2] J.L. LIONS
Sur l'approximation de la solution d'équations d'évolution couplées.
RENDICONTI DI MAT. 1 (1968) p.p.1-36.

[3] G.I. MARCHUK
Méthodes numériques en météorologie.
ARMAND COLIN PARIS (1971).

[4] R.D. RICHTMYER, K.W. MORTON
Différence Methods for initial-value problems.
Interscience Publishers (1967).

[5] R. TEMAM
Sur la stabilité et la convergence de la méthode des pas fractionnaires.
ANNALI DI MAT. ,(1968)

[6] N.N. YANENKO
Méthodes à pas fractionnaires.
ARMAND COLIN, PARIS (1968).

COMPUTATION OF LAMINAR RECIRCULATING FLOW BETWEEN SHROUDED ROTATING DISCS

A. D. Gosman and D. B. Spalding

Mechanical Engineering Dept., Imperial College, London

1. INTRODUCTION

1.1 Physical and mathematical features of the shrouded-disc flow.

The situation depicted in Fig.1 is representative of a class of flows which are encountered in rotating machinery such as gas turbines and large electrical generators. A fluid is confined within the cavity formed by a fixed cylindrical shroud and two rotating discs having a common axis and angular velocity Ω. A particular feature of the situation is that the ratio of the length L to the diameter D of the cavity is of the order of unity. As will be demonstrated shortly, in these circumstances the drag exerted by the shroud gives rise to axial variations in the swirl velocity of the fluid (which would otherwise be in a state of solid-body rotation); and the non-uniform centrifugal-force field so produced causes secondary flows, in the form of toroidal vortices, which further distort the swirl-velocity distribution. These features make particularly difficult the task of predicting the magnitude of the drag, which is the property of central interest to designers.

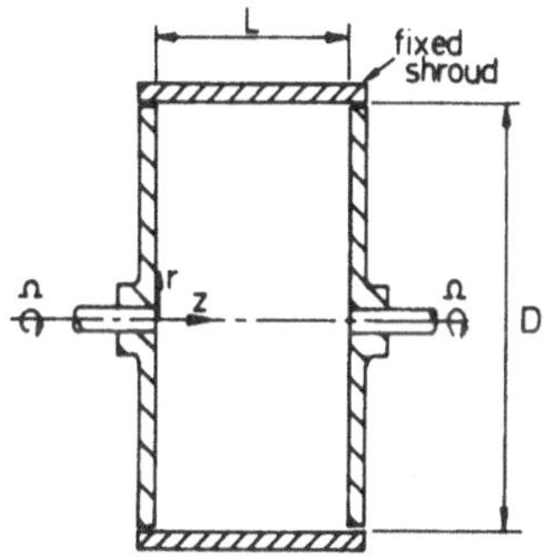

Fig.1 Illustration of the shrouded-disc problem.

The prediction task involves the solution of the set of partial differential equations which express the conservation laws of mass and momentum for a steady, two-dimensional, recirculating flow with swirl. In terms of the dependent variables vorticity ω and stream function ψ, these equations, for laminar flow with uniform viscosity ρ and density μ, are:

$$r^2[\frac{\partial}{\partial z}(\omega\frac{\partial\psi}{\partial r}) - \frac{\partial}{\partial r}(\omega\frac{\partial\phi}{\partial z})] - \mu[\frac{\partial}{\partial z}(r^3\frac{\partial\omega}{\partial z}) + \frac{\partial}{\partial r}(r^3\frac{\partial\omega}{\partial z})] - \frac{\rho}{r}\frac{\partial\phi^2}{\partial z} = 0 \quad ; \quad (1.1\text{-}1)$$

$$\frac{\partial}{\partial z}(\phi\frac{\partial\psi}{\partial r}) - \frac{\partial}{\partial r}(\phi\frac{\partial\psi}{\partial z}) - \mu[\frac{\partial}{\partial z}\{r^3\frac{\partial}{\partial z}(\frac{\phi}{r^2})\} + \frac{\partial}{\partial r}\{r^3\frac{\partial}{\partial r}(\frac{\phi}{r^2})\}] = 0 \quad ; \quad (1.1\text{-}2)$$

$$\frac{\partial}{\partial z}(\frac{1}{r}\frac{\partial\psi}{\partial z}) + \frac{\partial}{\partial r}(\frac{1}{r}\frac{\partial\psi}{\partial r}) + \rho r\omega = 0 \quad ; \quad (1.1\text{-}3)$$

where z and r are the axial and radial co-ordinates, and $\phi \equiv rV_\theta$. The variables ω and ψ are linked to the axial and radial velocities V_z and V_r by the definitions:

$$\left.\begin{aligned} \omega &\equiv (\partial V_r/\partial z - \partial V_z/\partial r)/r \quad ; \\ \partial\psi/\partial r &\equiv \rho V_z r \quad ; \\ \partial\psi/\partial z &\equiv -\rho V_r r \quad . \end{aligned}\right\} \quad (1.1\text{-}4)$$

The foregoing equations display the well-known properties of non-linearity and coupling, which disqualify them for analytical solution, and which also pose formidable obstacles to the application of numerical methods. The inter-linkage, it should be noted, is particularly strong in this equation set: thus ψ appears in each equation; while the swirl velocity $\emptyset$, a dependent variable of its own equation, is found also in the right-hand term of the vorticity equation. This term represents the source of vorticity due to the torque experienced by a fluid particle contained in an axially-varying centrifugal-force field. These interlinkages are of course nothing more than a reflection of the physical behaviour described earlier.

1.2 The nature of the present contribution. The present contribution is devoted to an account of the application to the above set of equations of a general finite-difference solution procedure, described by Gosman et al (1969). The emphasis in the present paper is on the computational, rather than the practical, aspects of the problem. Section 2.1 below outlines the main elements of the solution procedure as it was initially employed, and describes how the interlinkages mentioned above caused the procedure to become divergent beyond a limiting value of the swirl Reynolds number. Then, in section 2.2 we show that the convergence of the procedure may be considerably improved by use of two remedial measures, one of which is directed at the interaction between vorticity and stream function, and the other at the linkage between this pair and the swirl velocity. Finally, in section 2.3 we display examples of predictions obtained with the modified procedure.

2. THE PRESENT CONTRIBUTION

2.1 Application of the standard solution procedure to the problem. The procedure. The main features of the solution procedure which was employed in the initial stages of the calculations will now be described. A quick appreciation of these features may be obtained from inspection of the finite-difference equation for vorticity (the treatment of the other variables followed similar lines) given below: this should be considered in conjunction with Fig.2, which shows the grid notation.

$$\sum_{i=N,S,E,W} (C_i+D_i)(\omega_P-\omega_i)+S_\omega = 0; \quad (2.1\text{-}1)$$

where S_ω represents the source of vorticity, given by:

$$S_\omega \equiv \Delta r_{NS}\,\rho(\emptyset_E^2-\emptyset_W^2)/4r_P \; ; \qquad (2.1\text{-}2)$$

and the C_i and D_i are coefficients which express respectively the influence of convection and diffusion. For example:

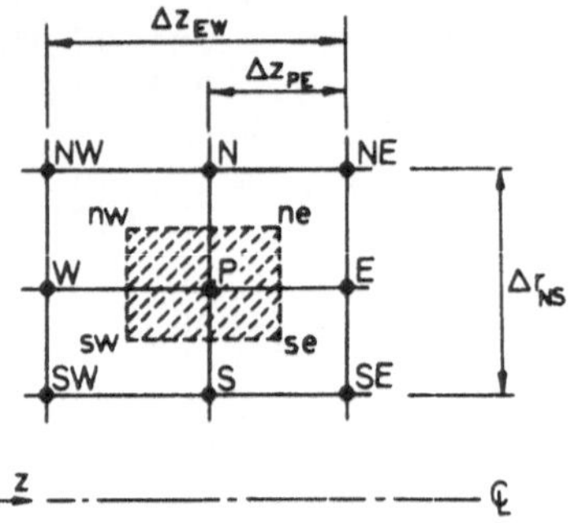

Fig. 2. Illustration of a portion of the grid. The boundaries of the control volume (shown hatched) which surrounds P lie midway between P and the neighbouring nodes.

$$C_E \equiv (|\Delta\psi_{ne-se}| - \Delta\psi_{ne-se})r_P^2/2 \quad ; \qquad (2.1\text{-}3)$$

where $\Delta\psi_{ne-se} \equiv (\psi_{NE}+\psi_N-\psi_{SE}-\psi_S)/4$, and:

$$D_E \equiv \mu r_P^3 \Delta r_{NS}/2\Delta z_{PE} \quad . \qquad (2.1\text{-}4)$$

Analogous expressions exist for the other coefficients. Inspection of the above relations will reveal that: (i) the difference scheme is a conservative one, i.e. the net flux across any of the boundaries of the control volume shown in Fig.2 will have the same value whether calculated from the difference equation for P, or from that for the relevent neighbouring node; and (ii) 'upwind' differences are employed for the convection terms. The motives for the incorporation of these features are now well-known: (i) is introduced for reasons of accuracy, while (ii) is employed so as to allow solutions to be obtained for large Reynolds numbers.*

Special difference equations are of course required for the boundaries of the domain of solution, i.e. the walls and the symmetry axis. ψ and ϕ assume known values on all boundaries. The wall vorticities are obtained from the no-slip condition, which we write as:

$$\omega_P = -[\frac{3(\psi_I - \psi_P)}{r_P n^2{}_I \rho} + \frac{\omega_I}{2}] \quad ; \qquad (2.1\text{-}5)$$

here I denotes the interior node adjacent to P, and n_I is the distance between P and I. The axis vorticities are calculated on the assumption of zero radial gradient. The origins of these formulae are explained by Gosman et al (1969).

The difference equations are solved by the Gauss-Seidel successive-substitution procedure. Each cycle of iteration involves the application of the procedure in turn to each of the three sets of equations for vorticity, stream function and swirl velocity, in the order of mention. This process is continued until the maximum change in any variable is less than 10^{-5} times a reference value. The solutions so obtained are regarded as acceptable if: (i) symmetry to within about 1% is obtained about a plane located mid-way between the discs; and (ii) the residual sources of the difference equations** have decreased monotonically to a small value during the iteration process.

Results. The calculations were performed for a length-to-diameter ratio 0.5, and for various values of the swirl Reynolds number Re_Ω ($\equiv D^2\Omega\rho/2\mu$). A mesh with 11 uniformly-spaced lines in each direction was employed unless stated otherwise.

These calculations revealed that the procedure became divergent for Re_Ω in excess of about 500. Moreover, it was found that straightforward remedial measures such as under-relaxation and refinement of the grid produced no appreciable improvement in stability. The failure of the procedure was surprising: for it had hitherto been successfully applied to several other swirling flows (see for example Pun and Spalding (1967)). These however were duct flows, whereas the present problem is of the closed-cavity variety. This distinction is an important one, for reasons which will be explained below.

2.2 The question of divergence. The causes. It turns out that there are two distinct causes of divergence: one is the interaction of vorticity and stream function in the presence of strong vorticity

*It should however be remembered that the diagonal dominance of the coefficient matrix which upwind differences procure will ensure unconditional stability only when the C_i's and S_ω are constants.

**The residual sources are defined as those values which are obtained when the estimates of ω, ψ and ϕ which prevail at the end of a cycle of iteration are substituted into the difference equations.

sources; and the other is associated with the coupling of the $\psi \sim \omega$ pair with the swirl velocity by way of the $\partial \phi^2/\partial z$ term. The mechanisms of instability are quite different, as are the remedial measures required; so we shall consider each case separately.

The interaction between vorticity and stream function. The mechanism of instability in this case may be illustrated by reference to the difference equation (2.1-1) for ω, and the analogous equation for ψ, which runs:

$$\sum_{i=N,S,E,W} F_i(\psi_P - \psi_i) - G_P\omega_P = 0 \quad ; \qquad (2.2\text{-}1)$$

where $G_P(\equiv r_P \Delta r_{NS} \Delta z_{EW}/4)$ is proportional to the area of the cell, and where, for example, F_E is defined as $\Delta r_{NS}/(\rho r_P \Delta z_{EP})$. The other F's have analogous definitions.

We consider the calculation of ω and ψ from these formulae, under conditions of large Re_Ω (and so small D_i), and large fixed vorticity sources. The procedure is first to compute ω_P from (2.1-1): here we note that the smaller are the C_i, i.e. the convective flow rates, the greater is the change in ω_P which results from a given flow perturbation. This change in ω_P causes the stream function to be altered, as is shown by equation (2.2-1): so the flow is further perturbed. Clearly the possibility of unstable interaction exists, even though the vorticity sources are fixed. Computer experiments, in which the fixed sources are produced by holding constant the ϕ values after about 50 cycles of iteration, confirm that this is indeed the case.

A remedy for this source of instability was developed from recognition that an important constraint on the ψ and ω fields is not fully-exploited in the standard solution procedure. The constraint in question is the circulation principle*, which is embodied in equation (2.2-1). In the standard procedure, the manner in which this relation is employed ensures only that the principle will be obeyed once the calculations have converged. The following paragraph describes how the procedure may be modified with advantage to ensure that the circulation condition is satisfied throughout the field, at all stages of the iteration process.

In the modified procedure, the initial values of ψ and ω are chosen so as to satisfy circulation: thereafter, the $\psi \sim \omega$ pair are adjusted in a single sweep of the field, in a special way. As each node is visited, changes are made to ω_P, ψ_P and the ω's at the four neighbouring nodes, such that the following conditions are satisfied: the conservation-of-ω principle for P; and the circulation principle for P,N,S,E and W**. The increments $\delta\psi$ and $\delta\omega$ which will satisfy these constraints may be deduced from equations (2.1-1) and (2.2-1).

Computer trials with the new 'multi-point circulation adjustment' (MPCA) procedure have revealed that it successfully suppresses the

---

*This states that $\int_A \omega dA = \oint v dS$, where A is an area in the $r \sim z$ plane, and v is the velocity along the curve S which encloses that area.

**This procedure is confined to the interior nodes of the grid; for the normal adjustment of the ω's on the boundaries already procures obedience to the circulation principle at these cells.

instabilities due to ψ-ω interactions: thus for fixed sources no difficulty was encountered in obtaining solutions at $Re_\Omega = 10^6$, while the limit for the standard procedure was $Re_\Omega \approx 500$. The improvement for the full problem, in which the sources vary from one iteration to the next, was however rather less, for the calculations diverged at $Re_\Omega = 10^3$. For the explanation of this behaviour, we must turn to the second kind of numerical instability.

The interaction with the swirl-velocity field. The role which the interlinkage between V_θ and the ψ-ω pair plays in inducing instability may be illustrated by reference to the especially-simple situation in which the shroud is presumed to rotate with the discs: in these circumstances the fluid will assume a state of solid-body rotation, with no secondary flow. This is a physically-stable situation, irrespective of the magnitude of Re_Ω. The solution procedure will however become unstable for large Re_Ω, as reference to Fig.2 and equations 2.1-1 and 2.2-1 will reveal: thus a small increase in V_θ at P will give rise to circulation in the clockwise direction about E, and in the opposite sense about W; so fluid will be caused to enter the control volume from below. The swirl velocity of this fluid is however lower than the value appropriate to solid-body rotation; so circulation will then occur in the opposite direction. For large Re_Ω the damping effect of viscosity is absent; hence an oscillatory motion will be established, and the physically-stable situation will never be regained. The existence of such a mechanism has been confirmed by computer experiments.

This source of divergence may be suppressed by under-relaxation on ϕ. The improvement which this brings is however modest: thus for the fixed-shroud problem, an under-relaxation factor of 0.15 raises the stability limit to $Re_\Omega \approx 3000$. Refinement of the grid has the effect of reducing the degree of under-relaxation required: for example, at the same Reynolds number, solutions for a 21 by 21 grid may be obtained without under-relaxation.

Unfortunately under-relaxation and grid refinement both increase computing times: for this reason, rather than attempt to map the stability limits for various combinations of grid and under-relaxation factor, efforts are currently under way to develop a remedy which does not exact this penalty.

Relevance to earlier studies. It is now possible to explain why the earlier applications of the procedure to swirling flows in ducts were not afflicted with problems of divergence. The reason is that the existence of forced convection increases the C_i's and so reduces the importance of the vorticity sources: indeed the greater is the flow rate, the less likely is divergence.

As a final comment on the question of stability, it my be remarked that problems of the kind described here may be expected in free-convection calculations, where temperature will play a role similar to that of swirl velocity.

2.3 Some predictions obtained with the modified procedure. Shown in Fig. 3 are predictions of the distributions of stream function, vorticity and swirl velocity, for $L/D = 0.5$ and $Re_\Omega = 1000$. A 21 x 21 uniform grid was employed, and the solution procedure incorporated the MPCA method. For these conditions, it was possible to over-relax by a factor of 1.2 the vorticity and swirl-velocity adjustments.

The results are displayed in the form of contour plots of ψ, ω and V_θ, for a domain bounded by the walls and the symmetry axis.

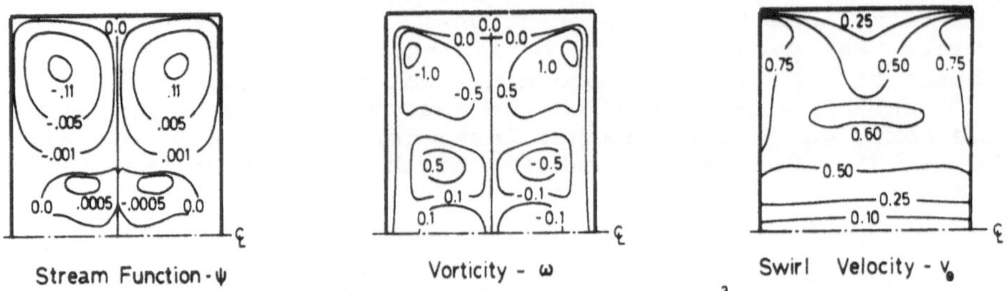

Fig 3. Predictions of ψ, ω and v_θ for $L/D = 0.5$, $Re_\Omega = 10^3$ and a 21x21 uniformly-spaced grid.

The swirl-velocity contours reveal that the fluid in the vicinity of the axis is very nearly in a state of solid-body rotation. However near the shroud, the action of shear and secondary motion gives rise to large axial gradients of V_θ, and so to large vorticity sources, as is evidenced by the vorticity contours. The stream-function plot reveals the existence of two pairs of centrifugally-driven toroidal vortices, of unequal sizes and strength. The upper pair are the larger and stronger of the two.

Predictions for higher and lower Reynolds numbers display the same general behaviour, although as Re_Ω is decreased, the small eddies eventually merge with the large ones.

3. CONCLUSIONS

The findings of the present study may be summarised as follows:

1. The application to the shrouded-disc problem of a solution procedure for steady two-dimensional flows has revealed that the procedure diverges beyond a limiting value of the swirl Reynolds number.
2. Two causes of divergence have been identified: one is the interaction between vorticity and stream function, in the presence of strong vorticity sources; and the other is the interaction between the swirl and secondary flows, by way of the vorticity sources.
3. The Reynolds-number limit has been increased by the following measures: modification of the procedure so as to ensure obedience to the circulation principle at all times; under-relaxation on the swirl velocity; and refinement of the grid. The suppression of divergence is however accompanied by an increase in computing time.
4. The existence of forced convection has a stabilising effect, which explains why previous applications of the procedure to swirling flows in ducts have been free from the divergence problems encountered here.

4. REFERENCES

1. Gosman, A.D., Pun, W.M., Runchal, A.K., Spalding, D.B. and Wolfshtein, M. (1969). Heat and Mass Transfer in Recirculating Flows. Academic Press, London and New York.
2. Pun, W.M. and Spalding, D.B. (1967). "A procedure for predicting the velocity and temperature distributions in a confined, steady, turbulent, gaseous, diffusion flame". Proc.Int.Astronautical Federation Mtd., Belgrade.

SOME SEMI-ANALYTICAL METHODS IN NUMERICAL FLUID DYNAMICS

B.W.Thompson
University of Melbourne, Australia.

In many problems it is possible to obtain inexact, but still (in some sense) leading approximations to the solution in analytic form. A decision then becomes available, whether to devise numerical methods for solving the equations directly as they stand, or whether to transform them in such a way as to take this analytic information into account. A method of solution which follows the latter course may be called "semi-analytical", and may well become obligatory if high accuracy is needed, or if the original unknowns are badly behaved. Examples involving both fast and slow flows will be discussed.

1. INTRODUCTION

The general object of mathematical research in fluid dynamics is to obtain numerical information about sets of parameters which typify, in some desired sense, the flows under consideration. Early research seems to have assumed that analytic estimates of parameters, involving clusters of e's, π's and particular values of special functions have greater intrinsic merit than simple decimal numbers truncated at some point beyond which their physical meaning becomes doubtful (having regard to the totality of physical influences neglected by the analysis). Few of us find this attitude convincing today: for example, in the Hele-Shaw cell calculation described below there occurs a parameter $\phi_1 = 744\pi^{-5}\zeta(5)$ where $\zeta(n)$ is the Riemann zeta-function, but most of us would surely prefer to see it as $\phi_1 \doteqdot 2{\cdot}521$.

If the end result of research is to be the specification of a certain number of parameters in numerical form, then the process by which they are determined may clearly contain as large a component of numerical processes as we please, even to the point of containing no explicit analysis at all: this, essentially, is the charter of Numerical Fluid Dynamics. There are, however, two powerful constraints on this apparent "carte blanche":

(i) Direct numerical methods in general assume considerable "smoothness" - even local analyticity is often an insufficient requirement-in the functions to be treated. If these functions vary too wildly, high accuracy can only be achieved by analytical transformations whereby the bad behaviour is made algebraically explicit, and the new unknowns are themselves "smooth";

(ii) Neither numeric nor analytic results can be considered valuable unless it has been possible to set realistic error bounds upon them, with respect to both the mathematical processes whereby they were determined and their accuracy as estimates of physical quantities in question. Unfortunately, at present the progress of numerical error analysis lags far behind the pace of algorithm development; this can preclude error assessment in a numerical determination.

Considerable ingenuity is often called for in devising transformations of the type described in remark (i); furthermore the subsequent numerical analysis can often carry us into fields well outside the repertoire of day-to-day fluid mechanical calculations. Nevertheless, for some reason it does not seem to be the convention to dwell on these matters in published work: indeed, all too frequently graphs and tables of values are appearing in the journals with no indication whatever of algorithms - let alone the error estimates - devised for their construction.

In the examples discussed below the reverse procedure will be followed. The physical details and results have all been, or are about to be published, but many of the computational details had to be suppressed. It seems worthwhile here to give the latter an airing.

2. SECONDARY FLOW IN A HELE-SHAW CELL

[Thompson (1968), Lee and Fung (1969), Gaydon and Shepherd (1964).]

The boundary layer equations for creeping flow past a circular cylinder in a Hele-Shaw cell may be reduced to a single biharmonic equation

$$\nabla^4\Psi = \Psi_{XXXX} + 2\Psi_{XXZZ} + \Psi_{ZZZZ} = 0 \tag{2.1}$$

for the stream function, subject to

$$\Psi = \Psi_Z = 0 \text{ at } Z = \pm 1, \quad \Psi_X = 0 \text{ at } X = 0, \quad \Psi \to 0 \text{ as } X \to \infty,$$

and

$$\Psi_Z = \tfrac{1}{2}\phi_1(1 - Z^2) - 8 \sum_{n=0}^{\infty} (-1)^n k_n^{-4} \cos k_n Z, \text{ at } x = 0 \tag{2.2}$$

where $k_n = (n + \frac{1}{2})\pi$ and $\phi_1 = 744\pi^{-5}\zeta(5) \doteqdot 2{\cdot}521$. The derivatives Ψ_X and Ψ_Z now specify the velocity components in the Z and X directions on the strip $0 \leq X < \infty$, $-1 \leq Z \leq 1$.

Gaydon and Shepherd (1964) have shown that there exists a rapidly converging eigenfunction expansion for Ψ in the form

$$\Psi = \sum_{0}^{\infty} C_r K_r^{-2} e^{K_r X} \{(1 + \cos 2K_r) \sin K_r Z - 2K_r Z \cos K_r Z\} \tag{2.3}$$

where K_r is the r-th solution ($\neq 0$) of $\sin 2K = 2K$ in the space $\frac{1}{2}\pi < \arg K < \pi$. However, the C_r are very hard to determine because, if we denote the quantity in braces by $\Phi_r(Z)$, then the set $\{\Phi_r(Z)\}$ are not orthogonal in any inner product space. Also, a least squares fitting of the C_r using (2.2) after truncating (2.3) is ill-conditioned.

Gaydon and Shepherd determine the C_r by expanding both the functions $\Phi_r(Z)$ and the boundary values in (2.2) in terms of the orthonormal set

$$Y_s(Z) = \{\sin \mu_s Z/\sin \mu_s - \sinh \mu_s Z/\sinh \mu_s\}/\sqrt{2} \tag{2.4}$$

$$(\tan \mu_s = \tanh \mu_s).$$

There then results an infinite set of linear simultaneous equations for the C_r whose coefficients depend only on the numbers $K_1, K_2, \ldots$ and $\mu_1, \mu_2, \ldots$. The μ_i were tabulated by Lord Rayleigh (1878), but they are not difficult to calculate anyway. The K_r are rather more difficult to find, but the first 10 of them were tabulated by Hillman and Salzer (1943), and apparently Gaydon and Shepherd confined themselves to this set. For this reason only the first 10 C_r could be determined, although the number of equations to which they could be fitted was arbitrary. They chose, in fact, to fit the C_r to the first 20 equations by least squares, and the point is worth making that at this stage other choices are not only possible but also may be more efficient. The limitation to the first 10 C_r is easily removed: I had in fact tabulated the first 30 of the K_r to 8 significant figures in 1959, and the basic algorithm is in Thompson (1967) together with an 11-figure determination of the first 20. My experience is that the direct solution of the first 20 equations, setting $C_r = 0$ $(r > 20)$, gives much better accuracy. This is discussed very fully in Thompson (1967).

3. THE BLAST WAVE

[Stewartson and Thompson (1968), Stewartson and Thompson (1970), Sedov (1959).]

To discuss the behaviour of a constant energy piston-generated flow of a perfect gas in a semi-infinite straight tube, assuming infinite downstream Mach number, one may set up an expansion in similarity-type variables:

$$X(t) = at^{2/3}[1+b_1t^{\beta_1}+\ldots],\quad P = \frac{4}{9}a^2t^{-2/3}[p_0(\eta) + b_1p_1(\eta)t^{\beta_1} +\ldots],$$

$$V = \frac{2}{3}at^{-1/3}[v_0(\eta) + b_1v_1(\eta)t^{\beta_1} +\ldots] \quad (\beta_1 < 0) \qquad (3.1)$$

where $\eta = \psi/X(t)$, ψ being the stream function, X the shock position, P the pressure and V the gas speed. Two crucial parameters are $P_0 = p_0(0)$ and $\underline{a}$ given by

$$(4/9)\ a^3 \int_0^1 [p_0{}^{1-1/\gamma}\ \eta^{-1/\gamma} + v_0{}^2]d\eta = 2. \qquad (3.2)$$

They are determined from the blast wave equations (Stewartson and Thompson, 1968)

$$p_0'(\eta) = \tfrac{1}{2}v_0 + \eta v_0'(\eta); \qquad v_0'(\eta) = \tfrac{1}{2}(1 - 1/\gamma)\eta^{-1/\gamma}p_0{}^{-1-1/\gamma}[p_0 + \eta p_0'(\eta)] \quad (3.3)$$

where $p_0(1) = v_0(1) = 1$ and γ is the adiabatic index. (3.3) has a known first integral

$$E(\eta) = [\eta v_0{}^2 + (\eta p_0)^{1-1/\gamma}]/(2p_0v_0) \equiv 1, \qquad (3.4)$$

in virtue of which (3.2) can be replaced by

$$(4/9)a^3 \int_0^1 p_0v_0/\eta\ d\eta = 1; \qquad (3.5)$$

furthermore (3.4) can be used as a check on the integration of (3.3). Prime interest in our work centred on the case $\gamma = 5/3$, for which one has

$$p_0 = P_0 + (9/28)P_0{}^{-3/5}\ \eta^{7/5} + O(\eta^{17/5}); \quad v_0 = \tfrac{1}{2}P_0{}^{-3/5}\eta^{2/5} + O(\eta^{12/5}). \qquad (3.6)$$

(3.6) indicates that v_0 will emerge very poorly from any numerical integration process near $\eta = 0$, but that the transformation $w = v\eta^{-2/5}$ gives a new unknown that is regular at $\eta = 0$. An estimate of P_0 can be found not only by direct integration of $p_0\ (\eta)$ but also by integrating the first equation (3.3), which gives

$$P_0 = \tfrac{1}{2}\int_0^1 v_0 d\eta\ .$$

This integrand is not well behaved near $\eta = 0$, but by (3.6) for small η

$$\tfrac{1}{2}\int_0^\eta v_0 d\eta = (5/14)\eta v_0(\eta) + O(\eta^{17/5}), \qquad (3.7\text{i})$$

and so the integration need not be carried right down to $\eta = 0$.

Similarly,
$$\int_0^\eta p_0v_0/\eta\ d\eta = (5/2)p_0(\eta)v_0(\eta) + O(\eta^{9/5}) \qquad (3.7\text{ii})$$

which saves us integrating (3.5) to $\eta = 0$.

Equations (3.3) were solved for p_0 and w by a standard fourth order Runge-Kutta process, together with the extra equations

$$u_1'(\eta) = -\tfrac{1}{2}v_0, \quad u_2'(\eta) = -p_0v_0/\eta, \quad u_1(1) = u_2(1) = 0 \qquad (3.8)$$

for determining P_0 and $\underline{a}$ respectively, using the steps 1·0(-0·025)0·5(-0·125)0·25 (-0·00625) ... down to $\eta \doteqdot 10^{-6}$. The quantity $E(\eta)$ was calculated at each step; the decreasing step-lengths are mainly for the benefit of (3.8). The maximum deviation of $E(\eta)$ from 1 was 2×10^{-6} at $\eta = 10^{-6}$. The correct value of P_0 (Sedov, 1959) is $2^{-2/3}(7/8)^{13/3} = 0{\cdot}353195287$. The computed value of $p_0(10^{-6})$ is

0·35319529, and by (3.8) we obtain $u_1(10^{-6}) = 0{\cdot}35319528$. Adding (3.7i) gives again 0·35319529. To compute $\underline{a}$ we have $u_2(10^{-6}) = 1{\cdot}20265024$; adding (3.7ii) gives 1·20587049, but even at the larger $\eta = 7 \times 10^{-6}$ we obtain $u_2(7 \times 10^{-6}) = 1{\cdot}19862274$, and adding (3.7ii) again gives

$$\int_0^1 p_0 v_0/\eta \, d\eta = 1{\cdot}20587049, \text{ whence } a = 1{\cdot}2311018.$$

An essentially similar error analysis governs the calculation of the functions $p_1(\eta)$ and $v_1(\eta)$ of (3.1) except that now there are genuine singularities at $\eta = 0$ and we solve for principal parts (Stewartson and Thompson, 1970).

4. ELECTROSTATIC PINCHING OF AN INCOMPRESSIBLE PLANE JET

[Michael and Thompson, to appear.]

In this problem an electrostatically perfectly conducting plane jet of incompressible fluid enters the space between parallel uniformly charged plates, parallel to the direction of -X. If $\eta(X)$ is the electrostatic pinching of the surface shape, then the following integral appears in the determination of η:

$$\eta_+(X) = \int_{-\infty}^{\infty} A_+(-\,ip)e^{-ipX}\, dp/(p \coth p\alpha + Bp^2), \tag{4.1}$$

where

$$H_+(z) = \int_{r_0}^{\infty} e^{-zt}\, t^{-z-1}\, dt/(1+t) \text{ and } r_0 + \log r_0 = 0, \tag{4.2}$$

A, α, B being physical constants. The function $H_+(z)$ as defined in (4.2) can be evaluated whenever Re $z \geq 0$. By writing it in the form

$$H_+(z) = \int_0^{\infty} (e^{-zt} - 1)t^{-z}dt \int_0^{\infty} [e^{-st} - e^{-s(t+1)}]ds + \int_0^{r_0} (e^{-zt} - 1)t^{-z}(\frac{1}{t} - \frac{1}{1+t})dt$$
$$+ \int_{r_0}^{\infty} t^{-z}(\frac{1}{t} - \frac{1}{t+1})dt$$

we can show that

$$H_+(z) = \pi \operatorname{cosec} \pi z \left\{ \sum_{n=1}^{\infty} z^{z+n}/\Gamma(z+n+1) - e^z \right\} + e^{r_0 z} \sum_{n=0}^{\infty} e_n(z)(-r_0)^n/(z-n) \tag{4.3}$$

where $e_n(z) = \sum_{r=0}^{n} z^r/r!$ This is bounded for all finite z and has an asymptotic expansion

$$H_+(z) \sim \frac{1}{(1+r_0)^2 z^2} - \frac{2r_0}{(1+r_0)^4 z} + O(z^{-3}). \tag{4.4}$$

Thus (4.1) can be evaluated by residues - which is an important simplification because a numerical integration of (4.1) carrying $(-ip)^{-ip}$ past $p = 0$ would be hard indeed to devise. The poles in (4.1) are all determined by the equation $p \coth p\alpha + Bp^2 = 0$, and are all pure imaginary.

Unfortunately, (4.3) is good for exposing the analytic behaviour of $H_+(z)$, but the boundedness property at integer values is deduced by delicate algebra which amply suggests that (4.3) is a poor algorithm for calculation.

(4.2) furnishes an alternative, and a new contour in t is all that is needed to handle negative z. Essentially what was done was to retain the series form for $|z| < 1$ and fit Chebyshev polynomials to $H_+(z)$ up until $|z| = 8$, beyond which point the asymptotic formula extended to $O(z^{-8})$ is adequate.

The Chebyshev polynomials were fitted by selective points collocation (Fox and Parker, 1968) which is a standard method, but the integration of (4.2) deserves a special word. Hartree (1958) suggests the use of a Gauss-Laguerre integration formula $\int_0^\infty e^{-x} f(x)dx = \sum_{i=1}^{n} w_i f(a_i)$ where a_i is the i-th zero of the n-th order Laguerre polynomial. This is bad advice when z is small, because the initial transformation x = zt gives an integral $z^{1+z} \int_{zr_0}^\infty e^{-x} x^{-z-1} dx/(x + z)$ and moves the pole of the integrand arbitrarily close to the contour as $z \to 0$. A more viable method is to split the range at t = 1 and obtain

$\int_1^\infty e^{-zt} t^{-z-1} dt/(1+t) = \int_0^1 e^{-z/u} u^z du/(1 + u)$. An adaptive quadrature is now possible on both ranges. The one used is due to my research student I.G.A. Robinson (unpublished) in which a 3-point Gaussian formula is used on a basic interval, this being trisected if need be to make the initial nodes the central nodes of the resulting 3 intervals, thus permitting us to re-use function values calculated at these points. At z = 2 Robinson's method gives $H_+(z) = 0{\cdot}164110697$; this is to be compared with a 15-th order Gauss-Laguerre value $H_+(2) = 0{\cdot}16408733$. At z = 1 matters are worse: Robinson's value is $H_+(1) = 0{\cdot}274576573$; the Gauss-Laguerre value is $0{\cdot}27235468$. For $z \ll 1$ Gauss-Laguerre is meaningless, but Robinson's values are uniformly reliable to 7 figures.

REFERENCES

Fox, L., and Parker, I.B., Chebyshev Polynomials in Numerical Analysis, (O.U.P., 1968) p.30.

Hartree, D.R., Numerical Analysis, (O.U.P., 1958) p.124.

Hillman, A.P., and Salzer, H.E., Phil.Mag. 34, 575 (1943).

Lee, J.S., and Fung, Y.C., J.Fluid Mech., 37, 657-670 (1969).

Michael, D.H., and Thompson, B.W., "Electrostatic Pinching of a Plane Incompressible Jet", (to appear).

Rayleigh, J.W.S., Theory of Sound, Vol. I., §174 (MacMillan, 1878).

Sedov, L.I., Similarity and Dimensional Methods in Mechanics, (Acad.Press, Inc., 1959) p.219.

Stewartson, K., and Thompson, B.W., Proc.Roy.Soc.Lond. A, 304, 255-273 (1968).

______________________________, Phys. of Fluids, 13, 227-236 (1970).

Thompson, B.W., "Complex Solutions of Equations z = f(z).", Comp.Lab.Rep., Melbourne (1959).

____________, Ph.D. Thesis, London, (1967) Ch. VI.

____________, J.Fluid Mech., 31, 379-395 (1968).

Gaydon, F.A., and Shepherd, W.M., Proc.Roy.Soc.Lond., A 281, 184-206 (1964).

LAMINAR FLOWS PAST A FLAT PLATE AT VARIOUS ANGLES OF ATTACK

Hans J. Lugt and Henry J. Haussling
Naval Ship Research and Development Center, Washington, D.C.

ABSTRACT

Numerical solutions are obtained by means of a stream function-vorticity formulation for laminar incompressible fluid flows past a flat plate at the angles of attack $\alpha = 0^{\circ}$, 45°, and 90° with Reynolds numbers Re = 30, 50, and 200. The cross-section of the plate is elliptic and the flow is assumed to be two-dimensional and time-dependent. Potential flow is selected as the initial condition. Asymptotic steady-state solutions have been obtained for the symmetric configurations $\alpha = 0^{\circ}$ and 90°, whereas the 45°-inclined plate always caused vortex shedding, even for Re as low as 30. The development of a Kármán vortex street for Re = 200, $\alpha = 45^{\circ}$ is presented in detail.

INTRODUCTION

A numerical study of flows around flat plates poses many more difficulties and demands much more computer time than such studies which have been performed for fluid motions past circular cylinders. This is due to the high surface curvature at the tips. The problem is discussed by Rimon and Lugt for disk-shaped bodies. The purpose of this study is twofold: to find methods to reduce the computer costs necessary for this type of calculation, and to study the characteristics of vortex shedding from inclined plates.

FORMULATION OF THE PROBLEM

The initial-boundary value problem for the Navier-Stokes equations is solved in the elliptic coordinate system (η, θ) which is defined by the transformation (see Fig. 1)

$$x + iy = a \cosh(\eta + i\theta), \quad a > 0, \; i^2 = -1 . \tag{1}$$

The equations of motion are formulated in terms of ψ, the dimensionless stream function, and ω, the dimensionless vorticity component normal to the (η, θ)-plane:

$$\frac{\partial \omega}{\partial t} + \frac{1}{h^2} \frac{\partial(\psi, \omega)}{\partial(\eta, \theta)} = \frac{2}{Re} \nabla^2 \omega , \tag{2}$$

$$\nabla^2 \psi = \omega . \tag{3}$$

Here, t and Re are the dimensionless time and the Reynolds number $Re = 2aU/\nu$, where a is the focal distance, ν the kinematic viscosity, and U the magnitude of the constant velocity at infinity. The characteristic length and velocity scales used in deriving Eqs. (2) and (3) are a and U. In particular, the del operator is made dimensionless by a. The coefficient h is defined by $h^2 = \cosh^2 \eta - \cos^2 \theta$. Boundary conditions are prescribed such that the velocity is zero at the body surface and approaches the constant $\vec{U}$ far from the body. The vector $\vec{U}$ is specified by α and U.

NUMERICAL TECHNIQUE

The infinite flow field is represented by a finite network of 75 x 60 = 4500 points in the (η, θ)-plane with the space increments $\Delta\eta = 0.05$, $\Delta\theta = 6^\circ$. Thus, the outer boundary is about 11 plate lengths removed from the body. The surface of the plate is located at $\eta = \eta_0$.

The numerical technique is similar to that used by Rimon. The linear part of Eq. (2) is approximated by the Dufort-Frankel scheme. The nonlinear terms are written in conservation form with the use of the velocity components corresponding to η and θ. Central differencing is applied to these terms. For the Poisson equation (3) the five-point difference approximation is used, and the resulting system of algebraic equations is solved by means of Gauss-Seidel line overrelaxation applied along lines of constant η. The overrelaxation factor is 1.82. The iteration process is halted after the k-th iteration if $|\nabla^2 \psi^k - \omega^k| < \epsilon$ at each grid point, where ϵ is of the order of 10^{-3}. The number of iterations varies with the magnitude of temporal change of the flow field. For Re = 200, $\alpha = 45^\circ$, $\eta_0 = 0.1$ the growth of the vortices attached behind the plate requires about 60 iterations per time step, whereas vortex separation needs about 130.

A one-sided difference scheme must be used to compute ω_0, the vorticity at the body surface. This approximation is very sensitive with regard to numerical stability. Of the various approximations tried, the following equation yields the best results:

$$\omega_0 = \frac{1}{4h_0^2(\Delta\eta)^2}(\psi_1 + 4\psi_2 - \psi_3) \; . \tag{4}$$

The subscripts 1, 2, and 3 refer to the grid points (η_1, θ), (η_2, θ), and (η_3, θ) with $\eta_1 = \eta_0 + \Delta\eta$, $\eta_2 = \eta_0 + 2\Delta\eta$, $\eta_3 = \eta_0 + 3\Delta\eta$. Formula (4) is derived from the Taylor series expansion of $\partial\psi/\partial\eta$ about (η_0, θ) evaluated at the points (η_1, θ) and (η_2, θ). The $\partial\psi/\partial\eta$-terms are replaced by central-difference terms.

The fact that the domain of numerical integration is bounded causes severe difficulties in prescribing the conditions at the outer boundary $\eta = \eta_\infty < \infty$. This boundary is almost circular and is divided in half. On the upstream half of the boundary the vorticity is specified as zero, and $\partial\psi/\partial\eta$ is expressed by the parallel-flow condition. Convection of vorticity across the downstream half of the boundary is allowed as proposed by Dawson and Marcus:

$$\frac{\partial\omega}{\partial t} + \frac{1}{U}(\vec{U} \cdot \nabla)\omega = 0 \; . \tag{5}$$

The second condition, which differs from that used by the above authors, is selected as the θ-component of

$$\frac{\partial\vec{v}}{\partial t} + \frac{1}{U}(\vec{U} \cdot \nabla)\vec{v} = 0 \; , \tag{6}$$

where $\vec{v}$ is the dimensionless velocity vector. The decision to use convection equations of this type is heuristic and based on flow photographs which reveal that the shapes of the vortices are well preserved over many cycles of the Kármán vortex street.

The vorticity ω is advanced in time at the interior points according to the finite-difference form of Eq. (2). Next, ω and $\partial\psi/\partial\eta$ are advanced on the downstream boundary points by means of Eqs. (5) and (6). The calculation of ψ follows with the aid of Eq. (3). The cycle concludes with the determination of ω_0 from Eq. (4).

The maximum stable time step Δt_{max}, beyond which instability occurs, is determined by increasing the time step until oscillations from one time step to another appear in the ω-values near the tips. The magnitude of Δt_{max} depends on other factors besides ω_0. It is a linear function of Re, at least for Re ≤ 200, and decreases rapidly as either $\Delta\eta$, $\Delta\theta$, or η_0 approaches zero (Lugt and Rimon). There is no dependence on α. Computer time is saved by enlarging η_0 as much as possible without the loss of the essential features of the flat plate (if η_0 is changed from 0.05 to 0.1, Δt_{max} increases fourfold). For Re = 200, $\eta_0 = 0.1$ the value of Δt_{max} is 0.005, smaller time steps being necessary near t = 0 when large vorticity gradients are present. For high surface curvature, that is for $\eta_0 < 0.1$, stability is improved if the grid system is shifted by $\Delta\theta/2$ in the θ-direction so that there are no grid points at the tips. This does not affect the overall accuracy of the solution and, in addition to improving the stability, makes possible the computation of the flow field when $\eta_0 = 0$. In this case the tips degenerate to singular body points and series expansions around the tips can be built into the numerical scheme. An account of this experience will be published in the future.

Accuracy considerations are similar to those for the flow past oblate spheroids (Rimon). Additional information on the quality of the solution is obtained by integrating $\partial\omega/\partial\eta$ over the body contour. The integral of this function over a segment of the surface is the pressure difference between the end points of the segment. Hence, the integration around the entire contour of the body should yield zero and is a sensitive test which must be passed to a certain degree by any accurate solution. For Re = 200, $\alpha = 45^0$, $\eta_0 = 0.1$ the value of this integral is always smaller than 3% of the difference between the maximum and minimum pressure at the surface. As discussed below other "confidence-building" tests are conducted by comparing results to those obtained with different methods.

RESULTS

Computations were carried out in double precision on an IBM 360-91 computer.

For $\alpha = 0^0$ the calculated drag coefficients approach to within a few percent the steady-state values of Dennis and Chang as $t \to \infty$. For $\alpha = 90^0$, Re = 50, $\eta_0 = 0.05$ a steady state is also approached. Fig. 2 shows the streamlines and equal-vorticity lines at t = 2.13. The corresponding drag coefficient is $C_D = \text{Drag}/\frac{\rho}{2} U^2 a \cosh \eta_0 = 2.8$. Here, ρ is the density of the fluid.

For $\alpha = 45^0$, $\eta_0 = 0.1$ an asymptotic steady state could not be obtained with Re between 30 and 200. Instead a Kármán vortex street develops. In this context it may be noted that Dumitrescu and Cazacu arrived at steady-state solutions for a flat plate between two parallel walls, which are two plate lengths apart, for $\alpha = 45^0$, Re = 15 and 25, $\eta_0 = 0$. Their method does not consider time-dependent motions.

In Fig. 3 the drag, lift, and moment coefficients C_D, C_L, and C_M are plotted versus time for the first two cycles with Re = 200, $\alpha = 45^0$, $\eta_0 = 0.1$. The quantities C_L and C_M are defined by C_L = -Lift/$\frac{\rho}{2}$ U^2 a cosh η_0 and C_M = -Torque/$\frac{\rho}{2}$ U^2 a^2 $\cosh^2$ η_0. The sudden start from potential flow forces C_D and C_L to be infinite and C_M to be zero at t=0. The lift is directed downward (in Figs. 4 and 5), and the torque always tends to turn the plate normal to the flow. The first two cycles are still in the transient state but already show the typical features of vortex shedding. The relation between the curves in Fig. 3 and the streamline patterns is discussed below.

Streamlines for Re = 200, $\alpha = 45^\circ$, $\eta_0 = 0.1$ are shown after the second cycle of periodic vortex shedding in Fig. 4. Notice the boundary of the region of integration on the left-hand side of the figure.

Fig. 5 is a picture sequence of streamlines and equal-vorticity lines exhibiting the first cycle in the development of a Kármán vortex street. The abrupt initial movement of the plate creates a starting vortex which is known from aerofoil theory to exist for high Reynolds number. This vortex is clearly visible in the streamlines as a "vortex wave" moving relative to the plate and in the equal-vorticity lines as a relative extremum. For Re = 30 the starting vortex is barely noticeable due to the dominance of frictional diffusion.

A comparison of Figs. 3 and 5 reveals that C_D attains its relative maxima when the vortices are about to separate from the plate, that is, when the vortices attached to the plate have gained their maximum strengths. The maxima are more pronounced when a vortex is separating from the rear of the leading tip than when a vortex is separating from the rear of the trailing tip. The shedding in the latter instance has almost no effect on C_L. The relative maxima of C_L and C_M have small phase differences with regard to those of C_D. The Strouhal number, 2an/U, for the second cycle is 0.23 (n = frequency).

All vortices which shed from the body dissipate in such a way that the relative extrema of their vorticity (which may be considered as vortex centers) decrease in their values approximately as 1/(t-const). This indicates that the vortices behave like decaying potential vortices when separated from the plate.

Detailed and additional results will be published in a forthcoming paper.

ACKNOWLEDGMENT

This study was performed under the Navy Independent Research Program, Task Area ZR0110101.

REFERENCES

Dawson, C. and Marcus, M., Proc. 1970 Heat Transfer and Fluid Mech. Inst., Stanford University Press, 323-338 (1970).

Dennis, S.C.R. and Chang, Gau-Zu, Physics of Fluids 12, Suppl. II, 88-93 (1969).

Dumitrescu, D. and Cazacu, M.D., ZAMM 50, 257-280 (1970).

Lugt, H.J. and Rimon, Y., Naval Ship Research and Development Center, Rep. No. 3306 (1970).

Rimon, Y., Physics of Fluids 12, Suppl. II, 65-75 (1969).

Rimon, Y. and Lugt, H.J., Physics of Fluids 12, 2465-2472 (1969).

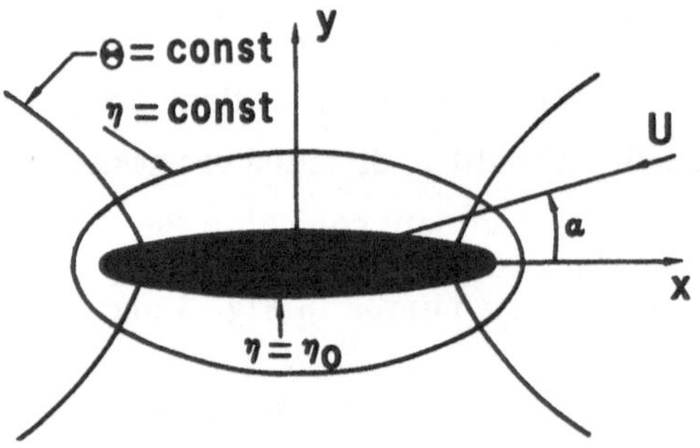

FIGURE 1:
ELLIPTIC COORDINATE SYSTEM AND DEFINITION OF ANGLE OF ATTACK

FIGURE 2:
STREAMLINES AND EQUAL-VORTICITY LINES FOR Re=50, α=90°, η_0=0.05 AT t=2.13 (ALMOST STEADY STATE)

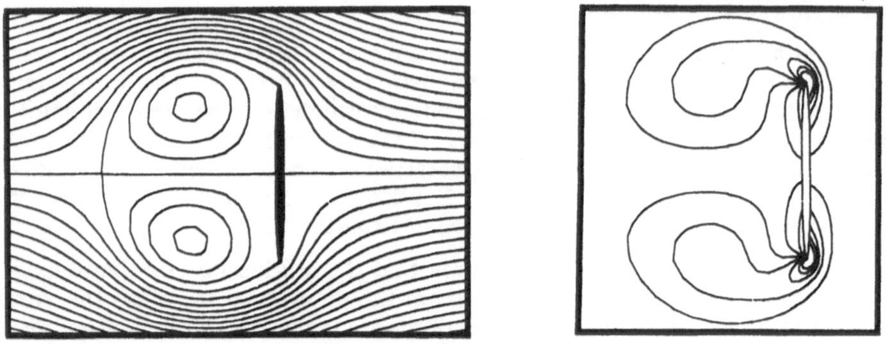

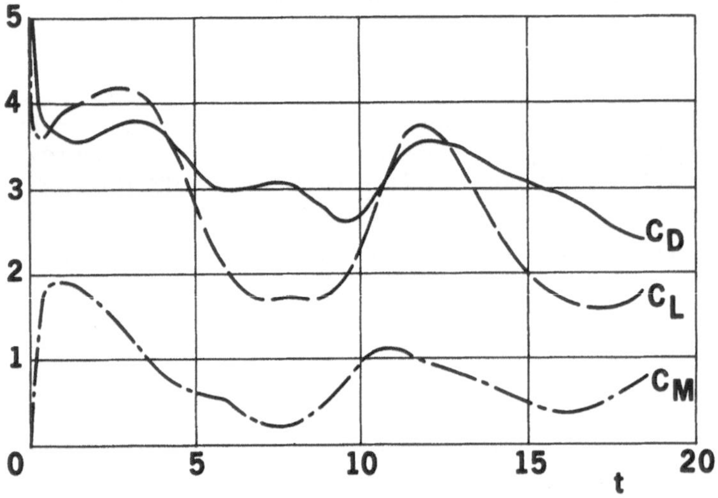

FIGURE 3:
DRAG, LIFT AND MOMENT COEFFICIENTS VERSUS TIME FOR Re=200, α=45°, η_0=0.1

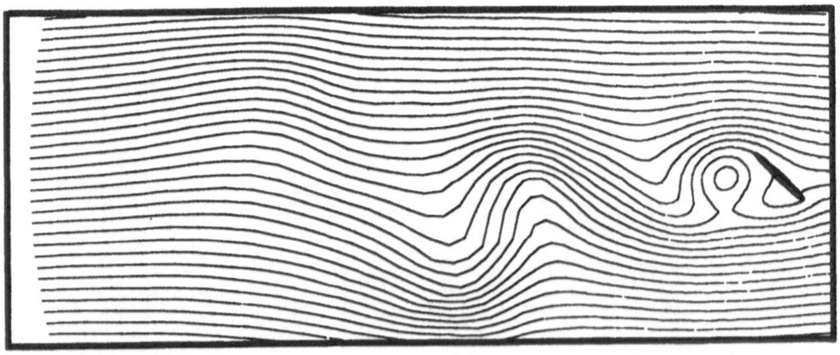

FIGURE 4:
STREAMLINES FOR Re=200, α=45°, η_0=0.1 AT t= 18.4

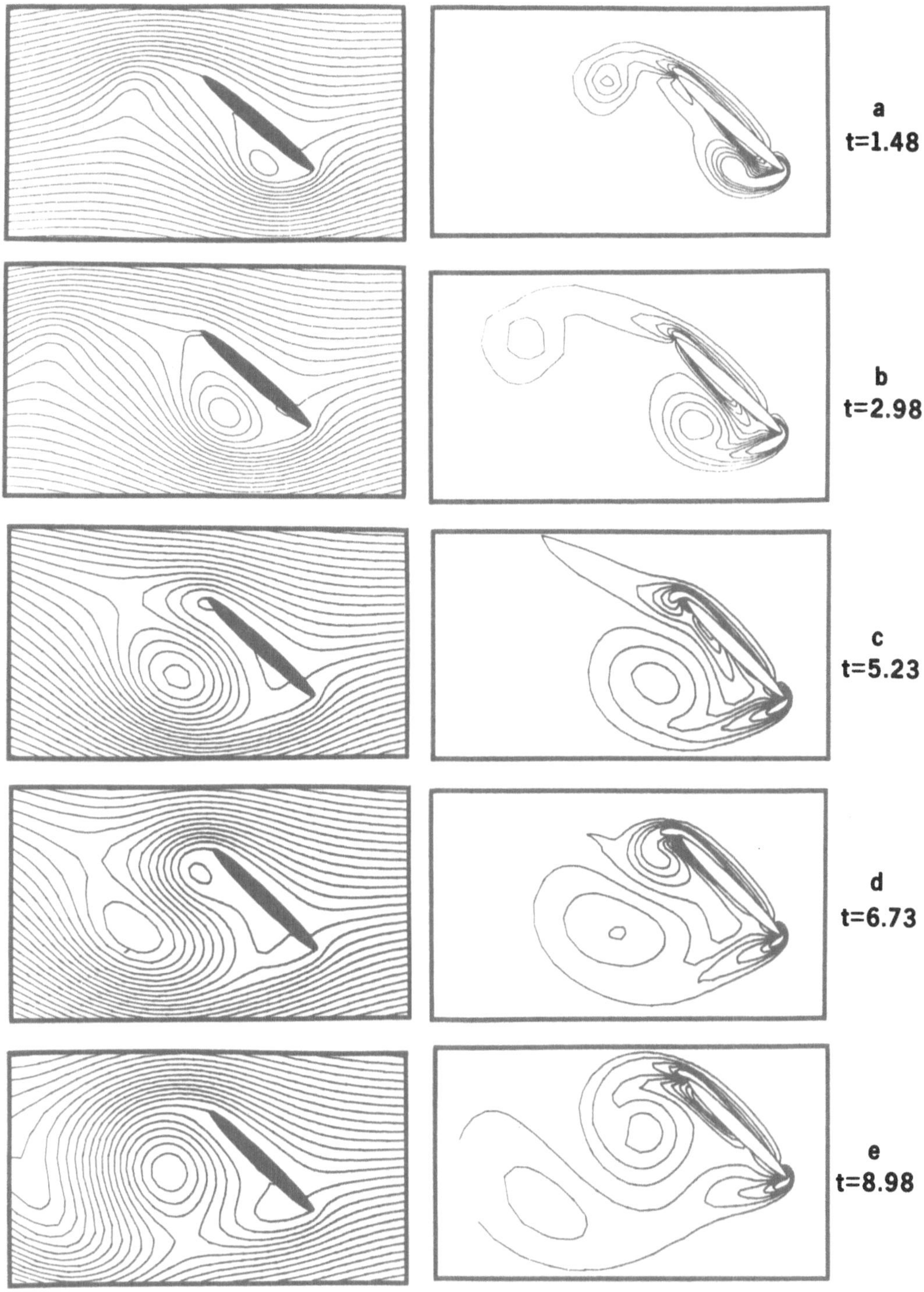

FIGURE 5: SEQUENCE OF STREAMLINES AND EQUAL-VORTICITY LINES FOR Re=200, α=45°, η_o=0.1 AT VARIOUS TIMES (POTENTIAL FLOW AT t=0)

INCLUSION OF TURBULENCE EFFECTS IN NUMERICAL FLUID DYNAMICS

Bart J. Daly and Francis H. Harlow*
Los Alamos Scientific Laboratory
University of California
Los Alamos, New Mexico

INTRODUCTION

If the time-dependent flow of a fluid is uniquely determined (except for arbitrary phase) by the macroscopic initial and boundary conditions, then we say that the flow is laminar. In contrast, if the flow features depend significantly on the microscopic influences (wall roughness, low amplitude vibrations, etc.), then the flow is turbulent.

To show this distinction, consider the flow of water past a circular cylinder. For very slow speeds, the flow development is uniquely determined by the magnitude of the Reynolds number, $Re \equiv ud/\nu$, in which u is the input flow speed, d is the diameter of the cylinder, and ν is the kinematic viscosity of the water. Since these are macroscopic parameters, this flow is called laminar. As the speed increases, a double vortex develops behind the cylinder, and when the Reynolds number exceeds about 40, the double vortex begins to oscillate. Except for an arbitrary phase, however, the oscillations are uniquely determined by the macroscopic parameters, u, d, and ν, and the flow is still completely laminar. As the Reynolds number continues to increase, the changing pattern maintains its laminar property until a stage is reached in which noticeable effects are observed that do not depend upon the macroscopic parameters. These spurious features are the result of microscopic perturbations, which for lower Reynolds numbers could never amplify to significance. They depend upon the details of cylinder roughness, slight mechanical vibrations and minor fluctuations of input water speed, and signify the onset of turbulence.

The effects of turbulence are usually of crucial importance in their influence on the mean flow. Since turbulent flows are much more commonly encountered than laminar ones, the subject of turbulence is accordingly of great importance. Especially we would like to be able to predict its occurrence and effects, and it is the purpose of this paper to describe one way in which this can be accomplished.

Specifically, we discuss here a transport-equation method, wherein the turbulent properties of the fluid are represented by a small number of moment functions, and their changes with time are described by differential transport equations. The detailed derivation of this theory has been presented elsewhere by Harlow and Hirt (1969); our purpose here is to show the manner in which the equations can be solved numerically for various specific instances in which analytical solutions cannot be

*This work was performed under the auspices of the United States Atomic Energy Commission.

attained. Especially, we wish to describe the incorporation of the turbulence transport equations into some of the current techniques for solving problems of time-dependent incompressible fluid flow in several space dimensions.

SOME FEATURES OF TURBULENCE TRANSPORT THEORY

The detailed structure of turbulence is exceedingly complicated. Our goal, however, is to describe the important properties of this fine-scale structure by means of a small number of variables, in the hope that specific details are unimportant except as they join to contribute to the required low-order moments.

This sort of approach, of course, is not without precedent. In gas dynamics the detailed motions of countless molecules would be hopeless to treat theoretically in detail. Nevertheless, their joint action is quite accurately represented in many circumstances of interest by such moment properties as density, pressure, heat energy, viscosity, and the coefficient of thermal conduction. Likewise, the variations of these quantities can be described by transport equations that represent the variations of mass, momentum and energy, supplemented by flux expressions for the higher moments that measure rates of molecular transport.

It is the success of such a gas-dynamical representation that encourages a similar model for turbulence dynamics and, indeed, furnishes some guidance in the derivations. Just as the gas dynamics equations are derived from a Liouville equation for the fluctuating molecular motions, so also are the turbulence transport equations derived by taking moments of the fluctuating Navier-Stokes equations. In both cases, higher-order flux moments must be approximated, so that neither type of transport equation is exact. The strongly non-linear turbulence equations require many more approximations for the higher moments than do their molecular counterpart. This means that our transport theory for turbulence requires a very extensive program of proof-testing. This is especially so because the theory is intended to be universal in its applicability, and accordingly must be capable of yielding meaningful results in a wide variety of circumstances.

In contrast to many other theories of turbulence, which are intended to have restricted applicability, the present transport equations must satisfy several invariance principles. These are

1. dimensional invariance: The theory must apply to any system of units,

2. Galilean invariance: The uniform translation of the reference frame must not alter the results,

3. tensor invariance: The equations must not be changed in physical content by rotation to another coordinate system,

4. problem invariance: The coefficients and functions in the transport equations must not depend upon the initial and boundary conditions of each specific application.

The first three of these have furnished useful guidance in the formulation of the various flux approximations, and have been relatively easy to satisfy. The fourth is more difficult, in that the burden of inaccuracy (e.g., from missing terms of higher order that might contribute to the results) can only fall upon the "universal" coefficients which, at this stage, still require a small amount of problem-dependent adjustment to achieve accurate results.

THE TRANSPORT EQUATIONS

For an incompressible fluid, the equations of fluid dynamics can be written

$$\frac{\partial u_j}{\partial x_j} = 0 \quad , \tag{1}$$

$$\frac{\partial u_i}{\partial t} + u_j \frac{\partial u_i}{\partial x_j} = - \frac{\partial \phi}{\partial x_i} + \frac{\partial}{\partial x_j} (\nu e_{ij} - R_{ij}) + g_i [1 + \beta (T - T_o)] \quad , \tag{2}$$

in which u_i and g_i are the components of velocity and gravity in the direction x_i, ϕ is the ratio of pressure to (constant) density, ν is the molecular kinematic viscosity, R_{ij} is the Reynolds stress (representing the turbulent diffusion of momentum), β is the volumetric expansion coefficient, T is the temperature, T_o is a reference temperature, and

$$e_{ij} \equiv \frac{\partial u_i}{\partial x_j} + \frac{\partial u_j}{\partial x_i} \quad .$$

Our central problem is to determine the transient behavior of R_{ij}, for which we have the transport equation of Daly and Harlow (1970):

$$\frac{\partial R_{ij}}{\partial t} + u_k \frac{\partial R_{ij}}{\partial x_k} = - R_{ik} \frac{\partial u_j}{\partial x_k} - R_{jk} \frac{\partial u_i}{\partial x_k} + \Phi_{ij}$$

$$+ \frac{\partial}{\partial x_k}\left(\frac{\alpha s^2 R_{k\ell}}{\nu\Delta} \frac{\partial R_{ij}}{\partial x_\ell}\right) - 4\nu D_{ij}$$

$$+ \theta \left[\frac{\partial}{\partial x_j}\left(\frac{s^2 q}{\nu\Delta} \frac{\partial R_{ik}}{\partial x_k}\right) + \frac{\partial}{\partial x_i}\left(\frac{s^2 q}{\nu\Delta} \frac{\partial R_{jk}}{\partial x_k}\right)\right]$$

$$+ \nu \frac{\partial^2 R_{ij}}{\partial x_k^{\,2}} + \frac{\tau s^2 \beta}{\lambda\Delta} (g_i R_{jk} + g_j R_{ik}) \frac{\partial T}{\partial x_k} \quad . \tag{3}$$

The additional symbols are $q \equiv R_{kk}/2$ (the turbulence energy density); $s \equiv$ a scale function (proportional to the integral scale); $D_{ij} \equiv$ Reynolds stress decay tensor; $\Phi_{ij} \equiv$ intercomponent coupling tensor; λ is the coefficient of themometric conductivity; while θ and τ are universal dimensionless constants with magnitudes near unity. Δ is a function of the turbulence intensity function, $\xi \equiv s(2q)^{1/2}/\nu$, which is proportional to the turbulence Reynolds numbers based on microscale; for $\xi << 5$, $\Delta \equiv 5$, while for $\xi >> 5$, $\Delta = \xi$. The scale function is currently taken to be

$$s \equiv (\Delta q/D_{kk})^{1/2} \quad , \tag{4}$$

while

$$\Phi_{ij} = - \frac{\omega\nu\Delta}{s^2} (R_{ij} - \frac{1}{3} R_{kk} \delta_{ij}) \tag{5}$$

plus other terms, as discussed by Daly and Harlow (1970), that need not concern us here. The coefficient, ω, is a function of the mean-flow strain rate:

$$\omega = 1 - \frac{s^2}{\nu\Delta q} R_{k\ell} \frac{\partial u_k}{\partial x_\ell} \quad .$$

The significance of the various terms in the Reynolds stress transport equation, Eq. (3), can be described as follows:

1. On the left are the Eulerian time derivative and a term representing the

transport of Reynolds stress by the mean motion of the fluid.

2. The first two terms on the right describe the creation of turbulence as a result of varying strain in the mean flow. Proportionality to the Reynolds stress itself shows that some "seeding" is necessary, in order that energy can be transferred from the mean motion to turbulent motion. There are no approximations in any of the terms discussed so far.

3. The role of Φ_{ij} is to accomplish the conservative (of energy) redistribution of R_{ij} among its various components. Equation (5) shows the term that contributes a tendency towards isotropy. In addition, there are near-wall terms of various possible forms that can describe the distortions of turbulence away from isotropy. These various terms are intended to represent contributions from several sources, including multiple velocity correlations and the correlations between pressure and rate-of-strain fluctuations.

4. The terms proportional to α and θ describe conservative diffusive processes, the former produced by convective properties of the interpenetrating turbulent eddy motions, and the latter by correlations between pressure and velocity fluctuations. The θ diffusion may effect some re-distribution among the components, while the α diffusion does not.

5. The decay term, $- 4\nu D_{ij}$, represents the loss of Reynolds stress as a result of the breakdown of large eddies into small ones and their subsequent dissipation by molecular viscosity. Despite the apparent dependence of this term upon the magnitude of the viscosity coefficient, that dependence disappears when the intensity of the turbulence is great. Physically, this means that the inertial properties of the fluid work to remove energy from the large eddies at a rate that is independent of the level of molecular viscosity, and to give that energy to the small eddies, from which any level of viscosity is capable of dissipating the energy at the same rate as it arrives. (This can happen because the scale of the small eddies automatically adjusts to the fact that the viscous dissipation rate depends upon both that scale and the magnitude of the viscosity coefficient.)

6. The other term in Eq. (3) explicitly containing ν is a purely diffusive term, which for intense turbulence is negligible in comparison with the α diffusion, but for weak turbulence (e.g., near a wall) can be a significant factor in the transport processes. The physical origin of the viscous diffusion terms arises from the influence of one eddy rubbing against another or against adjacent laminar fluid, thereby transporting energy through viscous stress.

7. Finally, for fluid flows in which there is a gradient in the mean temperature, the creation of turbulence from buoyancy effects is represented by the τ term. The physical origin of this comes from the fact that a local warm spot will tend to rise against gravity while a local cool spot will fall, in both cases producing an increase in the fluctuating velocity of the fluid.

There remains the question of the dissipation function, D_{ij}. The simplest approach is to relate D_{ij} to the scalar integral scale function, s, and to either specify the behavior of s from empirical (or other) considerations or else to write a single scalar scale equation. The advantage of this is that for many circumstances of interest, the dissipation of turbulence energy is a well-known function of the microscale, which in turn is closely related (in known fashion) to our integral scale function, s. In following such a procedure, we would write

$$D_{ij} \equiv \frac{\Delta}{2s^2} R_{ij} \tag{7}$$

and use the contraction of the D_{ij} equation, below, to find the scalar scale equation for $D \equiv D_{ii} \equiv \Delta q/s^2$. Indeed, this is the approach that has been used in all previous descriptions of our turbulence transport theory.

There is evidence, however, to suggest that the scalar proportionality in Eq. (7) between D_{ij} and R_{ij} is not correct for many circumstances of interest. Often D_{ij} is significantly more nearly isotropic than R_{ij}. To account for this, we have generalized the previously-scalar D equation to the following form:

$$\frac{\partial D_{ij}}{\partial t} + u_k \frac{\partial D_{ij}}{\partial x_k} = - g(\xi) \left(\frac{\partial u_i}{\partial x_k} D_{jk} + \frac{\partial u_j}{\partial x_k} D_{ik}\right)$$

$$+ \frac{\Delta}{2\nu s^2} \frac{\partial}{\partial x_k} \left(\frac{s^2 R_{k\ell}}{\Delta} \frac{\partial R_{ij}}{\partial x_\ell}\right) + \frac{\partial}{\partial x_k} \left(\frac{s^2}{\nu\Delta} R_{k\ell} \frac{\partial D_{ij}}{\partial x_\ell}\right)$$

$$+ \frac{\theta}{2\nu} \left[\frac{\partial}{\partial x_i} \left(q \frac{\partial R_{jk}}{\partial x_k}\right) + \frac{\partial}{\partial x_j} \left(q \frac{\partial R_{ik}}{\partial x_k}\right)\right]$$

$$+ \nu \frac{\partial^2 D_{ij}}{\partial x_k^2} - \frac{2\nu\Delta'}{s^2} D_{ij}$$

$$+ \frac{f(\xi)\beta}{2\lambda} \frac{\partial T}{\partial x_k} (g_i R_{jk} + g_j R_{ik})$$

$$- \frac{\omega'\nu\Delta}{s^2} (D_{ij} - \frac{1}{3} D_{kk} \delta_{ij})$$

$$- \frac{\omega''\nu\Delta}{s^2} (D_{ij} - \frac{D_{kk}}{2q} R_{ij}) \quad . \tag{8}$$

In addition to the previous symbols, $g(\xi)$ and $f(\xi)$ are slowly varying functions of the turbulence intensity function, ξ; Δ' is a function closely related to Δ and differing from it by no more than a factor of two; while ω' and ω'' are newly introduced coefficients that were not previously required for the scalar D equation. The ω' term is necessary in order to assure a continuing tendency to isotropy; it seems likely that ω' should exceed ω in magnitude, in order to assure the usually-observed greater isotropy of D_{ij} than of R_{ij}. The ω'' term is included speculatively in anticipation of the requirement for tying the variations of D_{ij} to those of R_{ij} as the components of the latter become small, so as to preclude the occurrence of finite Reynolds-stress decay rate for a vanishing component. The significance of the other terms in the D_{ij} equation is in each case similar to that of the corresponding term in the R_{ij} equation.

In view of the uncertainties of some of the terms in the full tensor D_{ij} equation, however, we recommend for now the use of the contracted version, in conjunction with Eq. (7).

APPLICATION TO NUMERICAL PROBLEM SOLVING

In almost all applications of interest, it is not possible to solve the turbulence transport equations analytically. An exception is the case of isotropic, homogeneous decay, in which case the results agree with experiments and with the analysis of other investigators.

The next order of complexity is the class of problems in which all quantities vary in one space dimension only. This is approximately true, for example, along the centerline of a wind tunnel with axisymmetric variations in cross sectional area. Solutions for such a case have been compared by Harlow and Romero (1969) with experimental results, with excellent agreement even for the anomalous examples involving very great contraction.

Another type of one-dimensional turbulent flow to which the equations have been applied is the passage of fluid through the space between two plates. This is a statistically steady situation for which there has also been satisfactory agreement with experiments.

In general, however, the greatest strength of a generalized turbulence transport theory is its potential applicability to non-steady flow problems in three space dimensions. Of great practical interest, for example, is the problem of air pollution dispersal in a varying, turbulent wind blowing over a city built on hills and adjacent to mountains.

To investigate such situations requires numerical solution techniques that can be applied with high-speed electronic computers. For the mean-flow equations, a variety of such techniques have been listed by Harlow (1969). These utilize one or another finite-difference approximation to the equations, together with a progression of time steps or cycles for the advancement of the changing configuration. To add the transient effects of turbulence, it is necessary to extend the computer program to include a finite-difference version of the turbulence equations. A previous description of this by Amsden and Harlow (1968) showed one way in which this could be accomplished; subsequent investigations have demonstrated that a somewhat more complicated formulation is required in order to overcome some of the problems of numerical instability. This is discussed in the following section. Here we wish to point out that the incorporation of the turbulence transport equations can also mitigate somewhat one of the most vexing of the numerical instability difficulties encountered in the solution of incompressible flow problems.

This difficulty is most clearly seen in the calculation of incompressible fluid flows at high Reynolds numbers. Numerical stability requires that the effective viscosity be large enough to damp out the fluctuations that otherwise tend to amplify from the effects of the convection terms. For low-Reynolds-number flows, this can be accomplished in various ways. The most desirable is to utilize the true viscosity of the fluid for this purpose, and to otherwise minimize the diffusive (second order) truncation errors. Alternatively the convective terms can be differenced in such a way as to introduce positive dissipation of a non-physical form, which is tolerable if the effects they produce on the mean flow are negligible.

For high Reynolds numbers, however, the molecular viscosity is much too small to be used for assuring numerical stability. Thus it is fortunate that the incorporation of turbulence transport equations introduces an added effective viscosity that can be many orders of magnitude greater than that produced by the molecular properties of the fluid. Accordingly, the calculation of turbulent flows at very high Reynolds numbers is substantially easier as a result of the mathematically stabilizing effect of this additional viscosity on the finite-difference equations.

One question that remains unresolved so far is whether or not the turbulence transport equations automatically accomplish the distinction described in the introduction between laminar and turbulent flows. Would a general numerical solution of the equations for the flow past a cylinder, for example, correctly treat the large-scale fluctuations as being part of the mean flow, while the small-scale fluctuations were properly represented by the turbulence moments? The question is closely related to the equally unresolved matter of the scale spectrum implied for the turbulence by the low-order transport approximation. Presumably, those parts of the flow that are not formed of the universal section of the scale distribution must indeed be part of the mean flow. Alternatively, if there is a double, superimposed form for the spectrum, then two sets of turbulence transport equations would be required, with coupling terms to show the interaction. Preliminary experiments with this latter have shown some success in describing the complicated turbulence in a wake with lateral intermittency.

In general, however, when the scale spectrum of a flow departs from a universal form appropriate for the turbulence transport equations, then the larger-scale parts of the motion will have to be treated by means of the mean-flow Navier-Stokes equations. An example of this is the problem of the buoyancy turbulence between two plates. For small temperature differences, the flow is entirely laminar, being composed of the Bénard cells. As the heating increases, turbulence is superimposed upon the mean flow, and the scale spectrum of the entire motion departs strongly from the universal form pertaining to that part which is the pure turbulence. Accordingly, the problem requires coupling of the mean flow equations with the turbulence transport equations, the latter containing both buoyancy and shear creation terms. Other examples of this type are the intermittent turbulent flows in boundary layers, jets and wakes.

NUMERICAL CONSIDERATIONS

The inclusion of turbulence transport effects in the finite-difference mean-flow equations requires only a trivial extension of existing techniques. Complications arise, however, in coupling the mean-flow effects back into finite difference versions of Eqs. (3) and (8). There are several reasons for these complications, including numerical stability limitations, resolution requirements and the difficulty of achieving either realistic initial turbulent conditions or the onset of turbulence from a previously laminar flow.

From a linear stability analysis one can estimate the effects of explicit time differencing of Eq. (3) and make comparisons with the mean flow stability limits. For example, if the term expressing the turbulent diffusion of R_{ij} is differenced explicitly, a stability limitation of the form,

$$\delta t \lessapprox (\text{mesh spacing})^2/\alpha\nu\Delta \quad , \tag{9}$$

is incurred. The corresponding requirement to avoid diffusional instability of the mean flow is

$$\delta t < (\text{mesh spacing})^2/2\nu \quad . \tag{10}$$

In calculations involving high intensity turbulence, the first limitation would be far more restrictive than the second. In such problems it would be advantageous to solve the turbulence equations in a time-advanced form so that these stability restrictions could be minimized.

A completely advanced-time implicit solution for Eqs. (2), (3), and (8) would be quite time-consuming and could be expected to result in a large truncation error accumulation. A time centered implicit version would reduce the truncation errors. In some cases, an even better approach is to use a two stage solution. In the first stage Eqs. (3) and (8) are solved explicitly to obtain an advanced-time estimate of R_{ij} and D_{ij}. Equations (1) and (2) are solved only once since they require the solution of a Poisson equation, but the R_{ij} term in Eq. (2) is time-centered using the advanced time estimate. The second stage of the solution involves the recalculation of Eqs. (3) and (8) in terms of time-centered values of u_i, R_{ij}, and D_{ij}. This completes one calculation cycle.

The mesh resolution needed to resolve high Reynolds number turbulent flows may be considerably finer than that which is usually required in low Reynolds number laminar flow studies. Thus, in calculations of turbulent flow between parallel plates by Daly and Harlow (1970), the minimum one-dimensional mesh resolution for which useful results could be obtained was 100 cells when the Reynolds number was twelve thousand; 300 cells were required at a Reynolds number of thirty thousand, and 500 cells were needed to calculate a flow at a Reynolds number of sixty thousand. If this fineness of resolution were needed in three-dimensional calculations, it would be extremely difficult to perform such numerical studies. However, the need for such fine resolution is ordinarily great only in regions of large mean flow gradients, such as occur in boundary layers. If these strong gradient regions are spatially fixed, then a variable mesh can be used to obtain fine resolution in those regions and poorer resolution elsewhere. In the parallel plate studies mentioned above, it was found that the minimum number of calculation cells could be reduced as much as 80% by changing from a uniform to a variable mesh. However, the time increment, Eq. (9), is limited by the smallest mesh spacing.

It sometimes can be difficult to arrive at appropriate initial conditions for turbulent flow calculations. This is especially true in cases where one wishes to examine the transient aspects of the flow, but it can even be a problem when one is seeking a steady solution. In the latter case the numerical analyst can be guided by his a priori knowledge of the steady configuration and choose his initial profiles to conform as closely as possible to that configuration in order to minimize the computation time. However, because of the extreme complexity of the turbulence equations and their strong coupling terms, it is not uncommon that these initial guesses lead rapidly to catastrophe, such as negative turbulence energies. Then much experimentation may be needed to find starting conditions that permit convergence to steady state.

The problem is more difficult in transient flows. In particular, it is not possible to examine the temporal flow development, from laminar flow to the onset of turbulence, and finally to fully developed turbulent flow. The reason is that the intermediate stages involve Reynolds numbers that are too high for numerical stability. Only if the initial stage of turbulence is rather well-developed can the transient calculations be expected to proceed without difficulty.

In the calculation of transient, initially turbulent flows, it is important that the starting conditions be physically "real" or that realistic configurations are attained rapidly in the computation. Tests for realism will require frequent comparison with experiment.

REFERENCES

Amsden, A. A., and Harlow, F. H., J. Comp. Phys. 3, 94 (1968).

Daly, B. J., and Harlow, F. H., Phys. Fluids (1970).

Harlow, F. H., "Numerical Methods for Fluid Dynamics, An Annotated Bibliography," Los Alamos Scientific Laboratory Report No. LA-4281 (1969).

Harlow, F. H., and Hirt, C. W., "Generalized Transport Theory of Anisotropic Turbulence," Los Alamos Scientific Laboratory Report No. LA-4086 (1969).

Harlow, F. H., and Romero, N. C., "Turbulence Distortion in a Nonuniform Tunnel," Los Alamos Scientific Laboratory Report No. LA-4247 (1969).

ACCURATE NUMERICAL METHODS FOR BOUNDARY LAYER FLOWS*

I. TWO DIMENSIONAL LAMINAR FLOWS

Herbert B. Keller

California Institute of Technology, Pasadena, California

Tuncer Cebeci

Douglas Aircraft Co., Long Beach, California

1. INTRODUCTION

A very simple and accurate numerical scheme which is applicable to quite general boundary layer flow problems has been devised. It has been tested extensively on laminar flows, turbulent flows (using the eddy viscosity and eddy conductivity formulations), wake flows and many other such problems. In the brief space alloted to us here we shall illustrate the method by showing its application in some detail to nonsimilar plane laminar incompressible boundary layers and in particular to the well known case of Howarth's flow [3].

The method is based on an accurate difference scheme which has been devised for boundary value problems for systems of ordinary differential equations [5], [6] and which has recently been applied to mixed problems for parabolic partial differential equations [7]. After writing the basic equations as a first order system the derivatives are approximated by centered difference quotients and averages centered at the midpoints of net rectangles or net segments. Arbitrary (non-uniform) meshes can be employed and second order accuracy is still retained. The nonlinear difference equations are solved by Newton's method using an efficient block-tridiagonal factorization technique. Richardson extrapolation is valid and yields two orders of magnitude reduction of the error per application.

2. THE DIFFERENCE SCHEME

A plane nonsimilar incompressible laminar boundary layer flow is determined by specifying the external velocity field $u_e(x)$. In terms of the Levy-Lees transformed variables (ξ, η) given by

$$d\xi = \mu\rho u_e dx\ , \quad d\eta = \rho u_e (2\xi)^{-\frac{1}{2}} dy$$

and a dimensionless stream function $f(\xi,\eta)$, such that $\psi(x,y) = (2\xi)^{\frac{1}{2}} f(\xi,\eta)$, the usual boundary layer equations become [2]:

$$\frac{\partial^3 f}{\partial \eta^3} + f\frac{\partial^2 f}{\partial \eta^2} + \beta(\xi)\left[1-\left(\frac{\partial f}{\partial \eta}\right)^2\right] = 2\xi\left[\frac{\partial f}{\partial \eta}\ \frac{\partial^2 f}{\partial \xi \partial \eta} - \frac{\partial^2 f}{\partial \eta^2}\ \frac{\partial f}{\partial \xi}\right]. \tag{2.1}$$

Here $\beta(\xi) \equiv (2\xi/u_e)\, du_e/d\xi$ is the pressure-gradient parameter. Appropriate boundary conditions are, for example:

$$\text{a) } f(\xi,o) = f_w(\xi) \equiv -\mu^{-1}(2\xi)^{-\frac{1}{2}}\int_o^{\xi} (^v w/u_e)d\xi; \quad \text{b) } \partial f(\xi,o)/\partial\eta = 0;$$
$$\text{c) } \partial f(\xi,\eta_\infty)/\partial\eta = 1. \tag{2.2}$$

*This work was supported in part by the U. S. Army Research Office, Durham, under Contract DAHC 04-68-0006.

We have allowed $v_w(x) \neq 0$ to simulate mass transfer at the wall. The edge of the boundary layer, at $\eta = \eta_\infty(\xi)$ say, could easily be determined throughout the course of the calculations. But we shall not dwell on this simple modification here and in fact we take $\eta_\infty \equiv \text{const.}$ To treat axisymmetric flows we need only modify the variable transformation slightly.

At $\xi = 0$ eq. (2.1) reduces to the well known Falkner-Skan equation:

$$\frac{\partial^3 f}{\partial \eta^3} + f\,\frac{\partial^2 f}{\partial \eta^2} + \beta_o\,\left[1 - \left(\frac{\partial f}{\partial \eta}\right)^2\right] = 0. \tag{2.3}$$

The solution of this equation, say $f^o(\eta)$, subject to (2.2) at $\xi = 0$ furnishes "initial" conditions $f(o,\eta) = f^o(\eta)$ for solving (2.1), (2.2) in $\xi > 0$, $0<\eta<\eta_\infty$. Many schemes have been devised for solving (2.3) (see for example [1]) and in fact a special case of our present procedure does this very efficiently.

We place an arbitrary rectangular net of points (ξ_n, η_j) on $\xi \geq 0$, $0 \leq \eta \leq \eta_\infty$ and use the notation:

$$\begin{aligned} &\text{a)}\quad \xi_o = 0;\ \xi_n = \xi_{n-1} + k_n,\ n=1,2,\cdots;\\ &\text{b)}\quad \eta_o = 0,\ \eta_j = \eta_{j-1} + h_j,\ j=1,2,\cdots,J;\ \eta_J = \eta_\infty. \end{aligned} \tag{2.4}$$

No additional restrictions need be placed on the meshwidths h_j and k_n. Now we write (2.1) as a first order system by introducing the new dependent variables $u(\xi,\eta)$ and $v(\xi,\eta)$ as follows:

$$\text{a)}\ \frac{\partial f}{\partial \eta} = u; \qquad \text{b)}\ \frac{\partial u}{\partial \eta} = v; \quad \text{c)}\ \frac{\partial v}{\partial \eta} = -fv - \beta[1-u^2] + 2\xi\left[u\frac{\partial u}{\partial \xi} - v\frac{\partial f}{\partial \xi}\right]. \tag{2.5}$$

The difference approximations to (2.5) are defined, for $1 \leq j \leq J$, by

$$\begin{aligned} &\text{a)}\quad h_j^{-1}(f_j^n - f_{j-1}^n) = u_{j-\frac{1}{2}}^n; \qquad \text{b)}\quad h_j^{-1}(u_j^n - u_{j-1}^n) = v_{j-\frac{1}{2}}^n;\\ &\text{c)}\quad h_j^{-1}\left(v_j^{n-\frac{1}{2}} - v_{j-1}^{n-\frac{1}{2}}\right) = -(fv)_{j-\frac{1}{2}}^{n-\frac{1}{2}} - \beta_{n-\frac{1}{2}}\left[1-(u^2)_{j-\frac{1}{2}}^{n-\frac{1}{2}}\right] + 2\xi_{n-\frac{1}{2}}\left[u_{j-\frac{1}{2}}^{n-\frac{1}{2}} k_n^{-1}\left(u_{j-\frac{1}{2}}^n - u_{j-\frac{1}{2}}^{n-1}\right) - v_{j-\frac{1}{2}}^{n-\frac{1}{2}} k_n^{-1}\left[f_{j-\frac{1}{2}}^n - f_{j-\frac{1}{2}}^{n-1}\right)\right]. \end{aligned} \tag{2.6}$$

Here we have introduced the notation, for averages and intermediate net points:

$$\xi_{n-\frac{1}{2}} \equiv \xi_n - k_n/2,\ \beta_{n-\frac{1}{2}} \equiv \beta(\xi_{n-\frac{1}{2}}),\ u_{j-\frac{1}{2}}^n \equiv (u_j^n + u_{j-1}^n)/2,$$

$$v_j^{n-\frac{1}{2}} \equiv (v_j^n + v_j^{n-1})/2,\ (fv)_{j-\frac{1}{2}}^{n-\frac{1}{2}} \equiv (f_j^n v_j^n + f_j^{n-1} v_j^{n-1} + f_{j-1}^n v_{j-1}^n + f_{j-1}^{n-1} v_{j-1}^{n-1})/4, \text{ etc.}$$

Note that (2.6a,b) are centered at $(\xi_n, \eta_{j-\frac{1}{2}})$ while (2.6c) is centered at $(\xi_{n-\frac{1}{2}}, \eta_{j-\frac{1}{2}})$. These difference equations are extremely compact, using at most values at four netpoints comprising a net rectangle. Alternative averages for the nonlinear terms could be employed provided the proper centering is maintained. We suggest a choice which reduces the computations needed to evaluate the iterates used to solve the difference equations.

The boundary conditions (2.2) become simply:

$$\text{a)}\quad f_o^n = f_w(\xi_n); \quad \text{b)}\quad u_0^n = 0; \quad \text{c)}\quad u_J^n = 1 \tag{2.7}$$

To generate initial data we retain (2.6a,b) with n=0 and simply alter (2.6c) by setting $\xi_{n-\frac{1}{2}} = 0$ and using superscripts n=0 rather than $n-\frac{1}{2}$ in the remaining terms. The resulting difference equations, are then easily solved by the scheme described below and accurate approximations to the solution of the Falkner-Skan equation are obtained.

3. SOLUTION OF THE DIFFERENCE EQUATIONS

We solve the nonlinear difference equations (2.6), (2.7) recursively starting with n = 0 (on $\xi=\xi_o=0$). In general when the solution is known on $\xi=\xi_{n-1}$ we next solve on the line $\xi=\xi_n$. In detail suppose $\{f_j^{n-1}, u_j^{n-1}, v_j^{n-1}\}_o^J$ are known. To simplify notation we now write: $\{f_j^n, u_j^n, v_j^n\} \equiv \{f_j, u_j, v_j\}$. With this notation in (2.6) we multiply (2.6a,b) by h_j and (2.6c) by $2h_j$ to get $1 \leq j \leq J$:

$$\text{a)}\ (f_j - f_{j-1}) - h_j u_{j-\frac{1}{2}} = 0; \qquad \text{b)}\ (u_j - u_{j-1}) - h_j v_{j-\frac{1}{2}} = 0; \tag{3.1}$$

$$\text{c)}\ (v_j - v_{j-1}) + h_j[(fv)_{j-\frac{1}{2}} - \beta_{n-\frac{1}{2}}(u^2)_{j-\frac{1}{2}}] - h_j\sigma_n[(u_{j-\frac{1}{2}})^2 - v_{j-\frac{1}{2}}(f_{j-\frac{1}{2}} - f^{n-1}_{j-\frac{1}{2}}) - v^{n-1}_{j-\frac{1}{2}} f_{j-\frac{1}{2}}] = S^{n-1}_{j-\frac{1}{2}}$$

Here we have used: $\sigma_n \equiv (2\xi_{n-\frac{1}{2}}/k_n)$ and

$$\text{d)}\ S^{n-1}_{j-\frac{1}{2}} \equiv -(v^{n-1}_j - v^{n-1}_{j-1}) - h_j[(fv)^{n-1}_{j-\frac{1}{2}} + \beta_{n-\frac{1}{2}}\{2-(u^2)^{n-1}_{j-\frac{1}{2}}\}] - h_j\sigma_n[(u^{n-1}_{j-\frac{1}{2}})^2 - v^{n-1}_{j-\frac{1}{2}} f^{n-1}_{j-\frac{1}{2}}]. \tag{3.1}$$

The nonlinear algebraic system (3.1) and (2.7) contain(3J+3) unknowns which we compute by means of Newton's method. The iterates are denoted by $\{f_j^{(i)}, u_j^{(i)}, v_j^{(i)}\}$ and we determine them by first writing:

$$f_j^{(i+1)} = f_j^{(i)} + \delta f_j^{(i)}, \quad u_j^{(i+1)} = u_j^{(i)} + \delta u_j^{(i)}, \quad v_j^{(i+1)} = v_j^{(i)} + \delta v_j^{(i)}. \tag{3.2}$$

Then we insert these expressions in place of $\{f_j, u_j, v_j\}$ in (3.1) and drop the quadratic terms in $\{\delta f_j^{(i)}, \delta u_j^{(i)}, \delta v_j^{(i)}\}$. The resulting linear system of equations can be written in vector-matrix form as:

$$R_j^{(i)} \boldsymbol{\delta}_j^{(i)} - L_j^{(i)} \boldsymbol{\delta}_{j-1}^{(i)} = \mathbf{r}_{j-\frac{1}{2}}^{(i)}, \quad 1 \leq j \leq J. \tag{3.3}$$

Here we have introduced the (column) vectors:

$$\boldsymbol{\delta}_j^{(i)} \equiv (\delta f_j^{(i)}, \delta u_j^{(i)}, \delta v_j^{(i)})^T, \quad \mathbf{r}_{j-\frac{1}{2}}^{(i)} \equiv (\beta_{j-\frac{1}{2}}^{(i)}, \alpha_{j-\frac{1}{2}}^{(i)}, \gamma_{j-\frac{1}{2}}^{(i)} + S_{j-\frac{1}{2}}^{n-1})^T \tag{3.4}$$

where:

$$\alpha_{j-\frac{1}{2}}^{(i)} \equiv -(f_j^{(i)} - f_{j-1}^{(i)}) + h_j u_{j-\frac{1}{2}}^{(i)}; \quad \beta_{j-\frac{1}{2}}^{(i)} \equiv -(u_j^{(i)} - u_{j-1}^{(i)}) + h_j v_{j-\frac{1}{2}}^{(i)};$$

$$\gamma_{j-\frac{1}{2}}^{(i)} \equiv -(v_j^{(i)} - v_{j-1}^{(i)}) - h_j[(f^{(i)}v^{(i)})_{j-\frac{1}{2}} - \beta_{n-\frac{1}{2}}(u^{(i)})^2_{j-\frac{1}{2}}]$$

$$- h_j\sigma_n[(u_{j-\frac{1}{2}}^{(i)})^2 - v_{j-\frac{1}{2}}^{(i)}(f_{j-\frac{1}{2}}^{(i)} - f_{j-\frac{1}{2}}^{n-1}) - v_{j-\frac{1}{2}}^{n-1} f_{j-\frac{1}{2}}^{(i)}].$$

The 3x3 matrices are:

$$R_j^{(i)} \equiv \begin{pmatrix} 0 & 1 & -h_{j/2} \\ 1 & -h_{j/2} & 0 \\ a_j^{(i)} & b_j^{(i)} & c_j^{(i)} \end{pmatrix}, \quad L_j^{(i)} \equiv \begin{pmatrix} 0 & 1 & h_{j/2} \\ 1 & h_{j/2} & 0 \\ \tilde{a}_j^{(i)} & \tilde{b}_j^{(i)} & \tilde{c}_j^{(i)} \end{pmatrix}; \tag{3.5}$$

where:

$$a_j^{(i)} \equiv h_j\,[\,v_j^{(i)}+\sigma_n(v_{j-\frac{1}{2}}^{(i)}+v_{j-\frac{1}{2}}^{n-1})/2],\quad \tilde{a}_j^{(i)} \equiv -h_j[v_{j-1}^{(i)}+\sigma_n(v_{j-\frac{1}{2}}^{(i)}+v_{j-\frac{1}{2}}^{n-1})/2],$$

$$b_j^{(i)} \equiv -h_j[\,\beta_{n-\frac{1}{2}}\,u_j^{(i)}+\sigma_n u_{j-\frac{1}{2}}^{(i)}\,],\quad \tilde{b}_j^{(i)} \equiv h_j\,[\,\beta_{n-\frac{1}{2}}\,u_{j-1}^{(i)}+\sigma_n\,u_{j-\frac{1}{2}}^{(i)}\,],\tag{3.5}$$

$$c_j^{(i)} \equiv 1+h_j\,\sigma_n\,(f_{j-\frac{1}{2}}^{(i)}-f_{j-\frac{1}{2}}^{n-1})/2,\quad \tilde{c}_j^{(i)} \equiv 1-h_j\,\sigma_n(f_{j-\frac{1}{2}}^{(i)}-f_{j-\frac{1}{2}}^{n-1})/2 .$$

We have intentionally interchanged the order in writing the first two equations of (3.1) in (3.3) for reasons explained below.

The boundary conditions (2.7) yield for our iteration scheme

$$\text{a) } \delta f_o^{(i)} = f_w(\xi_n)-f_o^{(i)},\quad \text{b) } \delta u_0^{(i)} = -u_0^{(i)},\quad \text{c) } \delta u_J^{(i)} = 1-u_J^{(i)} .\tag{3.6}$$

Clearly the right hand sides in each of these equations will vanish if the initial iterate $\{\delta f_j^{(o)},\ \delta u_j^{(o)},\ \delta v_j^{(i)}\}$ satisfies the correct boundary conditions. Assuming this to be the case the entire linear system (3.3) and (3.6) can be written in the block-matrix form.

$$\text{a) } \mathbb{A}^{(i)}\underset{\approx}{\delta}^{(i)}=\underset{\approx}{q}^{(i)};\quad \text{b) } \mathbb{A}^{(i)}\equiv\begin{pmatrix} \begin{pmatrix}1&0&0\\0&1&0\end{pmatrix} & & & \\ -L_1 & R_1 & & \\ & \ddots & \ddots & \\ & & -L_J & R_J \\ & & & (0\;1\;0)\end{pmatrix},\quad \underset{\approx}{\delta}^{(i)}\equiv\begin{pmatrix}\underset{\sim}{\delta}_o^{(i)}\\ \vdots\\ \underset{\sim}{\delta}_J^{(i)}\end{pmatrix},\quad \underset{\approx}{q}^{(i)}\equiv\begin{pmatrix}0\\0\\ \underset{\sim}{r}_{\frac{1}{2}}^{(i)}\\ \vdots\\ \underset{\sim}{r}_{J-\frac{1}{2}}^{(i)}\\0\end{pmatrix}.\tag{3.7}$$

The right hand sides in (3.6) replace the zeros in $\underset{\approx}{q}^{(i)}$ if they do not vanish.

We now make the important observation that $\mathbb{A}^{(i)}$ has block-tridiagonal form:

$$\mathbb{A}^{(i)}\equiv\begin{pmatrix}\ddots & \ddots & \ddots \\ B_j^{(i)} & A_j^{(i)} & C_j^{(i)} \\ \ddots & \ddots & \ddots\end{pmatrix}.$$

In particular the blocks are, from a comparison with (3.7b) and using (3.5):

$$\left.\begin{aligned} &A_o^{(i)} \equiv \begin{pmatrix}1&0&0\\0&1&0\\0&-1&-h_{j/2}\end{pmatrix};\ C_j^{(i)}\equiv\begin{pmatrix}0&0&0\\0&0&0\\0&1&-\frac{1}{2}h_{j+1}\end{pmatrix}\quad 0\leq j\leq J-1;\\ &A_J^{(i)}\equiv\begin{pmatrix}1&-h_{J/2}&0\\a_J^{(i)}&b_J^{(i)}&c_J^{(i)}\\0&1&0\end{pmatrix};\ B_j^{(i)}\equiv\begin{pmatrix}-1&-h_{j/2}&0\\-\tilde{a}_j^{(i)}&-\tilde{b}_j^{(i)}&-\tilde{c}_j^{(i)}\\0&0&0\end{pmatrix}\quad 1\leq j\leq J;\end{aligned}\right\}\ A_j^{(i)}\equiv\begin{pmatrix}1&-h_{j/2}&0\\a_j^{(i)}&b_j^{(i)}&c_j^{(i)}\\0&-1&-\frac{1}{2}h_{j+1}\end{pmatrix},\quad 1\leq j\leq J-1\tag{3.8}$$

With this form in (3.7a) we can solve the system by a more or less standard block-tridiagonal factorization procedure [4]. Specifically the matrices $D_j^{(i)}$ and $E_j^{(i)}$ are computed from the recursions:

$$\text{a)}\quad D_o^{(i)}=A_o^{(i)}\,;\quad E_j^{(i)}=(D_j^{(i)})^{-1}C_j^{(i)},0\leq j\leq J-1;\quad D_j^{(i)}=A_j^{(i)}-B_j^{(i)}E_{j-1}^{(i)}\,,\ 1\leq j\leq J. \qquad (3.9)$$

Then the intermediate vectors $\underset{\sim}{Z}_j^{(i)}\equiv(Z_{j,1}^{(i)}, Z_{j,2}^{(i)}, Z_{j,3}^{(i)})^T$ are computed from:

$$\text{b)}\quad \underset{\sim}{Z}_o^{(i)}=(D_o^{(i)})^{-1}\underset{\sim}{q}_o^{(i)};\quad \underset{\sim}{Z}_j^{(i)}=(D_j^{(i)})^{-1}[\underset{\sim}{q}_j^{(i)}-B_j^{(i)}Z_{j-1}^{(i)}]\,,\ 1\leq j\leq J. \qquad (3.9)$$

Finally the solution components $\underset{\sim}{\delta}_j^{(i)}$ are obtained as:

$$\text{c)}\quad \underset{\sim}{\delta}_J^{(i)}=\underset{\sim}{Z}_J^{(i)}\,;\quad \underset{\sim}{\delta}_j^{(i)}=\underset{\sim}{Z}_j^{(i)}-E_j^{(i)}\underset{\sim}{\delta}_{j+1}\,,\quad J-1\geq j\geq 0. \qquad (3.9)$$

The recursions (3.9a) require $A_o^{(i)}$ to be nonsingular. It is to achieve this that we have reordered the equations in writing the system as (3.3). Conditions to insure that all $D_j^{(i)}$ are nonsingular can be given.

To start the iterations it is most reasonable to use as an initial guess for the solution on $\xi=\xi_n$ the previously obtained solution on $\xi=\xi_{n-1}$; that is

$$f_j^{(o)}=f_j^{n-1}\,,\quad u_j^{(o)}=u_j^{n-1}\,,\quad v_j^{(o)}=v_j^{n-1}\,,\qquad 0\leq j\leq J. \qquad (3.10)$$

Of course the values $f_o^{(o)}$, $u_o^{(o)}$ and $u_J^{(o)}$ may need to be changed in order to satisfy (2.7) if these boundary conditions actually depend upon ξ.

4. ACCURACY, RICHARDSON EXTRAPOLATION & CONVERGENCE

The exact numerical solution of our difference equations (2.6), (2.7) is a second order accurate approximation. But since the local truncation error of this scheme has an expansion in powers of h_j^2 and k_n^2 and since the scheme is stable we can in fact show that the errors in approximation have asymptotic expansions of the form (see, for similar results, [7]):

$$f(\xi_n,\eta_j)-f_j^n=\sum_{\nu=o}^{m}\sum_{\mu=o}^{\nu} h^{2\mu}k^{2\nu-2\mu}\, e_{\nu\mu}(\xi_n,\eta_j)+\mathcal{O}(h^{2m+2}+\cdots+k^{2m+2}). \qquad (4.1)$$

Analogous expansions hold for the errors in $u(\xi,\eta)$ and $v(\xi,\eta)$. Here $h\equiv\max_j h_j$, $k=\max_n k_n$ and the functions $e_{\nu\mu}(\xi,\eta)$ depend upon the net only through two functions $\theta(\xi)$ and $\phi(\eta)$ which are such that:

$$h_j=\phi(\eta_{j-\frac{1}{2}})h\quad,\quad k_n=\theta(\xi_{n-\frac{1}{2}})k. \qquad (4.2)$$

On all nets which employ the same functions $\theta(\xi)$, $\phi(\eta)$ the expansions (4.1) hold with possibly different h and k values. Examples of such nets are, starting with an arbitrary initial net as in (2.4) and then: (i) dividing each ξ-<u>and</u> η - interval into r+1 equal subintervals for some r=1, 2, ⋯; (ii) dividing <u>only</u> the ξ- intervals into r+1 equal subintervals; (iii) reversing ξ and η in case (ii).

Let us denote by $\{f_j^n(r),\ a_j^n(r),\ v_j^n(r)\}$ the numerical solution on a net with maximal spacings $h^{(r)}$ and $k^{(r)}$. Now let two independent computations be made on nets related as in (ii) above with spacings $(h,k)=(h,k^{(o)})$ and $(h,k)=(h,k^{(1)})$. If we define: $\omega_{o1} \equiv k^{(1)2}/[k^{(o)2}-k^{(1)2}]$ and

$$\text{a)}\quad f_j^{n_o}(o,1) \equiv f_j^{n_o}(o) + \omega_{o1}\,[f_j^{n_o}(o)-f_j^{n_1}(1)],\ \text{etc;} \qquad (4.2)$$

where $\xi_{n_o}=\xi^*$ on the first net and $\xi_{n_1}=\xi^*$ on the second then from the above and (4.1) we find that:

$$\text{b)}\quad f(\xi^*,\eta_j) - f_j^{n_o}(o,1) = \mathcal{O}(h^2+k^4). \qquad (4.2)$$

Using a third net with $(h,k)=(h,k^{(2)})$, $\xi_{n_2}=\xi^*$ and:

$$\text{a)}\quad f_j^{n_o}(o,2) \equiv f_j^{n_o}(o) + \omega_{o2}\,[f_j^{n_o}(o)-f_j^{n_2}(2)]$$

$$\qquad (4.3)$$

$$\text{b)}\quad f_j^{n_o}(o,1,2) \equiv f_j^{n_o}(o,1) + \omega_{12}\,[f_j^{n_o}(o,1)-f_j^{n_o}(o,2)]$$

it follows that

$$\text{c)}\quad f(\xi^*,\eta_j) - f_j^{n_o}(o,1,2) = \mathcal{O}(h^2+k^6). \qquad (4.3)$$

Clearly the above procedure could be reversed, regarding ξ and η, to get $\mathcal{O}(h^6+k^2)$ order accuracy. More important this extrapolation can be used to get, say, $\mathcal{O}(h^6+k^6)$ accuracy if subdivisions of form i) are employed on three nets as then $\omega_{o1} \equiv h^{(1)2}/[h^{(o)2}-h^{(1)2}] = k^{(1)2}/[k^{(o)2}-k^{(1)2}]$, etc. Even higher order accuracy can be obtained by further extrapolations.

Since we do not obtain the exact solution of the difference equations the extrapolation procedure will only improve the accuracy provided the truncation errors are larger than the iteration errors (i.e. errors in approximating the exact numerical solution). However if the initial iteration error satisfies, say, $|f_j^n-f_j^{(o)}| = \mathcal{O}(h)$, then by the quadratic convergence of Newton's method it follows that $|f_j^n-f_j^{(i)}| = \mathcal{O}(h^{2^i})$. Thus if we wish to reduce the truncation error to $\mathcal{O}(h^{2m})$ it requires only about $i = 1+(\ln m/\ln 2)$ iterations. In practice we seldom require more than 2 or 3 iterations corresponding to m = 3 or 4 and two or three extrapolations (see Table I).

The proper use of these extrapolation procedures is to get numerical results of modest accuracy while only employing relatively crude nets. This results in very efficient computations (and quite simple codes) as we show in the example below.

5. COMPUTATIONS FOR HOWARTH'S FLOW

The choice $u_e(x) \equiv 1-x/8$ and $v_w(x) \equiv 0$ yields the case known as Howarth's flow. This flow has been computed extensively and accurately [2], [3], [8], so we employ it as a test case. As the basic net N_{oo} of 42 points we take:

a) $\xi_0=0,\ \xi_1=0.4,\ \xi_2=0.7,\ \xi_3=0.8,\ \xi_4=0.86, \xi_5=0.894;$

$$(5.1)$$

b) $\eta_0=0,\ \eta_1=1.0,\ \eta_2=2.0,\ \eta_3=3.0,\ \eta_4=4.0,\ \eta_5=5.0,\ \eta_6=6.0.$

We also compute on ten other nets $N_{r,s}$ for which each ξ-interval or η-interval in (5.1) is subdivided into (r+1) or (s+1) equal subintervals, respectively. The net $N_{r,s}$ contains (5r+6) (6s+7) points.

The most sensitive and most important quantity in such calculations is $\partial^2 f(\xi,o)/\partial\eta^2 = v(\xi,o)$ from which the stress on the wall is determined. Thus we shall examine only this quantity in our discussion of results. All other computed quantities behaved similarly. It should perhaps be stressed here that our scheme yields the same accuracy in $v\equiv\partial^2 f/\partial\eta^2$ as it does in f. Thus we do not lose any accuracy as would be the case if difference quotients had to be used to compute v from f.

We first show in Table I the quadratic convergence properties of Newton's method. Here the corrections $\delta v_o^{(i)}$ to v_o^n are given for each iterate at the six indicated values of $\xi=\xi_n$ on each of the two nets $N_{o,o}$ and $N_{9,9}$. All corrections were negative except those marked with an asterisk. The notation (-m) implies a factor 10^{-m}. Since the calculations were done in single precision on an I.B.M. 360/65 the roundoff errors prevented a full error reduction in many of the last iterates. The convergence criterion required $|\delta v_o^{(i)}| < 10^{-5}$ so that four significant decimal digits could be obtained if the truncation errors were sufficiently reduced.

In Table II we show v_o^n computed on four uniformly refined nets. Extrapolation was performed using: $N_{o,o}$ with $N_{1,1}$ to get the results under N(o,1): $N_{o,o}$ with $N_{2,2}$ to get N(o,2) and N(o,1) with N(o,2) to get N(o,1,2). These final extrapolated results have truncation errors which have been reduced to $\mathcal{O}(h_o^6+k_o^6)$. In the last column are v_o^n values computed using the relatively fine net $N_{9,9}$.

In Tables III and IV we show similar results on nets which have been subdivided only in the η and ξ intervals respectively. The indicated extrapolations are such that, for example, the column N(9; 0,1,2) in Table III has truncation error reduced to $\mathcal{O}\left(\left(\frac{h_o}{10}\right)^2+k_o^6\right)$. The final column N(9;9,19) in Table IV contains extrapolated values from the fine nets $N_{9,9}$ and $N_{9,19}$ given in Table II and are the most accurate values we have determined.

By comparing the first three columns in Tables II, III and IV we can see the relative effects of coarseness in the ξ-net or η-net spacings. It is rather apparent from these results that a fine η-spacing is more important, for this example, than a fine ξ-spacing. We also see that the results in N(0,1,2) are slightly more accurate [i.e. in better agreement with N(9;9,19)] than are the results from $N_{9,19}$. Since 489 points are used in the former and 6171 points are used in the latter, a reduction in computation <u>by a factor of 12.5 or more</u> has been achieved by means of the extrapolation procedure. This result is all the more impressive when it is realized that the computations in $N_{9,19}$ are at least as accurate as previous calculations [2], [3], [8] and much more efficient. (Detailed counts of the meshes and times employed in all these computations are not available.) Furthermore there is no difficulty whatsoever in using our scheme extremely close to the separation point in Howarth's flow ($\xi \doteq 0.901344$) but we do not report such results here.

TABLE I

Convergence of Newton iterates on two different nets.

ξ_n	$N_{o,o}$				$N_{9,9}$			
	i=0	i=1	i=2	i=3	i=0	i=1	i=2	i=3
0	*.203(0)	.289(-1)	.125(-2)	.255(-5)	*.165(0)	.271(-1)	.118(-2)	.241(-5)
.4	.127(0)	.392(-2)	.207(-4)	.123(-6)	.145(-1)	.181(-3)	.236(-7)	
.7	.129(0)	.914(-2)	.151(-3)	.259(-6)	.167(-1)	.473(-3)	.379(-6)	
.8	.589(-1)	.425(-2)	.457(-4)	.889(-6)	.763(-2)	.170(-3)	*.117(-6)	
.86	.432(-1)	.366(-2)	.475(-4)	.480(-6)	.649(-2)	.218(-3)	.175(-6)	
.894	.299(-1)	.276(-2)	.367(-4)	.204(-5)	.676(-2)	.632(-3)	.593(-5)	

TABLE II

Values of v_o^n on four nets,
uniformly refined in ξ and η, and three extrapolations.

ξ_n	$N_{o,o}$	$N_{1,1}$	$N_{2,2}$	N(o,1)	N(o,2)	N(o,1,2)	$N_{9,9}$
0	.5060647	.4789143	.4737529	.4698642	.4697139	.4695937	.4696975
.4	.3750923	.3534344	.3491818	.3462151	.3459430	.3457253	.346262
.7	.2368498	.2149650	.2106251	.2076701	.2073470	.2070985	.207481
.8	.1736072	.1486031	.1432982	.1402684	.1395096	.1389026	.139651
.86	.1266655	.0956397	.0886248	.0852987	.0838697	.0827312	.0834382
.894	.0939423	.0530091	.0416465	.0393647	.0351095	.0317053	.0317397
# pts.	42	143	304	185	346	489	3111

TABLE III

Values of v_o^n on four nets, refined only in η, and three extrapolations.

ξ_n	$N_{9,0}$	$N_{9,1}$	$N_{9,2}$	N(9;0,1)	N(9;0,2)	N(9;0,1,2)	$N_{9,19}$
0	.5060647	.4789143	.4737529	.4698642	.4697139	.4695937	.469694
.4	.377384	.353976	.3495226	.346140	.3460399	.3459598	.346021
.7	.238914	.2154687	.2108746	.2076536	.2073697	.2071426	.207227
.8	.175931	.1492218	.1437508	.1403189	.1397283	.1392558	.139343
.86	.1292425	.0963887	.089096	.0854374	.0840777	.0829899	.083004
.894	.0969247	.0540378	.0423506	.0409421	.0355288	.0311982	.030832
# pts	427	793	1159	1220	1586	2379	6171

TABLE IV

Values of v_o^n on three nets, refined only in ξ, and four extrapolations.

ξ_n	$N_{0,9}$	$N_{1,9}$	$N_{2,9}$	N(0,1;9)	N(0,2;9)	N(0,1,2;9)	N(9;9,19)
0	.4699748	.4699748	.4699748	.4699748	.4699748	.4699748	.4696013
.4	.3436177	.3457045	.3459175	.3464001	.3460478	.3460038	.3459413
.7	.2051605	.2069843	.2072321	.2075922	.2074303	.2074100	.2071427
.8	.1367188	.1390185	.1391890	.1397850	.1393254	.1392679	.13924
.86	.0798466	.0826372	.0829497	.0835674	.0831997	.0831538	.08286
.894	.0255880	.0304601	.0309323	.0320841	.0313100	.0312133	.0305307
# pts.	366	671	976	1037	1342	1708	9282

References

[1] Cebeci, T. and Keller, H. B., submitted to J. Comp. Phys.

[2] Cebeci, T. and Smith, A. M. O., Rep. No. DAC-67130, McDonnell Douglas (1968).

[3] Howarth, L., Proc. Roy. Soc. 164, p. 547 (1938).

[4] Isaacson, E. and Keller, H. B., Analysis of Numerical Methods, J. Wiley, New York, 1966.

[5] Keller, H. B., SIAM J. Num. Anal, 6, 8-30 (1969).

[6] Keller, H. B., Numerical Methods for Two Point Boundary-Value Problems, Blaisdell, Waltham, Mass, 1968.

[7] Keller, H. B. in Numerical Solution of Partial Differential Equations-II, B. Hubbard, Ed. Academic Press, New York, 1970.

[8] Smith, A. M. O. and Clutter, D. W., AIAA Jour. 1, 2002-2071 (1963).

THE REDUCTION OF THE STEFAN PROBLEM TO THE SOLUTION OF AN ORDINARY DIFFERENTIAL EQUATION

René De Vogelaere
Department of Mathematics
and Computer Center
University of California,
Berkeley

SUMMARY

The solution of the Stefan problem is reduced to the solution of a differential equation of order n. Computational evidence allows the conjecture of convergence to the boundary curve and to the solution of the problem when n tends to infinity, for all values of the time variable. The method generalizes.

1. INTRODUCTION

The Stefan problem [11], [5], [4] that we consider here is the solution $u(x, t)$ of the parabolic differential equation (1.1) with moving boundary $y(t)$ satisfying the boundary conditions (1.2) to (1.5):

(1.1) $u_{xx} - u_t = 0$ $\quad t > 0 \quad 0 < x < y(t)$

(1.2) $u_x(0, I) = -1$ $\quad t > 0$

(1.3) $u(y, I) = 0$ $\quad t \geq 0$

(1.4) $Dy = -u_x(y, I)$ $\quad t > 0$

(1.5) $y(0) = 0$

(I is the identity function, D the derivative operator, u_x is the partial derivative of u with respect to the first variable and u_t with respect to the second, u_{t^2} will also be used instead of u_{tt}, etc.)

The existence and uniqueness of the solution has been studied by Evans [3], Friedman [5][6], Sestini [10], Douglas [1], and others.

The problem has been replaced by the solution of an integral equation by Friedman [5, p. 216], and Evans, Isaacson, Macdonald [4] and given as the limit of problems with fixed boundary by Evans [3], and Friedman [5, p. 331]. A series solution is given by Evans, Isaacson and Macdonald [4], explicit difference methods have been studied by Trench [12], implicit difference methods by Douglas and Gallie [2]. Lanczos τ-methods by Wragg [13].

The exact solution in the case where (1.3) is replaced by

(1.2') $\quad u(0, I) = 1$

is given by Ruoff [9].

The starting point of this investigation is the result that partial derivatives of u on the moving boundary can be determined as a function of this boundary. The first partial derivatives are given by (1.3) and (1.4), the next two are contained in lemmas obtained by Osterby [8].

The solution is extrapolated to the t axis using Taylor expansion.

A first approximation to the moving boundary as a solution of a differential equation of order n, followed by an integration to get from Dy, y is given in section 4, formula (4.2). This approach can be used near the origin only.

Another approximation uses the boundary condition (2) for all values of t and leads to an approximation to the moving boundary by differential equations (5.1) of order $(n+1) \div 2$. The simplest case, which can be integrated gives an accuracy of

1% up to $t = 1$ and 6% up to $t = 10$.

The relation between the derivatives at the origin ot the exact boundary curves and these last approximations is furnished by Theorem 5.1.

2. SOLUTION ON THE BOUNDARY CURVE

Let us define

(2.1) $$f_{ij} = (-1)^i u_{x^i t^j}(y, I) , \qquad g_i = (-1)^i u_{x^i}(y, I)$$

(2.2) $$y_o = y , \qquad y_i = D^i y .$$

(The sign has been chosen to have only + signs in the expression for g_i .)

(2.1.) Theorem. The partial derivatives of the solution of (1.1) to (1.5) on the moving boundary are given by the recurrence relations (2.4), the starting values (2.5) and the relations (2.3).

Indeed, the differential equation (1.1) gives at once

(2.3) $$f_{ij} = f_{i+2,j-1} = f_{i+2j,0} = g_{i+2j} .$$

Taking the derivative of (2.1) gives

$$Df_{ij} = (-1)^i u_{x^{i+1}t^j}(y, I)y_1 + (-1)^i u_{x^i t^{j+1}}(y, I) \text{ or } Dg_{i+2j} = -g_{i+2j+1}y_1 + g_{i+2j+2}$$

hence the recurrence relation

(2.4) $$g_{k+2} = Dg_k + g_{k+1}y_1$$

while (1.3) and (1.4) give the first functions

(2.5) $$g_0 = 0 , \qquad g_1 = y_1 .$$

We obtain in succession:

$$u_{xx}(y, I) = g_2 = y_1^2$$

$$-u_{xxx}(y, I) = g_3 = y_2 + y_1^3$$

$$u_{xxxx}(y, I) = g_4 = 3y_1y_2 + y_1^4$$

$$-u_{xxxxx}(y, I) = g_5 = y_3 + 6y_2y_1^2 + y_1^5$$

$$u_{x^6}(y, I) = g_6 = 4y_1y_3 + 3y_2^2 + 10y_1^3y_2 + y_1^6$$

$$-u_{x^7}(y, I) = g_7 = y_4 + 10y_1^2y_3 + 15y_1y_2^2 + 15y_1^4y_2 + y_1^7$$

(2.6) $$u_{x^8}(y, I) = g_8 = 5y_1y_4 + 10y_2y_3 + 20y_1^3y_3 + 45y_1^2y_2^2 + 21y_1^5y_2 + y_1^8$$

$$-u_{x^9}(y, I) = g_9 = y_5 + 15y_1^2y_4 + 60y_1y_2y_3 + 35y_1^4y_3 + 15y_2^3 + 105y_1^3y_2^2 + 28y_1^6y_2 + y_1^9$$

$$u_{x^{10}}(y, I) = g_{10} = 6y_1y_5 + 15y_2y_4 + 35y_1^3y_4 + 10y_3^2 + 210y_1^2y_2y_3 + 56y_1^5y_3 + 105y_1y_2^3 + 210y_1^4y_2^2 + 36y_1^7y_2 + y_1^{10}$$

$$-u_{x^{11}}(y, I) = g_{11} = y_6 + 21y_1^2y_5 + 105y_1y_2y_4 + 70y_1^4 + 70y_1y_3^2 + 105y_2^2y_3 + 560y_1^3y_2y_3 + 84y_1^6y_3 + 420y_1^2y_2^3 + 378y_1^5y_2^2 + 45y_1^8y_2 + y_1^{11} .$$

3. SOLUTION ON THE t AXIS

The function y and its derivatives y_i are unknown. The only conditions not yet used are (1.2) and (1.5). Taking derivatives we can write a set of equivalent equations ($\delta_{i,j}$ is the Kronecker symbol).

$$-u_{xt^i}(0, I) = \delta_{i,0} . \tag{3.1}$$

Using Taylor expansion:

$$-u_{xt^i}(0, I) = - [u_{xt^i}(y, I) - yu_{x^2t^i}(y, I) + \frac{y^2}{2} u_{x^3t^i}(y, I) + \cdots]$$

we get

$$\delta_{i,0} = g_{2i+1} + yg_{2i+2} + \frac{y^2}{2} g_{2i+3} + \frac{y^3}{3!} g_{2i+4} + \cdots \tag{3.2}$$

These equations will determine, for $y = 0$, the values of the derivatives of the boundary curve at the origin; for $i = 0,1,2,3,4,5,$ we get in succession [4]

$$\begin{aligned} y_1(0) &= 1 \\ y_2(0) &= -1 \\ y_3(0) &= 5 \\ y_4(0) &= -51 \\ y_5(0) &= 827 \\ y_6(0) &= -18961 \\ y_7(0) &= 574357 . \end{aligned} \tag{3.3}$$

This suggests the problem of determining the radius of convergence of the function y expanded around the origin.

4. A FIRST APPROXIMATION TO THE MOVING BOUNDARY

The problem

$$y_3 = Dy_1 + y_1^3 = 0 \qquad\qquad y_1(0) = 1$$

has the solution

$$y_1 = (2I + 1)^{-\frac{1}{2}}$$

therefore

$$\int_0^t y_1 = y(t) = (2t + 1)^{\frac{1}{2}} - 1 . \tag{4.1}$$

This is precisely a lower bound for the boundary curve.

Moreover, $g_{2n+1} = 0$ has solutions of the form $y = AI^{\frac{1}{2}}$. For instance for g_5, because

$$y_i = A(\tfrac{1}{2})(-\tfrac{1}{2}) \dots (\frac{3 - 2i}{2})I^{\frac{1}{2}-i} ,$$

we have

$$\frac{3}{8}A - \frac{3}{8}A^3 + \frac{1}{32}A^5 = 0$$

which gives

$$A = 0 \quad\text{and}\quad A = \pm (6 + \sqrt{24})^{\frac{1}{2}} .$$

This suggests that an approximation to the boundary curve could be obtained by approximating it by

(4.2) $$-u_{xt^n}(y, I) = g_{2n+1} = 0$$

where the initial conditions are determined using (3.3).

This approach is extrapolatory in character because we use the conditions (1.2) or (3.2) only near the origin. We therefore must expect that $u_x(0, t)$ becomes soon much different from -1 when t increases.

In fact we obtain

n	y(0.2)	y(0.4)
1	0.18322	0.34164
2	0.18489	0.35108
3	0.18441	0.34628
4	0.18462	0.35009
5	0.18450	0.34612
∞	0.18454	

The last result is obtained by extrapolation using

$$y(0.2) = A \cdot C^{5-n} + B \quad \text{for } n = 3, 4, 5 .$$

The function u can be obtained from the first terms of Taylor expansion:

(4.3) $$u(x, t) = (y_0(t) - x)g_1(t) + \tfrac{1}{2}(y_0(t) - x)^2 g_2(t) + \cdots .$$

5. A SECOND APPROXIMATION TO THE MOVING BOUNDARY

Instead of using (1.2) near the origin we can use (1.2) globally and approximate the moving boundary by the solution of the differential equation

(5.1) $$1 = g_1 + yg_2 + \frac{y^2}{2!} g_3 + \cdots + \frac{y^{n-1}}{(n-1)!} g_n$$

which is of order $(n + 1) \div 2$.

We should use another notation of y for instance $\bar{y}^{(n)}$ or $\bar{y}$; we will only do so occasionally.

When $n = 2$, we obtain

$$1 = g_1 + yg_2$$

or

(5.2) $$yy_1^2 + y_1 = 1 , \quad \text{giving } y_1(0) = 1 , \quad \text{because of (1.5)}$$

or

(5.3) $$Dy = (-1 + \sqrt{1 + 4y})/(2y) , \quad y(0) = 0 .$$

Equation (5.2) can be integrated by the Legendre transformation [7, I.479 and A4.20] which gives the implicit solution

(5.4) $$t = -\frac{1}{6} - \frac{u^2}{2} + \frac{2}{3}u^3 , \; y(t) = -u + u^2, \; y_1(t) = Dy(t) = \frac{1}{u} , \quad u \in [1, \infty) .$$

In particular, for $u = 1.15926$, $t = .2$, $y(.2) = .18462$, $Dy(.2) = .86262$. (We can also compute t by $t = \frac{1}{6}(u + y(t) - 1) + \frac{2}{3}uy(t))$.

Moreover

$$y_2(t) = -y_1^3(t)/(1 + 2y(t)y_1(t)) = \frac{-1}{u^3(2u - 1)}$$

(5.5) $$y_3(t) = \frac{-(5y_1^2(t)y_2(t) + 2y(t)y_2^2(t))}{1 + 2y(t)y_1(t)} = \frac{8u - 3}{u^5(2u - 1)^3}$$

$$y_4(t) = \frac{-(12y_1(t)y_2^2(t) + 7y_1^2(t)y_3(t) + 6y(t)y_2(t)y_3(t))}{1 + 2y(t)y_1(t)} =$$

$$= -\frac{112u^2 - 80u + 15}{u^7(2u-1)^5} .$$

In particular

(5.6) $\bar{y}_2(0) = -1 , \quad \bar{y}_3(0) = 5 , \quad \bar{y}_4(0) = -47 , \ldots .$

We will now examine the case $n = 3$, the extension to $n > 3$ is straightforward, we will give associated results when relevant.

$$1 = g_1 + yg_2 + \frac{y^2}{2} g_3$$

gives

(5.7) $y_2 = Dy_1 = -y_1^3 + 2(1-y_1 - yy_2^2)/y^2 .$

The values of the derivatives $\bar{y}_i(0)$ at the origin are obtained by successive differentiation of

$$y^2y_2 + y^2y_1^3 + 2y_1 + 2yy_1^2 - 2 = 0$$

which gives

(5.8) $\bar{y}_1(0) = 1 , \quad \bar{y}_2(0) = -1 , \quad \bar{y}_3(0) = 5 , \quad \bar{y}_4(0) = -53 , \quad \ldots .$

We notice that the first derivative in error is the 4th one and that the error is 4%.

It is of interest to observe that contrary to the case of equation (5.3) we cannot use a Runge-Kutta method to start, even if we use $y_2(0) = -1$. Indeed in the first step of the Runge-Kutta method, if we write $z_i \equiv y_i(0)$, we approximate

$$y_0(\alpha h) \text{ by } z_0 + \alpha h z_1 , \quad y_1(\alpha h) \text{ by } z_1 + \alpha h z_2$$

and if we substitute in (5.7) we get

$$k_1(h) = (Dy_1)(\alpha h) = -(z_1+\alpha h z_2)^3 + \frac{2(1-z_1-\alpha h z_2-(z_0+\alpha h z_1)(z_1+\alpha h z_2)^2)}{(z_0 + \alpha h z_1)^2} = -1+4+O(h)$$

while

$$(Dy_1)(0) = -1 .$$

Therefore $\lim_{h\to 0} k_1(h) = 3 \neq -1$ and the method will fail. We therefore will start using the results for $n = 2$ at $t = .1$ or $t = .2$ say.

5.1. Theorem. The derivatives at the origin of the regular solution of (5.1) are related to those of the solution (3.3) by

(5.9) $\bar{y}_i^{(n)}(0) = y_i^{(n)}(0) , \quad i = 0, \ldots , 2(n \div 2) + 1$

(5.10) $\bar{y}_{n+1}^{(n)}(0) = y_{n+1}^{(n)}(0) + g_{n+1}(0)$ if n is odd

(5.11) $\bar{y}_{n+2}^{(n)}(0) = y_{n+2}^{(n)}(0) - n \quad g_{n+2}(0)$ if n is even.

If we take the derivative of (5.1) we get

$$0 = Dg_1 + y_1g_2 + yDg_2 + yy_1g_3 + \frac{y^2}{2} Dg_3 + \cdots + \frac{y^{n-1}}{(n-1)!} Dg_n .$$

Because of (2.4), this gives

$$0 = g_3 + yg_4 + \cdots + \frac{y^{n-1}}{(n-1)!} g_{n+2} - \frac{y^{n-1}}{(n-1)!} (y_1 g_{n+1}) .$$

By successive differentiation, we have

$$0 = g_5 + yg_6 + \cdots + \frac{y^{n-1}}{(n-1)!} g_{n+4} - \frac{y^{n-1}}{(n-1)!} [y_1 g_{n+3} + D(y_1 g_{n+1})] - \frac{y^{n-2}}{(n-2)!} y_1^2 g_{n+1}$$

$$0 = g_7 + yg_8 + \cdots + \frac{y^{n-1}}{(n-1)!} g_{n+6} - \frac{y^{n-1}}{(n-1)!} [y_1 g_{n+5} + D(y_1 g_{n+3} + D(y_1 g_{n+1}))]$$

$$- \frac{y^{n-2}}{(n-2)!} [y_1^2 g_{n+3} + y_1 D(y_1 g_{n+1}) + D(y_1^2 g_{n+1})] - \frac{y^{n-3}}{(n-3)!} y_1^3 g_{n+1}$$

.....

$$(5.12) \qquad 0 = g_{2p+1} + \cdots - \frac{y^{n-p+1}}{(n-p+1)!} [y_1^{p-1} g_{n+3} + \sum_{q=0}^{p-2} y_1^q D(y_1^{p-q+1} g_{n+1})] - \frac{y^{n-p}}{(n-p)!} y_1^p g_{n+1},$$

from which follows that the derivatives at 0 of the solution of (5.1) satisfy

$$g_3(0) = 0, \quad g_5(0) = 0, \ \ldots \ , \ g_{2n-1}(0) = 0 \ , \quad g_{2n+1}(0) = g_{n+1}(0) \ , \ \ldots \ .$$

If we compare with (3.2) and (3.3) and observe that $g_{n+1}(0) = 0$ when n is even, we get (5.9) and (5.10).

When n is even and $p = n+1$, (5.12) gives

$$g_{2n+3}(0) = \sum_{q=0}^{n-1} y_1^q D(y_1^{n-q} g_{n+1})(0) = n \, Dg_{n+1}(0) = -n \, g_{n+2}(0) .$$

In particular, we obtain the results of (5.6) and (5.8)

$$\bar{y}_4^{(2)}(0) = -47 \ , \quad \bar{y}_4^{(3)}(0) = -53 \ ,$$

as well as

$$y_6^{(4)}(0) = -19031 \ , \quad \bar{y}_6^{(5)}(0) = -18947 \ .$$

Just for comparison we also state the

5.2. Theorem. The derivatives at the origin of the solution $\bar{y}$ of (4.2) with $\bar{y}^{(n)}(0) = y_i(0)$, $i = 0,\ldots,n$, satisfy, for $n > 0$, $\bar{y}_{n+1}^{(n)}(0) = \bar{y}_{n+1}(0)$ and

$$g_{2n+3}(0) = g_{2n+2}(0), \ g_{2n+5}(0) = 2(g_{2n+4}(0) - g_{2n+2}(0)) \ ,$$

$$g_{2n+7}(0) = 3g_{2n+6}(0) - 8g_{2n+4}(0) + (9 + \bar{y}_3(0))g_{2n+2}(0) \ , \ \ldots \ .$$

For instance, if $n = 1$, $\bar{y}_3(0) = 3$, $\bar{y}_4(0) = -15$, $\bar{y}_5(0) = 105$, indeed very different from the exact values (3.3).

6. RESULTS

We give now summary tables which give, to within 1 unit of the last decimal place, the results of the integration for various values of n in (5.1). For $n > 2$, the integration is started at 0.1 using the values for function and derivatives obtained with $n = 2$.

TABLE 6.1. y(t), the moving boundary.

t\n	2	3	4	5	6
2	1.383	1.356	1.361	1.362	1.362
4	2.403	2.314	2.328	2.333	2.333
6	3.289	3.126	3.146	3.157	3.157
8	4.094	3.850	3.874	3.892	3.893
10	4.843	4.514	4.541	4.566	4.568

TABLE 6.2. u(0,t) the temperature at the fixed boundary.

t\n	2	3	4	5	6
2	1.081	0.993	1.009	1.043	1.043
4	1.766	1.513	1.543	1.658	1.658
6	2.335	1.899	1.934	2.146	2.149
8	2.839	2.214	2.249	2.564	2.572
10	3.300	2.484	2.516	2.937	2.951

The integration was performed using the Runge-Kutta method with Zonneveld's algorithm to control automatically the step of integration [14].

7. GENERALIZATION

The method generalizes. Indeed the recurrence relation depends only on the differential equation; if the boundary conditions at the moving boundary are given by explicitly differentiable functions which may depend on y and its derivatives, we can construct the sequence of g_n. The condition at the fixed boundary gives the first member of (5.1). From (1.5) we derive the derivatives of the boundary curve at the origin. It is less straightforward to do the same when $y(0) > 0$ and $u(x, 0)$ is given in $[0, y(0)]$.

8. ACKNOWLEDGMENT

This work was done while on leave at the "Mathematisch Instituut" and the "Instituut voor Toegepaste Wiskunde" of the University of Amsterdam. I want to thank all my colleagues there for their hospitality and for providing me an assistant P. Hollenberg who did the required programming.

9. REFERENCES

[1] Douglas, J., Proc. Amer. Math.Soc.,8, 402-408 (1957).

[2] Douglas, J. and Gallie, J.M., Duke Math. J., 22, 557-571 (1955).

[3] Evans, G.W., Quart. Appl. Math., 9, 185-193 (1951)

[4] Evans, G.W., Isaacson, E. and MacDonald, J.K.L. Quart. Appl. Math., 8, 312-319 (1950)

[5] Friedman, A., Partial differential equations of parabolic type, Prentice-Hall, Englewood Cliffs, N.J., (1964)

[6] Friedman, A., J. Math. and Mech., 9, 885-904 (1960).

[7] Kamke, E., Differentialgleichungen lösungsmethoden und lösungen, Akad. Verlag, Leipzig (1943)

[8] Osterby, O., Thesis, Univ. of California, Berkeley.(In preparation)

[9] Ruoff, A.L., Quart. Appl. Math., 16, 197-201 (1958)

[10] Sestini, G., Annali Mat. Pura Appl., 56, 193-207 (1961)

[11] Stefan, J., Sitzungsberichte Akad. der Wiss., Math.-Natuurw., Vienna., 98, 965-983 (1889)

[12] Trench, W.F., J. Soc. Indust. Appl. Math., 7, 184-204 (1959).

[13] Wragg, A. Comp. J., 9, 106-109 (1966).

[14] Zonneveld, J.A., Mathematical Center Tracts, Math. Centrum, Amsterdam, 8, 1-110 (1964).

DISCRETIZATION OF BOUNDARY CONDITIONS ON MOVING DISCONTINUITIES

Czeslaw P. Kentzer
Purdue University, Lafayette, Indiana USA

I. INTRODUCTION

Improper approximations of boundary conditions are known to cause serious numerical errors, see, e.g., Kreiss and Lundqvist, Moretti, and Osher. The difficulties are compounded in gasdynamics where internal boundaries between regions of continuous flow may exist in the form of discontinuities. Often, boundary conditions take the form of nonlinear relations. We shall propose here a procedure for imposing boundary conditions on partial derivatives of the dependent variables rather than on the variables themselves. Such conditions are exact and linear in derivatives, may be used with analytical or numerical methods of integration, and allow a formal passage to the limit of a stationary flow. The problem of discretization of boundary conditions is then reduced to a well-known problem of discretization of partial differential equations.

II. COMPATIBILITY CONDITIONS ON MOVING BOUNDARIES

Following Moretti, we shall formulate gasdynamic boundary conditions using the theory of characteristics. For the sake of generality, we shall consider time-dependent flows. Discontinuities in derivatives of the dependent variables are allowed in regions of continuous flow on characteristic surfaces. Compatibility conditions on characteristics are discussed by, e.g., Rusanov. Conditions of compatibility satisfied by the dependent variables on moving discontinuities were derived by Green and Naghdi from the principles of objectivity and without the use of Newton's laws of motion. Compatibility conditions for partial derivatives, on surfaces of discontinuity in the dependent variables, were obtained by Thomas. Conditions required of the partial derivatives on discontinuities in the dependent variables and on characteristics may be imposed simultaneously at intersections of such surfaces. This principle was used by Whitham to derive an approximate law for a one-dimensional shock motion; the same principle will be used in the sequel to derive partial differential equations applicable to points on moving boundaries.

We shall refer the gaseous medium to an inertial frame of Cartesian coordinates (x,y,z) and consider velocity components u, v, w, density ρ, and pressure p as functions of position and time t. Assuming the existence of an arbitrary differentiable transformation of (x,y,z) into a new coordinate system (ξ,η,ζ), we shall consider a moving surface $\Sigma(t)$ given by an equation of the form $\zeta=\zeta(\xi,\eta,t)$. Let $R(t)$ be the immediate neighborhood of $\Sigma(t)$ in which the surface $\Sigma(t)$ possesses a tangent plane, and let $\Sigma(t)$ divide $R(t)$, at time t, into two regions $R_1(t)$ and $R_2(t)$ obtained by excluding the points on the surface $\Sigma(t)$ from its neighborhood $R(t)$. Let $f_i(x,y,z,t)$, $i=1,2,\dots,I$, be continuous differentiable functions in $R_1(t)$ and $R_2(t)$, and such that f_i and its partial derivatives of first order approach finite limiting values as the surface $\Sigma(t)$ is approached from either side. We shall denote by the second subscript $(\ )_{i1}$ or $(\ )_{i2}$ the limiting values of f_i as $\Sigma(t)$ is approached from $R_1(t)$ or $R_2(t)$, respectively.

We may introduce now J surface parameters $\sigma_j(\xi,\eta,t)$, $j=1,2,..,J$, which are assumed to be continuous and differentiable along $\Sigma(t)$ in $R(t)$. Suppose further that we are given K boundary conditions on $\Sigma(t)$ in the form of K continuous, differentiable, algebraic relations between f_{i1}, f_{i2}, and σ_j, written symbolically as

$$A_k(f_{i1},f_{i2},\sigma_j) = 0. \tag{1}$$

A directional derivative of A_k, taken along an arbitrary curve $\Gamma(t)$, which lies on $\Sigma(t)$ and is given by $\xi=\xi(s)$, $\eta=\eta(s)$, $t=t(s)$, $\zeta=\zeta[\xi(s),\eta(s),t(s)]$, is

$$dA_k/ds = (dA_k/ds)_t + (dA_k/dt)_{\xi\eta}\cdot(dt/ds) = 0, \tag{2}$$

where $(dA/ds)_t = A_{k,f_{i1}}[f_{i1,\zeta}(\zeta_{,\xi}\xi_{,s}+\zeta_{,\eta}\eta_{,s})+f_{i1,\xi}\xi_{,s}+f_{i1,\eta}\eta_{,s}]$

$$+A_{k,f_{i2}}[f_{i2,\zeta}(\zeta_{,\xi}\xi_{,s}+\zeta_{,\eta}\eta_{,s})+f_{i2,\xi}\xi_{,s}+f_{i2,\eta}\eta_{,s}]$$

$$+A_{k,\sigma_j}(\sigma_{j,\xi}\xi_{,s}+\sigma_{j,\eta}\eta_{,s}) = 0 \text{ if } dt/ds = 0 \tag{3}$$

is an inner derivative (taken in the t=constant hyperplane), and where

$$(dA_k/dt)_{\xi\eta}=A_{k,f_{i1}}(f_{i1,\zeta}\zeta_{,t}+f_{i1,t})+A_{k,f_{i2}}(f_{i2,\zeta}\zeta_{,t}+f_{i2,t})+A_{k,\sigma_j}\sigma_{j,t}=0 \tag{4}$$

is an outer derivative (taken in a hyperplane normal to t=constant plane) which vanishes when $d\xi/ds=d\eta/ds=0$. Repeated indices denote summation, and a comma denotes differentiation. Since the direction of differentiation is arbitrary, one may use anyone of the Eqs. (2), (3), or (4), but not simultaneously. We shall use Eq. (4) for time-dependent problems and Eq. (3) for stationary problems.

III. GENERAL DISCUSSION OF BOUNDARY CONDITIONS

Let the number of primary flow variables f_i be I, then, in the time-dependent case, the field equations of gasdynamics form a system of I hyperbolic equations. The system admits I independent characteristic relations at all interior points. We shall consider cases where there exist only N_1 pertinent characteristic relations on $\Sigma(t)$ in $R_1(t)$, and N_2 relations in $R_2(t)$, $N_1 \leq I$, $N_2 \leq I$. In the time-dependent case we shall take the temporal derivatives of the flow variables f_{i1} and f_{i2} and the temporal derivatives of the surface parameters σ_j as unknowns. Let N be the total number of unknown derivatives. Then we must have $N=N_1+N_2+K$ where K=number of boundary conditions. Thus we have the rule:

> the linear algebraic system for the N unknown temporal derivatives is constructed by writing N_1 linearly independent characteristic relations evaluated on $\Sigma(t)$ from the $R_1(t)$ side, N_2 characteristic relations from the $R_2(t)$ side, and K differential boundary conditions in the form of outer derivatives, Eq. (4); in the stationary case Eq. (3) is used to form a system of N equations linear in spatial derivatives.

The following types of boundary conditions occur frequently in gasdynamics.

Impermeable Wall. We are interested only in one side of $\Sigma(t)$, say $R_2(t)$. The number of independent characteristic relations is $N_2=4$ (one on a wave surface, and three on a stream surface). There are five flow variables, I=5, and three surface parameters: U_w=wall velocity and two direction cosines of the unit normal $\vec{n}$. The direction cosines and their derivatives may be determined from geometry leaving U_w as the only unknown surface parameter. Thus J=1. The number of unknowns is N=I+J=6, and we need two boundary conditions. The kinematic boundary condition is $A_1=\vec{n}\cdot\vec{q}-U_w=0$, where $\vec{q}$=velocity vector.

Case (a). Stationary Wall. We add $A_2=U_w=0$, so that K=2 and the number of equations is $N=N_2+K=6$. This may be reduced to five by elimination of U_w. Taking the outer derivative of A_1 and adding the four characteristic relations yields a linear system of five equations for the temporal derivatives of u, v, w, ρ, and p.

Case (b). Moving Wall. Set $A_2=U_w-F(\xi,\eta,t)=0$, where F is a prescribed function, continuous and differentiable at least once. Substituting F for U_w and taking the outer derivative of A_1 we again have five independent linear equations in temporal derivatives of u, v, w, ρ, and p.

Case (c). Prescribed Pressure Distribution. In the time-dependent version of the classical inverse problem we take $A_2 = p(\xi,\eta,\zeta,t)-P(\xi,\eta,t)=0$, where P is an arbitrary, continuous, and differentiable function. Adding the outer derivatives of A_1 and A_2 to the four characteristic relations completes the system for the temporal derivatives of u, v, w, ρ, p, and U_w. Coordinates of the wall supporting a given pressure may be followed in time with the help of the wall velocity $U_w(\eta,\xi,t)$.

Moving Shock Wave. Let the flow across $\Sigma(t)$ be from $R_1(t)$ to $R_2(t)$. Relative to $\Sigma(t)$, the conditions are supersonic in $R_1(t)$ and the five field equations determine a unique solution there independently of the motion of $\Sigma(t)$. There remain five unknowns in $R_2(t)$ and one shock parameter, U_s=shock velocity. Thus the number of unknowns is six. The flow in $R_2(t)$, the high pressure side, is subsonic relative to $\Sigma(t)$, so that there exists only one pertinent characteristic relation. Appending to it five Rankine-Hugoniot conditions, differentiated in the outward direction, Eq. (4), we obtain a system of six linear equations. It suffices to solve it for the shock acceleration, $\partial U_s/\partial t$. Timewise integration yields shock velocity and position. The former determines the flow in $R_2(t)$ uniquely through the shock jump conditions.

Contact Surface. The continuity of pressure and of normal components of velocity across the surface may be written as

$$A_1 = \vec{n}\cdot\vec{q}_1 - U_c = 0, \; A_2 = \vec{n}\cdot\vec{q}_2 - U_c = 0, \; A_3 = p_2 - p_1 = 0,$$

where U_c=velocity of the contact surface. Since the surface $\Sigma(t)$ is a stream surface, there are, on each side of $\Sigma(t)$, three characteristic relations of the stream type and one of the wave type. Thus N_1=N_2=4, and the number of flow variables is 2I=10. With U_c as the only unknown surface parameter, we have N=11 unknowns. The number of available equations is $N=N_1+N_2+K=11$, where the three (K=3) boundary conditions are differentiated in the outward direction, Eq. (4).

IV. EXAMPLES OF BOUNDARY CONDITIONS IN TWO SPATIAL DIMENSIONS

First, we will give below general forms of the characteristic conditions as derived by Rusanov and specialized to two spatial dimensions. Let $\vec{N}$ be a normal to a characteristic surface in the time-space domain, and let its projection onto the physical space, $\vec{n}$, be a unit vector. Then $\vec{n}$ is completely specified by one angle. Let ϕ be the inclination of the unit vector $\vec{n}$, measured counterclockwise from the positive x axis.

We shall consider two types of characteristic surfaces, the wave surface (trajectory of acoustic disturbances) and the stream surface (particle trajectory). A characteristic compatibility relation, which holds on wave surfaces and is evaluated in a plane tangent to the wave surface and normal to the vector $\vec{N}$ with a spatial component $\vec{n}$, may be written using Rusanov's Eq. (2.18):

$$p_t+(u+a\cos\phi)p_x+(v+a\sin\phi)p_y+\rho a[u_t\cos\phi+(u\cos\phi+a)u_x+vu_y\cos\phi]$$
$$+\rho a[v_t\sin\phi+uv_x\sin\phi+(v\sin\phi+a)v_y] = \rho a(X\cos\phi+Y\sin\phi), \qquad (5)$$

where a=adiabatic speed of sound, and X,Y,=x and y components of a body force per unit mass. On a stream surface we shall use the principle of constancy of particle entropy, Rusanov's Eq. (2.23):

$$a^2(\rho_t + u\rho_x + v\rho_y) - (p_t + up_x + vp_y) = 0, \qquad (6)$$

and the projection of the momentum equation, Rusanov's Eq. (2.24):

$$(u_t+uu_x+vu_y)\sin\phi-(v_t+uv_x+vv_y)\cos\phi+(p_x\sin\phi-p_y\cos\phi)/\rho = X\sin\phi-Y\cos\phi. \qquad (7)$$

Since the number of primary flow variables is four, and since we must have at least one boundary condition on $\Sigma(t)$, the above three equations will be used no more than once on a given side of $\Sigma(t)$. We may agree to choose $\vec{n}$ to be parallel to a unit normal to $\Sigma(t)$ pointing from $R_2(t)$ to $R_1(t)$ and inclined at an angle ϕ. Consequently, Eqs. (5)-(7) will require $\phi_1=\phi+\pi$ when evaluated in $R_1(t)$ and $\phi_2=\phi$ when evaluated in $R_2(t)$.

For the purpose of illustration we shall adopt the polar coordinates (r,θ) as the auxiliary coordinates (ξ,η), and we shall denote the moving boundary $\Sigma(t)$ by $r=r(\theta,t)$. Since ϕ, the inclination of the normal to $\Sigma(t)$, is determined by the geometry of the boundary, its time derivative must be compatible with the motion of $\Sigma(t)$. Referring to Fig. 1, we have on $\Sigma(t)$

$$\partial r(\theta,t)/\partial t = U/\cos(\theta-\phi), \; \partial r(\theta,t)/\partial\theta = r(\theta,t)\tan(\theta-\phi).$$

By cross-differentiation and elimination of the cross-derivative, one obtains

$$\phi_t = \{[U\sin(\theta-\phi)]\phi_\theta - [\cos(\theta-\phi)]U_\theta\}/r(\theta,t). \qquad (8)$$

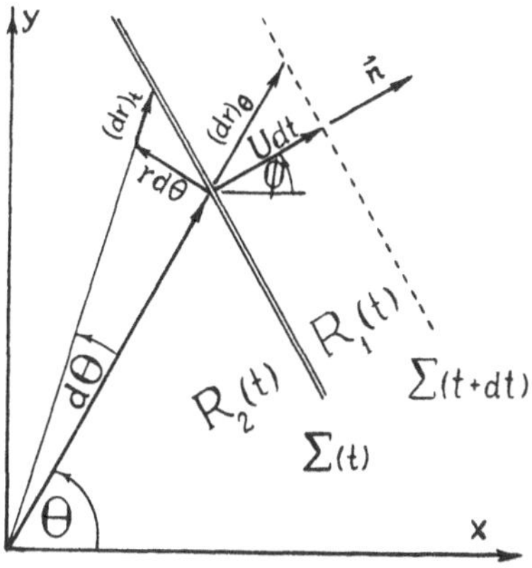

Fig. 1

Notation for Two-Dimensional Flow, $\xi=r, \eta=\theta$

In what follows, the time derivatives of $r(\theta,t)$ and of $\phi(\theta,t)$ on $\Sigma(t)$ will be considered as known at time t.

Impermeable Wall. Since only one side of the wall is of interest, say the $R_2(t)$ side, we omit the subscript 2. The first boundary condition on the wall becomes

$$A_1 = \vec{n}\cdot\vec{q}-U_w = u\cos\phi+v\sin\phi-U_w = 0. \quad (9)$$

For stationary wall, Case (a), we set $A_2=U_w=0$ and differentiate Eq. (9):

$$u_t\cos\phi + v_t\sin\phi = 0. \quad (10)$$

Let $u = q\sin\phi, \quad v = -q\cos\phi, \quad (11)$

where q=magnitude of fluid velocity at the wall, positive when directed at an angle $(\phi-\frac{1}{2}\pi)$ with respect to the x axis. Then $u_t=q_t\sin\phi, v_t=-q_t\cos\phi$, and the condition (10) is satisfied identically. Further, the number of unknowns is reduced by one and Eqs. (5)-(7) may be solved explicitly for q_t, p_t, and ρ_t:

$$q_t = (uv_x + vv_y + p_y/\rho-Y)\cos\phi - (uu_x + vu_y + p_x/\rho-X)\sin\phi, \quad (12)$$

$$p_t = -(u+a\cos\phi)p_x-(v+a\sin\phi)p_y-\rho a[(uu_x+vu_y-X)\cos\phi+(uv_x+vv_y-Y)\sin\phi+a(u_x+v_y)], \quad (13)$$

$$\rho_t = -(u\rho_x + v\rho_y) + (up_x + vp_y)/a^2 + p_t/a^2. \quad (14)$$

Eqs. (12)-(14) are exact and nonsingular except when the surface t=constant is characteristically oriented. Eqs. (12)-(14) may be used to replace the system of equations of gasdynamics at mesh points located on stationary walls. The velocity components u and v are expressible in terms of q by Eq. (11).

In the steady case we set all temporal derivatives in Eqs. (5)-(7) equal to zero and take the inner derivative of the boundary condition (9). Using Eq. (3) we obtain

$$(u_r r_\theta+u_\theta)\cos\phi+(v_r r_\theta+v_\theta)\sin\phi+(v\cos\phi-u\sin\phi)\phi_\theta = 0. \quad (15)$$

Again, the substitution of relations (11) into Eq. (15) satisfies the boundary condition (15), and we may treat q, ρ, and p as the only unknowns. Eqs. (5)-(7) provide us with a system of three partial differential equations which hold on the boundary. As may be easily shown, the system becomes singular when q=0. This singularity is of a saddle type indicating the existence of a branch point, so that two surfaces, along which wall boundary conditions hold, may intersect in the steady case only at a stagnation point unless a discontinuous solution is allowed.

For moving wall, Case (b), we use the second boundary condition $A_2=U_w-F(\theta,t)=0$, so that the wall velocity U_w may be considered as known. Substituting $F(\theta,t)$ for U_w in (9) and differentiating according to Eq. (4) we obtain

$$[u_t+(u_x\cos\theta+u_y\sin\theta)r_t]\cos\phi+[v_t+(v_x\cos\theta+v_y\sin\theta)r_t]\sin\phi+(v\cos\phi-u\sin\phi)\phi_t-F_t = 0. \quad (16)$$

The derivatives r_t and ϕ_t in Eq. (16) are considered known. Solving the characteristic relations (5)-(7) simultaneously with Eq. (16) one obtains

$$u_t = -(uu_x+vu_y+p_x/\rho-X)\sin^2\phi+(uv_x+vv_y+p_y/\rho-Y)\sin\phi\cos\phi+G\cos\phi,$$

$$v_t = (uu_x+vu_y+p_x/\rho-X)\sin\phi\cos\phi-(uv_x+vv_y+p_y/\rho-Y)\cos^2\phi+G\sin\phi,$$

$$p_t = -[(u+a\cos\phi)p_x + (v+a\sin\phi)p_y]$$
$$-\rho a[G+(u_x+v_y)a+(uu_x+vu_y-X)\cos\phi+(uv_x+vv_y-Y)\sin\phi],$$

$$\rho_t = -(u\rho_x+v\rho_y) + (up_x+vp_y)/a^2+p_t/a^2,$$

where $G=F_t+(u\sin\phi-v\cos\phi)\phi_t-[(u_x\cos\theta+u_y\sin\theta)\cos\phi+(v_x\cos\theta+v_y\sin\theta)\sin\phi]F/\cos(\theta-\phi)$.

The above four equations become singular only when the t=constant plane is characteristically oriented.

When the gas pressure is prescribed as a function of time and position on the wall, Case (c), the second boundary condition $A_2 = p[r_w(\theta,t),\theta,t] - P(\theta,t) = 0$ takes the differential form

$$p_t + (p_x\cos\theta + p_y\sin\theta)r_t - P_t = 0, \tag{17}$$

so that p_t may be considered as known and given by Eq. (17). The first boundary condition, $A_1=0$, takes the form

$$\partial U_w/\partial t = u_t\cos\phi + v_t\sin\phi - H, \tag{18}$$

where $H = (u\sin\phi-v\cos\phi)\phi_t - [u_x\cos\theta+u_y\sin\theta)\cos\phi+(v_x\cos\theta+v_y\sin\theta)\sin\phi]U_w/\cos(\theta-\phi)$. Then ρ_t is available from Eq. (6) in terms of p_t, and we only need to solve Eqs. (5) and (7) simultaneously for u_t and v_t. The result is

$$u_t = -\cos\phi[p_t+(u+a\cos\phi)p_x+(v+a\sin\phi)p_y]/(\rho a)+\sin\phi(p_y\cos\phi-p_x\sin\phi)/\rho$$
$$-(uu_x+vu_y-X)-a(u_x+v_y)\cos\phi,$$

$$v_t = -\sin\phi[p_t+(u+a\cos\phi)p_x+(v+a\sin\phi)p_y]/(\rho a)-\cos\phi(p_y\cos\phi-p_x\sin\phi)/\rho$$
$$-(uv_x+vv_y-Y)-a(u_x+v_y)\sin\phi.$$

The wall acceleration is then available from Eq. (18). The wall velocity, $U_w(\theta,t)$, and the radial coordinate, $r_w(\theta,t)$, may be obtained by a numerical integration subject to given initial conditions.

Moving Shock Wave. With the upstream conditions in $R_1(t)$ assumed known, calculated independently of the motion of the shock, we may write the Rankine-Hugoniot jump conditions in the form

$$A_1 = u_2\cos\phi+v_2\sin\phi-U_s+\{(\gamma-1)(U_s-q_{n1})+2\gamma p_1/[\rho_1(U_s-q_{n1})]\}/(\gamma+1) = 0, \tag{19a}$$

$$A_2 = (v_2-v_1)\cos\phi - (u_2-u_1)\sin\phi = 0, \tag{19b}$$

$$A_3 = p_2 + [(\gamma-1)p_1 - 2\rho_1(U_s-q_{n1})^2]/(\gamma+1) = 0, \tag{19c}$$

$$A_4 = \rho_2-\rho_1(U_s-q_{n1})^2(\gamma+1)/[(\gamma-1)(U_s-q_{n1})^2 + 2\gamma p_1/\rho_1] = 0, \tag{19d}$$

where $q_{n1} = u_1\cos\phi + v_1\sin\phi$ = component of the upstream velocity normal to the shock. Since ϕ_t is available from Eq. (8), we have only one unknown shock parameter, viz. $\sigma_1 = U_s$. Differentiating the boundary conditions (19a)-(19d) according to Eq. (4) and solving for the temporal derivatives of u_2, v_2, and p_2(that of ρ_2 is not needed), one may substitute them directly into the characteristic relation (5) evaluated in $R_2(t)$. Eq. (5) will contain only one unknown temporal derivative, namely the shock acceleration $\partial U_s/\partial t$. Solving for it gives

$$\frac{\partial U_s}{\partial t} = -\frac{1}{C_1}\left\{C_2\frac{\partial p_2}{\partial x} + C_3\frac{\partial p_2}{\partial y} + \frac{1}{\gamma+1}\left[C_4\frac{dq_{n1}}{dt} + C_5\frac{d\rho_1}{dt} + C_6\frac{dp_1}{dt}\right]\right.$$
$$\left.+\rho_2 a_2\left[C_7\frac{\partial u_2}{\partial x} + C_8\frac{\partial u_2}{\partial y} + C_9\frac{\partial v_2}{\partial x} + C_{10}\frac{\partial v_2}{\partial y} + C_{11}\frac{\partial\phi}{\partial t} + X\cos\phi + Y\sin\phi\right]\right\}, \tag{20}$$

where

$$C_1 = 2\{2\rho_1(U_s-q_{n1})+\rho_2 a_2[1+a_1^2/(U_s-q_{n1})^2]\}/(\gamma+1),$$

$$C_2 = u_2 + a_2\cos\phi - U_s\cos\theta/\cos(\theta-\phi), \quad C_3 = v_2 + a_2\sin\phi - U_s\sin\theta/\cos(\theta-\phi),$$

$$C_4 = (\gamma-1)\rho_2 a_2 - 2a_1^2\rho_2 a_2/(U_s-q_{n1})^2-4\rho_1(U_s-q_{n1}),$$

$$C_5 = 2(U_s-q_{n1})^2 + 2a_1^2\rho_2 a_2/[\rho_1(U_s-q_{n1})], \quad C_6 = -(\gamma-1) - 2\gamma\rho_2 a_2/[\rho_1(U_s-q_{n1})],$$

$$C_7 = u_2\cos\phi + a_2 - U_s\cos\theta\cos\phi/\cos(\theta-\phi), \quad C_8 = v_2\cos\phi - U_s\sin\theta\cos\phi/\cos(\theta-\phi),$$

$$C_9 = u_2\sin\phi - U_s\sin\phi\cos\theta/\cos(\theta-\phi), \quad C_{10} = v_2\sin\phi + a_2 - U_s\sin\theta\sin\phi/\cos(\theta-\phi),$$

$$C_{11} = u_2\sin\phi - v_2\cos\phi,$$

and where $d(\ )/dt = \partial(\ )/\partial t + U_s[\cos\theta\ \ \partial(\ )/\partial x + \sin\theta\ \partial(\ )/\partial y]/\cos(\theta-\phi)$.

It is worth noting that $U_s-q_{n1}\geq a_1$, so that the coefficient C_1 is positive definite. Thus the expression for the shock acceleration, Eq. (20), is nonsingular (even when the shock and the characteristic are parallel) except when $\rho_1=0$ (expansion into vacuum) or when $\cos(\theta-\phi)=0$, but then the shock is tangent to the radial line θ=constant, and the auxiliary coordinates (r,θ) are inappropriate. A change of coordinates removes this singularity.

Eq. (20) simplifies considerably when the upstream conditions are uniform and time-independent or when intrinsic coordinates are used. Coordinates other than polar coordinates may be introduced into (20) by an appropriate transformation of (r,θ).

After the shock velocity, U_s, is integrated numerically to the next time level, the boundary values are obtainable from the shock conditions (19a)-(19d) in terms of U_s and the upstream flow.

Contact Discontinuity. The two-dimensional time-dependent case calls for a simultaneous solution of nine equations linear in the temporal derivatives of u_1, u_2, v_1, v_2, ρ_1, ρ_2, p_1, p_2, and U_c, the velocity of the contact surface. It is possible to reduce the system to a coupled system of four partial differential equations linear in the temporal derivatives of the velocity components. For brevity, we shall omit the explicit expressions for the derivatives of the unknowns. We will mention as a point of interest that the determinant of the system is equal to $(\rho_1 a_1+\rho_2 a_2)$ and, consequently, a unique solution exists provided that the densities do not vanish on both sides of the discontinuity simultaneously. This would occur only if the Mach numbers on both sides of the surface would become infinite.

V. CONCLUSIONS

Compatibility conditions on characteristics and on boundaries, solved simultaneously for the temporal derivatives of the unknowns, afford convenient means of imposing boundary conditions on a numerical solution. Singularities appear whenever data are given on a characteristically oriented surface. Discretization of boundary conditions is reduced to the problem of discretization of partial differential equations. Thus stability, accuracy, and convergence may be studied by familiar methods. The technique proposed here was applied with success in the subsonic, transonic, and supersonic case by the author. It was observed that the technique is insensitive to approximations of the spatial derivatives.

REFERENCES

Green, A. E., and Naghdi, P. M., Int. J. Engng. Sci. 2, 621-624, (1965).
Kentzer, C. P., AIAA 8th Aerospace Sci. Meeting, paper No. 70-45 (1970).
Kreiss, H. O., and Lundqvist, E., Math. Comp., 22, 1-12 (1968).
Moretti, G., Phys. Fluids, Supplement II, 12, Part II, 13-20 (1969).
Osher, S., Math. Comp., 23, 567-572 (1969).
Rusanov, V. V., Zh. Vych. Mat. Mat. Fiz., 3, 508-527 (1963).
Thomas, T. Y., Int. J. Engng. Sci., 4, 207-233 (1966).
Whitham, G. B., J. Fluid Mech., 4, 337-360 (1958).

"THE VISCOUS FLOW AROUND A CIRCULAR CYLINDER"

P.J. ZANDBERGEN
TWENTE UNIVERSITY OF TECHNOLOGY, THE NETHERLANDS

1. INTRODUCTION

The problem of the viscous flow around a circular cylinder has already received large attention, therefore it might seem to be superfluous to present again a paper concerned with the same problem.

However, notwithstanding the many investigations reported so far, it cannot be said that the problem has been solved to any degree of satisfaction, neither fundamentally nor practically.

With a view to obtain more insight in these problems, at Twente University of Technology a group was formed by Mr. R.W. de Vries, Mr. H.Q.J. Meershoek, Mr. E.H. Derks and myself to attack these problems along a broad front.

Several schemes, some rather simple, were tried or tried again in order to obtain a set of overlapping results. As may be obvious some of these schemes failed while others were partially successful. In nearly all methods a trigonometric representation of the vorticity and the stream function was used, thereby reducing the partial differential equations to a set of ordinary differential equations.

At the moment the investigations continue and all that can be given now is a survey of the schemes tried and of the preliminary results obtained thereby so far.

It is only fair to say that an important stimulus to these investigations was formed by a paper of Son and Hanratty (ref. 3), since the results presented there left much to question.

2. FORMULATION OF THE PROBLEM

The partial differential equations governing the flow of a viscous fluid in two dimensions can be written in terms of the vorticity ζ and the stream function ψ, as follows

$$\frac{\partial\zeta}{\partial t} + \frac{1}{r}\left(\frac{\partial\psi}{\partial r}\frac{\partial\zeta}{\partial\theta} - \frac{\partial\psi}{\partial\theta}\frac{\partial\zeta}{\partial r}\right) = \frac{2}{R}\left[\frac{\partial^2\zeta}{\partial r^2} + \frac{1}{r}\frac{\partial\zeta}{\partial r} + \frac{1}{r^2}\frac{\partial^2\zeta}{\partial\theta^2}\right] \qquad (2.1)$$

$$\zeta = \frac{\partial^2\psi}{\partial r^2} + \frac{1}{r}\frac{\partial\psi}{\partial r} + \frac{1}{r^2}\frac{\partial^2\psi}{\partial\theta^2} \qquad (2.2)$$

Here R is the Reynolds number referred to the diameter of the cylinder while r and θ are cylindrical coordinates, where θ is measured from the rear of the cylinder. The boundary conditions take the following form

$$\psi = \frac{d\psi}{dr} = 0 \qquad \text{for } r = 1 \qquad (2.3)\text{a}$$

$$\psi = \zeta = 0 \qquad \text{for } \theta = 0,\pi \qquad (2.3)\text{b}$$

$$\zeta \to 0 \text{ and } \psi \to -r\sin\theta \text{ for } r \to \infty \qquad (2.3)\text{c}$$

In the case of a numerical solution based on finite difference methods, often use is made of the following transformation.

$$r = e^{\pi\xi} \qquad \eta = \theta/\pi \tag{2.4}$$

This has the advantage that the right hand side of eqs (2.1) and (2.2) reduce to the Euclidean laplace operator, while the main advantage is that more points are situated in the vicinity of the cylinder, i.e. there where the change in the physical quantities is largest.

3. INVESTIGATION OF DIFFERENT SCHEMES

In this chapter the various schemes and the results thereby obtained will be discussed in some detail. It should be emphasized that the results in all cases are only preliminary and that due to further investigations some of the tentative conclusions probably will be changed.

3.1 The truncated Stokes - Picard method

This is a time independant method, hence $\frac{\partial\zeta}{\partial t} \equiv 0$. Equations (2.1) and (2.2) are reduced to very simple linear partial differential equations by adopting the following scheme

$$\frac{R}{2r}\left(\frac{\partial\psi^k}{\partial r}\frac{\partial\zeta^k}{\partial\theta} - \frac{\partial\psi^k}{\partial\theta}\frac{\partial\zeta^k}{\partial r}\right) = \left[\frac{\partial^2\zeta^{k+1}}{\partial r^2} + \frac{1}{r}\frac{\partial\zeta^{k+1}}{\partial r} + \frac{1}{r^2}\frac{\partial^2\zeta^{k+1}}{\partial\theta^2}\right] \tag{3.1.1}$$

$$\zeta^{k+1} = \frac{\partial^2\psi^{k+1}}{\partial r^2} + \frac{1}{r}\frac{\partial\psi^{k+1}}{\partial r} + \frac{1}{r^2}\frac{\partial^2\zeta^{k+1}}{\partial\theta^2} \tag{3.1.2}$$

This system of partial differential equations is transformed into a set of ordinary differential equations by using the following representation

$$\zeta = \sum_{n=1}^{\infty} \zeta_n \sin n\theta \tag{3.1.3}$$

$$\psi = \sum_{n=1}^{\infty} \psi_n \sin n\theta \tag{3.1.4}$$

By this representation the symmetry condition (2.3)b is fulfilled automatically. The left hand side of eq. (3.1.1) is represented by

$$\sum_{n=1}^{\infty} f_n^k \sin n\theta \tag{3.1.5}$$

where f_n^k depends on all the functions ζ_n and ψ_n .

The sytem of the equations now becomes

$$f_n^k = \frac{d^2\zeta_n^{k+1}}{dr^2} + \frac{1}{r}\frac{d\zeta_n^{k+1}}{dr} - \frac{n^2}{r^2}\zeta_n^{k+1} \tag{3.1.6}$$

$$\zeta_n^{k+1} = \frac{d^2\psi_n^{k+1}}{dr^2} + \frac{1}{r}\frac{d\psi_n^{k+1}}{dr} - \frac{n^2}{r^2}\psi_n^{k+1} \tag{3.1.7}$$

For $n \geqslant 2$ these equations can be solved easily analytically by using the boundary conditions (2.3)a and (2.3)c. However for $n = 1$, a solution fulfilling all the boundary conditions is impossible, which is of course caused by the well known Stokes phenomenon. There are a number of possibilities to circumvent this difficulty. We will not go into any detail, since all the possibilities tried did not lead to success. We continue, however, our effort on this method, since the closely related method of Underwood (ref. 5) did lead to successful results, although a strong dependence on the starting profile was observed.

3.2 Approaches based on linearization

This time independent method has been used in fact by many investigators. The approaches of Thom (ref. 4) and of Apelt (ref. 1) belong to this category, although due to many special devices to improve the convergence, it is very difficult to see what is really the crucial point in these methods.

3.2.1 Direct linearization

A very simple and straight forward method is to linearize the left hand side of eq. (2.1) as follows $\left(\frac{\partial \zeta}{\partial t} \equiv 0\right)$

$$-\frac{R}{2r}\left(\frac{\partial \psi^k}{\partial r}\frac{\partial \zeta^k}{\partial \theta} - \frac{\partial \psi^k}{\partial \theta}\frac{\partial \zeta^k}{\partial r}\right) + \frac{R}{2r}\left(\frac{\partial \psi^{k+1}}{\partial r}\frac{\partial \zeta^k}{\partial \theta} + \frac{\partial \psi^k}{\partial r}\frac{\partial \zeta^{k+1}}{\partial \theta} - \frac{\partial \psi^{k+1}}{\partial \theta}\frac{\partial \zeta^k}{\partial r} - \frac{\partial \psi^k}{\partial \theta}\frac{\partial \zeta^{k+1}}{\partial r}\right) \qquad (3.2.1)$$

while in the right hand side and in eq. (2.2) the index k+1 is used.

The boundary conditions in this case are taken to be

$$\zeta = 0 \qquad \text{and} \qquad \psi = -r\sin\theta \qquad \text{for } r = r_{max} \qquad (3.2.2)$$

while the usual conditions hold for $r=1$. The linearized system has been calculated by use of the normal 5-point star in the (ξ,η) system. As starting condition is taken

$$\psi = \left(r - \frac{3}{r} + \frac{2}{r^2}\right)\sin\theta$$

$$\zeta = \frac{6}{r^4}\sin\theta$$

For several values of the Reynolds number R and for $r_{max} = 111,3$ this method of direct linearization was tried, so far only for 30 steps in the θ direction and 50 steps in the r direction. It turns out that the method in its present form is convergent for values of R up to 24. For values of R of 32 and 40 the method is divergent.

3.2.2 Linearized Truncation mathod

A new and rather unexpected result was obtained when this same method of linearization was used on the set of ordinary differential equations resulting from the substitution of the expressions (3.1.3) and (3.1.4). In this case we get a block tridiagonal matrix for the unknown functions ζ_n and ψ_n.
It turns out that this system is convergent for a much larger class of Reynolds numbers.

Up to now it has been shown that the process is convergent for Reynolds numbers of 150 and less.

If N is the maximum number of ψ_n and ζ_n functions taken into account, and h is the stepwidth in ξ direction the following cases have been considered.

R = 2 , 24 , 40	N = 8	h = 0,03	ξ_{max} = 1,5	(r_{max} = 111,3)
R = 100 , 150	N = 8,16	h = 0,03	ξ_{max} = 1,5	

In fig. 1 the flow field has been given for R = 24 as compared to the one calculated by direct linearization. In fig. 2 the vorticity distribution along the body surface has been compared. Whereas the flow fields agree rather well, rather large discrepancies occur in the vorticity distribution. Comparison with results from the literature for R = 40 indicates that the vorticity values obtained so far with the linearized truncation method are too large. This point is investigated further.

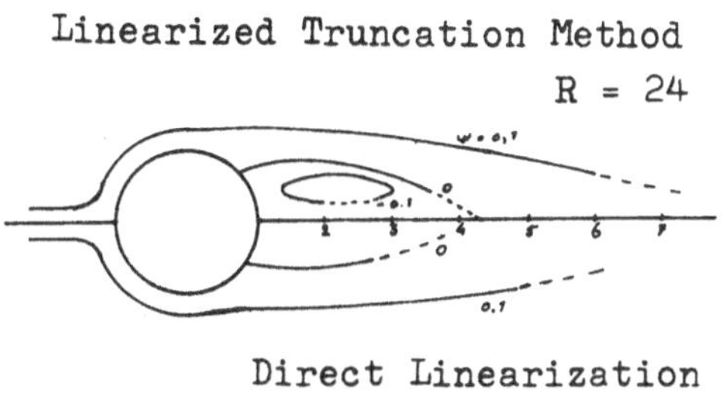

Flow Field for R = 24

Fig. 1

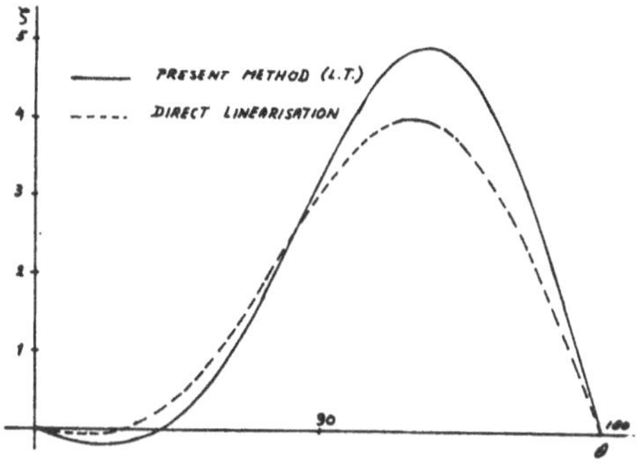

Comparison of Vorticity Distribution for R = 24

Fig. 2

In fig. 3 preliminary results for the case R = 100 are given for N = 16. They do not agree quite well, with N = 8 since it is evident from the computed results that in general the higher coefficients in the ζ and ψ series are not going to zero very fast.

The Linearized Truncation Method

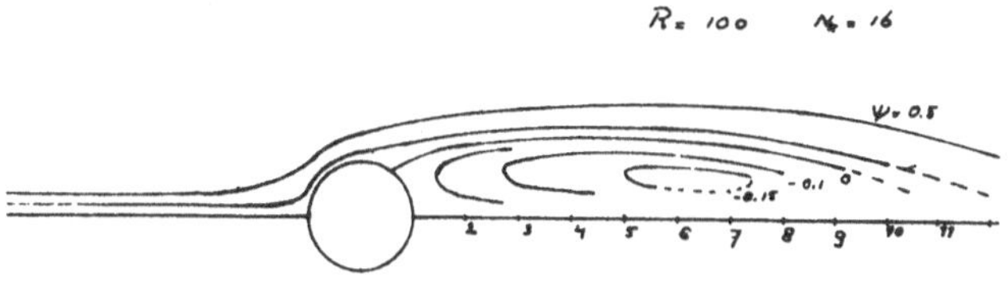

Flow Field for R = 100

Fig. 3

3.3 The truncated Oseen - Picard method

In the methods discussed so far one of the very weak points is the treatment of the boundary conditions at infinity. Oseen was the first to notice this point and to remedy the situation by retaining the term $U_\infty \frac{\partial\zeta}{\partial x}$ in the inertia terms. He thus was able to obtain an approximate analytical solution.

Without going into any detail we can state that the problem can be formulated now in terms of the reduced vorticity $\bar{\zeta}$ and the reduced stream function ψ^* as follows

$$- 2\beta^2\bar{\zeta}\left(\frac{1}{r}\frac{\partial\psi^*}{\partial\theta}\cos\theta + \frac{\partial\psi^*}{\partial r}\sin\theta\right) - \frac{2\beta}{r}\left(\frac{\partial\psi^*}{\partial\theta}\frac{\partial\bar{\zeta}}{\partial r} - \frac{\partial\psi^*}{\partial r}\frac{\partial\bar{\zeta}}{\partial\theta}\right) =$$

$$\frac{\partial^2\bar{\zeta}}{\partial r^2} + \frac{1}{r}\frac{\partial\bar{\zeta}}{\partial r} + \frac{1}{r^2}\frac{\partial^2\bar{\zeta}}{\partial\theta^2} - \beta^2\bar{\zeta} \qquad (3.3.1)$$

$$e^{\beta r\cos\theta}\,\bar{\zeta} = \frac{\partial^2\psi^*}{\partial r^2} + \frac{1}{r}\frac{\partial\psi^*}{\partial r} + \frac{1}{r^2}\frac{\partial^2\psi^*}{\partial\theta^2} \qquad (3.3.2)$$

Here $\beta = \frac{R}{4}$, $\zeta = e^{\beta r\cos\theta}\,\bar{\zeta}$, $\psi = - r\sin\theta + \psi^*$ (3.3.3)

The Oseen equations are obtained by setting the left hand side of eq. (3.3.1) equal to zero.

In the method proposed here, we again use the trigonometric representations (3.1.3) and (3.1.4) but now for the functions $\bar{\zeta}$ and ψ^*.

Analogous to eq. (3.1.5) the left hand side of eq. (3.3.1) can be written as a sine series, with amplitude functions $\bar{f}_n$.

The iteration scheme adopted leads to the following equation for $\bar{\zeta}_n^{k+1}$

$$\frac{d^2\bar{\zeta}_n^{k+1}}{dr^2} + \frac{1}{r}\frac{d\bar{\zeta}_n^{k+1}}{dr} - \left(\frac{n^2}{r^2} + \beta^2\right)\bar{\zeta}_n^{k+1} = \bar{f}_n^{\,k} \qquad (3.3.4)$$

where the term $\bar{f}_n^{\,k}$ is calculated from the previous values of $\bar{\zeta}_n^{\,k}$, ψ_n^{*k} , etc.

The solution of $\bar{\zeta}_n$ fulfilling the conditions at infinity reads

$$\bar{\zeta}_n^{k+1} = A_n^{k+1}K_n(\beta r) + I_n(\beta r)\int_\infty^r \xi\,\bar{f}_n^{\,k}(\zeta)\,K_n(\beta\xi)d\xi - K_n(\beta r)\int_1^r \xi\bar{f}_n^{\,k}(\zeta)I_n(\beta\xi)d\xi$$

$$= A_n^{k+1}K_n(\beta r) + F_n^{k+1}(n) \qquad (3.3.5)$$

The equation for ψ_n^* takes the following form

$$\frac{d^2\psi_n^{*k+1}}{dr^2} + \frac{1}{r}\frac{d\psi_n^{*k+1}}{dr} - \frac{n^2}{r^2}\psi_n^{*k+1} = g_n^{k+1}(r) \qquad (3.3.6)$$

The function g_n^{k+1} is given by

$$\sum_{m=1}^{n} A_m^{k+1}\{I_{n-m}(\beta r) - I_{n+m}(\beta r)\}\,K_m(\beta r) + \sum_{m=n+1}^{\infty} A_m^{k+1}\{I_{m-n}(\beta r) - I_{m+n}(\beta r)\}\,K_m(\beta r)\} +$$

$$+ \sum_{m=1}^{n} F_m^{k+1}\{I_{n-m}(\beta r) - I_{m+n}(\beta r)\} + \sum_{m=n+1}^{\infty} F_m^{k+1}\{I_{m-n}(\beta r) - I_{m+n}(\beta r)\} \qquad (3.3.7)$$

It is readily proved that the solution of eq. (3.3.6) can be written as

$$\psi_n^{*k+1} = \frac{r^n}{2n} \int_\infty^r \frac{g_n^{k+1}(\xi)}{\xi^{n-1}} d\xi - \frac{1}{2nr^n} \int_1^r \xi^{n+1} g_n^{k+1}(\xi) d\xi \qquad (3.3.8)$$

The integrals involving the coefficients A_m can be obtained in closed form (see also ref. 2). By using the boundary conditions at the surface of the cylinder, a system of linear algebraic equations is obtained for the unknown quantities A_n. The matrix of coefficients of this system is independent of the index k, only the right hand side changes. Once the coefficients A_n are known the full solution can be calculated. So far convergence of the method has been obtained for R = 2 and N = 5. As will be clear the first iteration provides the truncated solution to the full Oseen equations. One of the great advantages of the method is provided by the fact that the numerical methods involved, are normal integration procedures. Further investigations with regard to accuracy, rate of convergence with respect to Reynolds number and with respect to the number of terms taken into account are performed at the moment.

REFERENCES

1. Apelt, C.J. The Steady Flow of a Viscous Fluid past a Circular Cylinder at Reynolds Numbers 40 and 44. A.R.C. Report and Memoranda No. 3175, 1961.
2. Dennis, S.C.R., and Shimshoni, M. The Steady Flow of a Viscous Fluid past a Circular Cylinder. A.R.C. Current paper No. 797.
3. Son, J.S., and Hanratty, Th.J. Numerical Solution for the Flow around a Cylinder at Reynolds Numbers of 40, 200 and 500. J. Fluid Mech. (1969), vol 35, part 2.
4. Thom, A. The Flow past Circular Cylinders at Low Speeds. Proc. Roy. Soc. A 141, 1933.
5. Underwood, R.L. Calculation of Incompressible Flow past a Circular Cylinder at Moderate Reynolds Numbers. J. Fluid Mech. (1969), vol 37, part 1.

A NUMERICAL STUDY OF BUOYANCY DRIVEN FLOWS IN ROOM ENCLOSURES

Jacob E. Fromm

IBM Research Laboratory
Monterey and Cottle Roads
San Jose, California 95114

The objective of the present work is to extend the capability of finite difference solution of buoyant flows to ranges of the parameters which exist in the environment of living quarters. Consider a long hallway of rectangular cross section with a constant high temperature T_1 applied at the left wall and a constant low temperature T_0 applied at the right wall. Let the ceiling and floor have a constant linear gradient of temperature between the wall values. We may characterize the flow of air in the specified geometry by two parameters, namely the Rayleigh number $A=\alpha g(T_1-T_0)d^3/\kappa\nu$ and the Prandtl number $\sigma=\nu/\kappa$. α, κ and ν are taken as constants of the flow, representing fluid volume expansion coefficient, thermometric conductivity and kinematic viscosity respectively. d is the distance between the walls of the enclosure and g is the gravitational constant. Solutions may be obtained over a wide range of A and σ. Generally σ~1.0 and A small enough to correspond to creeping flows ($A\lesssim 10^4$) provides for cases where finite difference methods are most reliable. Agreement is sought between analytical and numerical solution for such highly viscous flows and solutions by Poots have been used in this connection. With increased values of A, no analytical solutions are available, but experimental heat transfer measurements can be used as a guide. Power law data comparisons are adequate up to $A\sim 3\cdot 10^8$ with $\sigma\sim 1.0$. Beyond this range experimental heat transfer measurements are no longer available but some authors show the transition to turbulence in this range through flow visualization (Torrance, Rochett et al). Many authors regard the range below this transition as turbulent but recent experiments and the present numerical results suggest the contrary.

We have intentionally sought to obtain solutions to high Rayleigh number flows without recourse to modeling with mean flow equations. In this context, we cannot play a numerical limitation against a physical one. Computational noise must be overcome without artificial or eddy viscosity. Of course, the method itself must not be one that damps short waves beyond what is necessary for representation on a finite grid. While it is expected that finite representation of turbulence is still impossible on present day computers, it should be noted that the capabilities of properly designed algorithms have not yet been fully exploited.

We seek solutions to the system of equations

$$\frac{\partial T}{\partial t} + \frac{\partial uT}{\partial x} + \frac{\partial vT}{\partial y} = \kappa\nabla^2 T \quad , \tag{1}$$

$$\frac{\partial \omega}{\partial t} + \frac{\partial u\omega}{\partial t} + \frac{\partial v\omega}{\partial y} = \nu\nabla^2\omega + \alpha g\,\frac{\partial T}{\partial x} \quad , \tag{2}$$

and $$\nabla^2\psi = -\omega \quad , \tag{3}$$

where $$u = \frac{\partial\psi}{\partial y} \quad , \quad v = \frac{-\partial\psi}{\partial x} \quad . \tag{4}$$

T is the temperature, ω the vorticity and ψ the streamfunction. u and v are the velocities in the horizontal x direction and the vertical y direction respectively. Numerically, we replace the indicated derivatives by finite differences and computation proceeds in the order of the given equations. Assuming all variables are known at time n we first update $T^n_{i,j} = T(n\Delta t,\, i\Delta x,\, j\Delta y)$ by a convective contribution to obtain $\tilde{T}_{i,j}$. The latter is then updated by a conduction contribution to obtain $T^{n+1}_{i,j}$. A similar procedure is applied to ω with the additional intervening step in which $T^{n+1}_{i,j}$ is used in providing the buoyancy update to ω. These successive updates provide for maximum range of stability of the various parts of the explicit difference procedure. For example, the criteria $\nu\Delta t/\Delta x^2<1/4$ is applicable even in the creeping flow range. A more stringent criteria would apply if successive updates were not used. In the case of the buoyancy contribution, the use of old values($T^n_{i,j}$) would lead to instability.

Standard difference methods were used for the linear parts of the time dependent equations and for equation (3) (see Fromm). For solution of equation (3) we made use of Buneman's programs which give a direct solution accurate to computer roundoff.

Vorticity boundary updates for no-slip boundary conditions were obtained following solution of equation (3). The usual procedure of making the tangential velocity go to zero by linear variation at the wall was applied. The use of Fourier transform methods for this calculation did not appreciably change the results. Although the boundary values of vorticity became larger, the additional drag at the walls quickly compensated to give essentially the same local conditions.

The crucial nonlinear (convection) calculations were performed with new programs. Several difference forms were used for various ranges of $\alpha=u\Delta t/\Delta x$ and $\beta=v\Delta t/\Delta y$. The reason for using several methods dependent upon local flow velocity was that lagging phase errors could not be tolerated. Also, leading phase errors had to be held to a minimum and a sharp cut off of the highest representable wave number had to be provided. Fourth order methods were used and the notion of combining upstream and central space approximations to minimize phase error (see Fromm) was employed. Alternate steps were calculated with a central approximation followed by an upstream approximation rather than combining the methods for each time step. This saved computation time without losing the benefits of the low phase error computation.

Unfortunately, the combined upstream and central space approximation method has leading phase errors only for $|\alpha|$ or $|\beta| > 0.5$. For $|\alpha|$ or $|\beta| < 0.5$ we must accept a larger phase error and use simply upstream approximations. For lower speeds of flow this is not serious because the adverse noise effects of larger phase error are not manifested in such a way as to upset the solution. This is perhaps because short wave variations and local high speed flows go hand in hand. With lower speeds (longer waves) the phase properties are very good for most difference methods.

In the preceding, the use of fourth order methods in two forms dependent on local flow speed was discussed. Both forms require upstream testing (in one case only every other time step). Since the upstream approximation requires additional mesh point information, flow away from boundaries requires us to first return to a central approximation and then even to drop down to a second order central form. Fortunately flow away from a boundary is best for central approximations because phase error noise is upstream (lagging) or in this instance, outside of the region of computation. Flow toward the wall is possible with upstream approximations and here again the phase error effects are ideally suited so they do not occur inside the fluid region.

In second order methods, the property of inherent damping is of concern but flows at right angles away from a wall are slow. No serious loss of amplitude is expected from this use of second order methods. Surprisingly for our particular choice of methods, it was also necessary to use an upstream second order method for $|\alpha|$ or $|\beta|$ less than 0.09. This was because of an anomoly in the upstream fourth order method that gave a slightly lagging phase error for short waves at low speeds. For flows of such relatively slow speeds the damping characteristics of second order are again of little consequence. The important aspect of this, however, is that it illustrated the necessity for leading phase properties even to this detail!

Our choice of methods dependent on α and β was based on current limited experience of stability characteristics of various alternatives. Clearly, a broader experience could have led to more ideal amplitude and phase properties over the entire range. It is doubtful that any single difference method could provide all the necessary features required in the range of $|\alpha|$ and $|\beta| \leq 1$. Fourth order upstream methods can provide stability up to $|\alpha|$ and $|\beta| \leq 2$ but unfortunately the added range involves lagging phase errors again at the most crucial velocities.

The complicated procedures outlined above are costly in computer time compared with earlier methods but the simpler methods are limited to $A \lesssim 10^6$ for the problem under consideration. The growth in complexity is unfortunately not only because of variability of methods as a function of α and β. It is not generally known that finite difference methods are not always isotropic in behavior. In a greater part of the literature where nonlinear approximations have been applied, there has been no consideration given to the fact that the characteristics of the approximation may differ radically if the flow is diagonal to mesh lines or in the direction of mesh lines in a two dimensional calculation. "Time splitting" is the expedient that is usually used to provide both stability and isotropic behavior for the type of difference methods here considered. The "time splitting" procedure is that of using updated convection values in one coordinate direction in evaluating updates in the second coordinate direction. Without "time splitting" one can have instability in the flow direction and damping normal to the flow direction. Obviously such effects are compensating only in maintaining some semblance of overall stability not in providing accurate results. The longitudinal instability is revealed by formal stability analysis but the transverse damping is not. The transverse damping is probably present in many of the difference methods that are of neutral stability with the exception of Arakawa's mixed schemes which appear to have similar properties to techniques involving "time splitting."

Conservation may be achieved with "time splitting" as Crowley has observed, but unfortunately inconsistencies occur in velocity specifications. This is revealed only by expansion of the two dimensional convection approximation into a single time step. For example, we find that $u_{i-1/2,j}$, and $v_{i,j-1/2}$ can appear in the same higher order term. The consequences of this are that squared vorticity is not conserved. The manifestation is a moderately fast unstable growth of vortex strength in regions of concentrated vorticity. Alternation of the "time splitting" sequence has no effect in reducing this growth.

To overcome the inconsistency in velocity specification we have explicitly programmed all cross derivative terms that are implied in "time splitting." By this means, one can separate higher order correction terms and divide them appropriately into component parts that will remove the inconsistencies. For fourth order approximation, this involves 42 terms out to eighth order in time. We shall here give only the expanded form for second order central computation. We write for example

$$\tilde{T}_{i,j} = T^n_{i,j} + \frac{\Delta t}{\Delta x} (F^n_{i-1/2,j} - F^n_{i+1/2,j}) + \frac{\Delta t}{\Delta y} (F^n_{i,j-1/2} - F^n_{i,j+1/2}) \quad , \qquad (5)$$

recalling that $\tilde{T}$ represents an update that is yet to be modified by conduction before obtaining T^{n+1}. Here F is the temperature flux yet to be defined.

Now in the second order central approximation let F imply $F^n_{i-1/2,j}$, α imply $\alpha^n_{i-1/2,j}$, and β imply $\beta^n_{i-1/2,j}$.

Also let

$$S_o = (T_{i-1,j} + T_{i,j}) \quad , \quad D_o = (T_{i-1,j} - T_{i,j}) \quad ,$$

$$S_2 = (T_{i-1,j+1} + T_{i,j+1}) \quad , \quad D_2 = (T_{i-1,j+1} - T_{i,j+1}) \quad ,$$

$$S_4 + (T_{i-1,j-1} + T_{i,j-1}) \quad , \quad D_4 = (T_{i-1,j-1} - T_{i,j-1}) . \qquad (6)$$

then

$$\frac{\Delta t}{\Delta x} F = \frac{\alpha}{2} S_0 + \frac{\alpha\beta}{8} (S_4 - S_2) + \frac{\alpha\beta^2}{12} (S_4 - 2S_0 + S_2)$$
$$+ \frac{\alpha^2}{2} D_o + \frac{\alpha^2\beta}{6} (D_4 - D_2) + \frac{\alpha^2\beta^2}{8} (D_4 - 2D_o + D_2) \qquad (7)$$

No provisions are made for higher order approximations of α and β (i.e. u and v) both because this information cannot be obtained without reverting to the primative equations and because the required values are averages over the local domain that is effective within the stability limit. Thus the approximation (equations 5-7) follows from taking

$$\frac{\partial T}{\partial t} = - \frac{\partial uT}{\partial x} - \frac{\partial vT}{\partial y} \quad .$$

and letting

$$\tilde{T} = T^n + \Delta t \frac{\partial T}{\partial t} + \frac{\Delta t^2}{2} \frac{\partial^2 T}{\partial t} + \frac{\Delta t^3}{3!} \frac{\partial^3 T}{\partial t^3} \quad \frac{\Delta t^4}{4!} \frac{\partial^4 T}{\partial t^4} \quad .$$

All velocity time derivatives and all higher derivatives of temperature not occurring in the equivalent one step expansion of the "time splitting" form are deleted. We have not given the remaining u flux term of (5) nor the v flux terms. These may be readily inferred from (6) and (7). Computation is carried out by adding the contribution (7) to the mesh point value $T^n_{i,j}$ and subtracting it from $T^n_{i-1,j}$. Clearly, the first moment of this equation is conserved and the procedure of equation (6) and (7) either conserves higher moments or closely approximates such conservation.

It would be possible to devise an implicit system along the same lines as described above to gain overall neutral stability. Unfortunately, without the possibility of using "time splitting" the solution in implicit form would become very complicated. Also neutral stability means no cut off of the $2\Delta x$ wave which is always stationary in finite differences. An artificial cut off would have to be provided. There is no known means for providing a sharp artificial cut off that does not introduce other complications. A similar problem occurs with Arakawa's methods where leap-frog time differencing gives neutral stability. The phase properties for time centered methods are difficult to obtain, making it improbably that such systems will be designed to have purely leading phase properties.

Computational noise related to phase distortion can be particularly bad in the application we are considering here. The buoyancy driven flow along the vertical walls acts like a jet impinging on a solid surface. Also circulating horizontal flow brings fluid directly to the high or low temperature walls. With lagging phase errors, the flow into the hot wall will cause false low temperatures to occur right adjacent to the wall. Clearly, these will have a detrimental effect upon the solution. With leading phases properties, the short waves are precursers of the flow. They do not arise at all at the wall and otherwise move into quiescent regions. They are removed by the inherent cut off of the method with no detectable influence upon the general flow.

The effects of non-conservation of second moments in a "time splitting" calculation are such as to produce hot or cold spots in the temperature fields that will exceed the fixed boundary extremes. In the present calculations this never occurs regardless of the violence of the circulations.

The computations here were all carried out on a 65x65 evenly spaced grid. This provided for use of available direct Poisson solvers and avoided the concern of even order tructation errors. While boundary layers become very thin their resolution is at least as good as that of small circulations occuring elsewhere in the mesh.

We have carried out a parameter study starting from $A = 10^4$ to link our results with previously obtained solutions. Compared with results by Poots, there are differences of the order of 2.0 percent in heat transfer values and in dominant Fourier coefficients in the temperature expansion. Because both methods are approximate, this degree of agreement is regarded as extremely good. In Table I we give the heat transfer measurements for the series of runs which have been completed to the present time. These values may be made dimensionless by multiplying by 0.0056492 $\sqrt{A\sigma}$. For $10^4 \leq A \leq 10^8$ the dimensionless heat transfer (Nusselt number N) obeys the relation $N = 0.0695\ A^{1/3}$ for $\sigma = 1$. This is in good agreement with data given by Jacob for similar experiments. The sharp rise following this range is consistent with observation in the sense of transition to turbulence, but no experimental heat transfer values are available for comparison. Current thoughts on the matter are that after the transition a power law behavior of less than 0.5 must again come into effect but the transition is necessary to provide observed heat transfer increases with transition to turbulence.

TABLE I

A	σ	Heat transfer (Calculated at the cold wall)
$1\cdot10^4$	1.0	3.10
$1\cdot10^5$	1.0	1.92
$1\cdot10^6$	10.0	.394
$1\cdot10^6$	1.0	1.23
$1\cdot10^6$	0.1	3.25
$1\cdot10^7$	1.0	0.81
$1\cdot10^7$	0.1	2.16
$1\cdot10^8$	10.0	0.20
$1\cdot10^8$	1.0	0.52
$1\cdot10^8$	0.1	~100
$3\cdot10^8$	1.0	0.47
$3\cdot10^8$	1.0	~33
$1\cdot10^9$	10.0	0.25
$1\cdot10^9$	1.0	~25
$1\cdot10^{10}$	10.0	~2.5

The given figures illustrate late time solutions in the range of transition. Five cases are shown by isotherm plots of the flows. In all cases these are just the last solution computed in the development from an initial state of no flow, with the interior temperatures of the region set to the mean at the surfaces. In all examples shown, the flow is highly unsteady, but with vastly different flow behavior. Unsteadyness, after the early transient phases, is observed at $A = 10^7$. It probably first occurs close to $A = 10^6$. Our first illustrated solution is for $A = 10^7$, $\sigma = 1.0$. Here one has a slow gradient in temperature in the vertical in the central region with rapid horizontal gradients at the vertical walls. There is little corner activity so that the thermal and velocity boundary layers both show only a small tendency toward separation. The flow separation that does occur, occurs on the horizontal surfaces downstream from the jet type flow that inpinges on the horizontal surfaces and initiates small plumes when the separation extends beyond the center of these surfaces. This, of course, is because of buoyant reinforcement of the separated flow and is associated with the change in temperature at the surface as specified.

At $A = 10^8$, $\sigma = 1.0$ the flow separation is followed by counter circulations that become strong independently of buoyant reinforcement. The flow separation causes a diagonal rebounding of the impinging jets. Farther along the horizontal surfaces the buoyant reinforcement then causes these circulations to become as strong as the main flow for short periods of time. The circulations in turn create waves in the vertical boundary layers as they grow and subside. In the central region there is a weak positive temperature gradient in both the vertical and horizontal. The latter is too weak to be important in terms of buoyant reduction in strength of the main flow tendency.

Surprisingly the case $A = 10^9$, $\sigma = 10.0$ is less irregular in the sense of producing large counter circulations. The diagonal rebounding of the jets is very dominant in the flow patterns but the buoyant plumes which develop are more like the case $A = 10^7$, $\sigma = 1.0$. This follows from the smaller conductivity and shows that small κ limits the plume activity but small ν causes stronger flow separation.

In complete contrast the case $A = 10^7$, $\sigma = 0.1$ shows no direct rebounding of the vertical jets. The potential for producing separated flow at the corners, where the jets strike the horizontal surfaces, has been reduced because of the larger viscosity. However, the potential for buoyant plumes to develop has increased. These plumes roll up to some extent and break out of the corners where they developed. The flow is predominantly a single large circulation with activity in the lower left and upper

$A = 10^7$, $\sigma = 1.0$

right corners.

A strong similarity exists between the low Prandtl number flow at $A = 10^7$ and the last case of $A = 10^9$, $\sigma = 1$. That is, the dominant behavior is a single large circulation. In contrast, however, we are now in the transition region so that the heat transfer has increased dramatically. The flow is characterized by activity in all four corners. Plume roll up causes activity in the central region but this is relatively small compared to the corner activity. Like the $A = 10^7$, $\sigma = 0.1$ case the central region may almost be regarded as isothermal. The flow is somewhat eccentric so that a bulge in the main circular flow can cause sudden increased strength in boundary generated vorticity. Such an occurance provides the

$A = 10^8$, $\sigma = 1.0$

$A = 10^9$, $\sigma = 10.0$

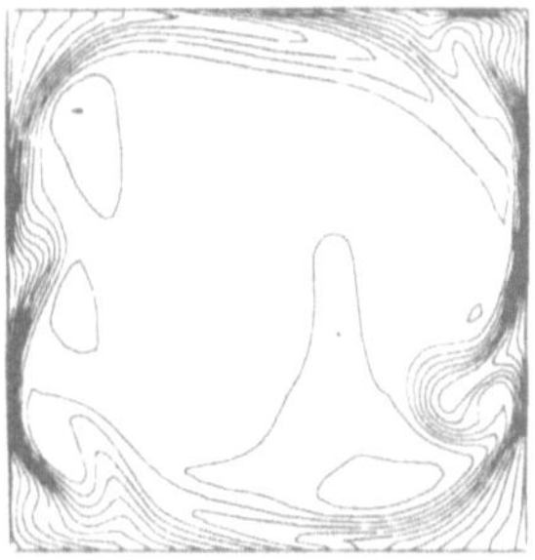

$A = 10^7$, $\sigma = 0.1$

proper mechanism for a plume to develop in the lower left or upper right corners. These plumes roll up, and pass swiftly along the vertical walls. This is followed by a swirling flow in the upper left or lower right corners exceeding that normally present. The boundary layer may here again roll up and be swiftly transferred along the horizontal boundaries. The occurances are random except for this tendency of the rolled up boundary layers to be passed along from corner to corner. The swift transfer of the rolled up boundary layers along the walls and the scouring of the surfaces by this process is characteristic of the presence of intermittent turbulence.

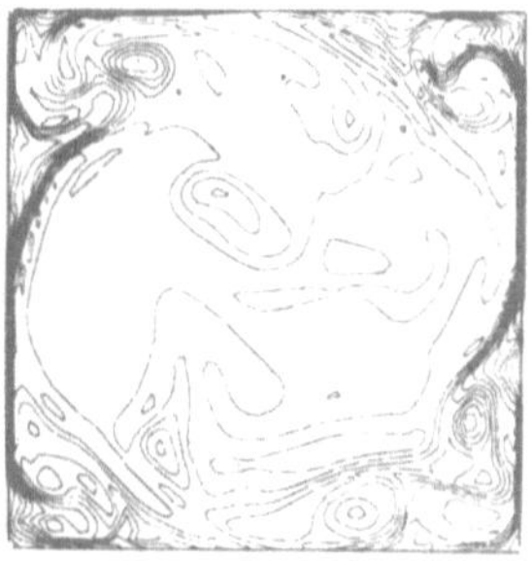

$A = 10^9$, $\sigma = 1.0$

To further investigate this transition region, we have taken the calculation for $A = 10^8$, $\sigma = 1.0$ and modified the parameters to give $A = 3 \cdot 10^8$, $\sigma = 1.0$. The cube root dependence emerged with flow unlike that at $A = 10^9$, $\sigma = 1.0$. However, the latter run was also modified to give $A = 3 \cdot 10^8$, $\sigma = 1$. Here the behavior remained like $A = 10^9$ so that two distinct behaviors were obtained for the same values of the parameters. In other words, a hysteresis effect was noted. A factor of 60 difference in the heat transfer was obtained and the time scale of the flow showed a factor of 120 in difference. Much work still needs to be done to analyse this transitional behavior. In spite of the violence in the transitional flow, the temperature field remains well behaved. Since the same numerical programs are used for the vorticity calculation there is confidence that meaningful behavior is being obtained. The tendency of the nonlinear approximations is more toward a loss of energy because of damping of the shortest waves. The buoyancy term also is slightly damped. These characteristics of the approximations should preclude difficulty with aliasing as we now understand it.

REFERENCES

Arakawa, A., J. Comput. Phys. 1, 119-143 (1966)

Buneman, O., Stanford University Institute of Plasma Research, SUIPR Report No. 294 (1969)

Crowley, W.P., Monthly Weather Rev. 96, 1-11 (1968)

Fromm, J.E., Phys. Fluids Suppl. II, 12, 3-12 and 113-119 (1969)

Jacob, M., Heat Transfer, Vol. I, Chap. 25 (Wiley, 1949)

Torrance, K.E., Orloff, L. and Rockett, J.A., J Fluid Mech. 36, 21-31 (1969)

Poots, G., Quart. J. Mech. App. Math., 11, 257-273 (1958)

Session III

Boundary Layers

R. Timman, Chairman

THE STRUCTURE OF THE LAMINAR BOUNDARY LAYER UNDER A POTENTIAL VORTEX

O. R. BURGGRAF AND K. STEWARTSON*
Department of Aeronautical and Astronautical Engineering
The Ohio State University, Columbus, Ohio

The structure of the laminar boundary layer induced by a potential vortex with axis normal to a fixed plane has been studied by a number of scientists, beginning with Taylor (1950). This flow problem is of interest because of its occurrence in vortex chambers, and more remotely as a simplified model for dust devils and tornadoes. Several idealizations must be made to represent a real flow of this type by a laminar boundary layer beneath a potential vortex. We consider a semi-infinite vortex maintained in an incompressible steady flow by appropriate conditions at infinity. Compressibility effects limit the radius from the axis to which the present theory may penetrate, while the laminar-flow requirement places a restriction on the Reynolds number.

The specific problem considered is that of a potential vortex whose axis coincides with that of a fixed disc of radius a, on which the boundary layer originating at the edge spreads inward toward the axis. The object of the study is to elucidate the structure of the boundary layer, particularly its behavior as $r \to 0$, where r is radius normalized with respect to the disc radius a. Earlier work on this problem has been reviewed by Rott and Lewellen (1966). Briefly the situation is that no similarity solution of the usual type exists for this problem, leaving the question open as to how the boundary layer terminates as r decreases. Indeed, Goldshtik (1960) showed that the full Navier-Stokes equations for this problem have no similarity solution if the Reynolds number Γ/ν exceeds 8, where $2\pi\Gamma$ is the circulation of the potential vortex and ν the kinematic viscosity. These results have been confirmed numerically by Kidd and Farris (1968) and by Schwiderski (1969) who found solutions up to $\Gamma/\nu = 5.5$, but not beyond.

Momentum-integral analyses for the boundary layer on the disc have been carried out by Taylor (1950), by Cooke (1952), and by Mack (1962). Although these analyses gave results even at $r = 0$, they must be questioned because of the non-existence of the similarity solution that would be expected to yield the terminal flow at $r = 0$.

To resolve this anomaly, a numerical integration was undertaken with the hope of either revealing the existence of some sort of singularity near $r = 0$, as suggested by Brown and Stewartson (1969), or otherwise elucidating the structure of the terminal profiles. Such calculations had been carried out earlier by Anderson (1966) and by Cooke (1966), but their calculations did not extend to sufficiently small r to permit recognition of the terminal profile. The present finite-difference computations, based on the steady-flow boundary-layer equations, were started at the edge of the disc ($r = 1$) and continued inward to $r = 0.03$. The results showed the possibility of a singularity in the solution to be remote for any $r > 0$, and the numerical results suggested the form of the terminal profile to be a double-scaled structure, for which a double asymptotic expansion was formulated.

The finite-difference calculations were carried out in the independent variables $\xi = -\log r$ and $\zeta = z/r(1-r)^{1/4}$, where $az(\nu/\Gamma)^{1/2}$

*Permanent Address: Department of Mathematics, University College London, Gower Street W.C. 1, London, England.

denotes distance above the disc. Use of the variable ξ ensured that increasingly smaller radial step-size was used as $r \to 0$ in the event of a singularity there. In the ξ-variable, the factor $(1-r)^{1/4}$ provides the proper initial behavior near $r = 1$, and the factor $1/r$ is the scaling that allows the local flow structure near $r = 0$ to develop independent of conditions at $r = 1$. The parabolic partial differential equations of the boundary layer were replaced by a set of implicit difference equations, much like the well-known Crank-Nicolson type. At a given ξ-station, the difference equations were solved by Gaussian elimination; since the matrix of coefficients is tridiagonal, this requires on the order of 6N multiplications and divisions after the matrix elements are formed, where N is the number of points taken across the mesh. Owing to the non-linearity of the equations of motion, the matrix elements themselves vary with the dependent variables. Hence the solution was iterated at each ξ-station, with the matrix elements at each iteration evaluated using the solution obtained in the preceding iteration. Usually convergence to a difference of 10^{-5} between successive iterations was obtained after 10 to 15 iterations, with 100 to 200 mesh points across the boundary layer and a ξ-step size of 0.0125.

Several difficulties were encountered in carrying out the calculations described here, and these are discussed in some detail in the complete paper (Burggraf, Stewartson, and Belcher, 1971), hereafter denoted by B.S.B.. In brief, the velocity component w perpendicular to the disc was found to be extraordinarily sensitive to inaccuracies in the computation. If the ξ-mesh was too thin at a given ξ-station, w would plunge to zero and, following sign reversal, the computation became unstable. Moreover, owing to the extremely thick boundary layer (in the ξ variable) that developed as $r \to 0$, slight oscillations in the radial component u, within the tolerance of accuracy of 10^{-5}, caused pronounced oscillations in w. These were reduced an order of magnitude by setting u and v equal to their edge values at any point at which the converged solution was within 10^{-5} of the edge value.

In carrying out the computation to small r, the solution was observed to develop a double structure, with radial flow near the surface tending to a fixed profile in terms of the similarity variable z/r, and outer profiles for u and v rapidly growing in thickness in terms of z/r. Accuracy was maintained for small r by use of a double-mesh structure: an inner mesh with $\Delta\zeta$ of 0.1 for $\zeta < 10$ and an outer mesh in which $\Delta\zeta$ was increased stepwise as needed to maintain approximately constant the number of points in the mesh. The accuracy of the computation was checked step-by-step by comparing the velocity gradient at the wall as computed by direct differencing and by a momentum-integral calculation using the computed velocity profiles. This comparison indicated an accuracy of about 1.5% at worst, and progressively better for $r \to 0$.

The observation of a double structure in the numerical solution suggested the possibility of a two-layer asymptotic expansion for the terminal behavior of the solution of the boundary-layer equations at $r = 0$. Such an expansion was carried out and was found to be in very good agreement with the numerical solution for $r < 0.06$. The details of the analysis are given in the complete paper (B.S.B.). In brief, the inner region, next to the disc, is described by the similarity variable z/r. The flow there is dominated by a radial flow toward the axis under the strong pressure gradient. The azimuthal velocity in this region, though large, is small compared with the radial component. The terminal structure of the radial flow in the inner region is independent of the previous history of the boundary layer, and the mass flux therein is zero at $r = 0$. On the other hand, the terminal structure of the tangential flow is an eigenfunction of the form $rv = r^{\lambda}\gamma(\eta)$ where $\eta = z/r$ and λ is a member of a set of positive eigenvalues, the lowest equal to 0.6797. Thus rv is known to within a multiplicative constant, which is determined by the previous history of the flow. In the outer layer, the similarity variable is z; the flow there is inviscid to first order, with the speed of the fluid independent of z. The structure of the outer layer depends entirely on the earlier history of the boundary layer and the mass flux steadily increases as $r \to 0$, as had been suggested by Moore (1956). The terminal velocity profiles in the outer layer were evaluated from the numerical solution, which was shown to be consistent with four terms of the outer asymptotic expansion. For details of this analysis, see the complete paper (B.S.B.).

A preliminary investigation was made of the more general vortex $v \propto r^{-n}$. On the premise that a multiple-structure terminal boundary layer exists for $n < 1$, with circulation in the inner layer given by $r^n v = r^{\lambda}\gamma(\eta)$, where $\eta = z/r^{(n+1)/2}$, the lowest eigenvalue λ_1 was computed as a function of n. The results show that λ_1 varies continuously from 0.6797 for $n = 1$ to zero for $n = 0.1217$. At this point the azimuthal velocity is of the same order as the radial velocity, and the multiple structure reduces to the conventional similarity boundary layer, as computed by King and Lewellen (1964). More recent computations based on the time-dependent boundary layer equations show that the inner structure described above for the r^{-n} vortex-boundary layer does develop in the steady-state limit. However, the outer region is considerably more complicated than the structure for the potential vortex.

REFERENCES

Anderson, O. L., Ph.D. Thesis, Rensselaer Polytechnic Institute, E. Windsor Hill, Conn. (1966).

Brown, S., and Stewartson, K., Ann. Rev. Fluid Mech. 1 (1969).

Burggraf, O. R., Stewartson, K., and Belcher, R., Phys. Fluids 14, to be published (1971).

Cooke, J. C., J. Aero. Sci. 19, 486-490 (1952).

Cooke, J. C., Royal Aircraft Establishment, T. R. 66 128 (1966).

Goldshtik, M. A., Prik. i. Mat. (English Translation) 24, 913-929 (1960).

Kidd, G. J., and Farris, G. J., J. App. Mech. 35, 209-215 (1968).

King, W., and Lewellen, W., Phys. Fluids 7, 1674-1680 (1964).

Mack, L., Jet Prop. Lab., Pasadena, Calif., T. R. 32-224 (1962).

Rott, N., and Lewellen, W. S., Prog. Aero. Sci. 7, 111-144, Ed. D. Kuchemann (Pergamon Press, New York, 1966).

Schwiderksi, E. W., J. Appl. Mech. 36, 614-619 (1969).

Taylor, G. I., J. Mech. Appl. Math. 3, 129-139 (1950).

EXACT NUMERICAL SOLUTIONS FOR THREE-DIMENSIONAL BOUNDARY LAYERS

Egon Krause and Ernst Heinrich Hirschel

DFVLR-Institut für Angewandte Gasdynamik, Porz-Wahn, Germany

BASIC EQUATIONS

For orthogonal curvilinear coordinates Prandtl's boundary-layer equations read

Continuity equation

(1) $$\frac{\partial}{\partial \xi}(\rho h_2 u) + \frac{\partial}{\partial \eta}(\rho h_1 v) + \frac{\partial}{\partial \zeta}(\rho h_1 h_2 w) = 0$$

Tangential momentum equation, ξ-direction

(2) $$\rho u \frac{\partial u}{h_1 \partial \xi} + \rho v \frac{\partial u}{h_2 \partial \eta} + \rho w \frac{\partial u}{\partial \zeta} - K_2 \rho u v + K_1 \rho v^2 = -\frac{\partial p}{h_1 \partial \xi} + \frac{\partial}{\partial \zeta}\left(\mu \frac{\partial u}{\partial \zeta}\right)$$

Tangential momentum equation, η-direction

(3) $$\rho u \frac{\partial v}{h_1 \partial \xi} + \rho v \frac{\partial v}{h_2 \partial \eta} + \rho w \frac{\partial v}{\partial \zeta} - K_1 \rho u v + K_2 \rho u^2 = -\frac{\partial p}{h_2 \partial \eta} + \frac{\partial}{\partial \zeta}\left(\mu \frac{\partial v}{\partial \zeta}\right)$$

Energy equation

(4) $$\rho u \frac{\partial T}{h_1 \partial \xi} + \rho v \frac{\partial T}{h_2 \partial \eta} + \rho w \frac{\partial T}{\partial \zeta} = \frac{E_\infty}{c_p}\left[u \frac{\partial p}{h_1 \partial \xi} + v \frac{\partial p}{h_2 \partial \eta} + \mu\left(\frac{\partial u}{\partial \zeta}\right)^2 + \mu\left(\frac{\partial v}{\partial \zeta}\right)^2\right] + \frac{1}{Pr_\infty c_p}\frac{\partial}{\partial \zeta}\left(k \frac{\partial T}{\partial \zeta}\right) .$$

The quantities u, v and w are the components of the velocity vector in the directions of the coordinates ξ, η, and ζ; the length parameters h_1 and h_2 depend on ξ and η only, and the local curvatures of the curves ξ = constant and η = constant are denoted by K_1 and K_2; they are defined below. All other quantities have the usual meaning.

(5) $$K_1 = -\frac{1}{h_1 h_2}\frac{\partial h_2}{\partial \xi} , \qquad K_2 = -\frac{1}{h_1 h_2}\frac{\partial h_1}{\partial \eta} .$$

The flow variables are introduced in a nondimensionalized form, each of them being referenced to its freestream value, only the pressure is normalized by $\rho^*_\infty u^{*2}_\infty$. The coordinates are nondimensionalized by a characteristic length of the body, L, and the normal coordinate and the normal velocity component are streched by the square root of the freestream Reynolds number:

(6) $$\zeta = (\zeta^*/L)\left(\sqrt{Re_\infty}\right), \quad w = (w^*/u^*_\infty)\left(\sqrt{Re_\infty}\right)$$

The starred quantities denote the flow variables in physical dimensions. The above equations are supplemented by the equation of state and relations defining the transport coefficients and the specific heat in terms of the temperature T:

(7) $$\rho = p/(p_\infty T), \; \mu = f_1(T), \; k = f_2(T), \; c_p = f_3(T)$$

The partial differential equations (1) - (4) are subject to the following boundary conditions: At the outer edge of the boundary layer the tangential velocity components and the temperature approach asymptotically their corresponding values of the inviscid flow.

(8) $$\lim_{\zeta \to \infty}\left[u(\xi,\eta,\zeta) = u_e(\xi,\eta), \; v(\xi,\eta,\zeta) = v_e(\xi,\eta), \; T(\xi,\eta,\zeta) = T_e(\xi,\eta)\right]$$

On the surface of the body the no-slip condition requires the tangential velocity components to vanish identically, while the normal velocity component may differ from zero when mass is transferred normally through the surface. Then, $w(\xi,\eta,0)$ must be prescribed in terms of ξ and η. The last boundary condition concerns the temperature. To make the problem definite, either the wall temperature or its first derivative in the direction normal to the wall must be prescribed. Thus,

(9) $$u(\xi,\eta,0) \equiv v(\xi,\eta,0) \equiv 0, \quad w(\xi,\eta,0) = 0 \text{ or } F_1(\xi,\eta)$$

(10) $$T(\xi,\eta,0) = F_2(\xi,\eta), \text{ or } \frac{\partial T}{\partial \eta}(\xi,\eta,0) = F_3(\xi,\eta).$$

The functions $F_1(\xi,\eta)$, $F_2(\xi,\eta)$ and $F_3(\xi,\eta)$ must be of $0(1)$ such that the boundary-layer concept remains valid. Since the system (2) - (4) is of first order in the ξ- and η directions, initial conditions must be prescribed for the tangential velocity components u and v and the temperature T. For three-dimensional flows the initial data have to be specified on a surface $E(\lambda,\zeta)$, being normal to the surface of the body. The coordinate λ is the trace of $E(\lambda,\zeta)$ on the surface of the body, and λ must be known in terms of the coordinates ξ and η. The initial conditions can then be written in the following way:

(11) $$\lambda = g(\xi,\eta): \; u(\lambda,\zeta) = F_4(\lambda,\zeta), \; v(\lambda,\zeta) = F_5(\lambda,\zeta), \; T(\lambda,\zeta) = F_6(\lambda,\zeta)$$

FINITE DIFFERENCE SOLUTION

An implicit finite difference method is used to solve the above system of equations. Since the differential equations (2) - (4) are of the same type the difference equations can be written in the form of a vector equation

(12) $$A_1 W^{n+1}_{l+1,m+1} + A_2 W^{n}_{l+1,m+1} + A_3 W^{n-1}_{l+1,m+1} + B_1 = 0$$
$$1 \leqq n, \; n+1, \ldots, N-1.$$

The subscripts l and m and the superscript n designate the coordinates ξ, η, and ζ respectively of a point P inside of the boundary layer and l+1, m+1, and n+1 those of a neighbouring point, being an increment $\Delta\xi$, $\Delta\eta$ and $\Delta\zeta$ away from P in the positive direction of the coordinates; N gives the maximum number of mesh points across the boundary layer; W is a vector with components u, v, and T. The general solution of Eq. (12) is given in terms of Richtmyer's algorithm:

(13) $$W^{n}_{l+1,m+1} = A_4 W^{n+1}_{l+1,m+1} + B_2$$

The matrix A_4 and the vector B_2 are immediately obtained from (12) when $W^{n=1}_{l+1,m+1}$ is replaced by the lower boundary conditions (9), (10). The upper boundary condition is satisfied by introducing an error bound ε and requiring that

(14) $$| W^{N-1}_{l+1,m+1} - W_{e\,l+1,m+1} | \leqq \varepsilon .$$

If Eq. (14) is not satisfied, N is repeatedly increased by one until the velocity and temperature differences of the last step are within the prescribed error bound. The elements of the matrices A_1 - A_3 and the components of the vector B_1 are given here for orthogonal rectilinear coordinates; the superscripts denote the matrix to which the elements belong:

(15) $${}^1A_{11} = {}^1A_{22}, \quad {}^2A_{11} = {}^2A_{22}, \quad {}^3A_{11} = {}^3A_{22},$$

(16) $${}^1A_{13} = -{}^3A_{13}, \quad {}^1A_{23} = -{}^3A_{23}, \quad {}^1A_{31} = -{}^3A_{31}, \quad {}^1A_{32} = -{}^3A_{32},$$

(17) $${}^1A_{12} = {}^1A_{21} = {}^2A_{13} = {}^2A_{21} = {}^2A_{23} = {}^3A_{12} = {}^3A_{21} = 0,$$

(18) $${}^1A_{11} = \frac{[\rho w]}{4\Delta z} - \frac{[\mu]}{2\Delta z^2} - \frac{1}{4\Delta z}\left[\frac{\partial \mu}{\partial T}\right]\frac{\partial T}{\partial z}(l,m,n), \qquad {}^2A_{11} = \frac{[\rho u]}{2\Delta x} + \frac{[\rho v]}{2\Delta y} + \frac{[\mu]}{\Delta z^2},$$

(18) $$^{3}A_{11} = -\frac{[\rho w]}{4\Delta z} - \frac{[\mu]}{2\Delta z^2} + \frac{1}{4\Delta z}\left[\frac{\partial\mu}{\partial T}\right]\frac{\partial T}{\partial z}(l,m,n), \qquad {}^{1}A_{13} = -\frac{1}{4\Delta z}\left[\frac{\partial\mu}{\partial T}\right]\frac{\partial u}{\partial z}(l,m,n),$$

$$B_1 = \frac{1}{2}[\rho u]\frac{\partial u}{\partial x}(l+\tfrac{1}{2},m,n) - \frac{[\rho u]}{2\Delta x}u(l,m+1,n) + \frac{1}{2}[\rho v]\frac{\partial u}{\partial y}(l,m+\tfrac{1}{2},n) - \frac{[\rho v]}{2\Delta y}u(l+1,m,n) +$$

$$+\frac{1}{2}[\rho w]\frac{\partial u}{\partial z}(l,m,n) - \frac{1}{2}[\mu]\frac{\partial^2 u}{\partial z^2}(l,m,n) + \frac{\partial p}{\partial x}(l+\tfrac{1}{2},m+\tfrac{1}{2},n).$$

(19) $$^{1}A_{23} = -\frac{1}{4\Delta z}\left[\frac{\partial\mu}{\partial T}\right]\frac{\partial u}{\partial z}(l,m,n),$$

$$B_2 = \frac{1}{2}[\rho u]\frac{\partial v}{\partial x}(l+\tfrac{1}{2},m,n) - \frac{[\rho u]}{2\Delta x}v(l,m+1,n) + \frac{1}{2}[\rho v]\frac{\partial v}{\partial y}(l,m+\tfrac{1}{2},n) - \frac{[\rho v]}{2\Delta y}u(l+1,m,n) +$$

$$+\frac{1}{2}[\rho w]\frac{\partial v}{\partial z}(l,m,n) - \frac{1}{2}[\mu]\frac{\partial^2 v}{\partial z^2}(l,m,n) + \frac{\partial p}{\partial y}(l+\tfrac{1}{2},m+\tfrac{1}{2},n).$$

(20) $$^{1}A_{31} = -\frac{E_\infty}{2\Delta z}\left[\frac{\mu}{c_p}\right]\frac{\partial u}{\partial z}(l,m,n), \qquad {}^{1}A_{32} = -\frac{E_\infty}{2\Delta z}\left[\frac{\mu}{c_p}\right]\frac{\partial v}{\partial z}(l,m,n),$$

$$^{2}A_{31} = -\frac{E_\infty}{4}\left[\frac{1}{c_p}\right]\frac{\partial p}{\partial x}(l+\tfrac{1}{2},m+\tfrac{1}{2},n), \qquad {}^{2}A_{32} = -\frac{E_\infty}{4}\left[\frac{1}{c_p}\right]\frac{\partial p}{\partial y}(l+\tfrac{1}{2},m+\tfrac{1}{2},n),$$

$$^{1}A_{33} = \frac{[\rho w]}{4\Delta z} - \frac{1}{2\Delta z^2}\frac{1}{Pr_\infty}\left[\frac{k}{c_p}\right] - \frac{1}{2\Delta z}\frac{1}{Pr_\infty}\left[\frac{1}{c_p}\frac{\partial k}{\partial T}\right]\frac{\partial T}{\partial z}(l,m,n),$$

$$^{2}A_{33} = \frac{1}{\Delta z^2}\frac{1}{Pr_\infty}\left[\frac{k}{c_p}\right] + \frac{[\rho u]}{2\Delta x} + \frac{[\rho v]}{2\Delta y},$$

$$^{3}A_{33} = -\frac{[\rho w]}{4\Delta z} - \frac{1}{2\Delta z^2}\frac{1}{Pr_\infty}\left[\frac{k}{c_p}\right] + \frac{1}{2\Delta z}\frac{1}{Pr_\infty}\left[\frac{1}{c_p}\frac{\partial k}{\partial T}\right]\frac{\partial T}{\partial z}(l,m,n),$$

$$B_3 = \frac{[\rho u]}{2}\frac{\partial T}{\partial x}(l+\tfrac{1}{2},m,n) - \frac{[\rho u]}{2\Delta x}T(l,m+1,n) - \frac{1}{2Pr_\infty}\left[\frac{k}{c_p}\right]\frac{\partial^2 T}{\partial z^2}(l,m,n) +$$

$$+\frac{[\rho v]}{2}\frac{\partial T}{\partial y}(l,m+\tfrac{1}{2},n) - \frac{[\rho v]}{2\Delta y}T(l+1,m,n) + \frac{[\rho w]}{2}\frac{\partial T}{\partial z}(l,m,n) -$$

$$-\frac{E_\infty}{4}\left[\frac{1}{c_p}\right]\frac{\partial p}{\partial x}(l+\tfrac{1}{2},m+\tfrac{1}{2},n)\left\{u(l,m,n) + u(l+1,m,n) + u(l,m+1,n)\right\} -$$

$$-\frac{E_\infty}{4}\left[\frac{1}{c_p}\right]\frac{\partial p}{\partial y}(l+\tfrac{1}{2},m+\tfrac{1}{2},n)\left\{v(l,m,n) + v(l+1,m,n) + v(l,m+1,n)\right\}$$

The matrix elements and vector components are listed in the order in which they appear in the momentum equations and in the energy equation. The terms in the square brackets denote the coefficients of the derivatives in the differential equations. They are defined as

(21) $$[H] = \alpha H(l+1,m+1,n) + \beta\left\{H(l,m,n) + H(l,m+1,n) + H(l+1,m,n)\right\}.$$

Eq. (21) involves the function evaluated at the four corner points of the mesh cell. It is sufficient to express the mean value through the functions evaluated at those corner points where the derivatives normal to the wall are incorporated in the scheme. The weighting factors α and β are zero and one third in the zeroth iteration and one fourth in the first and all following iterations. Further details concerning the difference equations are given by Krause, Hirschel, and Bothmann.

The normal velocity component is determind from the continuity equation by using the iteration method of Krause. For three-dimensional boundary layers Eq. (1) becomes when central difference quotients are used:

$$(\rho w)^{n}_{l+1/2,\, m+1/2} = (\rho w)^{n-1}_{l+1/2,\, m+1/2} - \left[\frac{\partial \rho u}{\partial x} + \frac{\partial \rho v}{\partial y}\right]\Delta z + 0(\Delta z^3). \tag{22}$$

REMARKS ON THE STABILITY OF THE DIFFERENCE EQUATIONS

Although the implicit numerical solution is unconditionally stable for two-dimensional flows a stability analysis of the difference equations revealed that for three-dimensional flows the solution is only conditionally stable. The stable region depends on the difference molecule used in the integration. Since the convective terms involve two directions in surfaces parallel to the surface of the body the domain of dependence condition (Courant-Friedrichs-Levy-condition) has to be satisfied in three dimensions. For this reason the integration must follow the direction of the characteristic lines of the convective operator such that the numerical domain of dependence of the difference scheme always includes the domain of dependence of the differential equations (Isaacson and Keller). For the boundary-layer equations in three dimensions the characteristic lines of the convective operator are given by the traces of the streamlines in surfaces parallel to the surface of the body. Although the direction of the streamlines may change substantially due to cross-flow components, stable difference schemes can be constructed which allow the velocity vector to turn as much as 135 degrees. Within the stable region the numerical solution is as accurate as for two-dimensional flows. Good agreement is obtained when the numerical results are compared with those of the similar solutions (Fig. 1). The maximum turning angle of the difference scheme for which the matrix elements and vector components are given is 90 degrees with respect to the positiv x-axis.

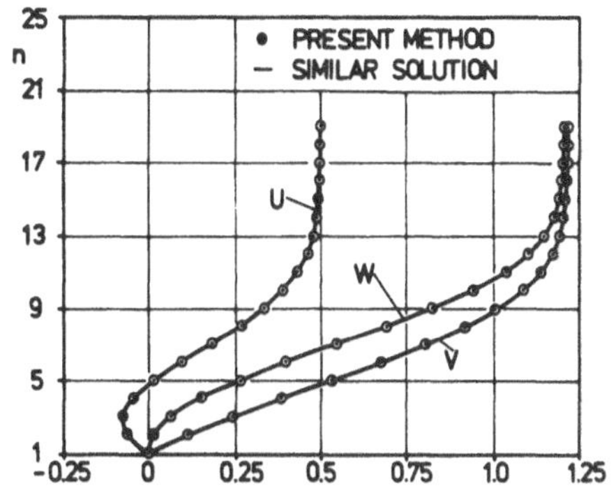

Fig. 1 Comparison with similar solution.

APPLICATION OF THE METHOD TO FLOW PROBLEMS

Some results obtained by machine calculation for an incompressible flow over a flat plate with different initial- and boundary conditions are given here. It was assumed that a three-dimensional boundary layer is generated by the presence of inviscid cross-flow component v, being linearly proportional to the x-coordinate. The tangentional velocity component u was kept constant, and the similar solutions of Yohner and Hansen were used to generate the initial data for the velocity. Two problems were investigated: In the first problem, a prescribed area of the wall is heated and the remainder of the surface is either insulated or kept at constant temperature T_w. The second problem deals with an enforced change of momentum such as may occur when a probe is immerged in the boundary layer. The wake of the probe then causes a defect in the origionally undisturbed velocity profile. On the other hand, momentum may be added at constant pressure through a jet tangential to the surface of the plate creating a velocity overshoot. The major results of the numerical integration are shown in Figs. 2 - 6. The leading edge of the plate coincides with the y-axis as can be seen in Fig. 2. For the inviscid velocity components u = 1 and v = x the streamlines form parabolas as indicated in the lower part of the Fig. If T_w is kept constant everywhere the lines of constant shearing stress are straight lines parallel to the y-axis. The shearing stress exhibits a minimum of about $\tau_w = 0.97$ at $x = 0.25$ and increases then with x because of the growing cross-flow component. If T_w is raised in the area hatched in Fig. 2 τ_w changes as indicated. The maximum temperature was twice as high as the wall temperature and fell off sinoidally towards the edge of the heated region. The shearing stress is increased by some forty percent in the immediate neighbourhood of the temperature maximum, but outside of the heated area only small changes are observed. The situation is quite different when the wall outside of the heated region is insulated. For this calculation the heat flux q_w was described instead of the temperature T_w. The same sinoidal law was used and the maximum value was chosen in such a way that the maximum temperature was about the same as before.

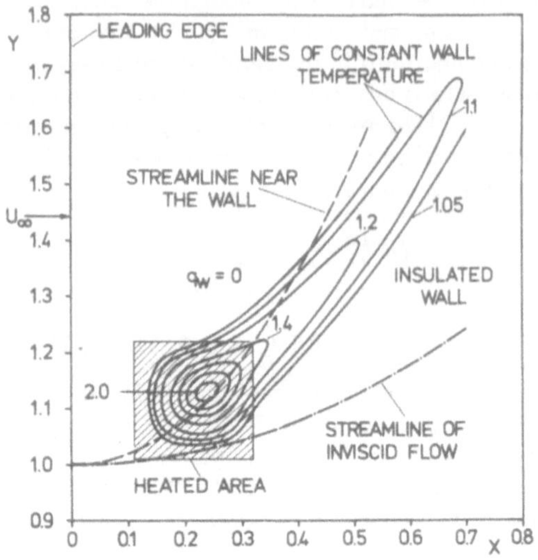

Fig. 2 Lines of constant shearing stress for partially heated wall.

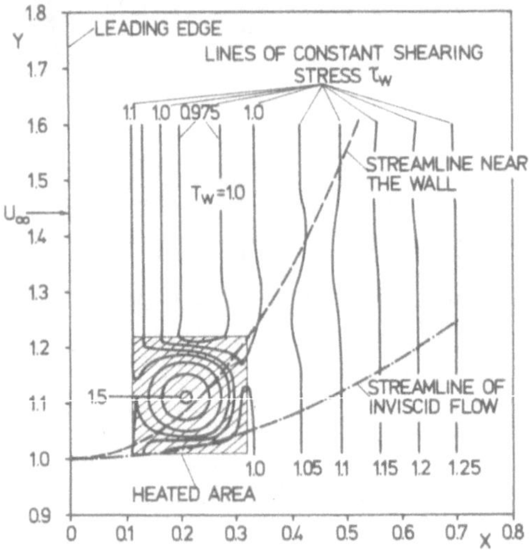

Fig. 3 Lines of constant wall temperature for partially heated wall.

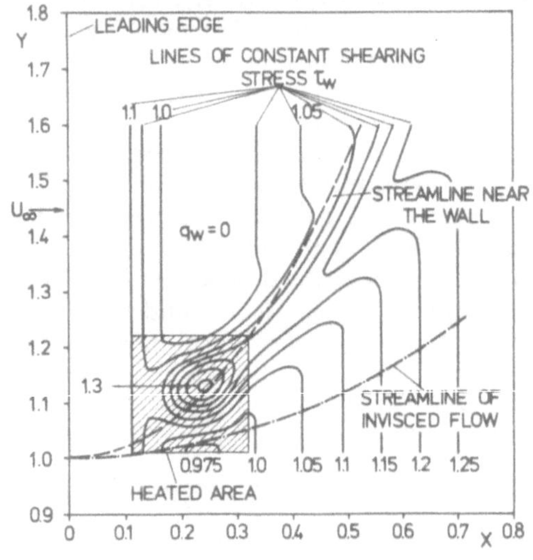

Fig. 4 Lines of constant shearing stress for partially heated wall.

Fig. 3 shows the lines of constant wall temperature. The heat is convected in the direction of the flow, which approximately follows the diagonal joining the lower left with the upper right corner of the plate. The shearing stress at the wall (Fig. 4) differs now markedly from its origional value in the region of increased wall temperature. This happens inspite of the fact that the tangential velocity gradients in the direction normal to the wall are lower now because of the higher viscosity.

In the second problem the effect of disturbed initial velocity profiles on the boundary layer was investigated. A velocity defect was superimposed on the origional profile, and, for the second run a velocity overshoot. The maximum amplitude of the disturbance was changed in a double sine distribution for the y- and z-direction. The boundary conditions were those of the undisturbed flow. The shearing stress at the wall is shown in Figs. 5 and 6.

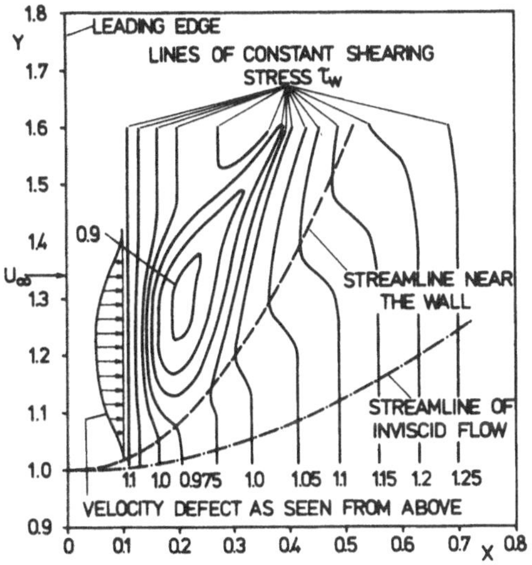

Fig. 5 Lines of constant shearing stress of the wake flow.

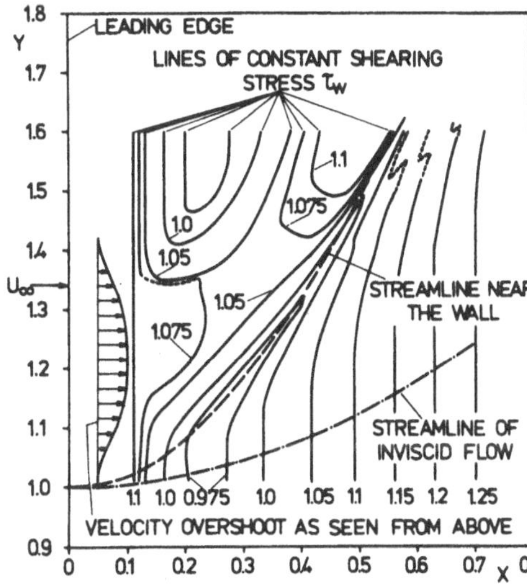

Fig. 6 Lines of constant shearing stress of the jet flow.

For the velocity defect, τ_w is decreased because of the momentum defect of the oncoming flow. The velocity overshoot generates a plateau in the shearing stress distribution, which is completely different from that of the unperturbed flow. Further downstream the local gradients in the x-direction become so large that for the step sizes chosen the calculation became inaccurate and vibrations were observed. However, the results confirm that the present method may be used to solve the complete initial-boundary value problem for three-dimensional boundary layers as long as the gradients in the direction of the tangential velocity components remain compatible with the boundary layer assumptions.

REFERENCES

Isaacson, E. and Keller, H. B.: Analysis of numerical methods. John Wiley a. Sons, Inc. New York, London, Sidney, 1966

Krause, E., Hirschel, E. H., and Bothmann, Th.: ZAMM Sonderheft 48, H. 8, T205 (1968); Fachtagung Aerodynamik, 03-1-03-49 Berlin (1968); AIAA-J. 7, 367-369 (1969)

Krause, E.: AIAA-J. 5, 1231-1237 (1967)

HIGHER ORDER EFFECTS IN LAMINAR BOUNDARY LAYER THEORY FOR CURVED SURFACES*

R. A. Wagstaff
Assistant Professor of Mechanical Engineering
San Diego State College, San Diego, Calif.

and

S. S. Lee
Associate Professor of Mechanical Engineering
University of Miami, Miami, Fla.

The flight of an aircraft and the lifting of a ship on hydrofoils are phenomena made possible by the flow of fluid over curved surfaces. The same principle is utilized to keep a child's kite aloft in a breeze and to provide lift and forward motion for some types of ground cushion vehicles. To predict performance and to achieve an optimum or even a reasonably acceptable design it is necessary to obtain verious boundary layer parameters. In the case of the ship and the airplane, for example, a knowledge of the shearing stress and pressure distribution on the curved lifting surfaces or wings is necessary in order to calculate the lift and drag. Boundary layer parameters are determined by the flow conditions and the curvature of the surface. It is therefore important to include the higher order terms of curvature in the mathematical model and be able to solve the resulting complex system of equations.

In the past the most popular solutions to these higher order equations have been similarity solutions. In such a solution a highly specialized pressure distribution and surface curvature are usually assumed facilitating transformation of the governing system of partial differential equations into a solvable system of ordinary differential equations. Murphy (1952) obtained a similarity solution to a system of governing equations which include some higher order terms. His analysis is limited to surfaces which have curvature K proportional to the reciprocal of the square root of the distance downstream of the leading edge $K \sim s^{-1/2}$ and for flows having a zero pressure gradient. The zero pressure gradient assumed automatically eliminates the possibility of considering types of flow for which separation is known to occur. Massey and Clayton (1965) also considered surfaces for which similarity solutions are possible. One of their examples is flow over a circular cylinder. They approximate the potential velocity on the surface by keeping only the first term in a Taylors Series expansion for the sine function. As a result, their analysis is good only near the leading stagnation point. Narasimha and Ojah (1967) considered similarity solutions to higher governing equations for surfaces which are an extension to the Falkner-Skan family. Taulbee and Patel (1968) considered surfaces which have curvature proportional to the velocity potential. This led to similarity solutions for flow over surfaces or thru concerning channels having the shape of a logarithmic spiral. In many cases the similarity solutions have been exact solutions. However, due to the limitations and restrictions placed on flow conditions and surface curvature distributions to make the similarity transformations possible, the solutions are of little practical value.

What is needed now is a solution to the higher order system which is valid throughout the boundary layer between attachment and separation and free from the restrictions and limitations required for a similarity solution. Most important, it must be for an arbitrary curvature distribution and must be able to predict, with reasonable accuracy, pertinent boundary layer parameters such as shearing stress, boundary layer thickness and the velocity and pressure distributions. To obtain such a solution is the objective of the present study. Results for flow over a circular and an elliptic cylinder will be obtained and compared to existing experimental and theoretical data. The circular cylinder

*This work was sponsored by the National Science Foundation under a Research Initiation Grant (GK-3617).

has been chosen because it has been a popular subject for many researchers. The elliptic cylinder was chosen for its variable surface curvature distribution.

EQUATIONS OF MOTION AND BOUNDARY CONDITIONS

The problem considered is the laminar steady flow of an incompressible fluid over a two dimensional curvilinear body. The range of the Reynolds number Re is in the order of $1/\delta^2$ where δ is the boundary layer thickness made dimensionless with respect to the free stream velocity U_∞ and some characteristic length of the body. Only moderate curvature K and moderate change in curvature along the surface $\partial K/\partial s$ are considered. Performing an order of magnitude analysis on the Navier-Stokes equations in curvilinear coordinates (with coordinates s along the surface and n normal to the surface) and keeping all terms of order one and δ in the momentum equation for the direction tangent to the surface (s - momentum equation) and the continuity equation and the terms of order one in the momentum equation in the direction normal to the surface (n momentum equation) yields the following system of governing equations:

s - momentum: $$\bar{u}\frac{\partial \bar{u}}{\partial s} + (1+Kn)\bar{v}\frac{\partial \bar{u}}{\partial n} + K\bar{u}\bar{v} = -\frac{1}{\rho}\frac{\partial p}{\partial s} + \nu\left[(1+Kn)\frac{\partial^2 \bar{u}}{\partial n^2} + K\frac{\partial \bar{u}}{\partial n}\right] \quad (1)$$

n - momentum: $$K\bar{u}^2 = \frac{1}{\rho}\frac{\partial P}{\partial n} \quad (2)$$

continuity: $$\frac{\partial \bar{u}}{\partial s} + \frac{\partial}{\partial n}\left[(1+Kn)\bar{v}\right] = 0 \quad (3)$$

The boundary conditions are: $\bar{u}=0$, $\bar{v}=0$ at $n=0$ and $\bar{u}=U(s)$ at $n=\infty$. $\bar{u}, s, \bar{v}, n, K, \rho, P, \nu$ and U are the dimensional velocity and space components tangent and normal to the surface, the curvature, density, pressure, kinematic viscosity and velocity along the boundary layer boundary respectively.

METHOD OF SOLUTION

The Method of Integral Relations (designated by MIR here after) will be used to solve this problem. In the past the use of this method has been restricted to the solution of first order (Prandtl boundary layer) equations. An extension to MIR will be proposed, in the present study, by which it can be utilized to solve systems of equations which include higher order terms, i.e. the system of equations (1), (2) and (3).

CHANGE OF VARIABLES AND FIRST INTEGRATION

The following transformation due to A.A. Dorodnitsyn is used.

$$u=\frac{\bar{u}}{U},\quad v=\frac{\bar{v}}{U\nu^{1/2}},\quad \zeta=\int_0^s U ds \quad \text{and} \quad \eta=\frac{Un}{\nu^{1/2}} \quad (4)$$

Substituting the transformation (4) into equations (1), (2) and (3) and using Bernoulli's equation along the boundary layer give:

$$u\frac{\partial u}{\partial \zeta} + \frac{\dot{U}}{U}u^2 + \frac{\dot{U}}{U}\zeta\frac{\partial u}{\partial \eta} + v\frac{\partial u}{\partial \eta} + \frac{K\nu^{1/2}}{U}\left[\eta\frac{\partial u}{\partial \eta} + Uv\right] = \frac{1}{\rho U^2}\frac{\partial P}{\partial \zeta} - \frac{\dot{U}}{\rho U^3}\eta\frac{\partial P}{\partial \eta}$$
$$+ \frac{\partial^2 u}{\partial \eta^2} + \frac{K\nu^{1/2}}{U}\left[\eta\frac{\partial^2 u}{\partial \eta^2} + \frac{\partial u}{\partial \eta}\right] \quad (5)$$

$$K u^2 = \frac{1}{\rho\nu^{1/2} U}\frac{\partial P}{\partial \eta} \quad (6)$$

$$\frac{\partial u}{\partial \zeta} + \frac{\dot{U}}{U} u + \frac{\dot{U}}{U} \eta \frac{\partial u}{\partial \eta} + \frac{\partial v}{\partial \eta} + \frac{K\nu^{1/2}}{U} \cdot v \left[v + \eta \frac{\partial v}{\partial \eta}\right] = 0 \tag{7}$$

The dot denotes differentiation with respect to ζ.

An additional useful relation is obtained by taking the derivative of equation (2) with respect to s and applying the transformation (4) giving:

$$U^3\dot{K}u^2 + 2KU^3 u\frac{\partial u}{\partial \zeta} + 2KU^2\dot{U}u^2 + 2KU^2\dot{U}\eta u\frac{\partial u}{\partial \eta} = \frac{U\dot{U}}{\rho\nu^{1/2}}\left[\frac{\partial P}{\partial \eta} + \eta\frac{\partial^2 P}{\partial \eta^2}\right]$$

$$+ \frac{U^2}{\rho\nu^{1/2}}\frac{\partial^2 P}{\partial \zeta \partial \eta} \tag{8}$$

Now an interpolation function f(u) is introduced which has the properties that along the outer boundary of the boundary layer: $f(1) = f'(1) = 0$. Along the body $f(0) = 1$. Next f times (7) plus f times (5), f times (6) and $\frac{f}{U^2}$ times (8) are integrated in the n direction, and the variable of integration is changed to u. After some manipulation the following three integral relations are obtained:

$$\frac{d}{d\zeta}\int_0^1 f(u)u\theta du = -\frac{\dot{U}}{U}\int_0^1 f'(u)u^2\theta du - \frac{f'(o)}{\theta_o} - \int_0^1 \frac{f''(u)}{\theta} du + \frac{1}{\rho U^2}\int_0^1 f'(u)\frac{\partial P}{\partial \zeta}\theta du$$

$$-\frac{\dot{U}}{\rho U^3}\int_0^1 f'(u)\eta\frac{\partial P}{\partial \eta}\theta du - \frac{K\nu^{1/2}}{U}\int_0^1 f'(u)uv\theta du - \frac{K\nu^{1/2}}{U}\int_0^1 \frac{f''(u)}{\theta}\eta du \tag{9}$$

$$K\int_0^1 f(u)u^2\theta du = \frac{1}{\rho\nu^{1/2}} U\int_0^1 f(u)\frac{\partial P}{\partial \eta}\theta du \tag{10}$$

$$\frac{d}{d\zeta}\left[UK\int_0^1 \theta f(u)u^2 du\right] - KU\int_0^1 \theta f'(u)u^2\frac{\partial u}{\partial \zeta}du - \dot{K}U\int_0^1 f'(u)\eta u^2 du =$$

$$-\frac{1}{\rho\nu^{1/2}}\frac{\dot{U}}{U}\int_0^1 f'(u)\eta\frac{\partial P}{\partial \eta}du - \frac{1}{\rho\nu^{1/2}}\left(\frac{\partial P}{\partial \zeta}\right)_o - \frac{1}{\rho\nu^{1/2}}\int_0^1 f'(u)\frac{\partial P}{\partial \zeta}du \tag{11}$$

where $\theta = \frac{1}{\frac{\partial u}{\partial \eta}}$ and a zero subscripted variable denotes its value along the surface.

Including the higher order terms in the governing equations (1) and (3) has introduced into equations (9) and (11) the variables η, $\partial u/\partial\zeta$ & v which are not present when considering only first order terms. Relationships for these in terms of u and ζ can be obtained in the following manner:

$$\text{since } \theta = \frac{1}{\partial u/\partial \eta} \quad \text{then } d\eta = \theta\, du \tag{12}$$

$$\text{and integrating gives:} \quad \eta = \int_0^\eta d\eta = \int_0^u \theta du \tag{13}$$

$\frac{\partial u}{\partial \zeta}$ can be obtained by taking the differential of (13) with respect to ζ:

$$\frac{d\eta}{d\zeta} = 0 = \frac{d}{d\zeta}\int_0^u \theta du \tag{14}$$

Integrating the first order continuity equation gives the following relationship for v:

$$v = -\frac{\dot{U}}{U}\eta u - \int_0^u \frac{\partial u}{\partial \zeta}\theta\, du \tag{15}$$

Relationships for θ and P in terms of u and ζ are assumed to meet boundary conditions along the body surface and along the dividing strips. Once θ is chosen the relationships for η, v and $\frac{\partial u}{\partial \zeta}$ can be determined. These along with the chosen relationships for f and P are substituted into equations (9), (10) and (11) and integrated giving an approximating system of nonlinear ordinary differential equations to be solved numerically. Since space is limited only the equations for the second approximation will be given. For the first and third see Wagstaff (1970). The following relationships for f, θ and P were assumed for the second approximation with the resulting values of η, $\frac{\partial u}{\partial \zeta}$ and v being obtained:

$$f_{21} = (1-u)^2, \quad f_{22} = (1-u)^3, \quad \theta = \frac{1}{1-u}[\theta_0(1-2u) + \theta_1 u], \quad \frac{1}{\theta} = (1-u)\left[\frac{1-2u}{\theta_0} + \frac{4u}{\theta_1}\right]$$

$$P = P_e + (1-u)(3P_o - 4P_1 + P_e) - (1-u^2)(2P_o - 4P_1 + 2P_e)$$

$$\eta = (\theta_0 - \theta_1)\ln(1-u) + (2\theta_0 - \theta_1)u, \quad \frac{\partial u}{\partial \zeta} = -\frac{1}{\theta}[(\dot{\theta}_0 - \dot{\theta}_1)\cdot \ln(1-u) + (2\dot{\theta}_0 - \dot{\theta}_1 u)]$$

$$v = -\frac{\dot{U}}{U}\eta u + (\dot{\theta}_1 - \dot{\theta}_0)[(1-u)\ln(1-u) + u] + (2\dot{\theta}_0 - \dot{\theta}_1)\frac{u^2}{2}$$

Substituting these relationships into equations (9), (10) and (11) yields the following approximating system of ordinary differential equations:

$$\sum_{L=0}^{1} \dot{\theta}_L \left\{C(1+5L,I) - \frac{K\nu^{1/2}}{U} \sum_{n=0}^{3} C(2+n+5L,I)\frac{\theta_0^{3-n}\,\theta_1^{n}}{\theta_0\,\theta_1}\right\} =$$

$$\sum_{L=0}^{1} \left\{C(11+L,I)\frac{\dot{U}}{U}\theta_L + C(13+L,I)\frac{1}{\theta_L} + \frac{K\nu^{1/2}}{U} C(15+L,I)\frac{\theta_0\theta_1}{\theta_L^{2}}\right\}$$

$$+ \frac{K\nu^{1/2}}{U}\left\{C(17,I) + \sum_{n=0}^{4} C(n+18,I)\frac{\theta_0^{4-n}\,\theta_1^{n}}{\theta_0\,\theta_1}\right\} + \frac{K\dot{\nu}^{1/2}}{U}\sum_{n=0}^{2} C(n+23,I)\,\theta_1^{n}\,\theta_0^{2-n} \qquad (16)$$

The values of C(J, I), J = 1,2,...,25 and I = 1,2 are given in Table I.

TABLE I: Coefficients for Approximating System of Equations 16

	J=1	J=2		J=1	J=2
C(1,J)	0.00000	-0.04166	C(14,J)	-1.33333	3.66666
C(2,J)	0.00888	-0.08658	C(15,J)	0.05555	0.09705
C(3,J)	-0.09786	-0.20107	C(16,J)	-0.22222	0.86113
C(4,J)	1.41237	1.24067	C(17,J)	-0.50000	0.70751
C(5,J)	0.26542	-0.18668	C(18,J)	-0.49778	-0.08802
C(6,J)	0.08333	-0.00000	C(19,J)	-6.32330	-1.55634
C(7,J)	0.73777	0.10468	C(20,J)	8.12101	5.31480
C(8,J)	0.36041	0.92410	C(21,J)	3.10193	-2.32835
C(9,J)	-1.36272	-2.37031	C(22,J)	0.18669	0.13089
C(10,J)	-0.18669	0.13089	C(23,J)	0.60000	0.16250
C(11,J)	-0.33333	1.04166	C(24,J)	-0.83889	-0.33889
C(12,J)	-0.50000	0.37500	C(25,J)	-0.32778	0.23472
C(13,J)	1.66666	-3.33333			

After relationships for $\dot{U}/U$, K and $\dot{K}$ in terms of u and ζ (Wagstaff, 1970) are obtained and initial conditions calculated equation (16) can be integrated numerically on a digital computer completing the solution. The conditions necessary to initialize the integration process downstream of the singularity at the leading edge of the elliptic cylinder calculated in a manner described by Liu (1962) are: $\theta = 1.152\ \zeta_o$ and $\theta_1 = 1.190\ \zeta_o^{1/2}$.

RESULTS AND DISCUSSION

The shearing stress at the surface of an elliptic cylinder of a semi-major to semi-minor axis ratio of 4 with major axis parallel to a uniform stream is given in figure 1. Here the results obtained from the solution to the system of governing equations (1), (2) and (3) by the second approximation of MIR are compared to the results obtained by Schlichting (1955). The agreement between the two curves obtained from the solution of the Prandtl boundary layer equations is reasonably good except near the point of vanishing shearing stress. Including the higher order terms of curvature reduces the calculated value of shearing stress along the surface, the effect increasing with decreasing Reynolds number.

The velocity profiles obtained from the same three solutions are compared in figure 2. Again the agreement between the first order profiles is reasonably good considering that both solutions are approximate. The deviation in shape of the MIR profiles from that obtained by Schlichting (1955) corresponding to a value of s/l' of .74 (l' is half the circumference) results from closeness to separation, i.e. when θ_o in the equations approaches infinity. The profiles corresponding to the solution to the higher order system indicate that the curvature terms have the effect of reducing the velocity for a given distance normal to the surface. Hence, the boundary layer thickness increases with decreasing Reynolds number.

The pressure profiles in figure 3 are for the same higher order solution to the ellipse problem discussed earlier. The authors know of no such profiles to which these curves can be compared. It is interesting to note that the pressure reduction across the boundary layer increases as the separation point is approached, a trend suggested by the experimental evidence of Parsons and Wallen (1930).

The present analysis is valid in the region which starts immediately downstream of the forward stagnation point and ends a small distance upstream of the separation point. Near and downstream from separation an analysis such as discussed by Catherall and Mangler (1966) could be used. In their work they specify the displacement thickness in the

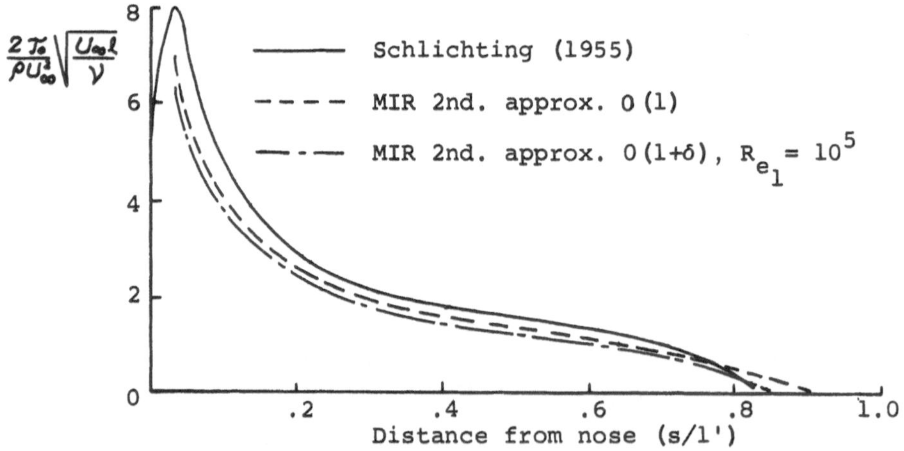

Fig. 1 Shearing stress at the wall over an a/b = 4 elliptic cylinder.

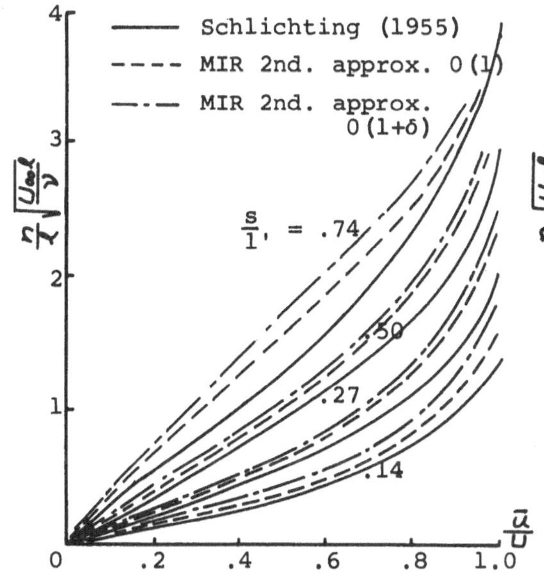

Fig. 2 Vel. profiles for a/b=4 elliptic cylinder at various dist.

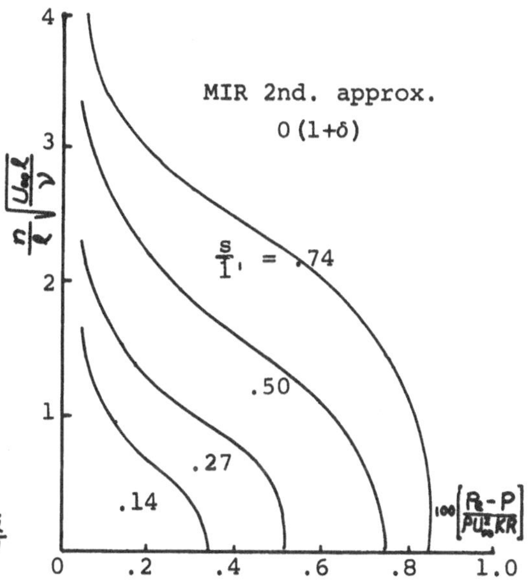

Fig. 3 Pressure distribution across b. l. for elliptic cyl.

region near and downstream of separation rather than the potential velocity distribution. Such a method eliminates the singularity in the governing equations resulting from the vanishing skin friction at the separation point.

The authors do not endorse the use of the present analysis in the neighborhood or downstream from the separation point. However, it was used to locate the separation point merely as an additional example of the effect of the higher order terms of curvature on the solution to the mathematical system of equations.

For circular cylinder, the separation point given by Schlichting (1955) was 108.8° while the MIR obtained 152°, 138°, 119° and 112° for first, second, third and fourth approximations respectively. When higher order terms were included in first and second approximations, the values became 126° and 106° for $R_e = 10^4$. Using experimental pressure data of Parsons and Wallen (1930) to calculate U/U term in equation (16), the separation point predicted by MIR was within one degree of error. This indicates that the error of MIR prediction is mainly due to the inaccuracy of the potential pressure used. In the case of some aerofoils, for example, the potential pressure distribution is nearly equal to the experimental measurement. The present analysis would then be expected to give good predictions up to a small distance upstream from the separation point.

REFERENCES

Catherall, D. & Mangler, K.W., *J. Fluid Mech.* 26, 163 (1966)
Liu, Shen-Tsuan, *J. of Computing Math. & Math. Phys.* V.2, No.4 (1962)
Massey, B.S. & Clayton, B.R. *Trans ASME, J. Basic Engr.* 483 (1965)
Murphy, J.S. *J. Aero.Sci.* 20,338 (1953)
Narasimha, R. & Ojha, S.K. *J. Fluid Mech.* 29, 187 (1967)
Parsons, J.F. & Wallen J.A. *NACA TN 354* (1930)
Schlichting, H. *Boundary Layer Theory*, 4th ed. N.Y.: McGraw-Hill (1960)
Taulbee, D.B. & Patel, H.N. *J. AIAA* 6, 1808 (1968)
Wagstaff, R.A. Ph.D. thesis, Univ. of Miami, C.G. Fla. (1970)

FINITE-DIFFERENCE METHODS FOR SOLVING THE BOUNDARY LAYER EQUATIONS WITH SECOND-ORDER ACCURACY

Frederick G. Blottner
Sandia Laboratories, Albuquerque, New Mexico

ABSTRACT

A number of second-order accurate finite-difference schemes are investigated for a simplified equation representative of reacting flows and a new method is proposed. The boundary layer equations for a binary gas mixture with a finite reaction rate are solved with this scheme and with the implicit, Crank-Nicolson, Predictor-Corrector and Petukhov methods. For very small step-sizes and for the problems investigated, the second-order finite-difference methods offer an advantage over the first-order implicit scheme.

INTRODUCTION

The present investigation is concerned with numerical methods for solving boundary layer equations with chemical reactions and applicable to cases with a large number of chemical species (≈ 25). A finite difference procedure has been developed previously by Blottner (1970A) for this type of flow. The method is used without iteration at each step and is of first-order accuracy. The same type of problem has also been investigated by Shnol' (1968) where an implicit finite-difference scheme is used with an iteration procedure at each step. The objective of the present study is to determine if the previous finite-difference scheme can be improved by going to a second-order procedure. As a part of this study, a secondary result is a comparison of several finite-difference methods. A more complete version of this paper is given in Blottner (1970B).

Since implicit finite-difference procedures have proved to be the better approach for the boundary layer equations, only the implicit procedures will be considered. Finite-difference procedures of second-order accuracy are readily obtained if initial conditions at two profiles across the boundary layer are employed. This type of procedure requires that the derivatives with respect to x (distance along the body surface) be smooth. For a sphere-cone body at the juncture point where there is a discontinuity of surface curvature, the derivatives with respect to x become very large. Therefore, a procedure with two lines of initial information would not be appropriate for this type of problem. Also, such a procedure is awkward to use since it cannot be started from a single initial profile of data. For boundary layer flows with finite chemical reactions, the evaluation of the chemical production term at the unknown grid point greatly reduces the occurrence of unstable solutions. This suggests that an implicit finite-difference scheme where the equations are evaluated at the unknown grid points is the desirable method when there are chemical reactions. When an implicit scheme of the Crank-Nicolson form is employed, unstable solutions can occur for boundary conditions which involve derivatives and for non-linear problems. Also, errors introduced will oscillate and damp-out slowly and this can occur downstream of the junction point on a sphere-cone.

ACCURACY OF METHODS

A number of finite-difference schemes are considered for solving parabolic partial differential equations of the form

$$a_4 \frac{\partial W}{\partial x} = L[W] = a_o \frac{\partial^2 W}{\partial y^2} + a_1 \frac{\partial W}{\partial y} + a_2 W + a_3 \qquad (1)$$

The following derivative in equation (1) is written as

$$\left(\frac{\partial W}{\partial x}\right)_{m+\Theta\lambda} = \left(W_{m+\lambda} - W_m\right)/(\lambda\Delta x) + (\Theta-\tfrac{1}{2})\lambda\Delta x\left(\frac{\partial^2 W}{\partial x^2}\right)_{m+\Theta\lambda} + O(\Delta x^2) \qquad (2)$$

and the function W or η-derivatives can be evaluated at $(m + \Theta\lambda)$ as

$$W_{m+\Theta\lambda} = \Theta W_{m+\lambda} + (1-\Theta)\, W_m + O(\Delta x^2) \qquad (3)$$

The value of λ determines the point the dependent variable is being evaluated with a value of ½ to 1 to indicate either a half or a full step ahead. The value of Θ varies from 0 to 1 as the scheme changes from explicit to fully implicit. If the second term on the right side of equation (2) is zero or is evaluated, the method will be second-order accurate.

To investigate the characteristics of finite-difference schemes, it is worthwhile to pick a simplified equation to study. Since it is the chemical production terms in the conservation of species equations that cause the greatest difficulty in solving a reacting boundary layer flow, the right side of equation (1) is represented as

$$L[W] = (P - QW)/\epsilon \quad \text{and } a_4 = 1 \tag{4}$$

The quantities ϵ and Q are assumed positive to make the equation inherently stable. We have neglected the y-derivatives to avoid a partial differential equation and this is reasonable when we are concerned with the effects of the production terms. In general, one has a system of equations that are nonlinear; therefore, any results obtained for equation (1) can be at best only approximate.

The exact solution to equation (1) with right side given by relation (4) is

$$W(x_m) = c\ \Lambda^m + P/Q \qquad \text{where } x_m = m\Delta x \text{ and } \Lambda = \exp(-\Delta xQ/\epsilon). \tag{5}$$

When equation (4) is solved with the various finite-difference schemes, the difference relations can be reduced to the form

$$W_{m+1} = \Lambda\ W_m + \varphi \tag{6}$$

where Λ and φ are a function of the difference scheme being used. The solution of all of the difference equations are the same form as equation (5). Therefore, the exact solution and the finite-difference solution will differ in how accurately Λ represents the exponential given in (5).

The variation of the Λ's for several finite-difference methods are compared to the exact solution in Figure 1. For small values of $\Delta xQ/\epsilon$, the second-order methods are more accurate than the first-order as expected. However, at large values of $\Delta xQ/\epsilon$, the order of the method does not give any indication of the accuracy of the method. The Euler and Runge-Kutta (explicit) methods become unstable for $\Delta xQ/\epsilon > 2$, and the solution W becomes large and diverges away from the exact solution. All of the other methods tend to converge to the exact result as the solution proceeds. When $\Delta xQ/\epsilon \to \infty$, the exact solution for Λ approaches 0. For all of the stable methods $\Lambda \to 0$ except for the Crank-Nicolson or Predictor-Corrector which approaches -1. This indicates why these methods might slowly damp-out an error with an oscillating value.

From the foregoing discussion, one would like the finite-difference procedure to have the following properties:

1. At least second-order accuracy for small step-sizes.
2. Correct asymptotic behavior for large step-sizes.
3. Balance between accuracy and computation time.

From the present accuracy considerations, the Petukhov (1966) method appears to be the best procedure. It has the disadvantage of requiring four solutions of the equations to obtain the solution ahead one step. Also, the asymptotic behavior is not completely satisfactory since negative values of Λ can occur. For a reacting flow this could lead to negative mass fraction of species, which is undesirable and physically impossible.

Are there other methods which are better than those already considered? First it should be noticed that there are several types of finite-difference schemes of second-order accuracy and they are described below:

1. Symmetric - For these methods the x-derivative is evaluated at the mid-point ($\Theta = \frac{1}{2}$) and has the desired accuracy if the right-side of equations (1) is evaluated with W at $m + \frac{1}{2}$. For nonlinear equations the determination of W at $m + \frac{1}{2}$ requires a prediction technique or an iteration scheme. The Crank-Nicolson and Predictor-Corrector methods are of this type. The present Runge-Kutta method can be interpreted as this type of method also.
2. Successive Substitutions - In these methods for each step taken, successive solutions are obtained with previous results employed in the subsequent

solutions. The solutions are combined such that the truncation error is made as small as possible. For second-order methods, two solutions are usually required. The general explicit Runge-Kutta method is of this type and an analogous general implicit process has also been developed for ordinary differential equations by Rosenbrock (1963).

3. Implicit with Truncation Corrections - For these methods $\Theta \neq \frac{1}{2}$ and the second derivative in equation (2) must be evaluated to obtain the desired accuracy. This derivative can be evaluated as in the Petukhov method. The use of explicit methods with truncation correction are not considered suitable.

It does not appear to be possible to improve the symmetric schemes without going to complex methods which have only a slight improvement in the variation of Λ. The method of successive substitutions of the explicit form are not stable for large values of $\Delta xQ/\varepsilon$ and are considered inappropriate. The implicit forms have been considered and the approach of Rosenbrock (1963) is followed. The variation of Λ for this case is given in Figure 2. Although this method has the correct asymptotic value, Λ is significantly negative for a large range of values of $\Delta xQ/\varepsilon$. With the lack of accuracy of this method, it does not seem desirable to extend this method to partial differential equations.

Several implicit methods with truncation correction were considered. The value of Λ for a two-step method which uses the governing equation to evaluate the second derivative is given in Figure 2. The scheme with the best properties evaluates the second derivative in Equation 2 by taking two steps forward to $(m+\lambda)$ and $(m + 2\lambda)$. The derivative is determined from

$$W'' = \frac{\partial^2 W}{\partial \xi^2} = (W_{m+2\lambda} - 2W_{m+\lambda} + W_m)/(\lambda\Delta\xi)^2 = (L[W_{m+2\lambda}] - L[W_{m+\lambda}])/(\lambda\Delta\xi) \qquad (7)$$

Then this method starts again with one step forward to obtain the final result at $(m+1)$. The value of $\Lambda \to 0$ as $\Delta xQ/\varepsilon \to \infty$, but can become negative depending on the value of λ. If $\lambda = 1/\sqrt{2}$, then Λ will be positive for any value of $\Delta xQ/\varepsilon$ and has the variation shown in Figure 2 for this three-step method. From the present investigation, this appears to be the most promising of the new methods considered.

FINITE-DIFFERENCE FORMULATION

The multi-component boundary layer equations with finite reaction rates are written in the form given in Blottner (1970A) where the equations are transformed into similarity variables. The conservation equations for momentum, energy and species are of the form of equation (1) and the a's can be a function of any of the dependent variables. It is advantageous to uncouple the governing equations when a large number of chemical species are being considered, and this is accomplished by assuming the dependent variables other than W appearing in the a's are known. These are determined from a previous profile or from iterated results. In order to evaluate the a's with the dependent variables at the proper location, an iteration of the solution is required if a second-order accurate result is desired.* If the original governing equation has a term involving W multiplying $\partial W/\partial \xi$ as occurs in the momentum equation, the equation is divided by W. If this is not done, the first term in equation (1) can never be made second-order accurate without an iteration process. For the various finite-difference schemes being investigated, the governing equations (1) must be evaluated at various locations which are denoted as $(m+\lambda\Theta)$. The dependent variables are known at (m) and unknown at $(m + \lambda)$. Before the derivatives are replaced with difference quotients, the equations are linearized. The details of the linearization technique are given in Blottner (1970B) and the governing equation (1) becomes

$$\frac{1}{\eta_e^2}\frac{\partial^2 W}{\partial \eta^2} + \frac{\alpha_1}{\eta_e}\frac{\partial W}{\partial \eta} + \alpha_2 W + \alpha_3 + \alpha_4\, 2\xi \frac{\partial W}{\partial \xi} = 0 \qquad (8)$$

The coefficients α for the momentum, energy and species equations are similar to those given in Blottner (1970A) and W is $f' = u/u_e$, $\theta = T/T_e$ or C_i for the governing

*If coupling is allowed between the equations, second-order accurate difference schemes can be obtained without iteration.

equations respectively. Although this equation looks almost the same as equation (1) the α's should be a weaker function of W due to the linearization. Some terms in the α's are evaluated at (m) while others must be evaluated at $(m + \Theta\lambda)$ due to the uncoupling of the governing equations. The continuity equation does not require linearization and is

$$2\xi \frac{\partial f'}{\partial \xi} + \frac{1}{\eta_e}\frac{\partial V}{\partial \eta} + f'\sigma = 0 \quad \text{with } \sigma = 1 + \frac{2\xi}{\eta_e}\frac{d\eta_e}{d\xi} \qquad (9)$$

The derivatives in equation (8) are replaced with relation (2) and centered differences in the η-direction. The derivatives and W are evaluated at $(m + \Theta\lambda)$ with the use of the corresponding quantities at $(m + \lambda)$ and (m) as shown in equation (3). The resulting difference equations have W's unknown at $(m + \lambda)$ and are of the tridiagonal type which is solved with the standard procedure. The continuity equation (9) is evaluated at $(m + \Theta\lambda,\ n - \frac{1}{2})$ and written in finite-difference form to evaluate V at $(m + \Theta\lambda, n)$.

The schemes that have been used to solve the equations are the following: implicit, Crank-Nicolson, Predictor-Corrector, Petukhov and the new method.

COMPARISON OF METHODS AND DISCUSSION

Two test cases have been used to investigate the accuracy and computing time of the various schemes. These cases are the following and the reason they are used is indicated:

1. Flat Plate - This example gives a large change in the chemical composition as the gas dissociates downstream.
2. Hyperboloid - For this example the gas is close to chemical equilibrium in the stagnation point region and is a good test of the stability of the various schemes.

For both cases the binary gas is taken as oxygen with simplified gas properties and is similar to that employed by Blottner (1964). Also for both cases the freestream conditions correspond to an altitude of 100K ft. and the surface is catalytic. For the flat plate the freestream velocity is 25 kfps with a wall temperature of 1177.2°R while for the hyperboloid the freestream velocity is 20 kfps with a wall temperature of 2520°R. In both examples 30 grid points are used across the boundary layer unless indicated otherwise.

A computer code has been developed for solving the boundary layer equations for the binary gas model. The five finite-difference methods which were mentioned previously have been combined into one computer program. General subroutines which are valid for all the methods are employed and for each method the logic is varied to correspond to the procedure being employed. Therefore, all of the methods experience the same quality of programming with the less complicated methods requiring perhaps additional computations.

For the flat plate case the maximum value of the mass fraction of atoms (CAMAX) at 10 ft. downstream from the leading edge is used to evaluate the accuracy of the solution to this problem. The mass fraction of atoms is zero at both the wall and edge of the boundary layer and also at the leading edge. As the gas flows along the flat plate the gas dissociates and the mass fraction of atoms increases toward an equilibrium value downstream. The variation of CAMAX is given in Figure 3 for several methods when the step-size is large (Δx = 1 ft.). Also shown in this figure is the solution for Δx very small and it is considered the exact solution. These results indicate the second-order methods are more accurate than the first-order implicit scheme when the step-sizes are the same. For the "exact" solution the number of points across the boundary layer was held constant and solutions were extrapolated to the case where $\Delta x = 0$. The effect of also varying the step-size $\Delta\eta$ has been investigated with the predictor-corrector method and the results are presented in Figure 4. As the step-sizes are made smaller, the solution appears to be converging to a unique value. The effects of the step-size $\Delta\eta$ will be neglected in the subsequent discussion since the methods we are investigating use the same second-order accurate difference relations for the η derivatives. It is the effect of the step-size Δx on the accuracy of the various methods which is of concern. Of course, for any desired accuracy there is some optimum choice of the two step-sizes to minimize the computing time.

The effect of varying Δx on the value of CAMAX at x = 10 ft. for the various methods is presented in Figure 5. As $\Delta x \to 0$, all of the methods converge to the same value for CAMAX as they should. The second-order methods are more accurate than the first-order implicit method. The computer results for the new method do not have quadratic convergence and the reason for this is being investigated. When $\Delta x = 0$, the value of CAMAX is extrapolated to the value indicated in Figure 5 and this is considered the exact value. In Figure 6 the percent error in the maximum mass fraction as a function of central processing time required on the CDC 6600 computer is presented for the various methods. For this problem the predictor-corrector method is the best method as it requires the minimum computer time for a desired accuracy. The first-order implicit method requires significantly more computer time than the second-order methods for the step-sizes investigated.

For the flat plate problem the effect of performing more than one iteration in the Crank-Nicolson and Petukhov methods was investigated. The results of this study are given in Figure 7 where it can be seen that iterations beyond the first is not necessary for the Crank-Nicolson scheme. However, for the Petukhov method, several iterations are required to make the scheme second-order accurate. These results use the first form of equation (7) to evaluate W" and the second form should require fewer iterations and is being investigated.

The second test case employed is the hyperboloid and the Stanton number at $x = 3$ nose radii is used as the parameter to judge the accuracy of the solution. The Stanton number is determined with $\Delta x = 0$ and used as the exact value to determine the error. The percent error for the various methods and the computer time required is given in Figure 8. In this case, the first-order implicit method is more competitive with the second-order methods, but the percent error is generally larger. The Crank-Nicolson and Predictor-Corrector methods have switched position when compared to the previous case.

In summary, for small step-sizes and for the problems investigated; the second-order finite-difference methods offer an advantage over the first-order implicit scheme. For solutions with approximately 10% error, the implicit scheme is competitive with the other methods. The Crank-Nicolson and Predictor-Corrector methods seem to be the best for the cases considered but these methods have marginal stability properties which limit their usefulness in general. The Petukhov method could compare more favorably with the Crank-Nicolson and Predictor-Corrector methods if the number of iterations is reduced by going to the second form of equation (7) to evaluate W". The new method investigated has not proved as successful as was anticipated from the analysis of the linear ordinary differential equation. Whether this results from programming difficulties or other reasons is being investigated.

ACKNOWLEDGEMENTS

The author expresses his appreciation to Margaret Johnson of Computer Application, Inc., for writing the computer code and for her assistance in obtaining the numerical results. The help of Molly Ellis in obtaining the results is also acknowledged.

REFERENCES

1. Blottner, F. G., AIAA J., Vol. 2, No. 2, 232-240 (1964).
2. Blottner, F. G., AIAA J., Vol. 8, No. 2, 193-205 (1970A).
3. Blottner, F. G., Sandia Corp. Research Report to be published (1970B).
4. Petukhov, I. V., Soviet J. of Computational Mathematics and Mathematical Physics, Vol. 6, No. 6, 1019-1028 (1966).
5. Rosenbrock, H. H., Computer Journal, Vol. 5, 329-330 (1963).
6. Shnol', E. E., Soviet J. of Computational Mathematics and Mathematical Physics, Vol. 8, No. 5, 1063-1075, (1968) English Translation: Sandia Laboratories SC-T-69-1014, March 1969.

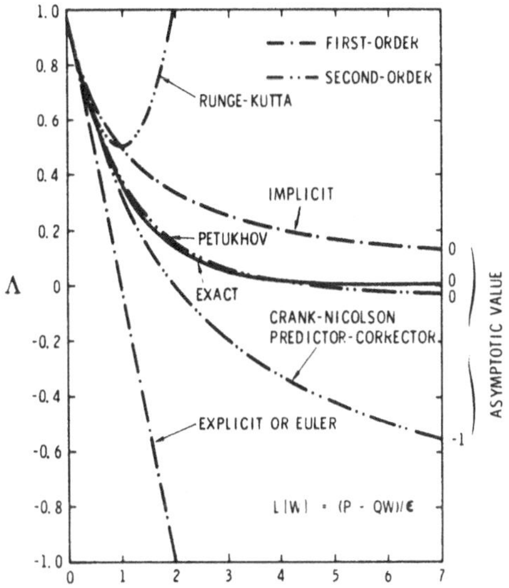

FIG. 1 - INDICATION OF THE ACCURACY OF VARIOUS METHODS

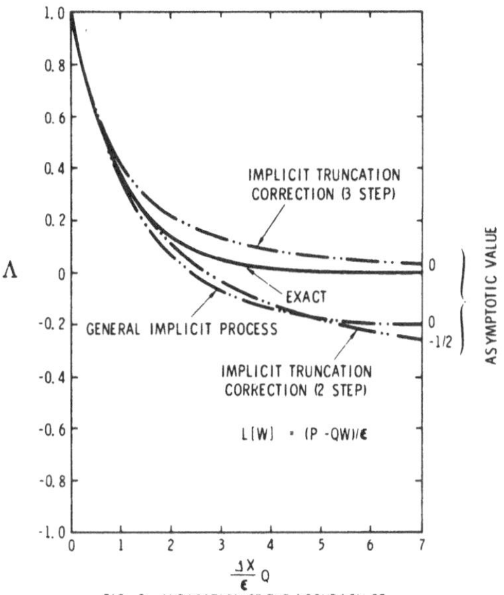

FIG. 2 - INDICATION OF THE ACCURACY OF SEVERAL SECOND-ORDER METHODS

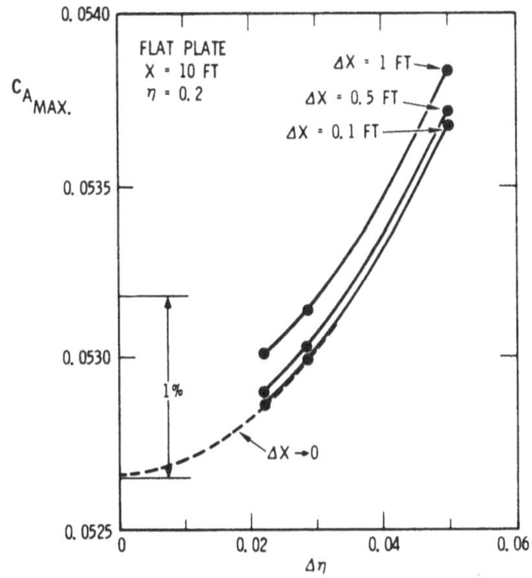

FIG. 3. EFFECT OF STEP-SIZES ON THE ACCURACY OF THE MAXIMUM MASS FRACTION OF ATOMS

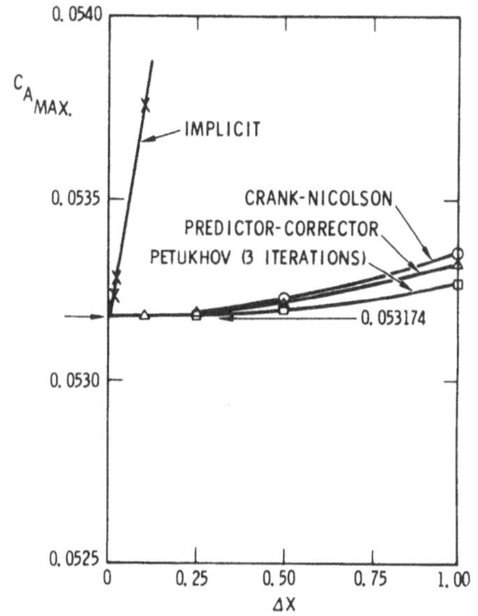

FIG. 4. EFFECT OF STEP-SIZE ΔX ON THE ACCURACY OF THE MAXIMUM MASS FRACTION OF ATOMS

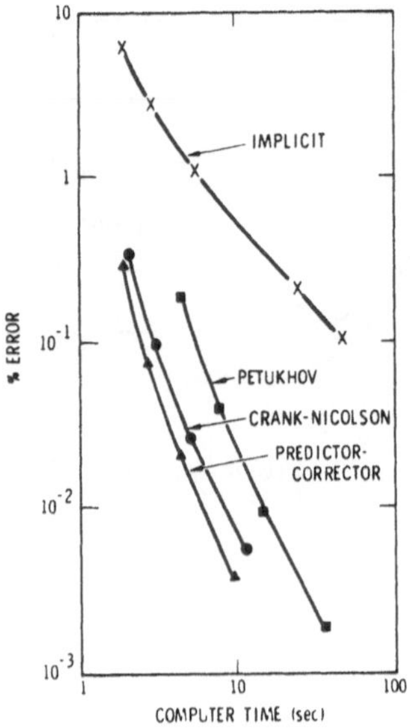

FIG. 5. ABSOLUTE ERROR IN MAXIMUM MASS FRACTION OF ATOMS AT X = 10 FT FOR A FLAT PLATE

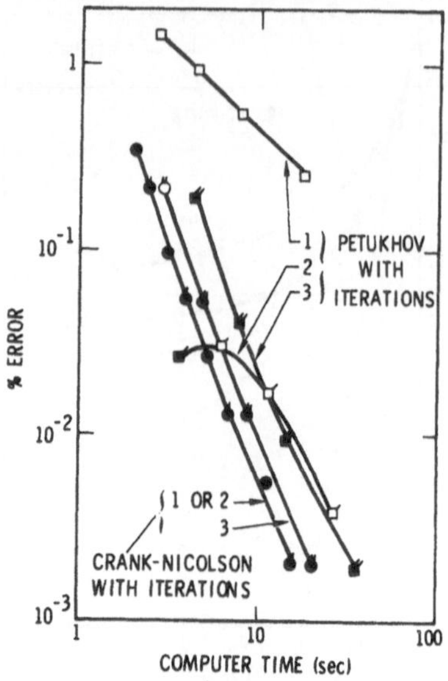

FIG. 6. EFFECT OF ITERATIONS ON THE ABSOLUTE ERROR IN MAXIMUM MASS FRACTION OF ATOMS AT X = 10 FT FOR A FLAT PLATE

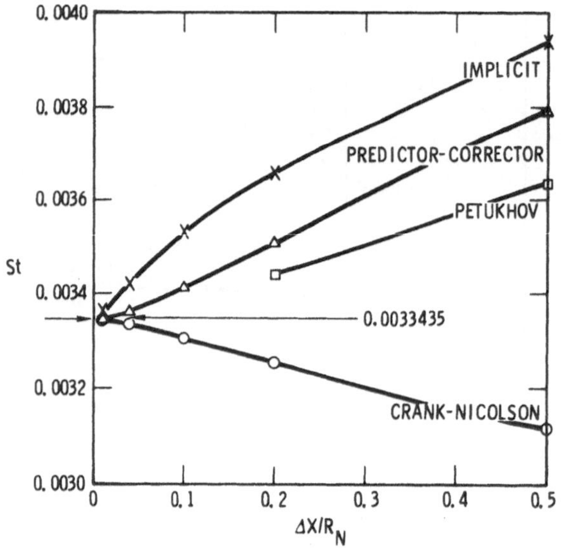

FIG. 7. EFFECT OF STEP-SIZE OF ΔX ON THE ACCURACY OF THE STANTON NUMBER AT $X/R_N = 3$

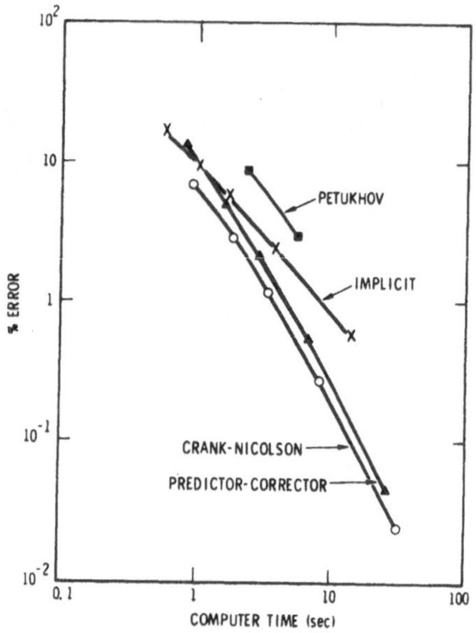

FIG. 8. ABSOLUTE ERROR IN STANTON NUMBER AT $X/R_N = 3$ FOR HYPERBOLOID

NUMERICAL SOLUTION OF THE INTERACTION OF A SHOCK WAVE WITH A LAMINAR BOUNDARY LAYER

Robert W. MacCormack
Ames Research Center, NASA
Moffett Field, Calif. 94035

INTRODUCTION

Currently many problems of fluid dynamics are being attacked by finite difference techniques. In the numerical solution of any given problem, several questions are raised concerning the accuracy or quality of the computed solution. Unfortunately too few of these questions have answers. Accuracy depends on the order of accuracy of the numerical method (consistency), numerical stability, matching of both the numerical and physical domains of dependence (to be discussed herein), and on simulating numerically all of the physically significant processes in the flow field. In addition to accuracy, the numerical analyst is also concerned with obtaining the solution as efficiently as possible. The purpose of this paper is to examine the numerical behavior of an efficient Lax-Wendroff difference technique of second order accuracy now being used to solve a wide range of problems in fluid dynamics (MacCormack, 1969; Kutler, 1969). In particular, this technique will be modified to meet the specific demands required for the solution of the interaction of a shock wave with a laminar boundary layer. The modified techniques, obtained through the concept of splitting, may however themselves have a much wider application in the solution of problems in fluid dynamics.

GENERAL NUMERICAL CONSIDERATIONS

The time-dependent Navier-Stokes equations, in two dimensions, neglecting body forces and heat sources, may be written in vector form as

$$\frac{\partial U}{\partial t} + \frac{\partial F}{\partial x} + \frac{\partial G}{\partial y} = 0 \tag{1}$$

where $U = \begin{pmatrix} \rho \\ \rho u \\ \rho v \\ e \end{pmatrix}$, $F = \begin{pmatrix} \rho u \\ \rho u^2 + \sigma_x \\ \rho uv + \tau_{xy} \\ (e + \sigma_x) u + \tau_{yx} v + k\frac{\partial T}{\partial x} \end{pmatrix}$, $G = \begin{pmatrix} \rho v \\ \rho uv + \tau_{yx} \\ \rho v^2 + \sigma_y \\ (e + \sigma_y) v + \tau_{xy} u + k\frac{\partial T}{\partial y} \end{pmatrix}$

$$\sigma_x = p - \lambda\left(\frac{\partial u}{\partial x} + \frac{\partial v}{\partial y}\right) - 2\mu\frac{\partial u}{\partial x}, \quad \tau_{xy} = \tau_{yx} = -\mu\left(\frac{\partial u}{\partial y} + \frac{\partial v}{\partial x}\right) \text{ and } \sigma_y = p - \lambda\left(\frac{\partial u}{\partial x} + \frac{\partial v}{\partial y}\right) - 2\mu\frac{\partial v}{\partial y}$$

for density ρ, x and y velocity components u and v, viscosity coefficients λ and μ, total energy per unit volume e, specific internal energy ϵ, coefficient of heat conductivity k, and temperature T. Finally, the pressure p is related to ϵ and ρ by an equation of state, $p(\epsilon,\rho)$, where $\epsilon = e/\rho - (u^2 + v^2)/2$.

A two-step difference method of second order accuracy (MacCormack, 1969) devised to solve Eq. (1) can be defined by

$$\left.\begin{aligned} U_{i,j}^{\overline{n+1}} &= U_{i,j}^n - \frac{\Delta t}{\Delta x}\left(F_{i+1,j}^n - F_{i,j}^n\right) - \frac{\Delta t}{\Delta y}\left(G_{i,j+1}^n - G_{i,j}^n\right) \\ U_{i,j}^{n+1} &= \frac{1}{2}\left\{U_{i,j}^n + U_{i,j}^{\overline{n+1}} - \frac{\Delta t}{\Delta x}\left(F_{i,j}^{\overline{n+1}} - F_{i-1,j}^{\overline{n+1}}\right) - \frac{\Delta t}{\Delta y}\left(G_{i,j}^{\overline{n+1}} - G_{i,j-1}^{\overline{n+1}}\right)\right\} \end{aligned}\right\} \tag{2}$$

where $F_{i,j}^n$ and $G_{i,j}^n$ equal $F(U_{i,j}^n)$ and $G(U_{i,j}^n)$. The subscripts refer to a spacial mesh of points (x_i, y_j) with spacing Δx and Δy, and the superscripts refer to times $t = n\Delta t$ where Δt is the time increment that the solution is advanced during each cycle of Eqs. (2). The method first obtains an approximate value, $U_{i,j}^{\overline{n+1}}$, at each point using two forward differences to approximate the two spacial derivatives. The approximate solution is then used in the second equation, using two backward differences, to obtain the new accepted value $U_{i,j}^{n+1}$. The above difference method is only one of four methods of essentially the same form. For example, if instead of first using two forward spacial differences and then two backward differences, the reverse procedure could be followed or one forward and one backward difference could be followed by corresponding backward and forward differences. The variant defined by Eqs. (2) will be used here only to illustrate the numerical behavior common to all four and it should be used to solve Eq. (1) with caution, as will become clear later in the section on accuracy and stability.

For the analysis of the numerical behavior, in particular, stability, of the difference equations, only the inviscid non-heat-conducting equations will be treated in detail. Both the viscous and heat conduction terms, physically and also numerically (if their magnitudes in the differenced equations are not too great), tend to damp out the high frequency components of the solution, which are normally the ones which cause numerical instability. Thus, it is expected that the inclusion of these terms in the analysis would only enhance stability.

Domain of Numerical Dependence

To advance the solution in time by Δt at one mesh point (x_i,y_j), using Eqs. (2), requires knowledge of only seven neighboring points. This "star" of points is illustrated in Fig. 1. The points marked A are primary in that information at these points more strongly modifies the solution at (x_i,y_j) during time Δt than does that at points B. The three other variants of Eqs. (2) have similar stars; the only differences are in the locations of the points B. The seven points define the numerical domain of dependence. The first requirement of any method is that its numerical domain include the physical domain of dependence. Clearly, if this is violated, the numerical scheme does not have in hand all the data necessary to advance the solution in time. On the other hand, there are two reasons why the numerical domain should not be much larger than the physical domain. First, to obtain an accurate numerical solution at a given point which "sees" much more data with time than the true solution, a rather severe demand is made of the numerical method to ignore or give little weight to the extraneous data. Second, the computation time spent in processing this unneeded data is costly. If the more restrictive domain, the primary domain defined by the points A (cross-hatched in Fig. 1) is taken as the effective numerical domain, the above requirement is met if

$$\Delta t \le (|u|/\Delta x + |v|/\Delta y + c\sqrt{1/\Delta x^2 + 1/\Delta y^2})^{-1} \tag{3}$$

where c is the local adiabatic speed of sound. This condition is usually called the Courant-Friedrich-Lewy (C.F.L.) condition.

Accuracy and Stability

The numerical stability of methods of this type, namely, those of the Lax-Wendroff class, cannot presently be completely analyzed in the general nonlinear form. The most successful attempt to date is to first linearize the set of differential Eqs. (1) and then to study the amplification of Fourier components of the solution by the difference method applied to the linearized set. The new set is then

$$\frac{\partial U}{\partial t} + J_F\frac{\partial U}{\partial x} + J_G\frac{\partial U}{\partial y} = 0 \tag{4}$$

where J_F and J_G are the Jacobian matrices of F and G with respect to U and are considered to be constant. This set of equations approximates Eqs. (1) locally, and difference methods found to be unstable for it can be expected to be unstable for the general nonlinear case. Two conditions inherent in such an analysis are: (a) the boundary conditions have no effect on stability, and (b) the exact solution to Eq. (1) is smooth. The latter condition allows the matrices J_F and J_G to be treated as constants (locally).

The amplification matrix of the difference Eqs. (2) applied to Eqs. (4) for a single Fourier component of the solution, $W(t)\exp[i(k_1x+k_2y)]$, becomes

$$\mathcal{G} = I - i\Delta t\left(\frac{J_F}{\Delta x}\sin\xi + \frac{J_G}{\Delta y}\sin\eta\right) - \frac{1}{2}\Delta t^2\left(\frac{J_F}{\Delta x}(1-e^{-i\xi}) + \frac{J_G}{\Delta y}(1-e^{-i\eta})\right)\left(\frac{J_F}{\Delta x}(1-e^{i\xi}) + \frac{J_G}{\Delta y}(1-e^{i\eta})\right)$$

where $\xi = k_1\Delta x$ and $\eta = k_2\Delta y$. For ξ and $\eta << 1$, $\mathcal{G} = \exp[-i\Delta t(J_F\xi/\Delta x + J_G\eta/\Delta y)]$ modulo terms of third order in ξ and η. The exact solution of Eqs. (4) for the above Fourier component is $\exp[-it(k_1J_F + k_2J_G)]\exp[i(k_1x + k_2y)]W(0)$. Hence, the exact amplification of the solution from $t = 0$ to $t = \Delta t$ is $\exp[-it(k_1J_F + k_2J_G)]$ which equals $\mathcal{G}$ modulo terms of third order in ξ and η. Thus the difference method is shown to be of second order accuracy.

In the limiting process $\Delta t \to 0$, stability is assured if all the eigenvalues, λ_i, of $\mathcal{G}$ satisfy the von Neumann condition

$$|\lambda_i| \le 1 + O(\Delta t) \tag{5}$$

The eigenvalues of $\mathcal{G}$ are invariant under a similarity transformation. There exists a transformation, S, such that $\mathcal{G}' = S^{-1}\mathcal{G}S = I + K - K^* - 2K^*K$,[1] where

$$K = (\Delta t/2)\left((1 - e^{i\xi})A/\Delta x + (1 - e^{i\eta})\,B/\Delta y\right)$$

$$A = \begin{pmatrix} u & c & 0 & 0 \\ c & u & 0 & 0 \\ 0 & 0 & u & 0 \\ 0 & 0 & 0 & u \end{pmatrix} \text{ and } B = \begin{pmatrix} v & 0 & c & 0 \\ 0 & v & 0 & 0 \\ c & 0 & v & 0 \\ 0 & 0 & 0 & v \end{pmatrix}$$

It can be shown that $|(w,\mathcal{G}'w)|^2 \le 1 + 4\,||Kw||^4$ where w is any unit vector, $(w,\mathcal{G}'w)$ is the inner product of w with $\mathcal{G}'w$ and $||Kw|| = (Kw,Kw)^{1/2}$; hence the eigenvalues of $\mathcal{G}$ can be made to satisfy (5) as $\Delta t \to 0$ if $\Delta t^3/\Delta x^4$ is held fixed. It can be shown that the bound on $|(w.\mathcal{G}'w)|^2$ is a least upper bound (i.e., the value is achieved by $(w,\mathcal{G}'w)$) by considering $v = -(\Delta y/\Delta x)u$, $\xi = -\eta$ and $w = (1,0,0,0)^T$. Then $|(w,\mathcal{G}'w)|^2 = 1 + 4(\sin\xi\,\Delta t u/\Delta x)^4$. The fixing of $\Delta t^3/\Delta x^4$ as $\Delta t \to 0$ is much more restrictive than the condition imposed by the C.F.L. condition (Eq. (3)). It is a necessary condition near the unfavorable velocity directions $v = -u\Delta y/\Delta x$. The other three variants also have unfavorable directions; however they are not all the same. The amplification matrix of each has in general different eigenvalues and vectors for each Fourier component of the solution; hence a single component would not in general maximize $|(w,\mathcal{G}w)|$ for each variant.

It was conjectured that if the four variants followed one another cyclically during the numerical calculation, the condition on Δt would be close to that of the C.F.L. condition. The difference method is then in terms of the permutted subscripts ii and jj and the mod function[2]

$$\left.\begin{aligned}
ii &= \bmod(n,2) \\
jj &= \bmod(n - ii,4)/2 \\
\overline{U^{n+1}_{i,j}} &= U^n_{i,j} - \frac{\Delta t}{\Delta x}\left(F^n_{ii,j} - F^n_{ii-1,j}\right) - \frac{\Delta t}{\Delta y}\left(G^n_{i,jj} - G^n_{i,jj-1}\right) \\
ii &= \bmod(ii + 1,2) \\
jj &= \bmod(jj + 1,2) \\
U^{n+1}_{i,j} &= \frac{1}{2}\left\{U^n_{i,j} + \overline{U^{n+1}_{i,j}} - \frac{\Delta t}{\Delta x}\left(\overline{F^{n+1}_{ii,j}} - \overline{F^{n+1}_{ii-1,j}}\right) - \frac{\Delta t}{\Delta y}\left(\overline{G^{n+1}_{i,jj}} - \overline{G^{n+1}_{i,jj-1}}\right)\right\}
\end{aligned}\right\} \tag{6}$$

It has not yet been possible to assess the validity of the conjecture analytically. Instead the eigenvalue least upper bound for several flow directions will be approximated using a power method and a simple numerical test problem. Consider a region of uniform flow containing a square array of mesh points, $U^o_{i,j}$ = constant, i, j = 1,2, . . . , N. Using the similarity transformation S on Eq. (4), we have

$$\frac{\partial Z}{\partial t} + A\frac{\partial Z}{\partial x} + B\frac{\partial Z}{\partial y} = 0 \tag{7}$$

where $Z = S^{-1}U$. The exact solution is $Z(t) = S^{-1}U(t) = S^{-1}U^o = Z^o$. Let L denote the numerical operator which obtains U^{n+1} from U^n according to Eqs. (6) and also where the inflow boundary point values are fixed and the outflow boundary point values are calculated using backward or, more precisely, upstream differencing. In the absence of roundoff error, since the technique defined by Eqs. (6) is consistent (of second order), $Z^{n+1} = LZ^n = (L)^nZ^o = Z^o$. The numerical solution is exact. Now let δ^o be a perturbation to Z^o rich in Fourier components which can be supported by the mesh. With $Z^o + \delta^o$ as an initial condition, by

[1]The superscripts * and T denote complex conjugate and transpose.

[2]The function mod (x,y) is defined as x - [x/y]y where [x/y] is the integral part of x/y.

linearity, both the set of differential Eqs. (7) and the difference equations reduce to

$$\frac{\partial \delta}{\partial t} + A\frac{\partial \delta}{\partial x} + B\frac{\partial \delta}{\partial y} = 0 \qquad \text{and} \qquad \delta^{n+1} = L\delta^{n} = (L)^{n}\delta^{o}$$

As earlier in this section, the magnitude of the exact amplification of each component of δ is unity. On the other hand, if all the eigenvalues of the amplification matrix $\mathcal{G}_L$ associated with the operator L are less than or equal to one in magnitude $\|\delta^{n+1}\| \leq \|\delta^{o}\|$; otherwise $\|\delta^{n+1}\|$ will grow exponentially with n. The numerical results for Mach number 2.0, a mesh of 20 x 20 points, $\delta^{o}_{i,j} = (0,0,0,0)^T$ for $i \neq 10$ and $j \neq 10$ and $\delta^{o}_{10,10} = (0,1,0,1)^T$ are contained in Fig. 2 for $\Delta y/\Delta x = 1$ and for $\Delta y/\Delta x = 0.1$ (the form of $\delta^{o}_{10,10}$ is explained in footnote 3). Each point of the figures represents the result of a numerical solution, δ^{n}, and is plotted in polar coordinates with $c\Delta t/\Delta x$ as the radial coordinate and arctan v/u as the angular coordinate. Each solution was advanced from 40 to 400 time steps until it could be determined if $\|\delta^{n}\|$ remained bounded by unity (open symbols) or grew exponentially (closed symbols). The closed curve of each figure represents the C.F.L. condition, Eq. (3), with the equality sign. These results indicate that the C.F.L. condition is a sufficient condition for the stability of the difference Eqs. (6). Also, Fig. 2(a) shows that for the difference Eqs. (2), the associated amplification matrix has eigenvalues greater than one in magnitude for the velocity directions $3\pi/4$ and $7\pi/4$. However, it was observed that at arctan v/u = 0, $\pi/4$, $\pi/2$, π, $5\pi/4$ and $3\pi/2$ the associated eigenvalues of Eqs. (2) were bounded by unity if $c\Delta t/\Delta x$ satisfied the C.F.L. condition.

A Nonlinear Instability

Difference schemes shown to be stable by linear analysis may still experience numerical difficulties in the solution of nonlinear problems. Normally they appear where linear theory does not apply, in particular in regions where the exact solution is not smooth. However, even in smooth regions difficulties can occur. To demonstrate this, consider the equation $\frac{\partial \rho u}{\partial t} + \frac{\partial \rho u^2}{\partial x} = 0$, where ρ is constant, and the finite difference approximation to it,

$$u_i^{n+1} = u_i^n - (\Delta t/\Delta x)\left\{\left(\rho u^2\right)^n_{i+1/2} - \left(\rho u^2\right)^n_{i-1/2}\right\}$$

where $\left(\rho u^2\right)^n_{i+1/2}$ may be $\rho u^{n^2}_{i+1}$ (forward difference), or $\rho u^{n^2}_{i}$ (backward difference) or $\left(\rho u^{n^2}_{i} + \rho u^{n^2}_{i+1}\right)/2$ (central difference). Viewing the mesh as composed of cells allows the following interpretation of the difference equation. The change in ρu_i during time Δt is equal to the difference in momentum transported across the cell face located at $x = (i + 1/2)\Delta x$ from that at $x = (i - 1/2)\Delta x$. Let us consider only the effect of transport across the cell face at $x = (i + 1/2)\Delta x$ and suppose that both u_i^n and u_{i+1}^n are nonzero. There are four cases to consider: (a) $u_i^n, u_{i+1}^n < 0$; (b) $u_i^n, u_{i+1}^n > 0$; (c) $u_i^n > 0, u_{i+1}^n < 0$; and (d) $u_i^n < 0, u_{i+1}^n > 0$.

For case (a), cell i containing negative momentum receives $-(\Delta t/\Delta x)(\rho u^2)^n_{i+1/2}$ momentum per unit volume from cell i + 1, the "donor cell." The magnitude of momentum of the receiving cell is increased while that of the donor cell is decreased. Thus the difference equation is consistent with the physics of the flow at the cell face. The reduction in magnitude of the donor cell is a stabilizing influence. Case (b) is essentially the same as case (a) with the roles of cells i and i + 1 reversed. In case (c) both cells are donors and because $\rho u_i^n > 0$ and $\rho u_{i+1}^n < 0$, the magnitudes of momentum of both cells are reduced, again a stabilizing process. Case (d) exhibits an entirely different behavior. Both cells are receivers, increase their momentum magnitude by $(\Delta t/\Delta x)(\rho u^2)^n_{i+1/2}$ during time Δt and will again satisfy the requirements for case (d) for the next time step. Thus, this process is self-aggravating, a destabilizing influence. The difficulty arises because the difference equation loses information about the sign of u_{i+1}^n through squaring. For cell i the difference equation cannot distinguish case (d) from case (a) or for cell i + 1 case (d) from case (b). Thus the difference

---

[3]The fourth equation of Eqs. (7) is not coupled to the first three; hence it needs to be directly perturbed. The first three are coupled; hence a perturbation to any one will perturb the other two.

equation can violate the physics of the flow by allowing a quantity of negative momentum to be transported out of a cell, possibly containing only positive momentum, through a surface at which u vanishes. This numerical difficulty, when encountered can be remedied by appealing to the physics of the flow. The fluid velocity at the cell face can be approximated to second order accuracy by $(u_i^n + u_{i+1}^n)/2$ and if the term $(\rho u^2)_{i+1/2}^n$ is replaced in the difference equation by $\{(u_i^n + u_{i+1}^n)/2\}(\rho u)_{i+1/2}^n$ where $(\rho u)_{i+1/2}^n$ is defined as before, the order of accuracy of the difference equation is unchanged. The modification reduces the destabilizing influence because $\{(u_i^n + u_{i+1}^n)/2\}(\rho u)_{i+1/2}^n < (\rho u^2)_{i+1/2}^n$ and also is more consistent with the physics of the flow. This modification is employed when the conditions of case (d) occur by the methods described in this paper.

SPECIFIC NUMERICAL CONSIDERATIONS

In this section the ideas and techniques of the previous section will be tailored to meet the specific needs for the solution of the interaction of a shock wave with a laminar boundary layer on a flat plate sketched in Fig. 3. In the process some ideas are developed which may be very significant in the numerical solution of fluid dynamic problems in general.

As in any problem, the primary necessity is to choose Δx and Δy small enough so that: (1) good spatial resolution of the features in the flow field is attained; and (2) all the significant physical processes, for example, viscous shear, are exhibited without being largely influenced by truncation error. Unfortunately, it is impossible without knowing the exact solution and its derivatives to choose Δx and Δy so that these requirements are sure to be achieved.

This particular flow problem has characteristic lengths normal to the plate (y-direction), boundary layer thickness, and along the plate (x-direction), the distance from the leading edge to the incident shock, x_s, differing by several orders of magnitude. One approach to meet the above needs is to choose Δx and Δy so that all of the significant terms of the set of differential equations are all of nearly the same magnitude in the set of difference equations. For example, consider the transport term $(\partial\rho vu/\partial y)$ and the viscous stress term $\dfrac{\partial\mu\frac{\partial u}{\partial y}}{\partial y}$ of the x-direction momentum equation.[4] If we linearize and difference them, we have[5] $\frac{\rho v}{2\Delta y}(u_{j+1} - u_{j-1})$ and $(\mu/\Delta y^2)(u_{j+1} - 2u_j + u_{j-1})$. Equating the coefficients of these difference terms results in $\rho v\Delta y/\mu = 2$, a mesh Reynolds number, $Re_{\Delta y}$, of two. Comparing this number with the free-stream Reynolds number, Re_{x_s}, we obtain the estimate

$$\Delta y \approx \frac{Re_{\Delta y}}{Re_{x_s}}\frac{u_o}{v}x_s \approx \frac{2u_o x_s}{vRe_{x_s}} \tag{8}$$

where u_o is the free-stream velocity and it is assumed that the kinematic viscosity, μ/ρ, varies little from that of the free stream. Now estimating Δx, we, as before, linearize and difference the terms $\frac{\partial\rho u^2}{\partial x}$ and $\frac{\partial\rho vu}{\partial y}$. Equating the coefficients of the difference terms yields

$$\Delta x = \frac{u}{v}\Delta y \tag{9}$$

In using Eqs. (8) and (9) to calculate mesh spacing in a region in which viscous phenomena are important, velocities u and v characteristic of the region will also have to be estimated. It is probable that the "equal in magnitude" of the difference coefficients could be relaxed to "of the same order of magnitude" without the viscous terms being swamped by truncation

---

[4]The choice of the x-direction momentum equation is consistent with the usual boundary layer analysis. The y-direction momentum equation normally reduces in the boundary layer equations to $\frac{\partial p}{\partial y} = 0$.

[5]A central difference approximation is used here because the use of one forward and one backward difference by the techniques of this paper during each time step is effectively that of using a central difference.

error. For this reason and the assumption about kinematic viscosity Eqs. (8) and (9) are believed to be conservative. Nevertheless, in the boundary layer region Δy is expected to be much smaller than Δx.

At moderate Mach numbers, if $\Delta y \ll \Delta x$, the calculation for Δt using the C.F.L. Eq. (3) is dominated by small y-direction mesh spacing. If Eqs. (6) are used to advance the solution, during each Δt, the numerical domain increases in the x and y directions by Δx and Δy. The physical domain increases in the y-direction by $(|v| + c)\Delta t \approx \Delta y$. However, in the x-direction, the increase is only $(|u| + c)\Delta t \ll \Delta x$. Thus the numerical domain is much larger than the physical domain. However, it is possible to modify Eqs. (6) using the concept of splitting so that this difficulty is avoided.

<u>Splitting</u>

The concept of splitting is originally due to Peaceman and Rachford (1955) and is commonly known as the method of alternating directions. Since then, it has been widely used (Yanenko, 1969) to transform complex operators into a sequence of simpler ones. This concept will now be used to reduce the set of two dimensional Eqs. (6) into two sets of one-dimensional equations while maintaining second order accuracy.

Eqs. (6) can be split as follows:

$$\left.\begin{aligned}
ii &= \mathrm{mod}\,(n,2) \\
jj &= \mathrm{mod}\,(n - ii,4)/2 \\
\overline{U_{i,j}^{n+1/2}} &= U_{i,j}^{n} - \left(\frac{\Delta t}{\Delta y}\right)\left(G_{i,jj}^{n} - G_{i,jj-1}^{n}\right) \\
jj &= \mathrm{mod}\,(jj + 1,2) \\
U_{i,j}^{n+1/2} &= \frac{1}{2}\left\{U_{i,j}^{n} + \overline{U_{i,j}^{n+1/2}} - \frac{\Delta t}{\Delta y}\left(\overline{G_{i,jj}^{n+1/2}} - \overline{G_{i,jj-1}^{n+1/2}}\right)\right\} \\
\overline{U_{i,j}^{n+1}} &= U_{i,j}^{n+1/2} - \frac{\Delta t}{\Delta x}\left(F_{ii,j}^{n+1/2} - F_{ii-1,j}^{n+1/2}\right) \\
ii &= \mathrm{mod}\,(ii + 1,2) \\
U_{i,j}^{n+1} &= \frac{1}{2}\left\{U_{i,j}^{n+1/2} + \overline{U_{i,j}^{n+1}} - \frac{\Delta t}{\Delta x}\left(\overline{F_{ii,j}^{n+1}} - \overline{F_{ii-1,j}^{n+1}}\right)\right\}
\end{aligned}\right\} \quad (10)$$

Letting L_y denote the operator which obtains $U^{n+1/2}$ from U^n and L_x that which obtains U^{n+1} from $U^{n+1/2}$ we have schematically $U^{n+1} = L_xL_yU^n$. It can be shown that because of the noncommutativity of L_x and L_y that the numerical method L_xL_y is only of first order. However, it will now be shown that the method defined by $U^{n+2} = L_yL_xL_xL_yU^n$ is of second order accuracy.

The amplification matrix associated with L_x is $\mathcal{G}_x = I - i(\Delta t\, J_F/\Delta x)\sin\xi - (\Delta t\, J_F/\Delta x)^2 (1 - \cos\xi)$. The eigenvalues of J_F are the same as A, u, u, $u \pm c$ and those of $\mathcal{G}_x$ are less than or equal to unity in magnitude if $\Delta t \le \Delta t_x = \Delta x/(|u| + c)$. Similarly, $\mathcal{G}_y = I - i(\Delta t\, J_G/\Delta y)\sin\eta - (\Delta t\, J_G/\Delta y)^2(1 - \cos\eta)$ and its eigenvalues are less than or equal to one in magnitude if $\Delta t \le \Delta t_y = \Delta y/(|v| + c)$. Thus Eqs. (10) are stable if $\Delta t \le \min(\Delta t_y, \Delta t_x)$, since each component operator is then stable. Now for $\xi, \eta \ll 1$, $\mathcal{G}_y\mathcal{G}_x\mathcal{G}_x\mathcal{G}_y = \exp[-i2\Delta t(J_F\xi/\Delta x + J_G\eta/\Delta u)]$, the exact amplification of the solution during $2\Delta t$, modulo terms of third order in ξ and η. The extension to three dimensions, $L_zL_yL_xL_xL_yL_z$, is simple.

Now suppose that $\Delta y \ll \Delta x$ so that $\Delta t_y/\Delta t_x \ll 1$. Let $L_x(\Delta t_x)$ be L_x as before with $\Delta t = \Delta t_x$, $L_y(\Delta t_y)$ be similarly defined and M be the smallest even integer greater than $\Delta t_x/\Delta t_y$. Then the following method of second order accuracy advances the numerical solution Δt_x in time.

$$U^{n+1} = (L_y(\Delta t_x/M))^{M/2}\, L_x(\Delta t_x)\, (L_y(\Delta t_x/M))^{M/2}\, U^n$$

(In a chain of time steps this operator is . . . $L_y^M L_x L_y^M L_x$ . . . .) The advantages of this split technique are: (a) For M time step advances in the "y-direction" where Δt_y is at least as large as the maximum allowed by Eq. (3), the "x-direction" terms, $\frac{\partial F}{\partial x}$, need only be computed once, as compared with M times with Eqs. (6) and (b) the physical and numerical domains of dependence are matched more closely, since during time Δt_x both domains increase by $(|u| + c)\Delta t_x = \Delta x$ in the x-direction and in the y-direction the physical domain increases by $(|v| + c)\Delta t_x \approx (|v| + c)M\Delta t_y = M\Delta y$ which is the numerical domain increase. The main disadvantage for large M is that the solution is advanced in the y-direction many times nearly completely independent of the nature of the solution in the x-direction. Although this small amount of coupling becomes unimportant in the limit as Δx, $\Delta y \to 0$, at practical choices of Δx and Δy this need not be the case. This difficulty can be overcome while retaining the advantages of splitting as follows. First, the term $\frac{\partial F}{\partial x}$ will be calculated at time $(n + 1/2)\Delta t_x$ to second order accuracy by Eqs. (6) using $\Delta t = \Delta t_x$. $\left(\frac{\partial F}{\partial x}\right)_{i,j}^{n+1/2} = \frac{1}{2}\left(F_{ii,j}^n - F_{ii-1,j}^n + F_{ii,j}^{\overline{n+1}} - F_{ii-1,j}^{\overline{n+1}}\right)/\Delta x$. Let $L_{xy}(m, \Delta t)$ be the operator defined by

$$
\left.
\begin{aligned}
&jj = \mathrm{mod}\,(m,2)\\[4pt]
&U_{i,j}^{n+\frac{m+1}{M}} = U_{i,j}^{n+\frac{m}{M}} - \frac{\Delta t}{\Delta y}\left(G_{i,jj}^{n+\frac{m}{M}} - G_{i,jj-1}^{n+\frac{m}{M}}\right) - \Delta t\left(\frac{\partial F}{\partial x}\right)_{i,j}^{n+1/2}\\[4pt]
&jj = \mathrm{mod}\,(jj + 1,2)\\[4pt]
&U_{i,j}^{n+\frac{m+1}{M}} = \frac{1}{2}\left\{U_{i,j}^{n+\frac{m}{M}} + U_{i,j}^{\overline{n+\frac{m+1}{M}}} - \frac{\Delta t}{\Delta y}\left(G_{i,jj}^{\overline{n+\frac{m+1}{M}}} - G_{i,jj-1}^{\overline{n+\frac{m+1}{M}}}\right) - \Delta t\left(\frac{\partial F}{\partial x}\right)_{i,j}^{n+1/2}\right\}\\[4pt]
&m = 0, 1, \ldots, M-1
\end{aligned}
\right\} \quad (11)
$$

Then the solution U^n is advanced in time by Δt_x by the following sequence:

(1) Calculate $\left(\frac{\partial F}{\partial x}\right)_{i,j}^{n+1/2}$ all i,j (2) $U^{n+1} = \prod_{m=0}^{M-1} L_{xy}(m, \Delta t_x/M)\, U^n$

where Π denotes the product. It is not difficult to show, using Taylor series expansions, that the above technique is of second order accuracy. For stability, if we again consider the linear case, $U_{i,j}^{n+\frac{m+1}{M}} = L_{xy}(m, \Delta t_x/M)\, U_{i,j}^{n+\frac{m}{M}} = L_y(\Delta t_x/M) + \frac{\Delta t_x}{M} C_{i,j}^{n+1/2}$, where

$C_{i,j}^{n+1/2} = -\left(\frac{\partial F}{\partial x}\right)_{i,j}^{n+1/2} + \left(\Delta t_x J_G/(2M\Delta y)\right)\left\{\left(\frac{\partial F}{\partial x}\right)_{i,jj}^{n+1/2} - \left(\frac{\partial F}{\partial x}\right)_{i,jj-1}^{n+1/2}\right\}$. Since the magnitudes

of the eigenvalues of $L_y(\Delta t_x/M)$ are bounded by unity, those of $L_{xy}(m, \Delta t_x/M)$ are at worst bounded by $1 + O(\Delta t_x)$ (von Neuman condition Eq. (5)).

NUMERICAL RESULTS

To illustrate the concepts of the previous sections numerical solutions of the two-dimensional interaction of an oblique shock wave with a laminar boundary layer at a Mach number of 2.0 will be presented and compared with the experimental data of Hakkinen. Some characteristic features of this interaction are illustrated schematically in Fig. 3. Hakkinen's data were chosen because the boundary layer is laminar throughout the interaction region and the experimental data include both plate surface pressure and skin friction. Two test cases, one in which the incident shock was not strong enough to cause flow separation, and one with a shock sufficiently strong to cause separation (Fig. 3), were solved numerically.

The computational rectangular mesh covered an area of 8.636 cm x 1.270 cm including 7.874 cm of plate. The boundary conditions were: (a) the upstream boundary values were initially set for uniform flow and thereafter held fixed; (b) the values of the boundary opposite the plate were initially set using inviscid oblique shock wave theory such that a shock of given strength and point of incidence on the plate would be generated and thereafter were held fixed; (c) the downstream exit boundary values were set equal to those just upstream after each step; and (d) for the boundary containing the plate a fictitious row of cells together with "mirror symmetry" was employed; that is, the plate was located midway between the row of fictitious points, denoted by j = 1, and the first interior row of points,

$j = 2$ (located $\Delta y/2$ off the plate). The $\rho_{i,1} = \rho_{i,2}$, $u_{i,1} = -u_{i,2}$, $v_{i,1} = -v_{i,2}$, $T_{i,1} = T_{i,2}$ and $p_{i,1} = p_{i,2}$ except for those points upstream of the plate where $u_{i,1} = u_{i,2}$. Boundary condition (c) assumes uniform flow at the exit. This is good near the plate. Away from the plate the flow field is supersonic and errors made at the boundary will not propagate upstream. By the temperature and density boundary conditions (d) the plate is treated as an adiabatic wall. The initial condition of the interior points is uniform flow. The Sutherland viscosity law for air, with $\lambda = -\frac{2}{3}\mu$, the perfect gas equation of state, and a constant Prandtl number equal to 0.72 were used.

Unseparated Flow

For this case, the pressure ratio, p_f/p_o (final (after reflection) to initial), was 1.2, the distance from the leading edge to the incident shock, x_s, was 4.88 cm, and the free-stream Reynolds number, Re_{x_s}, was 2.84×10^5. From Eqs. (8) and (9) with v taken as $v_1/2$ and u/v as u_1/v_1, where u_1 and v_1 are the velocity components after the incident shock and away from the plate, the estimates $\Delta x = 0.0848$ cm and $\Delta y = 0.0024$ cm for the boundary layer region are obtained as a reference.

An initial calculation with a mesh of 34 x 12 cells, $\Delta x = 0.254$ cm, $\Delta y = 0.127$ cm (more than 50 times larger than the estimate), was made using Eqs. (6) with $\Delta t = 0.9$, the maximum allowed by Eq. (3) ($u = u_1$, $v = v_1$ and $c = c_1$) and allowed to run 128 time steps when there was little change occurring in the flow field (approximately 5 minutes in machine time on the IBM 360/67). Figure 4 compares the numerical results with Hakkinen's data, where the surface pressure is approximated by $p_{i,1}$ and the local coefficient of skin friction by $\{\mu_{i,1}(u_{i,2} - u_{i,1})/\Delta y\}/(\rho_o u_o^2/2)$. The results for skin friction are in conspicuously poor agreement with experiment. Although the full set of Navier-Stokes equations was differenced, the coarseness of the mesh allowed the viscous terms to be swamped by truncation and round-off error. The numerical solution was essentially that of inviscid flow and the shock wave angles, strengths, etc., were within a few percent of inviscid theory. The calculation was then repeated; this time with a mesh of 34 x 32 cells containing a fine mesh near the plate, large enough to contain the estimated boundary layer and a coarse mesh away from the plate (Fig. 5). In the fine mesh of 34 x 22 cells $\Delta x = 0.254$ and $\Delta y = 0.00635$ cm, (approximately three times that estimated by Eqs. (8) and (9), and in the coarse mesh $\Delta x = 0.254$ and $\Delta y = 0.127$ cm. The solution was advanced in time separately for each mesh. First the fine mesh was advanced in time by $\Delta t_x = 0.9\ \Delta x/(|u_o| + c_o)$ using Eqs. (11) with $M = 20$. The row of cells, $j = 22$, was used as a boundary for the fine mesh. Data for these points were obtained by linear interpolation between points of the inner fine mesh and outer coarse mesh after each $\Delta t_x/M$ time step. After the inner mesh has been advanced in time by Δt_x, Eqs. (10) with $\Delta t = \Delta t_x$ are used in second order fashion to advance the outer mesh. During the calculation of the inner mesh transport through and stress at the boundary common to both meshes were saved. Their net transport and stress were then used as a boundary condition for the outer flow field. Thus, mass, momentum, and energy were rigorously conserved within the overall mesh. The basic assumption in using this mesh is that the viscous terms are important only in the boundary layer (although their effects are important away from the plate as well), thus allowing the outer flow field to be treated as inviscid by use of a coarse mesh. The numerical results for this calculation are displayed in Fig. 6. Figure 6(a) illustrates the asymptotic convergence in time from the initial condition to steady state of a velocity profile at $x = 4.699$ cm (the column of mesh points just upstream of the shock). Figure 6(b) compares the numerical results after 256 Δt_x time steps (four hours of machine time) with the experimental measurements of surface pressure and skin friction. Figure 6(c) shows the streamlines in the boundary layer obtained from the velocity fields using a third order interpolation subroutine and plotted on a cathode ray display tube. They are initially 0.00953 cm apart. Also contained in the figure are u—velocity profiles at $x = 2.794, 5.080, 6.985$ cm.

Separated Flow

The experimental conditions for this case were essentially the same as those for the separated case, $x_s = 4.953$ cm, $Re_{x_s} = 2.96 \times 10^5$, except the incident shock was stronger, $p_f/p_o = 1.40$. As before, using Eqs. (8) and (9) the estimates $\Delta x = 0.0238$ and $\Delta y = 0.00127$ cm can be obtained as reference values.

Again, exactly as in the unseparated case the solution was advanced in time on a mesh containing a fine sub mesh where $\Delta x = 0.254$ and $\Delta y = 0.00635$ (approximately ten and five times that estimated) and a coarse sub mesh where $\Delta x = 0.254$ and $\Delta y = 0.127$ cm and allowed to run until there was little change occurring in the flow field (320 Δt_x time steps). The results are contained in Fig. 7. Although the computed separation point agrees well with the experimental point, the length of the separated region is underpredicted and the characteristic plateau in the experimental pressure profile is not found in the numerical data. Two related probable causes for this are: (a) there were too few points in the separated region (four cells) to provide good spatial resolution, and (b) Δx is more than ten times that estimated from Eq. (9). The mesh was further refined by halving Δx. The new mesh of 38 x 32 cells began 2.54 cm downstream of the leading edge and covered an area of 4.826×1.270 cm^2. The new time increment Δt_x was half that of the former and M equalled ten. The upstream boundary condition was obtained from the previous calculation and thereafter held fixed. The initial condition was the previous converged solution with second order interpolation used to define values at the additional points. It was observed that the solution on the new mesh, initially out of equilibrium, converged asymptotically again in about 256 additional time steps. The new results are shown in Fig. 8. The streamlines of Fig. 8(c) are initially 0.00953 cm apart and the u—velocity profiles are located at x = 2.794, 5.080, 5.715, and 6.985 cm.

CONCLUDING REMARKS

1. Although the conditions of consistency and stability are sufficient for convergence of the numerical solution to the exact solution for well-posed linear problems as mesh and time increments tend to zero, there is no guarantee that this will occur in nonlinear problems. The numerical analyst must at present be content with assuming the same behavior for the nonlinear case. The test cases in the applications section support this. It is felt that mesh spacings nearer those estimated by Eqs. (8) and (9) would cause the remaining disparity between the numerical and experimental results (i.e., the plateau in the pressure profile of Fig. 8) to be reduced. Nevertheless, good agreement in general has been achieved.

2. The techniques developed from the concept of splitting are expected to have much more general application than just those with severe differences in coordinate mesh spacing. Not only do they exhibit more flexibility to allow better matching of dependence domains, but since $\min(\Delta x/(|u| + c), \Delta y/(|v| + c)) \geq (|u|/\Delta x + |v|/\Delta y + c\sqrt{1/\Delta x^2 + 1/\Delta y^2})^{-1}$ their time step increments can be larger than the unsplit techniques. There are however two questions to their unrestricted use: (a) Although there is no loss in order of accuracy, are they still less accurate for practical choice of Δx and Δy; and (b) although the solution can proceed at larger time steps, is the computation time per step correspondingly greater also? To partially answer these questions the calculation which produced the data contained in Fig. 5 obtained by the unsplit Eqs. (6) was recalculated using the split Eqs. (10) for the same simple 34 x 12 mesh. The steady-state solutions were nearly identical (Fig. 9). The time step ratio, $\Delta t_x/\Delta t$, equalled 1.45 where Δt_x was taken as $0.9\Delta x/(|u_0| + c_0)$ and the computation time per step ratio, $t_{unsplit}/t_{split}$, was 0.924. Thus for the case computed here the computation time to advance the solution to a given time by the split technique is 0.75 that of the unsplit technique.

REFERENCES

Hakkinen, R. J.; Greber, I.; Trilling, L.; and Abarbanel, S. S.: The Interaction of an Oblique Shock Wave with a Laminar Boundary Layer. NASA Memo 2-18-59W, 1959.

Kutler, P.: Application of Selected Finite Difference Techniques to the Solution of Conical Flow Problems. Ph.D. thesis, Iowa State Univ., 1969.

MacCormack, R. W.: The Effect of Viscosity in Hypervelocity Impact Cratering. AIAA Paper No. 69-354, 1969.

Peaceman, D. W.; and Rachford, H. H., Jr.: The Numerical Solution of Parabolic and Elliptic Differential Equations. SIAM Jour., vol. 3, 1955, pp. 28-41.

Yanenko, N. N.: The Method of Fractional Steps for Numerical Solution of the Problems of Mechanics of Continuous Media. Fluid Dynamics Transactions, vol. 4, 1969, pp. 135-147. Institute of Fundamental Technical Research, Polish Academy of Science, Warsaw.

○ SOLUTION BY EQUATIONS (6)
□ SOLUTION BY EQUATIONS (2)
— $C\Delta t/\Delta x$ FROM EQUATION (3) WITH EQUALITY SIGN

OPEN SYMBOLS MAX $|\lambda_i| \leq 1$
SOLID SYMBOLS MAX $|\lambda_i| > 1$

(r, θ)

(a) $\Delta x = \Delta y$

(b) $\Delta x = 10\ \Delta y$

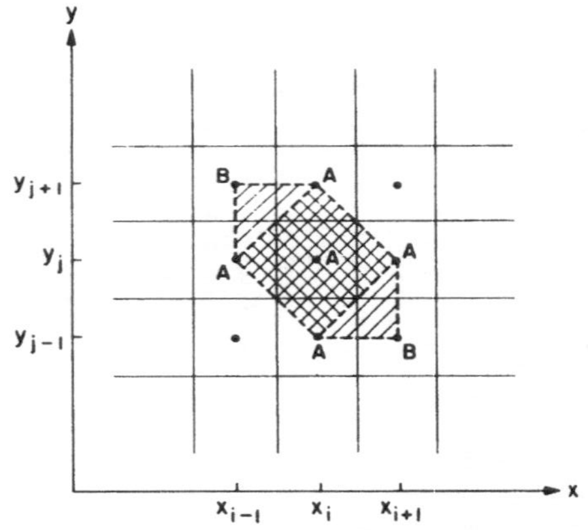

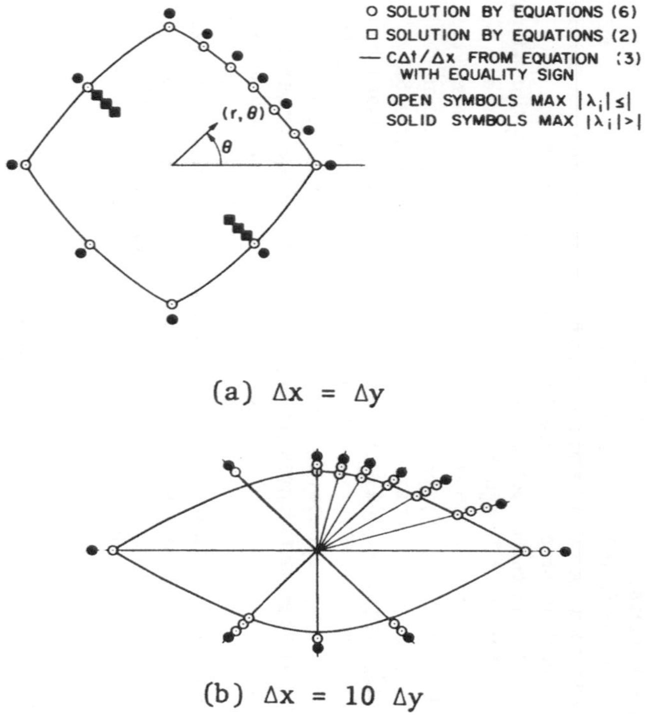

Fig. 2. Stability domain, $c\ \Delta t/\Delta x$ vs. $\theta = \arctan(v/u)$ at Mach 2.0.

Fig. 1. Star of mesh points involved in a single application of Eqs. (2).

INCIDENT SHOCK

EXPANSION

COMPRESSION WAVES

COMPRESSION WAVES

REATTACHMENT

SEPARATION

BOUNDARY LAYER

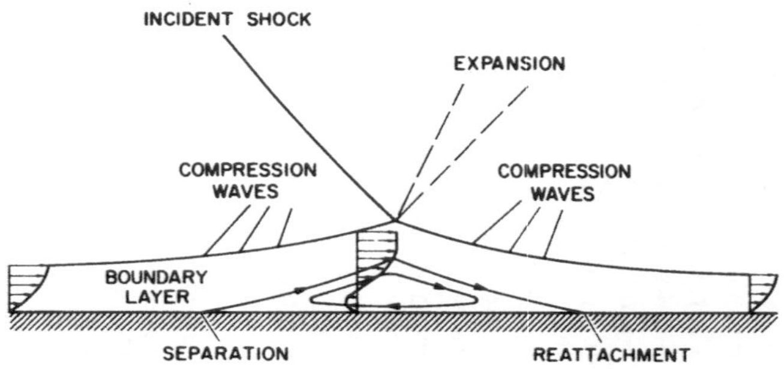

Fig. 3. Sketch of shock boundary layer interaction.

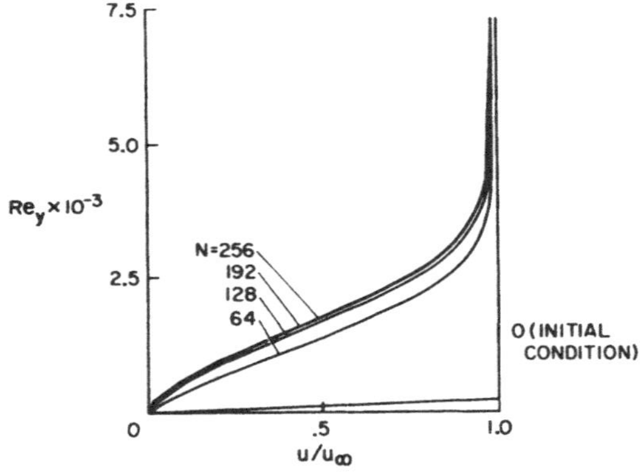

(a) Development of velocity profiles at $Re_x = 2.74 \times 10^5$ with time.

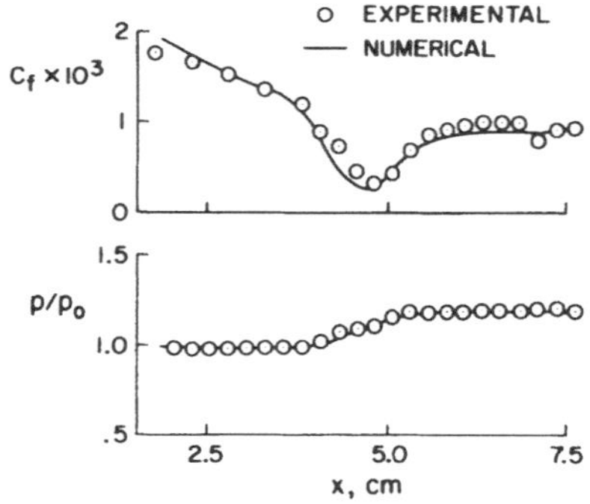

(b) Skin-friction coefficient and surface pressure.

Fig. 6. Numerical results with fine inner mesh coupled to coarse outer mesh; $p_f/p_o = 1.2$.

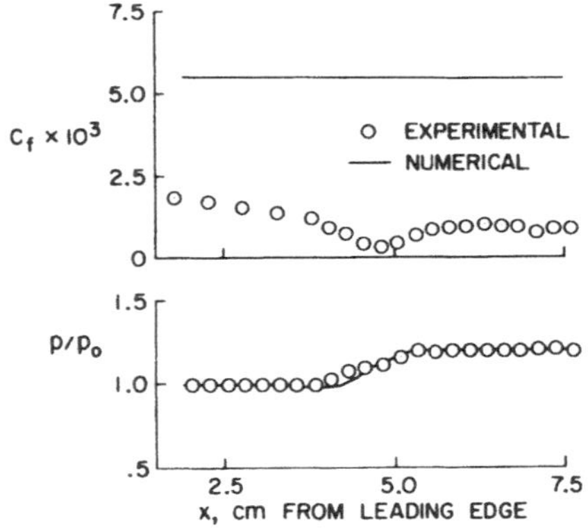

Fig. 4. Comparison of numerical and experimental results; $p_f/p_o = 1.2$, $\Delta x = 0.254$ and $\Delta y = 0.127$ cm.

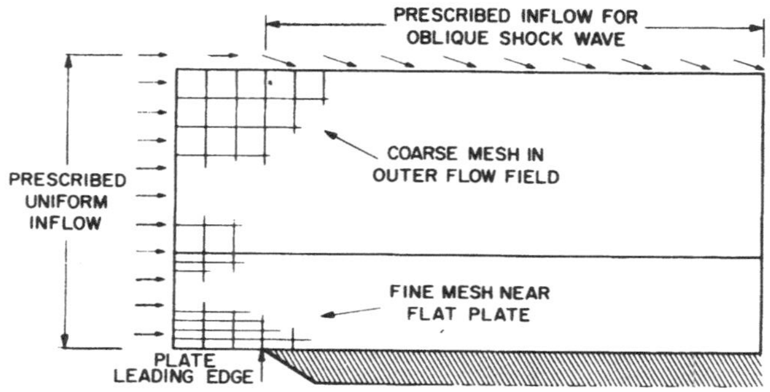

Fig. 5. Mesh setup.

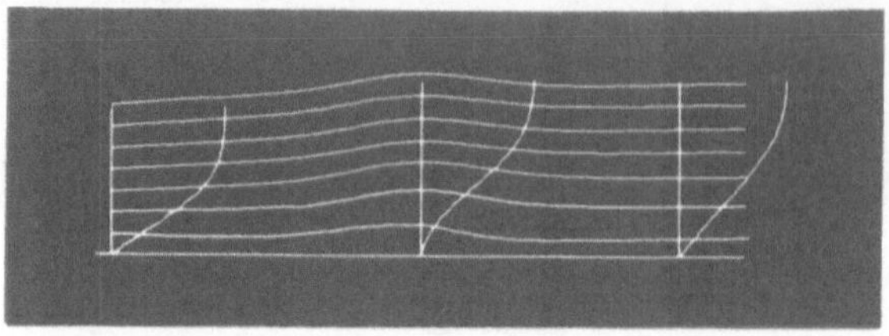

(c) Streamlines and velocity profiles.

Fig. 6. Concluded.

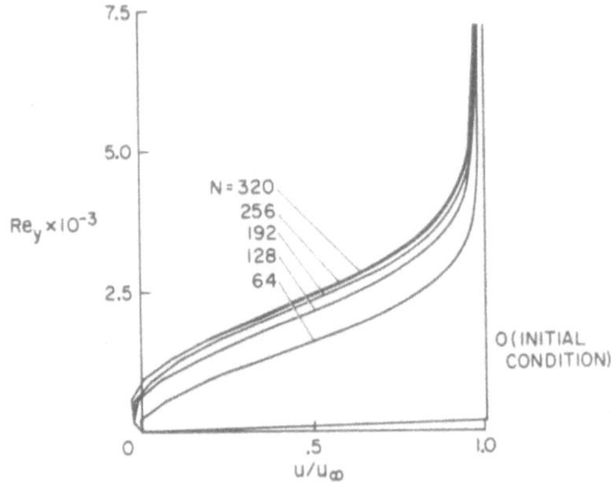

(a) Development of velocity profiles at $Re_x = 2.79\times10^5$ with time.

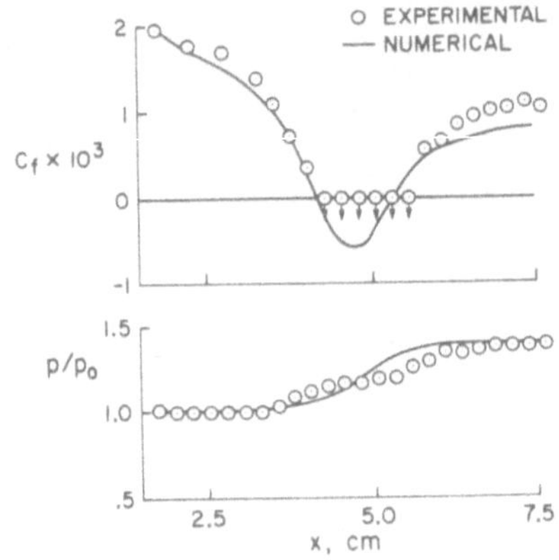

(b) Skin-friction coefficient and surface pressure.

Fig. 7. Numerical results for $p_f/p_o = 1.4$; $\Delta x = 0.254$ and $\Delta y = 0.00635$ cm in fine mesh, $\Delta x = 0.254$ and $\Delta y = 0.127$ cm in coarse mesh.

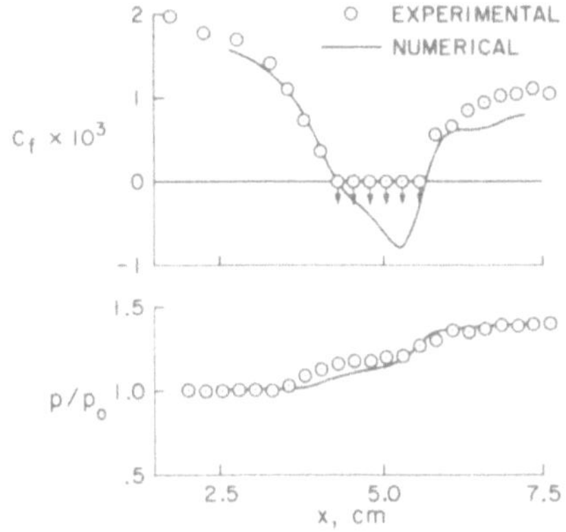

(a) Skin-friction coefficient and surface pressure.

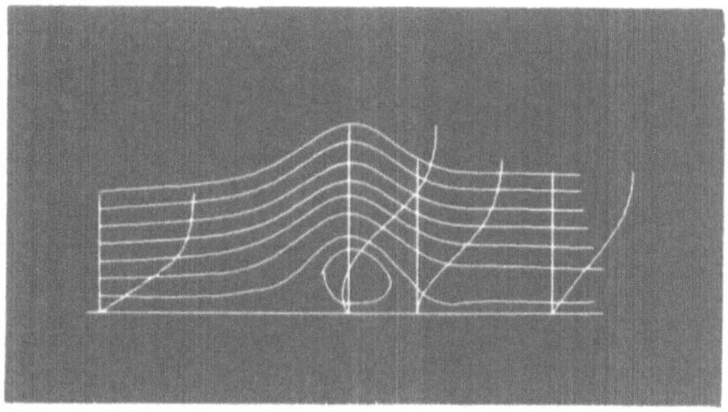

(b) Streamlines and velocity profiles.

Fig. 8. Numerical results for $p_f/p_o = 1.4$; $\Delta x = 0.127$ and $\Delta y = 0.00635$ cm in fine mesh, $\Delta x = 0.127$ and $\Delta y = 0.127$ cm in coarse mesh.

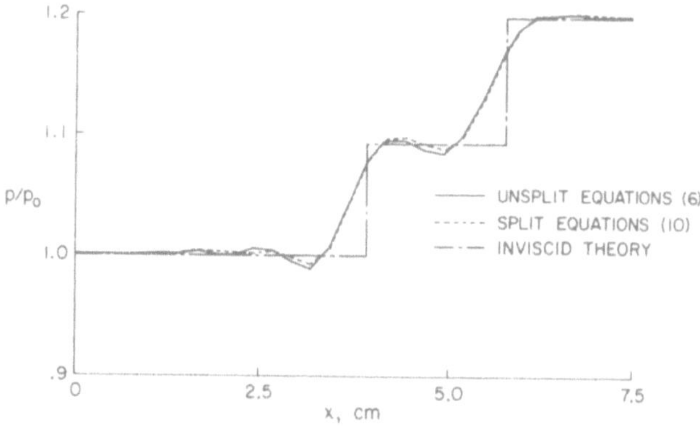

Fig. 9. Comparison of pressure profiles at $y = 0.5715$ cm from plate.

A NUMERICAL SOLUTION OF THE PLANAR SUPERSONIC NEAR-WAKE WITH ITS ERROR ANALYSIS +

B. B. ROSS* and S. I. CHENG**

Princeton University
Princeton, New Jersey

The planar flow field of a low-Reynolds-number viscous wake imbedded in a supersonic flow has been studied by Allen and Cheng and by Roache and Mueller using finite difference calculations based on the full Navier-Stokes equations. The present paper gives a critical study of the influence of boundary conditions in such a problem while extending the work of Allen and Cheng to include variable viscosity and base injection. Details are given in Ross' thesis.

The configuration of the computational field is illustrated in figure 1. We will denote the base height $\overline{OA}$ by H. Inflow across the boundary BC includes an inviscid, irrotational supersonic flow together with a Blasius-type boundary layer and its vertical displacement velocity. Minor modification of the conservative, two-step difference algorithm is needed to maintain the semi-explicit nature of the difference scheme for the variable viscosity equations. A uniform Cartesian grid is used with $H/\Delta x = 19.5$ and $\Delta x/\Delta y = 1.5$. Density, energy, velocity and pressure are nondimensionalized with respect to their inviscid inflow values on BC (denoted by subscript 1). The distribution of flow properties is displayed by means of contour plots. The values of contours are integral multiples of the "step" parameter given in each figure (with the exception of stream function whose contours are equally spaced along BC). Dashed contours denote negative values. The locations of maxima and minima are designated by points in the figures with their values and coordinates (x/H, y/H) given in the figure titles.

THE INFLUENCE OF BOUNDARY CONDITIONS

Cheng (1969) and Moretti discussed the importance of boundary condition errors in finite-difference computations. The boundary conditions provided by the differential formulation of an external flow problem are often not sufficient to permit the complete formulation of a difference model over a finite field (Cheng, 1970). So in the present problem, we must provide extraneous boundary conditions on outer boundaries CD and DE to approximate the flow behavior locally. We must therefore expect local errors to result from our improper choice of boundary conditions. On the other hand, inner boundary conditions on wall OAB and symmetry line OE have counterparts in the differential problem to help guide us in their difference formulation. However, computational errors are still committed along these boundaries due to discretization. Near the sharp corner A, such errors are significant.

Since we cannot evaluate the boundary errors without an exact solution, we can only determine the rate of spatial decay of boundary errors by means of numerical experiments. Such error decay depends not only on the nature of the boundary condition, but also on the nature of the solution near the boundary. Hence, while the diffusion model of Burgers' equation shows boundary disturbances decaying very slowly, (Cheng, 1969), the one-dimensional wave equation model exhibits a more rapid exponential decay (Cheng, 1970). We will consider here the extent to which such inferences regarding boundary error decay may apply in a multidimensional problem such as the planar near-wake. To do this, we alter the location or the method of treatment of a particular boundary condition and then note the resulting flow property changes in the overall solution by means of contour mappings (figures 5-12).

+This research was supported by the Aeronautical Research Laboratories, Wright-Patterson Air Force Base, under contract AF33615-70C-1244 with computations also supported by NASA grant Ns G(T)-38 as well as NSF grants GJ-34 and GU-3157.

*NSF Graduate Fellow, Department of Aerospace and Mechanical Sciences.

**Professor, Department of Aerospace and Mechanical Sciences.

Tests were made on the adiabatic-wall wake solution shown in figures 2-4, where Re_{H1}, M_1, and inflow boundary layer height δ/H are given in the figure titles. The disturbance behavior of density R and x-velocity component U were found to be typical and will thus be used in the discussion below.

Solid Wall Boundary:- The flow around the corner A poses serious computational difficulties because of the large errors created locally and the corner's location upstream of the near-wake region. In the present solution, static pressure changes over Δx upstream of the corner are comparable with those across the wake shock. Hence, the truncation error is certainly not small in this region, and in fact, the continuum formulation is questionable locally. We wish to determine experimentally how the resulting error propagates over the rest of the field. Instead of applying an arbitrary disturbance at the corner, we chose to alter the corner flow by changing the method of treating the wall boundary conditions. This serves not only to produce a reasonably small corner disturbance, with little effect on boundary behavior elsewhere on the wall, but also shows the sensitivity of the near-wake solution to the given choice of wall boundary conditions. Parallel calculations were performed, one using a simple reflection boundary condition and the other using the more realistic boundary condition employed by Allen and Cheng. In both cases, wall boundary points occur within the fluid a half space increment off the wall. As shown in figures 5 and 6, the maximum changes occur near the corner where gradients are largest. The greatest discrepancy (.0387 U_1) in velocity component U occurs just below the corner; the "effective corner" is more rounded for the reflection case than for the Allen-Cheng case. The velocity disturbance decays exponentially along downstream radial lines with the slowest decrease being 25% of the maximum value at $x/H = 0.5$ and 8% at $x/H = 2.0$. Density shows the greatest discrepancy (.0074 R_1) on the top wall with very rapid decay of the disturbance away from the corner. The expansion wave and wake shock regions show only slight density variations.

Top and Rear Boundaries:- The ratios h/H and L/H, as well as the treatment of boundary conditions along CD and DE, were chosen so that outer boundary influence on the near-wake region is smaller than that from other sources such as the corner. In order to determine the extent of such outer boundary influence, a number of calculations were performed using several different combinations of h/H and L/H listed below:

	L/H	h/H	L_1/H
Reference Case	7.58	2.97	1.81
Case A	6.04	2.26	1.81
Case B	4.04	2.26	1.81
Case C	5.04	1.85	1.81

Density R and velocity U differences, taken relative to the large-field reference calculation, are shown in figures 7-12.

A simple wave condition was used along CD, but some waves are still reflected from the top boundary as shown in figures 7-9. In case B, these waves pass harmlessly out the rear boundary, while in cases A and C they interact with the shock (the accumulation of contour lines) and alter the viscous wake immediately below. The extent to which such disturbances can alter the viscous region is shown by the U differences in figures 10-12. In case C, where h/H is the smallest of the three, the error waves are strongest, and the influence extends considerably farther upstream. Note that the density difference contours in figures 7 and 8 are nearly identical for $x/H \leq 4$ while the U difference contours in figures 10 and 11 are decidedly different. This change in velocity disturbances must then be attributed to the change in errors created at the effective outflow boundary $x/H = 4$.

The three-point extrapolation boundary condition used along the boundary DE can cause significant errors because this boundary terminates the recompression shock and the viscous wake. The most striking aspect of the density difference figures near the outflow boundary is the density change due to the slight upwards displacement of the shock relative to its position in the reference calculation (vertical displacement is less than 0.003H in all cases). The absence of any apparent upstream influence in the supersonic inviscid region above the shock (compare figures 7 and 8) may be due either to the lack of upstream numerical propagation of disturbances or else to a rapid exponential decay of boundary errors in the hyperbolic region (Kreiss and Lundqvist; Cheng, 1970). In the viscous region below the shock, the contour lines of U differences in figure 11 are primarily due to boundary errors on DE. The slow upstream decay of these errors is similar to the error behavior of the one-dimensional diffusion model studied by Cheng (1969).

In summary, so long as computational stability is maintained, the difference between corner boundary conditions seems to be relatively unimportant under the considerations of local numerical uncertainties and the rapid decay of errors away from the corner. The reflection wall boundary condition, being the simpler of the two tested, has thus been adopted in the calculations which follow. To avoid extensive viscous wake alteration from waves reflected off CD, we require $H/L \gtrsim 0.4$ and $h/H \gtrsim 2$. (for $\delta/H<0.6$). Under these restrictions, the slow upstream decay of errors from the outflow boundary DE makes the strategy of extending the outflow boundary quite costly. An estimate of the error of the extrapolation boundary condition by Cheng (1969, equation 16) suggests that the error in the near-wake region due to boundary DE is less than 1% for $L/H>4$. This should be compared with the residual error from the corner A which may well be of the order of a few percent.

SOME PHYSICAL RESULTS

Ross gives detailed results of wake calculations using the present model for different Reynolds numbers, Mach numbers, wall heat transfer conditions, and base injection conditions. The inviscid inflow Mach number M_1 and Reynolds number Re_{H1} are varied from $Re_{H1} = 1000$ at $M_1 = 3.0$ to $Re_{H1} = 4000$ at $M_1 = 4.0$. The inflow boundary layer thickness δ/H varies from .38 to .56. Composite mappings of pressure, sonic line, and selected streamlines are shown in figures 13 and 14 for two different cases which were chosen with nearly identical rear stagnation point locations. The flow features are seen to be similar to those calculated with fixed viscosity by Allen and Cheng. In fact, fixed and variable viscosity results were found to correlate well if the fixed viscosity value is based on the mean temperature across the wake shear layer. An inviscid region could not be easily discerned below the wake shock for the short-field calculations of figures 13 and 14; however, the viscous wake edge could be clearly distinguished below the shock for $x/H>5$ in the long-field calculation of figure 15. Each of the wake flows illustrated shows the pressure variations in the near-wake portion of the viscous wake to be of comparable magnitudes in the x and y directions.

A series of calculations were also performed using the flow parameters of figure 13 but with small fluid injection distributed uniformly over the base wall. Figures 16 and 17 show streamline mappings for two different injection rates (non-dimensionalized by the inviscid inflow conditions along BC). The recirculation eddy was found to persist up to 0.6% injection rate. Figure 18 compares the pressure contours for flow without injection and with 0.8% injection. The presence of injectant fluid along the wake center-line is seen to displace and weaken the recompression shock; flow upstream of the trailing edge of the expansion fan is unchanged.

REFERENCES

Allen, J.S., and Cheng, S.I., Physics of Fluids 13,37-52 (1970); also, Allen, J.S., Ph.D. Thesis, Princeton University (1968).

Cheng, S.I., Physics of Fluids, Supplement II, 34-41 (1969).

Cheng, S.I., AIAA Preprint No. 70-2 (1970); final manuscript to appear in AIAA J.

Kreiss, H.O., and Lundqvist, E., Math. Comp. 22,1-12 (1968)

Moretti, G., Physics of Fluids, Supplement II, 13-20 (1969).

Roache, P.J., and Mueller, T.J., AIAA J.8 530-538 (1970).

Ross, B.B., Ph.D. Thesis, Princeton University (Dec., 1970)

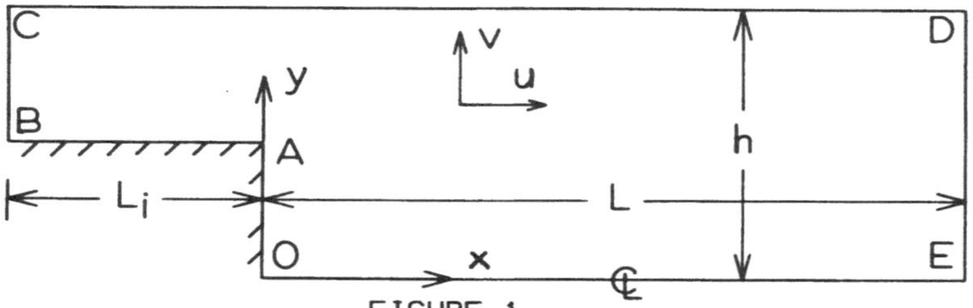

FIGURE 1

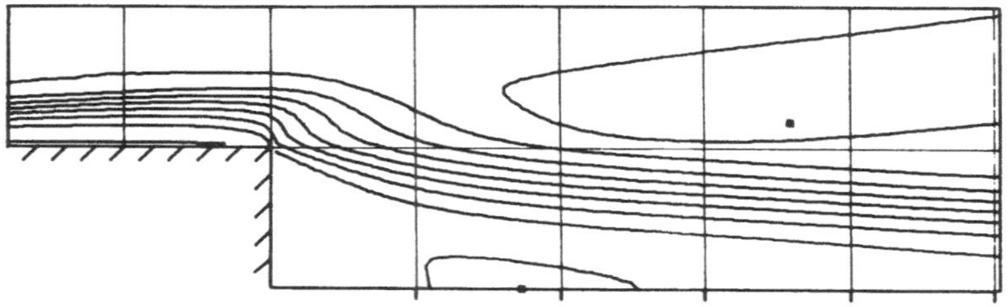

FIG. 2: INTERNAL ENERGY E

RE = 1250., MACH = 3.00, B.L.H. = 0.410, ADIABATIC WALL
MAX(1.73, 0.0) = 2.793, MIN(3.58, 1.18) = 0.644, STEP = 0.252

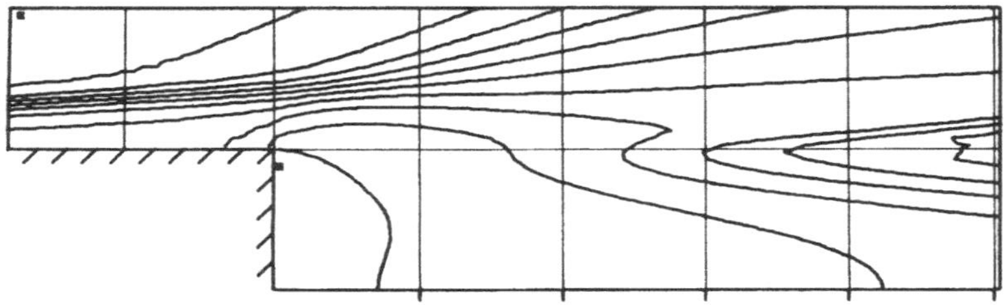

FIG. 3: DENSITY R

RE = 1250., MACH = 3.00, B.L.H. = 0.410, ADIABATIC WALL
MAX(-1.73, 1.95) = 1.004, MIN(0.04, 0.87) = 0.033, STEP = 0.099

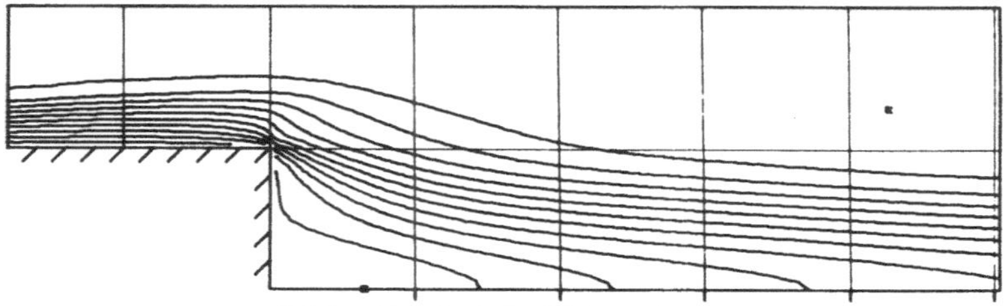

FIG. 4: X-VELOCITY U

RE = 1250., MACH = 3.00, B.L.H. = 0.410, ADIABATIC WALL
MAX(4.27, 1.28) = 1.062, MIN(0.65, 0.0) =-0.066, STEP = 0.099

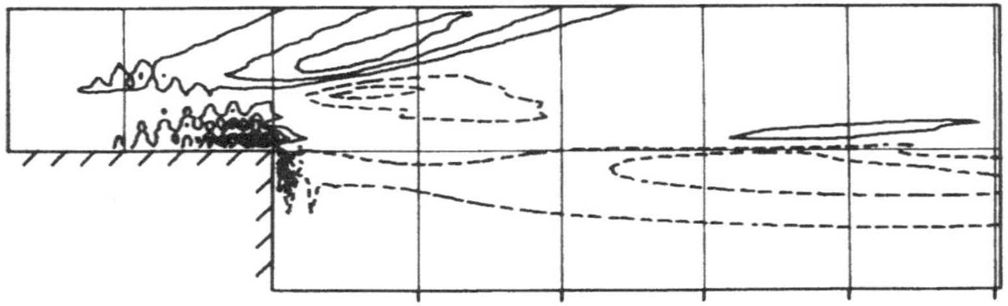

FIG. 5: DENSITY DIFFERENCE

RE = 1250., MACH = 3.00, B.L.H. = 0.410, ADIABATIC WALL
MAX(-0.12, 1.08) = 0.0074,MIN(0.12, 0.82) =-0.0044, STEP = 0.0010

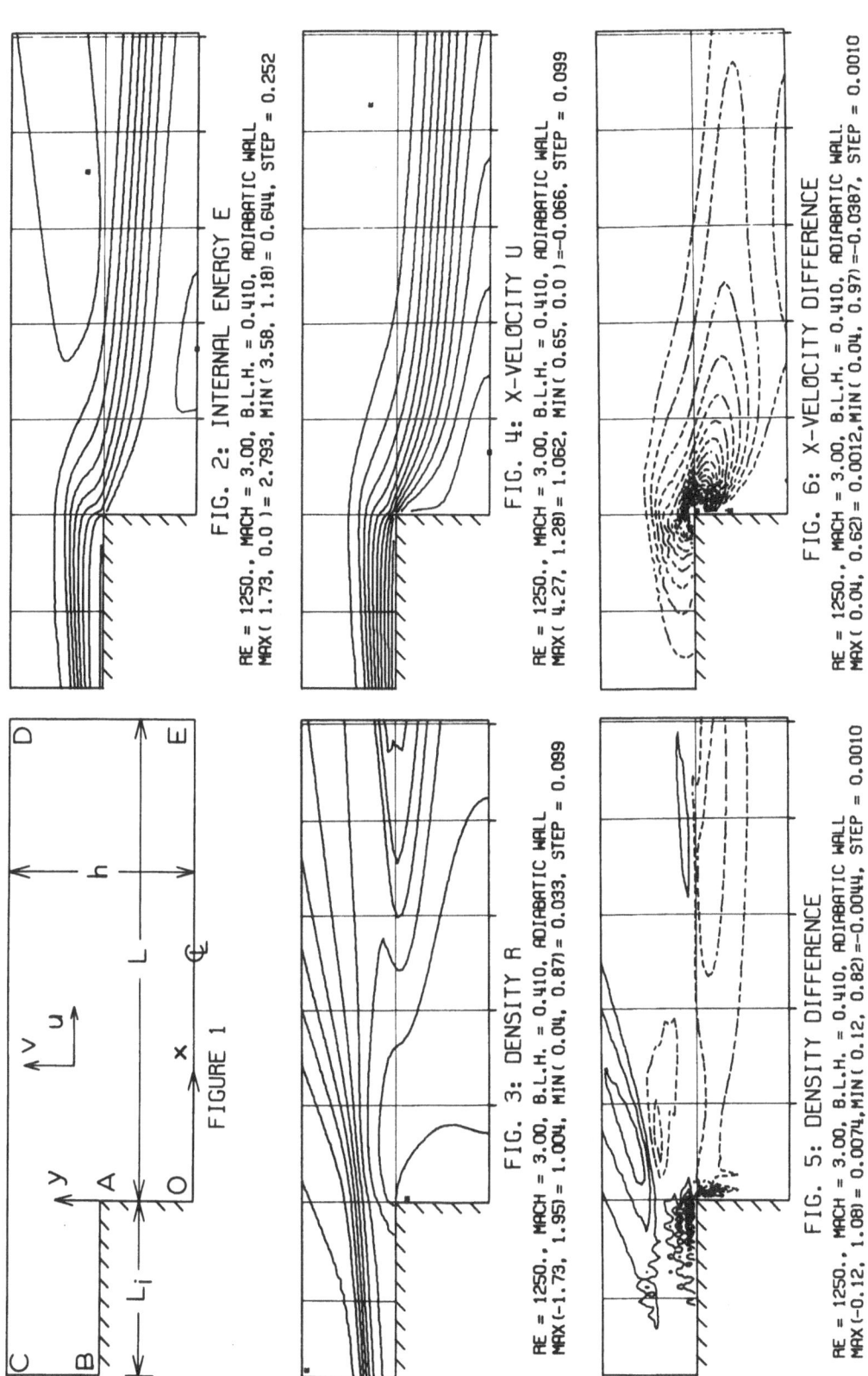

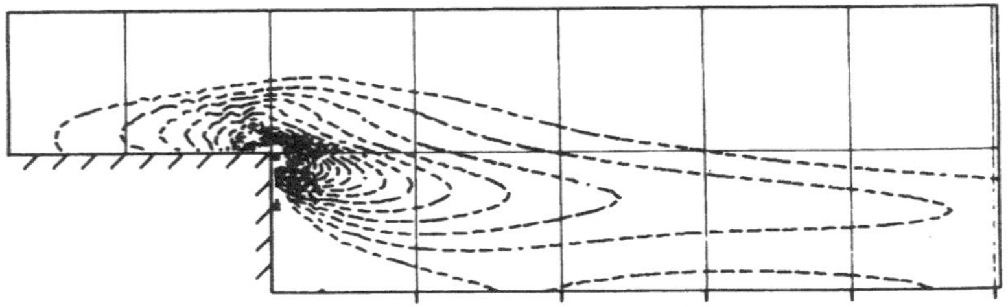

FIG. 6: X-VELOCITY DIFFERENCE

RE = 1250., MACH = 3.00, B.L.H. = 0.410, ADIABATIC WALL
MAX(0.04, 0.62) = 0.0012,MIN(0.04, 0.97) =-0.0387, STEP = 0.0010

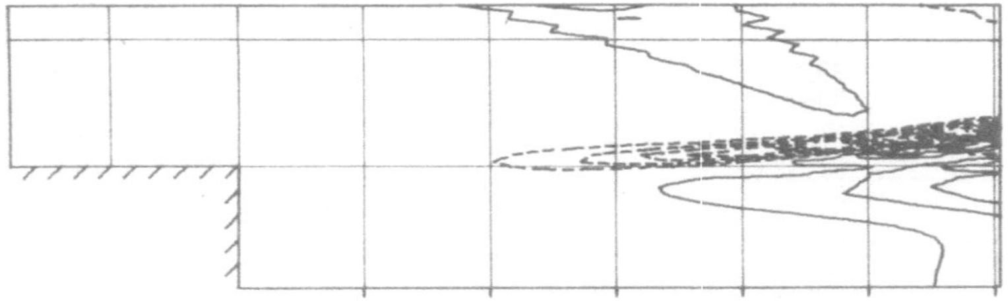

FIG. 7: DENSITY DIFF. (CASE A)

RE = 1250., MACH = 3.00, B.L.H. = 0.410, ADIABATIC WALL
MAX (5.65, 1.13) = 0.0021, MIN (5.96, 1.28) =-0.0037, STEP = 0.0005

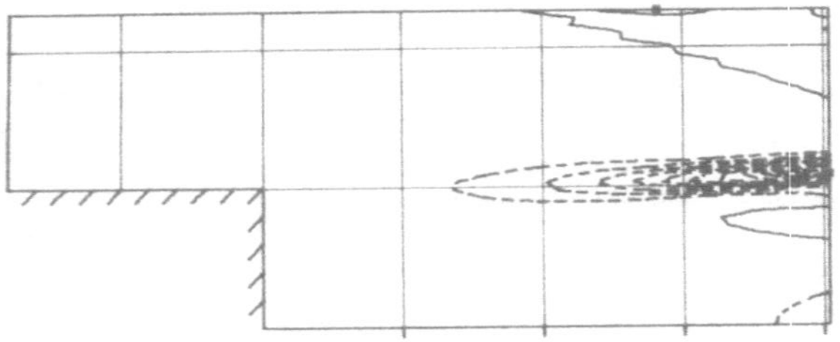

FIG. 8: DENSITY DIFF. (CASE B)

RE = 1250., MACH = 3.00, B.L.H. = 0.410, ADIABATIC WALL
MAX (2.81, 2.26) = 0.0012, MIN (4.04, 1.08) =-0.0060, STEP = 0.0005

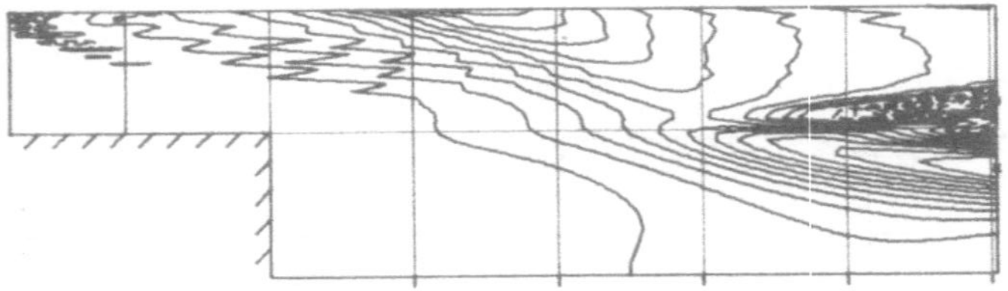

FIG. 9: DENSITY DIFF. (CASE C)

RE = 1250., MACH = 3.00, B.L.H. = 0.410, ADIABATIC WALL
MAX (5.04, 0.72) = 0.0063, MIN (5.04, 1.18) =-0.0057, STEP = 0.0005

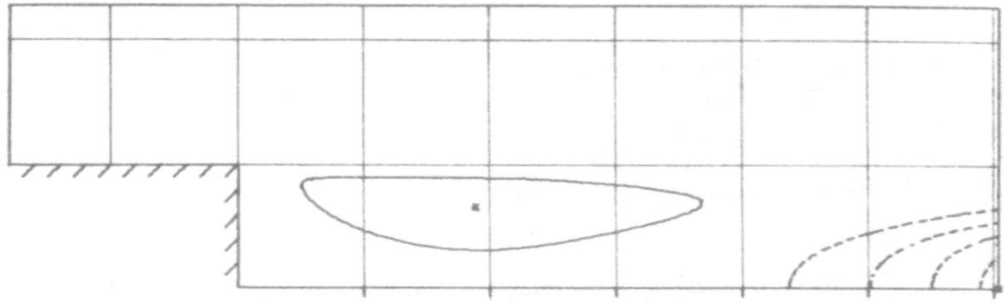

FIG. 10: X-VELOCITY DIFF. (CASE A)

RE = 1250., MACH = 3.00, B.L.H. = 0.410, ADIABATIC WALL
MAX (1.88, 0.67) = 0.0008, MIN (6.04, 0.0) =-0.0022, STEP = 0.0005

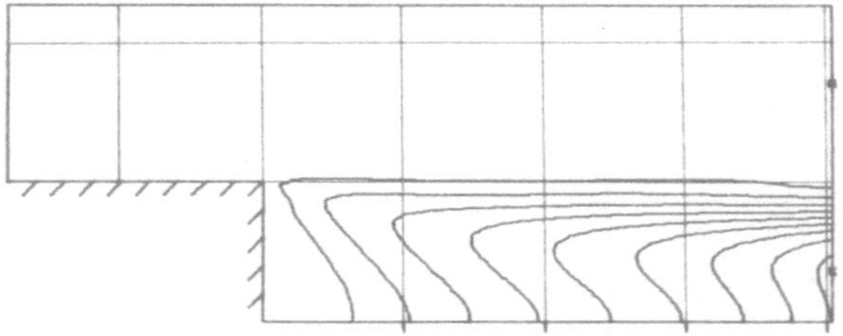

FIG. 11: X-VELOCITY DIFF. (CASE B)

RE = 1250., MACH = 3.00, B.L.H. = 0.410, ADIABATIC WALL
MAX (4.04, 0.36) = 0.0047, MIN (4.04, 1.69) =-0.0002, STEP = 0.0005

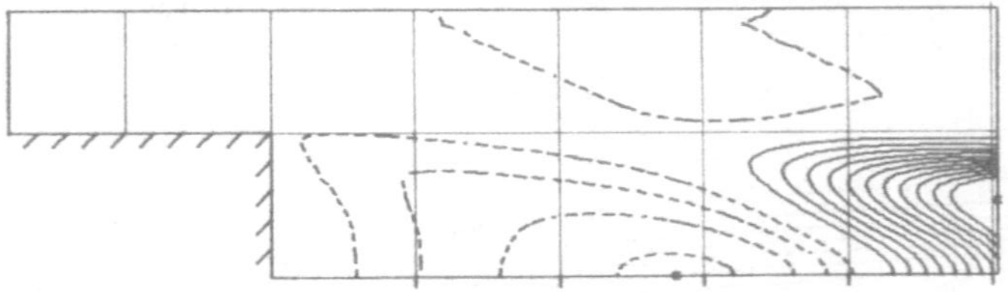

FIG. 12: X-VELOCITY DIFF. (CASE C)

RE = 1250., MACH = 3.00, B.L.H. = 0.410, ADIABATIC WALL
MAX (5.04, 0.51) = 0.0069, MIN (2.81, 0.0) =-0.0021, STEP = 0.0005

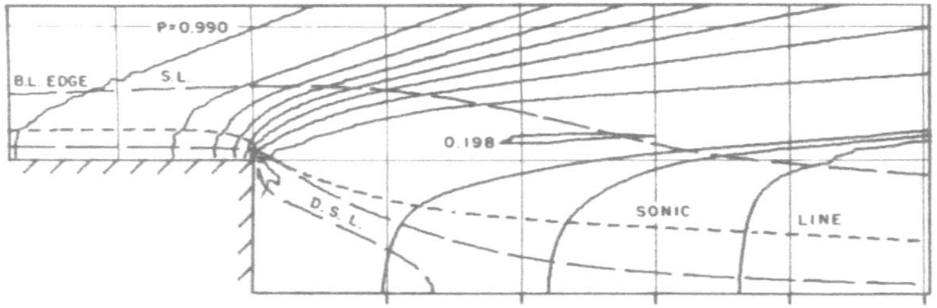

FIGURE 13

RE = 1500., MACH = 3.25, B.L.H. = 0.498, ADIABATIC WALL

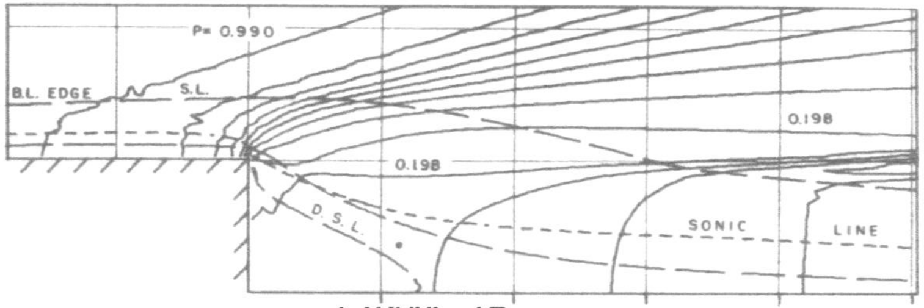

FIGURE 14

RE = 3000., MACH = 3.75, B.L.H. = 0.410, ADIABATIC WALL

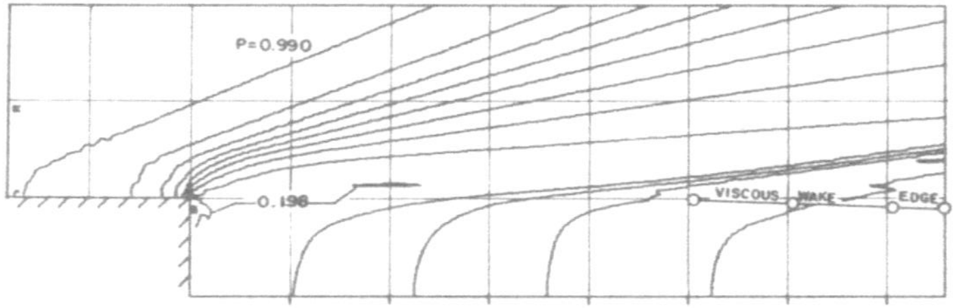

FIGURE 15

RE = 1250., MACH = 3.00, B.L.H. = 0.410, ADIABATIC WALL
MAX(-1.73, 1.90) = 1.004, MIN(0.04, 0.87) = 0.091, STEP = 0.099

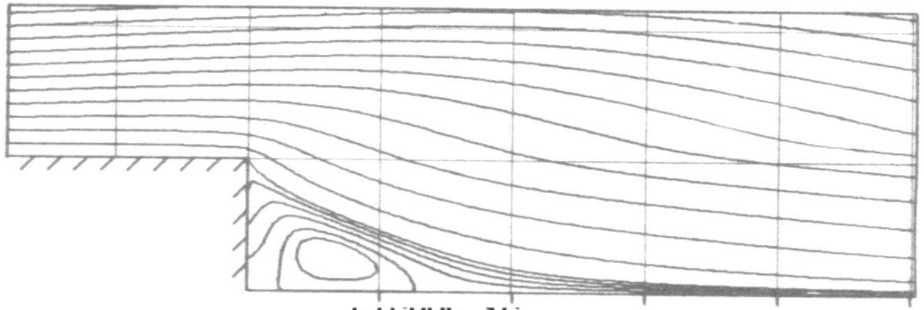

FIGURE 16

RE = 1500., MACH = 3.25, B.L.H. = 0.498, ADIABATIC WALL
BASE INJECTION MASS FLUX = 0.0020

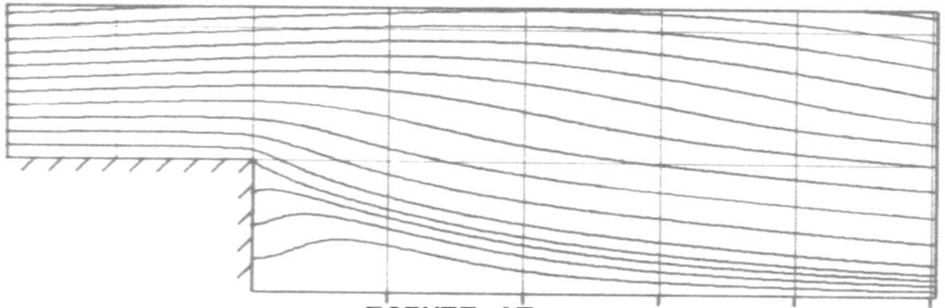

FIGURE 17

RE = 1500., MACH = 3.25, B.L.H. = 0.498, ADIABATIC WALL
BASE INJECTION MASS FLUX = 0.0080

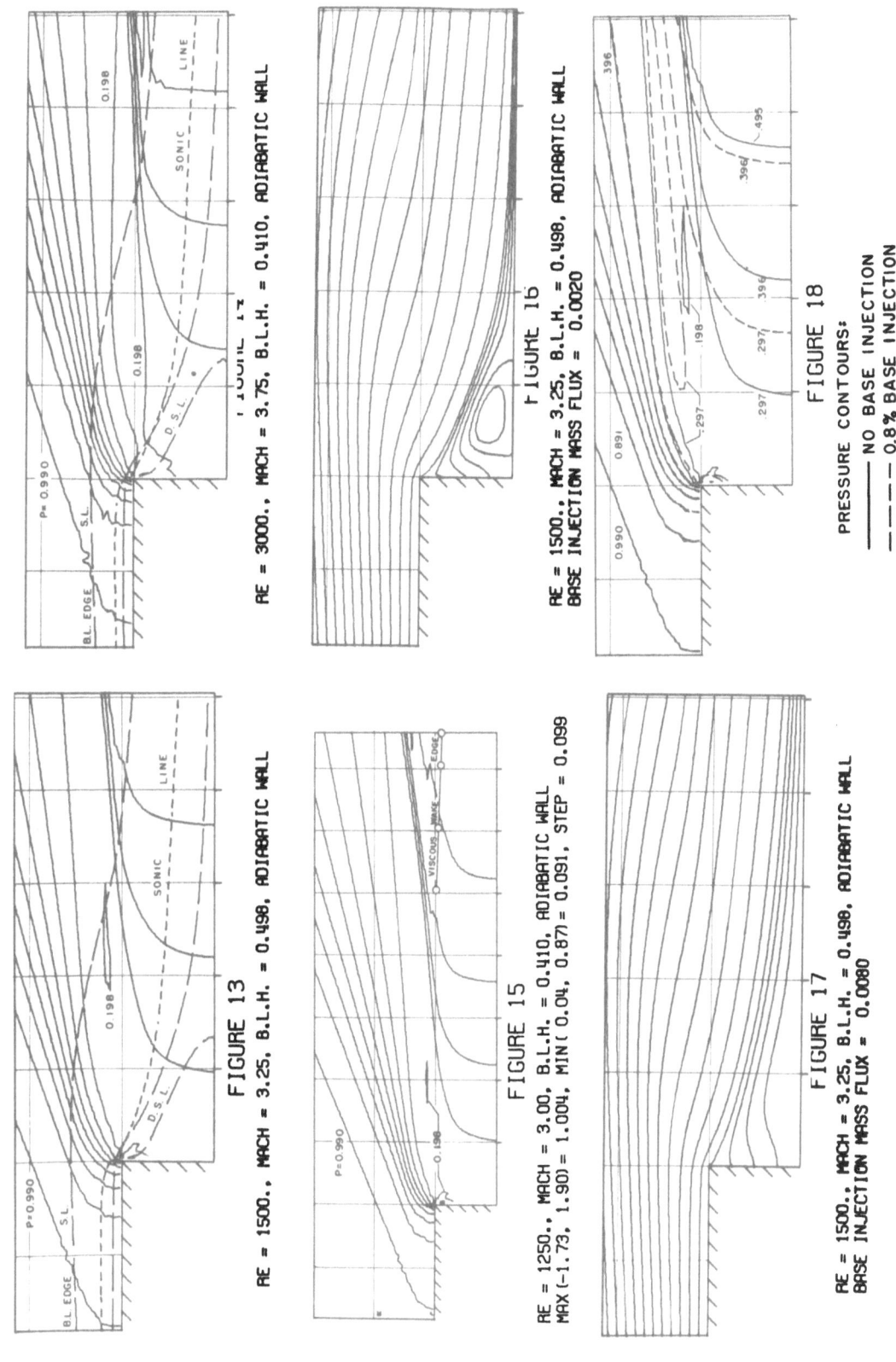

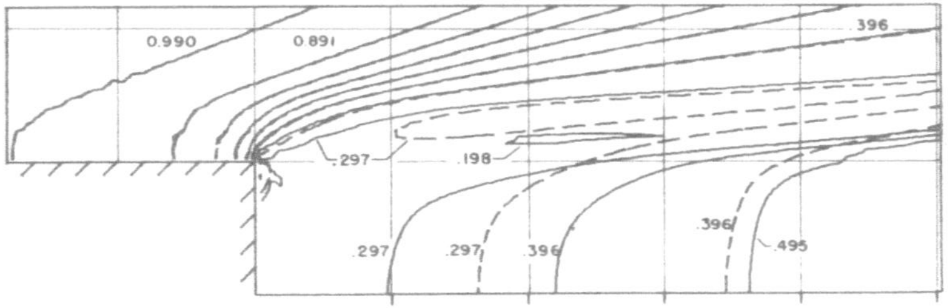

FIGURE 18

PRESSURE CONTOURS:
——— NO BASE INJECTION
– – – – 0.8% BASE INJECTION

BOUNDARY-LAYER SEPARATION AT A FREE STREAMLINE
FINITE DIFFERENCE CALCULATIONS

R.C. Ackerberg
Polytechnic Institute of Brooklyn
Graduate Center, Farmingdale, New York

This paper is concerned with the boundary layer flow just upstream of the sharp trailing edge of a body which has a free streamline attached to the edge. These problems are unusual because the pressure distribution in the potential flow exhibits a singularity which appears in the boundary-layer equations as a forcing term. It is of interest to know how this singularity influences the boundary-layer motion and the skin friction near the edge. The work presented here compares a numerical solution of the boundary-layer equations for the flow over a finite flat plate set perpendicular to a uniform stream with an asymptotic solution obtained by Ackerberg[1]. The potential motion outside the boundary-layer is assumed to be of the Kirchhoff-Rayleigh type with free streamlines attached at the sharp edges.

The asymptotic solution of the boundary-layer equations is valid just upstream of the trailing edge and near the wall; it predicts that (1) the skin friction at the edge is singular to first-order and proportional to the inverse eighth-power of the distance from the edge; (2) the proportionality factor for the first-order term is independent of the boundary-layer flow upstream and depends only on the potential flow solution; (3) the upstream boundary-layer flow influences the skin friction at the second-order by means of an eigenfunction with an unknown multiplicative constant; (4) the streamwise velocity component at separation is non-analytic near the wall and the first term in its series expansion is proportional to the two-thirds power of the distance from the wall.

An explicit finite difference technique was used to solve the boundary-layer equations in Mises variables. This method was introduced by Mitchell and Thomson[4], and has been used by Ackerberg[2] to study the development of the boundary-layer motion in a thin film flowing along a vertical plate. The new features which enter our calculation are as follows: (1) The separation at the free streamline occurs with an extremely favorable pressure gradient which is singular at the edge; thus, the small step size of the explicit technique is useful in suppressing the truncation errors. (2) An asymptotic formula derived by Brown and Stewartson[3] is incorporated into the calculation of the velocity at the outermost point of the boundary layer. (3) A spurious singularity enters the numerical solution as a result of truncating a series expansion, valid near the wall, which is used to calculate the skin friction.

The unknown multiplicative constant in the asymptotic expansion is evaluated by requiring the skin friction to agree with a numerical value close to the free streamline. Using this value, the velocity profiles, computed from the asymptotic expansion, are in excellent agreement with the numerical solution.

This work was supported by the U.S. Army Research Office-Durham under Contract DA-31-124-ARO-D-444.

REFERENCES

1. Ackerberg, R.C. "Boundary-Layer Separation at a Free Streamline. Part I. Two-Dimensional Flow". To be published in J. Fluid Mech. 1970.
2. Ackerberg, R.C. Phys. Fluids, 11, 1270-1291, (1968).
3. Brown, S.N. and Stewartson, K. J. Fluid Mech., 23, 673-687, (1965).
4. Mitchell, A.R. and Thomson, J.Y. Z. angew. Math. Phys., 9, 26-37, (1958).

COMPUTATIONAL MESHES FOR BOUNDARY LAYER PROBLEMS

Glyn O. Roberts

Courant Institute of Mathematical Sciences, New York

INTRODUCTION

Frequently the solution to a system of differential equations in a region has boundary layer structure, with functions varying on a very small length scale, the great majority of this variation being in thin layers against the boundaries. It is then inefficient to represent the function numerically by its values on a computational mesh with uniform spacing. The very fine spacing required in the layers is unnecessarily accurate away from the boundaries.

Computational meshes with non-uniform spacing can take two forms. In the first, here called the interface method, the region is divided into two or more parts, with different uniform mesh spacings, and suitable representations of the equations at the interface mesh points are found. In the second, the mesh spacing varies continuously. The description and illustration of this second method is given below for one-dimensional problems; the application in more dimensions is discussed in the concluding section.

THE INDEPENDENT VARIABLE CHANGE

Suppose that the single independent variable x lies in the range $|x| \leq a$, and that functions of x vary rapidly in layers with thickness of order d on the boundaries. The method consists in a change of variable to y, where $1 \leq y \leq n$, with the monotonic function y(x) chosen to increase rapidly in the thin boundary layers and more slowly in the interior. With a good choice of y(x), the unknown functions do not have boundary layer structure at all when regarded as functions of y, and they can be represented by their values at the uniformly-spaced integer values of y, with derivatives represented by central differences, with second (or fourth) order accuracy. Thus

$$\frac{dw}{dx} = \frac{dy}{dx}\frac{dw}{dy}\,,\quad \frac{d^2w}{dx^2} = \left(\frac{dy}{dx}\right)^2 \frac{d^2w}{dy^2} + \frac{d^2y}{dx^2}\frac{dw}{dy}\,, \tag{1}$$

where $\frac{dw}{dy} = \frac{1}{2}\{w(y+1) - w(y-1)\} + O\left(\frac{d^3w}{dy^3}\right)$,

$$\frac{d^2w}{dy^2} = w(y+1) - 2w(y) + w(y-1) + O\left(\frac{d^4w}{dy^4}\right)\,. \tag{2}$$

Further, for integrals,

$$\int_{-a}^{a} w\, dx = \int_{1}^{n} (w \frac{dx}{dy}) dy \ , \qquad (3)$$

$$\int_{1}^{n} f\, dy = \sum_{2}^{n-1} f(y) + \tfrac{1}{2}\{f(1) + f(n)\} + O\{f'(n) - f'(1)\}.$$

If the scale on which w(y) varies is everywhere greater than unity, the higher derivatives are small, and convergence can be expected as n increases.

CHOICE OF y(x)

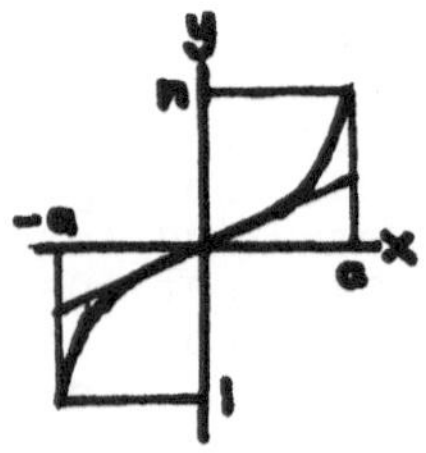

The functional form required is shown in the Figure. The function y(x) increases linearly in the interior, and much more rapidly, on a length scale of order d, in the boundary layers. A natural way of achieving this form is given by

$$y = \frac{n+1}{2} + \frac{n-1}{2}\{(1-\alpha)\frac{x}{a} + \alpha \frac{f(\frac{a-x}{d}) - f(\frac{a+x}{d})}{f(0)-f(\frac{2a}{d})}\} \ , \qquad (4)$$

where $0 < \alpha < 1$ and α is the proportion of the mesh points which particularly represent the boundary layer. Here f(z) is a supplied function of positive z, with f(0) = 1, f'(0) = -1, and with f(z) and all its derivatives tending monotonically to zero at infinity. The four functions

$$f = e^{-z},\ f = \frac{1}{1+z}\ ,\ f = \frac{1}{1 + \log(1+z)}\ ,$$

$$f = \frac{1}{1+\log\{1+\log(1+z)\}}\ , \qquad (5),(6),(7),(8)$$

were actually tested. A FORTRAN subroutine was written to set up a computional mesh according to equation (4), with n, α, a and d as input, giving as output the values of x, $\frac{dy}{dx}$ and $\frac{d^2y}{dx^2}$ at integer values of y. Equations (1) to (3) can then be used to solve the problem.

The choice (4) has two disadvantages. First, there is often more than one boundary layer scale d, and if one scale is well represented using a particular y(x), the other one is not. For one-dimensional problems a simple generalization of equation (4), with a sum over α_i and d_i, is sufficient, but in more than one dimension the boundary layer scale can vary continuously along a boundary and may be much larger near a corner. The second disadvantage is that the coefficients $\frac{dy}{dx}$ and $\frac{d^2y}{dx^2}$ vary sharply, on a small length scale, at the edge of the boundary layer. The natural definition of the length scale L of variation of $\frac{dy}{dx}$ is

$$L^{-1} = \left| \frac{d}{dy}\{\log(\frac{dy}{dx})\} \right| .$$

With $n = 41$, $\alpha = .75$, $a = 1$, and the indicated d values the peak values of L^{-1} for the four function choices are shown in table 1.

TABLE 1

Equation	(5)	(6)	(7)	(8)
$d=10^{-2}$	3.1	.74	.59	.64
$d=10^{-5}$	1600	18	3.5	3.2

The successive function choices (6), (7) and (8) clearly improve the situation considerably, and will be adequate for many problems. But there is another choice for the function y(x) which successfully deals with both these disadvantages.

FINAL CHOICE OF y(x)

Clearly the mesh spacing at $x = 0$ should be greater than at $|x| = a$ by a factor $\frac{a}{d}$, so that $\log(\frac{dy}{dx})$ increases by $\log(\frac{a}{d})$ as $|x|$ increases from zero to a. Write

$$\frac{d}{dy}(\log \frac{dy}{dx}) \propto x$$

so that $\frac{dy}{dx}$ varies smoothly and L^{-1} remains small. Then using the condition above,

$$y = \frac{n+1}{2} + \frac{n-1}{2} \log(\frac{b+x}{b-x})/\log(\frac{b+a}{b-a}), \qquad (9)$$

$$b^2 = a^2/(1 - \frac{d}{a}). \qquad (10)$$

For this final choice, $\frac{dy}{dx}$ is proportional to $\frac{1}{b^2-x^2}$, and thus the mesh spacing is proportional to b^2-x^2, and all boundary layer scales greater than d are represented equally well. Thus both disadvantages of the function choice (4) are removed for the choice (9); which effectively puts more mesh points between the boundary layer and the interior, and slightly fewer in the boundary layer and in the interior.

It may be remarked that with equation (10) replaced for small d by $b = a + \frac{d}{2}$, equation (9) is equivalent to equation (4) with $\alpha = 1$ and $f(z) = \log(z+ \frac{1}{2})$.

TWO APPLICATIONS

Two linear second-order ordinary differential equations are now proposed, analytically soluble, representing Ekman layers and thermal convection layers respectively, with boundary layer thickness of order k. Numerical solutions were obtained using the interface method, with a proportion α of the mesh points in boundary regions of thickness d, and using the five mesh choices in this paper. The second order central differences (2) were used, and the resulting equations solved by Gaussion elimination. The k values 10^{-2} and 10^{-5} were used, with $\frac{d}{k} = \frac{1}{2}$, 1, 2 and 4. The quantity α was always 0.75, since the solutions were constant in the interior. The values of n used were 11, 21, 41, 81, 161 and 321, and the convergence with increasing n for each problem was studied. For each problem and mesh there is a scalar result S which for fixed k, d and α is a function of the number m = n - 1 of grid intervals with the form,

$$S(m) = S + \sum_{j=1}^{\infty} S_j \, m^{-2j} \,, \tag{12}$$

because of the second-order finite differences used. Interpolation with a range of values of m was then used to obtain successively better approximations to S, the analytic solution.

EKMAN LAYER TEST

The horizontal fluid layer $|x| \leq \frac{1}{2}$ is contained between walls rotating with angular velocity $\underline{\Omega} = (\Omega,0,0)$. With respect to axes rotating at Ω, the steady uniform pressure gradient $\nabla p = (0,0,-2\rho\Omega)$ maintains a steady motion $\underline{u} = (0,v(x),w(x))$. With zero viscosity μ, $\underline{u} = (0,1,0)$. This flow adjusts to the no-slip condition at $|x| = \frac{1}{2}$ in Ekman layers. Then

$$k^2 \frac{d^2 f}{dx^2} = 2i(f-1), \tag{13}$$

where $k^2 = \mu/\rho\Omega$ and $f = v + iw$; $f = 0$ for $|x| = \frac{1}{2}$.

The analytic solution is

$$\begin{aligned} f &= 1 - \cosh\{\frac{(1+i)x}{k}\}/\cosh\{\frac{1+i}{2k}\} \\ &= 1 - \exp\{-\frac{(1+i)(\frac{1}{2}-|x|)}{k}\} \quad \text{for small } k. \end{aligned} \tag{14}$$

The length scale in the Ekman boundary layers is clearly of order k, with layers at $|x| = \frac{1}{2}$. The Ekman flux is defined as

$$\begin{aligned} Q &= \int_{-\frac{1}{2}}^{\frac{1}{2}} (f-1)dx \\ &= -k(1-i) \quad \text{for small } k. \end{aligned} \tag{15}$$

The numerical and analytic solutions of equation (13) are compared for the different cases in table 2 below. The numbers shown are negative decimal exponents, so that the figure 4 indicates a result below 10^{-4}. Four results are shown for each case. The upper row gives the modulus of the difference between the analytic and the interpolated numerical f values at $x = -\frac{1}{2} + k$, for n = 41 and 321 respectively. The third result is the relative error $|(Q(m) - Q)/Q|$ of the numerical Ekman flux found using equations (15) and (3), for m = n - 1 = 40. The fourth result is the relative error of the Ekman flux determined from equation (12), using m = 320, 160 and 80.

TABLE 2

Equation k	d/k	(5)	(6)	(7)	(8)	(11)	Interface
10^{-2}	$\frac{1}{2}$	1 3	1 4	1 3	1 3	1 3	-1 0
		1 2	2 7	2 8	2 8	2 10	0 0
	1	2 4	2 3	1 3	1 3	1 3	0 1
		2 5	2 10	2 10	2 9	2 9	0 1
	2	2 4	2 3	1 3	2 3	1 3	1 1
		2 8	2 10	2 10	2 11	1 9	1 2
	4	1 3	1 3	1 3	1 3	1 3	1 3
		1 9	1 9	1 9	2 10	1 8	1 4
10^{-5}	$\frac{1}{2}$	1 2	2 3	1 3	1 3	0 3	-1 -1
		1 2	2 6	2 6	2 6	1 7	-1 -1
	1	2 4	2 4	1 3	1 3	0 2	-1 0
		2 3	2 8	2 8	2 7	1 7	0 0
	2	2 3	2 4	1 3	1 3	0 2	1 1
		2 5	2 8	2 8	2 8	0 6	1 1
	4	1 3	2 3	2 3	1 3	0 2	1 2
		1 8	1 8	1 8	1 8	0 5	1 3

It may be seen that equations (6), (7) and (8) give the best results, then equations (11) and (5), and finally, and a long way behind, the interface method. Equation (5) is inferior because of the sudden changes in $\frac{dy}{dx}$, indicated in table 1. Equation (11) gives poor results because there are fewer mesh points in the boundary layer and because with this equation (13) accurate representation at the interior is not important.

CONVECTION LAYER TEST

The differential equation problem

$$\frac{d\theta}{dx} = k\,\frac{d^2\theta}{dx^2}\,, \qquad \theta(0) = 0,\ \theta(1) = 1\,, \qquad (15)$$

represents heat convection and conduction in a pipe of unit length with unit flow, input temperature zero, temperature unity maintained at the outlet, and small thermal diffusivity k. Heat can only diffuse back a little way, and the analytic solution,

$$\theta = e^{\frac{x-1}{k}} \quad \text{for small k,}$$

has a boundary layer at x = 1 with thickness of order k. A parameter of interest is the heat excess

$$H = \int_0^1 \theta \, dx$$
$$= k \quad \text{for small k.} \qquad (16)$$

Here there is only one boundary layer, and the quantities $\frac{n+1}{2}$ and $\frac{n-1}{2}$ in equations (4) and (9) must be replaced by the quantities 1 and n - 1, so that the range $0 \leq x \leq 1$ becomes the range $1 \leq y \leq n$.

The numerical and analytic solutions of equation (15) are compared in table 3. With $k \approx 0$, equation (15) with central differences used implies that $\theta(y) \approx 0$ for odd values of y (since it is zero for y = 1), while the values at even numbers are arbitrary except for coupling in the boundary layer where there are more grid points. The value of θ at y = 2 is therefore important, indicating the extent of this separation of the two sets of mesh points. Again four negative decimal exponents are shown in the table for each case. The first row gives $\theta(2)$ for n = 41 and 321 respectively; the second gives the relative error $|(H(m)-H)/H|$ for m = 40, and the relative error of the heat excess H determined from equation (12) using m = 320, 160 and 80.

TABLE 3

Equation k	d/k	(5)	(6)	(7)	(8)	(11)	Interface
10^{-2}	$\frac{1}{2}$	2 50	5 48	7 46	7 46	13 45	1 50
		0 2	1 8	1 9	1 9	2 11	-1 7
	1	3 49	7 47	9 46	8 46	15 45	1 50
		1 4	2 10	2 11	2 10	2 11	0 7
	2	5 49	10 46	11 45	10 45	17 44	2 49
		3 8	2 10	2 10	2 10	2 11	0 8
	4	8 48	12 45	13 45	12 45	20 44	3 49
		2 10	2 11	2 10	2 10	2 12	1 9
10^{-5}	$\frac{1}{2}$	0 1	2 10	3 18	3 17	5 50	1 1
		-4 -4	-2 1	-2 3	-1 2	1 9	-6 -5
	1	1 2	4 14	5 22	4 20	5 52	1 0
		-4 -2	-1 5	0 4	0 4	1 8	-5 -5
	2	2 4	8 20	5 22	5 25	6 55	0 0
		-2 0	2 6	0 6	0 5	1 9	-5 -5
	4	5 9	8 27	7 33	6 30	6 58	-1 -1
		1 5	2 9	2 9	1 8	1 9	-4 -4

For this test equation (11) gives the best results in practically every case. Next come equations (6), (7) and (8), then equation (5), and worst again the interface method. The tables do not show up the fact that using the interface method for $k = 10^{-2}$ and $n = 161$ and 321, the heat excess is almost exact. This can be confirmed analytically; the errors in the approximations (2) and (3) turn out to cancel identically if $\theta(y)$ is small at $y = 2$. In any case, the superiority of equation (11), even for a case with the solution constant in the interior, is clearly shown.

CONCLUSION

It has been demonstrated that the method described is versatile and simple to apply, and that it can give very accurate results for one-dimensional problems where a uniformly spaced grid will not give meaningful answers at all.

The method can easily be extended to higher order equations, by differentiating equation (1) successively with respect to x, and evaluating the additional derivatives

$\frac{d^3y}{dx^3}$, $\frac{d^4y}{dx^4}$, etc., at integer values of y.

For partial differential equations applying to rectangular regions in two or more dimensions, the different cartesian coordinates can be transformed separately, using different n, d and α for each, as necessary. As stated above, it is best to use equation (11) rather than (5), (6), (7) or (8), because of difficulties at the corner. For partial differential equations applying in a cylinder or sphere the method can still be applied if the solution has boundary layer structure. Here the radial coordinate should be transformed as the variable x was for the thermal convection test above.

ACKNOWLEDGEMENT

This work was done while the author was working with Professor P.H. Roberts as a Senior Research Associate in the Department of Applied Mathematics at the University of Newcastle upon Tyne, with the research grant GR/3/425 from the Natural Environment Research Council.

ERROR BOUNDS IN BOUNDARY LAYER THEORY

Karl L. E. NICKEL, Universitaet Karlsruhe/GERMANY and Mathematics Research Center, University of Wisconsin, Madison/Wisc. USA *)

ABSTRACT

The two dimensional steady laminar boundary layer of an incompressible medium is treated. The paper consists of two sections. In the first section a priori error bounds for the unknown solution of the Prandtl equation are given. Included are bounds for the wall shear stress, for the displacement thickness and for the momentum loss thickness; also an a priori upstream bound for the separation point is evaluated. In the second section for a given approximate solution error bounds are computed with the aid of the theory of parabolic inequalities. The bounds converge toward zero if the residual error vanishes.

1. The Prandtl Equation

Consider a laminar boundary layer at a wall of an incompressible medium of the kinematic viscosity $\nu > 0$ which is in steady two dimensional motion. The Prandtl differential equation is written as

$$Pu = 0 . \tag{1}$$

Here $u = u(x,y)$ denotes the velocity component parallel to the wall, x is the arc length in the wall direction, y is the distance perpendicular to the wall. P is the Prandtl differential operator, defined as $(u_x := \partial u/\partial x,\ u_y := \partial u/\partial y, \dots)$

$$Pu := uu_x - u_y \int_0^y u_x(x,t)dt - UU' - \nu u_{yy} ,$$

where $U = U(x)$ denotes the velocity component in x-direction of the outer flow. The initial and boundary conditions for a solid immovable wall are:

$$\begin{aligned} x = 0: &\quad u(0,y) = \hat{u}(y) && \text{for } 0 \le y < \infty \\ y = 0: &\quad u(x,0) = 0 && \\ y = \infty: &\quad u(x,\infty) = U(x) && \text{for } 0 \le x < x_1 . \end{aligned} \tag{2}$$

The domain G of the flow is defined as $G := \{x,y \mid 0 < x < x_1,\ 0 < y < \infty\}$ with the closure $\bar{G}$; the "parabolic" boundary of G — where boundary values are prescribed by (2) — is $\Gamma := \{x = 0,\ 0 < y < \infty\} \cup \{y = 0,\ 0 \le x < x_1\} \cup \{"y = \infty",\ 0 \le x < x_1\}$.

2. The Problem

The purpose of this paper is to find admissible functions $\underline{w}(x,y)$ and $\bar{w}(x,y)$ such that

$$\underline{w}(x,y) \le u(x,y) \le \bar{w}(x,y) \quad \text{on } \bar{G} \tag{3}$$

for any solution $u(x,y)$ of (1), (2). In words: The explicitly constructed functions $\underline{w}$ resp. $\bar{w}$ are pointwise lower resp. upper bounds for the (in general) unknown solution $u(x,y)$ of (1), (2). There are two ways to do this:

---

*) Sponsored by the United States Army under Contract No.: DA-31-124-ARO-D-462.

a) A priori bounds: No information about $u(x,y)$ is known. The bounds $\underline{w}$, $\overline{w}$ obtained are therefore in general not very close to $u(x,y)$.

b) A posteriori bounds: An approximation $\tilde{u}(x,y)$ to the solution $u(x,y)$ is known on $\overline{G}$. From the given data of $\tilde{u}$, functions $\underline{\varepsilon}(x,y)$, $[\overline{\varepsilon}(x,y)]$ are derived such that $\underline{w} := \tilde{u} + \underline{\varepsilon}$ $[\overline{w} := \tilde{u} + \overline{\varepsilon}]$ is a lower [upper] bound, i.e. such that on $\overline{G}$

$$\underline{\varepsilon}(x,y) \leq u(x,y) - \tilde{u}(x,y) \leq \overline{\varepsilon}(x,y) . \tag{4}$$

Important: Only certain properties of $\tilde{u}$ are used for the construction of the functions ε, such as bounds for the residual error $P\tilde{u}$. Hence (4) is true for the whole class of all approximate solutions $\tilde{u}$ with the same properties. In what follows all bounds $\underline{\varepsilon}, \overline{\varepsilon}$ have the property $\underline{\varepsilon}, \overline{\varepsilon} \to 0$ if $P\tilde{u} \to 0$ on $\overline{G}$, hence the bounds are "close" when the residual error $P\tilde{u}$ is "small".

Let (3) be fulfilled. Then bounds for the wall shear stress $\tau_0(x)$, for the displacement thickness $\delta^*(x)$ and for the momentum loss thickness $\vartheta(x)$ are given in $0 \leq x < x_1$ by

$$\underline{w}_y(x,0) \leq \tau_0(x) := u_y(x,0) \leq \overline{w}_y(x,0) , \text{ 1)}$$

$$\int_0^\infty (1-\overline{w}/U)dy \leq \delta^*(x) := \int_0^\infty (1-u/U)dy \leq \int_0^\infty (1-\underline{w}/U)dy ,$$

$$\int_0^\infty (\underline{w}/U)(1-\overline{w}/U)dy \leq \vartheta(x) := \int_0^\infty (u/U)(1-u/U)dy \leq \int_0^\infty (\overline{w}/U)(1-\underline{w}/U)\, dy .$$

3. Basic Theorem

Let $U(x) > 0$ in $0 < x < x_1$. The function $w(x,y)$ is called admissible on $\overline{G}$, if $0 < w(x,y)/U(x) < 1$ and $w_y(x,y) \geq 0$ for $0 \leq x < x_1$, $0 < y < \infty$ and if $w(x,y)/U(x)$ is sufficiently smooth on $\overline{G}$ (see [4]). For three admissible functions $\underline{w}(x,y)$, $u(x,y)$, $\overline{w}(x,y)$ the following theorem holds:

Theorem (see [4]):

> If $\underline{w} \leq u \leq \overline{w}$ on Γ and if $P\underline{w} \leq Pu = 0 \leq P\overline{w}$ in G, then the inequality (3) is true on $\overline{G}$.

Most of the following results can be proven with the aid of this theorem. For some of them however transformations of the Prandtl equation are convenient such as the v. Mises transformation [1], a similarity transformation [4] or the Crocco transformation [1]. In what follows bounds $\underline{w}$ and $\overline{w}$ for the unknown solution u are given in the form (3) or (4). These inequalities have the following meaning: it is assumed that they are true for the initial profile $x = 0$, $0 \leq y \leq \infty$. This is a condition for $\hat{u}(y)$ and for certain free constants called $\delta(0)$, $\varepsilon(0)$, ... Then it follows that they are true also on the whole region $\overline{G}$.

4. A Priori Bounds

Let

$$\delta(x) := \sqrt{\delta^2(0)U^2(0) + 2\nu \int_0^x U(t)dt} \;/\; U(x)$$

1) If $\underline{w}(x,0) = u(x,0) = \overline{w}(x,0) = 0$.

be a boundary layer thickness and let

$$\beta(x):=\delta^2(x)\, U'(x)/\nu$$

be the Pohlhausen parameter.

4.1. Similar Solutions as Bounds

Let $f(y,\lambda)$ be a solution of the Falkner-Skan equation $f'''+ff''+\lambda(1-f'^2)=0$ ($f':=\partial f/\partial y, \dots$) under the boundary conditions $f(0,\lambda)=f'(0,\lambda)=0$, $f'(\infty,\lambda)=1$. Then on $\overline{G}$

$$\boxed{U(x)\, f'(y/\delta(x),\beta_1) \le u(x,y) \le U(x)\, f'(y/\delta(x),\beta_2)}$$

as long as $\beta_1 \le \beta(x) \le \beta_2$. Hence it follows on $0 \le x \le x_1$

$$\boxed{U(x) f''(0,\beta_1)/\delta(x) \le \tau_0(x):=u_y(x,0) \le U(x)\, f''(0,\beta_2)/\delta(x) \quad ,}$$

$$\boxed{\delta(x) \lim_{t\to\infty} (t-f(t,\beta_2)) \le \delta^*(x):=\int_0^\infty (1-u/U)dy \le \delta(x) \lim_{t\to\infty}(t-f(t,\beta_1))}$$

(for the values of $\lim_{t\to\infty}(t-f(t,\lambda))$ see [4])

$$\boxed{\delta(x)\int_0^\infty f'(t,\beta_1)(1-f'(t,\beta_2))dt \le \vartheta(x) \le \delta(x)\int_0^\infty f'(t,\beta_2)(1-f'(t,\beta_1))dt \quad .}$$

<u>Separation</u>: The parameter $\beta_0:=-0.198\,838\dots$ corresponds to the "separation profile" with $f''(0,\beta_0)=0$. Hence separation cannot occur as long as $\beta(x)\ge\beta_0$.
<u>Example</u> (oscillating flow): $U(x):=1+2\varepsilon \sin\log x$ gives $\beta(x)\ge\beta_0$ for <u>all</u> $0\le x<\infty$ — i.e. no separation at all — as long as

$$\varepsilon \le [(1-\beta_0)-\sqrt{(1-\beta_0)^2+(\sqrt{2}-\beta_0)}]/2(\sqrt{2}-\beta_0) \ge 0.0442.$$

Note: $\overline{\lim_{x\to 0}}\; U'(x)=+\infty!$

4.2. Explicit A Priori Bounds

Let $g'(y):=1-12s/(s^2+10s+1)$, $s:=\exp(\sqrt{3}\,y/2)$ and

$h'(y):=1-\exp(-y^2/2-\sqrt{1+2\beta_3}\,y)$. Then on $\bar{G}$

$$\boxed{U(x)\, g'(\sqrt{2/3}\, y/\delta(x)) \le u(x,y) \le U(x)\, h'(y/\delta(x)) \quad ,}$$

as long as $-1/12 \le \beta(x) \le \beta_3$. Hence it follows on $0\le x\le x_1$

$$\boxed{0 \le \tau_0(x):=u_y(x,0) \le U(x)\sqrt{1+2\beta_3}/\delta(x) \quad ,}$$

$$\boxed{\int_0^\infty \exp(-t^2/2-\sqrt{1+2\beta_3}\,t)dt \le \delta^*(x)/\delta(x) \le \sqrt{3}\,\log(\sqrt{6}+2)/(\sqrt{6}-2) = 3.970\,608\dots .}$$

5. A Posteriori Bounds

In what follows it is assumed that the approximate solution $\tilde{u}(x,y)$ is an admissible function which fulfills the boundary conditions (2). Let $\tilde{\tau}_0(x) := \tilde{u}_y(x,0)$. The auxiliary functions $\rho(x)$, $\varepsilon(x)$, ... will not be the same from one case to another, but they will not be distinguished since no confusion should arise.

5.1 Let $|P\tilde{u}| \le \rho(x)$, $\varepsilon(x) := \varepsilon(0) + 2\int_0^x \rho(t)dt$. Then on $\bar{G}$

$$-\sqrt{\varepsilon(x)} \le \frac{-\varepsilon(x)}{\tilde{u} + {}_*\sqrt{\tilde{u}^2 - \varepsilon}}\,{}^{2)} \le u - \tilde{u} \le \frac{\varepsilon(x)}{\tilde{u} + \sqrt{\tilde{u}^2 + \varepsilon}} \le \sqrt{\varepsilon(x)},$$

if $\tilde{u}_{yy} \le 0$ ("hence"[3] $U' \ge 0$).

5.2 Let $|P\tilde{u}/\tilde{u}| \le \rho(x)$, $\varepsilon(x) := \varepsilon(0) + \int_0^x \rho(t)dt$. Then on $\bar{G}$

$$|u - \tilde{u}| \le \varepsilon(x), \quad \text{if } \tilde{u}_{yy} \le 0 \text{ and } U'(x) \ge 0 .$$

5.3 Let $\tilde{\eta}(x,y) := \int_0^y \tilde{u}(x,t)dt$, $|P\tilde{u}/\tilde{\eta}| \le \rho(x)$, $\varepsilon(x) := \varepsilon(0) + 2\int_0^x \rho(t)dt$.

Then on $\bar{G}$

$$-\sqrt{\varepsilon(x)\,\tilde{\eta}(x)} \le \frac{-\varepsilon\tilde{\eta}}{\tilde{u} + {}_*\sqrt{\tilde{u}^2 - \varepsilon\tilde{\eta}}}\,{}^{2)} \le u - \tilde{u} \le \frac{\varepsilon\tilde{\eta}}{\tilde{u} + \sqrt{\tilde{u}^2 + \varepsilon\tilde{\eta}}} \le \sqrt{\varepsilon(x)\,\tilde{\eta}(x)},$$

if $\tilde{u}_{yy} \le 0$ ("hence"[3] $U' \ge 0$). For $y \to 0$ it follows that

$$|\tau_0(x) - \tilde{\tau}_0(x)| \le \sqrt{\varepsilon(x)\,\tilde{\tau}_0(x)/2}\,.$$

5.4 Let $\tilde{\eta}(x,y) := \int_0^y \tilde{u}(x,t)dt$, $|P\tilde{u}/\tilde{u}\sqrt{\tilde{\eta}}| \le \rho(x)$, $\varepsilon(x) := \varepsilon(0) + 2\int_0^x \rho(t)dt$.

Then on $\bar{G}$ $\quad |u - \tilde{u}| \le \varepsilon(x)\sqrt{\tilde{\eta}(x,y)}$, if $\tilde{u}_{yy} \le 0$ and $U'(x) \ge 0$.

Hence $\quad |\tau_0 - \tilde{\tau}_0| \le \varepsilon(x)\sqrt{\tilde{\tau}_0(x)/2}\,.$

5.5 Let $|P\tilde{u}/\tilde{u}^2| \le \rho(x)$, $R(x) := \exp\int_0^x \rho(t)dt$, $\underline{\varepsilon} \le (u(0,y) - \tilde{u}(0,y)) \le \bar{\varepsilon}$.

Then on $\bar{G}$

$$(1+\underline{\varepsilon})R^{-1}(x) - 1 \le (u - \tilde{u})/\tilde{u} \le (1+\bar{\varepsilon})\,R(x) - 1\,,$$

2) ${}_*\sqrt{x} = \sqrt{x}$ for $x \ge 0$, ${}_*\sqrt{x} := 0$ for $x < 0$.

3) If $U' \ge 0$ and $u_{yy}(0,y) \le 0$ then $u_{yy} \le 0$ on $\bar{G}$, see [1].

if $\tilde{u}_{yy} \leq 0$ and $U'(x) \geq 0$. For $y \to 0$ it follows that

$$\boxed{(1+\underline{\varepsilon})R^{-1}(x) - 1 \leq (\tau_0 - \tilde{\tau}_0)/\tilde{\tau}_0 \leq (1+\bar{\varepsilon})R(x) - 1\ .}$$

5.6 Let $\tilde{\eta}(x,y) := \int_0^y \tilde{u}(x,t)dt$, $\varphi(y) := y \exp(-\sqrt[3]{y})$,

$$|P\tilde{u}/\varphi(\tilde{u})| \leq \rho(x),\ [UU'/\tilde{u}^3 - \nu\,\varphi''(\tilde{\eta})/\varphi(\tilde{\eta})]\ U(x) \geq \alpha(x)\ ,$$

$$A(x) := \exp \int_0^x \alpha(t)dt,\ \varepsilon(x) := A^{-1}(x)\,[\varepsilon(0) + \frac{1}{2}\int_0^x \rho(t)A(t)dt]\ .$$

Then on $\bar{G}$ $\boxed{|u - \tilde{u}| \leq \sqrt{\varepsilon(x)\,\varphi(\tilde{\eta}(x,y))}\ .}$ For $y \to 0$ it follows that

$$\boxed{|\tau_0 - \tilde{\tau}| \leq \sqrt{\varepsilon(x)\,\tilde{\tau}_0(x)/2}\ .}$$

5.7 Let y_0 be such that $\tilde{u}_{yy}(x,y) \leq 0$ and $\tilde{u}(x,y) \geq U(x)/2$ for $y \geq y_0$, let $|P\tilde{u}/(U^2 - \tilde{u}^2)| \leq \rho(x)$, $R(x) := \exp \int_0^x \rho(t)dt$ and let be for $x = 0$: $\underline{\varepsilon}(U^2 - \tilde{u}^2) \leq u^2 - \tilde{u}^2 \leq \bar{\varepsilon}(U^2 - \tilde{u}^2)$. Then on $\bar{G}$, if $y \geq y_0$:

$$\boxed{2\underline{\varepsilon} - 2(1-\underline{\varepsilon})(R^2(x) - 1) \leq (u - \tilde{u})/(U - \tilde{u}) = 2\bar{\varepsilon} + 2(1-\bar{\varepsilon})(1 - R^{-2}(x))\ .}$$

5.8 Let $\tilde{\psi}(x,y) := \tilde{u}^2 - \tilde{u}_y \int_0^y \tilde{u}(x,t)dt$ and let $\tilde{\psi}(x,y) > 0$ in G. Assume further that $\underline{\rho}(x) \leq P\tilde{u}/\tilde{\psi} \leq \bar{\rho}(x)$, $\tilde{u}\tilde{u}_x - \frac{3}{2}\tilde{u}_y \int_0^y u_x(x,t)dt \geq -\alpha(x)$ and let $A(x) :=$ $\exp 2\int_0^x \alpha(t)dt$, $\bar{\varepsilon}(x) := A(x)\,[\bar{\varepsilon}(0) + 2\int_0^x \bar{\rho}(t)A^{-1}(t)dt]$, $\underline{\varepsilon}(x) := A(x)[\underline{\varepsilon}(0) - 2\int_0^x \underline{\rho}(t)A^{-1}(t)dt]$. Then on $\bar{G}$

$$\boxed{\sqrt{1+\underline{\varepsilon}(x)}\,\tilde{u}(x, y/\sqrt[4]{1+\underline{\varepsilon}(x)}) \leq u(x,y) \leq \sqrt{1+\bar{\varepsilon}(x)}\,\tilde{u}(x, y/\sqrt[4]{1+\bar{\varepsilon}(x)})\ ,}$$

hence for $y \to 0$ it follows that on $0 \leq x \leq x_1$

$$\boxed{\sqrt[4]{1+\underline{\varepsilon}(x)} - 1 \leq (\tau_0(x) - \tilde{\tau}_0(x))/\tilde{\tau}_0(x) \leq \sqrt[4]{1+\bar{\varepsilon}(x)} - 1\ .}$$

6. Outlook

Similar a priori and a posteriori bounds can — with the aid of the Crocco transformation (see [1]) — be constructed also for the shear stress $u_y(x,y)$. Since the theorem of Section 3 is valid also for a moving wall and/or suction and blowing through the wall the above results can be extended to these cases. Also the case of unsteady flow can be treated, since an analog of the basic theorem is valid in the nonstationary case too (see [2]). The same is true for two- or more-component flows (see [3]). However the extension to the case of three-dimensional boundary layers or to the case of a compressible medium offers more difficulties, since at the present time nothing comparable to the above basic theorem is known.

The proofs of the results given above will be published elesewhere together with the mathematical backgound.

7. References

[1] Nickel, K.: Einige Eigenschaften von Loesungen der Prandtlschen Grenzschichtdifferentialgleichungen. Archive Rat. Mech. Analysis 2, 1-31 (1958).

[2] Nickel, K.: Ein Eindeutigkeitssatz fuer instationaere Grenzschichten I and II. Math. Zeitschr. 74, 209-220 (1960) and 83, 1-7 (1964).

[3] Nickel, K.: Parabolic equations with application to boundary layer theory. Partial Differential Equations and Continuum Mechanics, Madison, Wisconsin 1961, 319-330.

[4] Nickel, K.: Eine einfache Abschaetzung fuer Grenzschichten. Ing.-Archiv 31, 85-100 (1962).

Session IV

Flow Field Calculations

O. M. Belotserkovskii, Chairmann

TIME-DEPENDENT CALCULATION METHOD FOR TRANSONIC NOZZLE FLOWS

Pierre LAVAL

Office National d'Etudes et de Recherches Aérospatiales (ONERA) - 92 Châtillon, France

INTRODUCTION

The calculation of mixed flow in a converging-diverging axisymmetric nozzle is of great practical interest in the case when the ratio of throat radius of curvature to the nozzle radius at the throat R/r is small. This problem can be solved either by analytical methods (HOPKINS and HILL 1966, KLIEGEL and LEVINE 1969) or by numerical time-dependent methods (SAUNDERS 1966, MIGDAL, KLEIN and MORETTI 1969, IVANOV 1969).

In the present work, a time dependent method is used to calculate mixed two-dimensional or axisymmetric flows in a nozzle with or without central body. The equations of motion are written in terms of transformed spatial variables and solved by means of an explicit second-order scheme. The results will show that the method proposed here allows to approach the limiting case of the conical converging nozzle.

A different time-dependent method was also set up in which the independent variables are the images of the streamlines and of their orthogonal trajectories (P. CARRIERE 1968, P. LAVAL 1968). However, for the problem considered here, this method proved to be much more complicated and lead to much longer computing times than the first one, and it will not be presented here. A number of details which cannot be given here will be available in a more extensive paper to be published soon (P. LAVAL 1970).

I - GENERAL EQUATIONS

In the general case of axisymmetric or of two-dimensional nozzle with central body, let x and y represent the non-dimensional axial and radial or transversal coordinates, and let $y = y_s(x)$ and $y = y_i(x)$ be the equations of the upper and lower walls. The reference length is such that $y_s(x) - y_i(x) = 1$ at the throat. The transformed coordinates are defined by the equations :

$$(1) \qquad y \longrightarrow Y = \frac{y - y_i(x)}{y_s(x) - y_i(x)} \, ,$$

$$(2) \qquad x \longrightarrow X = f(x) = \frac{f_1(x) + e^{k(\bar{x}-1)} f_2(x)}{1 + e^{k(\bar{x}-1)}} \, ,$$

where $\bar{x} = \frac{x}{x_0}$, x_0 is the throat abscissa, $f_1(x) = x_0\left(\frac{1}{2}\bar{x} + k_1 \bar{x}^4\right)$, $f_2(x) = x_0\left(2\bar{x} + k_2 \bar{x}^2\right)$.

The following conditions are imposed at the throat : $f(x_0) = x_0$, $f''(x_0) = 0$. The parameters k , k_1 and k_2 can thus be expressed as functions of the only parameter $p_0 = f'(x_0)$.

The transformation from y to Y is made in order to obtain a rectangular domain (Fig. 1). The transformed axial variable X is used in the case of classical nozzles with very small throat radius of curvature ($0.1 \leqslant \frac{R}{r} < 0.5$). It allows one to stretch the throat region where large gradients occur (Fig. 2). When $\frac{R}{r} \geqslant 0.5$, the stretching is not used $(X = x)$.

The coordinates transformation being regular, it is possible (ANDERSON, LAPIDUS) to write the time-dependent equations of motion in conservative form in the new variables X , Y , t :

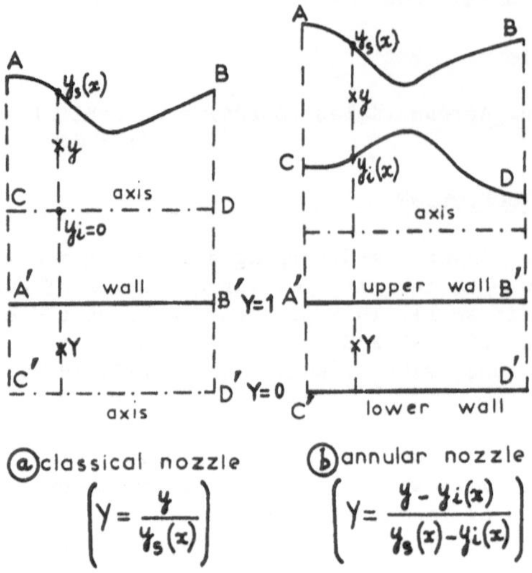

FIGURE 1

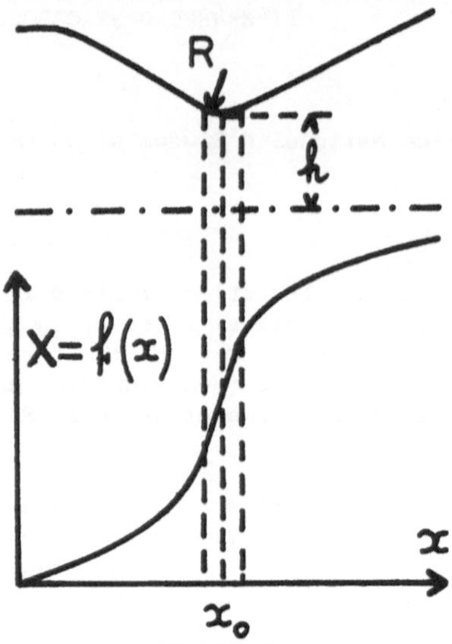

FIGURE 2

Stretching for $0{,}1 \leq \frac{R}{h} < 0{,}5$

$$\frac{\partial U}{\partial t} + \frac{\partial F}{\partial X} + \frac{\partial G}{\partial Y} = H . \tag{3}$$

The components of the vector U are :

$$U_1 = \rho y^{\varepsilon} r \ , \ U_2 = \rho V \cos\theta \, y^{\varepsilon} r \ , \ U_3 = \rho V \sin\theta \, y^{\varepsilon} r \ , \ U_4 = E y^{\varepsilon} r \ , \tag{4}$$

where $r(x) = y_s(x) - y_i(x)$, θ is the angle of the velocity vector with the x-axis, $E = p/(\gamma-1) + 1/2\, \rho V^2$ is total energy for a perfect gas. The parameter ε is equal to 0 for plane flow and to 1 for axisymmetric flow.

The components of the vectors F , G and H are given in terms of the components of U and of geometrical variables such that $y_i' = tg\,\theta_i(x)$, $r'(x) = tg\,\theta_s(x) - tg\,\theta_i(x)$, $f'(X)$ and $f''(X)$:

$$F = F(U, f') \ , \ G = G(U, Y, r, y_i', r') \ , \ H = H(U, f''/f', y) . \tag{5}$$

II - FINITE-DIFFERENCE SCHEME

A rectangular net of points is defined in the X , Y plane with the coordinates lines X_j $(j = 0, 2, 4, \dots, j_{max})$ and Y_ℓ $(\ell = 0, 2, 4, \dots, \ell_{max})$. Let $U_{j,\ell}^{n}$ be the value of U at the point $X_j = j\Delta X$, $Y_\ell = \ell \Delta Y$ and at time $t^n = n\Delta t$, with $n = n_0, n_0+2, \dots, n_{max}$.

We use the following explicit, second order scheme :

$$U_{j,\ell}^{n+1} = \frac{1}{4}\left(U_{j+1,\ell}^{n} + U_{j-1,\ell}^{n} + U_{j,\ell+1}^{n} + U_{j,\ell-1}^{n}\right) - \Delta t\left(\frac{F_{j+1,\ell}^{n} - F_{j-1,\ell}^{n}}{2\Delta X} + \frac{G_{j,\ell+1}^{n} - G_{j,\ell-1}^{n}}{2\Delta Y}\right) + \frac{1}{4}\Delta t\left(H_{j+1,\ell}^{n} + H_{j-1,\ell}^{n} + H_{j,\ell+1}^{n} + H_{j,\ell-1}^{n}\right) . \tag{6}$$

$$U_{j,\ell}^{n+2} = U_{j,\ell}^{n} - \Delta t\left(\frac{F_{j+1,\ell}^{n+1} - F_{j-1,\ell}^{n+1}}{\Delta X} + \frac{G_{j,\ell+1}^{n+1} - G_{j,\ell-1}^{n+1}}{\Delta Y}\right) + 2\Delta t\, H_{j,\ell}^{n+1} + Q_{j,\ell}^{n} . \tag{7}$$

The vectors $F_{j,l}^{n+1}$, $G_{j,l}^{n+1}$ and $H_{j,l}^{n+1}$ are determined through equations (5).

A pseudo-viscosity term is added to the right-hand side of (7) :

(8) $$Q = \chi \frac{\Delta t}{2\Delta X}\left(\delta_X\left(q_X \delta_X U\right) + \delta_Y\left(q_Y \delta_Y U\right)\right),$$

which is of the type defined by BURSTEIN.

One knows (BURSTEIN, RICHTMYER) that it is necessary to use a pseudo-viscosity term in a second-order scheme in order to avoid non-linear instabilities. The matrices q_X and q_Y are given by the expressions (LAPIDUS) :

(9) $$\left(q_X\right)_{j,l} = \left| u_{j+1,l} - u_{j-1,l}\right| I \quad , \quad \left(q_Y\right)_{j,l} = \left| v_{j,l+1} - v_{j,l-1}\right| I ,$$

where u and v are the velocity components and I is the unit matrix.

The study of the stability was made for plane flows and the results are given in (P. LAVAL, 1970).

For classical nozzles, preliminary numerical calculations have lead to define the parameter χ which appears in (8) as a linear function of Y :

(10) $$\chi = \chi_{max} Y ,$$

where χ_{max} , the value of χ at the wall, is of order unity.

III - INITIAL AND BOUNDARY CONDITIONS

For a given nozzle, the value of θ is known on the upper $(Y=1)$ and lower $(Y=0)$ boundaries (Fig. 3).

The initial values of p , ρ and V are obtained from the one-dimensional approximation, the angle θ being calculated by a linear interpolation : $\theta = Y\theta_s(X) + (1-Y)\theta_i(X)$.

For the flow values at the upper and lower boundaries, a parabolic extrapolation is used, taking into account the relation $U_3 = U_2 \, tg\,\theta(X)$. When the lower boundary is the nozzle axis, the flow symmetry is taken care of by a parabolic interpolation.

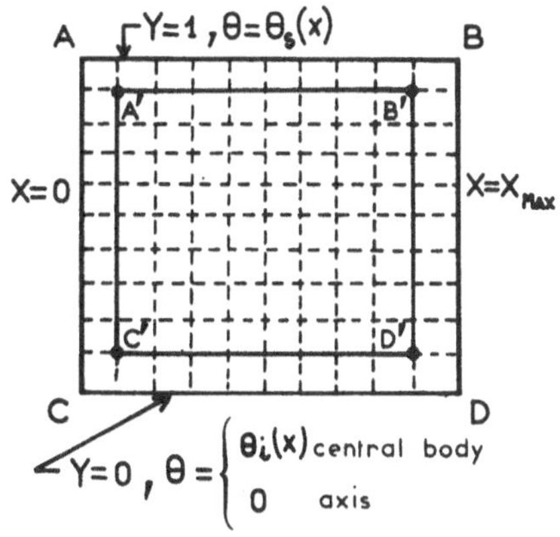

FIGURE 3

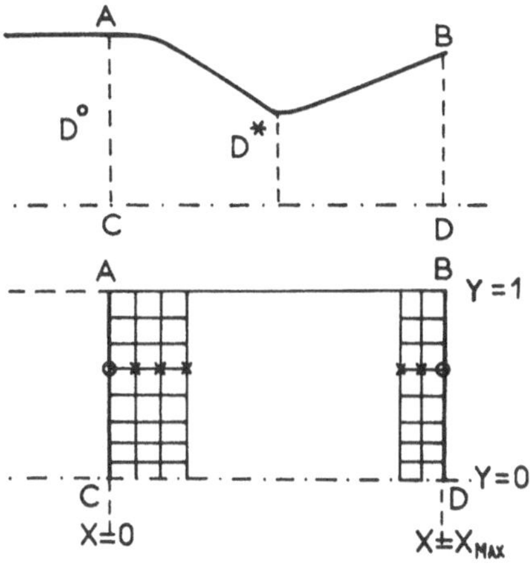

FIGURE 4

Time-dependent conditions are imposed at the entrance section AC by calculating the components of U by a parabolic extrapolation. The component U_2 is then corrected by writing that the mass flow rate D° through the entrance section at time $t+\Delta t$ is equal to the mass flow rate through the throat section at time t (Fig. 4).

At the exit section BD, values are extrapolated linearly (Fig.4).

IV - RESULTS

Calculations have been carried out on a 360/50 IBM computer, for $\gamma = 1.4$.

IV.1 - Axisymmetric nozzle with small throat radius of curvature ($R/h = 0.625$)

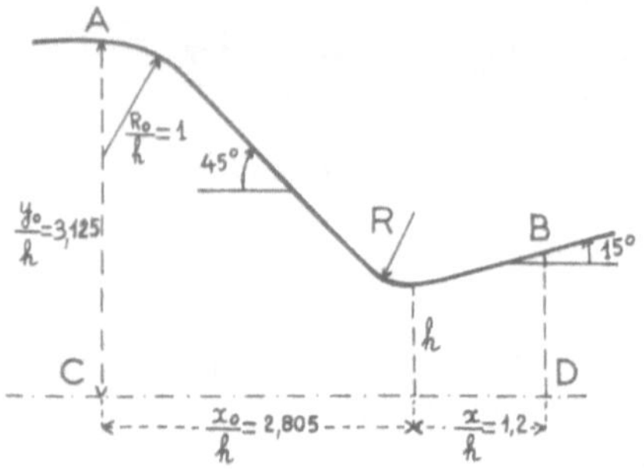

FIGURE 5 : Axisymmetric nozzle ($R/h = 0,625$)

This nozzle which is made of a 45° convergent and a 15° divergent (Fig.5) has been studied by various authors (CUFFEL, KLIEGEL, PROZAN, MIGDAL).

The results have been obtained with a net of 61 x 21 points (x,y) and with $X_{max} = 1.5$, the stretching of the axial coordinate being not used.

Steady state is considered to be obtained at the throat when the mass flow rate change at each iteration becomes smaller than $\Delta t^2 (\sim 2.10^{-5})$. This occurs after 398 iterations (Table I). The corresponding values of the Mach number are 1.3935 at the wall (Fig. 6) and 0.8 on the axis (Fig. 7). These values can be compared with CUFFEL's experimental values : $M_{wall} = 1.4$ and $M_{axis} = 0.8$.

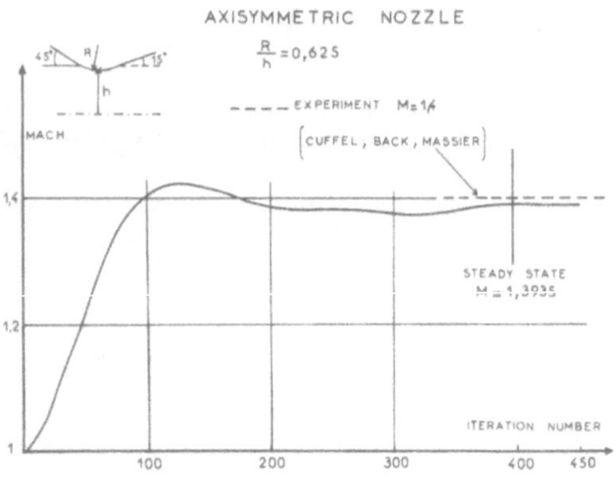

FIGURE 6 : Throat section Mach number on wall vs. time.

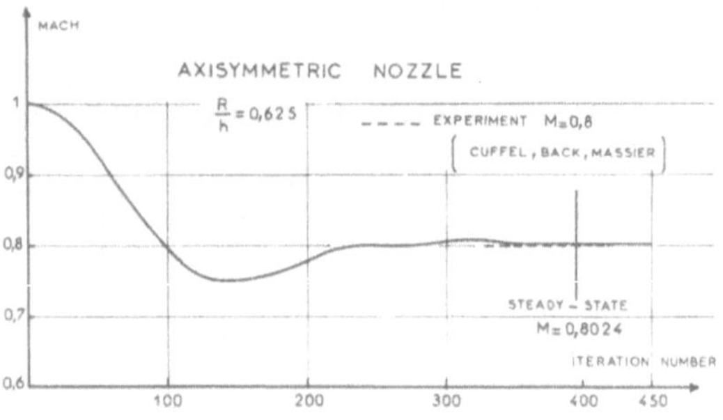

FIGURE 7 : Throat section Mach number on axis vs time.

TABLE I - Discharge coefficient C_D

N	C_D
392	0.9849025
393	0.9850113
394	0.9850981
395	0.9851634
396	0.9852066
397	0.9852279
398	0.9852274

KLIEGEL & LEVINE	0.982
CUFFEL, BACK & MASSIER	0.985
PROZAN (SAUNDERS method)	0.990
Present method	0.985227

The values of the discharge coefficient C_D calculated at the throat section for the last seven iterations are given in Table I, where N is the iteration number. The asymptotic value is compared with values obtained by other methods.

IV.2 - Axisymmetric nozzle with very small throat radius of curvature (R/ℓ = 0.1)

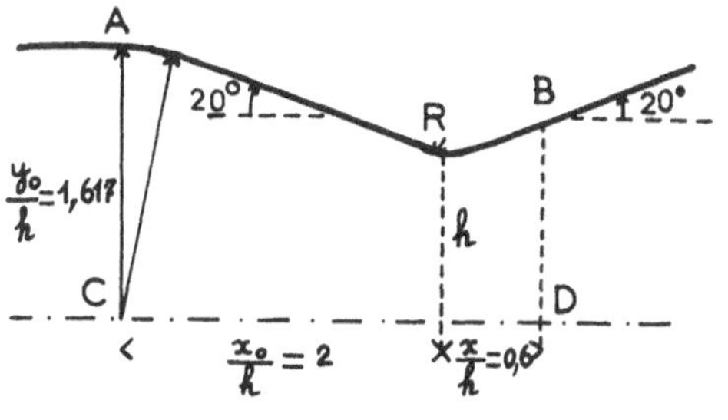

FIGURE 8 : Axisymmetric nozzle ($R/\ell = 0,1$)

Convergent and divergent angles are both equal to 20° (Fig. 8).

Stretching of the axial coordinate is used with ρ_o = 3.4. A net of 68 x 23 points (x,y) is used, with X_{max} = 1.15. Table II gives the discharge coefficient at the throat. One sees that after 278 iterations C_D changes by less than $\Delta t^2 (\sim 10^{-5})$ and one can consider that the asymptotic value of C_D is 0.968988.

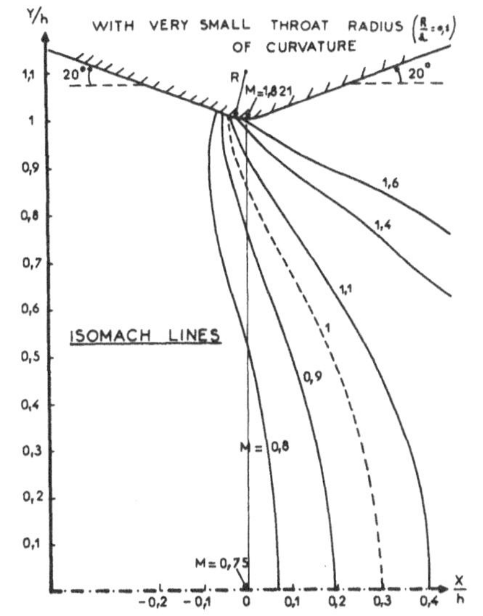

FIGURE 9 : Axisymmetric nozzle with very small throat radius of curvature ($\frac{R}{\ell}$ = 0.1)

TABLE II - Discharge coefficient

N	C_D
275	0.9690063
276	0.9689981
277	0.9689925
278	0.9689889
279	0.9689884
280	0.9689896
300	0.969543
325	0.970453
350	0.969428
375	0.969079

From Figure 9 it can be seen that large variations of Mach number occur, particularly at the wall, on the rectilinear part of which the sonic point is located, very close to the circular part, whereas the Mach number is equal to 1.821 at the throat section.

IV.3 - Axisymmetric nozzle with a 20° convergent and varying R/ℓ ratios

To study the approach to the limiting case of a conical converging nozzle, calculations have been made for three values of R/ℓ : 0.25 - 0.5 and 0.8, besides the value R/ℓ = 0.1 (IV.2). The results will be compared with experiments carried out at O.N.E.R.A. (SOLIGNAC) for a 20° conical converging nozzle.

Excellent agreement is found between theory and experiment concerning the location of the sonic line (Fig. 10). Extrapolating the numerical results relative to the 4 values of R/ℓ, one obtains C_D = 0.968 for R/ℓ = 0 (Fig. 11). The difference between this value and the experimental result C_D = 0.971 can be explained partly by the difficulty of the measurements in the neighbourhood of the angular point.

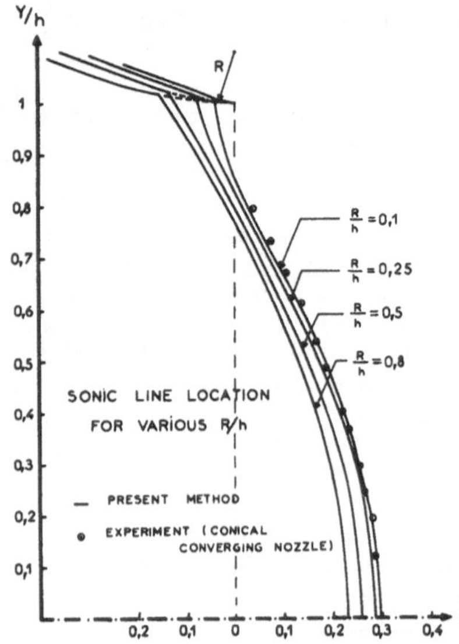

FIGURE 10 : Axisymmetric nozzle 20° convergent.

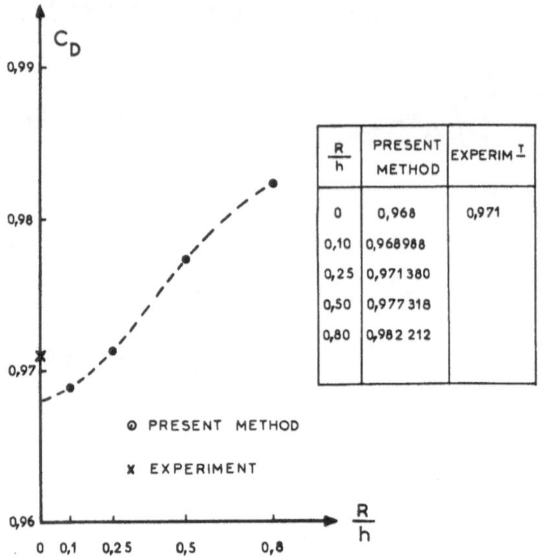

R/h	PRESENT METHOD	EXPERIM.T
0	0,968	0,971
0,10	0,968988	
0,25	0,971380	
0,50	0,977318	
0,80	0,982212	

FIGURE 11 : Axisymmetric nozzle with 20° convergent. Discharge coefficient vs. R/ℓ.

CONCLUSION

The calculations which have been presented, and particularly the excellent agreement which has been found between experimental and numerical results, show that the proposed method is well suited for calculating transonic nozzle flows. This method presents the following particular features :

- conditions in the entrance section are time-dependent;
- the parameter χ which appears in the pseudo-viscosity term is defined as a function of the transformed coordinate y (equ. 10);
- a stretching of the axial coordinate (equ. 2) allows one to treat the case of very small throat radius of curvature and to approach the case of a conical converging nozzle.
- the method is applicable also to annular nozzles (P. LAVAL, 1970).

REFERENCES

ANDERSON, J.L., PREISER, S., and RUBIN, E.L., J. of computational phys., 2, No 3, pp. 279-287 (1968).
BURSTEIN, S.Z., J. of computational physics, 1, No 2, pp. 198-222 (1966).
CARRIERE, P., C.R.Ac.Sc., t. 266 A, pp. 1015-1018 (1968).
CUFFEL, R.F., BACK, L.H., and MASSIER, P.F., AIAA J., 7, No 7, pp. 1364-1366 (1969).
HOPKINS, D.F., and HILL, D.E., AIAA J., 4, No 8, pp. 1337-1343 (1966).
IVANOV, M., and KRAIKO, A.H., Mekh. jidkosty i gaza (in Russian), No 5, pp.77-83 (1969).
KLIEGEL, J.R., and LEVINE, J.N., AIAA J., 7, No 7, pp. 1375-1378 (1969).
LAPIDUS, A., J. of computational phys., 2, No 2, pp. 154-177 (1967).
LAVAL, P., C.R.Ac.Sc., t. 267 A, pp. 754-756 (1968).
LAVAL, P., N.T.ONERA, to be published (1970).
MIGDAL, D., KLEIN, K., and MORETTI, G., AIAA J., 7, No 2, pp. 372-373 (1969).
PROZAN, R.J.*
RICHTMYER, R.D., and MORTON, K.W., Tracts in Mathematics, second edition, No 4, Interscience publishers (1967).
SAUNDERS, L.M., BSVD - P 66 - TN - 001 (1966).
SOLIGNAC, J.L., La Recherche Aérospatiale, to be published (1970)

* Private communication to CUFFEL, R.F., above reference.

GENERALIZED RELAXATION METHODS APPLIED TO PROBLEMS IN TRANSONIC FLOW

Joseph L. Steger and Harvard Lomax

Ames Research Center, NASA
Moffett Field, Calif., 94035

INTRODUCTION

The purpose of this paper is to discuss generalized relaxation and show how it applies to transonic flow problems. Two methods for calculating steady-state transonic flows have been used: time dependent integrations in which time is allowed to become sufficiently large, and generalized relaxation procedures. The time dependent methods (see, e.g., Magnus & Yoshihara, Sills, and Singleton) are represented by

$$\frac{d\vec{u}}{dt} = \vec{F}(\vec{u}) + \vec{\Delta}(\vec{u}) \qquad (1)$$

where the form of $\vec{u}$ depends upon the physics of the problem, the form of $\vec{F}$ depends upon both the physics and the differencing scheme chosen, and $\vec{\Delta}$ is an arbitrary vector whose magnitude is in some sense much less than that of $\vec{F}$. If the vector $\vec{\Delta}$ is used, its function is to control the stability and convergence of the computations. It is related to the concept popularly termed "artificial viscosity."

On the other hand, generalized relaxation schemes can be represented by

$$\frac{d}{d\tau}\vec{g}(\vec{w}) = \vec{f}(\vec{w}) + \vec{\delta}(\vec{w}) \qquad (2)$$

where the elements of $\vec{g}$ and $\vec{f}$ depend upon the physical variables and the differencing scheme, but in an unlimited number of possible ways, the condition on them being that when all elements of $d\vec{g}/d\tau$ are zero, $\vec{u}$ can be constructed from $\vec{w}$ such that $\vec{F}(\vec{u}) + \vec{\Delta}(\vec{u}) = 0$. The vector $\vec{\delta}$ may be used for smoothing. If so, its magnitude must be less than the truncation error involved in the construction of $\vec{f}$. The independent variable τ is not related to a physical coordinate and is introduced for the sole purpose of coupling classical relaxation theory to the theory of ordinary differential equations. The theory that results from this coupling is what we refer to as generalized relaxation. The components of $\vec{w}$ are assumed to be continuous functions of τ throughout the iteration path, leading from an initial estimate to the final stationary solution. If the integration is carried out numerically, $\vec{w}_n$ refers to the value of $\vec{w}$ at $\tau = n\Delta\tau$ along this path. More specific descriptions of methods represented by Eq. (2) are given below. The objective here is to indicate that for the computation of steady-state problems, generalized relaxation procedures have the potential for much more rapid convergence than do time dependent ones. It should also be pointed out that Eq. (2) may be derived from partial differential equations that are fundamentally elliptic, parabolic, or hyperbolic. That relaxation techniques can be treated as a subset of the numerical solution of ordinary differential equations is certainly not new (see, e.g., Forsythe and Wasow, p. 263). However, development of these techniques from this point of view is, at the very least, not popular. A brief outline of such a development follows.

GENERALIZED RELAXATION

It is sufficiently general for the purpose of this paper to let $d\vec{g}/d\tau = H^{-1}\, d\vec{w}/d\tau$ where H^{-1} is some nonsingular matrix. Further, for the sake of analyzing convergence, $\vec{f}$ is approximated by $A_n\vec{w} - \vec{c}_n$ where $\vec{c}_n = A_n\vec{w}_n - \vec{f}_n$ and A is the Jacobian of $\vec{f}(\vec{w})$. In this way the rate of convergence of a method in the vicinity of the point $n\Delta\tau$ on the iteration path is seen to be governed by the equation

$$\vec{w}' \equiv \frac{d\vec{w}}{d\tau} = H(A_n\vec{w} - \vec{c}_n + \vec{\delta}) \qquad (3)$$

Solutions of Eq. (3) depend upon the eigenvalues σ_m of the associated matrix HA_n which is referred to as the σ matrix. Clearly, the solutions converge if the real part of each σ_m is less than 0, and the rate of convergence depends upon the value of $\exp(\sigma_m\Delta\tau)$. It is at this point that the theory on the numerical integration of ordinary differential equations and the theory on classical relaxation combine to form the theory of generalized relaxation.

The analysis of numerical methods for integrating ordinary differential equations depends on the relation between the eigenvalues λ_m of the difference equations and the eigenvalues σ_m of the differential equations. Conditions under which there is a one to one correspondence between these eigenvalues are known (see, e.g., Lomax). The relation $\lambda_m = \lambda(\sigma_m \Delta\tau)$ determines the accuracy and stability of the numerical method. It is accurate insofar as $\lambda_m \cong \exp(\sigma_m \Delta\tau)$ and stable insofar as $|\lambda_m| < 1$ for values of $|\sigma_m \Delta\tau|$ that are as large as possible. When the relation applies to a method being used to integrate the iteration path of a generalized relaxation scheme, however, the accuracy condition simplifies to the requirement that $\lambda(0) = 1$ for all m; and the stability condition is replaced by an emphasis on the rate of convergence which, in turn, is optimized when $|\lambda_m| << 1$ or $= 0$ for every $\sigma_m \Delta\tau$.

In this report the theory referred to as classical relaxation satisfies the following conditions: First, it pertains to a set of difference equations that are formed by central differencing a set of partial differential equations. Second, the σ matrix associated with the set of difference equations so formed in "nearly" diagonally dominant and definite. Third, the solutions of these difference equations are found by applying a matrix iterative analysis such as that developed by Varga. In contrast to this, the theory referred to as general relaxation pertains to a set of difference equations that result from any differencing scheme, that have an arbitrary associated σ matrix, and that are solved as a set of ordinary differential equations as described above.

Regardless of the form of relaxation studied, the σ eigenvalues play a fundamental role. In actual application to nonlinear problems with a large number of mesh points it would be extremely difficult to compute both the Jacobian matrix and the corresponding eigenvalues. However, on both linear and nonlinear problems having a relatively small number of mesh points, this computation has been found to be both practicable and valuable. In such cases the overall convergence of differencing schemes applied at interior and boundary points can be analyzed and used for optimizing convergence. In any case it is clear that if one can estimate the σ eigenvalues from the form of the σ matrix, then one can make a rational choice from a variety of differencing formulas available for both the space and τ derivatives. An example illustrating these statements is presented at the end of this paper.

FORMS OF THE TRANSONIC EQUATIONS

In our study of transonic flow fields we have considered the following five well-known formulations of the steady-state, two-dimensional, inviscid, isentropic equations

$$\left.\begin{aligned} \psi_{xx} + \psi_{yy} - \rho^{-1}(\psi_x \rho_x + \psi_y \rho_y) &= 0 \\ (\rho_{st}^{-1}\rho)^{\gamma-1} + (\gamma - 1)(2a_{st}^2 \rho^2)^{-1}(\psi_x^2 + \psi_y^2) &= 1 \end{aligned}\right\} \tag{4}$$

$$\left.\begin{aligned} \phi_{xx} + \phi_{yy} + \rho^{-1}(\phi_x \rho_x + \phi_y \rho_y) &= 0 \\ (\rho_{st}^{-1}\rho) &= [1 - (\gamma - 1)(2a_{st}^2)^{-1}(\phi_x^2 + \phi_y^2)]^{1/(\gamma-1)} \end{aligned}\right\} \tag{5}$$

$$[1 - M_\infty^2 - U_\infty^{-1} M_\infty^2 (\gamma + 1)\phi_x]\phi_{xx} + \phi_{yy} = 0 \tag{6}$$

$$\left.\begin{aligned} (a^2 - \phi_x^2)\phi_{xx} + (a^2 - \phi_y^2)\phi_{yy} - 2\phi_x\phi_y\phi_{xy} &= 0 \\ a^2 = a_\infty^2 - 0.5(\gamma - 1)(\phi_x^2 + \phi_y^2 - U_\infty^2) & \end{aligned}\right\} \tag{7}$$

$$\left.\begin{aligned}(\rho u)_x + (\rho v)_y &= 0\\ u_y - v_x &= 0\\ (\rho_{st}^{-1}\rho) &= [1 - (\gamma - 1)(2a_{st}^2)^{-1}(u^2 + v^2)]^{1/(\gamma-1)}\end{aligned}\right\} \qquad (8)$$

The isentropic assumption was made to simplify both the computations and the boundary conditions. Its effect on the pressure jump across a normal shock is shown in Fig. 1 where the appropriate weak solution (in the sense introduced by Lax) of Eqs. (8) is compared with the Rankine-Hugoniot relation. The approximation is considered to be a good one if the local Mach number does not exceed about 1.3.

The above equations divide into two classes; one class has second-order derivatives of derived (ψ and ϕ) variables, and the other has only first-order derivatives of the primitive variables. A large body of literature is available for designing ways to relax the former class, while the only published relaxation procedure for supercritical flows (known to the authors) that applies to the latter class is that of steepest descent (Babaev and Prozan) – a technique that does not appear to have any potential for rapid convergence.

METHODS DERIVED FROM CLASSICAL RELAXATION

Equations (4) (see Emmons, Katsanis, and Sells) are probably the most efficient for subcritical flows since these equations are the only ones with Dirichlet boundary conditions. However, their use for transonic flow problems is hampered by the fact that the density is a double valued function of the mass flow parameter ($\psi_x^2 + \psi_y^2$), and a saddle point exists at the sonic line. Although some computations were made using Eqs. (4), they were eventually abandoned.

Equations (5) are also suitable for the analysis of subsonic flows and have been used by Howell & Spong. In transonic problems Eqs. (5) present no difficulties at the sonic line or in double valuedness. They were, therefore, used by the authors to compute the flow about a biconvex airfoil at a supercritical Mach number using thin-airfoil boundary conditions. Here it was found that when central differencing is applied to both the x and y derivatives, a symmetrical solution results even for supercritical flows. This is illustrated in Fig. 2. The same effect was found to be true for all calculations of airfoils with fore and aft symmetry. There are indications that such solutions will diverge, but at a very slow rate. Backward difference formulas for the first-order derivatives in the x direction were also tried for these equations. A shocklike solution did form in these cases but, as seen in Fig. 2, the result was not satisfactory. This particular approach was abandoned in lieu of the developments described below.

Numerical solutions for the velocity potential given by Eqs. (7) were computed for a biconvex airfoil in a manner similar to that used by Murman & Cole for the small perturbation Eqs. (6). In the subsonic portion of the flow, second-order central differencing was used, while in the supersonic region all x derivatives were differenced backward. The system so formulated was iterated by the method of successive overrelaxation. The results are shown in Fig. 3 where they are compared with results computed by the method of Murman & Cole on the same size grid. Taking the uniform free-stream as an initial guess, the solution shown for Eqs. (7) converged in about 600 iterations for a 40 by 60 mesh.

A METHOD DERIVED FROM GENERALIZED RELAXATION

At the present state of development, it is probably unwise to rely on any one numerical method to solve a practical transonic flow problem. Studies of transonic airfoil problems are being carried out, therefore, paralleling those problems discussed above but based on Eqs. (8). Since Eqs. (8) have no second-order derivative terms, and therefore, do not "naturally" difference to form diagonally dominant σ matrices, their analysis can be said to fall into the realm of generalized relaxation. Such an analysis deserves some discussion. First Eqs. (8) were recast for relaxation into the form

$$\left.\begin{aligned} u' &= u_y - v_x + \delta(u) \\ G' &= (\rho u)_x + G_y + \delta(G) \\ v' &= G - \rho v \\ \rho' &= (\rho_{st}^{-1}\rho) - [1 - (\gamma - 1)(2a_{st})^{-1}(u^2 + v^2)]^{1/(\gamma-1)} \end{aligned}\right\} \tag{9}$$

which is an example of the use of $\vec{w}$ in Eq. (2). Note that $G \to \rho v$ as $\tau \to \infty$. A variety of forward, central, and backward differencing combinations were considered. For example, the G_y derivative was replaced by alternating forward and backward differencing schemes according to the equation

$$\left.\begin{aligned} (G_y)_{i,j} &= (-3G_{i,j} + 4G_{i+1,j} - G_{i+2,j})/2\Delta y \\ (G_y)_{i+1,j} &= (3G_{i+1,j} - 4G_{i,j} + G_{i-1,j})/2\Delta y \end{aligned}\right\} \quad i = 2, 4, 6 \ldots \tag{10}$$

and the iteration derivative G' was evaluated using the interchange

$$\left.\begin{aligned} (G')_{i,j} &= [(\rho u)_x + G_y]_{i+1,j} + \delta_{i,j} \\ (G')_{i+1,j} &= -[(\rho u)_x + G_y]_{i,j} + \delta_{i+1,j} \end{aligned}\right\} \tag{11}$$

Equations (11) are examples of the use of H in Eq. (3). The fact that such intermingled sets of differencing formulas can be used is clear. The reasons for settling on a particular combination are suggested by the resulting form of the σ matrix. Thus, if the Jacobian matrix is ordered in successive blocks of u, G, v, and ρ, then u_y and G_y can only contribute to diagonal blocks, while v_x and $(\rho u)_x$ can only contribute to off-diagonal blocks. Then strictly forward differencing of u_y causes the diagonal block it affects to be upper triangular, and the differencing of G_y and G' specified above causes the diagonal block they effect to be definite and diagonally dominant. These were key issues in making the method convergent.

Equations (8) were first used to solve subcritical problems applying thin-airfoil boundary conditions. With the experience thus gained, the method was extended to supercritical flows. Results for a biconvex airfoil are shown in Fig. 4. The computations for the results shown required about 600 iterations under the same conditions that apply to the solution in Fig. 3. On the basis of the solutions shown in these figures, we can say that second-order, convergent, iterative differencing techniques can be constructed from either Eqs. (7) or (8) to solve transonic flow problems.

BIBLIOGRAPHY

Babaev, D. A. *AIAA J.* 1, 2224-2231 (1963).
Emmons, H. W. *NACA TN 1746* (1948).
Forsythe, G. E. and Wasow, W. R. *Finite-Difference Methods for Partial Differential Equations.* John Wiley and Sons (1960).
Howell, R. H. and Spong, E. D. *AIAA J.* 7, 1392-1393 (1969).
Katsanis, T. *NASA SP–228* (1969).
Knechtel, E. D. *NASA TN D–15* (1959)
Lax, P. D. *Comuns. Pure Appl. Math.* VII, 159-193 (1954).
Lomax, H. *NASA TN D–4547* (1968).
Magnus, R. and Yoshihara, H. *AIAA Paper 70–47* (1970).
Murman, E. M. and Cole, J. D. *AIAA Paper 70–188* (1970).
Prozan, R. J. *Lockheed Missiles and Space Co.*, Rep. LMSC/HREC D162177 (*NASA Contract NAS 7–743*) (1970).
Sells, C. C. L. *Royal Aircraft Establishment*, Tech. Rep. 67146 (1967).
Sills, J. A. *General Dynamics*, Fort Worth, Res. Rep. ERR-FW-806 (1968).
Singleton, R. E. *AGARD* CP 35 (1968).
Varga, R. S. *Matrix Iterative Analysis*. Prentice-Hall, Inc., Englewood Cliffs, N. Y. (1962).

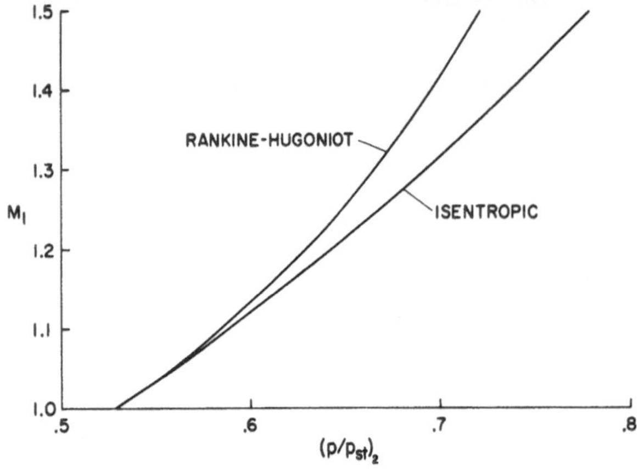

Fig. 1.– Isentropic weak solution compared to the Rankine-Hugoniot normal shock relation for a perfect gas.

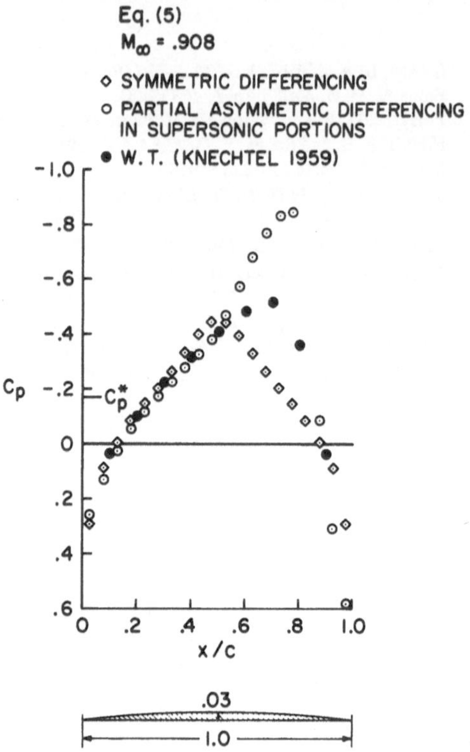

Fig. 2.– Pressure coefficient along chord of biconvex airfoil using ϕ, ρ as dependent variables.

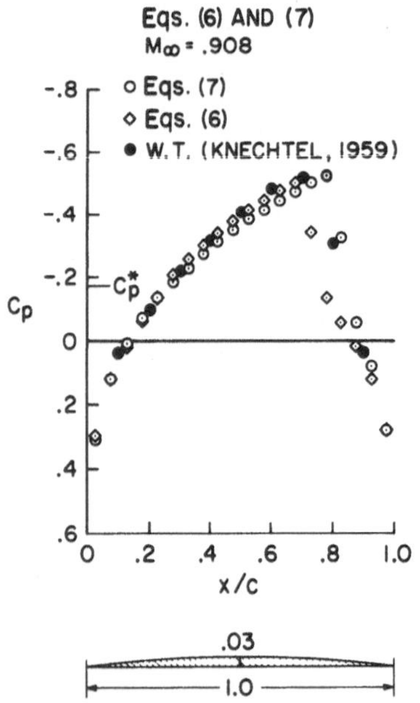

Fig. 3.– Pressure coefficient along chord of biconvex airfoil using ϕ, a as dependent variables.

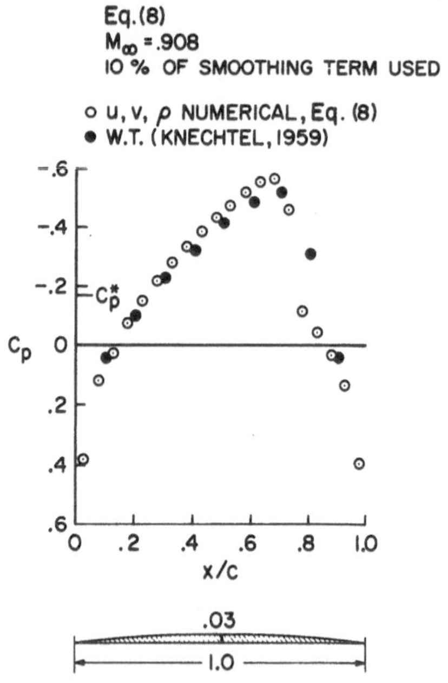

Fig. 4.– Pressure coefficient along chord of biconvex airfoil using u, v, ρ as dependent variables.

SOLUTION OF THE TRANSONIC POTENTIAL EQUATION USING A MIXED FINITE DIFFERENCE SYSTEM

E. M. Murman and J. A. Krupp
Boeing Scientific Research Laboratories, Seattle, Washington

I. INTRODUCTION

Transonic flows are a common occurrence in compressible flow problems. Some examples where the solution of a transonic problem is required include flows through nozzle throats, around airfoils, wings, rotor and compressor blades, past blunt bodies, and near weak shock waves which must terminate. Such flow patterns are characterized by the presence of adjacent subsonic and supersonic regions which may be separated on one boundary by a shock wave. The governing partial differential equations are non-linear, of mixed hyperbolic-elliptic type, and admit discontinuities. The problem may be either an initial or boundary value problem or a combination thereof, depending on the particular situation.

In this paper we report some numerical solutions of an equation which describes small disturbance transonic flows. The transonic potential equation

$$[K-(\gamma+1)\phi_x]\phi_{xx} + \phi_{\tilde{y}\tilde{y}} = 0 \qquad (1)$$

models the flow past thin airfoils in a nearly sonic freestream (Murman and Cole, 1970; Spreiter, 1952; Cole and Messiter, 1957). The constant K is a similarity parameter defined by

$$K = (1-M_\infty^2)/\left(g(M_\infty)\delta\right)^{2/3} . \qquad (2)$$

Flows with different values of M_∞, the freestream Mach number, and δ, the airfoil thickness ratio, are similar if they have the same value for K. An arbitrary function $g(M_\infty)$ where $g(M_\infty) \to 1$ as $M_\infty \to 1$ is not defined by the systematic derivation leading to Eq. 1. The definition $g(M_\infty) = M_\infty^2$ is customarily adopted (Spreiter, 1952). The specific heat ratio of the gas γ is a constant, $\tilde{y} = y\delta^{1/3}$ is a transformed coordinate, and ϕ is a perturbation velocity potential. The airfoil shape is described by the equation $y = \delta F(x)$. The non-linear term $\phi_x\phi_{xx}$ allows Eq. 1 to change type and is responsible for the formation of shock waves. A transonic pressure coefficient

$$\overline{C}_p(x,\tilde{y};K) = -2\phi_x \qquad (3)$$

is a universal function for a given body shape function. It is related to the usual pressure coefficient by

$$C_p(x,\tilde{y}) = \left(\frac{\delta^2}{g(M_\infty)}\right)^{1/3} \overline{C}_p(x,\tilde{y};K) . \qquad (4)$$

The boundary value problem for a non-lifting symmetric airfoil in a slightly subsonic freestream is outlined in Fig. 1. The mean surface approximation is applicable so that the airfoil boundary condition $\phi_{\tilde{y}} = F'(x)$ is applied on the line segment $|x|\leq 1$, $\tilde{y} = 0$. By symmetry, $\phi_{\tilde{y}} = 0$ for $|x|>1$, $\tilde{y} = 0$. On the other boundaries, ϕ is specified by a far field formula (Murman and Cole, 1970). Equation 1 is to be solved numerically inside the domain.

During the past year we have developed a new computational procedure to solve the transonic potential equation using a mixed finite difference system. Equation 1 is approximated by either a hyperbolic, parabolic, or elliptic difference equation depending upon its local behavior at each mesh point. The difference equations are solved iteratively using a line relaxation procedure. The sonic line and shock waves evolve naturally during the course of the solution.

It should be mentioned that the small disturbance flows past wings (Spreiter, 1952), bodies of revolution (Cole and Messiter, 1957), and at a caustic (Hayes, 1968) are described by differential equations similar to Eq. 1. The caustic problem will be mentioned further in Section IV.

II. NUMERICAL ANALYSIS

The numerical solution of mixed elliptic-hyperbolic differential equations is a subject which has received little attention. Emmons (1944) used relaxation methods to solve the inviscid, compressible, two-dimensional stream function equations for transonic nozzle and airfoil flows with imbedded shock waves. His method required that locally normal shocks be fitted into the flow. The method has not been extended to other problems, presumably due to difficulties with convergence. Chu (1959) formulated a type insensitive finite difference method for solving certain linear systems which change type. His method is too specialized for general use. Filippov (1957) reported a method for solving the Tricomi equation by a finite difference method. The present work was first reported by Murman and Cole (1970). Since then, the numerical procedures have been further refined.

It is well known that separate finite difference procedures are required to solve elliptic and hyperbolic differential equations [e.g., Isaacson and Keller (1966)]. The derivatives in elliptic equations are replaced by centered finite difference formulae. The difference equations with the boundary conditions are solved by inverting a large matrix usually by some iterative technique such as relaxation. For hyperbolic equations, the derivatives in the time-like direction must be replaced by one-sided finite difference formulae. The domain of dependence of the hyperbolic equations is accounted for by using either an implicit difference formulation or an explicit formulation with a restriction on the step size in the time-like direction. The difference equations are solved iteratively by marching forward in time, starting with given initial conditions and incorporating any boundary conditions. These disparate features are incorporated in the mixed finite difference system.

To be specific let us assume that Eq. 1 is to be solved in a rectangular domain with the boundary conditions given in Fig. 1. We select uniform mesh spacing in the x and $\tilde{y}$ directions; $x = i\Delta x$, $\tilde{y} = j\Delta\tilde{y}$; $i,j = 1,2,\cdots$. In the actual computations variable mesh spacing is used but there are no essential changes in the methodology. From physical considerations, we know that the time-like direction in the hyperbolic regions is the positive x coordinate.

At each mesh point, the local velocity must be computed to determine if the flow is subsonic, sonic, or supersonic. The test formula using a centered difference for ϕ_x

$$V(i) = \left(K-(\gamma+1)\phi_x\right)_i = K-(\gamma+1)\,\frac{\phi_{i+1}-\phi_{i-1}}{2\Delta x} \tag{5}$$

was found to be best after some numerical experimentation. If $V(i) > 0$, the flow is subsonic and the x-derivatives are approximated by the <u>ELLIPTIC</u> difference formula

$$\left(K-(\gamma+1)\phi_x\right)\phi_{xx} = \left[K-(\gamma+1)\,\frac{\phi_{i+1}-\phi_{i-1}}{2\Delta x}\right]\frac{(\phi_{i+1}-2\phi_i+\phi_{i-1})}{(\Delta x)^2}\quad . \tag{6a}$$

If $V(i) < 0$ the flow is supersonic and the <u>HYPERBOLIC</u> difference formula is

$$\left(K-(\gamma+1)\phi_x\right)\phi_{xx} = \left[K-(\gamma+1)\,\frac{\phi_i-\phi_{i-2}}{2\Delta x}\right]\frac{(\phi_i-2\phi_{i-1}+\phi_{i-2})}{(\Delta x)^2}\quad . \tag{6b}$$

By comparing Eqs. 5 and 6b, we see that an anomaly can arise when $V(i) < 0$ but $V(i-1) > 0$. If this happens, the coefficient of the ϕ_{xx} term in Eq. 1, when written in difference form, is positive rather than negative as it should be for a hyperbolic equation. This will occur whenever the flow accelerates from subsonic to supersonic velocities between mesh points i-1 and i. To resolve this discrepancy, we introduce a third PARABOLIC formula.

$$\left(K-(\gamma+1)\phi_x\right)\phi_{xx} = 0 \tag{6c}$$

whenever $V(i) < 0$ and $V(i-1) > 0$ or $V(i) = 0$.

Equations 6 were actually written for the x derivatives expressed in divergence (or conservation) form

$$(K\phi_x - \frac{\gamma+1}{2}\phi_x^2)_x ,$$

a procedure which has been shown by many authors to be superior when shock waves are present. The equations have been factored to clearly show the relationship between the derivatives and the difference formulae. It should be noted that Eq. 6b is written in implicit form to avoid any stability restriction on the allowable size for Δx. If an explicit formulation were written for the hyperbolic region, the Courant-Friedrichs-Lewy stability criterion (Isaacson and Keller, 1966) would require $\Delta x \rightarrow 0$ at points where $V(i) \rightarrow 0$, an intolerable restriction.

The $\tilde{y}$ derivatives are replaced everywhere by the centered formula

$$\phi_{\tilde{y}\tilde{y}} = \frac{\phi_{j+1} - 2\phi_j + \phi_{j-1}}{(\Delta\tilde{y})^2} \tag{7a}$$

except at the boundary $\tilde{y} = 0$ $(j = 1)$. At this point, the boundary condition on $\phi_{\tilde{y}}$ is incorporated by writing either

$$\phi_{\tilde{y}\tilde{y}} = \frac{1}{\Delta\tilde{y}}\left[\frac{\phi_2 - \phi_1}{\Delta\tilde{y}} - (\phi_{\tilde{y}})_{\tilde{y}=0}\right] \tag{7b}$$

or

$$\phi_{\tilde{y}\tilde{y}} = \frac{2}{\Delta\tilde{y}}\left[\frac{\phi_2 - \phi_1}{\Delta\tilde{y}} - (\phi_{\tilde{y}})_{\tilde{y}=0}\right] \tag{7c}$$

depending on whether the mesh point $j = 1$ is placed a half mesh step off the boundary or on the boundary, respectively. Both versions have been used successfully.

Equations 6 and 7 written for each mesh point comprise the finite difference system. The resulting set of non-linear algebraic equations are solved in an iterative fashion using line relaxation. Values of ϕ are solved for along a vertical line (x = constant). Each vertical line is then successively relaxed by marching forward in the positive x direction. The latest values of ϕ are always used as they become available. At each stage of the iteration process, the local velocity is tested (Eq. 5) and the appropriate difference system is selected. The sonic line and shock wave evolve naturally as the iterations proceed. The process is terminated when the solution converges to a final answer. We found that the best way to determine convergence is to calculate the pressures on the airfoil surface and require that they do not change substantially (say 0.1%) during 10 iterations. Absolute convergence was checked for several cases by approaching the solution with initial guesses from "above" and "below". The double integral needed in the boundary condition for ϕ (Fig. 1) was recomputed every 5 iterations using the trapezoidal rule.

The line relaxation algorithm used is similar to the Gauss- Seidel line relaxation process (Isaacson and Keller, 1966) except that the set of equations on each vertical line is non-linear. Newton's method (Isaacson and Keller, 1966) is used to solve these non-linear equations. Convergence for Newton's method is guaranteed provided that the initial guess for ϕ is "close enough" to the solution. Successive line over relation was tried but the computations diverged due to the presence of the hyperbolic region.

The computations are started by solving for a subcritical case (typically $K = 5$) using any initial choice for ϕ. A series of calculations are then performed for successively lower values of K. The initial guess for ϕ at each K is taken as the solution of the proceeding value of K. Typical step sizes in K may go as 5, 3, 2.5, 2.0, 1.8, 1.6, 1.4, but these intervals vary with the particular airfoil geometry. This process is needed to satisfy the "close enough" constraint on Newton's method. Thus, a sequence of solutions corresponding to different Mach numbers or thickness ratios given by Eq. 2 are obtained for each airfoil geometry. The iterations at each K value do not need to be carried to convergence before proceeding to the next K.

III. SHOCK STRUCTURE

Any finite difference system which approximates Eq. 1 must possess the correct properties for admitting shock waves in the solution. Much work has been done to identify what these properties are for purely hyperbolic systems (e.g., Richtmeyer and Morton, 1967). In general, the finite difference system must be dissipative and must approximate a differential equation (or system of differential equations) which is written in divergence form. No previous work has been done to determine the necessary properties of mixed difference systems. In this section, we examine the features of Eqs. 6 and 7 which pertain to shock wave structure. As was already noted in Section II, the present difference equations have been written for the divergence form of Eq. 1. The analysis of Eq. 1 is simplified in details if a transformed potential ψ

$$\psi = -Kx + (\gamma+1)\phi$$

is introduced to give

$$-\psi_x\psi_{xx} + \psi_{\tilde{y}\tilde{y}} = 0 \quad . \tag{8}$$

Using a Taylor series expansion, one can show that the finite difference approximation to Eq. 8 represents the original differential equation plus added truncation terms, assuming, of course, that the derivatives are all continuous. The resulting equations for the hyperbolic and elliptic difference systems are, respectively,

$$-\psi_x\psi_{xx} + \psi_{\tilde{y}\tilde{y}} + \Delta x(\psi_x\psi_{xx})_x = 0\left((\Delta x)^2, (\Delta\tilde{y})^2\right) \tag{9}$$

$$-\psi_x\psi_{xx} + \psi_{\tilde{y}\tilde{y}} - \frac{(\Delta x)^2}{24}\left((\psi_{xx})^2 + 2\psi_x\psi_{xxx}\right)_x + \frac{(\Delta\tilde{y})^2}{12}\psi_{\tilde{y}\tilde{y}\tilde{y}\tilde{y}} = 0\left((\Delta x)^4, (\Delta\tilde{y})^4\right) \tag{10}$$

The form of these equations is examined to ascertain if they permit a solution which has a shock wave like structure.

Equation 9 is similar to the viscous transonic (VT) equation. If the truncation error term is locally linearized by setting $\psi_x = u = \text{constant} > 0$, one obtains the VT equation

$$\psi_x\psi_{xx} - \psi_{\tilde{y}\tilde{y}} = u\Delta x\, \psi_{xxx} \quad . \tag{11}$$

This equation is parabolic and models transonic shock wave structure. The physical viscosity in the VT equation is replaced by the numerical viscosity in Eq. 11 proportional to $u\Delta x$. Thus, we conclude that Eq. 9 is dissipative when the flow is supersonic ($\psi_x > 0$) and contains a solution with shock wave like structure.

To test the above conclusion, computations of an oblique shock wave with supersonic downstream velocities were carried out. One result, given in Fig. 2, shows an oblique shock wave with the correct jump in velocity and a thickness of 6-8 mesh points. Other results show that as the shock strength increases, and the non-linear steepening of the shock profile overcomes the dissipation effects, the shock thickness decreases.

If a finite difference system could be found with a numerical viscosity proportional to $(\Delta x)^2$, a thinner oblique shock wave would result. The second order system introduced in a previous report (Murman and Cole, 1970) is dispersive rather than dissipative, and was found to be inferior in actual computations. It would be useful to have a dissipative second order hyperbolic system.

For a shock wave with a subsonic downstream velocity, the difference equation is changed midway through the shock. The form of Eq. 10 may be determined by examining the highest order derivatives, the conclusion is that Eq. 10 is elliptic when the flow is subsonic ($\psi_x < 0$). Conceptually one may explain the shock wave structure as an exponential decay of velocity in the supersonic region, governed by the parabolic Eq. 9, followed by an algebraic decay in the subsonic portion governed by the elliptic Eq. 10.

In the limit of a normal shock, Eq. 8 reduces to the simple ordinary differential equation,

$$\psi_x \psi_{xx} = 0 \tag{12}$$

whose solution $\psi_x = \pm c$ gives a uniform flow with a shock. It is instructive to examine the finite difference equations corresponding to Eq. 12.

Elliptic
$$\frac{\psi_{i+1} - \psi_{i-1}}{2\Delta x} \cdot \frac{\psi_{i+1} - 2\psi_i + \psi_{i-1}}{(\Delta x)^2} = 0$$

Hyperbolic
$$\frac{\psi_i - \psi_{i\ 2}}{2\Delta x} \cdot \frac{\psi_i - 2\psi_{i-1} + \psi_{i-2}}{(\Delta x)^2} = 0 .$$

Each equation has two solutions

Elliptic
$$\frac{\psi_{i+1} - \psi_i}{\Delta x} = \pm \frac{\psi_i - \psi_{i-1}}{\Delta x}$$

Hyperbolic
$$\frac{\psi_i - \psi_{i-1}}{\Delta x} = \pm \frac{\psi_{i-1} - \psi_{i-2}}{\Delta x} .$$

The plus sign corresponds to a uniform flow while the minus sign represents an abrupt jump in the velocity. Thus, for this simple case, it is clear that the finite difference equations allow a discontinuity in ψ_x between two mesh points. The airfoil calculations presented in the next section contain nearly normal shock waves with thicknesses of 3-5 mesh points.

The above discussion does not completely explain shockwave formation and structure in the mixed finite difference system. However, it has provided insight into understanding the computed results. One question which must be raised is should one examine the partial differential equation including the truncation error terms, or the finite difference equations themselves, or both, to understand shock wave structure?

IV. RESULTS

A series of transonic airfoils shapes have been computed by Korn (1969), each one of which is designed to be shock free at a specific freestream Mach number. We have computed the flow for one of these airfoils using the above methods. Figure 3 shows a comparison between the present results and Korn's solution for the shock free (design point) case. The agreement is excellent. Figure 4 shows solutions for three off-design conditions. The solution of $K = 1.50$ contains a shock wave while the case at $K = 1.91$ does not have a shock wave strong enough to be seen in the results. We also have computed solutions for the flow past the NACA 4 digit series of airfoils (0006, 0012, etc.) for various K values. Some results are shown in Fig. 5. As $K \to 0$ $M_\infty \to 1$), the supersonic region increases in size and the shockwave strengthens. Results for circular arc airfoils and a Nieuwland airfoil were previously reported by Murman and Cole (1970). A comparison of the computed and experimental pressure jump across the shockwave is contained in that report.

The above calculations were done with a 100 x 41 mesh point network with 50 mesh points along the chord. Typically 50-100 iterations are needed at each K value before proceeding to the next. For K values where an accurate answer is desired, the calculations must run an additional 400-500 iterations. The total computing time for one K value is about 40 minutes on an IBM 360/44 (equivalent to about 5 min. on a CDC 6600).

In Section 1, it was pointed out that an arbitrary function $g(M_\infty)$ enters into the definition of K. It is possible to select a definition for $g(M_\infty)$ by directly comparing the solutions of the small disturbance equations with the Korn and Nieuwland solutions of the exact inviscid equations. The choice

$$g(M_\infty) = M_\infty^{3/2} \tag{13}$$

gives excellent results. This definition was compared with the criteria (Spreiter, 1952) used to select the customary choice $g(M_\infty) = M_\infty^2$ and Eq. 13 was found to give equally good results. Equation 13 has been used in Eq. 4 for computing the C_p values shown in Fig. 3.

The numerical methods of Sec. II have been applied to another transonic flow problem. A caustic is formed when the normal velocity of a weak oblique shock wave traveling through a stratified medium approaches the local speed of sound. The incoming wave reflects from the caustic with a local amplification of the wave strength. The solution of the equation modeling the caustic problem (Hayes, 1968)

$$(\eta+\phi_\xi)\phi_{\xi\xi} - \phi_{\eta\eta} = 0 \tag{14}$$

has been computed by Prof. A. Richard Seebass and the present authors. A typical result showing the sonic line and characteristics is given in Fig. 6. Boundary conditions for the elliptic region and upstream side of the hyperbolic domain were obtained from the linear version of Eq. 14. A characteristic boundary condition was applied on the downstream hyperbolic boundary to permit the reflected waves to pass out of the region. These results will be discussed in detail in a later publication.

V. CONCLUSIONS

A new method has been presented for numerically solving the transonic potential equation including problems with imbedded shock waves. Elementary concepts of finite difference techniques are combined to form a mixed finite difference system. Solutions have been presented for flows past four transonic airfoils and for the structure of a caustic. The method is accurate, straight forward, and relatively fast.

The authors wish to acknowledge many valuable discussions with Professors A. Richard Seebass and Julian D. Cole.

REFERENCES

Chu, C. K., Ph.D. Thesis, New York University (1959)

Cole, J. D. and Messiter, A. F., Zeit. ang. Math. u. Physik, 8(1), 1-25 (1957)

Emmons, H. W., NACA TN 932 (1944) [See also NACA TN 1003 (1945) and NACA TN 1746 (1948)]

Filippov, A. G., Isv. Akad. Nauk SSSR ser mat 21, 73-88 (1957)

Hayes, W. D., NASA SP-180, 165-171 (1968)

Isaacson, E. and Keller, H. B., Analysis of Numerical Methods, John Wiley & Sons, New York (1966)

Korn, D. G., Courant Inst. of Math. Sci., NYO-1480-125 (1969)

Murman, E. M. and Cole, J. D., AIAA Paper 70-188 (1970); also to appear in AIAA J. 8(12) (1970)

Richtmeyer, R. D. and Morton, K. W., Difference Methods for Initial Value Problems, Interscience Publishers, Second Ed. (1967)

Sichel, M., Phys. of Fluids, 6(5), 633-662 (1963)

Spreiter, J. R., NACA Rep. 1153 (1952)

Far Field Solution
Doublet

$$\phi = \frac{1}{2\pi\sqrt{K}} \left\{ 2\int_{-1}^{1} F(x)dx + \frac{\gamma+1}{2} \iint_{-\infty}^{\infty} \phi_x^2 \, dx d\tilde{y} \right\} \frac{x}{x^2 + K\tilde{y}^2}$$

Near Field
Finite Difference Solution

$\phi_{\tilde{y}} = 0$ $\phi_{\tilde{y}} = F'(x)$ $\phi_{\tilde{y}} = 0$

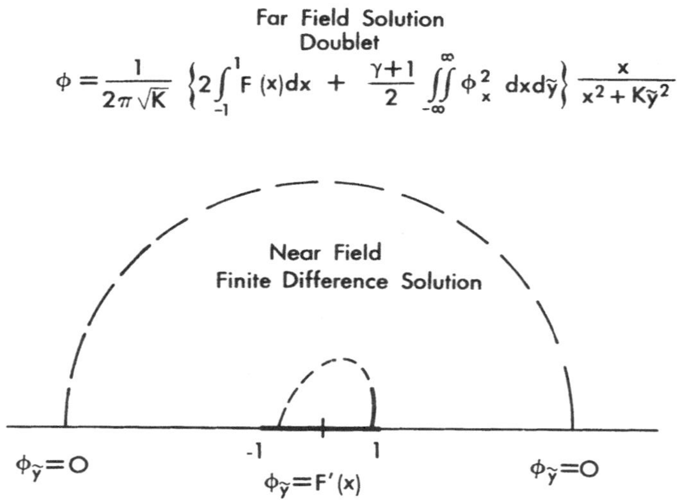

Fig. 1. Transonic boundary value problem.
Non-lifting symmetric airfoil, $M_\infty < 1$.

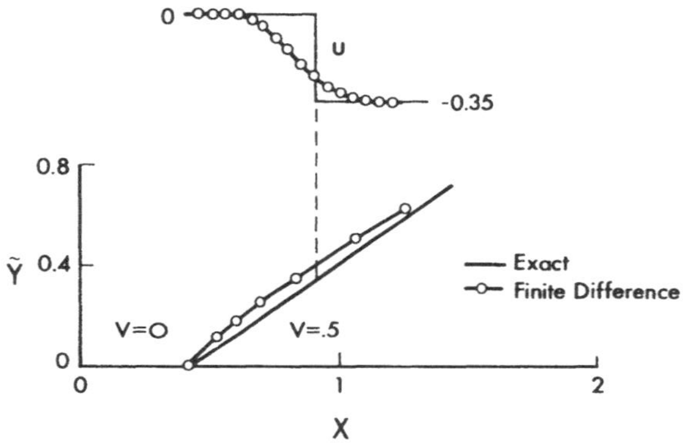

Fig. 2. Location and strength of oblique
shock wave, K = - 2.4.

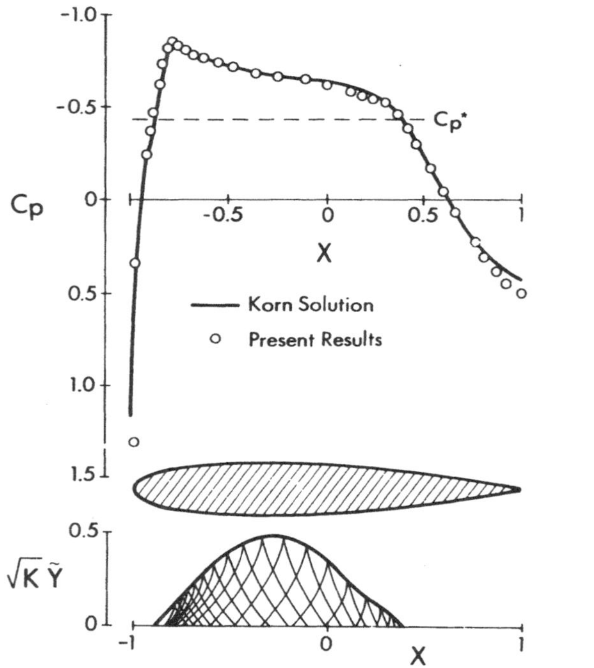

Fig. 3. Korn airfoil. Comparison of
exact solution ($M_\infty = .8$, $\delta = .128$) with
present calculations, K = 1.77.

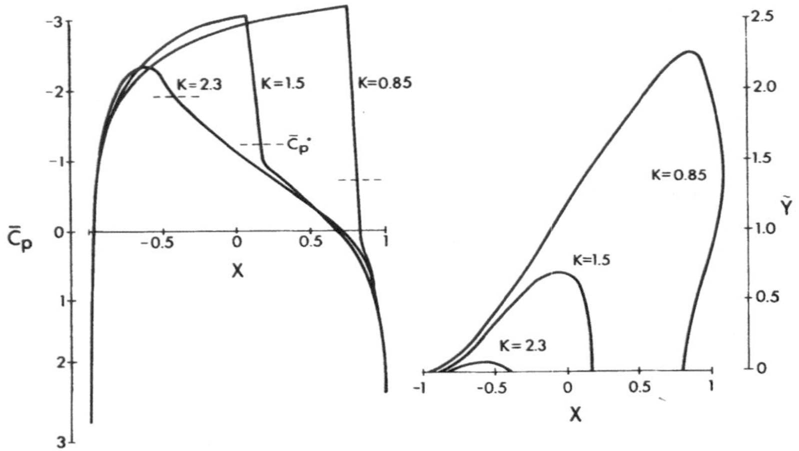

Fig. 5. NACA 4-digit series airfoil pressure distributions and sonic lines.

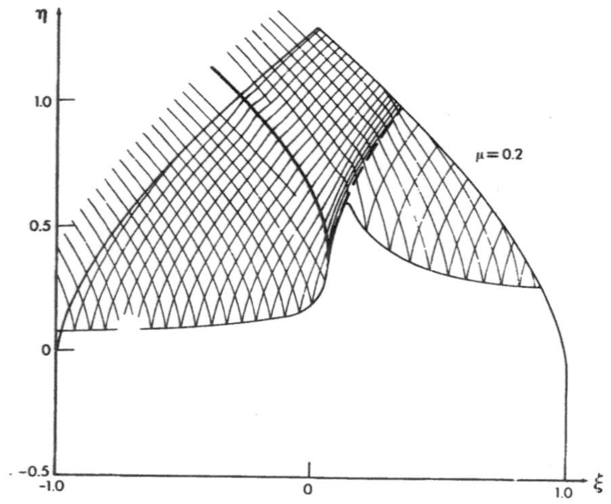

Fig. 6. Solution for a caustic.

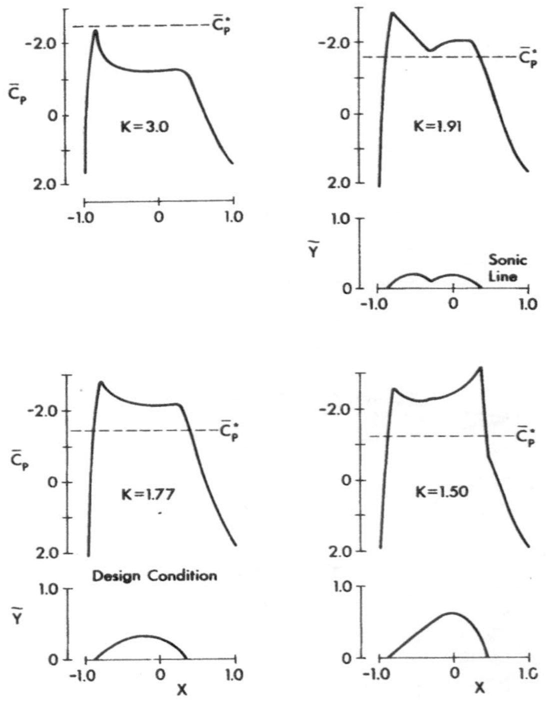

Fig. 4. Korn airfoil. Off design sonic lines and pressure distributions.

THE CALCULATION OF HIGH SUBSONIC FLOW PAST BODIES BY THE METHOD OF INTEGRAL RELATIONS

Maurice Holt
University of California, Berkeley

and

Bruce S. Masson
Northrop Corporate Laboratories, Hawthorne, California

INTRODUCTION

The problem of flow past two dimensional bodies at high subsonic speeds has recently received renewed attention. Nieuwland (1) has generated a number of flow fields past symmetrical shapes by the hodograph method, all of which are shock free. Magnus and Yoshihara (2) calculated high subsonic flows by introducing the time as a third independent variable and integrating the unsteady flow equations by a forward marching finite difference method. Cole and Murmann (3) calculated steady flow past thin airfoils using an iterative finite difference method to solve the transsonic small disturbance equations. They generate results for a Nieuwland airfoil which agree very well with those obtained by the hodograph method. For other shapes they calculate unsymmetrical flows with local supersonic regions terminated by weak shock waves.

In the present paper the problem of subsonic plane flow is investigated by the Method of Integral Relations. This method was successfully applied by Chushkin (4) to calculate subcritical flow past ellipses and airfoil profiles and also to sonic flow past such shapes (5). In applying the method to flows with local supersonic regions no restriction is imposed on the thickness of the body, since the full non-linear equations of motion can be treated just as easily as the small disturbance equations. The method is a development of that used by Holt (6) to determine trans-sonic flow through nozzles.

To simplify the algebra the present analysis is carried out for the circular cylinder, although its extension to more general (symmetrical) shapes presents no difficulties in principle. The equations of motion are written in polar coordinates centered on the cylinder. The equations (continuity and irrotationality) are written in divergence form and integrated in the radial direction across strips bounded by circles concentric with the body. The unknown integrands are represented as poly-nomials in the reciprocal of the radial distance with coefficients depending on the angular coordinate. The number of strips used depends on the order of the approxima-tion; here we carry out the second approximation with three strips. Conditions on the outer radius are determined from the solutions of the Prandtl-Glauert equations for a circular cylinder (containing an unknown dipole strength). On the cylinder the condition of zero radial velocity is satisfied.

One of the unknowns is the transverse velocity on the circular cylinder. The ordinary differential equation to determine this has a saddle point singularity at the forward sonic point, in the supercritical case. There are two unknown parameters in the second approximation. The values of these are determined by two conditions, firstly that the integral curve along the body is regular near the saddle point, and secondly, that the flow is symmetrical.

Once the regular solution on the forward part of the cylinder has been found it is continued into the supersonic region over the rear surface of the cylinder. It is terminated at a shock point, downstream of which the integral curve leads to a stagnation point at the rear of the cylinder. The location of this shock point is unique and is analogous to that in a Laval nozzle in the off design condition.

The approximation is carried out for a range of free stream Mach numbers between 0.05 and 0.45. At the lower limit the solution agrees very well with the incompres-sible results. At supercritical Mach numbers all the flows generated have local supersonic regions terminated by shock waves and in no case was it possible to produce shock free symmetrical flows by this method.

The flow is assumed to be irrotational except at the shock. The entropy changes across the shock on the body boundary. Entropy changes on other strip boundaries are not considered since the free stream Mach number does not exceed a value for which the sonic region crosses the middle strip boundary.

2. MATHEMATICAL FORMULATION

The equations of motion for plane, compressible flow are referred to polar coordinates based on the center of the cylinder

$$(r\tau u)_r + (\tau v)_\theta = 0 \tag{1}$$

$$(rv)_r - u_\theta = 0 \tag{2}$$

where

$$\tau = (1 - u^2 - v^2)^{1/(\gamma-1)} \tag{3}$$

and u,v are the velocity components in the polar coordinate system r,θ. The quantity τ is the scaled density.

The boundary conditions correspond to those of a uniform free stream at $r = \infty$.

$$u = - U_\infty \cos \theta \tag{4}$$

$$v = U_\infty \sin \theta \tag{5}$$

Also, the normal velocity component vanishes at the body surface

$$\frac{1}{r_o} \frac{dr_o}{d\theta} = \frac{u}{v} \tag{6}$$

The Rankine-Hugoniot relations will be assumed to hold across shock waves forming the internal boundaries.

An approximate solution to (2), (3) is obtained by application of the Method of Integral Relations. The flow field is divided into three strips in the radial direction, Fig. 1. The innermost strip lies between the body $r_o(\theta)$ and an arbitrary radius r_2. The middle strip lies between r_2 and an outer radius r_1, while the last strip lying beyond r_1 will be eliminated. It is assumed that r_1 can be chosen large enough so that the flow on it is given by a two term multipole expansion. Denoting the variable on this strip boundary by the subscript 1, it is seen that they can all be determined by differentiation of the potential

$$\phi = - U_\infty r_1 \cos \theta - \frac{D}{2\pi \lambda^{1/2} r_1} \left(\frac{\cos \theta}{\cos^2 \theta + \lambda \sin^2 \theta} \right) \tag{7}$$

where

$$\lambda = 1 - M_\infty^2 ,$$

and D is a doublet strength to be determined during the course of the solution. The potential (7) follows from the assumption that the flow perturbations on $r = r_1$ are small and subsonic, and therefore satisfy the Prandtl Glauert equation

$$\lambda \phi_{xx} + \phi_{yy} = 0$$

for which ϕ is the two term multipole expansion in which the transverse doublet strength has been neglected.

On the two inner strip boundaries the flow variables are determined by ordinary differential equations. These are derived from the integral relations resulting from the integration of Eqs. (2) and (3) with respect to r across the middle and inner strips, respectively, namely

$$\frac{d}{d\theta}\int_{r_2}^{r_1} t\,dr + r_1h_1 - r_2h_2 = 0 \tag{8}$$

$$\frac{d}{d\theta}\int_{r_2}^{r_1} u\,dr - r_1v_1 + r_2v_2 = 0 \tag{9}$$

$$\frac{d}{d\theta}\int_{r_o}^{r_2} t\,dr + r_2h_2 = 0 \tag{10}$$

$$\frac{d}{d\theta}\int_{r_o}^{r_2} u\,dr + u_o\frac{dr_o}{d\theta} - r_2v_2 + r_ov_o = 0 \tag{11}$$

The surface boundary condition has been employed to simplify Eq. (10).

The integral relations (8) through (11) are now simplified to ordinary differential equations by specifying the dependence of the variables t,u on the radial coordinate r. The interpolating functions are taken to be polynomials in 1/r. The leading terms will be $1/r^2$ to match the far field variation, while a $1/r^3$ term is added for near field modification. The 1/r term is omitted in the present formulation in conformity with the behavior for incompressible flow past a circular cylinder without circulation. We write

$$t = t^{(0)} + t^{(1)}/r^2 + t^{(2)}/r^3 \tag{12}$$

$$u = u^{(0)} + u^{(1)}/r^2 + u^{(2)}/r^3 \tag{13}$$

The coefficients $t^{(k)}$, $u^{(k)}$ can be expressed in terms of the strip values t_k, u_k, k = 0,1,2.

When these expressions are then substituted in the integrands of Eqs. (8) - (11) we find, after some reduction, the following system of ordinary differential equations,

$$Z_{01}t_0' + Z_{11}t_1' + Z_{21}t_2' + r_2h_2 = 0 \tag{14}$$

$$Z_{02}t_0' + Z_{12}t_1' + Z_{22}t_2' + r_1h_1 - r_2h_2 = 0 \tag{15}$$

$$Y_{01}u_0' + Y_{11}u_1' + Y_{21}u_2' + u_0r_0' - r_2v_2 + r_0v_0 = 0 \tag{16}$$

$$Y_{02}u_0' + Y_{12}u_1' + Y_{22}u_2' - r_1v_1 + r_2v_2 = 0 \tag{17}$$

where the coefficients Y_{ij}, Z_{ij} are known functions of r_o, r_1 and r_2, and primes denote differentiation with respect to θ.

It is convenient to introduce two velocity gradient equations for v_0, v_2 in terms of t,u and their derivatives. These equations are identities based on the derivatives of the definitions of t,

$$\frac{dv_k}{d\theta} = \frac{d}{d\theta}\left(\frac{t_k}{\tau_k}\right) = \frac{E_k}{F_k} \tag{18}$$

where k = 0, 2 , respectively.

The specific forms of E_k and F_k are

$k = 0$

$$E_o = C_*^2 \tau_o^{\gamma-2} t_o' + \frac{2}{\gamma+1} u_o v_o^2 \left(\frac{r_o'}{r_o}\right)' \quad (19)$$

$$F_o = C_*^2 (1-u_o^2) - v_o^2 - \frac{2}{\gamma+1} u_o v_o \frac{r_o'}{r_o} \quad (20)$$

$k = 2$

$$E_2 = C_*^2 \tau_2^{\gamma-2} t_2' + \frac{2}{\gamma+1} u_2 v_2 u_2' \quad (21)$$

$$F_2 = C_*^2 (1-u_2^2) - v_2^2 \quad (22)$$

where

$$C_*^2 = \frac{\gamma-1}{\gamma+1} \quad .$$

At this point in the development there are six unknowns u_k, t_k and v_k, $k = 0,2$, which are described by four integral relations, two velocity gradient equations and one surface boundary condition

$$\frac{r_o'}{r_o} = \frac{u_o}{v_o} \quad (23)$$

for a total of seven equations. One equation must have its status lowered. It is natural to permit u_o to be a multiple of v_o as required by Eq. (23), in which case u_2 must be determined by one linear combination of Eqs. (16) and (17). However, the doublet strength D of Eq. (7) enters as a parameter in the problem, requiring that the second linear combination of (16) and (17) be a parametric condition.

Before summarizing the formulations we note that the sum of Eqs. (14) and (15) provides an integral condition on a linear combination of t_o and t_2 in terms of the values on the outer boundary $r = r_1$. Since values of r_1 are given by the asymptotic expression which contains the correct symmetry, it is seen that boundary values applied to either t_k apply to both. The two boundary conditions

$$v_o(0) = v_o(\pi) = 0 \quad (24)$$

imply the four additional conditions

$$v_2(0) = v_2(\pi) = 0$$

$$u_o(0) = u_o(\pi) = 0$$

The basic dependent variables are therefore taken as u_2, v_o, the remaining variables being connected by known algebraic relationships.

The formulation consists of the use of a pair of nonlinear first order differential equations for v_o and u_2

$$\frac{dv_o}{d\theta} = \frac{E_o(v_o,u_2,D,\theta)}{F_o(v_o,u_2,D,\theta)} \quad (25)$$

$$\frac{du_2}{d\theta} = G_2(v_o,u_2,D,\theta) \quad (26)$$

Equation (25) is given explicitly by Eq. (18), while Eq. (26) is the sum of Eqs. (16) and (17). For transsonic cases Eq. (25) has a critical point, θ_*, at which $F_\rho = 0$. To pass smoothly through this saddle point E_0 must vanish simultaneously, which provides an additional condition. However, in this case a shock wave may possibly terminate the supersonic region at an angle θ_s, which provides the additional parameter that is required.

The boundary and subsidiary conditions are:

(1) stagnation points at $\theta = 0,\pi$; $v_0(0) = v_0(\pi) = 0$

(2) regularity at the sonic saddle point, θ_*

(3) Rankine-Hugoniot shock at θ_s on $r = r_0(\theta)$

(4) Compatibility of $u_2(\pi)$, $u_2(0)$ with Eq. (16) above.

The four conditions listed here permit a three parameter iteration to be performed for θ_*, θ_s and D, and determine, in addition, the unknowns themselves u_2, v_0.

Our numerical procedure is to integrate Eqs. (25) and (26) numerically by a 4th order Runge-Kutta procedure. The initial conditions are

$$v_0(0) = 0 \quad , \quad u_2(0) = a$$

and an initial value of D. The value of D is changed until an integral curve is found which passes smoothly through the saddle point at θ_*. In reality the approach to θ_* can only be carried out to some arbitrary degree, and then a polynomial extrapolation performed to jump to the other side of the saddle point. A quartic has been found sufficient for this purpose. The integration of the differential equations is then resumed and continued into the supersonic region. A value of θ_s is assumed and the shock relations are used to provide a jump to subsonic conditions from which the integration is restarted and carried to $\theta = \pi$. The parameter θ_s is then changed until a stagnation point is found at $\theta = \pi$. The value of a is then changed and the entire integration repeated until condition (4) is satisfied.

3. RESULTS AND DISCUSSION

The method is applied to calculate flow past a circular cylinder for a range of free stream Mach numbers between 0.05 and 0.45. Figure 2 shows results for the subcritical Mach number 0.37 and the almost incompressible case $M_\infty = 0.5$. The lowest Mach number curve shows excellent agreement with the analytical incompressible solution.

Figure 3 shows the surface velocity for free stream Mach numbers 0.35, 0.37, 0.39, 0.41, 0.43 and 0.45. The curves shown at the four higher Mach numbers are all unsymmetrical with flow accelerating over the rear cylinder surface until a shock wave is reached.

Up to the present time symmetrical flows (which are shock free) have been calculated up to a free stream Mach number of 0.37. On the other hand, unsymmetrical flows with local supersonic regions terminated by a shock, have been found for Mach numbers down to 0.39. Figure 4 shows the variation of velocity at the highest point of the cylinder with free stream Mach number, calculated both by the present method and by Simisaki (7). On the same graph is shown the variation of critical sound speed, C^*, with free stream Mach number M_∞. The intersection of this curve with the $v(\pi/2)$ curve determines the critical free stream Mach number. Extrapolation of the curve calculated in this paper indicates a critical Mach number close to 0.39. By comparison, use of Simisaki's surve indicates a critical Mach number of 0.40.

The precise value of the critical Mach number has still to be determined by the present method. For this purpose a separate calculation is required since both the subcritical and supercritical procedures need to be refined as the critical condition is approached and neither can be extended all the way to this condition.

Figure 5 shows the variation of stagnation point velocity gradient with free stream Mach number comparing the present results with a curve derived by the Jantzen-Rayleigh method. As expected, the expansion method overestimates the gradient as the critical Mach number is approached.

The present calculations are considered to give reliable values of surface velocity (and pressure) at supercritical Mach numbers between 0.41 and 0.45 and at subcritical Mach numbers below 0.35. In the intermediate range a three or four strip approximation is needed for accurate results and the iteration procedure should also be refined in this sensitive region. These extensions of the method are now being carried out. In addition, the formulation is geing generalized to apply to a wide class of airfoil profiles for flow both with and without circulation.

We wish to thank Mr. Gordon Duckworth for his assistance in the programming.

REFERENCES

1. Nieuwland, G. Y. Technical Report T-172, NLR, The Netherlands.

2. Magnus, R. and Yoshihara, H. General Dynamics-Convair Div., Rept. 1969 "Inviscid Transsonic Flow over Airfoils."

3. Cole, J. D. and Murman, E. M. Boeing Scientific Research Lab. Document D1-82-0943, 1970.

4. Chushkin, P. I. Vych. Mat. 2, 20-44, 1958.

5. Chushkin, P. I. Vych. Mat. 3, 99-110, 1958.

6. Holt, M. Symposium Transsonium (Ed. K. Oswatitsch), pp. 310-324, Springer-Verlag, 1962.

7. Simisaki, T. Bull. Univ. Osaka Prefecture A 4, 27-35, 1955.

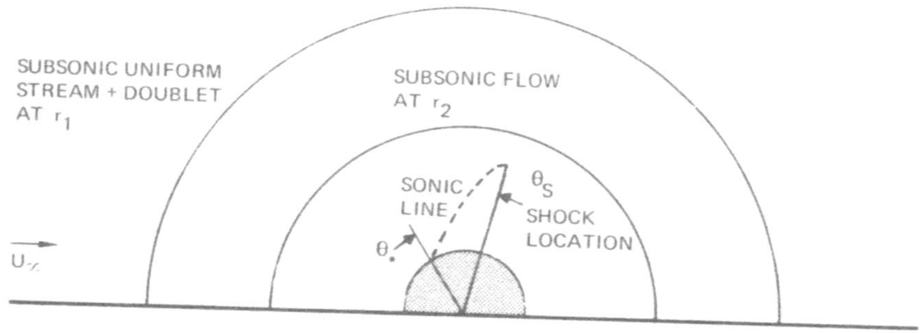

Fig. 1 Two strip MIR applied to transonic flow over a circular cylinder

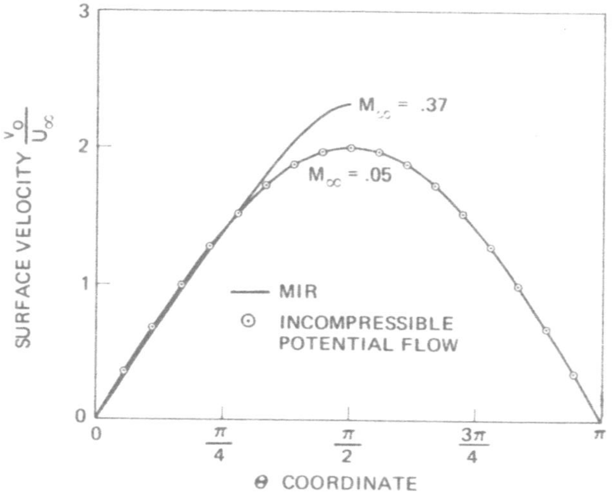

Fig. 2 Comparison of surface velocity for $M_\infty = .05$ with incompressible potential theory

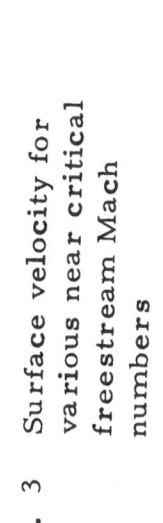

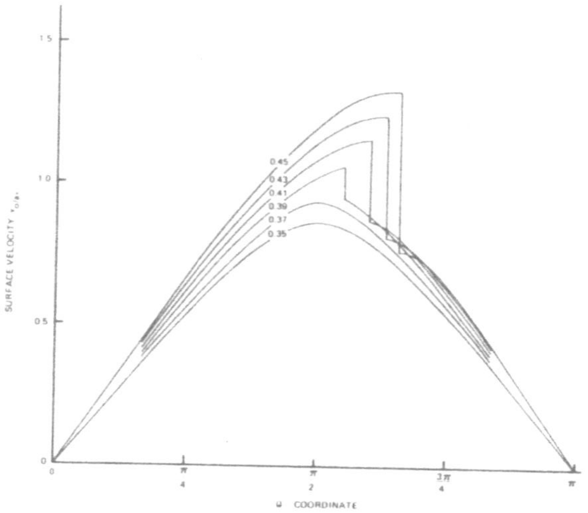

Fig. 3 Surface velocity for various near critical freestream Mach numbers

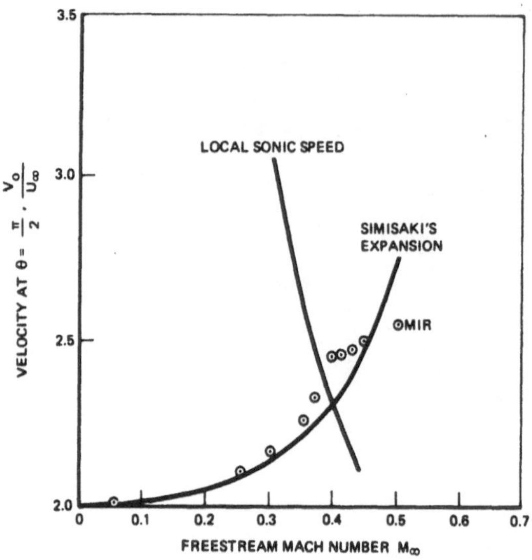

Fig. 4 Comparison of surface velocity at $\theta = \pi/2$ with Simisaki's five term expansion in M_∞^2

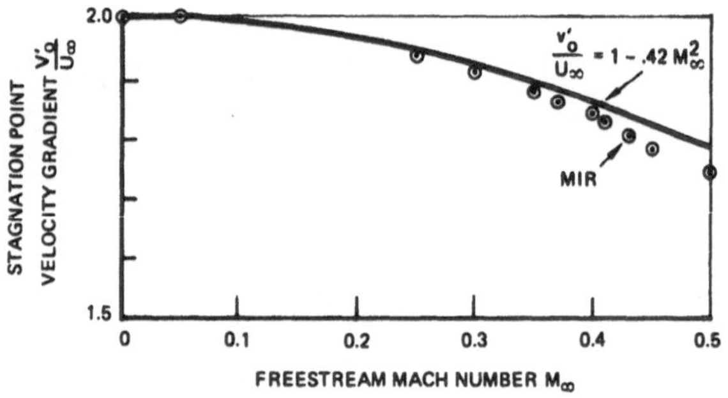

Fig. 5 Comparison of the stagnation point velocity gradient on a circular cylinder with the two term Rayliegh-Janzen Expansion for a circular cylinder

THE CALCULATION OF HYDRODYNAMIC FORCES WITH TANGENTIAL DISCONTINUITIES

B. D. Moiseenko and B. L. Rozhdestvenskii

Institute of Applied Mathematics
Academy of Sciences, U.S.S.R., Moscow

1. Introduction

We consider stationary flows of non-viscous fluid inside plane channels or axially symmetric nozzles. The flow can consist of several streams sliding over one another and along the walls of the channel. In the case of an electroconductive fluid we take into account the influence of an external magnetic field.

A typical longitudinal section of the channel (nozzle) and the coordinate system adopted are shown in Fig. 1.

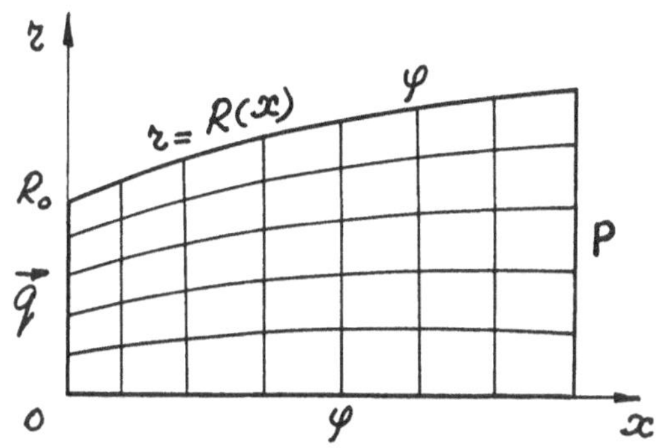

Fig. 1.

In the case of a plane channel, x and r are Cartesian coordinates; in the case of a nozzle--cylindrical coordinates. In the latter case the line r = 0 is the axis of the nozzle.

The velocity vector we denoted by $\vec{q}$ (x, r). If the magnitude $q_0(r)$ of the velocity vector in the entry section of the channel has some jumps, then the flow has tangential discontinuities on the streamlines going out from the points of discontinuity of $q_0(r)$.

In this paper we shall not consider the problems of instability of these tangential discontinuities, development of boundary layers and distortions.

We shall consider only steady solutions of the equations of motion of a non-viscous non-heat-conducting fluid

$$\rho(\vec{q}\ \nabla)\ \vec{q} + \nabla\ p = \frac{\sigma}{c^2}\ [[\vec{q}\ \vec{H}]\ \vec{H}] \qquad (1)$$

$$\mathrm{div}(\rho\ \vec{q}) = 0$$

p, ρ being pressure and density of fluid, respectively, and $\vec{H}$ a given external magnetic field strength.

In the case of a compressible fluid the equation of state should be added to these equations.

We consider now the case when ρ is constant along streamlines, but can change on different streamlines.

2. Choice of Variables

Cartesian components of velocity are not convenient for calculation of flows with tangential discontinuities. So as principal variables we choose the magnitude of the velocity vector q and the angle ϕ between the velocity vector and the x-axis. The angle ϕ is continuous everywhere in the flow.

By means of this choice of variables the equations (1) can be transformed into two equations

$$p_x - \rho q^2 \phi_r = \frac{\sigma}{c^2} [[\vec{q}\,\vec{H}]\,\vec{H}]_{(x)} + \frac{\nu\rho q^2}{r} \cos\phi \sin\phi$$

$$p_r + \rho q^2 \phi_x = \frac{\sigma}{c^2} [[\vec{q}\,\vec{H}]\,\vec{H}]_{(r)} + \frac{\nu\rho q^2}{r} \sin^2\phi \qquad (2)$$

containing only derivatives of continuous functions and the characteristic equation of system (1) - Bernoulli's equation:

$$\rho\, q\, q_e + p_e + \frac{\sigma}{c^2} \frac{[\vec{q}\,\vec{H}]^2}{q} = 0 \quad . \qquad (3)$$

Equation (3) contains the derivatives of the discontinuous function $q(x,r)$ only along the streamlines

$$q_e = \frac{dq}{d\ell} = (\vec{q}\,\nabla)\, q \qquad (4)$$

and q is continuous along streamlines.

In the equations (2) $\nu = 0$ in the case of a plane channel, and $\nu = 1$ for the axially symmetric flows.

3. Boundary Conditions

The formulation of boundary conditions for the system (2), (3) depends on the physical and technical requirements.

In this paper we are interested mostly in the methods of calculating flows with tangential discontinuities, so we apply certain boundary conditions, which are described below.

In the entry section of the channel $(x = 0)$ the velocity vector $\vec{q}\,(r)$ is given. Its magnitude $q_o(r)$ may be discontinuous at several points, but its direction is continuous. On the walls of the channel the velocity vector $\vec{q}$ is required to be directed along the wall, i.e., ϕ is given. In the exit section $(x = L)$ the pressure p is given.

We suppose that this boundary problem for system (1) is correctly posed. Naturally it is possible to consider others.

4. The Difference Scheme

The difference scheme we use is iterative. Let the approximation

$$\overset{(s)}{\phi}(x, r) \ , \ \overset{(s)}{p}(x, r) \ \text{ to } \ \phi \ , \ p$$

be known. For the calculation of the next iterative values

$$\overset{(s+1)}{\phi} \ , \ \overset{(s+1)}{p}$$

we use a new grid which is formed by intersections of the family of lines $x = x_i$ and the family of lines

$$r = \overset{(s+1)}{r_j}(x) \quad ,$$

where

$$\frac{d\overset{(s+1)}{r_j}}{dx} = \operatorname{tg} \overset{(s)}{\phi}(x, \overset{(s+1)}{r_j}(x)) \quad , \quad \overset{(s+1)}{r_j}(0) = r_j^{\,o}$$

Thus the lines $r = r_j^{(s+1)}(x)$ are approximations of streamlines outgoing from the points $r = r_j^o$ of the input section.

The intersection of these two families of lines

$$(x = x_i \quad , \quad r = r_j^{(s+1)}(x))$$

give us the points of the new grid.

The grid changes from one iteration to another. We call these iterations "outer iterations."

The values

$$p_{ij}^{(s+1)} \quad , \quad \phi_{ij}^{(s+1)}$$

are related to the full nodal points

$$(x_i, r_{ij}^{(s+1)})$$

of the grid and the values

$$q_{ij+1/2}^{(s+1)}$$

to the "middle" points

$$(x_i \ , \frac{1}{2}(r_{ij}^{(s+1)} + r_{ij+1}^{(s+1)}))$$

The new values

$$q_{ij+1/2}^{(s+1)}$$

are obtained from Bernoulli's equation (3) or directly from the equation of continuity

$$\operatorname{div}(\rho \vec{q}) = 0 \tag{5}$$

The difference analog of (5) is

$$q_{ij+1/2}^{(s+1)} (r_{ij+1}^{(s+1)\nu+1} - r_{ij}^{(s+1)\nu+1}) \cos \phi_{ij+1/2}^{(s)} = \text{const}_j \tag{6}$$

for each valid $i = 0, 1, 2, \ldots, N$

The values

$$p_{ij}^{(s+1)} \quad , \quad \phi_{ij}^{(s+1)}$$

are obtained from the difference equations which are formed from equations (2) by means of replacement of derivatives by finite differences.

For this replacement we use the approximative equalities

$$\frac{\partial f}{\partial x} = \frac{1}{S_{ij}} \begin{vmatrix} f_{i+1j} - f_{ij+1} & r_{i+1j} - r_{ij+1} \\ f_{i+1j+1} - f_{ij} & r_{i+1j+1} - r_{ij} \end{vmatrix}$$

$$\frac{\partial f}{\partial r} = \frac{1}{S_{ij}} \begin{vmatrix} x_{i+1} - x_i & f_{i+1j} - f_{ij+1} \\ x_{i+1} - x_i & f_{i+1j+1} - f_{ij} \end{vmatrix}$$

$$S_{ij} = \begin{vmatrix} x_{i+1} - x_i & r_{i+1j} - r_{ij+1} \\ x_{i+1} - x_i & r_{i+1j+1} - r_{ij} \end{vmatrix}$$

and obtain the difference equations:

$$A_1 \begin{pmatrix} p_{i+1j+1} - p_{ij} \\ p_{i+1j} - p_{ij+1} \end{pmatrix} + A_2 \begin{pmatrix} tg\,\phi_{i+1j+1} - tg\,\phi_{ij} \\ tg\,\phi_{i+1j} - tg\,\phi_{ij+1} \end{pmatrix} = F \qquad (7)$$

$$A_1 = \begin{pmatrix} r_{ij+1} - r_{i+1j} & r_{i+1j+1} - r_{ij} \\ x_{i+1} - x_i & x_i - x_{i+1} \end{pmatrix} \qquad (8)$$

$$A_2 = \begin{pmatrix} x_i - x_{i+1} & x_{i+1} - x_i \\ r_{ij+1} - r_{i+1j} & r_{i+1j+1} - r_{ij} \end{pmatrix} (\rho\, q^2 \cos^2 \phi)_{i+\frac{1}{2}\, j+\frac{1}{2}}$$

In system (7) p_{ij}, ϕ_{ij} denote the unknown variables

$p_{ij}^{(s+1)}$, $\phi_{ij}^{(s+1)}$.

In the equalities (8) r, q, ϕ denote

$r^{(s)}$, $q^{(s+1)}$, $\phi^{(s+1)}$.

The scheme we have obtained has second order accuracy.

5. Methods of Solution of Difference Equations (7)

These equations are linear in the p_{ij}, $tg\phi_{ij}$, and with the boundary conditions they determine uniquely all unknowns of p_{ij}, $tg\phi_{ij}$

Two method of solution have been used:

a) Method of successive approximations; the iteration of this method we call "inner iteration."

b) Modified method of orthogonal double sweep (method of exact solution of system (7)).

The method of successive approximations we used is similar to the method of simple iteration for the finite-difference approximation of Laplace's equation. Although this method appeared to be suitable in most cases, some restrictions are required for its convergence.

The second method is free from such restrictions, so we use now mostly this method.

6. The Calculation of Plane Flow

To test the accuracy of our procedures we consider the following example of plane irrotational flow with $H = 0$ in which the density ρ is constant along each streamline.

This flow can be described with the aid of the complex potential $F(z)$, an analytic function of the complex variable $z = x + ir$.

The streamlines (and the walls of the channel) are defined by the equation

$$\operatorname{Im} F(z) = \text{Const} \quad , \tag{9}$$

p, ρq^2, ϕ - by the formulas

$$\rho q^2 = |F'(z)|^2 \quad , \quad \phi = -\operatorname{Arg} F'(z) \quad , \quad p = B - \frac{\rho q^2}{2} \tag{10}$$

(B is a constant in Bernoulli's equation.)

The flow of two fluids with tangential discontinuity can be described by two complex potentials $F_1(z)$ and $F_2(z)$, which determine the common streamline, i.e., a line on which the two conditions

$$\operatorname{Im} F_1(z) = \text{Const}_1 \quad ; \quad \operatorname{Im} F_2(z) = \text{Const}_2 \tag{11}$$

are simultaneously satisfied. Besides, on this line, the condition of continuity of pressure

$$|F_1'(z)|^2 - |F_2'(z)|^2 = \text{Const} \tag{12}$$

has to be satisfied.

There exists a trivial case when each function $F_1(z)$ and $F_2(z)$ differs from the analytic continuation of the other only by a constant. In this case the quantity ρq^2 is continuous through the tangential discontinuity. We do not know any non-trivial example of analytic functions $F_1(z)$, $F_2(z)$ satisfying the conditions (11) and (12).

Because of this we have chosen for the test computations a flow with a continuous value ρq^2. This flow is defined by the following complex potential

$$F(z) = \frac{1}{2} (z - \lambda)^2$$

where λ is a real number.

In Fig. 2 the streamlines and the walls of the channel corresponding to this complex potential are shown. Figure 2 illustrates also the initial approximation to the streamlines. The computations were performed with both 10^2 and 19^2 grid points.

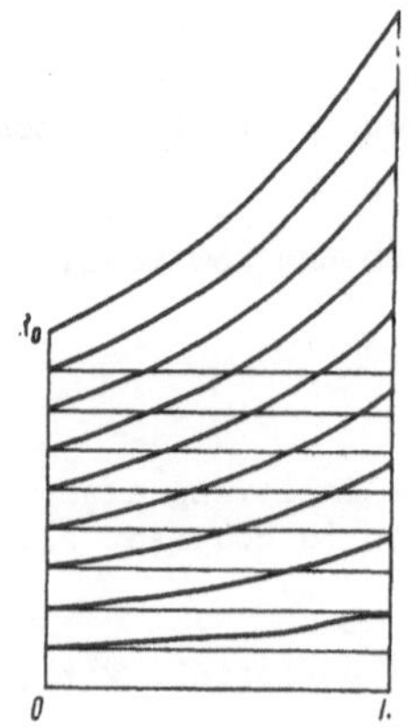

Fig. 2.

In Table I the computed values of pressure (in a single section) and streamline coordinates (in two sections) are presented and compared with the corresponding exact values. In this example the ratio of areas of entry and exit flow of the channel is 1.9.

TABLE I

The values of pressure, $x = 0.5$

exact	computed with 10^2 points	computed with 19^2 points
17.020	17.018	17.019
17.011	17.008	17.009
16.983	16.980	16.981
16.938	16.933	16.935
16.873	16.867	16.870
16.790	16.782	16.736
16.688	16.679	16.684
16.569	16.558	16.565
16.431	16.420	16.427
16.274	16.264	16.271

The values of coordinates

exact $x=0.5$	computed by 10^2 points	computed by 19^2 points	exact $x=0.9$	computed by 10^2 points	computed by 19^2 points
0.1357	0.1349	0.1353	0.19	0.1883	0.1892
0.2714	0.2697	0.2707	0.38	0.3767	0.3785
0.4171	0.4048	0.4061	0.57	0.5653	0.5678
0.5429	0.5399	0.5415	0.76	0.7542	0.7573
0.6786	0.6752	0.6770	0.95	0.9435	0.9470
0.8143	0.8107	0.8127	1.14	1.1334	1.1370
0.9500	0.9467	0.9485	1.33	1.3243	1.3275
1.0857	1.083	1.0846	1.52	1.5162	1.5184
1.2214	1.221	1.2214	1.71	1.7100	1.7100

It can be seen from this table that the accuracy is quite good.

7. The Calculation of Radial Outflow

This flow is defined by the law of mass conservation and the Bernoulli integral:

$$\rho q R^2 = K \quad ; \quad p + \frac{\rho q^2}{2} = B$$

where R is the distance from the center of outflow 0.

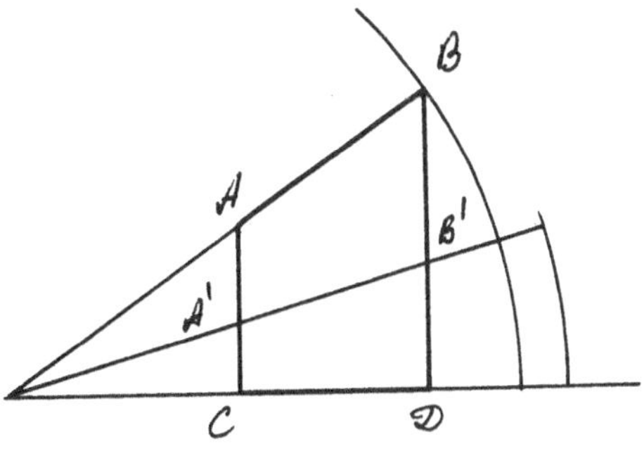

Fig. 3.

The test computations were performed in the part of the cone OBD contained by two planes AC and BD (see Fig. 3).

The tangential discontinuity was situated on the conical surface A'B'.

In Table II the computed values of the pressure (in two sections) are chosen and compared to the corresponding exact values. The number of points in the grid is 10^2, the angle of the cone OBD $\sim 10°$, K' = 18, B = 4.

TABLE II

Pressure x = 0.5		Coordinates			
		x = 0.5		x = 0.9	
exact	computed	exact	computed	exact	computed
2.920	2.930	0.1167	0.1179	0.13	0.1325
2.923	2.933	0.2333	0.2297	0.26	0.2530
2.930	2.939	0.3500	0.3441	0.39	0.3783
2.942	2.949	0.4667	0.4593	0.52	0.5055
2.958	2.963	0.5833	0.5795	0.65	0.6346
2.978	2.982	0.7000	0.6924	0.78	0.7654
3.002	3.004	0.8167	0.8104	0.91	0.8980
3.029	3.030	0.9333	0.9294	1.04	1.0326
3.059	3.060	1.0500	1.0500	1.17	1.1700
3.091	3.094				

The exact position of the streamlines as well as their initial approximations are shown in Fig. 4.

Although the initial approximations of the streamlines are very poor (the broken lines in Fig. 4), the outer iterations converge and give quite good results (see Table II).

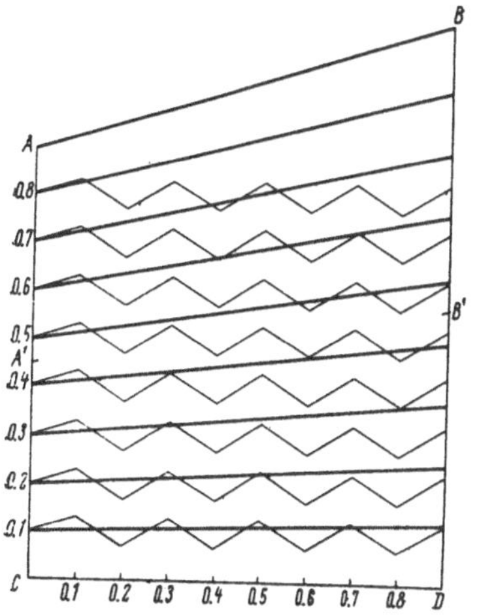

Fig. 4.

8. Calculation of Flow with Discontinuous ρq^2

We computed the flow with a single tangential discontinuity on which the quantity ρq^2 was discontinuous. The channel and streamlines are shown in Fig. 5. In this example the quantity ρq^2 changes in the entry section by a factor of 25.

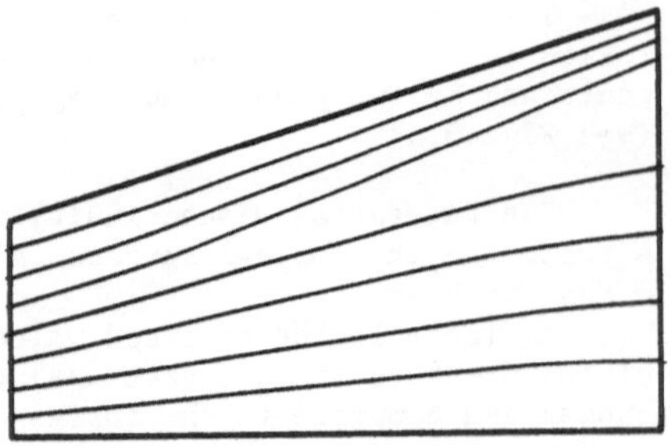

Fig. 5.

The most interesting effect observed in the computations is the compression of streams in the domain where ρq^2 is great and the expansion of streams where ρq^2 is small.

9. Conclusions

The performed computations have shown the suitability and sufficiently good accuracy of our procedures for the calculations of flows with tangential discontinuities.

We intend to proceed with calculations in cases of compressible electroconductive fluid.

10. Acknowledgement

We are thankful to Mrs. V. Sidorova for her help in programming and computations.

Reference

1. B. D. Moiseenko and B. L. Rozhdestvenskii, "The Numerical solution of stationary hydrodynamic equations with tangential discontinuities," Journal of Computational Mathematics and Mathematical Physics, No. 2, 10 (1970), pp. 499-505.

SOME RESULTS OF CALCULATIONS OF FLOWS AROUND CONICAL BODIES AT LARGE INCIDENCE ANGLES

A. P. Bazzhin

Physical Technical Institute, Moscow, U.S.S.R.

A series of calculations of flows around conical bodies at large incidence angles was made. The methods used allowed the construction of a solution in the major part of a flow field, with the exception of a region above the body. Some of the results obtained are presented in this paper. It is shown, in particular, that the increase of the angle of incidence results in changing the entropy distribution around a circular cone; the thin entropy layer near the surface is no longer observed at very large incidence angles. In accordance with Sychev's theory the normal force coefficient of various conical bodies may be presented as a function of two similarity parameters. The strength of the conical shock wave depends on the body geometry; in some cases the shock wave is expected to be of nonzero intensity all along its surface and remain closed in spite of large incidence angle.

- - - - -

1. Conical flows have been the subject of many papers. Most of them dealt with the flows around conical bodies at small or intermediate angles of incidence. These are, for instance, recent papers [1] - [3]. At the angles of incidence referred to the system of governing equations is of elliptic type and one has to construct a solution in the whole region between the body and the shock wave simultaneously. On the other hand, if the angles of incidence are large enough, the physical nature of flow past conical bodies is such that there is a possibility of dividing the problem into two separate problems, as is usually done when dealing with a blunt body problem in a supersonic stream.

2. Consider a circular cone (or - an arbitrary cone with a smooth cross section) placed in a supersonic stream of an inviscid gas at large angle of incidence. The paper deals only with angles of incidence when there is a closed region I, adjacent to the lower surface of the body, in which the system of conical flow equations is of elliptic type (Fig. 1a). The flow in this region was numerically calculated by the method of straight lines [5]. (We construct a steady flow field in the plane $X = 1$. The X axis of the cartesian coordinate system coincides with the axis of the body).

Further downstream there exists another region - II, where the system of governing equations is of hyperbolic type. The method of characteristics was applied to calculate the flow field in this region [6]. These two methods enable us to construct solutions in the major parts of flow fields and in some cases even to reach the upper plane of symmetry ($\omega = \pi$). The conditions of flow symmetry, however, are not satisfied, which means that in general internal shock waves must occur above the upper surface of the body.

It is clear, that the problem of gas flow above the upper surface of a conical body should be handled with consideration of all the effects due to viscosity. But our aim was much more modest, it was to find a solution for inviscid gas flow. In principle, however, from the results of the calculations of inviscid gas flow fields it is possible to give an approximate estimate of the size of flow field most affected by viscosity. Figure 1b shows the upper part of the calculated flow field around the circular cone with the semiangle $\theta_c = 20°$, the angle of incidence 40°, and $M_\infty = 7$. The boundary layer was assumed to separate in the plane with the circumferential angle $\omega = 140°$. The "boundary" of the separation zone was represented by a solid surface AC, with an angle of about 20° between this plane and the plane tangent to the cone on the corresponding generator A. The internal shock wave AB, appearing upstream of such a "separation zone," was calculated. This bounds a relatively narrow region, adjacent to the upper plane of symmetry, in which the flow is entirely determined by the effects of viscosity. In the rest of the flow field the viscosity effects are represented mainly by the existence of the boundary layer. And in this sense the results of numerical calculations of inviscid gas flows can be taken as valid in the major part of the whole flow field.

3. The numerical calculations of conical flows have been made within a fairly wide range of defining parameters. Elliptic cones with semiangles $\theta_c = 10°$, $15°$ and $20°$ and with the ratio of cross section axes $\delta = 1$, 2 and 3 have been considered. In most cases the angle of incidence varied from 30° up to 50°. The bulk of calculations have been made for supersonic free stream Mach number 7. Many variants of flow defined by different parameter combinations have also been calculated.

Here are some comments of the results obtained.

a) The entropy distribution in a flow field around the cones at large angles of incidence can be qualitatively different from the one at small incidence. Figure 2 shows the entropy profiles in various planes $\omega = \text{const}$ in the case of the circular cone with $\theta_c = 15°$ at $M_\infty = 7$ and angles of incidence $\alpha = 10°$, $30°$ and $50°$. (Data at $\alpha = 10°$ are borrowed from reference [1].) In this figure n is a normalized distance from the body surface $(n = 0)$ up to the shock wave $(n = 1)$ measured along the ray $\omega = \text{const}$ in the plane $X = 1$. It follows from these curves that there is a distinct qualitative difference between the entropy distributions at $\alpha = 10°$ and $\alpha = 50°$. At the angle of incidence 10° there exists the well-known thin layer of rapidly changing and large gradients of entropy near the body. On the other hand, there is no such layer at the angle of incidence 50°, and it is even possible to discuss the existence of a layer near the surface, where the entropy is nearly constant and its gradients are rather small. The entropy distribution at the angle of incidence 30° represents some intermediate state compared with the two former ones.

b) The fact that the calculations have been carried out within the wide range of parameters enables us to estimate the shock wave intensity under various flow conditions. From a series of experimental measurements of the shock wave shape the authors of reference [4] came to the conclusion that at large incidences the shock wave degenerates into the Mach wave, its intensity becomes zero and it is no longer a closed surface. This conclusion was made for the circular cone with $\theta_c = 9°$ at $M_\infty = 6.85$, $Re_\infty = 110000$ 1/sm and $\alpha = 15°$. (The measurements of the shock wave shape were performed in the plane $X = 185$ mm). A similar conclusion that the shock wave degenerates into a Mach wave in the plane of symmetry at the angle of incidence near 12° was made by the author of reference [3]. This conclusion was based upon the results of the calculations of the inviscid perfect gas flow past the circular cone with $\theta_c = 9°$ at $M_\infty = 7$ and incidence angles $\alpha = 5$, 7, 9, 10 and 11°.

Our results show that this conclusion is not the only one possible. To be more exact, it is valid for relatively slender cones or not very great Mach numbers of the free stream. With thicker cones or greater Mach numbers the shock wave might be expected to retain nonzero intensity all along its length and to be closed.

Figure 3 shows the calculated shock waves around the circular cone with the semiangle 20° at various angles of incidence up to 50°. The points C are traces of the characteristic from the cone vertex in the plane $X = 1$ at $\omega = \pi$. At the incidence angles 40° and 50° the flow fields have been calculated up to the planes with ω equal to 165° and 167°, correspondingly. The shock wave intensities in these planes are still not zero and will probably not be zero even in the plane of symmetry $\omega = \pi$.

Distributions of the shock wave intensities versus the circumferential angle ω at various angles of incidence are shown in Fig. 4. The ratio ρ/ρ_∞ is taken here as a measure of the shock wave intensity. These curves illustrate once again the probability for the shock waves to retain nonzero intensities all along their length at large angles of incidence. The two curves for $\alpha = 10°$ and $\alpha = 15°$ are borrowed from reference [1].

c) Circular cones represent a family of affinely related bodies. According to Sychev's theory [7], the aerodynamic coefficients of slender circular cones at large incidence angles and the parameters of flow fields around them must be functions of two similarity parameters, namely $K_1 = 2\ \mathrm{tg}\ \theta_c \cdot \mathrm{ctg}\ \alpha$ and $K_2 = M_\infty \sin \alpha$. In Fig. 5 the normal force coefficient, divided by $\sin^2 \alpha$, is represented versus K_1 parameter with K_2 equal to 3, 4 and 5. (A quadratic interpolation of calculated data was used to obtain the values of the coefficient C_n^* at the integer values of K_2, which seems to be a quite permissible procedure.) The dependence of the coefficient $C_n^* \equiv C_n/\sin^2 \alpha$ on those two similarity parameters K_1 and K_2 only is quite evident. Furthermore, this dependence includes also the cases when circular cones are rather thick (e.g., $\theta_c = 20°$, point V) or the free stream Mach number rather small ($M_\infty = 4$, point I). The points at $K_1 = 0$ correspond to the flows past the circular cylinder with the free stream velocities corresponding to the integer values 3, 4 and 5 of the parameter $K_2 \equiv M_\infty \sin \alpha$. (The corresponding values of C_n^* vary from 1.23 to 1.24 and are represented by a single point in Fig. 5.) The tangent force coefficient may be represented in a similar form. There is, however, greater scattering of data points and all the pattern is not as evident as in the previous case of the coefficient C_n^*.

d) The information about flow fields around elliptic cones, obtained by the two methods mentioned above, was used for heat-transfer calculations. Figure 6 shows the distribution of the heat-transfer rates along any cross section body contour of elliptic cones with various axes ratio δ. These data are due to the courtesy of V. A. Bashkin [8]). For the heat-transfer calculations the Prandtl number was assumed to be 0.7, the dynamic viscosity coefficient proportional to the temperature to the 0.76 power ($\mu \sim T^{0.76}$). It was also assumed that there was an intensive heat transfer at the surface of the body ($H_w/H_e = 0.05$, where H_w is the enthalpy at the surface of the body and H_e is the enthalpy at the outer edge of the boundary layer). The calculations were performed by the method of integral relations [9]. It can be seen from the figure, that the variation of δ results in qualitative alterations of the heat-transfer rates distribution along the body cross section. For some value of δ there occurs a minimum among the greatest local heat-transfer rates. The distribution of heat-transfer rates on the windward side of the body is most uniform in this optimum regime.

In conclusion the author would like to express his acknowledgement of the valuable help of his colleagues Miss J. F. Tchelysheva and Mrs. O. N. Trusova, who have performed all the calculations.

References

1. Babenko, K. I., et al. "Three dimensional flow of an ideal gas past smooth bodies," "Nauka." Moscow, 1964.
2. Moretti, G., "Inviscid flowfield about a pointed cone at an angle of attack," AIAA J., 1967, IV, v. 5, N 4, pp. 789-791.
3. Gonidou, R., "Écoulements supersoniques autour de cônes en incidence." La Recherche Aérospatiale, N 120, Sept. Oct. 1967.
4. Guffroy, B. Roux, J. Marcillat, R. Brun & J. Valensi, "Étude theorique et expérimentale de la couche limite autour d'un cone circulaire placé en incidence dans un courant hypersonique," AGARD Confrerence Proceedings N 30, May 1967.
5. Bazzhin, A. P. & Chelysheva, I. F., "Application of the method of straight lines to the calculation of flow past conical bodies at large angles of attack," Izv. Akad. Nauk SSSR M. Zh. G. No. 3, 1967.
6. Bazzhin, A. P., Trusova, O. N. & Chelysheva, I. F., "Calculation of flow of a perfect gas past elliptic cones at large angles of attack," Izv. An SSSR M. Zh. G. No. 4, 1968.

7. Sychev, V. V., "Three dimensional hypersonic flow of a gas past pointed bodies at large angles of attack," Prik. Mat. i Mekh. XXIV 2, 1960.
8. Bashkin, V. A., "Investigation of heat transfer on pointed elliptic cones in a supersonic stream at large angles of attack," Izv. Akad Nauk SSSR M. Zh. G. 1, 1969.
9. Bashkin, V. A., "Calculation of the equations of three dimensional laminar boundary layers by the method of integral relations," Zh. Vych. Mat. i Mat. Fiz. 8, 6, 1968.

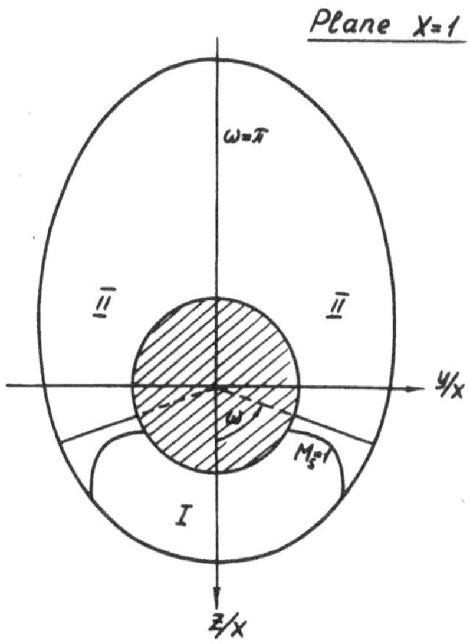

Fig. 1a

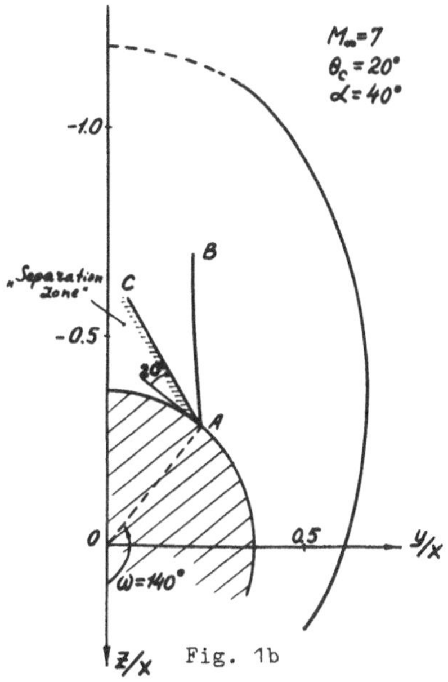

Fig. 1b

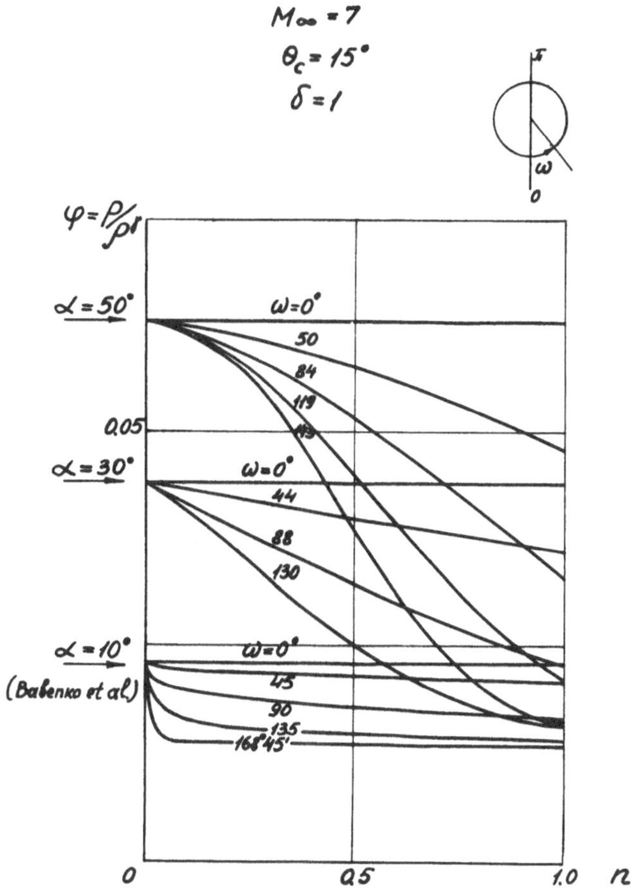

Fig. 2

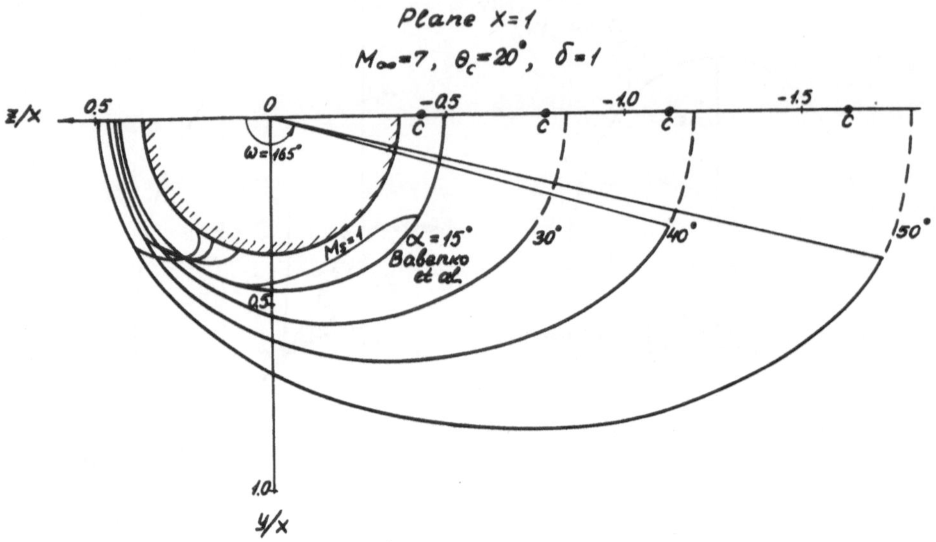

Plane X=1
$M_\infty=7$, $\theta_c=20°$, $\delta=1$
z/x
0.5
0
-0.5
-1.0
-1.5
C
C
C
C
ω=165°
$M_s=1$
α=15°
Babenko et al.
30°
40°
50°
0.5
1.0
y/x

Fig. 3

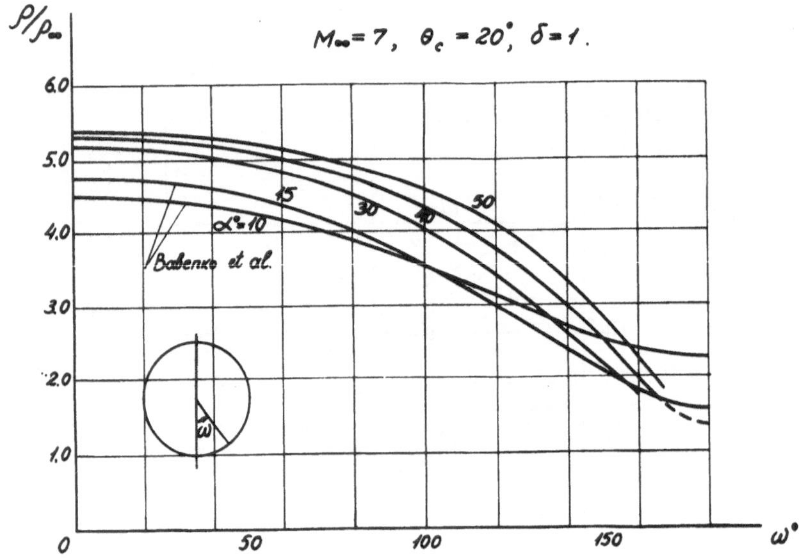

ρ/ρ∞
$M_\infty=7$, $\theta_c=20°$, $\delta=1$.
6.0
5.0
4.0
3.0
2.0
1.0
0
50
100
150
ω°
15
30
40
50
α°=10
Babenko et al.
ω

Fig. 4

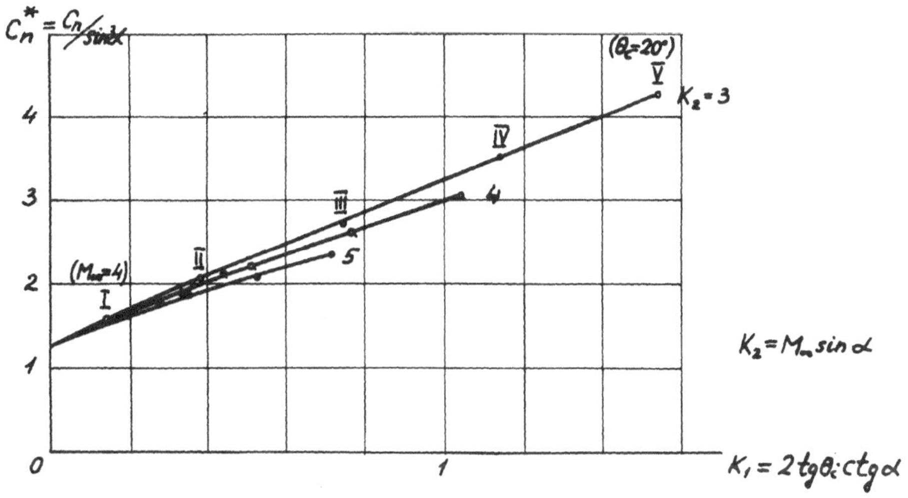
$C_n^* = C_n/\sin^2\alpha$
$(\theta_c = 20°)$
$K_2 = 3$
$(M_\infty = 4)$
I
II
III
IV
V
4
5
$K_2 = M_\infty \sin\alpha$
$K_1 = 2\,tg\theta_c\,ctg\alpha$
0
1
2
3
4

Fig. 5

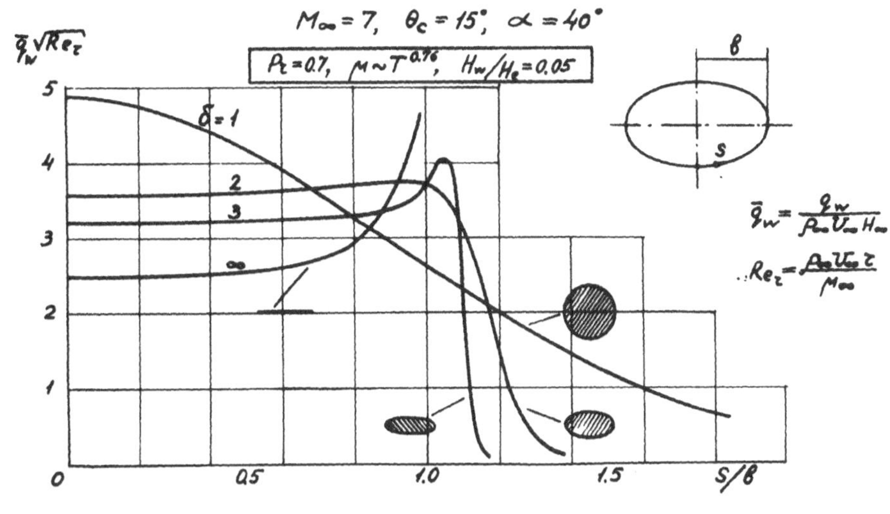
$M_\infty = 7, \quad \theta_c = 15°, \quad \alpha = 40°$
$Pr = 0.7, \quad \mu \sim T^{0.76}, \quad H_w/H_e = 0.05$
$\bar{q}_w\sqrt{Re_\tau}$
$\delta = 1$
2
3
∞
$\bar{q}_w = \frac{q_w}{\rho_\infty V_\infty H_\infty}$
$Re_\tau = \frac{\rho_\infty V_\infty \tau}{\mu_\infty}$
S/b
0.5
1.0
1.5

Fig. 6

THE COMPUTATION OF GENERAL PROBLEMS IN ONE DIMENSIONAL UNSTEADY FLOW BY THE METHOD OF CHARACTERISTICS

N. E. Hoskin and B. D. Lambourn

Atomic Weapons Research Establishment, Aldermaston, Berkshire, England

This paper describes the logical structure of a general purpose characteristic code RICSHAW which overcomes many of the difficulties that have previously discouraged the use of such codes. The code has the ability to consider multiple shocks, shock interactions and formation of shocks in compression waves in any number of regions using arbitrary equations of state. The relative advantages of characteristic and finite difference schemes are discussed and comparisons are given with calculations using a finite difference mesh code. It is shown that for similar usage of computer time the characteristic code gives smoother solutions, with much greater certainty of the position of shocks and other waves, and a physical insight into the wave phenomena that occur.

1. INTRODUCTION

The main phenomena occurring in one dimensional unsteady compressible fluid flow are controlled by waves. Discontinuities in flow parameters are propagated as shock waves and discontinuities in derivatives as boundaries of compression or rarefaction waves. Furthermore the number of waves multiplies because each wave is both transmitted and reflected at material interfaces.

Finite difference mesh methods of solving such problems avoid the difficulties of tracking waves by the addition of pseudo-viscous terms to smear shock discontinuities and the use of a finite mesh size which smears derivative discontinuities. The approach is very successful in assessing overall behaviour, can give considerable detail at the expense of increased computing time, but suffers from a number of disadvantages. These include the poor treatment of reflected rarefactions in shock/interface interactions, the frequent presence of small oscillations on the solution and perhaps most important the loss of the essential physics of wave motion.

In this paper we describe a general 1D code which uses the more fundamental method of characteristics. Shocks and their interactions are treated explicitly, and discontinuities in derivatives are automatically tracked since they propagate along characteristics. Characteristic methods are well known but previously have only been used to solve specific problems with few shocks, e.g. [1-6], since the logical complexity increases severely with the number of waves. The new code, RICSHAW, has a logical structure (section 2) which extends the method to track multiple shocks in a number of materials. The results of RICSHAW and mesh calculations are compared in section 3, and the two methods contrasted in section 4.

2. THE STRUCTURE OF RICSHAW

2.1 General

The equations of motion of 1D unsteady flow in characteristic form may be written

$$dp + \rho c\, du + \alpha\, \rho u c^2\, dt/R = 0 \quad \text{on} \quad da = +\,\rho c\, R^{\alpha}\, dt \tag{1}$$

$$dp - \rho c\, du + \alpha\, \rho u c^2\, dt/R = 0 \quad \text{on} \quad da = -\,\rho c\, R^{\alpha}\, dt \tag{2}$$

$$de + p dv = 0 \quad \text{on} \quad da/dt = 0 \tag{3}$$

where t, R, p, ρ, v, e, u, c are time, Eulerian coordinate, pressure, density, specific volume, specific internal energy, particle velocity and velocity of sound respectively; $\alpha = 0, 1, 2$, for plane, cylindrical and spherical geometry and a is a

Lagrangian mass coordinate defined by

$$da = \rho R^{\alpha} (dR - udt) \tag{4}$$

Equations (1) to (4) together with an equation of state of the form

$$p = p(\rho, e) \tag{5}$$

are sufficient to solve for the flow away from shocks. Shocks are treated as discontinuities across which the Rankine-Hugoniot conservation relations hold.

The numerical solution of (1) to (3) is performed by replacing the total differentials by differences and the coefficients (ρc etc.) by averages. The equations are solved by iteration.

In a characteristic code the solution is determined at the intersection points of two families of discrete positive and negative characteristics, and also at the intersections of characteristics with shocks, interfaces and boundaries. Thus there are four main types of mesh point, each being split into several cases requiring slightly different computational or logical procedure.

A typical characteristic mesh generated by RICSHAW is shown in Figure 1 in the space defined by time (t) and Lagrangian coordinate (a). The simple shock points (type 3) are calculated normally at equal intervals in a, while the negative characteristics (i.e. left facing in Figure 1) are generated either at the simple shock points or in rarefactions created at interactions. An example of this is illustrated in Figure 1 when the simple shock reaches the interface between the second and third materials. Positive characteristics are mainly generated by reflection of negative characteristics at an interface (or of course created in any right facing rarefactions which may arise at interactions). Characteristics of either type may be terminated within the mesh. The complex shock created at the interface between the first and second materials is clearly seen to be supersonic relative to the flow ahead so that the points labelled (2) in the previously calculated flow are made redundant by the calculation of the shock points (type 4) at the intersection of the shock and members of the opposite family of characteristics.

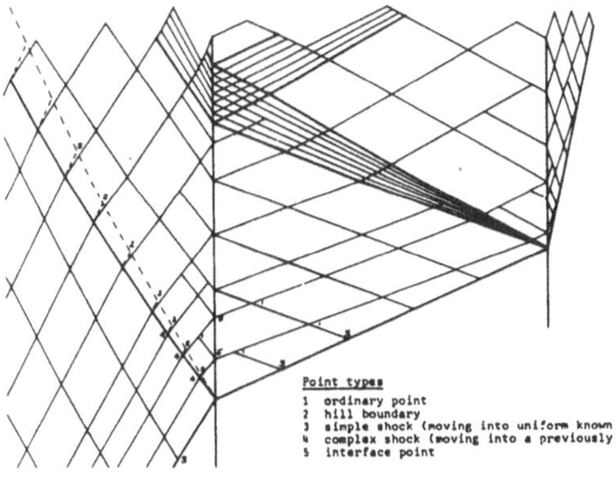

Figure 1 A typical characteristic mesh

The logic of a general purpose characteristic code must satisfy four conditions

(i) Addition and removal of points and characteristics must be simple.
(ii) Multiple shocks in multiple regions must be allowed for.
(iii) Conditions ahead of a shock must be calculated in advance.
(iv) All possible types of interaction must be solved in the correct time sequence.

In RICSHAW (i) is overcome by extending the concept of neighbours [2], and (ii) to (iv) by building the mesh up to zig-zag about constant time lines, by coping with multiple shocks as a series of single shocks by building mesh hills ahead of each shock and having a comprehensive interaction routine.

2.2 Neighbours

Each point is identified by being numbered in the order of computation (e.g. 12 in Figure 2) and is logically associated with its four neighbours (5, 14, 13 and 3). The point-numbers of these neighbours are stored as part of the solution of each point. Examination of these point-numbers, determining the presence or absence of neighbours, enables chains of connected points to be tracked through the mesh. In this way it is easy, for example, to determine the current chain of points (a zig-zag) along the top of the mesh in Figure 2. Points may easily be inserted or discarded by simple adjustments of the stores containing the point-numbers of neighbours.

2.3 Zig-Zags

Characteristic points cannot all be built up to the same time without losing many of the advantages of the method. Instead the mesh is built up until the latest points zig-zag about a constant time, which is increased in equal time steps. The basic requirements for a zig-zag to have been properly updated are that

(i) the latest shock, interface and boundary points must be above the current time

(ii) any ordinary point below the current time must have two neighbours above the time.

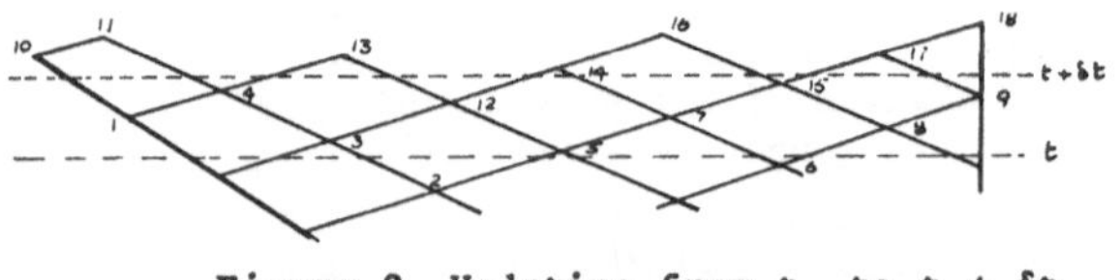

Figure 2 Updating from t to t + δt a zigzag bounded by a shock to the left and a piston to the right.

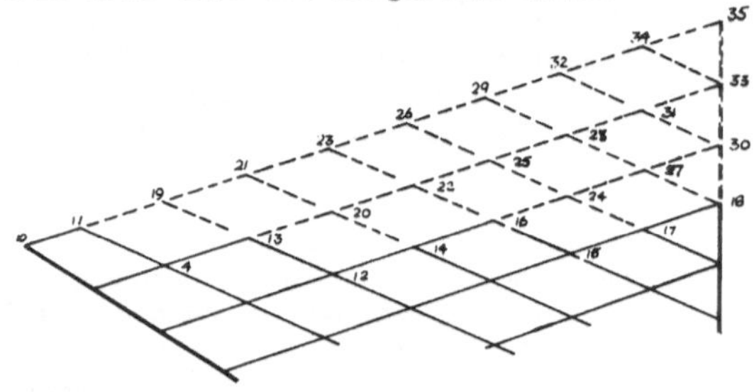

Figure 3 A hill constructed on the zigzag of Figure 2 showing the sequence of calculation

Figure 2 shows a zig-zag consisting of the points, 1, 4, 3, 2, 5, 7, 6, 8, 9 at time t, which has been correctly updated to the points 10, 11, 4, 13, 12, 14, 16, 15, 17, 18 at the new current time t + δt. The zig-zag is updated from the left, one negative characteristic at a time. Thus after the shock 10 and its associated ordinary point 11 have been calculated the points 4, 3, 2 are examined in turn until a point 2 is found having a right neighbour on a positive characteristic. Then ordinary points 12 and 13 are calculated until a point 4 is found on the negative characteristic which satisfies rule (ii).

2.4 Interactions and hills

An interaction occurs when a shock meets an interface or another shock, when two surfaces collide or when like characteristics run together and a shock is formed. The interaction is in two stages: (1) conditions are calculated at the instant immediately after the interaction, determining the nature of the reflected wave and ensuring that pressure and particle velocity are continuous across the interface; (2) conditions are calculated on the transmitted shock, reflected wave and interface one mesh distant from the interaction.

To solve for conditions at a shock point it is necessary to know the flow ahead of it. This could be computed at the same time as the shock point but this procedure has many practical difficulties and limitations. Instead, before the interaction calculation, "hills" of the characteristic mesh are constructed ahead of each potential shock. Conditions ahead of each shock point as the shock passes through the hill (as it must do because of the supersonic shock velocity) are then obtained by interpolation. Such a hill is shown in Figure 3 built on top of the zig-zag of Figure 2. A hill is built one characteristic at a time until, if possible, its peak has passed the next interface; for then it must contain the next interaction.

2.5 General calculational procedure

The calculation usually starts with an interaction (e.g. detonation of an explosive) in which a simple shock is produced. The mesh is constructed in a single expanding zig-zag until an interaction is detected. The zig-zag calculation is then temporarily suspended to build a hill and to do the interaction calculation. Afterwards the zig-zag procedure is continued. Since hills are built over a zig-zag the zig-zag may be split, so that the number of zig-zags increases with time. Conversely two zig-zags may be built up on two opposite sides of a hill and swamp it, coalescing into one zig-zag. The general mesh in RICSHAW therefore consists of a number of zig-zags, whose ends may be shocks, boundaries or hill edges, separated by hills. Figure 4 shows how a sequence of zig-zags and hills is constructed.

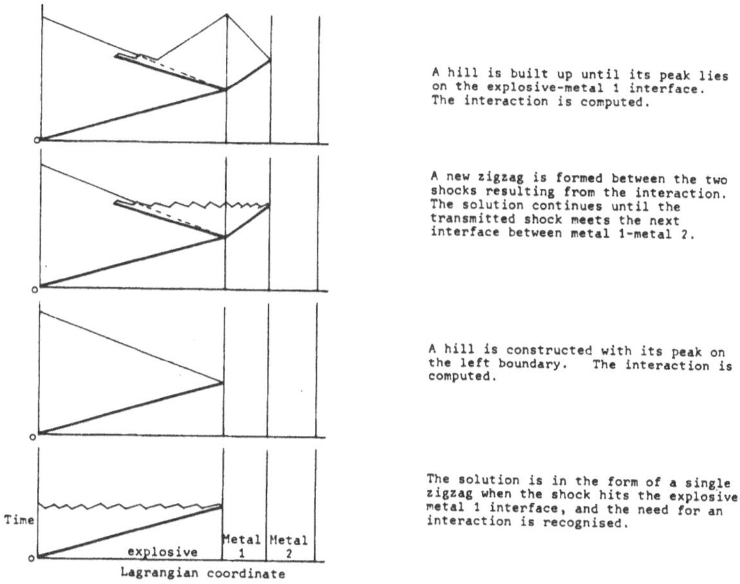
A hill is built up until its peak lies on the explosive-metal 1 interface. The interaction is computed.
A new zigzag is formed between the two shocks resulting from the interaction. The solution continues until the transmitted shock meets the next interface between metal 1-metal 2.
A hill is constructed with its peak on the left boundary. The interaction is computed.
The solution is in the form of a single zigzag when the shock hits the explosive-metal 1 interface, and the need for an interaction is recognised.
Time
o
explosive
Metal 1
Metal 2
Lagrangian coordinate

Figure 4 Construction of the solution by hills and zig-zags

Both characteristics and shocks are multiplied by repeated interactions. To keep the number of points to a minimum, shocks can be discarded if they become very weak, and characteristics are discarded if they are too close together and carry too little difference in pressure. It is also necessary to erase unwanted points when the storage becomes full. Erasure can be done most easily by cancelling the neighbour connections of the unwanted points and making point numbers and storage available for further use.

3. A TYPICAL CALCULATION

We consider the multiple shocks in a slab of polyethylene driven by an explosive through a thin stainless steel plate. Wright et al [5] describe the experiment and show that reasonable agreement was obtained between calculation and the experimental results. Figure 5 is a wave diagram for the problem in which are plotted shocks (×), interaces (+) and wave boundaries (•).

The reverberations in the stainless steel drive a succession of compression waves into the polyethylene each of which ultimately steepens into a shock. The first of these shocks catches up the primary shock in the polyethylene and strengthens it - the determination of this time of "catch-up" is an essential part of the experiment (and calculation).

Figure 6 shows as a full line a profile of pressure against distance through the three materials, at the time indicated in Figure 5. The various shocks and waves are clearly seen and may be easily identified with the consequences of the several interactions shown in Figure 5.

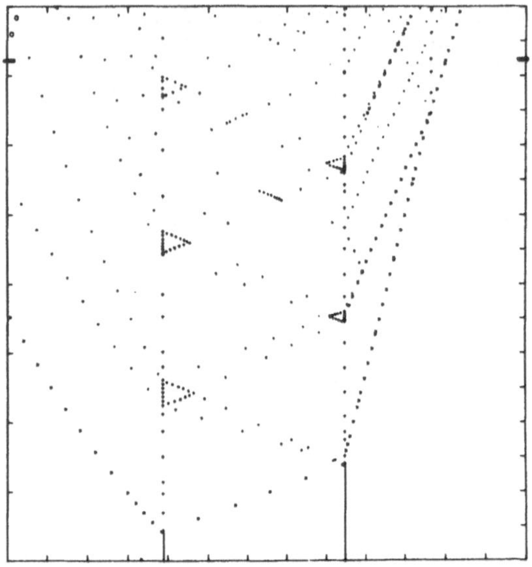

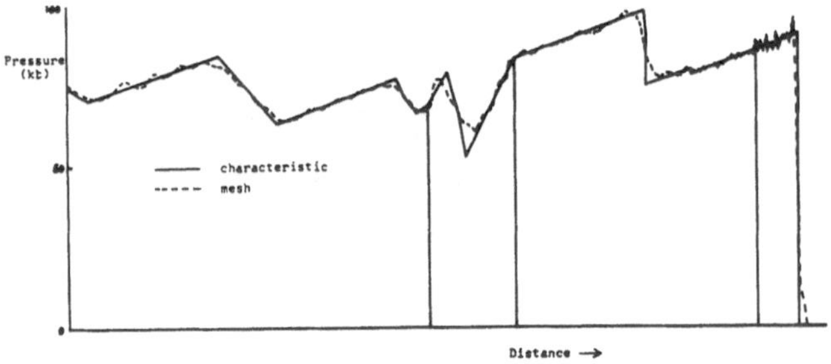

Figure 5 Wave diagram showing shocks, interfaces and wave boundaries.

Figure 6 Comparison between solutions obtained by characteristic and mesh codes.

The dotted line in figure 6 is the same profile calculated by a mesh code. This calculation used 470 points and took three times as long as the characteristic calculation (15 minutes instead of 5). It will be seen that there is a larger uncertainty in the shock positions and strengths. Smoother profiles can be obtained by using different artificial viscosities, but at the expense of further shock broadening and a corresponding increase in uncertainty.

The need for the large number of meshes used in the mesh code calculation arises because the problem includes a thin steel plate sandwiched between two much thicker materials. To obtain even the relatively poor resolution shown in the profile for the compression wave in the steel, it was necessary to have 20 mesh points in the steel. To preserve optimum mesh ratios across interfaces into the explosive and polyethlene, it was then necessary to use over 200 meshes in each of these regions.

4. COMPARISON OF CHARACTERISTIC AND MESH CODES

The respective advantages and disadvantages of mesh codes and characteristic codes are summarised in Table 1.

The general conclusion is that given the availability of both mesh and characteristic codes, one would use a mesh code for long time problems where the ultimate state of the materials is required but not the details leading to that state, and a characteristic code for short time problems where an understanding of detailed wave motion is required.

5. ACKNOWLEDGMENTS

We gratefully acknowledge the work of Mrs. Eileen Haddrell, Mr. D. Page and Mr. M. Tuck who carried out the programming of RICSHAW so efficiently.

TABLE 1

Property	Mesh Code	Characteristic Code
Advantages of a Mesh Code		
1. Logic	Simple	Complicated
2. Increase in difficulty due to multiple materials and variety of problems (e.g. shock formation)	Little	Great
3. Constant time profiles	Given continuously	Need interpolation
4. Probability of a new problem running at first attempt	Good	Fair
Advantages of a Characteristic Code		
1. Treatment of discontinuities and their interactions	Smeared, uncertainty in position	Treated explicitly
2. Profiles	Noisy	Smooth between discontinuities
3. Details of solution	Poorly defined	Good
4. Time step	Usually determined by stability of smallest mesh	Variable in space and time
5. Number of meshes within a material	Usually fixed. Need a minimum to let shocks form properly	Varied during a problem to give detail where needed
6. Utilization of a computer	Poor - many meshes required for accuracy	Optimum

6. REFERENCES

1. P. C. Chou, R. R. Karpp and S. L. Huang, "Numerical Calculation of Blast Waves by the Method of Characteristics", Journal AIAA, 5, 1967, pp 618-623.

2. A. E. Glennie, "Solution by Characteristics of the Equations of One-dimensional Unsteady Flow", Numerical Solution of Ordinary and Partial Differential Equations edited by L. Fox (Pergamon Press, London, 1962) pp 325-338.

3. N. E. Hoskin, "Solution by Characteristics of the Equations of One-dimensional Unsteady Flow", Methods in Computational Physics edited by B. Alder, S. Fernbach and M. Rotenberg (Academic Press, New York, 1964) pp 265-293.

4. B. D. Lambourn and J. E. Hartley, "The Calculation of the Hydrodynamic Behaviour of Plane One-dimensional Explosive/Metal Systems", Proceedings of Fourth Symposium (International) on Detonation, ACR-126 (Naval Ordnance Laboratory, 1965) pp 538-554.

5. P. W. Wright, G. Eden and B. D. Lambourn, "Behaviour of Various Materials under Multiple Shocking", Behaviour of Dense Media under High Dynamic Pressures (Dunod, Paris, 1968) pp 137-152.

6. A. I. Zhukov, No. 58 Tr. Mat. Inst. Attad. Nauk. SSSR.

STABILITY AND ACCURACY STUDIES ON A SECOND-ORDER METHOD OF CHARACTERISTICS SCHEME FOR THREE-DIMENSIONAL, STEADY, SUPERSONIC FLOW*

Victor H. Ransom**, H. Doyle Thompson†, and Joe D. Hoffman†

ABSTRACT

A new explicit method of characteristics (MOC) numerical scheme for three-dimensional steady flow has been developed which has second-order accuracy. A complete numerical algorithm has been developed for computing internal supersonic flows. A comprehensive stability analysis was conducted in which both the Courant-Friedrichs-Lewy (CFL) stability criterion and the von Neumann stability analysis were applied. Although necessary and sufficient criteria for stability exist only for linear difference equations and analytic initial data, experience has indicated that these same criteria are appropriate for nonlinear systems when applied locally to the linearized form of the equations. This thesis is supported by the results of the present research in which the nonlinear scheme was found to be stable only when the analysis of the linearized system indicated stability. The numerical scheme has been tested for order of accuracy using exact solutions for source flow and Prandtl-Meyer flow. The results of these tests have verified the second-order accuracy of the scheme. Additional tests using axisymmetric flows have shown that the accuracy of the scheme is comparable to that of a proven second-order two-dimensional MOC scheme.

INTRODUCTION

The equations of motion for a steady supersonic flow of an inviscid fluid in three independent space variables are a well established system of quasi-linear, first-order, hyperbolic partial differential equations. Although the number of dependent variables of the system depends upon the assumed nature of the flow, the mathematical character of the system does not and, consequently, neither does the theoretical method of solution. However, in practice the numerical algorithm and the associated computer program generally are developed for a specific system of equations.

The finite difference integration schemes which have been proposed for the solution of systems of hyperbolic partial differential equations fall into two categories: 1) schemes based on characteristic directions and 2) schemes based on ordinary coordinate directions. Variations of both approaches have been investigated both analytically and numerically; however, no one method has been shown to be superior for all problems. Numerical stability, accuracy, and time of computation are the factors of primary importance in evaluating a particular method. Secondary factors such as ease of programming and ease of incorporating boundary conditions are also important.

To a limited extent the numerical stability and accuracy of finite difference schemes can be investigated by theoretical methods. These methods provide certain necessary conditions which should be satisfied by any numerical scheme. In the application of the MOC to two independent variable problems, numerically stable schemes result and second-order accuracy is easily achieved. For three independent variable problems, stable schemes do not necessarily result and second-order accuracy is much more difficult to achieve. There is no way a priori to assure stability and a comprehensive investigation must be made for each scheme which is devised.

A number of investigators have proposed numerical schemes using the characteristic compatibility relations for three independent variable problems; however, only a few have developed complete numerical algorithms and obtained results. Good surveys of the various approaches which have been taken are given by Chushkin (1968), Fowell (1961), Strom (1965), and Thompson (1965). With the exception of the work of Butler (1960) all explicit schemes which have been proposed have had only first-order accuracy. Some of the schemes may be more accurate than others for a finite step size. However, all share the characteristic that the accumulated error approaches zero linearly with the step size, whereas with Butler's scheme, the accumulated error

---

* This research was sponsored by AFAPL, Wright-Patterson AFB, Ohio, Contract F33615-67-C-1068. The technical monitor was Lt. Gary Jungwirth.

** Manager, Thermal and Fluid Dyn. Sect.,Aerojet Liquid Rocket Co.,Sacramento, Calif.

† Associate Professor, School of Mechanical Eng., Purdue Univ., Lafayette, Ind.

approaches zero with the square of step size.

A numerical algorithm, and associated computer program, based on a modification of Butler's scheme have been developed for three-dimensional internal flow calculations by Ransom, et al. (1969) (1970). This paper is a summary of the numerical stability and accuracy studies that were made in the development of that method.

NUMERICAL STABILITY

The possibility of numerical instability is an ever present danger in any numerical scheme used to approximate systems of hyperbolic partial differential equations. Stability is solely a function of the numerical scheme and does not depend upon the differential system that it approximates. The stability of linear difference schemes has been studied extensively by a number of investigators and a good summary of this work as well as applications of the stability criteria to several different schemes for three-dimensional steady flow are given by Heie and Leigh (1964) (1965). Unfortunately, necessary and sufficient criteria for stability exist only for the case of linear equations and analytic initial data. However, although no formal proof exists, experience has shown that these same criteria when applied locally to the linearized form of the equations are appropriate for nonlinear systems. This hypothesis is supported by the results of the present research in which the nonlinear scheme proved to be stable only when the analysis of the linearized system indicated stability.

The two stability criteria which exist are the Courant-Friedrichs-Lewy (CFL) criterion and the von Neumann condition. The CFL criterion is a necessary condition which applies to both linear and nonlinear systems and states that the zone of dependence of the differential system must be embedded within the convex hull of the difference scheme. The relation between the differential zone of dependence and the convex hull of the difference scheme for three-dimensional supersonic flow is illustrated in Fig. 1. The CFL condition ensures that the speed of propogation of numerical disturbances, such as round off error, everywhere exceeds the speed of propagation of disturbances in the differential system (i.e., the speed of sound for compressible flow). Thus numerical disturbances diffuse throughout the network and do not accumulate.

The von Neumann condition is a stronger stability criterion which states that in order for a numerical scheme to be stable a finite limit must exist for the amplification of any Fourier component of the initial data. The criterion which must be satisfied in order to ensure this condition is $|\lambda_i|_{max} \leq 1 + O(\Delta x_1)$ where λ_i are the eigenvalues of the amplification matrix for the numerical scheme. The von Neumann condition is sufficient for stability of linear difference schemes only for the case of analytic initial data. However, the von Neumann condition has turned out to be sufficient for nonlinear, as well as linear, schemes which are known to have been investigated (Heigh and Leigh (1964)).

In the present analysis the CFL criterion is regarded as a necessary condition and it is satisfied in the nonlinear difference scheme by regulating the integration step size. The linearized difference scheme, with the CFL condition satisfied, is then tested for stability in the von Neumann sense. The von Neumann condition is also used to analyze the effect of various modifications on the numerical stability, but does not play an active role in the actual numerical calculations.

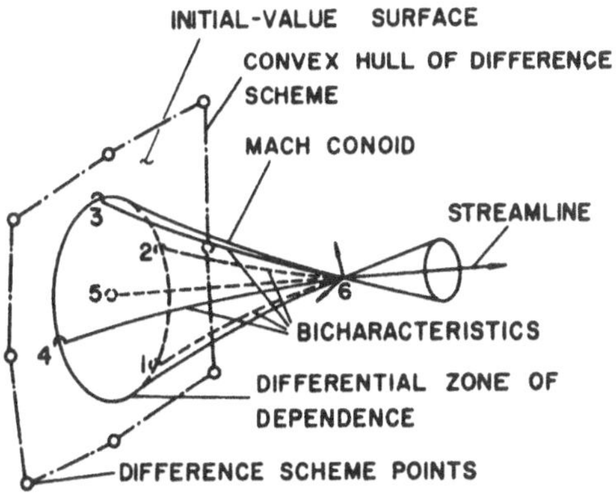

FIG. 1. Bicharacteristic Network For Computational Scheme

Linear Difference Equations

A system of linear difference equations is obtained from the system of nonlinear differential equations by application of small perturbation theory and replacement of the directional differential operators by first difference operators (Ransom, et al.

(1969)). The difference operators apply along the network of bicharacteristic segments illustrated in Fig. 1. In the linearization the scalar terms containing partial derivatives, $\alpha_i\alpha_j(\partial u_i/\partial x_j)$ and $\beta_i\beta_j(\partial u_i/\partial x_j)$, are treated as unknown quantities at the point under consideration. The resulting six linear equations are just sufficient to evaluate these two terms as well as the four dependent variables.

The specific form of the two scalar terms containing derivatives in the system of differential equations is a direct result of the scheme for maintaining second-order accuracy and the stability analysis must include consideration of these terms. The analysis is simplified if the terms containing derivatives are algebraically eliminated to reduce the system to four independent linear difference equations.

The final linearized difference equations for the selected network and coordinate orientation illustrated in Fig. 1 are

$$\tilde{p}(6)-\tilde{p}(1)+\bar{\rho}\bar{c}[\tilde{u}_2(6)-\tilde{u}_2(1)]=\tilde{p}(6)-\tilde{p}(3)-\bar{\rho}\bar{c}[\tilde{u}_2(6)-\tilde{u}_2(3)] \tag{1}$$

$$\tilde{p}(6)-\tilde{p}(2)+\bar{\rho}\bar{c}[\tilde{u}_3(6)-\tilde{u}_3(2)]=\tilde{p}(6)-\tilde{p}(4)-\bar{\rho}\bar{c}[\tilde{u}_3(6)-\tilde{u}_3(4)] \tag{2}$$

$$\tilde{p}(6)-\tilde{p}(5)=\tilde{p}(6)-\tilde{p}(1)+\tilde{p}(6)-\tilde{p}(2)+\bar{\rho}\bar{c}[\tilde{u}_2(6)-\tilde{u}_2(1)+\tilde{u}_3(6)-\tilde{u}_3(2)] \tag{3}$$

$$\tilde{p}(6)-\tilde{p}(5)=-\bar{\rho}\bar{u}_1[\tilde{u}_1(6)-\tilde{u}_1(5)] \tag{4}$$

where the bar denotes constant mean values and the tilde denotes a small variation.

Application of the von Neumann Condition

The analysis for stability in the von Neumann sense must include consideration of all the numerical operations of the basic integration scheme. In particular, the interpolation process for data in the initial-value surface must be included. It is assumed that the analytic solution of the system of linear difference equations can be obtained by separation of variables. Thus the general term of the Fourier representation of the solution is

$$\bar{U} = e^{i\pi M x_2/L}\, e^{i\pi N x_3/L}\, \bar{a}(x_1) \tag{5}$$

where $\bar{U}$ is the vector whose elements are the dependent variables of the problem, $\bar{a}(x_1)$ is a corresponding vector function of the direction of integration, x_2 and x_3 are the rectangular cartesian coordinates of the initial-value surface, L is a characteristic dimension, and M and N are the frequency factors for an arbitrary component of the solution.

For purposes of the analysis, data on the initial-value surface are assumed to be known at the points of a uniform rectangular grid in an x_2, x_3 plane with spacings Δx_2 and Δx_3 respectively. Thus the independent variables x_2 and x_3 in the general Fourier term are only permitted to have values which are integral multiples, m and n, of the grid spacings, Δx_2 and Δx_3. The values of the dependent variables at these points, given by Eq. (5), are thus

$$\bar{U} = \zeta^m \eta^n\, \bar{a}(x_1) \tag{6}$$

where ζ and η are defined as the complex quantities $\zeta = e^{i\pi M\Delta x_2/L}$; $\eta = e^{i\pi N\Delta x_3/L}$, and $x_2 = m\Delta x_2$; $x_3 = n\Delta x_3$ $(m,n = 0, \pm1, \pm2, \ldots)$.

In the numerical scheme the solution is advanced along the streamlines passing through each of the points of the network. The central streamline point and its eight nearest neighbors are used for local interpolation. The analysis is simplified, without loss of generality, if the central point of the local mesh is chosen such that $x_2(5)=x_3(5)=0$. Thus the nine points used for interpolation correspond to values for the integers $m = 0, \pm1$ and $n = 0, \pm1$. A second-order bivariate interpolating polynomial is fit to these nine points by a least squares method. The resulting polynomial has the general form

$$\bar{U} = (A_1 + A_2x_2 + A_3x_3 + A_4x_2x_3 + A_5x_2^2 + A_6x_3^2)\, \bar{a}(x_1) \tag{7}$$

where the coefficients are given by Ransom, et al. (1969, p. 47).

The dependent variables at the intersections of the four bicharacteristics with the initial-value surface must be evaluated by means of Eq. (7) since these intersections do not generally correspond to points of the initial-value surface network. On the other hand the streamline intersection coincides with a network point, and interpolation is not necessarily required. It was found however that whether or not interpolation is used at the streamline intersection has a significant effect on stability.

The system of difference equations, Eqs. (1)-(4), constitutes a recursion relation for the values of the dependent variables at point (6) in terms of values in the initial-value surface. For the case $x_2(5)=x_2(6)=x_3(5)=x_3(6)=0$, and when the interpolating polynomial, Eq. (7), is used to determine the values of the dependent variables at the points in the initial-value surface, the recursion relation has the form

$$\bar{U}[x_1(5) + \Delta x_1] = \bar{a}[x_1(5) + \Delta x_1] = A\, \bar{a}[x_1(5)] \qquad (8)$$

in which A, the amplification matrix, is a fourth-order matrix with nonzero coefficients A_{11}, A_{12}, A_{13}, A_{14}, A_{22}, A_{24}, A_{33}, A_{34}, A_{42}, A_{43} and A_{44}. The values of the nonzero coefficients are given by Ransom, et al. (1969, p. 48).

The bound for the eigenvalues of the matrix A was investigated numerically by choosing a characteristic length L and mesh spacings Δx_2 and Δx_3, consistent with physical problems of interest, and searching for the maximum eigenvalue within the range $0 \leq I \leq L/\Delta x$, $M \leq I$, and $N \leq I$.

The results of the eigenvalue analysis are shown plotted in Fig. 2 for three cases, the basic difference scheme without interpolation, interpolation except at the streamline point and interpolation at all points. When interpolation is not used the scheme is clearly unconditionally unstable since only eigenvalues greater than or equal to unity are obtained. The maximum absolute value of an eigenvalue, approximately 3, occurs for I equal to 10, which corresponds to a Fourier component having a wave length twice the mesh spacing. This result is not surprising since, when only five points in the initial-value surface are used (the four bicharacteristic intersections and the streamline intersection) the convex hull of the four outermost points is a square lying entirely within the circular differential zone of dependence. Thus, the CFL necessary condition for stability is not satisfied.

When interpolation based on nine points is used only at the bicharacteristic intersections, a marked improvement in stability characteristic results, but the maximum magnitude of the eigenvalues is still everywhere equal to or exceeds unity, see Fig. 2. Numerical tests of this scheme, using the nonlinear second-order algorithm, revealed that numerical instabilities did occur after 20 to 30 integration steps.

When interpolation is used at all five points in the initial-value surface a sufficiently stable scheme results, see Fig. 2. In this case eigenvalues greater than unity occur only for low frequency components and even these are only slightly greater than one at values for the frequency index of 1, 2, and 3, which correspond to Fourier components having wave lengths of 20, 10 and 7 times the grid spacing. Although the scheme cannot be judged unconditionally stable because of the amplification factors slightly greater than one, experience with the scheme has shown it to be highly stable. Numerical tests with this scheme have revealed no evidence of instability even after 120 integration steps and in the presence of severe disturbances which would result in shocks in a real flow.

The eigenvalues were also evaluated for reductions in the x_1 step size and for rotation of the network. Only for the case of zero x_1 step size were all eigenvalues less than or equal to unity. However, the numerical results which have been obtained using the nonlinear scheme clearly indicate that the scheme using interpolation at all points is sufficiently stable if the CFL condition is

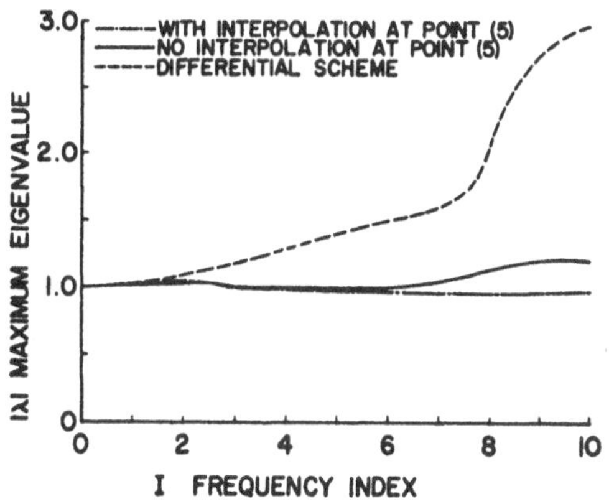

FIG. 2. Stability Results

satisfied.

In summary the results of the stability study did not indicate that any of the schemes investigated were unconditionally stable. However, the results dramatically illustrated the effect of the various modifications on the numerical stability. In particular the results showed the final scheme to be stable at the fundamental frequency (i.e., the frequency corresponding to a wave length twice the mesh spacing). The Fourier components corresponding to the fundamental frequency are the ones which are normally most amplified by an unstable scheme.

The results of this analysis, and subsequent numerical experience with the difference scheme, support the general thesis that the von Neumann condition is sufficient for stability of nonlinear schemes.

ACCURACY STUDIES

The absolute accuracy of a numerical scheme is difficult to establish without actually comparing numerical results with an exact solution. However, a desired order of accuracy, which governs how rapidly the solution converges with reduction of the step size, can be achieved by using consistent approximations throughout the numerical scheme. Throughout the development of the three-dimensional characteristic integration scheme only numerical relations accurate to at least second-order were used. The accuracy studies provided a means for experimentally verifying the order of the error for the resultant algorithm.

The more accurate second-order scheme permits the use of larger step sizes and thus fewer points need to be computed. In three independent variable problems, increasing the step size by a factor of two results in a reduction in the number of computed points by a factor of eight. Thus, a more accurate but more complex scheme may actually require less total computer time.

Comparison with Exact Solutions

To test the order and magnitude of the numerical error of the scheme, the numerical solution was compared with the exact solution for a spherical source flow and a Prandtl-Meyer or simple wave flow. Specialized computer programs were written which generated an exact initial-value surface and subsequently generated both the numerical and exact solutions at the points of each subsequent solution surface. The point spacing on the respective initial-value surfaces was successively halved in order to study the error behavior for reductions in step size. In addition to providing tests to determine the order of accuracy for the scheme, the comparisons with the exact solutions provided a quantitative way to readily evaluate the effect of new numerical innovations on the accuracy of the scheme.

A spherical source flow and a Prandtl-Meyer flow were used because of their three-dimensional geometrical character. In the Prandtl-Meyer flow, unlike the spherical source flow, the streamlines are curved and thus should provide a more severe test for the numerical scheme.

The local truncation error in the numerical scheme was assumed to be third order in step size; thus, for integration to a fixed point in the flow for which the

TABLE 1
Error Study For Source Flow and Prandtl-Meyer Flow, $M_I = 4.0$

Case	(1)	(2)	(3)
Relative Step Size	1	1/2	1/4
Accum. Error (%)			
Source Flow	.01040	.00210	.00050
P-M Flow	.00318	.000827	.000209
Ratio to Case (1)			
Source Flow	1	1/5	1/21
P-M Flow	1	1/4	1/15.5
Theoretical Ratio	1	1/4	1/16

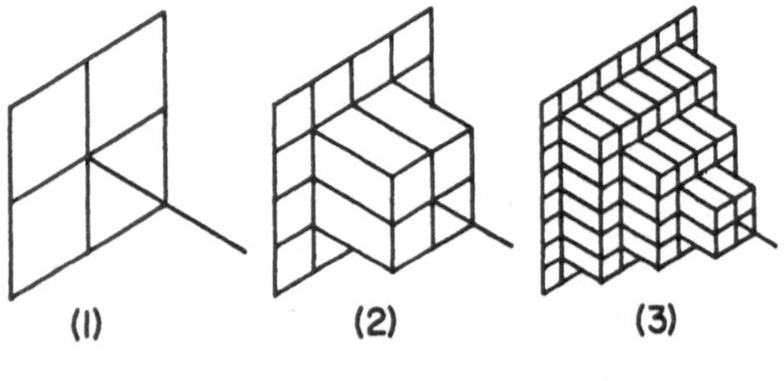

FIG. 3. Accuracy Study, Step Size Halving

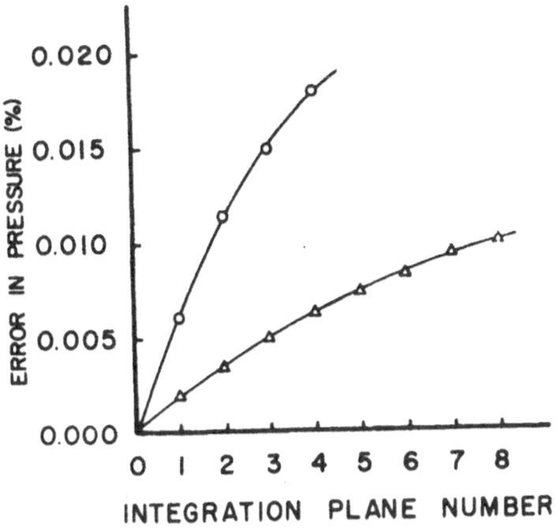

FIG. 4. Accuracy Study, Error Accumulation, O - Source Flow, Δ - Prandtl-Meyer Flow

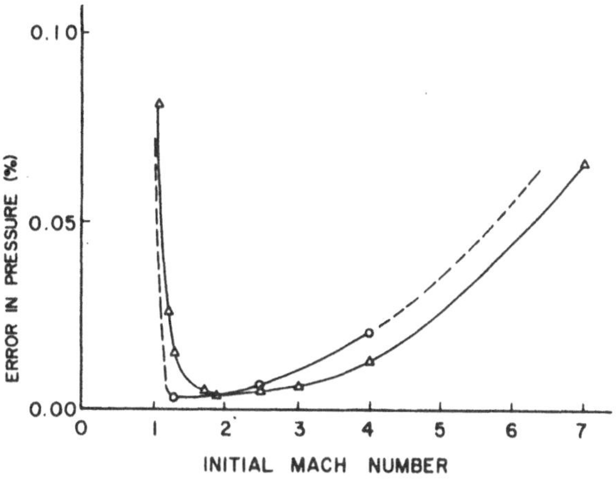

FIG. 5. Accuracy Study, Accumulated Error Vs. Initial Mach Number, O - Source Flow (4 planes), Δ - Prandtl-Meyer Flow (8 planes)

number of steps is of the order $(1/\Delta x_1)$, the accumulated error is second order in step size. The order of the actual error is established numerically by successively reducing the step size and comparing the ratio of accumulated errors to the ratio of step sizes raised to the assumed order of the error. This process is illustrated schematically in Fig. 3.

The static pressure was found to be the dependent variable which is computed least accurately and thus is the most sensitive error indicator. The accumulated error in pressure for three step sizes is tabulated in Table 1. These results are for an initial Mach number near 4.0 and a rectangular point network. The results confirm the second-order accuracy of the scheme for these initial conditions, since the ratios of the errors are in each case less than the ratio of the step sizes squared.

In Fig. 4 the absolute error along the central streamline is shown plotted versus the number of integration steps for both the source flow and the Prandtl-Meyer flow. Note that the rate of error increase monotonically decreases with the number of steps, thus indicating a bounded error characteristic.

The effect of the local Mach number on the error was also investigated using the rectangular network and the results are presented in Fig. 5. The high inaccuracies at a Mach number near unity are due, at least in part, to the fact that the equations approach a parabolic character, the family of wave surfaces degenerates to a single surface, and the numerical scheme degenerates. In addition, the gradients in velocity and pressure are greatest near a Mach number of unity, which also contributes to the increase in error.

A conical boundary and a circular network of points on the initial-value surface were used with source flow initial data to further test the accuracy of the overall scheme. This approach permitted the solution to be calculated beyond the zone of determinacy of the initial data. The results of this error study for an initial Mach number of 1.05, a 10 degree source, and for three step sizes are shown plotted as a function of axial length from the initial-value surface in Fig. 6 for a streamline at the boundary. The combination of a 10 degree source angle and a Mach number of 1.05 produces a flow having a gradient comparable to that which exists at the throat of a rocket nozzle. The results in Fig. 6 do not show the expected error reduction between the two largest step sizes, cases (1) and (2). However, the reduction between the two smaller step sizes, cases (2) and (3), does have the proper second-order characteristic. These results are not in conflict since the order of accuracy criteria only applies in the limit as step size becomes small such that the coefficient of the error term in the power series representation of the solution approaches a constant value. A somewhat different but analogous result was obtained for the relative error along the centerline of the flow.

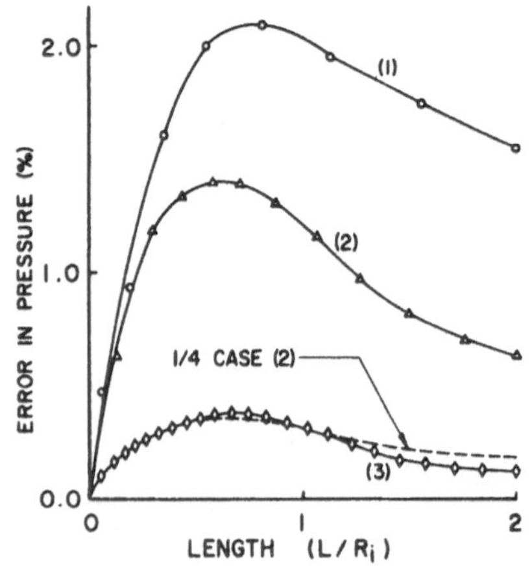

FIG. 6. Accuracy Study, $M_i = 1.05$, $\Delta x/R_i$: (1) - 0.333, (2) - 0.167, (3) - 0.083

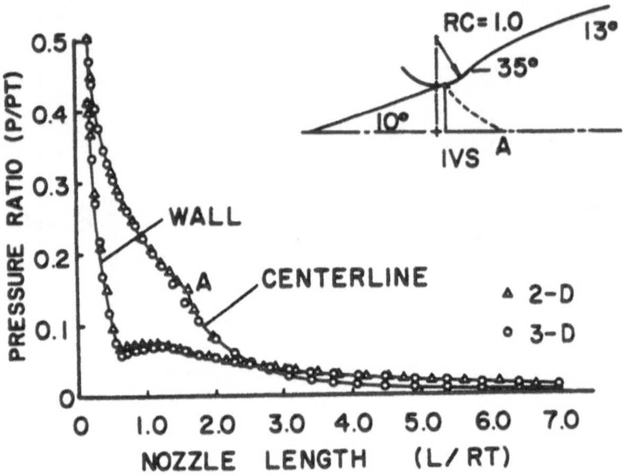

FIG. 7. Solution For Contoured Nozzle

The accuracy studies for Prandtl-Meyer flow revealed an additional phenomenon of some significance. When the initial-value surface included a portion of the uniform flow which preceeds the region of simple wave flow, a significant increase in error was noted at the junction of the two regions. This is due to the fact that the derivatives of the fluid properties are discontinuous at that junction. The interpolation scheme using second-order polynomials assumes continuity in derivatives up to second order and as a result the accuracy of interpolation drops to first order across the discontinuity.

Axisymmetric Results

As an additional test of the accuracy, the three-dimensional numerical results for two axisymmetric nozzles were compared with the numerical solution obtained using the two-dimensional method of characteristics. The two nozzles used for comparison were a conical nozzle having a 15 degree half-angle and a contoured axisymmetric nozzle. In both cases the initial data were generated using a 10 degree spherical source flow. The results of the comparison for the axisymmetric contoured nozzle are shown in Fig. 7. Both the nozzle wall and centerline pressures are plotted. The agreement is exceedingly good except at points of discontinuous rates of change in flow properties. The point on the nozzle axis at which the first expansion wave from the wall reaches the axis, point A, is such a point. The two-dimensional solution, which proceeds along characteristics and does not require interpolation, shows the true character of the solution, while the three-dimensional solution smooths out this point. This diffusive characteristic is inherent in any three-dimensional calculation scheme because to satisfy the CFL stability criterion it is necessary to interpolate on the initial-value surface. The results are very similar for the conical nozzle comparison.

REFERENCES

Butler, D. S., Proc. Roy. Soc. A 255, 232 (1960).

Chushkin, P. I., Chap. 3 Progress in Aeronautical Sciences Vol. 9, Ed. by D. Kuchemann, Pergamon Press (1968).

Fowell, L. R., IAS Paper 61-208-1902 (1961).

Heie, H., and Leigh, D. C., Rept. 705, Dept. of Aerospace and Mechanical Sciences, Princeton Univ. (1964).

Heie, H. and Leigh, D. C., AIAA J. 3, 1099 (1965).

Ransom, Victor H., Thompson, H. Doyle, and Hoffman, Joe D., Rept. AFAPL-TR-69-98, Vol. I, Aero Propulsion Laboratory, Wright-Patterson AFB, Ohio (1969).

Ransom, V. H., Cline, M. C., Hoffman, J. D., and Thompson, H. D., Rept. AFAPL-TR-69-98, Vol. II, Aero Propulsion Laboratory, Wright-Patterson AFB, Ohio (1970).

Strom, C. R., Dissertation, Univ. of Illinois (1965).

Thompson, H. D., Rept. TM-64-1, Jet Propulsion Center, Purdue University (1965).

SUBCRITICAL FLOWS OVER TWO DIMENSIONAL AIRFOILS BY A MULTISTRIP METHOD OF INTEGRAL RELATIONS

R. E. Melnik and D. C. Ives

Grumman Aerospace Corporation

1. INTRODUCTION

Recent attempts to improve the performance of aircraft at transonic speeds have led to a resurgence of interest in the subsonic airfoil problem, particularly in the supercritical regime. Most of the new approaches take advantage of the increased computing capacity of the modern digital computer to seek a direct numerical solution by finite difference techniques.

In the present investigation we explore the possibility of using Dorodnitsyn's method of integral relations (MIR) (see Belotserkovskii and Chushkin) to treat this classical nonlinear potential flow problem. We employ the scheme I version of the method, in which the strip boundaries are taken more or less parallel to the airfoil surface. We feel this scheme is best suited for airfoil type problems in which the rapid variations are along the airfoil surface. The chief advantage of MIR is that, because a relatively coarse mesh can be taken across the flow field, useful solutions can be obtained with less computer time then might be required by other methods. This approach offers the possibility of developing an economical calculation scheme for practical airfoil design.

Other applications of the MIR to subsonic potential flow problems were carried out many years ago by Chushkin, and more recently by Jones. Chushkin considered planar and axisymmetric flows and employed both the scheme I and II versions of the MIR. The only planar case treated by the scheme I version was for a symmetrical nonlifting Joukowski profile at a subcritical Mach number. Jones employed the scheme II version (method of lines) and has obtained accurate results only for flow over relatively thick ellipses.

In our study the MIR is programed for an arbitrary number of strips, and solutions using up to five strips have been obtained in order to investigate the convergence properties of the method. Although the technique is applicable to supercritical flows, the work reported here is restricted to the subcritical and critical cases only. The method is formulated for the general lifting airfoil, but numerical results have only been obtained to date for symmetrical Karman-Trefftz profiles at zero lift.

In the following section we outline the particular formulation of the MIR employed in this study and discuss some of the numerical considerations involved in integrating the resulting ordinary differential equations. In Section 3 we apply the method to a number of specific profiles, and in Section 4 we summarize our main conclusions and indicate the future direction of this work.

2. FORMULATION

Derivation of Integral Relations

We employ a curvilinear coordinate system defined by the conformal mapping of the region outside the airfoil to the region inside a unit circle, as illustrated in Fig. 1 on page 6. If the mapping is specified by an equation of the form

$$z \equiv x + iy = f(\zeta) \quad \text{where} \quad \zeta \equiv Re^{i\omega} \quad . \tag{2.1}$$

The metric $h(R,\omega)$ is given by

$$h(R,\omega) = |df/d\zeta| \tag{2.2}$$

where (x,y) are Cartesian coordinates in the original plane, $f(\zeta)$ is an analytic function (except at any boundary singularities) of the transformed coordinate, ζ, and (R,ω) are polar coordinates in the computational plane. This coordinate system transforms the infinite domain of the original problem to a finite region and smoothes the functions that are approximated in the MIR, particularly near the leading and trailing edges. The equations expressing conservation of mass and the condition of irrotationality can be written in the form

$$\partial/\partial R(\rho h q_r/R) - \partial/\partial\omega(\rho h q_\omega/R^2) = 0 \tag{2.3}$$

$$\partial/\partial R(h q_\omega/R) + \partial/\partial\omega(h q_r/R^2) = 0 \tag{2.4}$$

where

$$\rho = \rho_{stag}(1 - q_r^2 - q_\omega^2)^{1/\gamma-1} \quad , \tag{2.5}$$

(q_r, q_ω) are the velocity components in the (R,ω) directions, ρ is the density, ρ_{stag} the stagnation density, and γ the ratio of specific heats. All velocities are referred to the limiting thermodynamic velocity and densities to free stream density.

In accordance with the usual procedures of scheme I of the MIR, the computational region is divided into N concentric circular strips separated by the $N - 1$ lines, $R = R_j$ = constant. The equations of motion are successively integrated with respect to R between the airfoil surface $R = 1$ and each of the strip boundaries $R = R_j$ to generate the following 2N integral relations.

$$\frac{d}{d\omega}\left[\int_1^{R_j} \frac{\rho h q_\omega - q_{\omega o} - (\Gamma\beta/2\pi)R}{R^2}\, dR\right] = (1 - R_j)q_{\omega o} + S_j \tag{2.6}$$

$$\frac{d}{d\omega}\left[\int_1^{R_j} \frac{hq_r - q_{ro}}{R^2}\, dR\right] = (1 - R_j)q_{\omega o} + T_j \tag{2.7}$$

where the quantities $S_j(\omega)$ and $T_j(\omega)$ are functions of the strip values q_{rj}, $q_{\omega j}$ and the circulation Γ. The quantity M_∞ is the free stream Mach number, U_∞ is the free stream velocity, and β is equal to $\sqrt{1 - M_\infty^2}$. The circulation is related to the lift force L' by the Kutta-Joukowski theorem,

$$L' = \rho' U_\infty' \Gamma \tag{2.8}$$

(primes denote dimensional quantities), and $(q_{ro}, q_{\omega o})$ are the velocity components at infinity, which for a uniform stream at incidence δ are given by

$$q_{ro} = - U_\infty \cos(\omega - \delta) \quad , \quad q_{\omega o} = U_\infty \sin(\omega - \delta) \quad . \tag{2.9}$$

The integrals appearing in Eqs. (2.6) and (2.7) are evaluated by a quadrature formula based on a Lagrangian interpolation polynomial approximation of the integrands. This reduces the equations to a system of ordinary differential equations. We choose our profiles so that they: 1) are exact for incompressible flow, 2) are consistent with Ludford's asymptotic solution at infinity (except for certain logarithmic terms which arise in the lifting, compressible case), 3) satisfy the boundary condition $q_r = 0$ at the surface, and 4) satisfy certain wall compatability conditions derived from the exact partial differential equations. The profiles are assumed in the form

$$hq_r = \left[1 - R^2\right]\left[q_{ro} + R^2 \sum_{i=1}^{N+\nu_1} \alpha_i(\omega) R^{2i}\right] \tag{2.10a}$$

$$h\rho q_\omega = q_{\omega o} + (\Gamma\beta/2\pi)R + R^2 \sum_{i=1}^{N+\nu_2} \beta_i(\omega) R^{2i} \tag{2.10b}$$

where ν_1 and ν_2 are equal to either zero or 1 depending on the number of wall compatability conditions imposed. Three schemes labeled A, B, and C are considered. In scheme A no wall compatability conditions are employed and both ν_1 and ν_2 are zero. In scheme B we require that the exact continuity equation be satisfied at the wall and take $\nu_1 = 0$ and $\nu_2 = 1$. In scheme C we require that both the exact continuity equation and the condition of irrotationality be satisfied at the wall. In this case both ν_1 and ν_2 are equal to 1.

The ordinary differential equations obtained from the integral relations for all three schemes can be written in the form

$$\frac{d\alpha_i}{d\omega} + \nu_1 h_{io} \frac{d\Lambda}{d\omega} = \sum_{j=1}^{N} h_{ij} T_j(\omega) \qquad i = 1,2,\ldots,N \tag{2.11a}$$

$$\frac{d\beta_i}{d\omega} + \nu_2 f_{io} \frac{dm_b}{d\omega} = \sum_{j=1}^{N} f_{ij} S_j(\omega) \qquad i = 1,2,\ldots,N \tag{2.11b}$$

where h_{ij} and f_{ij} are a set of constants that depend only on the location of the strip boundaries. For schemes B and C these equations are supplemented by compatability conditions which are obtained by substituting the profiles given in Eqs. (2.10) into the exact partial differential equations and evaluating the result on the airfoil surface.

For scheme A the MIR reduces the problem to a system of 2N ordinary differential equations for 2N unknowns α_i, β_i. For scheme B the problem reduces to $2N + 1$ ordinary differential equations for the $\alpha N + 1$ unknowns α_i, β_i and m_b. For scheme C the problem reduces to $2N + 1$ ordinary differential equations plus one algebraic equation for the $2N + 2$ unknowns α_i, β_i, m_b, and Λ.

In the lifting case we require an additional condition to determine the circulation constant Γ. For smooth bodies without a sharp trailing edge, the circulation can be chosen arbitrarily. For bodies with a sharp trailing edge, we impose the Kutta condition which requires the trailing edge to be a stagnation point. The use of this condition in the $\rho h q_\omega$ profile leads to the following condition

$$\Gamma = -\left[2\pi/\beta\right]\left[q_{\omega o}(\omega_{stag}) + \sum_{i=1}^{N+\nu_2} \beta_i(\omega_{stag})\right] \tag{2.12}$$

<u>Numerical Solution of the Differential Equations</u>

In our procedure we treat the α_i's and β_i's as the principal dependent variables and integrate the equations in the form presented in this section. However, the forcing terms in these equations depend on the strip values of the velocity components q_r and q_ω. To carry out this scheme we must provide a procedure for calculating the velocity component q_ω from the mass flow ρq_ω and the normal component of velocity q_r. This is done by first using a change of variables to reduce the problem to one of inverting the one dimensional streamtube relations. The inversion is carried out in closed form using an accurate curve fit developed by Ives. This procedure kept the required algebraic manipulations within bounds and made multistrip applications feasible.

The differential equations to be integrated are inherently unstable, that is, if we linearize them about a given point the resulting linear equations possess real eigenvalues in plus and minus pairs.

Thus there is exponential growth in ω in either direction. The magnitude of the eigenvalues (and correspondingly the rate of growth of the linear perturbation solutions) increases rapidly with increasing N. This implies that the determination of the particular periodic solution of the differential equations which represents the solution for a given airfoil is improperly posed as an initial value problem in either direction of integration. This is, of course, consistent with the elliptic nature of the boundary value problem we are attempting to solve and is to be expected whenever we try to solve elliptic problems by the MIR. Accordingly, we employed boundary value techniques to obtain the solutions presented in this investigation. Although these techniques require more computing time than is generally required of initial value methods, they have been reliable and practical for the symmetric cases considered to date.

In the application of the MIR to potential flow problems based on the continuity and irrotationality equations, as in the present formulation, a curious decoupling of the resulting system of ordinary differential equations occurs. For example, one equation in schemes A and B and two in scheme C decouple from the full system governing the problem. As a result, the surface quantities are determined by a lower order system of equations involving fewer integral relations. This leads to lower accuracy then would otherwise be expected, particularly in the case of one strip which our experience indicates is generally poor in all our schemes. Chushkin, in an approach closely related to our scheme A, avoided these difficulties with decoupling by relaxing the boundary condition on q_r at infinity given by Eq. (2.9) and treated q_{ro} as a primary unknown in the problem. Although his solutions then have nonuniform flow at infinity his one and two strip results for the surface velocity distribution on a nonlifting Joukowski airfoil appear to be quite good. The usefulness of this approach for the lifting case has not yet been demonstrated, and there are indications of some difficulty for this case.

3. APPLICATIONS

We have carried out a number of calculations on a variety of symmetrical profiles at zero lift using schemes A, B, and C with various strip locations and up to five strips. In this section we present a few of these results to illustrate the effectiveness of using wall compatability conditions and certain special strip locations to improve the convergence of the MIR. Solutions with a uniform distribution of the strip boundaries and with strips distributed according to the zeros of the Chebyshev polynomial (e.g., scheme B - CBV) are presented. Although we have not been able to prove that the Chebyshev polynomial is the optimum polynomial for this problem, numerical experience has indicated that it must at least be near optimum. A small ($\sim$ 10%) shift of the strip boundaries from the Chebyshev locations results in a significant loss of accuracy. Other results are presented to demonstrate the applicability of the method to thin airfoils with small nose radii and finite trailing edge angles. In addition, we also present a highly accurate five strip solution for critical flow over a circular cylinder.

The solution for the surface pressure distribution on a 20 percent thick ellipse at $M_\infty = 0.70$ is given in Figs. 2 and 3. In Fig. 2 we compare the solutions obtained using scheme C and uniform strip spacing with the solution obtained using scheme B - CBV (i.e., Chebyshev strip locations) and also with a solution obtained by Sells employing a finite difference technique coupled with an iteration on the density field. The results indicate that five strips were required to obtain an accurate solution using scheme C but that only three strips were required to obtain a converged solution with scheme B - CBV (which used one less wall condition). The results in the figure show good agreement with Sells's solution.

In Fig. 3 solutions for the same geometry and Mach number obtained by using schemes A, B, and C with $N = 3$ are compared with the converged three strip scheme B - CBV solution. We recall that schemes A, B, and C employ zero, one, and two wall compatability conditions, respectively. The results in the figure clearly show the effectiveness of the wall conditions in improving the accuracy for fixed N. Comparison of the two scheme B solutions also shows the significant improvement in accuracy obtained by using Chebyshev spacing.

The effectiveness of the Chebyshev spacing shows up more dramatically for the thinner ellipse considered in Fig. 4. In this figure we compare solutions obtained for a 3 percent thick ellipse at $M_\infty = 0.86$ using scheme B with both uniform and Chebyshev spacing. The results employing uniform spacing are highly oscillatory and obviously divergent. The use of the Chebyshev spacing considerably improves the solution. Although the two strip Chebyshev experiences an overshoot near the nose, the three strip result is monotonic and well behaved. Unfortunately, four strip solutions have not as yet been obtained for this case so that we cannot be absolutely certain of the convergence of the Chebyshev solution.

In Fig. 5 we apply scheme B - CBV to a 6 percent thick circular arc airfoil at $M_\infty = 0.806$ and compare the results with the wind tunnel data of Ref. 5. The results for $N = 2$ to 4 clearly show the rapid convergence of the method for this case. As is usually the case, the one strip results are quite poor, and we do not show them. The agreement with the experiments is fairly good and can be improved considerably by employing a blockage correction to account for wall interference effects. On the basis of information in Ref. 5, a shift in

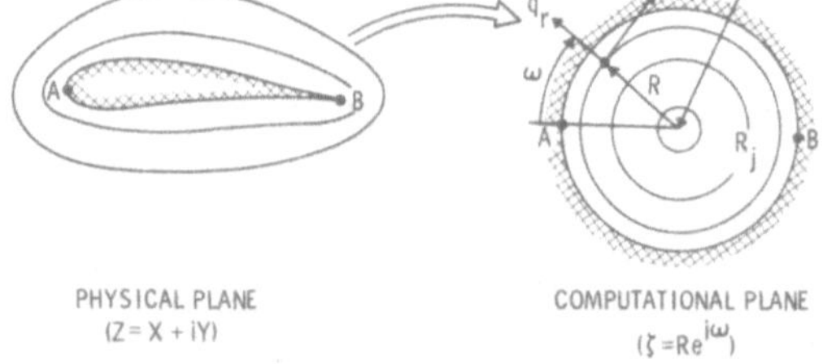

FIG. 1 COORDINATE SYSTEM

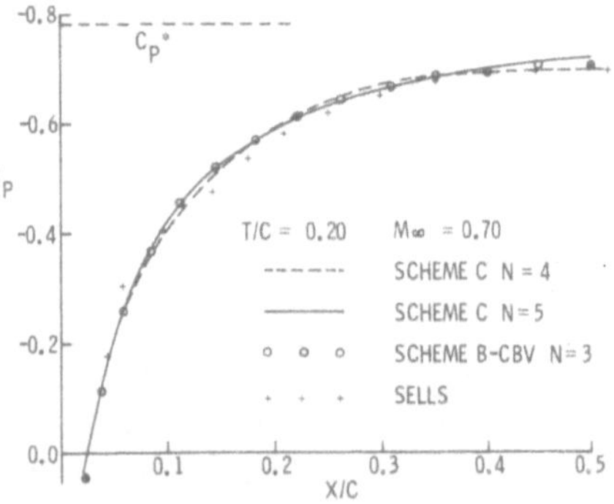

FIG. 2 PRESSURE DISTRIBUTION ON AN ELLIPSE

Mach number of -0.02 appears reasonable. Accordingly we recomputated our four strip results at a corrected Mach number of M = 0.786 and achieved excellent agreement. Since the symmetry of the data implies small boundary layer displacement effects, good agreement with an inviscid calculation is to be expected. Detailed examination of the behavior of the solutions near the stagnation points at the nose (and by symmetry at the trailing edge) indicates that the solution converges rapidly there also. This implies that the sharp edges are adequately treated by use of the conformal transformation.

In Fig. 6 we consider a 10 percent thick Joukowsky airfoil at $M_\infty = 0.660$ and compare the results with Chushkin's solution (although at a slightly different Mach number, M = 0.666). Our scheme B - CBV results have fully converged at N = 3. The one strip results are quite poor, and the two strip results have converged almost everywhere except for small errors near the suction peak. Our peak value closely agrees with Chushkin's so that the slight difference in free stream Mach number is probably insignificant. The discrepancy in trailing edge pressure is relatively large and is most likely due to the nonconformal coordinate system employed by Chushkin.

In Fig. 7 we present the results for a symmetric 12.7 percent thick Karman-Trefftz airfoil at $M_\infty = 0.7$ obtained by scheme B - CBV.

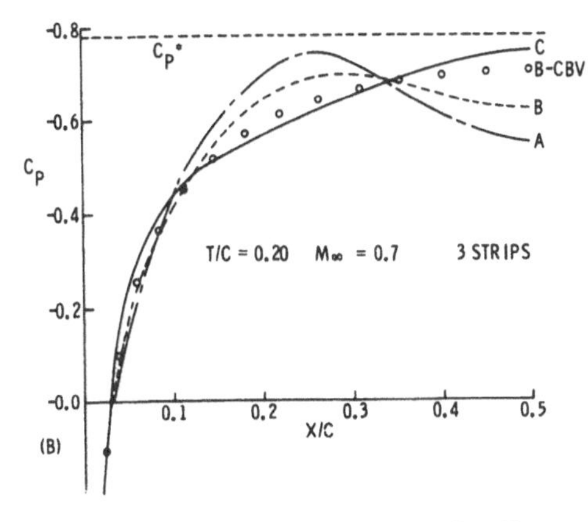

FIG. 3 PRESSURE DISTRIBUTION ON AN ELLIPSE

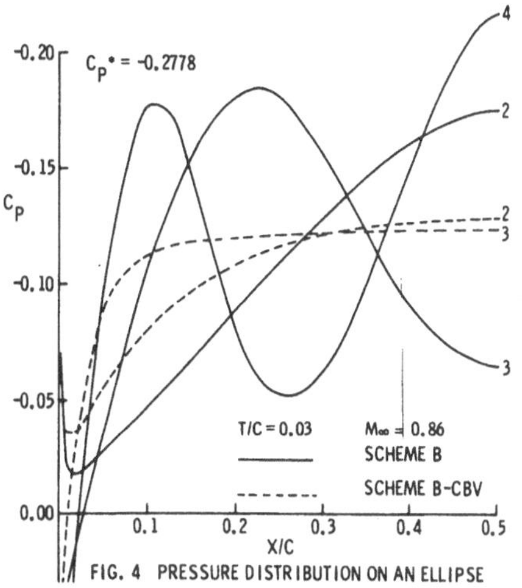

FIG. 4 PRESSURE DISTRIBUTION ON AN ELLIPSE

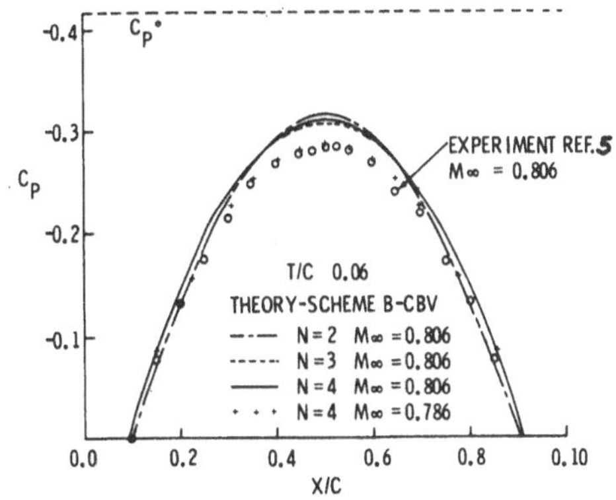

FIG. 5 PRESSURE DISTRIBUTION ON A CIRCULAR ARC PROFILE

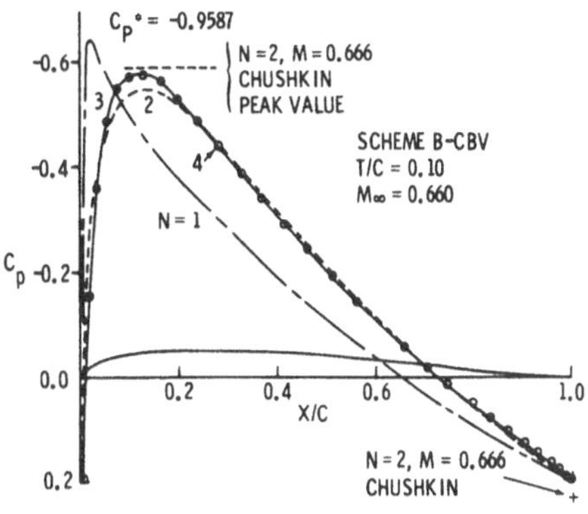

FIG. 6 PRESSURE DISTRIBUTION ON A JOUKOWSKY PROFILE

The one strip results are again poor. In this case the pressure peak is broader than on the Joukowsky profile, and the two strip results are somewhat better. A detailed examination again indicates that the solution converges rapidly in the vicinity of the front and rear stagnation points. Since this airfoil has a very small leading edge radius and a finite trailing edge angle, these results demonstrate the usefulness of the MIR for airfoil type geometries.

In Fig. 8 we employ scheme C with uniform spacing to obtain a solution for the interesting case of critical flow over a circular cylinder. The critical Mach number is determined in the course of the calculation by iterating the free stream Mach number. The MIR converged very rapidly in this case and the four and five strip results are virtually identical. We believe the critical Mach number is converged to the five strip value given in the table to within an error of ± 2 in the last digit. A five term Rayleigh-Janzen solution for the surface velocity obtained by Simisaki agrees quite well with our converged solution, and his estimate of $M_{crit} = 0.400$ is consistent with our more accurate solution. Professor Van Dyke recently informed us of the results obtained by Hoffman from a six term Rayleigh-Janzen solution giving $M_{crit} = 0.3983 \pm 0.0002$, which is in remarkable agreement with our solution.

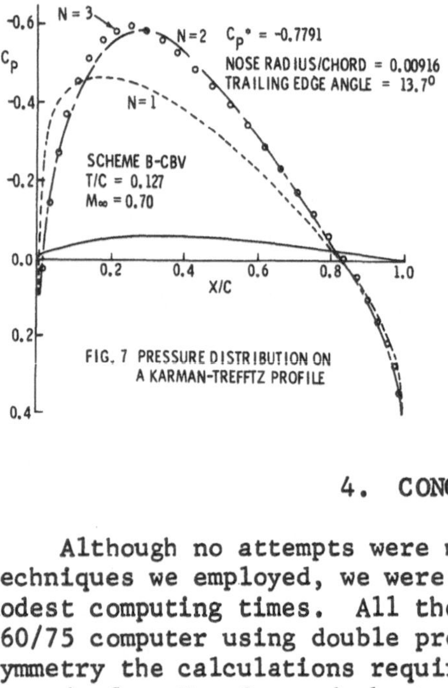

FIG. 7 PRESSURE DISTRIBUTION ON A KARMAN-TREFFTZ PROFILE

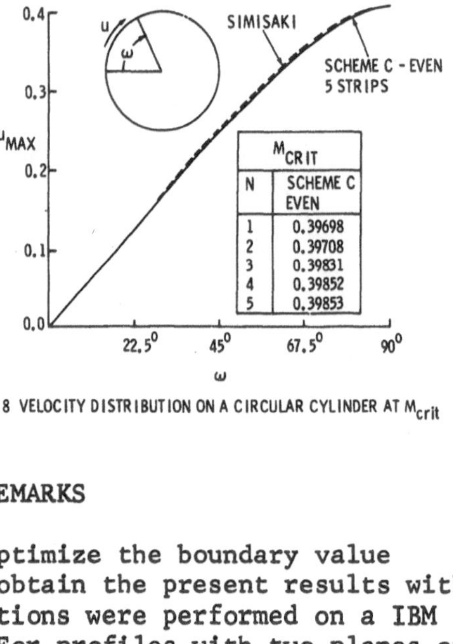

FIG. 8 VELOCITY DISTRIBUTION ON A CIRCULAR CYLINDER AT M_{crit}

4. CONCLUDING REMARKS

Although no attempts were made to optimize the boundary value techniques we employed, we were able to obtain the present results with modest computing times. All the calculations were performed on a IBM 360/75 computer using double precision. For profiles with two planes of symmetry the calculations required about 10 seconds for $N = 1$, 30 seconds for $N = 2$, and about a minute and a half for $N = 3$. With just one plane of symmetry the times were 40 seconds of $N = 1$, a minute and a half for $N = 2$, and about 4 minutes for $N = 3$. We believe these times can be considerably reduced by slightly modifying our program.

The present results seem to indicate that a two strip version using scheme C with Chebyshev spacing should be adequate for engineering purposes, and we are in the process of coding this scheme. We also plan to investigate the usefulness of Chushkin's procedure (i.e., relaxing one boundary condition at infinity) for the lifting case.

In the near future we plan to apply the present methods to lifting airfoils and to explore the possibility of extending the method to the supercritical regime.

5. REFERENCES

1. Belotserkovskii, O. M. and Chushkin, P. I., Basic Developments in Fluid Dynamics 1, 1-123 (ed. M. Holt), Academic Press (1965).

2. Chushkin, P. I., Sb Vych. Mat. 3, 99-110 (1958).

3. Ives, D. C., RM-452, Grumman Aerospace Corporation (1969).

4. Jones, D. J., Private Communication (1970).

5. Knechtel, E. D., NASA TN-D15 (1959).

6. Ludford, G. S. S., J. of Math. and Phy. 30, 131-139 (1951).

7. Sells, C. C. L., Proc. Roy. Soc. A 308, 377-401 (1968).

8. Simisaki, T., Bull. Univ. Osaka Prefecture A 4, 27-35 (1955).

9. Van Dyke, M. D., Private Communication (1970).

Session V

1. Shock Waves, 2. Turbulence

H. Cabannes, Chairman

ON THE CALCULATION OF GAS FLOWS WITH SECONDARY FLOATING SHOCKS

O. M. Belotserkovskii

Computing Center of the Academy of Sciences, U.S.S.R., Moscow

In certain regimes of supersonic flow about blunt bodies (cones, wedges, etc.) there arise, in the flow field behind the detached shock wave, secondary floating shocks (SFS), which have a significant effect on the aerodynamic characteristics of flying devices. Thus in the course of calculations of flow about profiles with a sharp corner [1] a solution with a floating shock in the region behind the main shock wave was obtained. This phenomenon was also noted earlier in experiments; its cause, however, was not particularly well understood. It was suggested by M. J. Lighthill, for example, that the shock was generated as the result of successive separation and reattachment of the boundary layer in the vicinity of the corner [2]. The reasons for the generation of SFS, the conditions under which they arise (body configuration, flight regime, etc.), and the properties of such flows have not been sufficiently studied to the present time.

We present here certain results of computations related to the generation of SFS. The computations were carried out by the aforementioned methods within the context of an ideal gas for both smooth blunt bodies and profiles with a sonic or supersonic corner [1].

1. Let us examine first those flow properties which lead to the generation of SFS.

In flow about blunt cones or wedges there occurs a rapid deceleration of the flow in passing from the blunt nose portion to the straight portion of the contour. This results in a crowding together of the characteristics and in the appearance of compression waves emanating from the body surface behind the corner. In those cases when the pressure gradient downstream from the characteristic bounding the region of influence of the blunt headform is large and the shock layer is sufficiently thick the compression waves develop into a shock wave of initially zero intensity and arising in the interior of the flow.

We shall define [2]

1. The shock wave -- the main shock wave arising in the uniform supersonic flow ahead of the body.
2. A secondary shock -- a shock in the region downstream from the main shock.
3. A floating shock -- a secondary shock on both sides of which the flow is supersonic. By analogy with characteristics we will call it a shock of the first (second) family if the tangent thereto is achieved by rotating the velocity vector through a positive (negative) acute angle. Shocks of different families do not connect smoothly with each other.

Every instance of spontaneous generation of shock waves is always associated with the intersection of characteristics of one family. In such flows there arises an entire region doubly covered by characteristics of one family. Furthermore, the cusp of the envelope of these characteristics marks the beginning of the shock wave. The cusp itself does not yet belong to the discontinuity. The derivatives there of velocity, density, pressure, etc., become finite, and hence the intensity of the SFS at this point is zero and a "real" shock wave arises only subsequently.

In practice it is not usually necessary to determine exactly the point at which the envelope has a cusp and to start out the SFS at zero intensity. The shock wave is considered instead to emanate from the point of intersection of those characteristics of one family which are followed in the computations.

It should be noted that SFS, in view of their relatively weak intensity, can be rigorously constructed numerically only by direct application of the method of characteristics.

Figure 1, taken from Ref. [3], shows examples of flow computations at $M_\infty = 100$; $\kappa = 1.4$ about truncated cones (semi apex angle $\omega_1 = 20°$ - Fig. 1a; $\omega_1 = 10°$ - Fig. 1b; $\omega_1 = 0°$ - Fig. 1c) with a common conical nose ($\omega_o = 54.5°$). The broken line ACB is the generator of the body of revolution; AD - shock wave; BD - the characteristic of the second family at which characteristics of the first family intersect. Th- region of intersection of characteristics (if such exists) is denoted by a circle with the letter N. In Fig. 1b the line SN is the last characteristic of the fan emanating from point C. The dot-dash curve in Figs. 1a and 1b represents the pressure on pointed cones with semi apex angles 20° and 10°, respectively, for the same Mach number, M_∞. Lengths are referred to the radius at the corner point; pressures to the stagnation pressure.

As can be seen from the curves, in the third case the SFS is generated inside the flow field, in the second, immediately behind the main shock wave, and in the first case there is no SFS generated in the shock layer. Furthermore, it should be noted that as the apex angle of the truncated cone ω_1 decreases, there is a corresponding increase in pressure gradient at the corner, and the shock layer becomes relatively thicker.

The calculations were performed here within the context of an ideal gas; the initial conditions for the method of characteristics were obtained from the exact solution for a pointed cone, insuring a high degree of accuracy in the computations. These examples show that the generation of SFS is not dependent on flow properties related to viscosity (i.e., separation of the boundary layer) and that the fact of their occurrence is not related to the use of "coarse" initial data for the numerical calculation of the supersonic region in the method of characteristics.

Thus the regimes of flow about blunt cones and wedges which are conducive to the development of a positive pressure gradient behind the corner and a sufficiently thick shock layer lead to the generation of SFA, the position of which is determined by the intersection of two neighboring characteristics of the same (first) family.

We shall now cite several examples.

In Fig. 2 are shown shock waves and longitudinal pressure distributions for plane ($\nu = 0$) and axisymmetric ($\nu = 1$) bodies of one configuration (circular nose, a corner at $x = 68°$, $\omega = 0$), while Fig. 3 shows pressure profiles along smooth circular cones (solid line) and wedges (dotted line) for various apex angles ω [4],[6].

From these figures it is apparent that, other things being equal (the flow regime, contour shape, etc.), the plane blunt body, in contrast to the axisymmetric (or a body with a sonic corner in contrast to a smooth one) introduces a greater disturbance into a supersonic flow, and hence the shock layer here will be thicker. Furthermore, in the case of blunt plates $\omega = 0$ and wedges $\omega \neq 0$ (plane flows), and also for bodies with a sonic corner (Fig. 4) there arises immediately after the blunt nose a zone of overexpansion, and a positive pressure gradient showing rapid increases with increasing ω obtains at all ω at the point of connection between the blunt nose and the rectilinear portion. It follows that, all other things being equal, more favorable conditions from the point of view of generation of SFS are realized in the plane two dimensional case or with bodies with a sonic corner.

Figure 5 shows the main shock wave and SFS which arise in the supersonic zone in the case of flow about cones, spherical headforms, and a sonic corner ($f = 20°$; $\kappa = 1.4$) [1]. Results are shown for cones with semi apex angles

(ω = - 5°, 0°, 10°) and incident Mach numbers $M_\infty = 4$ (the solid curve) and $M_\infty = 6$ (dotted) curve.

The intensity of a floating shock first rapidly increases and the, with increasing distance from the blunt nose, gradually decreases. In the cases cited above the angle through which the flow passing through a floating shock is deflected achieves a maximum of a few degress, while at a distance of some 30 to 40 nose radii this angle amounted to only a few seconds. V. F. Ivanov first computed tables of the locations of floating shocks [ ]. It is interesting to note that in the given cases for which floating shocks formed, the shape and location of the main shock waves (in the region lying ahead of their intersection with the secondary shocks) coincided (for identical values of incident Mach number) for various semi apex angles of the cone in the range $-10° \leq \omega \leq +10°$ (Fig. 6).

This property is explained, apparently, by the substantial extent of the region of influence of a blunt nose, which, for such regimes of flow, includes a relatively large portion of the main shock wave. Figure 7 illustrates the effect of the corner point on the position of the limiting characteristics.

An analogous situation regarding the region of influence is observed also with smooth blunt cones. Figure 8 shows the coordinates of the shock wave for smooth cones with spherical noses. Also shown are the coordinates at various ω of the point B, through which passes the characteristic bounding the region of influence of the nose.

Figure 9 shows examples of flow with SFS obtained by V. I. Kosarev for axisymmetric bodies with $\alpha = 0°$ and 5° ($\kappa = 1.4$; $M_\infty = 6$) insuring isentropic compression of the flow (smoothly joined cones: the fore-portion $\omega_0 = 10°$ and an aft-portion $\omega_1 = 20°$). Flow patterns (SFS are shown by dotted lines) and longitudinal pressure distributions are shown for the planes $\phi = 0, \pi/2, \pi$. Here, too, the generation of SFS is clearly seen when the conditions of a sufficiently thick shock layer and positive pressure gradient are present ($\alpha = 0°$; $\alpha = 5°$ at $\phi = 0, \pi/2$).

2. Next we will present the results of some analytic studies of the questions of generation of SFS. The existence of such shocks in the case of a flat plate with a leading wedge was first proved by A. A. Nikolskii under the assumption of constant entropy. E. G. Shifrin [2,5] considered the conditions under which secondary waves would be generated in plane rotational flows. Let us consider the results of this work.

It proved possible, for example, to examine certain reasons underlying the generation of a secondary shock in the case of flow about a convex profile with a detached shock wave. It is assumed that the form of the profile is such that the flow behind the shock wave is continuous and that in the minimal region of influence the entropy is a monotonically decreasing function of the stream function (the shock is convex).

The condition is found for the breakdown of continuous supersonic flow in the characteristic triangle ABC, based upon the profile and contiguous with the minimal region of influence (Fig. 10).

The following theorem is proved:

"If the profile is subjected to a continuous deformation, replacing part of its contour (located downstream from some point E) by the tangents to the profile at this point and moving it upstream, then, before the point E falls into the minimal region of influence, continuous supersonic flow in the maximal characteristic triangle, adjacent to the minimal region of influence, breaks down."

It is shown that in flow about an infinite blunt wedge with a detached shock wave, if its vertex angle is gradually increased (but so that the flow at infinity behind the shock wave always remains supersonic), then there will arise in the flow either a shock or a local subsonic zone of isentropically retarded gas (Fig. 10). An analogous result is obtained in the case of flow about a profile with a corner from which a sonic line originates.

The results of analytical studies agree very well with the results of computations cited in the preceding section for smooth bodies.

We shall restrict the proof of the preceding theorem to the case of a smooth convex profile in incident flows with $M_\infty < M_0(\kappa)$ (Fig. 10). Here the first type of minimal region of influence of the blunt nose is found, bounded in the shock layer by the characteristic of the first family AB.

Integrating the compatibility relations on the characteristics in the direction of increasing entropy from point A to the profile contour, we obtain [5]

$$\beta_B \geq \beta_A + I_1 = \beta_A + \frac{1}{2R\kappa} \int_{(AB)} \sin 2\alpha \, dS \quad ,$$
$$\beta_C \leq \beta_A - I_2 = \beta_A - \frac{1}{2R\kappa} \int_{(AC)} \sin 2\alpha \, dS \quad , \tag{A}$$

where $\alpha = \arcsin(1/\mu)$, β is the angle of inclination of the velocity vector to the axis of symmetry.

Since both integrals are positive there must exist on the profile contour a point D, lying between points A and C in which $\beta_D = \beta_A$.

We subject the profile to a continuous deformation, replacing part of its contour located downstream from some point E, by the tangent to the contour at this point and then moving the point E upstream from its aft-most position on the profile all the way to the point D. We shall show that just as soon as point E coincides with point D, either one of the aforementioned assumptions is contradicted or the minimal region of influence is reconstructed.

Indeed, we have

$$\varepsilon = \frac{1}{2R\kappa} \inf_{E \in [DF]} \int_{(AC)} \sin 2\alpha \, dS = \frac{1}{2R\kappa} \inf_{E \in [DC]} \int_{(AC)} \sin 2\alpha \, dS \geq$$
$$\geq \frac{1}{2R\kappa} \int_{(AP)} \sin 2\alpha \, dS$$

Here the point P is the point of intersection of characteristics of the second family AC with characteristics of the first family DP.

Let us now define the point G on the original profile by the equality

$$\beta_G = \beta_D - \varepsilon \quad .$$

In view of the convexity of the profile, point G lies between points D and C. If point E is located between points D and G, i.e., if

$$\beta_D \geq \beta_E > \beta_G$$

Then we obtain

$$\beta_C = \beta_E > \beta_G = \beta_D - \varepsilon = \beta_A - \frac{1}{2R\kappa} \inf_{E \in [DC]} \int_{(AC)} \sin 2\alpha \, dS \quad ,$$

i.e., $\beta_C > \beta_A - E$, which contradicts the second of the inequalities (A).

Thus, while with the original profile the problem of determining supersonic continuous flow in the characteristic triangle ABC from Cauchy data on the characteristic AB and the inpenetrability conditions on the profile BC had a solution in the large, this is no longer true for the profile deformed in the manner indicated.

In an analogous way the case $M_\infty > M_0(\kappa)$ can be examined, as can the case in which the sonic line starts from the profile corner.

Let us now examine [2] the flow of uniform supersonic flow about a profile with a corner (at the corner the angle is convex). We consider the case of low incident supersonic velocity, when changes in entropy on the shock wave can be neglected; we shall also take advantage of the possibility of making use of the transsonic approximation to the polar and the characteristics intersecting it.

Let us examine first the flow with attached shock wave about a profile with straight-line segments OA and AF with $\beta_0 = 0$. In Fig. 11 OB is the straight-line segment of the shock wave; AB, AC, AD are the straight-line segments of characteristics of the first family, BE is a characteristic of the second family.

Let us map the region behind the shock wave on the velocity hodograph plane $\eta,\beta(\eta = (\kappa+1)^{1/3}(\lambda-1)$, λ is the velocity coefficient, β is the angle of inclination of the velocity vector to the axis of symmetry, κ is the adiabatic index; the axes β,y are directed vertically upward, the axes λ,x horizontally to the right).

The mapping of the region OBEAO is the segment of the characteristic a_1a_2 of the second family $\beta = C - (2/3)\,\eta^{3/2}$; the point a_1 with coordinate β_1 lies on the shock polar $\beta = \Omega^{-1/2}(\eta_\infty+\eta)^{1/2}(\eta_\infty-\eta)$, the point a_2 lies on the η axis. The equations of the characteristics and shock polar are given in the transsonic approximation.

Theorem

"If in flow with an attached shock wave about a profile with straight line segments OA and AF the flow in that region behind the shock wave is everywhere supersonic, then it cannot be continuous."

Let us examine in the physical plane the fan of characteristics of the first family emanating from the corner point A (subsequently we shall denote this as the fan A). If the flow behind the shock wave is everywhere supersonic and continuous, then every characteristic of the fan either intersects the shock wave or extends to infinity. Corresponding to this in the $\eta\beta$ plane each characteristic of the first family passing through the characteristic a_1a_2 must end up either somewhere on the shock polar or at the point n, also lying on the shock polar and representing the mapping of the uniform rectilinear flow at infinity.

This represents a contradiction inasmuch as characteristics of the first family emanating from points of the segment ca_2 in the $\eta\beta$ plane cannot end up on the shock polar, consequently if the flow is to be supersonic a secondary shock must arise.

This theorem with appropriate changes in formulation can be extended also to the case of a detached shock wave.

It can be shown [2], that the beginning of the shock does not lie at the apex of the convex angle, but again in the region covered by the characteristics of the fan A. The floating shock is located downstream from its end point (beginning or end). The end of the shock of the first family (where it degenerates into a characteristic) lies at an infinite distance from the profile. If the shock of the first family intersects the main shock wave or another shock, then at the point of intersection its intensity does not vanish.

The proof of these properties of SFS is based on the one-to-one property of the mapping in the hodograph plane of the apex of the convex angle and its neighborhood covered by characteristics of the fan A.

We denote by Q_R the region bounded by the shock wave, the segment of profile OA, the floating shock, and the last characteristic of the fan A. In accordance with Theorems [2] the beginning of the shock lies either on this characteristic or an a characteristic of the first family located downstream of it. A study of the mapping of the region Q_R in the hodograph plane leads to the conclusion that the flow in the region Q_R does not depend on the shape of the profile over the segment AF. This property was observed also in the calculation [1]. In the flow about a profile with a straight-line segment AF the last characteristic of fan A represents a branch line.

And finally, we can establish the following theorem:

Theorem

"In flow about a profile with a rectilinear segment AF with $\beta_0 = 0$ the shock does not cross the main shock wave at a finite distance; apart from that, there are no other floating shocks of the first family.

In case there exists a secondary shock with flow about a profile with a rectilinear segment AF and $\beta_o > 0$, it must intersect the main shock wave. At the point of intersection of the main shock wave behind the secondary shock, $\beta < \beta_0$. The shock wave on the segment located downstream from the point of intersection with a secondary shock (or from the point of intersection with the last characteristic of the fan A, if there is no secondary shock) consists of an infinite number of segments with curvatures of opposite signs;* the oscillations in the angle of inclination of the shock wave decay with increasing distance from the profile."

This result also agrees well with numerical results for bodies with corners.

References

1. O. M. Belotserkovskii, A. B. Bulekbaev, M. M. Golomazov, V. G. Gruditskii, V. K. Luskin, V. F. Ivanov, Yu. P. Lun'kin, F. D. Popov, G. M. Ryabinkov, T. Ya. Timofeeva, A. I. Tolstikh, V. N. Forin, F. V. Shugaev. Flow past blunt bodies in supersonic flow; theoretical and experimental results, edited by O. M. Belotserkovskii. Publication of Computing Center AN SSSR Moscow 1966 (1st edition) 1967 (2nd edition revised and enlarged) NASA TT, F-453, 1967.

2. E. G. Shifrin. "Formation of floating shock waves for flow past profiles with broken contours," Prik. Mat. i Mekh. (to appear).

3. O. N. Katskova, I. N. Naumova, Yu. D. Shmiglevskii, N. P. Shulishnina. Experiments to calculate plane and axisymmetric supersonic flow of a gas by the method of characteristics. Publication of Computing Center AN SSSR Moscow 1961.

4. P. I. Chushkin. "Blunt bodies of simple form in supersonic flow of a gas," Prik. Mat. i Mekh. 5, 24, 927-930, 1960.

5. E. G. Shifrin. "On one condition for the breakdown of regions of continuous supersonic flow for streamline profiles with detached shock wave," Doklady AN SSSR. 4, 176, 797-800, 1967.

6. O. M. Belotserkovskii. "Flow past symmetric profiles with detached shock waves," Prik. Mat. i Mekh. 2, 2, 206-219, 1958.

*The property of oscillation of the angle of inclination of the main shock is found in connection with the theory of propagation of disturbances in the case of flow about wedge-shaped profiles.

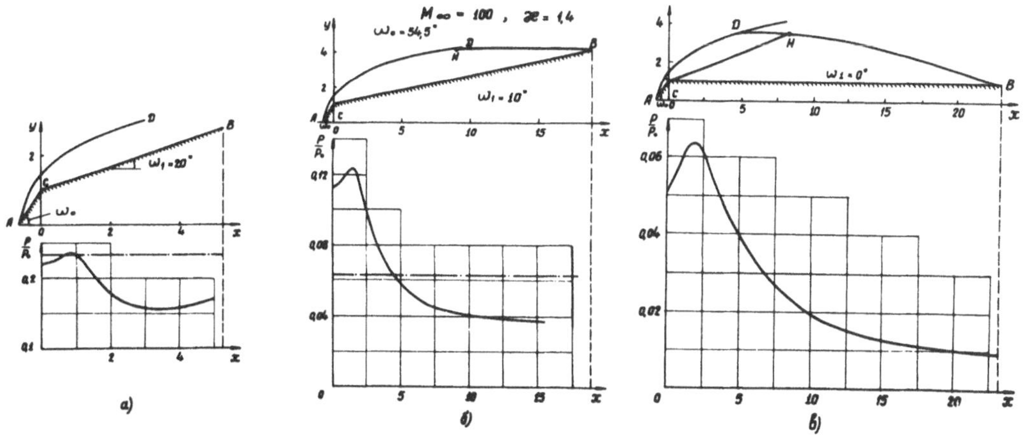

Fig. 1

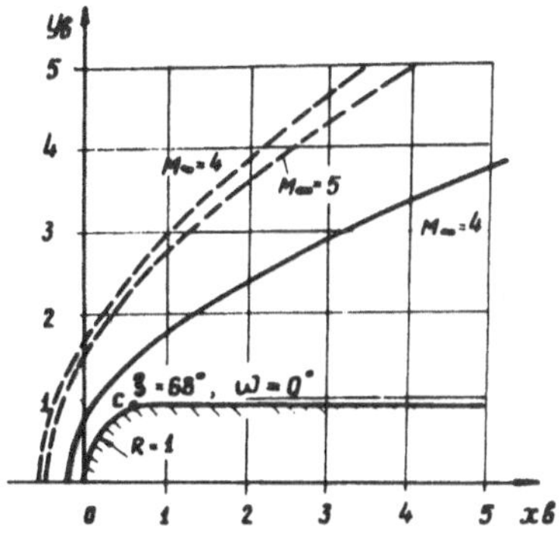

Fig. 2

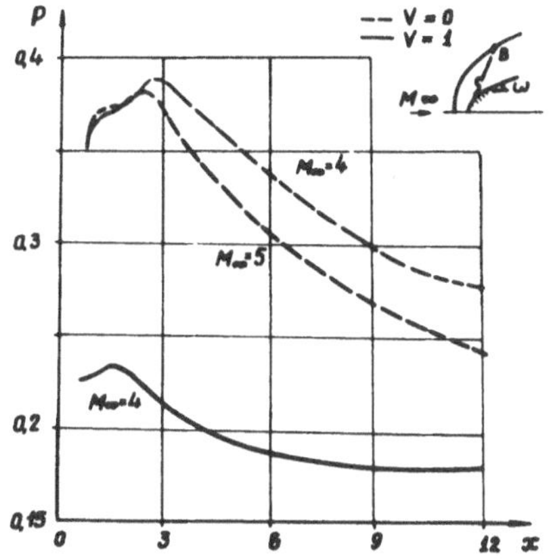

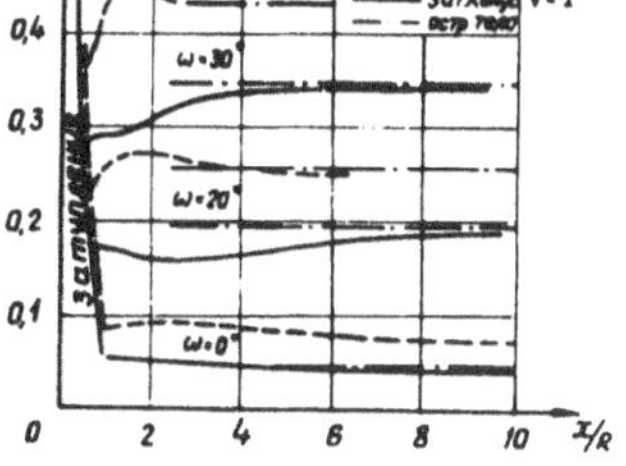

Fig. 3

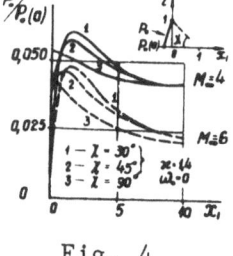

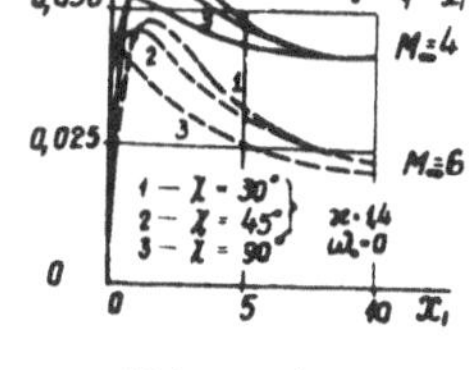

Fig. 4

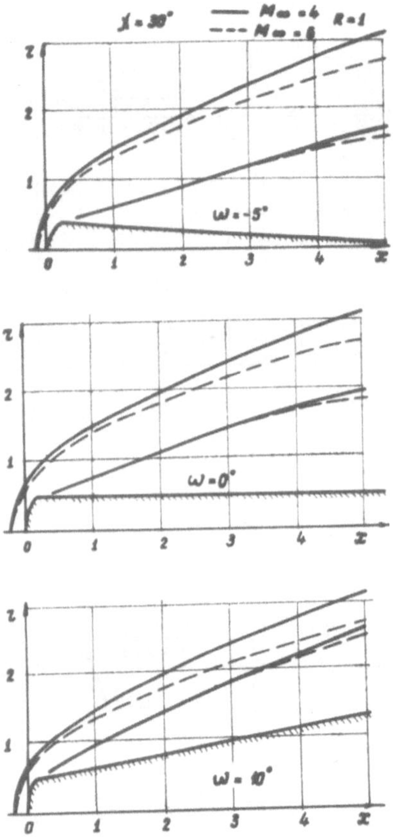

Fig. 5

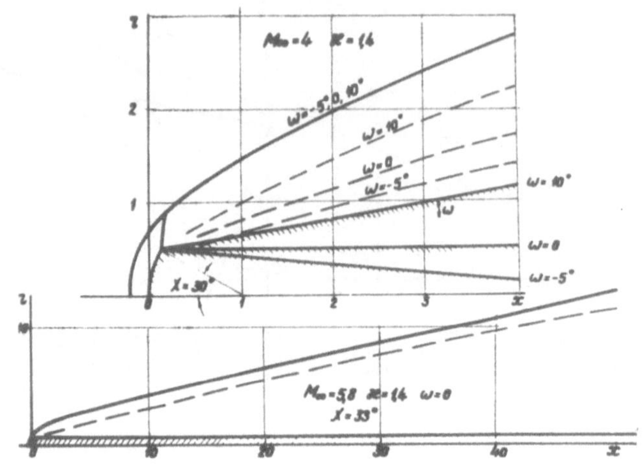

Fig. 6

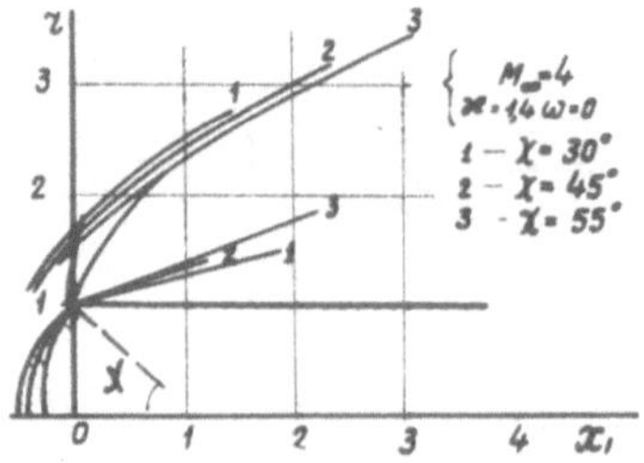

Fig. 7

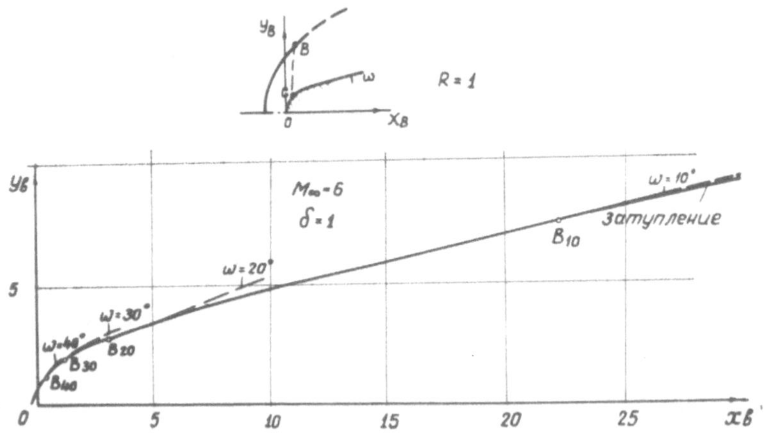

Fig. 8

$$M_\infty = 6, \quad \mathscr{æ} = 1,4$$
$$\alpha = 0°, 5°$$

$\alpha = 0°$ $\varphi = \pi$ $\alpha = 5°$

$\varphi = \frac{\pi}{2}$

$z = 2$

$\varphi = 0$

Fig. 9

$\beta_0 = 0$

$\eta = (K+1)^{1/3}(\lambda - 1)$

Fig. 11

$$I_2 = \frac{1}{2R\mathscr{æ}} \int\limits_{(AC)} \sin 2\alpha \, dS$$

$$\varepsilon = \ln I_2$$
$$\varepsilon \in [DF]$$

$M_\infty < M_0(\mathscr{æ})$

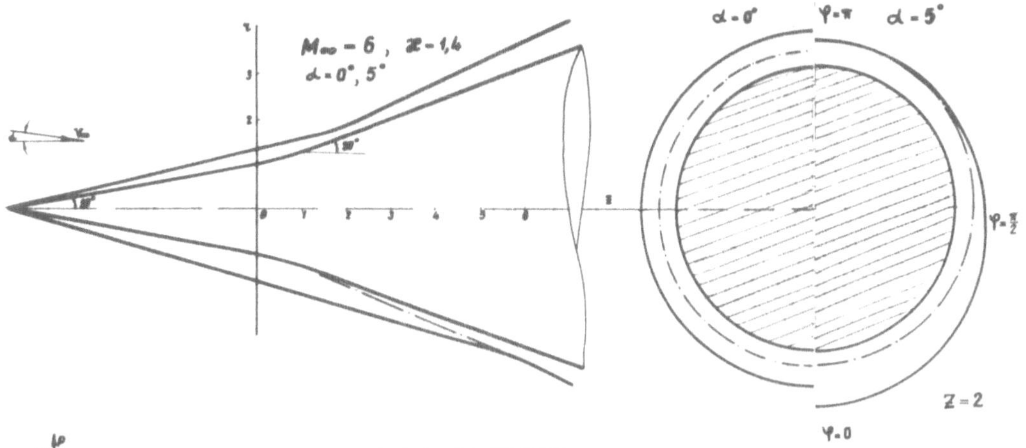

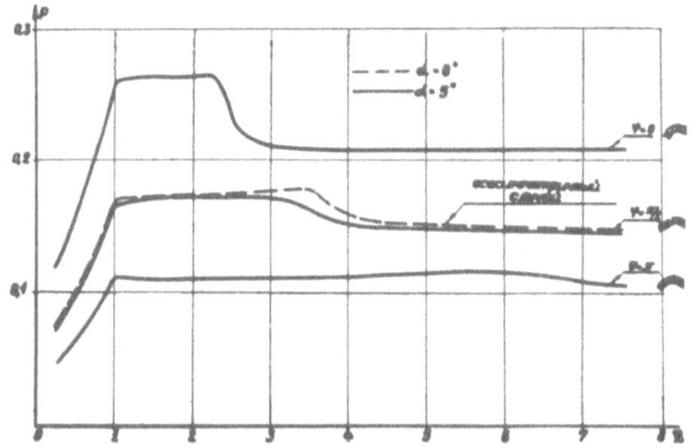

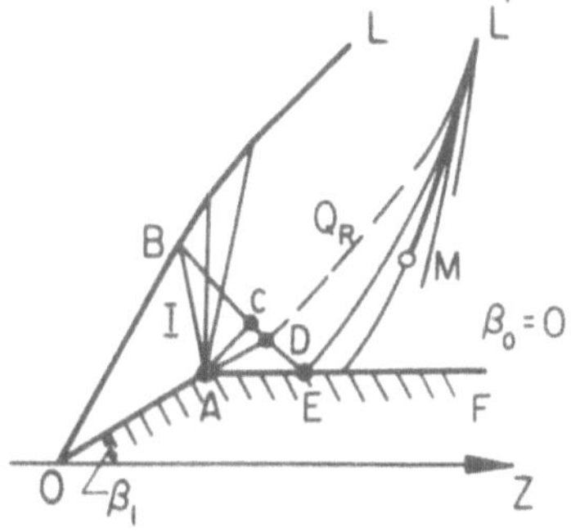

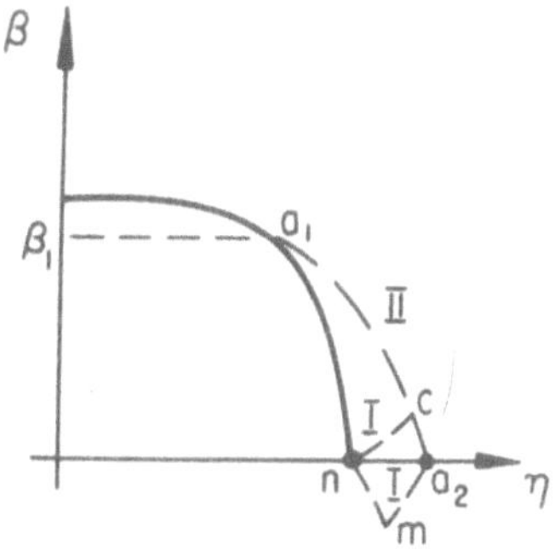

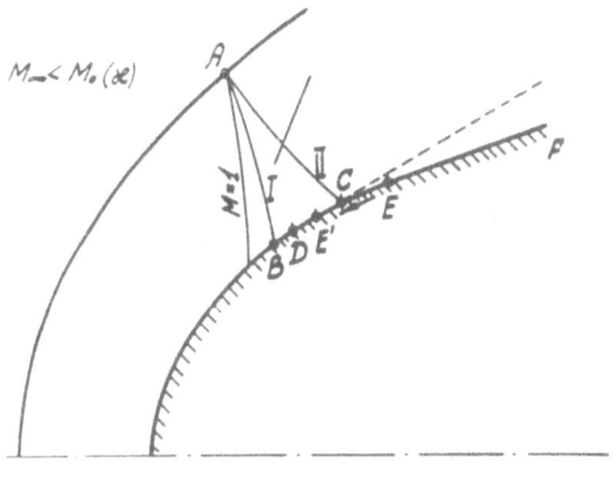

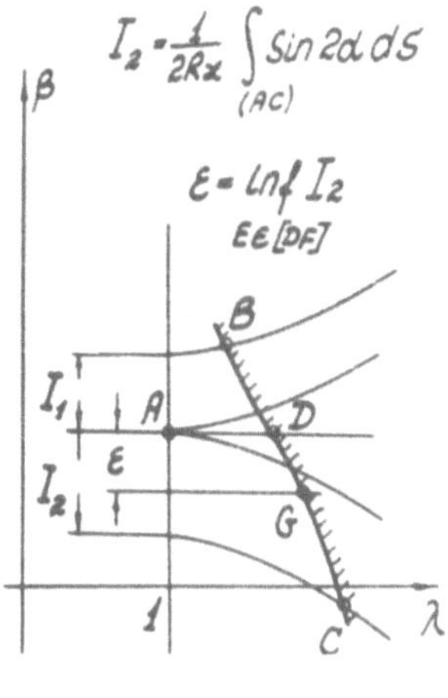

Fig. 10

MOTION OF A SHOCK WAVE THROUGH A NONUNIFORM FLUID

Richard Collins and Hsiang-Teh Chen
School of Engineering and Applied Science, University of California, Los Angeles

I. INTRODUCTION

In this paper, a general numerical method is developed for the diffraction of a shock wave of arbitrary strength as it propagates into a nonuniform fluid. Application may be made to a variety of problems in fluid dynamics in which a shock front interacts with a nonuniform fluid region, such as wave propagation in a stratified ocean, the motion of a shock front along a channel of varying fluid properties, and the rate of return of a perturbed wave to a stable equilibrium profile.

The method will be used to treat all of the above examples, but is developed first in its most general form for the case of shock wave propagation into a region consisting of two nonuniform semi-infinite regions separated by a contact discontinuity. The formulation then serves to treat any variant of the problem containing arbitrarily situated fluid discontinuities imbedded in a general nonuniform region. It must be stated at the outset that application is restricted to two-dimensional unsteady flows.

The present numerical scheme is useful mainly in its prediction of the shape and strength of the diffracting shock front with time.

II. MATHEMATICAL FORMULATION

We consider the propagation of an initially plane uniform shock wave into a region in which sound speed, density, pressure, and specific heat ratio may vary as general functions of the space coordinates x, y. In addition, this region is separated into two semi-infinite half planes by an interface, across which the sound speed, density and specific heat ratio are discontinuous. At time $t=0$, the uniform plane shock front reaches the contact discontinuity at $x=0$. A shock wave is transmitted, and in general, either a shock or rarefaction wave is reflected, according to whether the element of shock encounters a region of higher or lower density, respectively. The interaction of the incident shock wave with the contact discontinuity at $x = 0$ determines the initial strength of the transmitted wave which then propagates in the general nonuniform region $x > 0$, $-\infty < y < \infty$, for $t > 0$.

2.1 Initial Strength of the Transmitted Shock

Each element of the incident shock may be considered independently of adjacent elements during its interaction with the contact discontinuity, and is therefore governed by the usual one-dimensional laws. The strength of the transmitted wave may be determined as a function of the strength of the incident wave and the density, sound speed and specific heat ratio in the two fluid regions on either side of the shock element. The results for both cases of a reflected shock and a reflected rarefaction wave have been presented by Collins and Chen (1970).

2.2 A-M Relation

The motion of a shock front in a nonuniform fluid may be likened to that of shock elements in each of a series of adjacent one-dimensional "channels."

If a shock wave travels in a channel in which the sound speed, pressure, and specific heat ratio vary with distance along the channel, the relation between channel cross-sectional area A and shock Mach number M must be generalized. It will be shown that the modified relation is in general not integrable.

We consider the flow field in a channel at a small change in cross-sectional area.

Across the incident shock separating regions 1 and 2 in Fig. 2.1, the Rankine-Hugoniot relations relate the shock strength $z = p_2/p_1$ to the shock Mach number M, the fluid density ρ_2, velocity u_2 and flow Mach number M_2 behind the shock. The fluid ahead of the shock is initially at rest.

Across the area change 2-3, the continuity and energy equations for a polytropic gas may be combined to give the ratios of pressure, density, and fluid velocity respectively across the area change δA as

$$\frac{p_3}{p_2} = 1 - \frac{\gamma M_2^2}{M_2^2 - 1} \frac{\delta A}{A}, \tag{2.1}$$

$$\frac{\rho_3}{\rho_2} = 1 - \frac{M_2^2}{M_2^2 - 1} \frac{\delta A}{A} \tag{2.2}$$

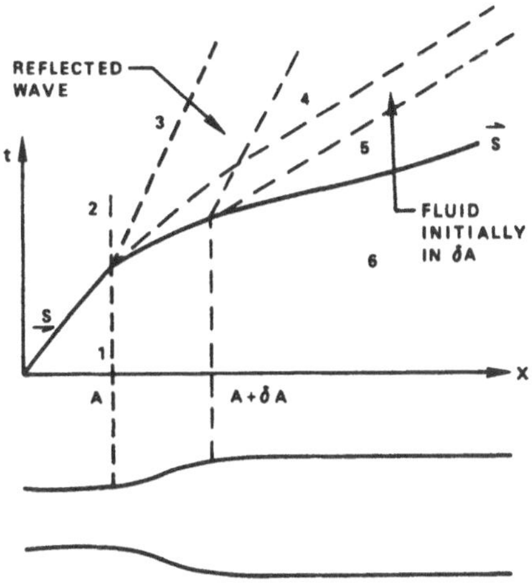

Figure 2.1. Flow in a Channel Element

$$\frac{u_3}{u_2} = 1 + \frac{1}{M_2^2 - 1} \frac{\delta A}{A} . \tag{2.3}$$

Across the simple expansion wave 3-4 the Riemann invariant

$$(\gamma - 1)\, u + 2a = \text{constant}, \tag{2.4}$$

yielding the pressure ratio

$$\frac{p_4}{p_3} = 1 - \gamma M_3 \left(\frac{u_4}{u_3} - 1 \right) \tag{2.5}$$

Across the contact discontinuity 4-5, the pressure and normal component of fluid velocity are continuous.

$$p_4 = p_5, \; u_4 = u_5 \tag{2.6}$$

In the undisturbed region 6, general nonuniformities in density, pressure and specific heat ratio are specified respectively as:

$$\rho_6 = \rho \left(1 + \frac{\delta \rho}{\rho} \right), \; p_6 = p \left(1 + \frac{\delta p}{p} \right), \; \gamma_6 = \gamma \left(1 + \frac{\delta \gamma}{\gamma} \right) \tag{2.7}$$

where the unsubscripted ρ, p and γ refer to region 1.

The Rankine-Hugoniot conditions for the transmitted shock are expressed in terms of ρ_6, p_6, γ_6.

Now the pressure ratio across the expansion wave 3-4 may be determined by combining Eqs. (2.1), (2.7), and the shock strengths $z = p_2/p_1$ and $z_6 = p_5/p_6$

$$\frac{p_4}{p_3} = \frac{p_5}{p_3} = \frac{p_5}{p_6} \cdot \frac{p_6}{p_1} \cdot \frac{p_1}{p_2} \cdot \frac{p_2}{p_3} \approx 1 + \frac{\delta z}{z} + \frac{\delta p}{p} + \frac{\gamma M_2^2}{M_2^2 - 1} \frac{\delta A}{A} \tag{2.8}$$

Similarly, using Eqs. (2.3), (2.6), and the Rankine-Hugoniot relations

$$\frac{u_4}{u_3} = \frac{u_5}{u_2} \cdot \frac{u_2}{u_3} = 1 + \frac{\delta z}{z-1} + \frac{da}{a} - \frac{1}{2} \cdot \frac{(\gamma+1)\delta z + (z+1)\delta\gamma}{(\gamma-1) + (\gamma+1) z} - \frac{1}{2}\frac{\delta\gamma}{\gamma} - \frac{1}{M_2^2 - 1} \frac{\delta A}{A} \tag{2.9}$$

When Eqs. (2.8) and (2.9) are substituted into Eq. (2.5), and it is recognized that M_2 and M_3 can differ only by order $\delta A/A$, as $\delta A \rightarrow 0$, the desired area-Mach number relation results:

$$\frac{dA}{A} = PdM + Qda + Rdp + Sd\gamma \tag{2.10}$$

where

$$T = -\left\{1 + \frac{2(M^2-1)}{[2\gamma M^2-(\gamma-1)]^{1/2}\,[(\gamma-1)M^2+2]^{1/2}}\right\}$$

$$P = T\cdot\left\{\frac{2M[(\gamma-1)M^2+2]^{1/2}}{(M^2-1)\cdot[2\gamma M^2-(\gamma-1)]^{1/2}} + \frac{M^2+1}{M(M^2-1)}\right\}, \quad Q = \frac{T}{a}$$

$$R = T\cdot\frac{[2\gamma M^2-(\gamma-1)]^{1/2}\,[(\gamma-1)M^2+2]^{1/2}}{2\gamma p(M^2-1)} \qquad S = T\cdot\left\{1 - \frac{M^2-1}{2M^2(\gamma+1)}\right\}$$

This relation is in general nonintegrable (for example $\partial P/\partial a \neq \partial Q/\partial M$). Thus the Mach number of the shock depends not only upon local conditions, but upon the complete "history" of the shock propagation in the channel.

2.3 Characteristic Relations

We consider a curvilinear coordinate systems formed of successive shock wave positions (t=constant), and the "rays" (β=const) orthogonal to them, as proposed by Whitham (1957). If an incremental length along the ray is measured by Udt, where U is the local shock velocity, and an elemental length $Ad\beta$ along the shock front, where A is the local cross-sectional area of the channel element between two adjacent rays, then the coordinates are related geometrically by

$$Ma\frac{\partial\theta}{\partial\beta} = \frac{\partial A}{\partial t}, \quad M\frac{\partial a}{\partial\beta} + a\frac{\partial M}{\partial\beta} = -A\frac{\partial\theta}{\partial t} \qquad \begin{matrix}(2.11)\\(2.12)\end{matrix}$$

where $M = U/a$ is the local shock Mach number based upon the sound speed immediately ahead of the shock. The derivative of cross-sectional area A occurring in Eq. (2.11) may be eliminated using the area-Mach number relation (2.10). The governing equations (2.11) and (2.12) then become

$$-PA\frac{\partial M}{\partial t} + Ma\frac{\partial\theta}{\partial\beta} = A\left\{Q\frac{\partial a}{\partial t} + R\frac{\partial p}{\partial t} + S\frac{\partial\gamma}{\partial t}\right\} \tag{2.13}$$

$$a\frac{\partial M}{\partial\beta} + A\frac{\partial\theta}{\partial t} = -M\frac{\partial a}{\partial\beta} \tag{2.14}$$

The characteristic directions are given by

$$\frac{d\beta}{dt} = \pm\frac{a}{A}\left(-\frac{M}{P}\right)^{1/2} \equiv \pm c \tag{2.15}$$

where c corresponds to the "shock-shock" velocity of the disturbance travelling on the main shock.

If gradients in the undisturbed region are expressed in Cartesian coordinates by the transformation

$$dy = aM\sin\theta\,dt + A\cos\theta\,d\beta$$

$$dx = aM\cos\theta\,dt - A\sin\theta\,d\beta$$

where θ is the inclination of a ray (β=const) with the horizontal (x-axis), then the characteristic relations valid on (2.15) become

$$d\theta \pm \frac{a}{Ac}\, dM = \pm\, Ac \left\{ Q\left(\cos\theta \frac{\partial a}{\partial x} + \sin\theta \frac{\partial a}{\partial y}\right) + R\left(\cos\theta \frac{\partial p}{\partial x} + \sin\theta \frac{\partial p}{\partial y}\right) + S\left(\cos\theta \frac{\partial \gamma}{\partial x} + \sin\theta \frac{\partial \gamma}{\partial y}\right)\right\} dt + M\left(\sin\theta \frac{\partial a}{\partial x} - \cos\theta \frac{\partial a}{\partial y}\right) dt \qquad (2.16)$$

where the (+) sign is taken along the C^+ characteristic $d\beta/dt = +C$, and the (-) sign along the C^- characteristic $d\beta/dt = -C$.

It was noted that Eq. (2.10) relating A and M is not integrable, implying that one cannot find a function $A = A(M)$. Fortuitously, A appears in the characteristic equations (2.16) only in combination with C, and from Eq. (2.15) it is clear that Ac is a function of M, and not of A. Hence the explicit dependence on A disappears from Eq. (2.16). In this respect, C may be regarded as an integrating factor, rendering the characteristic relations (2.16) integrable.

III. COMPUTATIONAL PROCEDURE AND RESULTS

The characteristic equations (2.15) and (2.16) are written in finite difference form in terms of points I and III on a C^- characteristic, and points II and III on an intersecting C^+ characteristic, for the special case in which fluid properties in the undisturbed region ahead of the shock are functions only of distance y from the interface.

Then, along C^- characteristics,

$$y_{III} - y_I = \bar{P}_{y^-} \cdot (t_{III} - t_I), \qquad x_{III} - x_I = \bar{P}_{x^-} \cdot (t_{III} - t_I) \qquad (3.1)$$

$$\theta_{III} - \theta_I = (M_{III} - M_I)/\bar{N}^- - \bar{P}_{M^-} (t_{III} - t_I) \qquad (3.2)$$

and along C^+ characteristics,

$$y_{III} - y_{II} = \bar{P}_{y^+} (t_{III} - t_{II}), \qquad x_{III} - x_{II} = \bar{P}_{x^+} \cdot (t_{III} - t_{II}) \qquad (3.3)$$

$$\theta_{III} - \theta_{II} = -(M_{III} - M_{II})/\bar{N}^+ - \bar{P}_{M^+} (t_{III} - t_{II}) \qquad (3.4)$$

where

$$P_y^{\pm} = (M \sin\theta \pm N \cos\theta)\, a(y), \qquad P_x^{\pm} = (M \cos\theta \mp N \sin\theta)\, a(y)$$

$$P_M^{\pm} = M\left(\frac{da}{dy}\right)\cos\theta \mp Na\left[Q\left(\frac{da}{dy}\right) + R\,\frac{dp}{dy} + S\,\frac{d\gamma}{dy}\right]\sin\theta, \quad N = \left(-\frac{M}{P}\right)^{1/2}$$

Suffices $\pm$ refer to the C^+ and C^- characteristics, while the overbar indicates an average value along the characteristic segment. Then, the numerical procedure described in Collins and Chen (1970) is employed, with some minor modifications to ensure numerical stability, to determine the flow field point III, in terms of the known points I and II. At each point III, the fluid properties are calculated using the specified nonuniform distributions of sound speed, pressure, and specific heat ratio. The computations proceed for constant time steps between successive shock positions. Accordingly each new shock position is easily identified as a locus of points III corresponding to the same value of time.

Three examples have been computed: a) the propagation of an initially plane shock past an air-water interface with gradients of sound speed in both the air and water, b) the distortion of an initially uniform shock through a channel of non-uniform fluid, and c) the stability of an initially perturbed shock as it propagates into a uniform medium at rest.

Fig. 3.1 shows the resulting evolution of shock shapes for shock propagation past an air-water interface. Small gradients in sound speed, density and pressure (maintained by a hydrostatic force field) are permitted both in the air and water. The shock profiles are very similar to those calculated in Collins and Chen (1970) for a uniform undisturbed region. The calculation is not restricted however to small gradients.

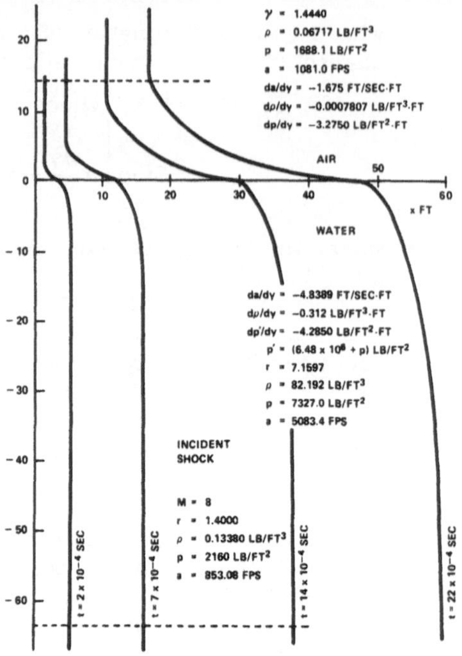

Figure 3.1. Shock Shapes for Air and Water

Examples b), and c), are presented particularly for comparison with analytical results obtained by Chester (1970) during the completion of this work. The closed-form linearized solution reported by Chester and Collins (1970) for the diffraction of a shock wave of arbitrary strength travelling into a region separated into two uniform half-planes of slightly differing sound speed by a contact discontinuity has been superposed to give the flow in a fluid "channel" whose properties vary only slightly from those of the adjacent fluid. An asymptotic solution for example b), valid for large times, would suggest that the maximum width of disturbance formed on the initially plane shock during diffraction in the channel should grow as log t. Fig. 3.2 indicates that this behavior is indeed reached for $t > 60 \times 10^{-4}$ seconds. The agreement is unexpectedly good, since for large times, the disturbance grows beyond the magnitude strictly appropriate to a linear theory. The evolution of the shock shape with time is shown in Fig. 3.3. It would appear that no steady limiting solution exists for this flow; rather the disturbance continues to grow under the influence of the differences in sound speed, the shock profiles maintaining a similar shape, but increasing in proportions.

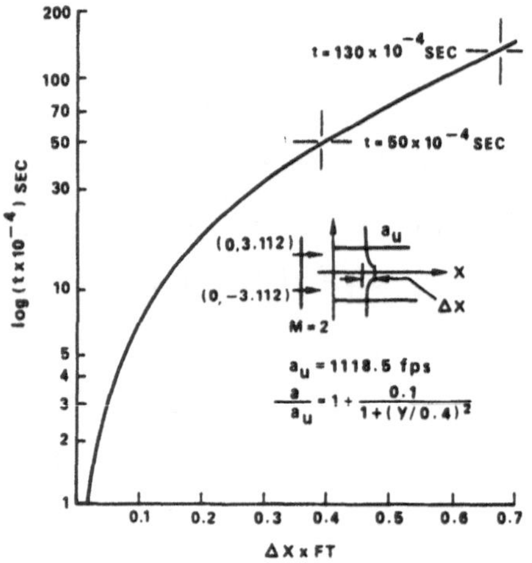

Figure 3.2. Amplitude of Shock Disturbance in Channel

In example c), we consider the stability of a plane shock which, after having been perturbed by passage through a zone of high nonuniformity, reenters a uniform fluid region. The linear theory predicts that after the shock has travelled a distance large compared with the dimensions of the nonuniform zone, the maximum width of the disturbance on the shock profile will decay inversely with time. Here, the linear theory is strictly appropriate. Fig. 3.4 confirms the prediction with excellent accuracy for times $t > 28 \times 10^{-4}$ second. The flattening of the shock profiles is evident in Fig. 3.5.

REFERENCES

Chester, W., Unpublished research notes (1970).

Chester, W., and Collins, R., Israel Journal of Techn. Vol. 8, No. 4, 345-351 (1970).

Collins, R. and Chen, H-T., J. Comput. Physics, Vol. 5, No. 3, 415-442 (1970).

Whitham, G.B., Journ. Fluid Mech. 2, 145-171 (1957).

ACKNOWLEDGMENT

This research was supported by the Office of Naval Research, under Contract N00014-69-A-0200-4013, NR089-005.

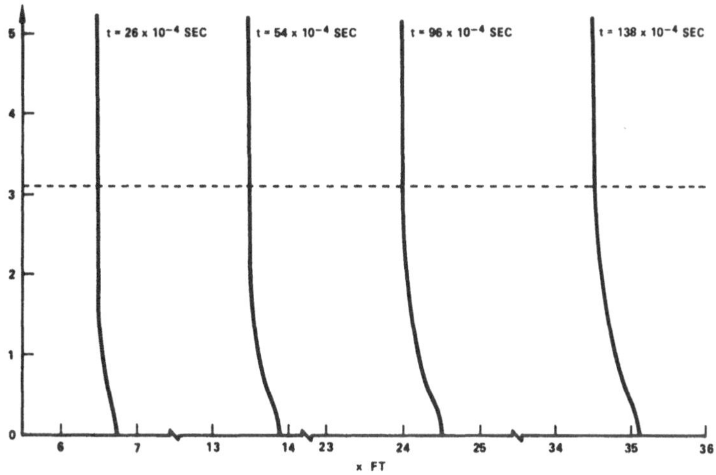

Figure 3.3. Shock Shapes in Fluid "Channel"

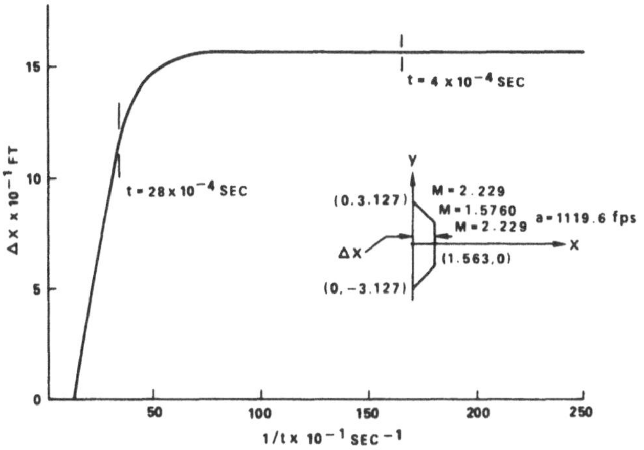

Figure 3.4. Amplitude of Shock Disturbance after Non-uniform Zone

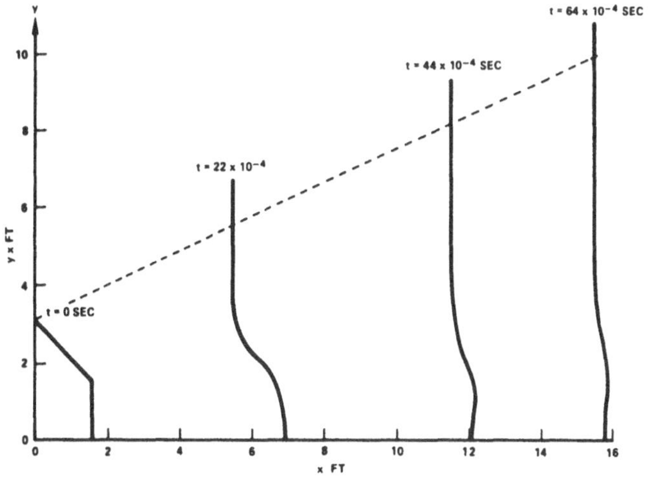

Figure 3.5. Shock Shapes after Passage through Non-uniform Zone

NON-LINEAR ANALYSIS OF THE SHOCK PROFILE

IN DIFFERENCE SCHEMES

V. V. Rusanov

Institute of Applied Mathematics, Academy of Sciences, U.S.S.R., Moscow

To calculate gas flows by difference schemes using continuous computations, i.e., without taking into consideration discontinuities, the character of behavior of the numerical solution near a discontinuity is very important. For the case of one equation with two independent variables such as $f_t + af_x = 0$ a number of authors have formulated the idea of a monotonic difference scheme which converts any monotonic initial data $f_m^n = f(x_m,t)$ into monotonic values f_m^{n+1}. It is easy to show (see 1) that the condition for the difference scheme to be monotonic is equivalent to the condition that all coefficients be positive in a formula expressing f_m^{n+1} in terms of f_m^n. Namely, with

$$f_m^{n+1} = \sum_k b_k \, f_{m+k}^n \tag{1}$$

then the scheme is monotonic if and only if $b_k \geq 0$ for all k.

The concept of monotonic behavior can be extended to a system of equations in the form $f_t + Af_x = 0$ where A is a constant matrix having real eigen values $\lambda_1,\ldots,\lambda_2$, and reducing to a diagonal form by means of matrix X. Substitution of the desired functions f_j by the characteristic combinations ψ_j, i.e., $\psi = X f$ decomposes the system into z independent equations for ψ_j:

$$\psi_{j,t} + \lambda_j \, \psi_{j,x} = 0$$

Let the difference scheme have the form (1) where b_k are matrices (each of them is reduced to diagonal form by means of matrix X). Then the substitution $\psi_m^n = X f_m^n$ decomposes the difference scheme into r independent equations

$$\psi_{j,m}^{n+1} = \sum_k \beta_{j,k} \, \psi_{j,m+k} \quad , \quad j = 1, 2, \ldots, r \tag{2}$$

It is natural to call the difference scheme monotonic if for all ψ_j the difference schemes are monotonic; in other words, if $\beta_{j,k} \geq 0$ for $j = 1, 2, \ldots, z$ and for all k.

Strictly speaking, all the preceding discussion is applied to equations having constant coefficients. When we construct difference schemes for nonlinear one-dimensional problems in gas dynamics the difference scheme must be monotonic when it is applied to the linear system. However, as is shown by numerical experiment, the monotonic character of the profile of a diffusion shock wave may be broken in spite of the fact that the scheme satisfies the monotonic condition. This example is shown in Fig. 1, where distributions of velocity u and pressure p are plotted for a strong shock wave in Lagrangian coordinates. The profiles were calculated by means of explicit three-point schemes (see formula 3 below), which are monotonic for linear equations. Thus the concept of a monotonic scheme, as was determined above, loses its meaning when it is applied to non-linear equations, and especially in the two-dimensional case. In this work another method for estimation of the quality of computation is considered. The idea of the method is based on an investigation of the profile of the shock which is really developed by the difference scheme itself.

I. <u>The Statement of the Problem</u>

Let us consider the non-linear hyperbolic system

$$(\partial f/\partial t) + [\partial F(f)/\partial x] = 0 \tag{3}$$

with the matrix $A = dF/df$ which has real and different eigen values $\lambda_j(f)$, $j = 1, 2, \ldots, r$ and

$$\lambda_1(f) > \lambda_2(f) > \ldots > \lambda_{r-1}(f) > \lambda_2(f) \tag{4}$$

We shall suppose that the matrix A can be reduced to diagonal form by means of matrix X.

Let us consider for the system (3) the difference scheme for continuous argument x:

$$\begin{aligned} f^{n+1}(x) &= f^n(x) - \kappa\{F[f^n(x+h)] - F[f^n(x-h)\}/2 + \\ &+ \frac{1}{2}\,\Omega[f^n(x+\frac{h}{2}),\kappa]\{f^n(x+h) - f^n(x)\} - \\ &- \frac{1}{2}\,\Omega[f^n(x-\frac{h}{2}),\kappa]\{f^n(x) - f^n(x-h)\} \end{aligned} \tag{5}$$

where $f^n(x) = f(x,n\tau)$, $\kappa = \tau/h$ and Ω is a matrix with real eigenvalues ω_1, $\omega_2, \ldots, \omega_r$ which could be reduced to diagonal form by means of the same matrix X. The stability condition for the scheme (5) is

$$\sigma_j^{\,2} \le \omega_j \le 1 \tag{6}$$

for all $j = 1, \ldots, r$, where $\sigma_j = \kappa\lambda_j$. If $|\sigma_j| \le \omega_j \le 1$ then scheme (5) is monotonic for linear equations. If $\omega_j = \sigma^2$ then the scheme (5) is of second order accuracy. Let $f(x,t)$ be the discontinuous solution of system (3) which describes the shock wave moving with velocity D and dividing the regions with constant values $f = f_-$ and $f = f_+$. Then

$$F(f_+) - F(f_-) = D(f_+ - f_-) \tag{7a}$$

and for certain k

$$\lambda_1(f_-) > \ldots > \lambda_k(f_-) > D > \lambda_k(f_+) > \ldots > \lambda_r(f_+) \tag{7b}$$

The function $f(x,t)$ depends only upon the combination $x - Dt = \xi$, namely,

$$f(x,t) \equiv g(\xi) = \begin{cases} f_- & \text{if } \xi < \xi_o \\ f_+ & \text{if } \xi > \xi_o \end{cases}$$

By analogy with this solution let us search for a quasi-stationary solution of the difference equations (5) which has the form $f^n(x) = g(\xi)$ where

$$\xi = (x - n\cdot D\tau)/h = x/h - n\delta \quad , \quad \delta = \kappa D$$

It is evident from (6) and (7b) that for a stable scheme $|\delta| < 1$. For $g(\xi)$ we obtain from (5) the equation:

$$g(\xi) = P\{g(\xi+\delta)\} \tag{8a}$$

with boundary conditions

$$\lim_{\xi\to\pm\infty} g(\xi) = f_\pm \tag{8b}$$

The operator P is determined by the formula

$$P\{g(\xi)\} = g(\xi) - \frac{\kappa}{2}\{F[g(\xi+1)] - F[g(\xi-1)]\} + \tag{9}$$
$$+ \frac{1}{2}\Omega[g(\xi + \frac{1}{2})]\{g(\xi+1) - g(\xi)\} - \frac{1}{2}\Omega[g(\xi - \frac{1}{2})]\{g(\xi) - g(\xi-1)\}$$

It was proposed by P. D. Lax [2] in the case $r = 1$ that there exists a unique solution of (8a) satisfying the conditions (8b) which could be obtained as a limit of the functional sequence $g^n(\xi)$ as n tends to infinity where

$$g^n(\xi) = f^n(\xi h + nD\tau) \tag{10}$$

From (5) and (10) we obtain the recurrence formula for $g^n(\xi)$:

$$g^{n+1}(\xi) = P\{g^n(\xi+\delta)\} \tag{11}$$

Now we can formulate the following questions:

1. Does a continuous vector-function $g(\xi)$ exist which satisfies equation (7a) and boundary conditions (7b)?
2. If such a function exists could it be obtained as a limit $g^n(\xi)$ when n tends to infinity?
3. Supposing the existence of $g(\xi)$ what are its properties, for example, is it monotonic or not?

2. The Necessary Condition for the Existence of the Limiting Solution

In this section we shall consider the first two questions and show that for the convergence of the sequence to the solution of equation (8) it is necessary that the initial function fulfill some special condition.

Let us consider the case of one equation and introduce the function

$$\alpha_{g^n}(\zeta) = (f_+ - f_-)^{-1} \sum_{m=-\infty}^{\infty} (\zeta + m + \frac{1}{2})\{g^n(\zeta + m + 1) - g^n(\zeta + m)\} \tag{12}$$

connected with $g^n(\xi)$.

If $g^n(\xi)$ tends sufficiently rapidly to the limiting eigenvalues (for example exponentially) the function $\alpha_{g^n}(\zeta)$ exists for any ζ and, as is easy to show, has a period equal to unity

$$\alpha_{g^n}(\zeta+1) = \alpha_{g^n}(\zeta) \tag{13}$$

To determine the connection between $\alpha_{g^{n+1}}(\zeta)$ and $\alpha_{g^n}(\zeta)$ let us write out (12) for $\alpha_{g^{n+1}}(\zeta)$ and substitute for $g^{n+1}(\xi)$ its expression in terms of $g^n(\xi)$ given by formula (11). After substituting the sums and changing some summation indices we obtain

$$\alpha_{g^{n+1}}(\zeta) = \alpha_{g^n}(\zeta+\delta) \tag{14}$$

For the limiting distribution $g(\xi)$ satisfying equation (8) we obtain

$$\alpha_g(\zeta) = \alpha_g(\zeta+\delta) \tag{15}$$

i.e., $\alpha_g(\zeta)$ has the period δ. As $\alpha_{g^{n+1}}(\zeta)$ differs from $\alpha_{g^n}(\zeta)$ only by the shift by δ then in order that the limiting function has the period δ it is necessary that $d_{g^o}(\zeta)$ have the period δ. On the other hand, all $\alpha_{g^n}(\zeta)$ have period equal to unity. So for the limit $g^n(\xi)$ to exist when δ is rational and equal to p/q it is necessary that $\alpha_{g^o}(\zeta)$ have the period $1/q$, and when δ is irrational $\alpha_{g^o}(\zeta)$ should be constant.

The existence of the limiting function when condition $\alpha_{g^o}(\zeta)$ = const is satisfied has been verified numerically by a large number of calculations for the various convex functions $F(f)$. The results obtained make it possible to justivy the following statement, which answers the first question.

If the function $F(f)$ satisfies the condition of convexity $F''(f) \geq 0$ while the initial data $g^o(\xi)$ are continuous and satisfy the conditions

$$g^o(\xi) = \begin{cases} f_- & \text{at } \xi \leq \xi_1 \\ f_+ & \text{at } \xi \geq \xi_2 \end{cases}, \qquad (16)$$

$$\alpha_{g^o}(\zeta) \equiv 0$$

then there exists $\lim g^n(\xi) = G(\xi)$ where $G(\xi)$ satisfies equation (8a), conditions (8b) and $\alpha_G(\zeta) \equiv 0$. The function $G(\xi)$ is determined uniquely by these conditions and does not depend upon ξ_1, ξ_2 and the values $g^o(\xi)$ in the interval $[\xi_1, \xi_2]$. Condition (16) can evidently be replaced by the condition that $g^o(\xi)$ approaches the limiting eigenvalues f_- and f_+ exponentially.

Let us note that from this statement the asymptotic behavior of $g^n(\xi)$ and $f^n(x)$ can be deduced for any initial data. If we put $x = x_m = mh$ in (5) then we obtain the difference scheme for a discrete mesh. For the values f_m^n at the nodes of this mesh the following asymptotic formula applies (for large n):

$$f_m^n \approx G[m - n\delta - \alpha_{f^o}(m)]$$

Numerical experiment shows that the values f_m^n, f_m^{n+1}, etc., appropriately shifted very well on a smooth curve.

In Fig. 2 the profile of the shock wave is shown at several values n for equation (1) with the function

$$F(f) \equiv Df + \ell(f-f^2) \quad , \quad f_- = 0 \quad \text{and} \quad f_+ = 1 \quad .$$

In this case

$$\lambda = F'(f) = D + \ell(1-2f) \quad , \quad \sigma = \delta + (1-2f)\eta$$

where

$$\delta = \kappa D \quad , \quad \eta = \kappa\ell \quad , \quad \sigma_- = \delta+\eta \quad , \quad \alpha_+ = \delta-\eta \quad .$$

The shock wave in Fig. 2 is calculated by a non-monotonic scheme of second order with $\omega = \sigma^2$, $\delta = 0.4$, $\eta = 0.1$.

The profile seems to be irregular and changing its shape on transition frome one time-step to another. But really the form of the profile is generated by the unique smooth curve $G(\xi)$ which moves along the mesh with the velocity D (see Fig. 3).

In the case of a system in which g consists of $r > 1$ components $g_j(\xi)$ the necessary condition for the existence of the limiting function $g(\xi)$ is the fulfillment of the analogous equalities:

$$\alpha_{g_j^o}(\zeta) = \alpha_{g_j^o} = \text{const} \quad , \quad j = 1, 2, \ldots, r \qquad (17)$$

separately for each component g_j where the $\alpha_{g_j^n}$ are determined by a formula which is analogous to (12), with the replacement $(f_{j,+} - f_{j,-})$ for $(f_+ - f_-)$. But here we find a circumstance which at first sight makes the existence of the limiting profile impossible. According to (14) at $\alpha_{g_j^n}$ = const, the values of the differences

$$\alpha_{g_j^n} - \alpha_{g_k^n} \equiv \alpha_{g_j^o} - \alpha_{g_k^o}$$

do not depend on n and are determined by initial data. As it is evident that these differences can change for various initial data the limiting profile cannot be unique. But detailed numerical investigation shows that in calculations at large n, of the variants having one and the same f_- and f_+ but different values $\alpha_{g_j^o} - \alpha_{g_k^o}$, in the neighborhood of a discontinuity, one and the same components $G_j(\xi)$ are developed. As $|\xi| \to \infty$ the differences $G_j(\xi) - f_{j,\pm}$ first of all approach zero, but then they begin to increase again. It is these disturbances "going to infinity" that compensate the change of differences

$$\alpha_{g_j^n} - \alpha_{g_k^n}$$

in the neighborhood of the discontinuity in comparison with those generated by the initial data. For this reason when ξ = const and $n \to \infty$ there exists convergence $g^n(\xi)$ to the vector-function $g(\xi)$ satisfying (8a) and (8b) and defined uniquely to within a shift with respect to ξ.

It was found by means of numerical experiments that the disturbances recede from a shock wave in separate groups with velocities equal to the characteristic values and in fact concentrate in the corresponding characteristic combinations.

3. Study of the Limiting Solution

Now in the assumption that there is a solution of equation (8a), let us clarify the asymptotic behavior of $G(\xi)$ as $|\xi| \to \infty$. We assume $G(\xi) = \hat{f}+\phi(\xi)$ where $\hat{f}$ denotes either f_- or f_+ depending upon the sign of ξ and correspondingly mark the function ϕ by indices. On the basis of (8b) $\phi(\xi) \to 0$ as $|\xi| \to \infty$. Substituting $G(\xi)$ into (8b) we obtain, after Taylor-series expansion and rejection of terms of order ϕ^2:

$$\begin{aligned}\phi(\xi) &= \phi(\xi+\delta) - \frac{\kappa A}{2}\{\phi(\xi+\delta+1) - \phi(\xi+\delta-1)\} + \\ &+ \frac{1}{2}\Omega\{\phi(\xi+\delta+1) - 2\phi(\xi+\delta) + \phi(\xi+\delta-1)\}\end{aligned} \tag{18}$$

The matrices A and Ω are calculated for $f = f_-$ and $f = f_+$, respectively. For the vector $\phi(\xi)$ we obtain a linear difference equation with constant coefficients. It may be simplified by taking advantage of the fact that A and Ω can be reduced to diagonal form simultaneously. Having multiplied (18) by X we obtain a reducible system of difference equations for the components of the vector $\psi = X\phi$:

$$\begin{aligned}\psi_j(\xi) &= \psi_j(\xi+\delta) - \frac{\sigma_j}{2}\{\psi_j(\xi+\delta+1) - \psi_j(\xi+\delta-1)\} + \\ &+ \frac{\omega_j}{2}\{\psi_j(\xi+\delta+1) - 2\psi_j(\xi+\delta) + \psi_j(\xi+\delta-1)\} \quad , \quad j = 1, 2, \ldots, r\end{aligned} \tag{19}$$

The components of the vector ψ simply represent the characteristic combinations of the system $\phi_t + A\phi_x = 0$ with the fixed matrix $A = A_-$ or $A = A_+$. We shall search for a solution of each of the equations (19) in the form $\psi_j = \mathrm{Re}\{C \exp(\mu\xi)\}$ where C is a real coefficient and μ is a real or complex number. Substituting this expression into (19) we obtain the equation for μ:

$$\frac{\omega_j - \sigma_j}{2}\ell^{2\mu} + (1 - \omega_j)\ell^{\mu} + \frac{\omega_j + \sigma_j}{2} = \ell^{(1-\delta)\mu} \tag{20}$$

As $\lim \psi_j(\xi) = 0$ then it is possible to assume that the asymptotic behavior of $\psi_j(\xi)$ will be determined as $\xi \to +\infty$ by the root of equation (20) $\mu_{j,+}$ with a maximum real part among the roots situated in the left half-plane, and as $\xi \to -\infty$ by the root $\mu_{j,-}$ with a minimum real part among the roots situated in the right half-plane. Setting

$$\mu_{j,\pm} = \pm (\ln r_{j,\pm} + i\, \theta_{j,\pm})$$

we find that the principal term $\psi_{j,\pm}$ as $|\xi| \to +\infty$ has the form

$$\mathrm{Re}\{C_{j,\pm}\, r_{j,\pm}^{|\xi|} \exp(i\, \theta_{j,\pm}\, |\xi|)\}$$

where $0 < r_{j,\pm} < 1$. Let $r_\pm = \max r_{j,\pm}$, then r_+ and r_- determine the rapidity with which $G(\xi)$ approaches its limiting values and the width of the shock region. The values θ_- and θ_+ determine the character of the approach. If $\theta_- = 0$ $(\theta_+ = 0)$ then $G(\xi)$ tends to f_- (f_+) monotonically and we shall say that in this case $G(\xi)$ is monotonic at $-\infty$ (at $+\infty$). If $\theta_\pm \neq 0$ the approach $G(\xi)$ to $f_\pm$ is oscillatory. The value $\Pi_\pm = \pi/\theta_\pm$ is one-half the period of the oscillations measured in units of h and $q_\pm = r_\pm^{\Pi_\pm}$
decreasing for one-half period. Numerical experiments show that in any case when θ_- or θ_+ is equal to zero the profile $G(\xi)$ is monotonic in a corresponding direction coming from the shock.

Let us note that although the statements given above are based on the considerations of the linear equation (19), nonlinearity of the initial equations (3) and (5) is taken into account because $\delta = \kappa D \neq \kappa\lambda_j = \sigma_j$. In other words, the velocity of displacement of the investigated steady state profile does not coincide with any velocity of the small disturbances propagating along characteristics.

4. Examples for Application of the Method

Let us consider as the first example the case of one equation with the function $F(f)$ satisfying the condition $F''(f) > 0$ and a shock wave moving to the right with the speed $D > 0$ so that $\lambda_- = F'(f_-) > D > F'(f_+) = \lambda_+$ and $\sigma_- > \delta > 0$, $\delta = \sigma_+$. Investigation of equation (20) shows that θ_+ is equal to zero, if $\omega_+ + \sigma_+ > 0$ and $\theta_+ \neq 0$ if $\sigma_+^2 < \omega_+ < -\sigma_+$ (in this case it is obvious $\sigma_+ < 0$). Hence it follows that the condition $\omega_+ > |\sigma_+|$ assures monotonic behavior of the limiting profile in front of the wave. If $\omega_+ < -\sigma_+$ then the linear condition of monotonic behavior is not fulfilled and it is possible that oscillations in the non-linear profile appear also. But the calculations show that the value r_+ in almost all cases is very small, 10^{-2} and even smaller. That is why non-monotonic behavior of the profile ahead of the wave cannot be found in practice except in certain special cases.

When $\xi \to -\infty$, i.e., behind the front of the wave the situation is different. As is seen from the analysis of (2) in the interval $[\sigma_-^2, \sigma_-]$, there is always a value $\omega_* = \omega_*(\sigma_-, \delta)$ such that $\theta_- = 0$ at $\omega_- \geq \omega_*$ and $\theta_- \neq 0$ at $\omega_- < \omega_*$. Knowing the function $\omega_*(\sigma, \delta)$ one can determine in every particular case (for given wave and grid) if the stationary solution is "monotonic at infinity" or not. For this case it is sufficient to find ω_-, σ_- and determine the sign of the difference $\omega_- - \omega_*(\sigma_-, \delta)$. The behavior will be monotonic if the difference is positive. The numerical experiments with one equation show that the scheme which is "monotonic at infinity" is monotonic in practice as the stationary profile is stabilized very quickly and monotonic behavior is broken only at the moment of the wave interaction.

The boundary of monotonic behavior of the limiting profile $\omega_- = \omega_*$ lies below the boundary of the strict monotonic scheme found according to the linear theory $(\omega_- = |\sigma_-|)$. It means that the limiting profile can be monotonic at infinity for a scheme which is not strictly monotonic from the linear point of view.

The examples of the limiting profile of the smoothed shock for one equation were given above for $\delta = 0.4$ (Figs. 2 and 3). The corresponding values

σ, ω, r, θ, Π, q, are given in Table I.

Table I.

	σ	ω	r	$\lvert\theta\rvert$	Π	q
-	0.5	0.25	0.76265	1.23092	2.55227	0.50079
+	0.3	0.09	0.17385	0	-	-

For practical purposes, the character of the approach of $G(\xi)$ to its limiting values obtained by means of computation as is shown in Fig. 3 agreed well with the theoretical one. For example, according to the results of calculations, we obtain, as $\xi \to -\infty$: $\Pi_- = 0.2550$, $q_- = 0.499$, which agrees well with the data in Table I.

As the second example we consider the shock wave for the gas dynamics equations in Lagrangian coordinates. In this case there are three eigenvalues $\lambda_1 = C$, $\lambda_2 = 0$, $\lambda_3 = -C$, which correspond to three characteristic velocities. The value $C = \sqrt{kp\rho}$ is the mass velocity of a sound, k is the isentropic exponent, p is a pressure, ρ is a density, and u is the gas velocity

If $D > 0$ is the velocity of a shock, then we have the inequalities

$$\begin{aligned} \sigma_{3,-} < \sigma_{2,-} &= 0 < \delta < \sigma_{1,-} \\ \sigma_{3,+} < \sigma_{2,+} &= 0 < \sigma_{1,+} < \delta \end{aligned} \tag{21}$$

From equation (20) we find just as for one equation, that $\theta_{j,+} = 0$ for all j if $\omega_{j,+} > -\sigma_{j,+}$ and all the components of $\psi(\xi)$ in this case approach their limiting values monotonically. Formally, violation of monotonic behavior may take place only in component $\psi_3(\xi)$ at $\omega_{3,+} + \sigma_{3,+} < 0$. But as has already been said, r_+ in this case is very small and that is why oscillations cannot be observed. As $\xi \to -\infty$ the behavior of the component $\psi_{1,-}(\xi)$ corresponding to the eigenvalue $\sigma_{1,-}$ and the characteristic arriving at the wave from the left is analogous to the behavior of the solution of one equation. Namely, if we denote $\omega_{*,1,-} = \omega_*(\sigma_{1,-},\delta)$ then $\theta_{1,-} = 0$ at $\omega_{1,-} \geq \omega_{*,1,-}$ and $\theta_{1,-} = 0$ at $\omega_{1,-} < \omega_{*,1,-}$. The greatest distinction of the system from one equation is to be seen in the existence of the characteristics having $\sigma_{2,-} = 0$ and $\sigma_{3,-} < 0$ departing from the shock wave into the region behind the front. The study of equation (20) shows that if $\sigma_- < \delta$ then for $\omega_- - \sigma_- > 0$ equation (20) has no positive roots for any ω and therefore $\theta_{2,-}$ and $\theta_{3,-}$ always differ from zero and the corresponding components $\psi_{2,-}$ and $\psi_{3,-}$ are always non-monotonic. Thus behind the front of the wave it is found that entropy and the Riemann invariant corresponding to the characteristic departing from the wave are non-monotonic. How important this non-monotonic behavior is depends upon the compared value $r_{j,-}$. In some cases $r_{3,-}$ and $r_{2,-}$ can be significantly greater than $r_{1,-}$, and in such case non-monotonic behavior of a "non-basic" characteristic combination can be the cause of the noticeable non-monotonic behavior of the component of the function $G(\xi)$. Let us note that equation (20) for irrational δ has an infinite number of roots located to the left as well as to the right of one-half plane of the complex variable μ. In addition, if $\delta > 0$ and $\sigma > \delta$ then among roots having $\mathrm{Re}\,\mu < 0$ there are the roots with a real part, the modulus of which is arbitrarily small. Among the roots with $\mathrm{Re}\,\mu > 0$ the real part is limited from below by a positive number. As $\delta > 0$ and $\sigma < \delta$ the situation is reversed, namely, there is an upper bound for the roots having $\mathrm{Re}\,\mu < 0$ and the roots with $\mathrm{Re}\,\mu > 0$ have real parts which are arbitrarily small. Hence apparently it follows that for $\sigma_{3,-} < \sigma_{2,-} = 0 < \delta, r_{3,-}$ and $r_{2,-}$ must be arbitrary close to unity and the asymptotic form of $G(\xi)$ as $\xi \to -\infty$ cannot be found by means of consideration of equation (20). But calculation shows that the asymptotic form of $G(\xi)$ is determined first of all by the roots with small $|\theta|$. The roots having the greater $|\theta|$ and exciting the high-frequency oscillations do not show their effect in the behavior of $G(\xi)$ even if the corresponding z are very close to unity.

Let us give some examples of calculations.

The distributions of pressure p and velocity u illustrated in Fig. 1 were obtained as the result of calculation of the shock wave moving with the velocity $D = 1$ in an immovable gas with pressure $p_+ = 0$ and density $\rho_+ = 1$. The calculation had been carried out by scheme (5) with the matrix $\Omega = \omega_1 \kappa C E$ where E is the unit matrix. The parameters of the scheme had the values $\kappa = 0.340168$, $\omega_1 = 1.05$, and therefore $\delta = 0.340168$, $\sigma_{1,-} = -\sigma_{3,-} = 0.9$, $\omega_{j,-} = \omega_- = \omega_1 \kappa C_+ = 0.945$.

Table II represents the values of r, $|\theta|$, Π, q corresponding to each characteristic behind the front.

Table II.

j	1	2	3
$r_{j,-}$	0.07705	0.47275	0.92548
$\lvert\theta\rvert_{j,-}$	0	4.65889	4.64833
$\Pi_{j,-}$	-	0.67432	0.67585
$q_{j,-}$	-	0.60339	0.94901

In Fig. 4 the components $\psi_j(\xi)$ are plotted for the profile given in Fig. 1. Non-monotonic behavior of the components $\psi_2(\xi)$ and $\psi_3(\xi)$ is clearly seen. The values $\Pi_{j,-}$ and $q_{j,-}$ were determined for $j = 2$ and $j = 3$ and the calculations are given in Table III.

Table III.

j	1	2	3
$\Pi_{j,-}$	-	0.677	0.676
$q_{j,-}$	-	0.594	0.946

Comparing Tables II and III we can find that the actual values Π and q for $j = 2,3$ coincide with their theoretical values up to the third decimal point. We could not estimate r_1 because of the smallness of this value. Let us note that the only possibility of decreasing the oscillations arising in ψ_2 and ψ_3 is by decreasing the time-step τ since $r_{2,-}$ and $r_{3,-}$ diminish when σ and δ are decreased.

References

1. S. K. Godunov. "Difference methods for the numerical calculation of discontinuous solutions of the equations of hydrodynamics," Mat. Shovnik 1959.

2. P. D. Lax. "Weak solutions of non-linear hyperbolic equations and their numerical computation," Comm. Pure Appl. Math. 1954, v. 7, N 1, 159-193.

3. N. N. Yanenko. Method of Fractional Steps for the Solution of Problems of Mathematical Physics in Several Variables, Nauka Novosibirsk 1967, to appear in English, Springer-Verlag 1971.

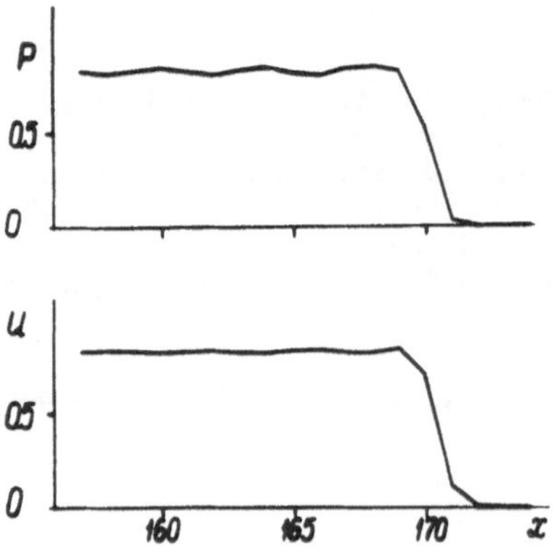

Fig. 1.

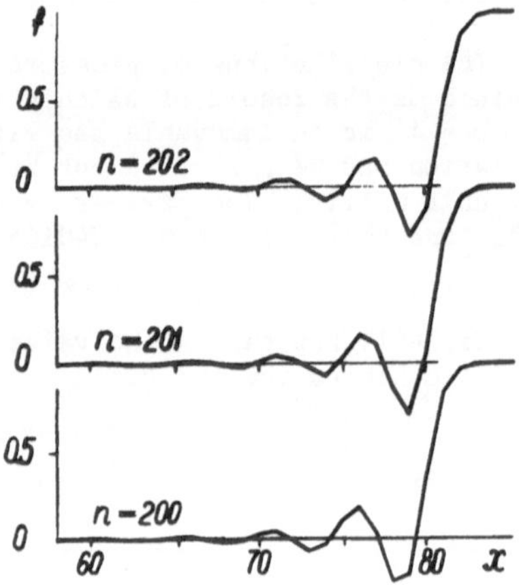

Fig. 2.

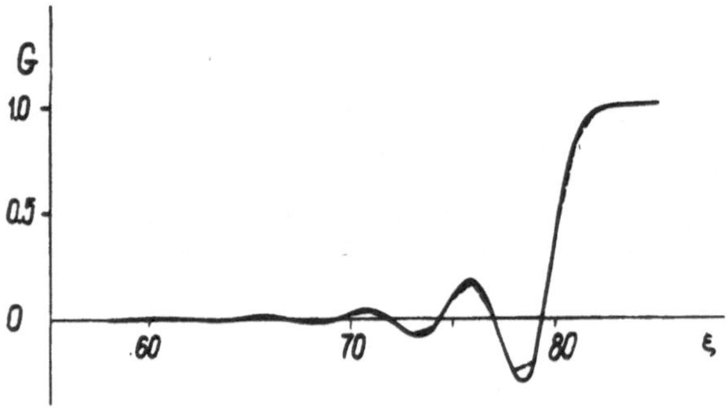

Fig. 3.

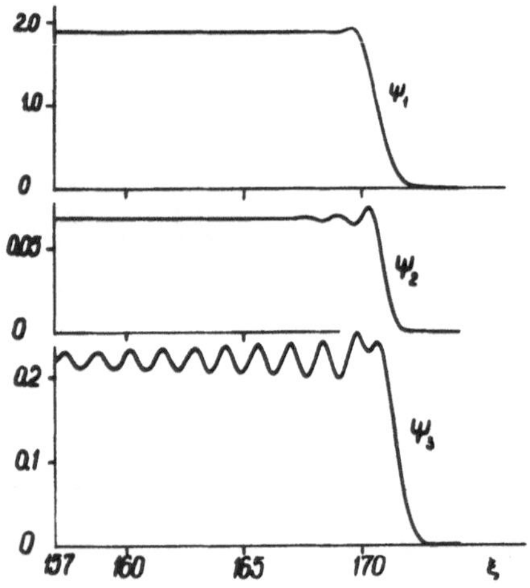

Fig. 4.

SHOCK WAVE STRUCTURE
BY SEVERAL NEW MODELED BOLTZMANN EQUATIONS

B. M. Segal and J. H. Ferziger, Stanford University, Stanford, California

INTRODUCTION

The purpose of this paper is to compare the structure of strong shock waves as predicted by several nonlinear kinetic models, with exact Monte Carlo solutions of the full Boltzmann equation (Bird, 1970). The four models considered are the well-known BGK (Bhatnagar, et al., 1954) and Ellipsoidal (Holway, 1966) models, and two new models developed by us: the Polynomial and Trimodal Gain Function (TGF) models. (Shakhov, 1968, has independently derived the Polynomial model.) Exact numerical solutions are obtained for the physical moments contained in each model. Previous workers (Chahine, et al., 1965; Anderson, et al., 1965) have obtained such BGK solutions, and discrete ordinates calculations for both the BGK and Ellipsoidal models have been reported (Giddens, et al., 1969).

I. FORMULATION OF THE MODEL EQUATIONS FOR A SHOCK WAVE

The full Boltzmann equation for a plane shock may be written:

$$c_x \frac{\partial f(\underline{c},x)}{\partial x} = G(f,f) - f \cdot L(f) \; ; \qquad \begin{array}{l} f(x=-\infty) = f_1^{(o)} \\ f(x=+\infty) = f_2^{(o)} \end{array} \tag{1}$$

where G and L are the gain and loss collision operators. Two of the six parameters of the endpoint Maxwellia $f_1^{(o)}$ and $f_2^{(o)}$ are used for normalization and the rest depend only on the Mach number and are related by the Rankine-Hugoniot conditions. The molecular velocity $\underline{c}$ has components c_x and c_t parallel and transverse to the flow axis x.

All the kinetic models considered here replace the right side of (1) by an operator of the form: $K(m_i) \cdot [\Psi(\underline{c},m_i) - f]$, where m_i are the following set of moments of f: density n, velocity u, kinetic temperature T, stress τ_{xx}, and heat flux q_x. The function $K \approx L(f)$ is assumed independent of $\underline{c}$ and represents a collision frequency, while $\Psi \approx G(f,f)/L(f)$ is termed the emission function. Each model specifies a particular K and Ψ as detailed below.

A closed set of nonlinear integral equations for the moments m_i is obtained as follows. For convenience K is absorbed into a new coordinate t defined by: $dt = K(x)dx$; among other advantages, shock profiles in the t frame tend to an invariant form at high Mach number. Now consider the five velocity moments of f, $\mu_i \triangleq \int f\psi_i d\underline{c}$ with $\{\psi_i\} = 1, c_x, c^2, c_x^2, c_x c^2$; these are related to the m_i introduced above as: $\mu_1 = n$, $\mu_2 = nu$, $\mu_3 = n(3RT + u^2)$, $\mu_4 = n(RT + u^2 + \tau_{xx})$, $\mu_5 = nu(5RT + u^2 + 2\tau_{xx}) + 2nq_x$. Because μ_2, μ_4 and μ_5 are flow invariants, and τ_{xx} and q_x both vanish at the endpoints, only μ_1 and μ_3 suffice to determine all the m_i. The equations for μ_1 and μ_3 are obtained from (1) after formal integration over t, yielding:

$$\begin{aligned} \mu_1(t) &= 2\pi \int_0^\infty c_t dc_t \int_{-\infty}^\infty dc_x \int_{-\infty}^\infty dt' \frac{\Psi(\underline{c},t')}{|c_x|} \exp - \left|\frac{t-t'}{c_x}\right| \\ \mu_3(t) - \mu_4 &= 2\pi \int_0^\infty c_t^3 dc_t \int_{-\infty}^\infty dc_x \int_{-\infty}^\infty dt' \frac{\Psi(\underline{c},t')}{|c_x|} \exp - \left|\frac{t-t'}{c_x}\right| \end{aligned} \tag{2}$$

For $\Psi(\underline{c})$ functions limited in form to Maxwellia multiplied by velocity polynomials, both the c_x and c_t integrals in (2) can be performed; the former lead to the functions defined by: $H_n(p,q) \triangleq \frac{1}{\sqrt{2\pi}} \int_0^\infty y^{n-2} \exp - [\frac{1}{2}(y-p)^2 + q/y]dy$. The pair of equations (2) can then be cast into the form used for solution:

$$\begin{aligned} n(t) &= \int_{-\infty}^\infty n(t')\mathcal{K}_n(n,T;\, t-t')dt' \\ n(t)T(t) &= \int_{-\infty}^\infty n(t')\mathcal{K}_T(n,T;\, t-t')dt' + \mathcal{S}(n,T;t) \end{aligned} \tag{3}$$

where the kernels $\mathcal{K}$ are sums of low order H_n functions, possessing a logarithmic singularity at $t' = t$.

II. MODEL CHOICES FOR K AND Ψ

The choices made for K and Ψ are presented in Table I.

TABLE I

Model	Ψ Function	K Func.	λ_{02}	λ_{11}	Pr
BGK	$\Psi = f^{(o)} \triangleq n\left(\frac{\beta}{\pi}\right)^{3/2} \exp - \beta(\underline{c} - u)^2 ;\ \beta \triangleq 1/2RT$	(p/μ)	-1	-1	1
Ellipsoidal	$\Psi = n\frac{\beta_x^{1/2}\beta_t}{\pi^{3/2}} \exp - [\beta_x(c_x - u)^2 + \beta_t c_t^2]$ with: $1/\beta_x = 2RT - \tau_{xx};\ 1/\beta_t = 2RT + \frac{1}{2}\tau_{xx}$.	$(2p/3\mu)$	-3/2	-1	2/3
Polynomial	$\Psi = f^{(o)}\{1 + 2(1+\lambda_{02})\tau_{xx}\beta^2(C_x^2 - C^2/3)$ $+ \frac{8}{5}(1 + \frac{2}{3}\lambda_{02})q_x\beta^2(\beta C^2 - 5/2)C_x\}$ with: $\underline{C} = (\underline{c} - u);\ \beta \triangleq 1/2RT$	$\frac{(p/\mu)}{\lvert\lambda_{02}\rvert}$	free	$\frac{2}{3}\lambda_{02}$	2/3
TGF	$\Psi = k(\nu)f_{MS} + (1 - k(\nu))f_3^{(o)}$ with: $\nu = (n - n_1)/(n_2 - n_1)$; $k(\nu)$ in Fig. 1 $f_{MS} = (1-\nu)f_1^{(o)} + \nu f_2^{(o)}$ $f_3^{(o)} = n(2\pi RT_3)^{-3/2} \exp - [(\underline{c} - u)^2/2RT_3]$ $T_3 = \frac{(T - T_{MS})}{(1-k)}$; $T_{MS} \triangleq \frac{1}{3Rn}\int C^2 f_{MS}\, d\underline{c}$	(p/μ)	-1	-1	1

The BGK model makes the simplest choice for Ψ, namely the local Maxwellian function $f^{(o)}$. This represents an isotropic emission of molecules in local equilibrium after collision. The Ellipsoidal and Polynomial Ψ functions allow small deviations from isotropy and equilibrium. Only the TGF Ψ specifically attempts to model the emission function expected in a shock wave. The assumption is made that the distribution function within the shock is of Mott-Smith form, f_{MS}, namely a linear combination of $f_1^{(o)}$ and $f_2^{(o)}$. The gain and loss integrals G_{MS} and L_{MS} can then be calculated in closed form for an intermolecular force law (quasi-Maxwellian) consistent with the modelling procedure. The method follows that used for hard spheres (Deshpande and Narasimha, 1969), but setting $b = (\sigma/g)\sin\psi$. The resulting function $(G/L)_{MS}$ is then approximated by the trimodal form shown for Ψ in Table I; this is done to retain the use of H_n kernel functions. This approximation obeys conservation requirements and closely reproduces the two most significant features of $(G/L)_{MS}$, namely the presence throughout the shock of a low-speed hot component with temperature T_2, and a cold beam-like component. The dominant effect of the hot component occurs near the cold side of the shock, and vice versa; this is reflected in the choice of the function $k(\nu)$ shown in Figure 1.

For each model, once Ψ is chosen the function K can be found in terms of scalar pressure p and viscosity $\mu(T)$ via the Chapman-Enskog treatment, yielding $K = (p/\mu)/|\lambda_{02}|$ and a Prandtl number $Pr = \lambda_{11}/\lambda_{02}$. Here λ_{02} and λ_{11} are the eigenvalues of the linearized model operators belonging to their eigenfunctions $(c_x^2 - c^2/3)$ and $c_x(5/2 - c^2)$, respectively. A major role of the eigenvalue λ_{02} is its appearance as a scale factor in the inverse transformation from the t frame back into the physical shock frame $x/\Lambda_1 = (1/\Lambda_1)\int_{t_o}^{t} dt/K(t) = (|\lambda_{02}|/\Lambda_1) \cdot \int_{t_o}^{t} dt\,\mu(T)/nkT$. (Here Λ_1 is the cold side Maxwell o mean free path for hard

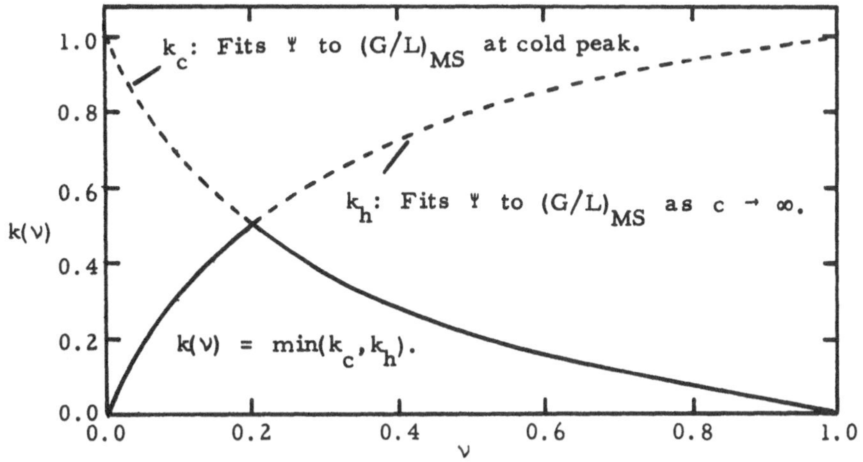

Figure 1. Choice of $k(\nu)$ for TGF model.

spherical molecules and is a normalizing constant.) As λ_{02} is determined by Ψ, its resulting value should influence the choice of Ψ for the shock problem. While a correctly shaped Ψ will lead to good profiles in terms of t, the profiles in terms of x/Λ_1 will be wrongly scaled unless $|\lambda_{02}|$ for that Ψ is also suitable. Conversely, as is found for example in the Ellipsoidal model, an incorrect Ψ yielding contracted t profiles can possess a high value of $|\lambda_{02}|$ which compensates by stretching out the x/Λ_1 profiles. (This is apparently a fortuitous result in the Ellipsoidal case, as its Ψ was designed using arguments unrelated to the shock problem.) This important effect has not been previously discussed to our knowledge: the Polynomial model allows it to be studied conveniently, by making the dependence of λ_{02} on Ψ adjustable and explicit. In the TGF model, λ_{02} is made independent of changes in Ψ, allowing the physical effects of modelling the Ψ function to be clearly demonstrated.

III. NUMERICAL METHODS

The methods are basically due to Anderson (1966), including some unpublished improvements communicated directly to us. The equations (3) are enforced at N grid points in the t variable, chosen so that differences of density between points are reasonably constant, with extra points added to deal with any special profile features such as a possible temperature overshoot. A compound M point Gaussian quadrature scheme replaces the integral operators, in conjunction with cubic spline interpolation between grid points. Interpolation is done on the small differences between the computed profiles and an analytical "backbone" profile, rather than directly; this increases accuracy. The quadrature scheme is tailored to the kernel shape using 6 zones: 3 for $t' < t$ and 3 for $t' > t$, centered about $t' = t$. The two outer Laguerre zones handle the exponential decay, the two inner zones are specially adapted to the logarithmic singularity, and two intermediate Legendre zones complete the range. In practice $N \sim 20$-30 and $M \sim 100$ are found to be optimal.

Computing the kernel functions $H_n(p,q)$ is the costliest operation even though power and asymptotic series methods are used for $n = 1$ to 3, and recurrence relations for $n > 3$. Therefore the H_{1-3} together with their arguments p and q are stored in core for all M and N for a single iteration. On all subsequent iterations, if the changes in p and q are found to be sufficiently small, a fast Taylor expansion is made for the new H_{1-3} about the old set at each (M,N). As convergence proceeds this Taylor mode quickly predominates and the iteration time falls by a factor of 3-5.

Solution is by iteration, but the basic Jacobi iteration is found to be either slowly convergent or weakly divergent. A complication exists: due to the translational invariance of plane shock profiles, iteration operators often cause small origin shifts which disguise the profile shape changes. This can be handled for a Jacobi type iteration by reshifting between sweeps and before combining iterates in the case of acceleration. A higher degree "extrapolation algorithm" due to Anderson (1965) incorporating this technique and forming linear combinations of Jacobi iterates which minimize a certain linearized residual, is found to produce rapid convergence. As an alternative, we have also developed a conceptually simpler "Relaxed Gauss-Seidel" iteration of similar power. A sweep is made from the cold to the hot side of the shock, re-evaluating interpolation coefficients as each of the N points is completed. We avoid cumulative errors from the shift phenomenon by first estimating its size by sampling for it near the shock center, and then accounting for it point by point during the sweep. A relaxation factor θ is incorporated so that each point is replaced by $(1-\theta)$.(its old value) + θ.(its unrelaxed new value). The optimal value of θ is found empirically to fall from about 1.2 far from convergence to about 0.8 in the final stages.

Typically 8-12 iterations by either method are needed to converge the profiles to good graphical accuracy (about 0.3%), starting from the analytical "backbone" profile, which is roughly BGK. Iteration times fall from about 45 sec. to about 8 sec., giving total times of about 3 to 4 minutes on the IBM 360/67.

IV. RESULTS

Density and temperature profiles are presented in Figure 2. Lacking a Monte Carlo temperature profile among Bird's published results, one due to Perlmutter (1969) was used; their density profiles are in close agreement. The Monte Carlo comparisons were made with Maxwell molecules as these correspond with the assumption of constant collision frequency made for all the models. The variation of viscosity with temperature was assumed to be linear for the models to complete this consistent approach. A Mach number of 10 was selected as representative of the range from about 5 to infinity. Within this range the models yield almost invariant profiles in terms of t, which can be shown to correspond to maximum density-gradient shock thicknesses in terms of x/Λ_1 which are directly proportional to Mach number. This is consistent with Monte Carlo results.

All the models give convergent solutions in much less time (about one tenth) than the Monte Carlo method. The BGK model predicts too late a rise of both n and T on the cold side, and too small a shock thickness. The Ellipsoidal model improves the T rise and the shock thickness due to its larger scaling factor, but at the price of distorting the profiles. Also a small ($\sim 1\%$) T overshoot appears on the hot side, unsupported by the Monte Carlo results; presumably this is because the Ellipsoidal Ψ is inappropriate to the shock problem. The Polynomial model with $\lambda_{02} = -3/2$ strongly resembles the Ellipsoidal model and is not plotted; that with $\lambda_{02} = -3/5$, the correct linearized value, corrects the T overshoot but its smaller scale gives too small a shock thickness. Only the TGF model and the Polynomial model with $\lambda_{02} = -1$ successfully predict the early rise of both n and T on the cold side, the absence of T overshoot, and the correct shock thickness.

The latter two models are thus found to be the most appropriate for predicting Maxwell molecule shock profiles. We feel that the success of the TGF model is the more significant result due to its greater physical content, but the versatility of the Polynomial model as λ_{02} is adjusted is also an interesting result.

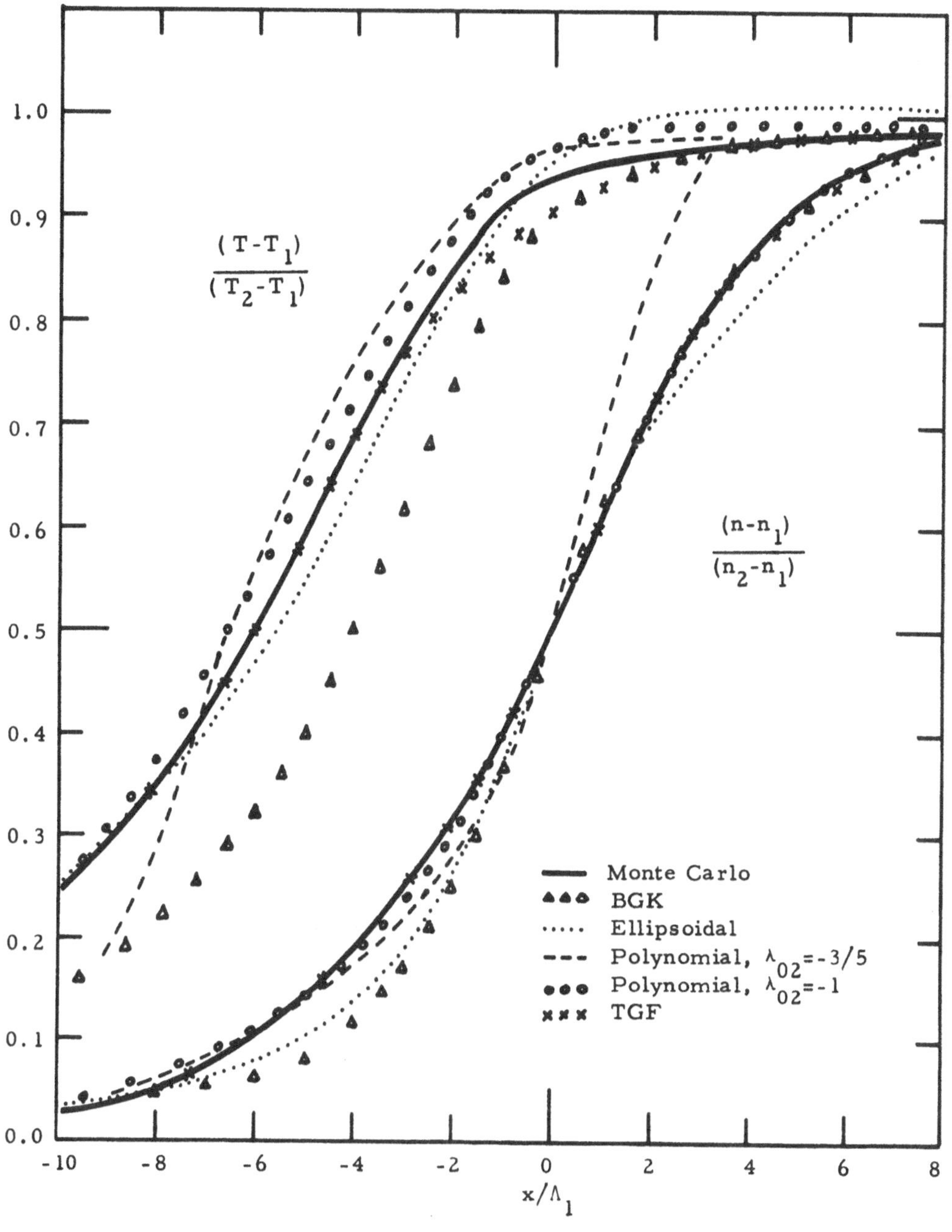

Figure 2: Reduced density and temperature profiles vs. x/Λ_1

ACKNOWLEDGMENTS

The authors wish to thank Prof. Donald G. Anderson of Harvard University for supplying details of his numerical methods, and Prof. G. A. Bird of the University of Sydney for clarifying certain Monte Carlo results.

This research has been supported by NSF Grant GK-4251 and by NASA Grant NGR-05-020-420.

REFERENCES

Anderson, D. G., and Macomber, H. K., in Rarefied Gas Dynamics, Suppl. 3, Academic Press, New York (1965).

Anderson, D. G., J. Fluid Mech., 25, 271 (1966).

Anderson, D. G., J. ACM, 12, 4, 547 (1965).

Bhatnagar, P. L., Gross, E. P., and Krook, M., Phys. Rev., 94, 511 (1954).

Bird, G. A., Phys. Fl., 13, 5, 1172 (1970).

Chahine, M. T., and Narasimha, R., in Rarefied Gas Dynamics, Suppl. 3, Academic Press, New York (1965).

Deshpande, S. M., and Narasimha, R., J. Fluid Mech., 36, 3, 545 (1969).

Giddens, D. P., Huang, A. B., and Young, Y. C., Proc. 20th Intl. Cong. Astronautics, Mar del Plata, Argentina (1969).

Holway, L. H., Phys. Fl., 9, 1658 (1966).

Perlmutter, M., in Rarefied Gas Dynamics, Suppl. 5, Academic Press, New York (1969).

Shakhov, E. M., Akad. Nauk. SSSR, Mekh. Zh. Gaza, 1, 156 (1968).

COMPUTATIONAL ASPECTS OF THE TURBULENCE PROBLEM*

Alexandre J. Chorin
Courant Institute of Mathematical Sciences, New York University

Introduction. Let $\underline{u}(\underline{x},t)$ be a solution of Euler's equations

$$\text{(1a,b)} \qquad \underline{u}_t + (\underline{u}\cdot\underline{\nabla})\underline{u} + \text{grad } p = 0 \ ; \qquad \text{div } \underline{u} = 0 \ ,$$

where an underline denotes a vector, p is the pressure and t is the time. Let the initial data $u(x,0)$ be smooth, for example in C_0^∞. $u(x,t)$ cannot be expected to remain smooth; there will appear surfaces and lines across which $u(\underline{x},t)$ can have no more than m bounded derivatives, m not necessarily an integer. One expects that the solutions of the Navier-Stokes equations for very small viscosity ν and far from boundaries will contain mollified forms of these singularities and that the presence of viscosity will affect expressions such as

$$\int u_i(\underline{x})u_j(\underline{x}+\underline{r})\, d\underline{x} \ , \qquad (u_i \text{ components of } \underline{u})$$

only for small separation $r = |\underline{r}|$, and affect the Fourier transform $\hat{u}(\underline{k},t)$ of $\underline{u}(x,t)$ only for large wave number $k = |\underline{k}|$. (For some results on the limiting process $\nu \to 0$, see Ebin and Marsden [4].) The study of the singularities of Euler's equations is thus equivalent to the study of the behavior of the Fourier transforms of the solutions of the Navier-Stokes equations as ν tends to zero and k tends to infinity (in that order), i.e. to the study of the inertial range behavior of $\hat{\underline{u}}$. (For a definition of the inertial range, see Batchelor [1].)

The Burgers equation

$$\text{(2)} \qquad u_t + uu_x = \nu u_{xx}$$

provides a striking illustration of these statements. As ν tends to zero, the solutions of (2) tend to weak solutions of

$$\text{(3)} \qquad u_t + uu_x = 0$$

(Hopf, [5]). The solutions of (3) contain jump discontinuities. The transform of a jump discontinuity is $O(k^{-1})$ for large k; the corresponding energy spectrum is $O(k^{-2})$. This is the well known form of the inertial range spectrum of Burgerlence (see e.g. Saffman [10]).

The aim of this paper is to describe how the singularities of equations (1) can be catalogued by numerical means. Conclusions will be drawn regarding the design of difference schemes for the study of turbulent flows.

A Numerical Method. One could attempt to determine the nature of singularities of Euler's equations by solving an initial value problem with smooth data and observing the subsequent flow, in the manner suggested by Taylor and Green [12] for the Navier-Stokes equations and pursued more recently by, among others, Jain [6], Orszag [9], and the author (unpublished).

*This work was partially supported by the U. S. Atomic Energy Commission, Contract AT(30-1)-1480.

The futility of such computations is illustrated by the fact that the best results obtained by the newest computers barely extend the hand computations of Taylor and Green, and are valid for times much too short and Reynolds numbers much too low to have a bearing on the problem of turbulence. The reasons for this dismal performance are suggested by the error estimates given by the author [2], and the object of this section is to explain how a meaningful study can be undertaken.

It is natural in this problem to use the variables $\hat{\underline{u}}$ and $\underline{k}$ rather than $\underline{u}$ and $\underline{x}$; as $\underline{u}$ loses smoothness, in a physical space computation one would have either to refine the mesh throughout the field, which is wasteful, or to refine it only in the region of the singularity, which is difficult to do. On the other hand, the behavior of $\hat{\underline{u}}$ for large k is determined by the least smooth part of the flow (see e.g. Lighthill [8]); in a k-space computation one merely keeps extending the range of the variables k. The advantages of Fourier variables have of course been observed in practice.

Since the solution of an initial value problem for $\hat{\underline{u}}$ is not practicable, and since there is no natural choice of initial values $\hat{\underline{u}}(k,0)$, one may consider solving a final value problem, i.e. guessing a function $\hat{u}(k,t)$ which could have arisen from <u>some</u> data $\hat{\underline{u}}(\underline{k},0)$ through a flow obeying the Euler equations. This is a more tractable problem. Furthermore, one is interested in the behavior of $\hat{u}$ for large k only; one therefore introduces variables $\underline{k}'$, which focus attention on the large-k region, through the scaling

$$\underline{k}' = \underline{k}/K \qquad K \text{ a large constant.}$$

Moderate non-zero values of k' will correspond to large k; since the total energy does not depend on K we find that $\hat{\underline{u}}(\underline{k}')$ becomes very large for small k'. Finally,

$$\frac{1}{K}\hat{\underline{u}}_t \to 0 \;;$$

in other words, the time scale characteristic of the inertial range is much shorter than the time scale of the energy-containing range (Batchelor [1], p. 104). We now take the Fourier transform of (1), scale k and drop the primes, and are led to the following problem: Find $\hat{\underline{u}}(\underline{k})$, with components $\hat{u}_\alpha(\underline{k})$, satisfying

$$(4a,b) \quad k_\delta\left(\delta_{\alpha\gamma} - \frac{k_\alpha k_\gamma}{k^2}\right)\int \hat{u}_\delta(\underline{k}-\underline{k}')\,\hat{u}_\gamma(\underline{k}')\,d\underline{k}' = 0 \;; \qquad k_\alpha \hat{u}_\alpha = 0 \;,$$

(the summation convention is in force), and such that

(i) $\hat{\underline{u}}(\underline{k})$ is the complex conjugate of $\hat{\underline{u}}(-\underline{k})$,

(ii) $\lim_{\varepsilon\to 0} |\hat{u}(\varepsilon)| = \infty$,

(iii) the integrals in (4a) have a meaning,

(iv) $\hat{\underline{u}}(k)$ is realizable, i.e. there exist initial functions $\hat{u}(k,0)$ which can give rise to $\hat{u}(k)$. For details, see Chorin [3]. (The analogous problem for the Burgers equation (2) is

$$ik\int \hat{u}(k')\,\hat{u}(k-k')\,dk' = 0 \;, \qquad \lim_{\varepsilon\to 0} |\hat{u}(\varepsilon)| = \infty,$$

and it can be readily verified that this equation has no realizable solution but

$$\hat{u} = C\,e^{iak}\,k^{-1} \;, \qquad C, a \text{ arbitrary real constants;}$$

i.e., the Fourier transform of a shock).

Vortex Formation and Turbulence Spectra. We now set out to solve the problem of the last section. Consider first the case of two space dimensions. To satisfy equation (4b) we must have

$$\hat{u}_1 = -ik_2\rho(\underline{k}) , \qquad \hat{u}_2 = ik_1\rho(\underline{k}) ;$$

ρ is a stream function. To construct ρ we note that a solution of (4) represents an equilibrium, in which the interactions enhancing $\hat{u}(k)$ and those depleting $\hat{u}(k)$ are in balance. (Once it is realized that the process of energy transfer to large k is identical to the process of formation of singularities, a resemblance to classical universal equilibrium theories becomes evident.) Evidence can be presented to the effect that ρ is of the form

$$\rho = C \exp (iak_1 + ibk_2) / H(\underline{k}) ,$$

where C, a, b are arbitrary constants and H is a homogeneous function of k which vanishes only when k = 0. (For the Burgers equation, an analogous argument leads to a form of $\hat{u}$ whose uniqueness can be proved.) With this choice of ρ, equation (4a) is satisfied for all $\underline{k}$ if it is satisfied on a closed curve surrounding the origin. Numerical experimentation shows that an appropriate choice of H is

$$H = k^{2\beta} .$$

The resulting velocity field $\hat{\underline{u}}$ is the Fourier transform of a vortex; we have reached the conclusion that the process of energy transfer from large to small eddies is identical to the process of vortex formation. There remains only the task of determining β. This is an easier undertaking than the solution of the original Green-Taylor problem. For each β, $\hat{u}$ is a definite function of k whose values need not be stored, and there is no opening for the error accumulation associated with the solution of an initial value problem. Some of the computations exhibited in Chorin [3] take into account the interaction of about 10^8 distinct Fourier components.

Given β, the energy spectrum is proportional to $k^{-\sigma}$, $\sigma = 4\beta - 3$. The fact that the energy is non-increasing yields $\sigma > 1$; conservation of vorticity by the Euler equations in two dimensions yields $\sigma > 3$, numerical computation yields the estimate $\sigma < 2$. Thus the solution obtained cannot be matched to an energy range flow in two dimensions unless vorticity sources are present. We may speculate that once the problem is reformulated to allow time dependence we will find the energy transfer still embodied in vortex-formation with spectra of the form

$$E = A(t)k^{-\sigma(t)} , \quad \sigma(t) = 3 + \varepsilon(t) , \quad \varepsilon(t) > 0 .$$

Such vortices and spectra have in fact been found by Zabuski and Deem [13].

One can show that the three-dimensional equations (4) admit only two-dimensional solutions. This two-dimensionality is by definition only local; the radius of curvature of the vortex lines is of the order of magnitude of the energy-containing scales, and the velocity field will be differentiable in a direction parallel to the vortex lines. In fact flow parallel to the vortex line ("vortex-stretching") is necessary for vorticity not to be conserved and for the appearance of spectra with $1 < \sigma < 3$.

The best estimate of σ I could obtain is $\sigma = 1.7 \pm 0.1$, in striking agreement with the Kolmogorov law. (The computation is performed by choosing various values of β and seeking to minimize the residual in equation (4a).) It should be stated however that precise determination of σ is difficult, for reasons which can be made explicit, and this estimate may not be free of a priori notions. Three-dimensional turbulence should thus be visualized as a jumble of vortex lines of random location, orientation and intensity, but with a velocity field in the normal plane having stream function proportional to $|\underline{r}|^{4/3}$ for small $|\underline{r}|$.

Applications. The arguments just presented have important implications for the theory of turbulence; in particular they indicate that there is no tendency towards energy equipartition among degrees of freedom even in the limit of vanishing viscosity. They also have implication for the design of difference schemes. As derivatives of the solution increase accuracy is lost (some manifestations of this effect are known as "aliasing"); the use of numerical "conservation laws" has been suggested to offset this phenomenon; it can be shown that some of the "conservation laws" found in the literature guarantee large errors rather than increased accuracy.

Practical applications being carried out include the following: Consider a fluid flow, with fairly large Reynolds number R; when R^{-1} is comparable to the truncation error, accuracy is lost; vortices may form for whose description the given mesh becomes increasingly more inadequate. However, a rough pointwise estimate of the truncation error can be obtained; as soon as the scheme is seen to be failing at some point, a vortex of appropriate structure and intensity is injected at that point. This process is readily carried out and should extend the range of validity of difference schemes in hydrodynamics by handling separately the difficult parts of the computations. Like the eddy coefficients of Smagorinski [11] and Leith [7], this process insures the correct behavior of the computed $\hat{u}$ for large k.

When the mesh width is much larger than $R^{-1/2}$ one has to fall back on eddy coefficients which merely remove vorticity from the grid. This can be justified on the assumption that the vortices are distributed with no preferred direction or location and thus that the velocity field to which they give rise is zero on the average. Under these conditions, the exact form of the eddy coefficient seems to be of little import.

Finally, it is useful to note that a random collection of vortices can provide an apt description of a flow field for use in numerical simulation of turbulent diffusion.

References

[1] Batchelor, G. K., The Theory of Homogeneous Turbulence, Cambridge University Press, (1960).

[2] Chorin, A. J., Math. Comp., 23, 341, (1969).

[3] Chorin, A. J., Inertial Range Flow and Turbulent Cascades. to appear, also AEC Report NYO-1480-135 (1969).

[4] Ebin, D. G. and Marsden, E., to appear, Arch. Rat. Mech. Anal.

[5] Hopf, E., Comm. Pure Appl. Math. 3, 201 (1950).

[6] Jain, P., Math. Research Center Report 751, Madison, Wisc. (1967).

[7] Leith, C. E., Proc. WMO/IUGG Symp. Num. Weather Prediction, Tokyo (1969).

[8] Lighthill, M. J., Fourier Analysis and Generalized Functions, Cambridge University Press (1959).

[9] Orszag, S., to appear.

[10] Saffman, P. G., in Topics in Nonlinear Physics, N. J. Zabuski (Ed.), Springer-Verlag, New York, (1968).

[11] Smagorinski, J., Monthly Weather Review, 91, 257 (1960).

[12] Taylor, G. I. and Green, A. E., Proc. Roy. Soc. A. 158, 499 (1937).

[13] Zabuski, N. J. and Deem, G., to appear.

TWO-DIMENSIONAL TURBULENCE AND ATMOSPHERIC PREDICTABILITY

C. E. Leith
National Center for Atmospheric Research*
Boulder, Colorado

TWO-DIMENSIONAL TURBULENCE

When a fluid is confined to flow in horizontal planes with vanishing horizontal divergence the flow is two-dimensional and satisfies a vorticity equation. One consequence is that in addition to the energy integral of three-dimensional inviscid flow there is also an enstrophy (half squared vorticity) integral. The barotropic models of the atmosphere, which have been moderately successful in describing the large scale motions, are essentially two-dimensional. Thus we may expect that the statistical behavior of the atmospheric large scales may be similarly approximated by the statistical behavior of ideal two-dimensional incompressible flows, that is, by two-dimensional turbulence.

An important characteristic of two-dimensional turbulence is the tendency for energy to be trapped in the low wavenumbers (Fjørtoft, 1953) but for enstrophy to be cascaded through an inertial spectral range to be dissipated at high wavenumber (Batchelor, 1969). The same sort of dimensional argument that predicts a -5/3 power energy-cascading inertial range for three-dimensional homogeneous isotropic turbulence predicts a -3 power enstrophy-cascading inertial range (Batchelor, 1969; Kraichnan, 1967; Leith, 1968) for two-dimensional homogeneous isotropic turbulence. Numerical simulation of two-dimensional turbulence first by Bray (Batchelor, 1969) and later with higher resolution by Lilly (1969) have indeed shown such enstrophy-cascading inertial ranges.

Recent observations indicate that the planetary-scale motions of the atmosphere also exhibit an enstrophy-cascading inertial range. In this paper will be described an eddy-damped Markovian approximation for two-dimensional turbulence. This approximation will be applied to atmospheric motions in order to predict the nonlinear transfer rate and rate of error growth associated with the observed energy spectra. In this way there are derived estimates of the predictability of the atmosphere and of the errors inherent in numerical models. Such estimates were obtained by Lorenz (1969) using a quasinormal approximation.

Turbulence deals with the statistical properties of the solutions of the Navier-Stokes equation which we shall use in its wavevector space form:

$$[d/dt + \nu k^2 - \alpha(k)]u_i(\underset{\sim}{k},t) = -\tfrac{1}{2}iP_{ijk}(\underset{\sim}{k})\sum_{\underset{\sim}{p}+\underset{\sim}{q}=\underset{\sim}{k}} u_j(\underset{\sim}{p},t)u_k(\underset{\sim}{q},t) \qquad (1)$$

with the incompressibility condition $k_i u_i(\underset{\sim}{k},t) = 0$ and with $P_{ijk}(\underset{\sim}{k}) = k_j P_{ik}(\underset{\sim}{k}) + k_k P_{ij}(\underset{\sim}{k})$, $P_{ij}(k) = \delta_{ij} - k_i k_j/k^2$, and $k = |\underset{\sim}{k}|$. We have included in (1) a specified instability function $\alpha(k)$ that can be used to simulate an energy source. Eq. (1) is for flows periodic of length L thus for a discrete $\underset{\sim}{k}$-space with area elements $V = (2\pi/L)^2$, but we shall consider the limit as $V\to 0$ in which $V\sum_{\underset{\sim}{p}} \to \int d\underset{\sim}{p}$.

The statistics of concern here are the single-time second and third moment tensors defined as follows:

$$U_{ij}(\underset{\sim}{k}) = \lim_{V\to 0} V^{-1}\langle u_i(\underset{\sim}{k})u_j(-\underset{\sim}{k})\rangle,$$

$$T_{ijk}(\underset{\sim}{k},\underset{\sim}{p},\underset{\sim}{q}) = \lim_{V\to 0} V^{-2}\langle u_i(\underset{\sim}{k})u_j(\underset{\sim}{p})u_k(\underset{\sim}{q})\rangle \text{ with } \underset{\sim}{k}+\underset{\sim}{p}+\underset{\sim}{q} = \underset{\sim}{0}.$$

---

*The National Center for Atmospheric Research is sponsored by the National Science Foundation.

By using the Navier-Stokes Eq. (1) we derive evolution equations for the second moment tensor in terms of the third and for the third in terms of the fourth:

$$[d/dt + 2\nu k^2 - 2\alpha(k)]U_{ij}(\underset{\sim}{k}) = -\tfrac{1}{2}\int d\underset{\sim}{p}[P_{i\ell m}(\underset{\sim}{k})T_{j\ell m}(-\underset{\sim}{k},-\underset{\sim}{p},-\underset{\sim}{q})$$
$$+ P_{j\ell m}(-\underset{\sim}{k})T_{i\ell m}(\underset{\sim}{k},\underset{\sim}{p},\underset{\sim}{q})] \qquad (2)$$

$$[d/dt + \nu(k^2 + p^2 + q^2) - \alpha(k) - \alpha(p) - \alpha(q)]T_{ijk}(\underset{\sim}{k},\underset{\sim}{p},\underset{\sim}{q}) =$$
$$-iR_{ijk}(\underset{\sim}{k},\underset{\sim}{p},\underset{\sim}{q}) + C_{ijk}(\underset{\sim}{k},\underset{\sim}{p},\underset{\sim}{q}) \qquad (3)$$

with $\underset{\sim}{k} + \underset{\sim}{p} + \underset{\sim}{q} = 0$. Here $C_{ijk}(\underset{\sim}{k},\underset{\sim}{p},\underset{\sim}{q})$ is a linear expression in fourth moment tensors and

$$R_{ijk}(\underset{\sim}{k},\underset{\sim}{p},\underset{\sim}{q}) = P_{i\ell m}(\underset{\sim}{k})U_{j\ell}(\underset{\sim}{p})U_{km}(\underset{\sim}{q}) + P_{j\ell m}(\underset{\sim}{p})U_{i\ell}(\underset{\sim}{k})U_{km}(\underset{\sim}{q}) + P_{k\ell m}(\underset{\sim}{q})U_{i\ell}(\underset{\sim}{k})U_{jm}(\underset{\sim}{p})$$

Closure of the moment equations at this level requires an approximation for $\underset{\approx}{C}$ in terms of, at most, $\underset{\approx}{U}$ and $\underset{\approx}{T}$. The quasinormal approximation, $\underset{\approx}{C} = \underset{\approx}{0}$, was discredited by Ogura (1962, 1963) who showed that it leads to negative energy spectra in both two and three dimensions. A better approximation is the eddy-damped quasinormal approximation (Orszag, 1970):

$$C_{ijk}(\underset{\sim}{k},\underset{\sim}{p},\underset{\sim}{q}) = -[\mu_e(\underset{\sim}{k}) + \mu_e(\underset{\sim}{p}) + \mu_e(\underset{\sim}{q})]T_{ijk}(\underset{\sim}{k},\underset{\sim}{p},\underset{\sim}{q})$$

where $\mu_e(k)$ is a positive eddy-damping rate intended to simulate the presumed destruction of triple correlations $\underset{\approx}{T}$ by the nongaussian random velocity field. We choose the eddy-damping rate to be proportional to the local or Lagrangian eddy rate, $\mu_e(\underset{\sim}{k}) = \gamma k^2[U_{ii}(\underset{\sim}{k})]^{\frac{1}{2}}$, where γ is a dimensionless coefficient to be determined from experience. The integration of Eq. (3) then leads to an expression for the third moment tensor $\underset{\approx}{T}$ as a time integral over past values of an integrand involving only second moment tensors $\underset{\approx}{U}$. The eddy-damped Markovian approximation, which we shall use, makes the further approximation of replacing in the time integral for $\underset{\approx}{T}$ the past values of $\underset{\approx}{U}$ by current values. At times large compared to the local eddy times initial conditions are irrelevant and our approximation leads to the equation

$$T_{ijk}(\underset{\sim}{k},\underset{\sim}{p},\underset{\sim}{q}) = -iD(\underset{\sim}{k},\underset{\sim}{p},\underset{\sim}{q})R_{ijk}(\underset{\sim}{k},\underset{\sim}{p},\underset{\sim}{q}) \qquad (4)$$

where the damping time $D(\underset{\sim}{k},\underset{\sim}{p},\underset{\sim}{q}) = [\mu(\underset{\sim}{k}) + \mu(\underset{\sim}{p}) + \mu(\underset{\sim}{q})]^{-1}$ with $\mu(\underset{\sim}{k}) = \mu_e(\underset{\sim}{k}) + \nu k^2 - \alpha(k)$. Eqs. (2) and (4) form a closed set that can be integrated to determine $\underset{\approx}{U}$ and $\underset{\approx}{T}$.

For isotropic turbulence $U_{ij}(\underset{\sim}{k}) = P_{ij}(\underset{\sim}{k})U(k)$ with the scalar function $U(k)$ determining the two-dimensional kinetic energy spectrum $E(k) = \pi k U(k)$. The eddy-damping rate becomes $\mu_e(k) = \gamma k^2[U(k)]^{\frac{1}{2}}$, and the damping time $D(k,p,q)$ is a function not of wavevectors but of the wavenumbers k, p, and q. The second moment equation of the eddy-damped Markovian approximation reduces to:

$$[d/dt + 2\nu k \ - 2\alpha(k)]U(k,t) = S(k,t) = 4\iint_{\Gamma(k)} \frac{B(k,p,q)}{\sin(p,q)} D(k,p,q,t)$$
$$\cdot\ [U(q,t)U(p,t) - U(q,t)U(k,t)]dpdq \qquad (5)$$

The integration is over that part $\Gamma(k)$ of the positive quadrant in the p,q plane for which triangles can be formed with sides k,p,q. The integrand includes a purely geometric factor

$$B(k,p,q) = -P_{ijk}(\underset{\sim}{k})P_{j\ell}(\underset{\sim}{q})P_{k\ell i}(\underset{\sim}{p}) = (k^2 - q^2)(p^2 - q^2)\sin^2(p,q)/k^2$$

Error may be defined (Kraichnan, 1970) in terms of the difference field $u_i^{(1)}(\underset{\sim}{k}) - u_i^{(2)}(\underset{\sim}{k})$ for an ensemble of pairs of velocity fields $u_i^{(1)}, u_i^{(2)}$ the first and second members each having the same statistical properties to which we apply the moment equations already derived. We are now concerned with the crossmoment tensor $W_{ij}(\underset{\sim}{k}) = \lim\,(V\to 0)V^{-1}\langle u_i^{(1)}(\underset{\sim}{k})u_j^{(2)}(-\underset{\sim}{k})\rangle$ in terms of which we define the error tensor $\Delta_{ij}(\underset{\sim}{k}) = U_{ij}(\underset{\sim}{k}) - W_{ij}(\underset{\sim}{k})$. For the isotropic case the error tensor becomes $\Delta_{ij}(k) = P_{ij}(k)\Delta(k)$. The eddy-damped Markovian approximation, in a derivation similar to that of an equation for $U_{ij}(\underset{\sim}{k})$, leads to an evolution equation for $W_{ij}(\underset{\sim}{k})$ and thus to one for $\Delta_{ij}(\underset{\sim}{k})$. In the isotropic case the error equation reduces to:

$$[d/dt + 2\nu k^2 - 2\alpha(k)]\Delta(k,t) = S'(k,t) = 4\iint_{\Gamma(k)} \frac{B(k,p,q)}{\sin(p,q)} D(k,p,q,t)$$

$$\cdot\,[\Delta(p,t)U(q,t) + \Delta(q,t)U(p,t) - \Delta(p,t)\Delta(q,t) - U(q,t)\Delta(k,t)]\,dp\,dq. \quad (6)$$

Note that if $\Delta(k) = 0$ then such an error-free condition persists. For total ignorance $\Delta(k) = U(k)$ and the Δ equation becomes the same as the U equation. In general Eq. (6) predicts the growth of small errors and is used in the determination of the predictability of the atmosphere.

The most complex and time consuming part of the numerical solution of the eddy-damped Markovian equations is the evaluation of the integrals over all triangles appearing in Eqs. (5) and (6). We use a finite approximation that satisfies all of the symmetry and integral properties of the continuous equations and thus itself serves as a consistent turbulence approximation. The wavenumber axis is divided into 4 logarithmic intervals per factor of 2. The mesh points k_ℓ, p_m, q_n are surrounded by blocks which subdivide the positive k,p,q octant. For each block we compute the fraction $v(\ell-m,\ell-n)$ of the volume corresponding to possible triangles. Integrals are approximated by a sum of mesh point values weighted by the surrounding triangle volume. The logarithmic subdivision and the scaling properties of $B(k,p,q)/\sin(p,q)$ lead to the integration approximation

$$4\iint_{\Gamma(k)} \frac{B(k,p,q)}{\sin(p,q)} f(k,p,q)\,dp\,dq \approx 2k_\ell^{4}\sum_{m,n} A(\ell-m,\ell-n)\,f(k_\ell,p_m,q_n)$$

where the $A(\ell-m,\ell-n)$ form a fixed two-index array of integration coefficients.

The computation requires at each time step the evaluation of three such sums

$$\sigma_\ell = 2k_\ell^{4}\sum_{m,n} A(\ell-m,\ell-n)\,D_{\ell mn}U_n$$

$$F_\ell = 2k_\ell^{4}\sum_{m,n} A(\ell-m,\ell-n)\,D_{\ell mn}U_mU_n$$

$$Y_\ell = 2k_\ell^{4}\sum_{m,n} A(\ell-m,\ell-n)\,D_{\ell mn}[\Delta_mU_n - \Delta_nU_m - \Delta_m\Delta_n]$$

in terms of which Eqs. (5) and (6) become

$$dU_\ell/dt = F_\ell - (\sigma_\ell + 2\nu k_\ell{}^2 - 2\alpha_\ell)U_\ell$$

$$d\Delta_\ell/dt = Y_\ell - (\sigma_\ell + 2\nu k_\ell{}^2 - 2\alpha_\ell)\Delta_\ell$$

A two step time integration scheme was used. The calculation for the atmospheric spectrum over a range of wavenumbers from 1 to 1024 with a time step Δt = 0.05 day took 1 minute of CDC 6600 time for 5 days. The truncation errors are comparable to the uncertainties in the turbulence approximation itself. The truncation does, however, make $A(m,n) = 0$ for sufficiently large $|m|$ or $|n|$ for $m \neq 0$ and thus the numerical approximation becomes local in the sense of Kolmogorov. The resulting error is diminished by the fact that the eddy damping dimensionless coefficient is chosen as $\gamma = 0.8$ to match observations.

ATMOSPHERIC PREDICTABILITY

In simulating the statistical properties of the atmosphere we are using a two-dimensional plane isotropic homogeneous turbulent analog of the actual spherical, non-isotropic, inhomogeneous random flow of the atmosphere, but observations indicate that for $k > 5$ rad^{-1} this may be satisfactory. (We measure k in planetary wavenumbers; 1 rad = 4.8×10^6 m. at 40° latitude.)

In order to maintain a statistically stationary spectrum it is necessary to simulate the atmospheric energy sources and sinks. We do this by choosing the instability function $\alpha(k)$ in such a way as to compensate in Eq. (5) for the nonlinear transfer S(k) computed for a function U(k) chosen to fit observations of the atmospheric energy spectrum. Such observations are usually of the one-dimensional energy spectrum $E_1(k_1) = 2\int_{k_1}^{\infty}(k/k_2)U(k)dk$ where $k^2 = k_1{}^2 + k_2{}^2$. In Fig. 1 is shown a comparison of the turbulence model spectrum and the atmospheric observations of Wiin-Nielsen (1967).

In Fig. 2 are shown the results of computations by Yang (1967) of the one-dimensional spectrum of nonlinear energy transfer $T_1(k_1) = 2\int_{k_1}^{\infty}(k/k_2)S(k)dk$ using

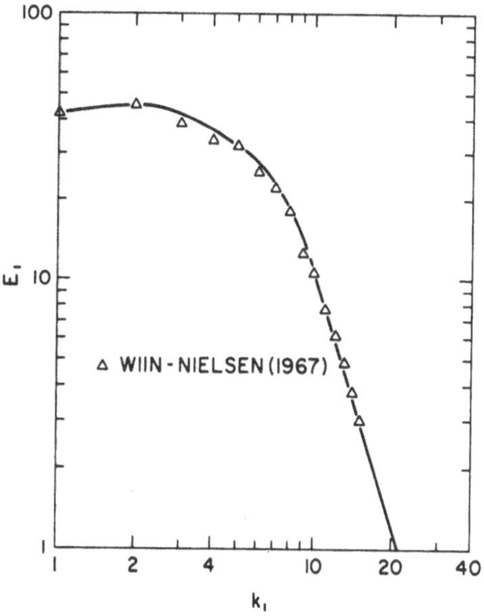

Fig. 1 One-dimensional spectrum for the atmosphere and the model

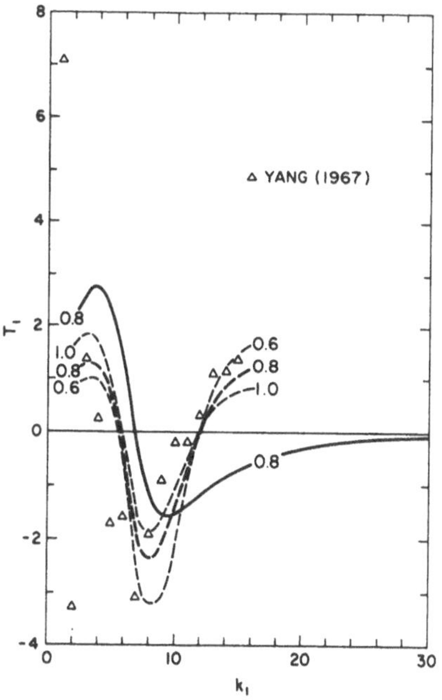

Fig. 2 Nonlinear transfer spectrum for the atmosphere and the model

the same data base as did Wiin-Nielsen and taking into account planetary wavenumbers with $1 \leq k_1 \leq 15$. Such a truncation is not possible in the model, but a truncation to wavenumbers with $1 \leq k \leq 17$ gives, for different values of γ, the dashed curves in Fig. 2. This comparison of the model prediction with observations is the basis for the choice of $\gamma = 0.8$. Without truncation the computed nonlinear transfer is shown as the solid curve.

The predictability of the atmosphere will depend on the magnitude of the initial error as determined by the resolution of the observing system and on the growth rate of error as determined by the equations of motion. We shall specify the resolution of an observing network as the smallest wavenumber k_o such that for it and all greater wavenumbers our ignorance of the observed initial state is complete so that $\Delta(k,0) = U(k)$, $k \geq k_o$. For $k < k_o$ we assume either $\Delta(k,0) = U(k_o)$ or $\Delta(k,0) = (k/k_o)U(k_o)$ or $\Delta(k,0) = 0$ to simulate, respectively, random instrumental error or alias error owing to contamination by unresolved scales $(k \geq k_o)$ or, finally, no error as an extreme ideal case.

In Fig. 3 is shown as solid lines labeled by days the evolution through 10 days of an alias error spectrum for $k_o = 128$ as computed with Eq. (6). There is also shown the evolution after 1 day of the random error case (dashed line) and the no error case (dotted line) showing the development of a characteristic error spectrum whose shape is not strongly dependent on the initial spectrum assumed.

In Fig. 4 the ratio of total error energy (defined as $\frac{1}{4}\langle[u^{(1)} - u^{(2)}]^2\rangle$) to total energy is shown as a function of time for different values of the resolution k_o for alias error (solid lines). For $k_o = 128$ there are also shown error growth curves for the random error case (dashed line) and the no error case (dotted line). Although there is some knowledge of large scales left after 10 days a practical definition of predictability can be based on the ratio 0.25 indicated.

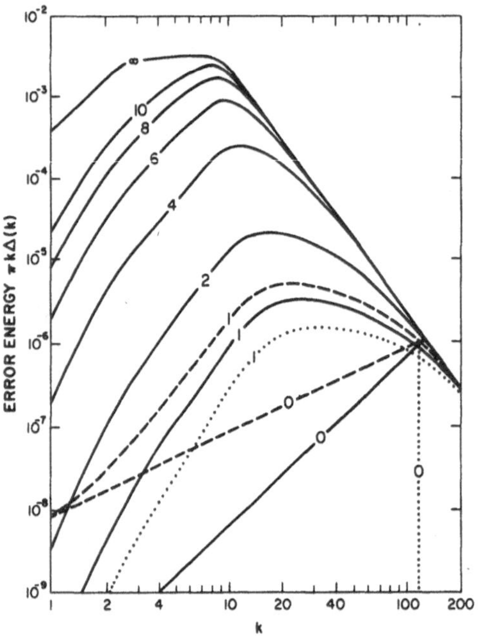

Fig. 3 Error energy spectra labeled by days after observation

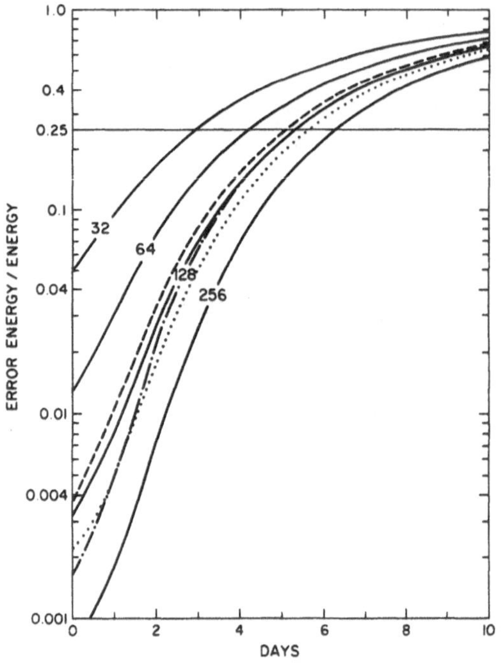

Fig. 4 Error growth curves for different observing resolutions

Numerical models of the atmosphere have themselves been used to determine the predictability of the atmosphere. These are severely truncated to wavenumbers $k \leq k_*$ and require an added eddy dissipation term to simulate mean nonlinear transfer into uncomputed wavenumbers $k > k_*$. To test the consequence of using such a model to determine error growth, the turbulence model error growth calculation was repeated with similar truncation. In Fig. 4 is shown as a dash-dot line the error energy growth curve for $k_o = 128$ in the alias error case but for truncation at $k_* = 128$. Thus we see that, even though the truncation is so severe that $k_* = k_o$, the error growth curve is not greatly changed. It would seem then that there is no inherent limitation in the ability of numerical models to determine the predictability of the atmosphere even if they cannot explicitly compute small scales of motion.

REFERENCES

Batchelor, G. K.: Phys. Fluids Suppl. II, 12, II-233 - II-239 (1969)

Fjørtoft, R.: Tellus, 5, 225-230 (1953)

Kraichnan, R. H.: Phys. Fluids, 10, 1417-1423 (1967)

________________: Phys. Fluids, 13, 569-575 (1970)

Leith, C. E.: Phys. Fluids, 11, 671-672 (1968)

Lilly, D. K.: Phys. Fluids Suppl. II, 12, II-240 - II-249 (1969)

Lorenz, E. N.: Tellus, 21, 289-307 (1969)

Ogura, Y.: Phys. Fluids, 5, 395-401 (1962)

________: J. Fluid Mech., 16, 33-40 (1963)

Orszag, S. A.: J. Fluid Mech., 41, 363-386 (1970)

Wiin-Nielsen, A.: Tellus, 19, 540-559 (1967)

Yang, C.-H.: Tech. Rept., Dept. of Meteorology and Oceanography, University of Michigan (1967)

A HEURISTIC APPROACH TO THREE-DIMENSIONAL BOUNDARY LAYERS

Roberto Vaglio-Laurin and Gabriel Miller

New York University

Bronx, New York

The formulation of numerical methods for general three-dimensional laminar boundary layer problems involves several factors, namely: the analysis of the mathematical nature of the governing equations; the proper description of the physical transport of mass, momentum and energy and the consistent definition of regions of influence that correctly represent the history of the boundary layer; the selection of suitable coordinate transformations and of stable convergent finite difference schemes. Among these factors only the first has proved amenable to a rigorous analysis, which classifies the three-dimensional boundary layer equations as essentially parabolic. All other factors have been examined only on heuristic grounds, often relying on numerical experiments[1,2,3].

An alternative approach to the investigation and solution of the problem consists in seeking heuristic and analytical models that are intimately related to the characteristics of the flow and, at the same time, are amenable to description by closed form solutions. In this vein, a two layer analysis of three-dimensional boundary layers was set forth in Refs. 4 and 5. The analysis relinquishes the quest for "exact" solutions in favor of the capability for broad use, for comprehensive interpretation and coordinated understanding of results, and for substantive reduction as well as simplification of the numerical aspects of the solution. A first satisfactory assessment of the two layer model, as applied to self-preserving hypersonic laminar flows near a plane of symmetry, has been presented in Ref. 5. Further favorable assessments, as well as the detailed statement of the method of solution for compressible three-dimensional nonsimilar laminar boundary layers, are set forth in the present paper. First, the basic features of the two layer model, and its implications for the numerical analysis, are briefly reviewed.[5] The solutions of outer and inner equations for general three-dimensional flows[5] are then presented, and the details of the numerical matching process are outlined. Finally the assessment of the model is extended by consideration of several recent applications and of their comparisons with available exact solutions; included are incompressible three-dimensional self-preserving flows, incompressible two-dimensional flows in the presence of adverse pressure gradients that produce separation, compressible three-dimensional flows.

THE TWO LAYER MODEL

In the heuristic formulation of the model[4,5] the three-dimensional boundary layer is viewed in a system of inviscid streamline coordinates x_i (Fig. 1). Two regions amenable to distinct approximate description are recognized, namely: 1) the outer portion of the boundary layer, where the streamwise momentum flux per unit area is comparable to the momentum flux in the external inviscid stream, while the crossflow velocity component remains much smaller than the streamwise component; 2) the inner portion of the layer adjacent to the wall, where the crossflow velocity component can become appreciable, but the inertial forces become small compared to the pressure and viscous forces. Although the flow in the outer region is viewed as two-dimensional, its history and development are governed by boundary (matching) conditions that depend upon integral three-dimensional effects in the inner region. The flow in the inner region is fully three-dimensional, but everywhere in local equilibrium because the inertial forces are negligible. The high momentum outer region plays the role of an equivalent solid plate with respect to the low momentum inner region; accordingly, the inner motion is Couette-like in the direction of the external flow, and Poiseuille-like in the direction transversal thereto.

Order of magnitude analysis consistent with the aforenoted characteristics, readily yields approximate statements of the equations valid in the two regions[4,5]. For nonsimilar flows, the outer region is described by linear, parabolic, partial differential equations in two independent variables (x_1 streamwise, x_3 normal to the body surface in the notation of Fig. 1); the inner region is always described by nonlinear ordinary differential equations. Three-dimensional effects are manifested in the inner region and in the matching conditions between inner and outer solutions.

In terms of the independent variables

$$\xi_1 = \int_0^{x_1} \rho_e \mu_e v_{1e} e_2^2 e_1 dx_1 \; ; \quad \xi_2 = x_2 \tag{1-a,b}$$

$$\Psi = \int_0^{x_3} \rho v_1 e_2 e_3 dx_3 + \int_0^{x_1} \frac{\partial}{\partial \xi_2} \left[\int_0^{\infty} \rho v_2 e_1 e_3 dx_3 \right] dx_1 \tag{1-c}$$

and of the dependent variables

$$\nu_1 = 1 - (v_1/v_{1e})^2 \; ; \quad \nu_2 = (v_2/v_{1e}) \; ; \quad g = 1 - (H/H_e) \tag{2-a,b,c}$$

where ρ denotes the density, μ the viscosity, H the stagnation enthalpy, v_i the velocity component along x_i, e_i the metric element of the coordinate system, and subscript e the flow properties evaluated at the outer edge of the boundary layer, the outer equations, for flows of a perfect gas with Pr = 1 and linear viscosity - temperature relation [Eqs. (22), (23) and (24) of Ref. 5], take the form

$$(\partial^2 \nu_1/\partial \Psi^2) - (\partial \nu_1/\partial \xi_1) = b_1(\xi_1)(\nu_1 - g) \tag{3-a}$$

$$(\partial^2 \nu_2/\partial \Psi^2) - (\partial \nu_2/\partial \xi_1) - c_2(\xi_1)\nu_2 = b_2(\xi_1)(\nu_1 - g) \tag{3-b}$$

$$(\partial^2 g/\partial \Psi^2) - (\partial g/\partial \xi_1) = 0 \tag{3-c}$$

with

$$b_1(\xi_1) = 2\left[1 - m(\xi_1)\right]^{-1} \left[\partial (\log v_{1e})/\partial \xi_1\right] \tag{4-a}$$

$$b_2(\xi_1) = (\rho_e \mu_e v_{1e} e_2^3)^{-1} \left[1 - m(\xi_1)\right]^{-1} \left[\partial (\log e_1)/\partial \xi_2\right] \tag{4-b}$$

$$c_2(\xi_1) = \left[\partial \log(e_1 v_{1e})/\partial \xi_1\right] \tag{4-c}$$

$$m(\xi_1) = v_{1e}^2/(2 H_e) \tag{4-d}$$

for the general streamtube ξ_2 = constant.

The closed form solutions satisfying the boundary conditions

$$\Psi \to \infty \qquad \nu_1 \to 0; \qquad \nu_2 \to 0; \qquad g \to 0 \tag{5-a,b,c}$$

$$\Psi = 0 \qquad \nu_1 = \nu_{10}(\xi_1) \; ; \quad \nu_2 = \nu_{20}(\xi_1); \; g = g_0(\xi_1) \tag{6-a,b,c}$$

with $\nu_{10}(\xi_1)$, $\nu_{20}(\xi_1)$, and $g_0(\xi_1)$ to be determined in the course of matching with the inner solutions, are

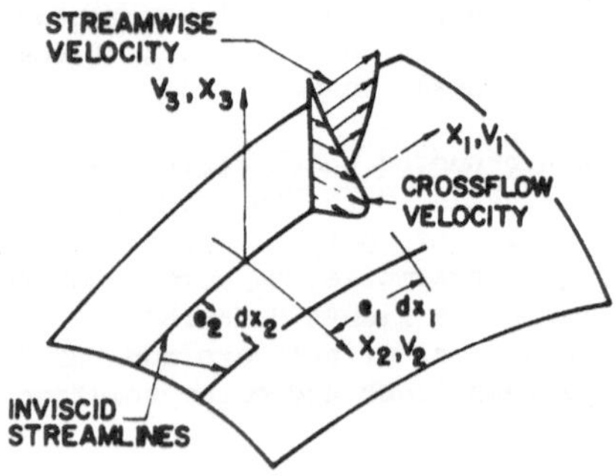

Fig. 1 Schematic Diagram of Streamline Coordinate System

$$\nu_1(\xi_1,\tilde{\Psi}) = g(\xi_1,\tilde{\Psi}) + \exp\left[-\int_0^{\xi_1} b_1 d\tau\right] \frac{\tilde{\Psi}}{2\sqrt{\pi}} \int_0^{\xi_1} \frac{[\nu_{10}(\tau)-g_0(\tau)]\ \exp(\int_0^{\tau} b_1 d\omega)\ d\tau}{(\xi_1-\tau)^{3/2}\exp[\tilde{\Psi}^2/4(\xi_1-\tau)]} \tag{7-a}$$

$$\nu_2(\xi_1,\tilde{\Psi}) = -\ I\ (\xi_1)\left[\nu_1(\xi_1,\tilde{\Psi}) - g(\xi_1,\tilde{\Psi})\right] \tag{7-b}$$

$$+ \exp\left(-\int_0^{\xi_1} c_2 d\tau\right)\frac{\tilde{\Psi}}{2\sqrt{\pi}}\int_0^{\xi_1} \frac{\left\{\nu_{20}(\tau) + I\ (\tau)[\nu_{10}(\tau)-g_0(\tau)]\right\}\exp\left(\int_0^{\tau} c_2 d\omega\right)\ d\tau}{(\xi_1-\tau)^{3/2}\exp[\tilde{\Psi}^2/4(\xi_1-\tau)]}$$

$$g(\xi_1,\tilde{\Psi}) = \frac{\tilde{\Psi}}{2\sqrt{\pi}}\int_0^{\xi_1} g_0(\tau)\ (\xi_1-\tau)^{-3/2}\exp\left[-\ \psi^2/4(\xi_1-\tau)\right]\ d\tau \tag{7-c}$$

where

$$I\ (\xi_1) = \exp\left[\int_0^{\xi_1}\left(b_1-c_2\right)\ d\tau\right]\ \int_0^{\xi_1} b_2\ \exp\left[-\int_0^{\tau}(b_1-c_2)\ d\omega\right]\ d\tau \tag{8}$$

In the region about

$$\tilde{\Psi}_1\ (\xi_1,\xi_2) = \int_0^{x_1}\frac{\partial}{\partial x_2}\left[\int_0^{\infty}\rho\ v_2\ e_1\ e_3\ dx_3\right]\ dx_1 \tag{9}$$

where inner and outer solutions for streamwise velocity and enthalpy are to be matched, the solutions (7a) and (7c) possess the behavior

$$\nu_1(\xi_1,\psi) = A_{v1}\left(\nu_{10},g_0,\tilde{\Psi}_1\right) + B_{v1}\left(\nu_{10},g_0,\tilde{\Psi}_1\right)\ \psi + \ldots = A_{v1}(\xi_1,\xi_2) + B_{v1}(\xi_1,\xi_2)\ \psi + \ldots \tag{10-a}$$

$$g(\xi_1,\psi) = A_H\left(g_0,\tilde{\Psi}_1\right) + B_H\left(g_0,\tilde{\Psi}_1\right)\ \psi + \ldots = A_H(\xi_1,\xi_2) + B_H(\xi_1,\xi_2)\ \psi + \ldots \tag{10-b}$$

where ψ is the streamwise mass flux

$$\psi = \int_0^{x_3} \rho\, v_1\, e_2\, e_3\, dx_3 \tag{11}$$

Similarly, in the region about

$$\tilde{\Psi}_2\ (\xi_1,\xi_2) = \tilde{\Psi}_1\ (\xi_1,\xi_2) + \int_0^{x_3 (v_1^{(i)}=1)} \rho^{(i)}\ v_1^{(i)}\ e_2\ e_3\ dx_3 \tag{12}$$

where crossflow matching must be performed, the solution (7b) possesses the behavior

$$\nu_2(\xi_1,\psi) = A_{v2}\ (\nu_{10},\nu_{20},g_0,\tilde{\Psi}_1,\tilde{\Psi}_2) + B_{v2}\ (\nu_{10},\nu_{20},g_0,\tilde{\Psi}_1,\tilde{\Psi}_2)\ \psi + \dots$$

$$= A_{v2}(\xi_1,\xi_2) + B_{v2}\ (\xi_1,\xi_2)\ \psi + \dots \tag{10-c}$$

The rather complicated functional dependence of A_{v1},B_{v1}, etc. on $\nu_{10}(\xi_1,\xi_2)$, $\nu_{20}\ (\xi_1,\xi_2)$, $g_0(\xi_1,\xi_2)$, $\tilde{\Psi}_1(\xi_1,\xi_2)$ and $\tilde{\Psi}_2(\xi_1,\xi_2)$ is extracted from (7a,b,c) in the step by step process of numerical matching described below.

Upon generalization to non-hypersonic conditions, the inner equations (18), (19), (20) of Ref. 5 take the form

$$(d^2\bar{v}_1/d\eta^2) = -\ 2\xi_1(1-m)^{-1}(\partial \log v_{1e}/\partial\xi_1)\left[\bar{H} - \bar{v}_1^2 - m\bar{v}_2^{\ 2}\right]$$

$$= -\ \hat{\beta}\ [\bar{H} - \bar{v}_1^{\ 2} - m\bar{v}_2^{\ 2}] \tag{13-a}$$

$$(d^2\bar{v}_2/d\eta^2) = 2\xi_1\left[(1-m)\ (\rho_e\mu_e v_{1e} e_2^3)\right]^{-1}\ (\partial \log e_1/\partial\xi_2)\left[\bar{H}-\bar{v}_1-m\bar{v}_2^{\ 2}\right]$$

$$= -\ \hat{\gamma}\ [\bar{H} - \bar{v}_1^{\ 2} - m\bar{v}_2^{\ 2}] \tag{13-b}$$

$$(d^2\bar{H}/d\eta^2) = 0 \tag{13-c}$$

in terms of the independent variable

$$\eta = (2\xi_1)^{-\frac{1}{2}}\ (e_2\ v_{1e}) \int_0^{x_3} \rho\ e_3\ dx_3 \tag{14}$$

and of the dependent variables

$$\bar{v}_1 = (v_1/v_{1e})\ \ ;\ \ \bar{v}_2 = (v_2/v_{1e})\ ;\ \ \bar{H} = (H/H_e) \tag{15}$$

Solutions must be determined subject to the boundary conditions

$$\eta = 0 \quad \bar{v}_1 = \bar{v}_2 = 0\ ; \quad \bar{H} = \bar{H}_w \tag{16-a}$$

$$\eta = \eta_o \quad \bar{v}_1 = 1\ ; \quad \bar{v}_2 = \bar{v}_{20}\ ; \quad \bar{H} = \bar{H}_o \tag{16-b}$$

with η_o, $\bar{v}_{20}$, and $\bar{H}_o$ to be defined in the course of matching. The solution of (13c) is readily obtained

$$\bar{H} = \bar{H}_w + (\bar{H}_o - \bar{H}_w)\ (\eta/\eta_o) \tag{17-c}$$

as is the solution of (13b) when $\hat{\beta} \neq 0$ and $\hat{\gamma} \neq 0$ viz.

$$\bar{v}_2 = (\hat{\gamma}/\hat{\beta})\,\bar{v}_1 + [\bar{v}_{20} - (\hat{\gamma}/\hat{\beta})]\,(\eta/\eta_o) \qquad (17\text{-b})$$

However, due to the nonlinear nature of (13a) when $\hat{\beta} \neq 0$, $\bar{v}_1$ can be determined in closed form only for incompressible flow (m=0) where

$$\eta = 2^{-\frac{1}{2}} \int_o^{\bar{v}_1} \left[\hat{\beta}\,\frac{\bar{v}_1^3}{3} - \hat{\beta}\,\bar{v}_1 + \tfrac{1}{2}\,(d\bar{v}_1/d\eta)_{\eta=0} \right]^{-\frac{1}{2}} d\bar{v}_1 \qquad (17\text{-d})$$

For compressible flow, closed form results can be obtained only if a linearizing assumption is introduced.* The approximation generally and successfully employed in all applications reported in this paper is

$$\bar{v}_1 = (\eta/\eta_o) + \tilde{v}_1 \qquad (17\text{-a})$$

with $\tilde{v}_1 << (\eta/\eta_o)$, so that only linear terms are retained in the equation for $\tilde{v}_1$, and $\tilde{v}_1(0) = \tilde{v}_1(\eta_o) = 0$. The equation for $\tilde{v}_1$ is then

$$(d^2\tilde{v}_1/d\eta^2) - k^3\,\eta\,\tilde{v}_1$$

$$= (\beta/\eta_o^2)\left[(1+m\,\bar{v}_{20}^2)\,\eta^2 - (\bar{H}_o - \bar{H}_w)\,\eta_o\eta - \bar{H}_w\,\eta_o^2\right] \qquad (18)$$

with $k = [(2/\eta_o)\,(\hat{\beta} + m\,\hat{\gamma}\,v_{20})]^{1/3}$. The solution satisfying the boundary conditions $\tilde{v}_1(0) = \tilde{v}_1(\eta_o) = 0$ is

$$\tilde{v}_1(\eta) = \Big\{ D\,(\eta_o, \bar{v}_{20}, \bar{H}_o) - (\pi\,c\hat{\beta})(k\eta_o)^{-2}\Big[1 + m\bar{v}_{20}^2 + \frac{k\eta_o}{\sqrt{3}}\,(\bar{H}_o - \bar{H}_w)\Big]\Big\}\Big\{A_i(k\eta) - 3^{-\frac{1}{2}}\,B_i(k\eta)\Big\}$$

$$+ \left[\beta/(k^3\eta_o)\right]\Big\{\pi k^2\eta_o\bar{H}_w\Big[A_i(k\eta)\int_o^{\eta} B_i(k\eta)\,d\eta - B_i(k\eta)\int_o^{\eta} A_i(k\eta)\,d\eta\Big]$$

$$+ (\bar{H}_o - \bar{H}_w)\left[1 - \tfrac{2}{3}\,\pi\,c\,B_i(k\eta)\right] - (1+m\,\bar{v}_{20}^2)\,(\eta/\eta_o)\Big\} \qquad (19)$$

where $A_i(k\eta)$, $B_i(k\eta)$ denote Airy functions, c = 0.77643 is a constant arising in the evaluation of quadratures like $\int_o^{\eta} \eta A_i(k\eta)d\eta$, and $D\,(\eta_o, \bar{v}_{20}, \bar{H}_o)$ is a quantity uniquely defined in terms of $\eta_o, \bar{v}_{20}, \bar{H}_o$.

In the region of matching ($\eta \rightarrow \eta_o$) the inner solutions (17a,b,c) and (19) possess the behavior

$$\bar{v}_1 \simeq 1 - \frac{\nu_1}{2} \simeq 1 + (2\xi_1)^{-\frac{1}{2}}\,(d\bar{v}_1/d\eta)_{\eta=\eta_o}\,(\psi - \psi_2) + \ldots \qquad (20\text{-a})$$

---

*If $\hat{\beta}=0, \bar{v}_1 = (\eta/\eta_o)$; however, (17b) may not be used. The solution of the nonlinear equation (13b) must then be sought by the linearizing approximation described here for equation (13a) and $\hat{\beta} \neq 0$.

$$\bar{v}_2 = \nu_2 \simeq \bar{v}_{20} + (2\xi_1)^{-\frac{1}{2}}(d\bar{v}_2/d\eta)_{\eta=\eta_o} \, (\psi - \psi_2) + \ldots \qquad (20\text{-b})$$

$$\bar{H} = 1 - g \simeq \bar{H}_o + (2\xi_1)^{-\frac{1}{2}} \, (\bar{H}_o - \bar{H}_w) \, \eta_o^{-1} \, (\psi - \psi_2) + \ldots \qquad (20\text{-c})$$

where

$$\psi_2 \, (\eta_o, \bar{v}_{20}, \bar{H}_o) = (2\xi_1)^{\frac{1}{2}} \int_o^{\eta_o} \bar{v}_1 \, d\eta \qquad (21)$$

Accordingly

$$\nu_1 = a_{v1}(\eta_o, \bar{v}_{20}, \bar{H}_o) + b_{v1}(\eta_o, \bar{v}_{20}, \bar{H}_o) \, \psi + \ldots \qquad (22\text{-a})$$

$$\nu_2 = a_{v2}(\eta_o, \bar{v}_{20}, \bar{H}_o) + b_{v2}(\eta_o, \bar{v}_{20}, \bar{H}_o) \, \psi + \ldots \qquad (22\text{-b})$$

$$g = a_H(\eta_o, \bar{v}_{20}, \bar{H}_o) + b_H \, (\eta_o, \bar{v}_{20}, \bar{H}_o) \, \psi + \ldots \qquad (22\text{-c})$$

In addition since

$$\int_o^{\infty} \rho v_2 \, e_1 \, e_3 \, dx_3 \simeq (2\xi_1)^{\frac{1}{2}} e_1 \, v_{1e} \int_o^{\eta_o} \bar{v}_2 d\eta = (2\xi_1)^{\frac{1}{2}} e_1 \, v_{1e} \, f_2(\eta_o, \bar{v}_{20}, \bar{H}_o) \qquad (23\text{-a})$$

the stream function Ψ_1 defined in (9) exhibits the dependence

$$\Psi_1 \, (\xi_1, \xi_2) = \left[\partial/\partial\xi_2\right] \left[\int_o^{x_1} f_1(\xi_1, \xi_2) \, f_2(\eta_o, \bar{v}_{20}, \bar{H}_o) \, dx_1\right] \qquad (23\text{-b})$$

The coefficients A_{v1}, B_{v1}, etc in (10a,b,c) are therefore functionally dependent on the parameters governing the inner solutions, viz., $A_{v2} = A_{v2} \, (\nu_{10}, \nu_{20}, g_o, \eta_o, \bar{v}_{20}, \bar{H}_o)$.

Matching of inner and outer solutions requires that the coefficients A_{v1}, B_{v1}, etc. be identically equal to the coefficients a_{v1}, b_{v1} etc in (22a,b,c) at each station (ξ_1, ξ_2) in the boundary layer. The requirement yields at each station six algebraic equations in the six unknowns ν_{10}, ν_{20}, g_o, η_o, $\bar{v}_{20}$, $\bar{H}_o$. Actually the matching process must recognize that, since the outer independent variable Ψ is used in an asymptotic sense, the relation (1c) must be generalized to the form

$$\Psi \, (\xi_1, \xi_2, x_3) = \psi(\xi_1, \xi_2, x_3) + \Psi_1 \, (\xi_1, \xi_2) + \psi^*(\xi_1, \xi_2) \qquad (24)$$

where $\psi^*(\xi_1, \xi_2)$ represents an a priori unknown shift of the zero stream surface. The introduction of ψ^* influences the overall solution insofar as it causes the behavior of the outer solution for $\Psi \to \Psi_1$ to be described by

$$\nu_1(\xi_1, \psi) = \left[A_{v1}(\nu_{10}, g_o, \Psi_1) + B_{v1}(\nu_{10}, g_o, \Psi_1) \, \psi^*\right] + B_{v1}(\nu_{10}, g_o, \Psi_1) \, \psi + \ldots$$

$$= A^*_{v1} \, (\nu_{10}, g_o, \Psi_1, \psi^*) + B_{v1}(\nu_{10}, g_o, \Psi_1) \, \psi + \ldots \qquad (25\text{-a})$$

and by analogous relations for ν_2 and g, instead of (10 a,b,c). Accordingly, the conditions $A^*_{v1} = a_{v1}$ etc. replace $A_{v1} = a_{v1}$ etc. in the matching process with evident influence on the magnitude of wall shear and heat transfer parameters. It is then apparent that the magnitude of ψ^* should be determined from integral conservation relations. In fact, three distinct functions ψ^* arise (say $\psi^*_{\nu_1}$, $\psi^*_{\nu_2}$, ψ^*_g), respectively connected with three equations (3a,b,c) and their solutions (9a,b,c). The three functions are determined by the requirement that the overall solution (composite of inner and outer) simultaneously satisfy conservation of streamwise momentum, of crossflow momentum, and of energy.* From this viewpoint, the two layer model takes the role of a mechanism for generating accurate streamwise velocity, crossflow velocity and stagnation enthalpy profiles parametrically dependent on the quantities $\psi^*_{\nu_1}$, $\psi^*_{\nu_2}$ and ψ^*_g, which, in turn, are determined by classical integral considerations. The favorable comparisons with exact solutions reported in Ref. 5 and here indicate that the mechanism is quite accurate. Indeed, satisfactory predictions are generated by the model even if the three functions ψ^* are assumed identical to each other, and a single integral conservation requirement is imposed. This is the case for the results reported in Ref. 5, where a single matching criteria in physical (instead of stream function) coordinates was employed instead of a single integral requirement.

The success in the profile generating function indicates that the two layer model provides the proper description of the mechanism and balances of physical transports within the boundary layer. Hence, the model may also be used to define the regions of influence that correctly represent the boundary layer history in three-dimensional situations. The solutions and the matching conditions presented above clearly indicate that the interplay between streamwise flow and crossflow is embodied in the function of integration $\tilde{\Psi}_1$ which, in turn, carries the flow history in the form stipulated by equation (23). The effects on the solutions are formally and conceptually equivalent to those of injection (or suction) at a solid boundary. Since $\tilde{\Psi}_1$ represents the cumulative divergence of integral (across the boundary layer) crossflow in specified control volumes (bound by inviscid flow stream surfaces), the domain of influence is not associated with selected limiting streamlines, but, instead, with fictitious average streamlines in the inner region that cause the correct integral crossflow divergence within the control volumes. Clearly in this context the evaluation of $\tilde{\Psi}_1$ depends on the determination of the entire boundary layer flow upstream of a considered station x_1= constant, consistent with the boundary conditions of the problem on surfaces x_2 = constant as well as on the surfaces x_3= 0 and $x_3 \to \infty$. However, a correct account of boundary layer history is reduced to a careful finite difference representation of the x_2 - derivatives of the integral inner region crossflow at each station x_1= constant. This represents the only requirement on the numerical formulation of the matching process. The problem is divorced from questions related to the nature of equation and solutions, and is confined to a standard finite difference fit for a function of a single variable.

The matching process is carried out step by step, proceeding from one station x_1= constant to the next $(x_1+\Delta x_1)$ = constant. In view of the complicated dependence of the coefficients $A_{\nu 1},\ldots,a_{\nu 1}\ldots$ on the unknown parameters $\nu_{10},\ldots,\eta_o,\ldots\psi^*_{\nu 1}\ldots$ the process is carried out numerically under the assumption that the coefficients are linearly dependent on perturbations $\delta\nu_{10},\ldots,\delta\eta_o,\ldots\delta\psi^*_{\nu 1},\ldots$ of the parameters from estimated values $\nu^{(e)}_{10},\ldots,\eta^{(e)}_o,\ldots,\psi^{*(e)}_{\nu 1},\ldots$ As an example, with the solution assumed known on x_1= constant, the coefficient $A_H(g_o,\tilde{\Psi}_1,\psi^*_g)$ at selected points $(x_1 + \Delta x_1, x_2)$ is computed from equation (7c) with $g_o(x_1+\Delta x_1,x_2)=g_o^{(e)}$, $\tilde{\Psi}_1(x_1+\Delta x_1,x_2)=\tilde{\Psi}_1^{(e)}$, $\psi^*_g(x_1+\Delta x_1,x_2) = \psi_g^{*(e)}$. The derivatives, e.g. $(\partial A_H/\partial\tilde{\Psi}_1)$ are evaluated by finite difference

$$(\partial A_H/\partial\tilde{\Psi}_1) = (1/\delta\tilde{\Psi}_1^{(e)})\left[A(g_o^{(e)}, \tilde{\Psi}_1^{(e)}+ \delta\tilde{\Psi}_1^{(e)}, \psi_g^{*(e)}) - A(g_o^{(e)}, \tilde{\Psi}_1^{(e)}, \psi_g^{*(e)})\right]$$

and the coefficient is cast in the form

*Conservation of mass is implicit in the use of the stream functions ψ and $\tilde{\Psi}$ as independent variables. Inclusion of the three ψ^* functions and of the three integral relations raises to nine the number of equations and unknowns to be considered at each station.

$$A_H(g_o,\tilde{\Psi}_1,\tilde{\Psi}_g^*) = A_H(g_o^{(e)},\tilde{\Psi}_1^{(e)},\tilde{\Psi}_g^{*(e)}) + (\partial A_H/\partial g_o)\,\delta g_o +$$

$$(\partial A_H/\partial \psi_g^*)\,\delta\psi_g^* + (\partial A_H/\partial\tilde{\Psi}_1)\left[\frac{\partial\tilde{\Psi}_1}{\partial(\partial\eta_o/\partial x_2)}\,\delta(\partial\eta_o/\partial x_2) + \right. \tag{26}$$

$$\left.\frac{\partial\tilde{\Psi}_1}{\partial(\partial\bar{v}_{20}/\partial x_2)}\,\delta\,(\partial\bar{v}_{20}/\partial x_2) + \frac{\partial\tilde{\Psi}_1}{\partial(\partial\bar{H}_o/\partial x_2)}\,\delta(\partial\bar{H}_o/\partial x_2)\right]$$

where the terms in square bracket express the variation of $\tilde{\Psi}_1$ as a function of the variations of the inner solution parameters and their derivatives. If finite difference approximations are stipulated for the variations of these derivatives, viz.

$$\delta(\partial\eta_o/\partial x_2) = (1/2\Delta x_2)\left[\delta\eta_o(x_2+\Delta x_2) - \delta\eta_o(x_2-\Delta x_2)\right]$$

and the linear forms (26) are substituted into the matching conditions

$$A_i = a_i \tag{27-a}$$

$$B_i = b_i \tag{27-b}$$

as well as into the integral conservation relations

$$F_i(\delta\nu_{10},\ldots,\delta\eta_o,\ldots,\delta\psi_{v1}^*\ldots) = 0 \tag{27-c}$$

(where $i = v_1, v_2, H$, and F_i denotes a linear function) nine linear equations in the unknowns $\delta\nu_{10}(x_2),\ldots,\ \delta\eta_o(x_2),\ \delta\eta_o(x_2-\Delta x_2),\ \delta\eta_o(x_2+\Delta x_2),\ldots,\ \delta\psi_{v1}^*(x_2),\ldots$ are obtained at each station $(x_1+\Delta x_1, x_2)$. Clearly a solution can be determined only if all stations on $(x_1+\Delta x_1)$ = constant are considered simultaneously. If crossflow transports of monemtum and energy are assumed to be confined to the inner region, the unknowns ν_{10}, ν_{20}, g_o, ψ_{v1}^*, ψ_{v2}^*, ψ_g^* can be eliminated a priori from (27-a,b); only three equations (27-c) in the unknowns $\delta\eta_o$, $\delta\bar{v}_{20}$, $\delta\bar{H}_o$ then remain to be solved for each station $(x_1+\Delta x_1, x_2)$. The matrix associated with the system of equations is nine - diagonal and, therefore, amenable to rapid and straightforward inversion.

Applications and Discussion

Several applications of the two layer model have recently been carried out with a view to broadening the initial assessment set forth in Ref. 5. Results and comparisons are reported here for several classes of flows, namely: a) incompressible three-dimensional laminar flows, b) incompressible two-dimensional laminar flows undergoing separation, and c) compressible three-dimensional nonsimilar laminar flows. The analysis of class a) flows, together with the earlier results for hypersonic problems[5], is aimed at establishing the effectiveness of the model over the entire spectrum of flight velocities. The study of class b) flows is mainly concerned with demonstrating the applicability of the model to nonsimilar situations. Finally, the investigation of class c) flows is directed toward the development and demonstration of a computer code for general applications.

In connection with the problems of class a) specific attention has been focused on flows over a planar surface (y=0) characterized by the inviscid velocity fields[6]

$$\underline{\text{Case I}} \qquad U = ax^n \qquad W = apx^m \tag{28}$$

$$\underline{\text{Case II}} \qquad U = ax^n z^{m-1} \qquad W = apx^{n-1} z^m \tag{29}$$

in catersian coordinated (x,y,z). In terms of the independent variable $\eta = y(U/\nu x)^{\frac{1}{2}}$, and of the dependent variables $F = \int_0^\eta (u/U)d\eta$, $G = \int_0^\eta (w/W)d\eta$, the equations describing the self-preserving flows are

$$\underline{\text{Case I}} \quad n(F'^2 - 1) - F''' = [(n+1)/2]\, FF'' \tag{30-a}$$

$$m(F'G' - 1) - G''' = [(n+1)/2]\, FG'' \tag{30-b}$$

$$\underline{\text{Case II}} \quad n(F'^2 - 1) + p\,(m-1)\,(F'G' - 1) - F''' = [(n+1)/2]\, FF'' + p[(m+1)/2]GF'' \tag{31-a}$$

$$pm(G'^2 - 1) + (n-1)(F'G' - 1) - G''' = p[(m+1)/2]\, GG'' + [(n+1)/2]FG'' \tag{31-b}$$

subject to the boundary conditions

$$\eta = 0, \; F = F' = G = G' = 0 \tag{32-a}$$

$$\eta \to \infty, \; F' \to 1, \quad G' \to 1 \tag{32-b}$$

Solution by the two layer approach proceeds along the following lines. The outer equations, readily obtained by linearization of (30-a,b) and (31-a,b), are cast in terms of the independent variable Υ and solved in terms of confluent hypergeometric functions, as described in Ref. 5. The inner equations, obtained by setting the right hand sides of (30-a,b) and (31-a,b) equal to zero, are nonlinear but amenable to approximate solution along the lines presented in the previous section [see eqs. (17-a), (18) and (19)]. Matching of inner and outer solutions is carried out with the realization that the stream functions F and G do not individually describe the streamwise and the crossflow mass fluxes at a general station in the flow.

For Case I the x-momentum equation (30-a) has the form appropriate to strictly two-dimensional flows; it can, therefore, be treated and solved first, without consideration of three-dimensional effects. The crossflow described by $[v_2/(U^2+W^2)^{\frac{1}{2}}] = (G'-F')\, g(x)$ is considered subsequently. The equation is obtained by linear combination of (30-a,b); the solution is determined by the process described in the previous section. Results generated by the two layer model are compared with exact predictions[6] for the wall shear parameters $F''(0)$, and $G''(0)$ in Fig. 2. Although many of the considered solutions are physically unrealistic[7], good agreement [within 3% for $F''(0)$ and 10% for $G''(0)$] is observed between approximate and exact predictions for this class of flows. The success with physically unrealistic solutions indicates that the physically motivated two layer model provides an adequate mathematical representation of the nonlinear boundary layer equations and their solutions. The satisfactory overall comparison confirms its applicability in the analysis of low speed flows.

In the analysis of Case II flows the two equations must be solved simultaneously. The function F is treated as the streamwise mass flux, and the function G as the crossflow, in considerations relating to the x-momentum equation (31-a); the roles of the two functions are inverted in considerations relating to equation (31-b). F and G = F + Constant are used as independent variables in the outer approximations to (31-a) and (31-b), respectively. The matching process is implemented as $F \to G^{(i)}$ [or $G \to F^{(i)}$] for (31-a) [or(31-b)], in accord with the criteria evolved for plane of symmetry problems[5]. A comparison between predictions of the two layer model and exact solutions for the shear parameters $F''(0)$ and $G''(0)$ (Fig. 3) indicates satisfactory agreement, within the limits noted for Case I. The conclusions about applicability and significance of the model are thus reiterated, at least for self-preserving flows.

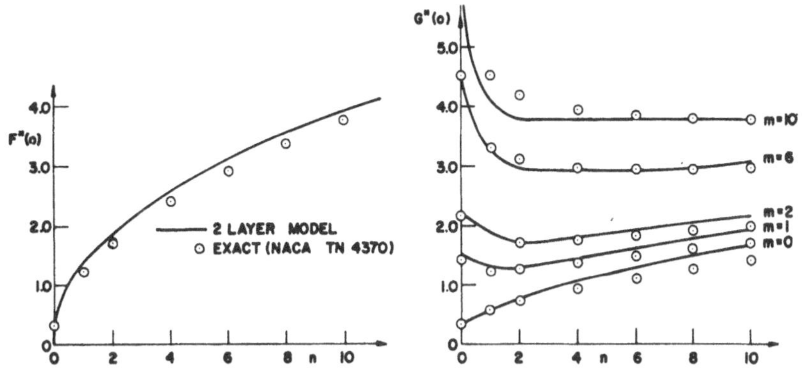

Fig. 2 Wall shear parameters for incompressible self-preserving flows, Case I

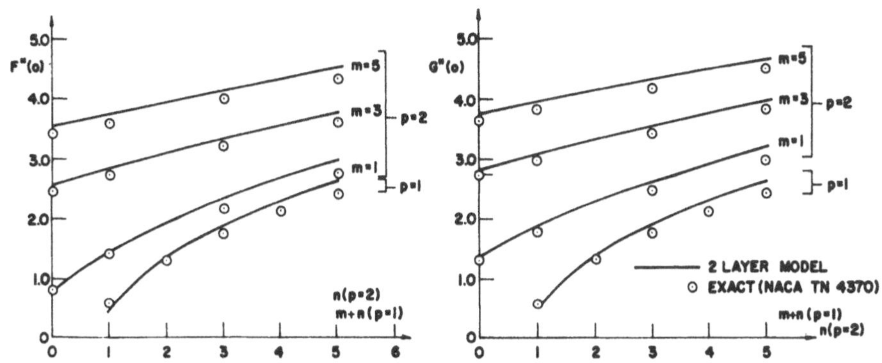

Fig. 3 Wall shear parameters for incompressible self-preserving flows, Case II

The questions posed by the analysis of nonsimilar problems are resolved by calculations and comparisons for incompressible two-dimensional laminar flows undergoing separation. The test can be approached with confidence as the present model represents a variation and extension of those originally proposed and employed by Von Karman - Millikan, Stratford and others for two-dimensional problems[8]. The point of view is supported by the results shown in Fig. 4 for the classical flows defined by the inviscid velocity distributions $U = U_o[1-(x/\ell)]$, $U = U_o[1-(x/\ell)^2]$, and $U = U_o[1-(x/\ell)^4]$. The streamwise distributions of skin friction [wall shear parameter $F''(0)$] and the positions of separation predicted by the method of the previous section (specialized to two-dimensional problems) are compared in the figure with the exact results[8] for the separation point. The agreement is reasonable, within 8% or better. Surprisingly the largest departures arise in the case of mildest adverse pressure gradient and longest run; sources of small cumulative errors should be checked accordingly.

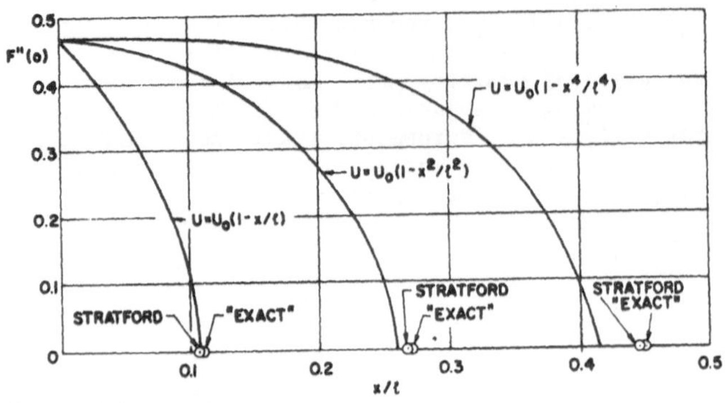

Fig. 4

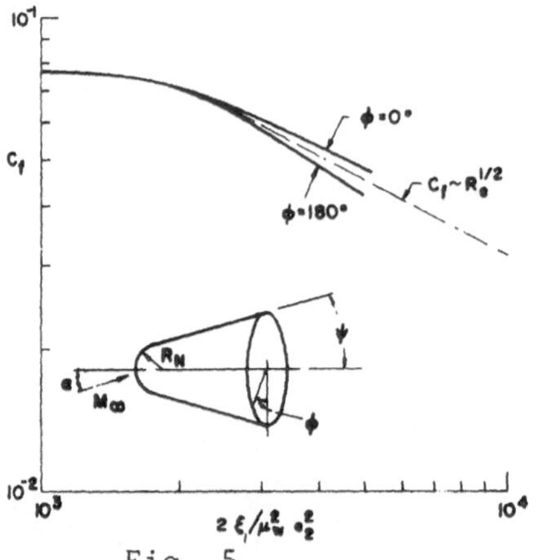

Fig. 5

The success of the two layer model in describing the various flows reported here and in Ref. 5 has motivated the development of a computer code which implements the analysis of three-dimensional nonsimilar flows and in particular, the numerical matching process outlined in the previous section. Some partial results obtained from this code are presented in Fig. 5. There is considered the three-dimensional laminar flow around a spherically capped, 15° half - angle cone at 15° incidence; the free stream is characterized by $M_\infty = 5$, $Re_{\infty R_N} = 10^4$ (R_N being the nose radius), the wall is assumed adiabatic, and entrainment of the rotational external flow into the boundary layer is neglected. The diagram in Fig. 5 exhibits the distributions of skin friction coefficient, $c_f = 2\tau_w/(\rho_e v_{1e}^2)$, along the windward and leeward generators for stations $(s/R_N) \leq 3$ from the stagnation point. The results are in reasonable agreement with predictions based on local similarity; however, a thorough assessment, involving comparisons with other numerical analyses[2] and with experiments, is still incomplete.

Although some aspects remain to be tested exhaustively, the results reported here and in Ref. 5 recommend the two layer model as a useful basis for extensive analysis and understanding of three-dimensional boundary layer porblems. On physical grounds the model provides the framework for comprehensive interpretation of results; on numerical grounds it represents a convenient computational method as well as a test for the evaluation of algorithms.

Acknowledgments

The paper is based on research sponsored by the U.S. Air Force, Office of Scientific Research under Grant AF-AFOSR-68-1551. The authors express their appreciation to M. Salita for providing the numerical results on incompressible three-dimensional self-preserving flows.

References

1) Der, J., Jr. and Raetz, G.S., Inst. Aero/Space Sci. Paper 62-70 (1962).

2) Der, J., Jr., AIAA Paper 69-138 (1969).

3) Dwyer, H.A., AIAA J., 6, p. 1336 (1968).

4) Vaglio-Laurin, R., in "Hypersonic Boundary Layers and Flow Fields", AGARD Conf. Proc. No. 30, Paper No. 25 (1968).

5) Vaglio-Laurin, R. and Miller, G., AIAA J., 8, (1970).

6) Yohner, P.L., and Hansen, A.G., NACA TN 4370 (1958).

7) Kovasznay, L.S.G., and Hall, M.G., AIAA J., 5, p. 2065 (1967).

8) Rosenhead, L., Ed., Laminar Boundary Layers, Ch. VI, Parts III and IV, pp. 318-331, Clarendon Press, Oxford (1963).

NUMERICAL SIMULATION OF TRANSITION AND TURBULENCE IN PLANE POISEUILLE FLOW

T. H. Gawain and G. D. O'Brien, Jr.
Naval Postgraduate School, Monterey, Calif.

1. THE PROBLEM

In a plane Poiseuille flow with constant flow rate, let Reynolds number R and other quantities be referred to channel half-width and mean velocity. Let x be the streamwise coordinate, y the transverse coordinate, and t time. Let U denote laminar velocity, ψ perturbation vorticity.

The following exact relations apply

$$U = \frac{3}{2}(1-y^2) \qquad U'' = -3 \tag{1-1}$$

$$\Gamma = \nabla^2 \psi \tag{1-2}$$

$$\Gamma_t = \frac{1}{R}\nabla^2 \Gamma - U\,\Gamma_x + U''\,\psi_x + Q \tag{1-3}$$

$$Q = \psi_y\,\Gamma_x - \psi_x\,\Gamma_y \tag{1-4}$$

$$\psi = 0 \text{ and } \psi_y = 0 \text{ at } y = \pm 1 \tag{1-5}$$

The perturbation is taken as periodic in x with arbitrary but large wave length.

For small perturbations, (1-3) reduces to the linear Orr-Sommerfeld equation. This has known solutions of the form

$$\psi = \varphi(y)\, e^{i(\alpha x+\beta t)} \tag{1-6}$$

For given values of R and α, these solutions exist only for certain characteristic values of the complex constant β. Only one of these roots represents an unstable, growing disturbance.

The linear case can be formulated as an algebraic eigenvalue problem and solved by means of the QR algorithm (O'Brien, Wilkenson). This method yields all the characteristic solutions, although somewhat less accurately than is possible with more restricted techniques.

The unstable linear solution provides a suitable initial condition for the numerical integration of the full nonlinear equations. The subsequent evolution of the perturbation can thus be simulated. That is the objective.

Note that the two dimensional turbulence considered here resembles three dimensional physical turbulence in some respects, but differs in others.

2. METHOD OF SOLUTION

A rectangular MxN grid is used. Indices k, j, and ℓ denote stations along x, y and t, respectively, and δx, δy, δt are the corresponding intervals.

Overbars denote time averages defined as follows

$$\overline{\Gamma}^{\ell}_{k,j} = \tfrac{1}{2}\left(\Gamma^{\ell}_{k,j} + \Gamma^{\ell+1}_{k,j}\right) \tag{2-1}$$

Symbols Δ_x, Δ_y, Δ_t denote ordinary first central differences, and $\overline{\Delta}_t$ denotes a forward difference.

The operators Δ^2 and $\overline{\Delta^2}$ are defined as follows

$$\Delta^2\Gamma^{\ell}_{k,j} = \frac{\Gamma^{\ell}_{k+1,j} - 2\Gamma^{\ell}_{k,j} + \Gamma^{\ell}_{k-1,j}}{(\delta x)^2} + \frac{\Gamma^{\ell}_{k,j+1} - 2\Gamma^{\ell}_{k,j} + \Gamma^{\ell}_{k,j-1}}{(\delta y)^2} \tag{2-2}$$

$$\overline{\Delta^2}\Gamma^{\ell}_{k,j} = \frac{\Gamma^{\ell}_{k+1,j} - (\Gamma^{\ell+1}_{k,j} + \Gamma^{\ell-1}_{k,j}) + \Gamma^{\ell}_{k-1,j}}{(\delta x)^2} + \frac{\Gamma^{\ell}_{k,j+1} - (\Gamma^{\ell+1}_{k,j} + \Gamma^{\ell-1}_{k,j}) + \Gamma^{\ell}_{k,j-1}}{(\delta y)^2} \tag{2-3}$$

With this notation, the finite difference expressions needed for the solution are summarized below, then discussed briefly. Indices are deleted whenever the context leaves the meaning clear. Thus

$$Q = Q^{\ell}_{k,j} = \frac{1}{3}\,[(\Delta_y\psi)(\Delta_x\Gamma) - (\Delta_x\psi)(\Delta_y\Gamma)] + \frac{1}{3}\,[\Delta_y(\psi\Delta_x\Gamma) - \Delta_x\,(\psi\Delta_y\Gamma)]$$

$$+ \frac{1}{3}\,[\Delta_x\,(\Gamma\Delta_y\psi) - \Delta_y\,(\Gamma\Delta_x\psi)] \tag{2-4}$$

$$\overline{\Delta}_t\Gamma = \frac{1}{R}\,\Delta^2\,\overline{\Gamma} - U\Delta_x\overline{\Gamma} + U''\Delta_x\overline{\psi} + \overline{Q} \tag{2-5}$$

$$\Delta_t\Gamma = \frac{1}{R}\,\overline{\Delta^2}\,\Gamma - U\Delta_x\Gamma + U''\Delta_x\psi + Q \tag{2-6}$$

$$\Delta^2\,\psi = \Gamma \tag{2-7}$$

$$\theta_{n,k} = \frac{2\pi(n-1)(k-1)}{M} \qquad n,k = 1,2,3\ldots M \tag{2-8}$$

$$G_{n,j} = \sum_{k=1}^{M} \Gamma_{k,j}\, e^{+i\theta_{n,k}} \qquad j = 2,3,4\ldots(N-1) \tag{2-9}$$

$$\alpha_n = 2\left\{1 + \left(\frac{\delta y}{\delta x}\right)^2 \left[1 - \cos\frac{2\pi(n-1)}{M}\right]\right\} \tag{2-10}$$

$$F_{n,j+1} - \alpha_n\, F_{n,j} + F_{n,j-1} = G_{n,j}(\delta y)^2 \tag{2-11}$$

$$\psi_{k,j} = \frac{1}{M}\sum_{n=1}^{M} F_{n,j}\, e^{-i\theta_{n,k}} \tag{2-12}$$

$$\psi_{k,1} = \psi_{k,N} = 0 \tag{2-13}$$

$$\Gamma_{k,1} = \frac{1}{2(\delta y)^2}\,[8\,\psi_{k,2} - \psi_{k,3}] \qquad \Gamma_{k,N} = \frac{1}{2(\delta y)^2}\,[8\,\psi_{k,N-1} - \psi_{k,N-2}] \tag{2-14}$$

$$\delta t = \frac{f\,\delta x\,\delta y}{(U+u)_0\delta y + v_0\delta x} \qquad f = 0.6 \tag{2-15}$$

Eq. (2-2) defines the ordinary Laplacian differential operator Δ^2, and (2-3) is the modified form $\overline{\Delta^2}$ of DuFort-Frankel. Eq. (2-4) denotes the quadratic vorticity transport approximation of Arakawa; this form conserves overall energy and mean square vorticity.

The calculation starts with a forward difference (2-5), but is continued with central differences (2-6) which are far more efficient. Eq. (2-6) requires the operator $\overline{\Delta^2}$ rather than Δ^2 for stability. Even so, it is subject to a possible weak instability associated with the central operator Δ_t. The stability is suppressed by reverting briefly to (2-5) at regular intervals.

Eq. (2-7) is the difference analogue of (1-2). It represents a set of simultaneous algebraic equations for the ψ's. Because of the many mesh points, conventional iteration methods are inadequate here. Instead, Eqs. (2-8) thru (2-12) provide a solution which is many times faster and far more accurate.

Eqs. (2-9) and (2-12) are computed by the fast Fourier algorithm of Cooley and Tukey. Eqs. (2-11) are tri-diagonal, so are easily solved. Sequence (2-8) thru (2-12) is done in double precision.

Spectral data such as the coefficients $G_{n,j}$ generated by this method are themselves of considerable interest. The method also ensures the required periodicity in x.

Eqs. (2-13) and (2-14) fix conditions at the walls, j=1 and j=N thru approximations of the same order as (2-7). They satisfy the wall conditions (1-5) implicitly.

Eq. (2-15) is a semi-empirical stability limit on δt. Velocity components $(U+u)_0$ and v_0 are the maximum absolute magnitudes over a large sample of the mesh points. The value f = 0.6 consistently produces satisfactory results. Values near unity may lead to catastrophic instability. The time step δt is readjusted according to (2-16) at regular intervals.

3. SOME TYPICAL RESULTS

Extensive results on the linear problem calculated with the QR algorithm are reported in O'Brien's dissertation. These include eigenvalues, eigenfunctions, and stability contours in the Rα plane. This work generally confirms and also extends earlier results obtained by Thomas, Grosch and others, by quite different methods.

Fig. 3-1 is an arbitrary but typical example. It shows the various eigenvalues of β in the complex plane for the indicated values of R and α.

O'Brien's linear stability calculations are summarized in Fig. 3-2. Using 201 grid points gives good agreement with published data. Reducing the number of grid points markedly narrows the region of apparent instability in the Rα plane. When 41 grid points are used, all evidence of linear instability is entirely lost!

O'Brien's calculations also confirm that if the non-linear terms in (2-5) and (2-6) be suppressed, the disturbance grows at precisely the constant exponential rate predicted by the linear theory, provided that the original perturbation is an accurate eigenfunction. Moreover, even if there are initial inaccuracies, these attenuate automatically during an initial transient, after which the flow develops precisely as indicated by the linear theory. These results inspire some confidence in the numerical model.

Fig. 3-3 summarizes some of O'Brien's calculations using the full non-linear solution. It shows the gradual change in dimensionless total kinetic energy for initial perturbations of three widely different amplitudes. The total kinetic energy includes not only the energy of turbulence, but also that of the mean flow. The three curves seem to tend toward a common equilibrium state. More extensive calculations are needed to confirm this trend.

A further energy break-down is given in Fig. 3-4. While the total remains substantially constant, there is a continual oscillatory exchange of energy between the mean flow and the turbulence. However, this oscillation attenuates as the system approaches equilibrium. Also, while this is not shown on the diagram, the data indicate that most of the turbulent energy is in the primary mode.

The calculation provides much detailed spectral information. Fig. 3-5 shows a typical example, the energy spectrum at arbitrary station y = $\pm$ 0.85. The computed points are reasonably consistent with an inverse cubic law, somewhat as suggested by Kraichnan.

Space limitations preclude a fuller discussion of results here, but the above examples illustrate the capabilities of the method.

4. REFERENCES

Arakawa, A. J. Comp. Phys., 1, 1, pp. 119-143 (1966)
Cooley, J. W., and Tukey, J. W., Math. Comp.,19, 90, pp. 297-301 (1965)
Grosch, C. E., and Salwen, H., J. Fluid Mech., 34, 1, pp. 177-205 (1968)
Kraichnan, R. H., Phys. Fluids, pp. 10, 7, (1967)
O'Brien, G. D., Jr., A Numerical Investigation of Finite Amplitude Disturbances in a Plane Poiseuille Flow, Ph.D. dissertation. Naval Postgraduate School, (1970)
Thomas, L. H., Phys. Rev., 91, 4, pp. 780, (1953)
Wilkinson, J. A., The Algebraic Eigenvalue Problem, pp. 485-569, Clarendon Press (1965)

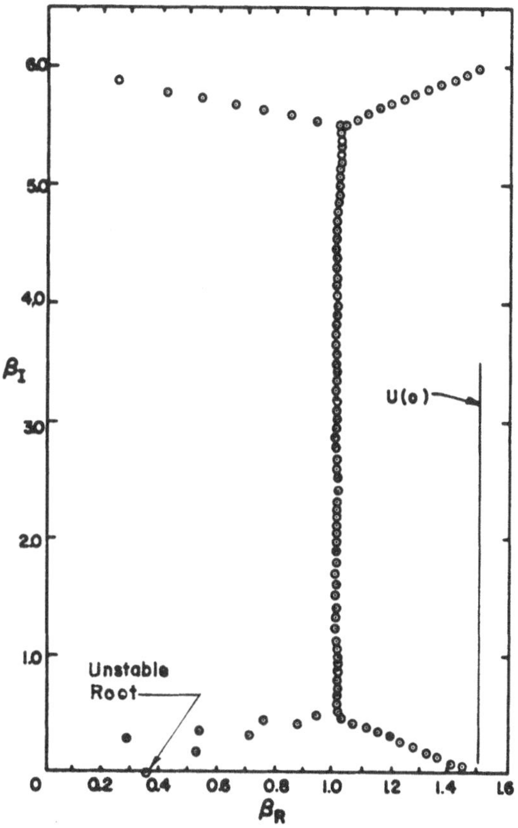

Fig. 3-1 Calculated Eigenvalues of Orr-Sommerfeld Equation. Symmetric Modes.
R = 6667 α = 1 N = 201

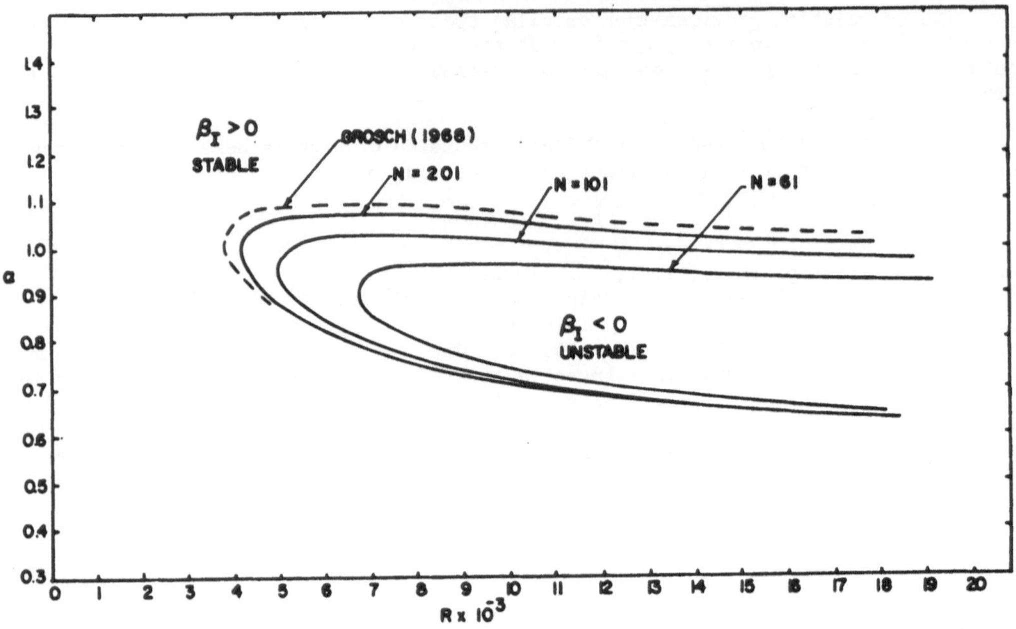

Fig. 3-2 Effect of Mesh Size on Apparent Stability Boundaries

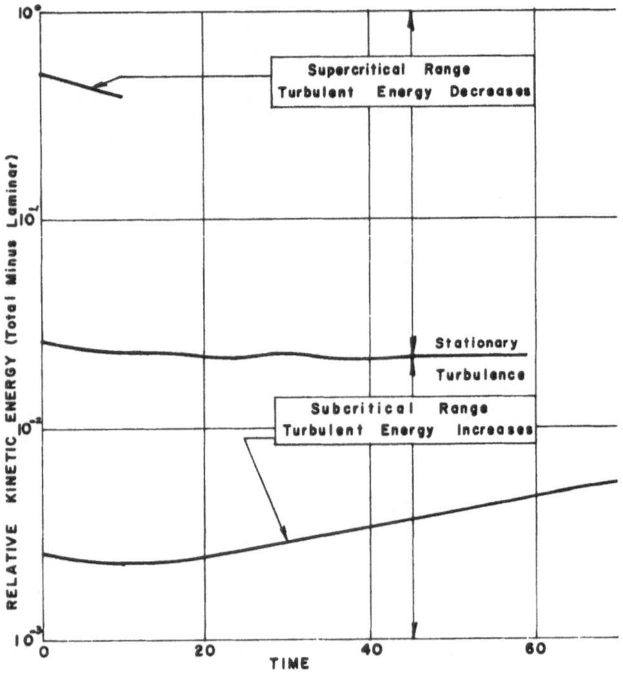

Fig. 3-3 Relative Kinetic Energy versus Time for Three Initial Perturbation Amplitudes

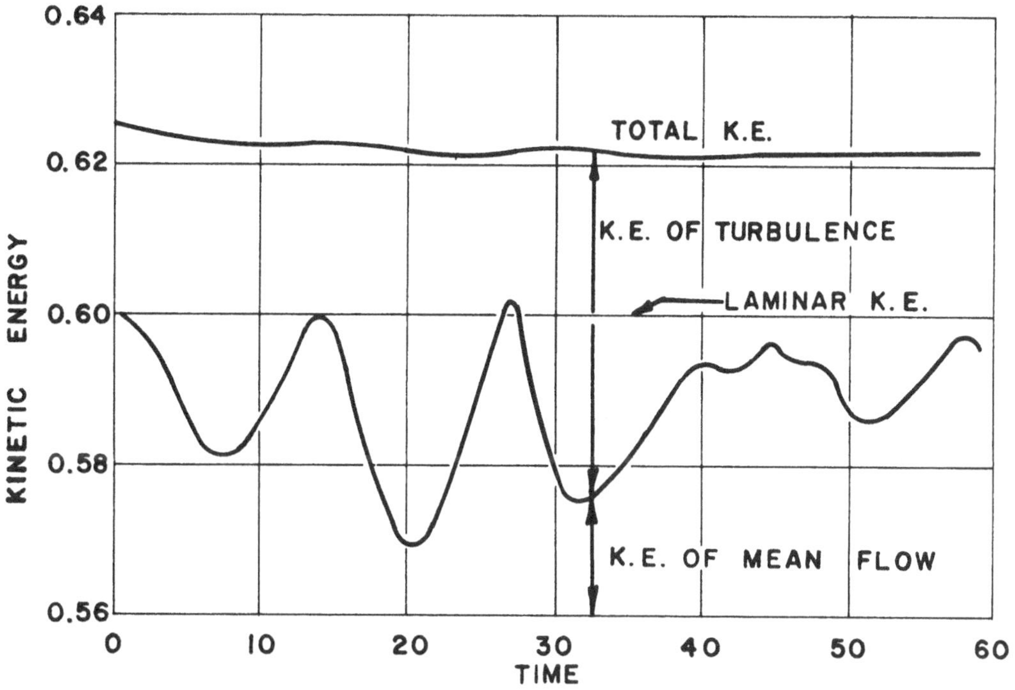

Fig. 3-4 Kinetic Energy of Turbulence and of Mean Flow versus Time

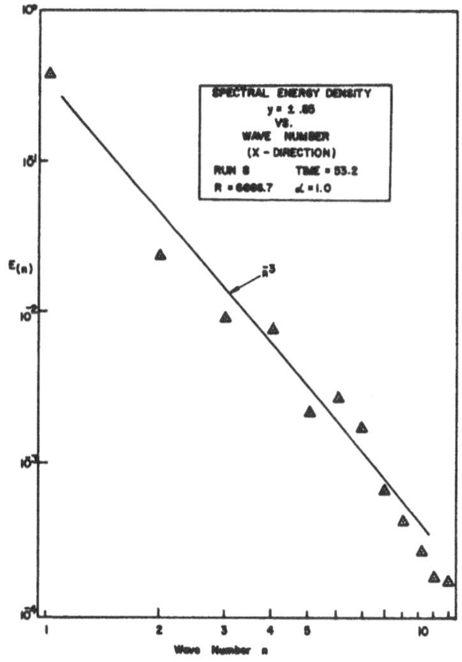

Fig. 3-5 Typical Energy Spectrum Near Wall

HEURISTIC ANALYSIS
OF
CONVECTIVE FINITE DIFFERENCE TECHNIQUES+

L. D. Tyler
Dynamic Analysis Division
Sandia Laboratory, Albuquerque, New Mexico

Abstract

A heuristic analysis is made of the Lax, Rusanov, Lax-Wendroff two-step and MacCormack techniques to determine truncation, stability and utility conditions for practical application of the methods. A simple artificial viscosity is introduced for compression region in the second order techniques, which reduced the overshoot at shock waves to less than 0.2% of the waves' defined strength.

INTRODUCTION

In the past several years a number of finite difference techniques have been developed which may be used to solve transient inviscid flows involving propagating shock waves. In this paper the Lax, Rusanov, Lax-Wendroff two-step and MacCormack techniques are examined using the heuristic approach presented by Hirt. From this analysis a better understanding is gained of the different techniques, particularly practical conditions for utilizing the techniques. These practical conditions are referred to as utility conditions. Because of the desirability of second order accurate schemes for nondissipative processes, such as rarefactions waves, the second order techniques are modified to eliminate the undesirable overshoot problems at shock fronts. An artificial viscosity model is introduced for compression process which greatly reduces and eliminates, in some cases, the overshoot present at the shock wave.

HEURISTIC ANALYSIS

A simple analysis is made of the four techniques utilizing a procedure recently used by Hirt[1] to study the stability of difference methods. The analysis has been used to determine consistency conditions for partial differential equations by Richtmyer and Morton[9]. The analysis in this paper is applied to the difference equations for the simple linear convective equation

$$\frac{\partial f}{\partial t} + c\,\frac{\partial f}{\partial x} = 0 \qquad (1)$$

where c is a positive constant. The difference equations for the four techniques applied to equation (1) are given in Table I. The Courant number is defined as

TABLE I

Lax[5]	$f_j^{n+1} = f_j^n - \frac{\sigma}{2}\left(f_{j+1}^n - f_{j-1}^n\right) + \frac{1}{2}\left(f_{j+1}^n - 2f_j^n + f_{j-1}^n\right)$
Rusanov[10]	$f_j^{n+1} = f_j^n - \frac{\sigma}{2}\left(f_{j+1}^n - f_{j-1}^n\right) + \frac{\omega\sigma}{2}\left(f_{j+1}^n - 2f_j^n + f_{j-1}^n\right)$
Lax-Wendroff Two-Step;[8] MacCormack[7]	$f_j^{n+1} = f_j^n - \frac{\sigma}{2}\left(f_{j+1}^n - f_{j-1}^n\right) + \frac{\sigma^2}{2}\left(f_{j+1}^n - 2f_j^n + f_{j-1}^n\right)$

+This work was supported by the U.S. Atomic Energy Commission.

$\sigma = c\Delta t/\Delta x$. The Lax technique has been written in a slightly different form to reflect a similarity between the difference equation. Both the Lax-Wendroff two-step and MacCormack methods reduce to the same difference equation for the linear differential equation, which is the same as the difference equation for the Lax-Wendroff single step technique. An examination of the difference equations in Table I shows only a difference of coefficients in the last term on the right hand side of the equations. This similarity might allow the misinterpretation that the partial differential equation being solved has the form

$$\frac{\partial f}{\partial t} + c\,\frac{\partial f}{\partial x} = A\,\frac{\partial^2 f}{\partial x^2} \qquad (2)$$

where A is a function of Δt, Δx and a constant. When an examination of this kind is made it may give the wrong equation or the wrong function for the coefficient A. The proper form of the differential equation will be one of the results of the following analysis.

The essence of the analysis is to expand the dependent variable of the difference equation in both the time and space variable in Taylor series about the point $j\Delta x$, $n\Delta t$ and substitute the series into the difference equation. The Taylor series for the variable f_j^{n+1} has the form

$$f_j^{n+1} = f_j^n + \Delta t\,\frac{\partial f}{\partial t} + \frac{(\Delta t)^2 c^2}{2!}\,\frac{\partial^2 f}{\partial x^2} + \frac{(\Delta t)^3 c^3}{3!}\,\frac{\partial^3 f}{\partial x^3} + \frac{(\Delta t)^4 c^4}{4!}\,\frac{\partial^4 f}{\partial x^4} - O(\Delta t^5)$$

where the higher order time derivatives have been replaced by space derivatives using equation (1). Substitution of this series and the series for f_{j+1}^n and f_{j-1}^n into the difference equations in Table I give partial differential equations which have the general form

$$\frac{\partial f}{\partial t} + c\,\frac{\partial f}{\partial x} = T \qquad (3)$$

where the function T represents the higher order terms of the substituted series. The T functions for the difference equations of Table I are given in Table II.

TABLE II

Lax

$$\frac{c(\Delta x)}{2!}\left(\frac{1}{\sigma} - \sigma\right)\frac{\partial^2 f}{\partial x^2} - \frac{c(\Delta x)^2}{3!}(1 - \sigma^2)\frac{\partial^3 f}{\partial x^3} + \frac{c(\Delta x)^3}{4!}\left(\frac{1}{\sigma} - \sigma^3\right)\frac{\partial^4 f}{\partial x^4} + O(\Delta t^4) - O(\Delta x^4)$$

Rusanov

$$\frac{c(\Delta x)}{2!}(\omega - \sigma)\frac{\partial^2 f}{\partial x^2} - \frac{c(\Delta x)^2}{3!}(1 - \sigma^2)\frac{\partial^3 f}{\partial x^3} + \frac{c(\Delta x)^3}{4!}(\omega - \sigma^3)\frac{\partial^4 f}{\partial x^4} + O(\Delta t^4) - O(\Delta x^4)$$

Lax Wendroff 2-S
MacCormack

$$-\frac{c(\Delta x)^2}{3!}(1 - \sigma^2)\frac{\partial^3 f}{\partial x^3} + \frac{c(\Delta x)^3}{4!}(\sigma - \sigma^3)\frac{\partial^4 f}{\partial x^4} + O(\Delta t^4) - O(\Delta x^4)$$

The solution of equation (3) is the same as the numerical solution of the difference equation at the point $j\Delta x$, $n\Delta t$. The T function represents the difference between the original differential equation (1) and the partial differential equation which the finite difference equation actually solves. The two differential equations become identical only if the T functions vanish. A comparison of equations (2) and (3) shows that the coefficient A would have the wrong form for the Lax and Rusanov techniques if higher order terms are neglected and equation (2) is not the partial differential equation being solved by the second order techniques. In the following analysis the T function is inspected to determine truncation, stability and utility conditions with the goal of determining the conditions needed to minimize the T function for practical application.

The benefit of examining the T function for truncation error comes when an "ad hoc" difference formula has been used in which an explicit truncation error is not defined. For the Lax equation the truncation error is $O(\Delta t)$, if $\Delta x = O(\Delta t)$, which shows the method to be first order accurate in both space and time. The Rusanov equation is an "ad hoc" example of differencing because of the parameter ω. If $\omega = 1/\sigma$, the Rusanov equation reduces to the Lax and is first order accurate in time and space. If $\omega = O(\Delta x)$ the method is first order accurate in time but second order accurate in space. If $\omega = \sigma$, the Rusanov equation reduces to the Lax-Wendroff two-step and MacCormack equation. The equation is second order accurate in time and space.

To determine the stability condition using this analysis, Hirt[1] has shown that the coefficient on the lowest even order derivative must be positive for a stable solution to exist. Examination of the Lax and Rusanov T functions shows the lowest even order derivative to be second order, allowing a large amount of numerical diffusion for a stable solution. Since c and Δx are positive quantities, the terms in the parenthesis must be positive. Therefore, the stability condition for the Lax and Rusanov are given to be $\sigma < 1$ and $\sigma < \omega$, respectively. The Lax condition is the same as obtained by a linear Fourier stability analysis. The Rusanov condition is less restrictive than the Fourier analysis which gives $\sigma \leq 1$ and $\sigma \leq \omega \leq 1/\sigma$. A few simple numerical experiments show that the condition given by the T function is necessary but not sufficient for stability and a Fourier analysis is needed to determine limits for the parameter ω. The lowest even order derivative in the Lax-Wendroff and MacCormack T function is fourth order which indicates weak numerical diffusion. The stability condition for the two techniques is the $\sigma < 1$, which is consistent with the Fourier analysis.

When a computation is performed it is desirable for the T function to vanish or be very small. The conditions for this to occur are called utility conditions. Utility conditions are defined as the conditions by which the T function may vanish or be reduced to a small value when the finite difference equations are used in a stable computation. Theoretical studies of the T function have been made as Δt vanishes which give consistency conditions. These are not utility conditions since the limit of $\Delta t \to 0$ may be approached but not realized in a computation. Examination of the coefficients of the derivatives in the T function of Table II reveal two utility conditions. One of the utility conditions is to reduce the space increment Δx to a very small value. The quantity Δx is usually normalized by a characteristic length L. For a multidimensional computation the condition $\Delta x = O(0.1L)$ is very common. To reduce Δx by one order of magnitude may be almost impossible because of the storage requirements or computational speed of a given computer. The second utility condition is not as dependent on the computer as the nature of the mathematics. An inspection of the Courant number, within the limits of stability, reveals that a value of $\sigma = O(1)$ reduces the coefficients in all of the T functions in Table II. This condition is not too surprising since the characteristic solution of equation (1) is $\frac{dt}{dx}c = 1$. A combination of these conditions may allow a reasonable reduction of the T function. Consider $\Delta x = 0.1L$ and $\sigma = 0.95$ the coefficient for the lowest terms becomes $10^{-3}\frac{Lc}{2!}$ for the Lax equations, $0.5 \times 10^{-3}\frac{Lc}{2!}$ for Rusanov with $\omega = 1.0$ since $0.95 \leq \omega \leq 1.05$, and $0.975 \times 10^{-3}\frac{Lc}{2!}$ for the second order techniques.

NUMERICAL RESULTS

In this section the utility conditions are used to obtain finite difference solutions for one dimensional transient compressible flow phenomena. All results are for the conservation form of the mass, momentum and energy equations in Eulerian coordinates using an ideal gas equation of state. The results were obtained using a Courant number of 0.95 and an ω = 1.0 for the Rusanov method. The shock thickness for a plane Mach 3 shock wave using the Lax and Rusanov techniques was 8Δx and 3 to 4Δx, respectively, with no overshoot at the shock front. The results for the second order techniques are shown in Figure 1. The Lax-Wendroff method gave a shock thickness of 6Δx through the shock to the uniform state behind with 3Δx at the front. The overshoot in pressure was a minimum of 16% and maximum of 20%. The permuted indices

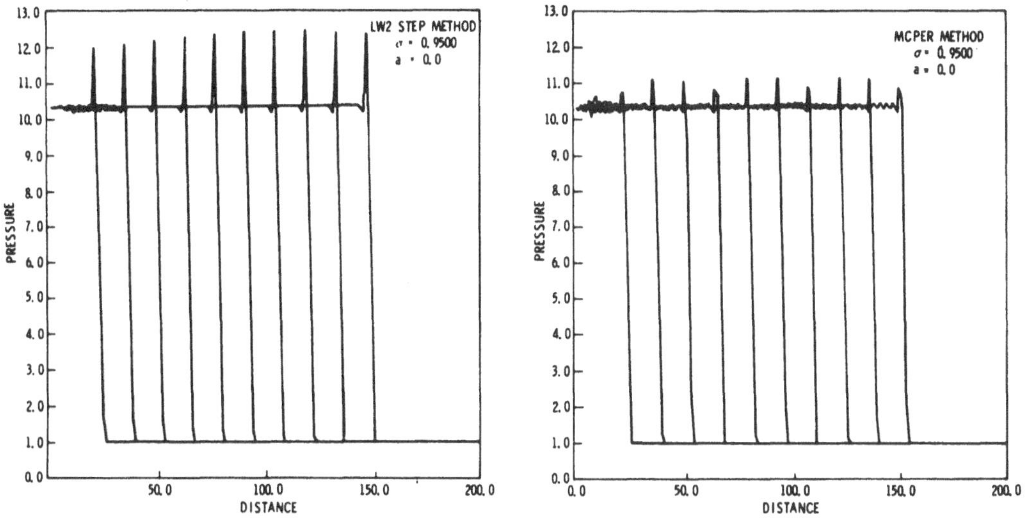

FIGURE 1

version of the MacCormack method was used for all calculations and gave a shock thickness of 6Δx across the shock to the uniform flow with 3Δx at the front. The pressure overshoot ranged from a minimum of 4% to a maximum of 8%. Lax and Wendroff[6] and Lapidus[4] have introduced artificial dissipation into the Lax-Wendroff two-step method in an attempt to eliminate the overshoot. The dissipation was added to all the governing flow equations. This approach allowed the overshoot to be reduced but not eliminated. It was found in this paper that using the concept of an artificial viscosity as defined by Von Neumann and Richtmyer[11] the overshoot is greatly reduced and in some cases eliminated for the second methods. A linear artificial viscosity is defined as

$$q = a\ \Delta x \rho(|u| + c)\frac{\partial u}{\partial x}$$

where a is a constant and ρ is density with u and c, respectively, the local velocity and sound speed and added to the pressure terms in the momentum and energy equations for compression processes only. This artificial viscosity is a linear combination of Landshoff[3] and Evans and Harlow[2] artificial viscosity. The results using this artificial viscosity are shown in Figure 2. The shock thickness for the Lax-Wendroff two-step and MacCormack results are, respectively, 2 to 3Δx and 3 to 4Δx with a maximum overshoot of 0.18%. A shock tube problem was computed to determine the effectiveness of the difference techniques on other phenomena and to test the effect of the artificial viscosity on these phenomena. The diaphragm pressure ratio was 20 for an isothermal initial condition. Both the Lax and Rusanov methods gave shock characteristics comparable to the plane wave case, but the rarefaction wave and contact discontinuity were diffused to the extent the phenomena was not definable. The results for the second order techniques are shown in Figure 3*. The Lax-Wendroff and MacCormack results had overshoots of 23% and 7%, respectively, with a shock thickness of 4 to 5Δx. Good agreement exists between the exact and numerical solutions for the rarefaction wave except for a 2% undershoot at the tail in the MacCormack solution.

*To start the MacCormack calculation the diaphragm pressure ratio was defined across 2Δx and a backward-forward computation was done for the first 5 time cycles.

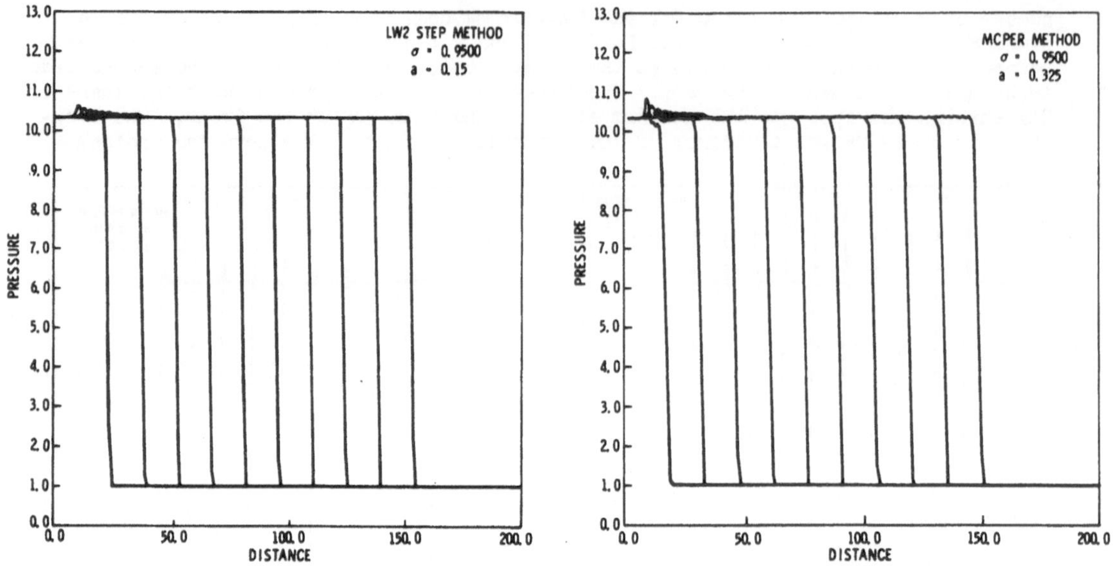

FIGURE 2

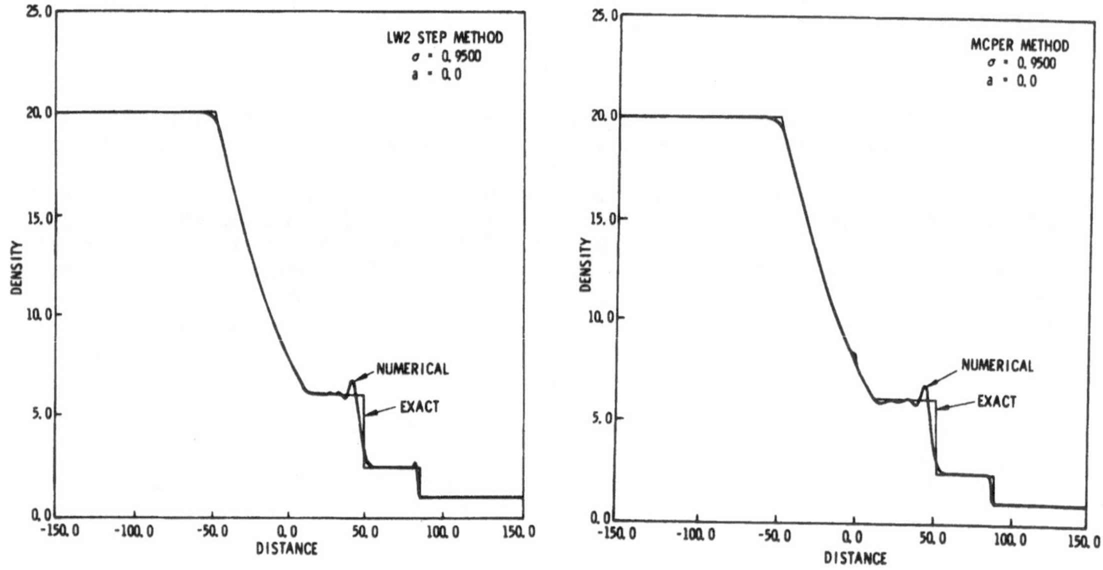

FIGURE 3

The results using the artificial viscosity gave no overshoot at the shock wave as shown in Figure 4 and the shock thickness was 2 to $3\Delta x$ for both methods. The rarefaction is well defined with a pressure undershoot of 2% at the tail in both solutions. The contact discontinuity is better defined than for the first order techniques but is still quite smeared.

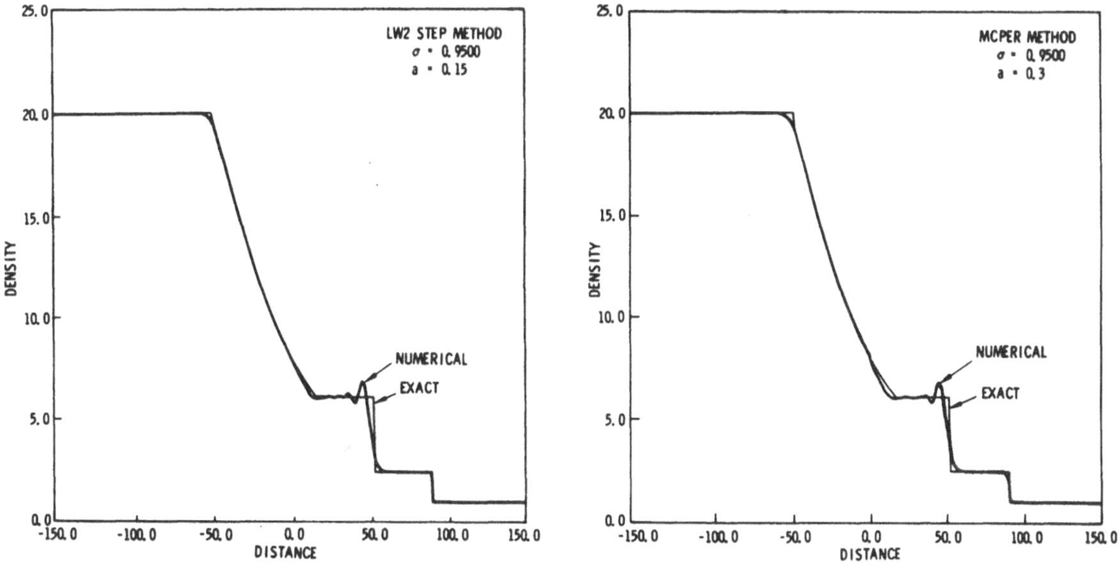

FIGURE 4

CONCLUSIONS

A heuristic analysis has been performed which allows the definition of truncation error for "ad hoc" difference methods and utility conditions to be determined to give better computational results. An artificial viscosity has been presented which eliminates the undesirable overshoot at shock fronts for second order difference techniques.

BIBLIOGRAPHY

1. Hirt, C. W., J of Computation Physics 2, 339-355 (1968)
2. Evans, M. and Harlow, F., LA-2139 (1957)
3. Landshoff, R., LA-1930 (1955)
4. Lapidus, A., J of Computation Physics 2, 154-177 (1967)
5. Lax, P., Comm Pure Appl Math 7, 159-193 (1954)
6. Lax, P. and Wendroff, B., Comm Pure Appl Math 13, 217-237 (1960)
7. MacCormack, R. W., AIAA Hypervelocity Impact Conference, Paper 69-354 (1969)
8. Richtmyer, R. D., NCAR Tech Note 63-2 (1962)
9. Richtmyer, R. D. and Morton, K. W., Difference Method for Initial Value Problems (1967)
10. Rusanov, V. V., USSR Computational Math and Math Physics 2, (1962)
11. Von Neumann. J. and Richtmyer, R. D., J of Applied Physics 21, 232-237 (1950)

THE CONICAL WING IN HYPERSONIC FLOW

A. L. Gonor, V. I. Lapygin and N. A. Ostapenko
Institute of Mechanics, Moscow State University, U.S.S.R.

Supersonic and hypersonic flow past a V-shaped conical wing with a shock attached to the leading edge is considered.

It is observed in experiments [1 - 4] that in the windward part of flow about V-shaped wings a complex system of shocks exists. No methods for calculation of such flows are available. Only a few exact solutions of the inverse problem are known when, for special geometric forms of the wing, some simple regimes of flow exist [5 - 7].

Two approaches are developed for solution of the general problem. In the first a time-dependent method is used in which the difference scheme contains an artificial viscosity and the role of time is played by one of the space coordinates with respect to which the stationary system of the equations is hyperbolic.

In the second approach an analytical method is used; one developed recently for investigation of hypersonic flow past tringular wings with supersonic leading edges [8,9].

NUMERICAL METHOD

1. Basic equations and difference scheme. Let us consider flow about a V-shaped wing set at an angle of attack (Fig. 1). The coordinates used are

$$\xi = \ln x \quad , \quad \eta = y/x \quad , \quad \zeta = z/x$$

The equations of motion in difference form are

$$\frac{\partial f}{\partial \xi} + \frac{\partial}{\partial \eta}(F^y - \eta f) + \frac{\partial}{\partial \zeta}(F^z - \zeta f) + 2f = 0$$

$$f = \begin{Bmatrix} \Omega \\ R \\ S \\ T \\ E \end{Bmatrix} , \quad F^y = \begin{Bmatrix} S/u \\ S \\ R-\Omega u+S^2/\Omega u \\ TS/\Omega u \\ ES/\Omega u \end{Bmatrix} , \quad F^z = \begin{Bmatrix} T/u \\ T \\ TS/\Omega u \\ R-\Omega u+T^2/\Omega u \\ ET/\Omega u \end{Bmatrix} \qquad (1.1)$$

$$\Omega = \rho u \quad , \quad R = P + \rho u^2 \quad , \quad S = \rho uv \quad , \quad T = \rho uw$$

$$E = \rho u(e + P/\rho + (u^2+v^2+w^2)/2)$$

u, v, w - the components of velocity along the axis x, y, z; e - the internal energy of gas. The system (1.1) is in dimensionless form. The velocity components are referred to the velocity of the undisturbed flow, the pressure P - to the dynamic pressure, the density ρ - to the density of undisturbed flow.

Consider a vector

$$\phi = \begin{Bmatrix} u \\ v \\ w \\ p \\ \rho \end{Bmatrix}$$

which is connected with the components of the vector f by

$$u = kR/(k+1)\Omega + \{k^2R^2/(k+1)^2\Omega^2 - 2(k-1)[E/\Omega - (S^2+T^2)/2\Omega]/(k+1)\}^{1/2}$$

$$v = S/\Omega \quad , \quad w = T/\Omega \quad , \quad P = R - \Omega u \quad , \quad \rho = \Omega/u \tag{1.2}$$

The x-axis is chosen in such a way that $u > a$, where $a = \sqrt{(kP/\rho)}$ is the local velocity of sound.

The boundary conditions:

On the wing surface $\psi(\xi, \eta, \zeta) = \eta - g(\zeta) = 0$ the solid boundary condition is

$$(v - u\eta) - (w - \zeta u)g' = 0 \tag{1.3}$$

On the plane of symmetry $\zeta = 0$

$$w = 0 \tag{1.4}$$

In the undisturbed flow

$$u = \cos\alpha \quad , \quad v = -\sin\alpha \quad , \quad w = 0 \quad , \quad P = \frac{1}{kM_\infty^2} \quad , \quad \rho = 1 \tag{1.5}$$

The problem (1.1), (1.3) - (1.5) is solved by the time-dependent method where the space coordinate ξ plays the role of the time. In the calculations Rusanov's difference scheme [10] was used.

Denote

$$\Delta\xi = \tau \quad , \quad \Delta\zeta = h_1 \quad , \quad \Delta\eta = h_2 \quad , \quad h = \sqrt{h_1^2+h_2^2}$$

$$h_1 = h\cos\chi \quad , \quad h_2 = h\sin\chi \quad , \quad \chi_i = \frac{\tau}{h_i} \quad , \quad \chi = \sqrt{\chi_1^2+\chi_2^2}$$

The value of a quantity A at the mesh point with coordinates $(n\tau, \ell h_2, mh_1)$ is denoted by $A^n_{\ell,m}$

We consider the case when the surfaces of the V-shaped wing are plane and the line intersection of the wing with the plane $x = \text{const}$ goes through the mesh points along diagonals of the mesh cells (Fig. 2).

The system of equations (1.1) is approximated in accordance with [10]

$$f^{n+1}_{\ell,m} = f^n_{\ell,m} - \frac{\chi_2}{2}\{(F^y-\eta f)_{\ell+1,m} - (F^y-\eta f)_{\ell-1,m}\}^n - \frac{\chi_1}{2}\{(F^z-\zeta f)_{\ell,m+1} -$$

$$- (F^z-\zeta f)_{\ell,m-1}\}^n + \frac{1}{2}\{\phi^y_{\ell+\frac{1}{2},m} - \phi^y_{\ell-\frac{1}{2},m} + \phi^z_{\ell,m+\frac{1}{2}} - \phi^z_{\ell,m-\frac{1}{2}}\} - 2\tau^n f^n_{\ell,m}$$

$$\phi^y_{\ell+\frac{1}{2},m} = \frac{1}{2}(\beta_{\ell+1,m}+\beta_{\ell,m})(f_{\ell+1,m}-f_{\ell,m}) \ , \ \beta_{\ell,m} = \chi\omega\sigma_{\ell,m}\cos^2\chi$$

$$\phi^z_{\ell,m+\frac{1}{2}} = \frac{1}{2}(\alpha_{\ell,m+1}+\alpha_{\ell,m})(f_{\ell,m+1}-f_{\ell,m}) \ , \ \alpha_{\ell,m} = \chi\omega\sigma_{\ell,m}\sin^2\chi$$

$$\sigma_{\ell,m} = \{[u(v^2+w^2)^{\frac{1}{2}} + a(u^2+v^2+w^2-a^2)^{\frac{1}{2}}]/(u^2-a^2)\}_{\ell,m} \ , \ \omega = \text{const}$$

The stability analysis of the difference scheme was carried out by Fourier's method. The stability condition is

$$\sigma_o < \omega < (1-\tau)/\sigma_o$$

$$\sigma_o = \max_{\ell,m} \sigma_{\ell,m} \quad , \quad \sigma_o < 1$$

2. <u>The calculation formulae</u>. 1) The point O_1 (Fig. 2) of the plane of symmetry is also on the wing surface. It follows from (1.3) and (1.4) that $T_{o,o} = S_{0,0} = 0$ and at this point it is necessary to find the components of the vector f, R and E only. To derive the calculation formula the following new system of coordinates is used (Fig. 2)

$$\xi = \xi_1 \quad , \quad \eta = \eta_1 + \zeta \sin\chi \quad , \quad \zeta = \zeta_1 \cos\chi$$

The formula is

$$\frac{1}{\chi^n}[f_{o,o}^{n+1} - (1-2\tau^n)f_{o,o}^n] = -\,\mathrm{tg}\,\chi(F^z-\zeta f)_{1,1}^n - \cos\chi(F^y-\eta f)_{1,o}^n +$$

$$+ \sin\chi\, F_{1,0}^{zn} + \frac{\omega\cos^2\chi\sin^2\chi}{4}(\sigma_{1,1}-\sigma_{o,o})^n(f_{1,1}-f_{o,o})^n$$

2) A point of the plane of symmetry $\zeta = 0$. From the boundary condition (1.4) it follows that $T_{\ell,o} = 0$ and here all components of the vector f other than T should be found.

$$\frac{1}{\chi^n}[f_{\ell,o}^{n+1} - (1-2\tau^n)f_{\ell,o}^n] = -\sin\chi(F^z-\zeta f)_{\ell,1}^n - \frac{1}{2}\cos\chi\{(F^y-\eta f)_{\ell+1,o} -$$

$$- (F^y-\eta f)_{\ell-1,0}\}^n + \frac{1}{2}\omega\sin^2\chi(\sigma_{\ell,o}+\sigma_{\ell,1})(f_{\ell,1}-f_{\ell,o})^n +$$

$$+ \frac{1}{4}\omega\cos^2\chi\{(\sigma_{\ell+1,o}+\sigma_{\ell,o})(f_{\ell+1,o}-f_{\ell,o})-(\sigma_{\ell,o}+\sigma_{\ell-1,o})(f_{\ell,o}-f_{\ell-1,o})\}^n$$

3) A point on the wing surface. In the system of coordinates $_1$, $, n$ (Fig. 2)

$$\xi_1 = \xi \quad , \quad \rho = \zeta\cos\chi + \eta\sin\chi \quad , \quad n = -\zeta\,\mathrm{xin}\,\chi + \eta\cos\chi$$

the calculation formula is

$$\frac{1}{\chi^n}[\tilde{f}_{\ell,\ell}^{n+1} - \tilde{f}_{\ell,\ell}^n(1-2\tau^n)] = -\frac{1}{2}\sin\chi\cos\chi\{(\tilde{F}^z-\tilde{\zeta}\tilde{f})_{\ell+1,\ell+1} - (\tilde{F}^z-\tilde{\zeta}\tilde{f})_{\ell-1,\ell-1}\}^n -$$

$$- \{(\tilde{F}^y-\tilde{\eta}\tilde{f})_{\ell,\ell-1}\sin^2\chi + (\tilde{F}^y-\tilde{\eta}\tilde{f})_{\ell+1,\ell}\cos^2\chi\}^n +$$

$$+ \frac{1}{4}\omega\sin^2\chi\cos^2\chi\{(\sigma_{\ell+1,\ell+1}+\sigma_{\ell,\ell})(\tilde{f}_{\ell+1,\ell+1}-\tilde{f}_{\ell,\ell}) +$$

$$+ (\sigma_{\ell,\ell}+\sigma_{\ell-1,\ell-1})(\tilde{f}_{\ell,\ell}-\tilde{f}_{\ell-1,\ell-1})\}^n$$

All quantities marked by the sign ($\tilde{\ }$) are calculated by using the above formulae for corresponding quantities without such a sign but with v, w, η, ζ replaced by

$\tilde{v} = - w \sin \chi + v \cos \chi \quad , \quad \tilde{w} = w \cos \chi + v \sin \chi$

$\tilde{\eta} = - \zeta \sin \chi + \eta \cos \chi \quad , \quad \tilde{\zeta} = \zeta \cos \chi + \eta \sin \chi$

The third component of the vector $\tilde{f}_{\ell,\ell}$ is not calculated as it follows from (1.3) that $\tilde{S}_{\ell,\ell} \equiv 0$.

When $f_{\ell,\ell}$ is known the vector $f_{\ell,\ell}$ is found by formulae

$\Omega_{\ell,\ell} + \tilde{\Omega}_{\ell,\ell} \ , \quad R_{\ell,\ell} = \tilde{R}_{\ell,\ell} \ , \quad S_{\ell,\ell} = \tilde{T}_{\ell,\ell} \sin \chi \ , \quad T_{\ell,\ell} = \tilde{T}_{\ell,\ell} \cos \chi \ ,$

$E_{\ell,\ell} = \tilde{E}_{\ell,\ell}$

4) An internal point of the flow region.

$$\frac{1}{\chi^n}\,[f^{n+1}_{\ell,m} - (1-2\tau^n) f^n_{\ell,m}] = -\frac{1}{2} \sin \chi\{(F^z-\zeta f)_{\ell,m+1} - (F^z-\zeta f)_{\ell,m-1}\}^n -$$

$$-\frac{1}{2} \cos \chi\{(F^y-\eta f)_{\ell+1,m} - (F^y-\eta f)_{\ell-1,m}\}^n +$$

$$+\frac{1}{4}\,\omega \sin^2 \chi\{(\sigma_{\ell,m+1}+\sigma_{\ell,m})(f_{\ell,m+1}-f_{\ell,m}) - (\sigma_{\ell,m}+\sigma_{\ell,m-1})(f_{\ell,m}-f_{\ell,m-1})\}^n +$$

$$+\frac{1}{4}\,\omega \cos^2 \chi\{(\sigma_{\ell+1,m}+\sigma_{\ell,m})(f_{\ell+1,m}-f_{\ell,m}) - (\sigma_{\ell,m}+\sigma_{\ell-1,m})(f_{\ell,m}-f_{\ell-1,m})\}^n$$

The calculated region O_1DCB is shown in Fig. 2. On the boundary BC all quantities correspond to the undisturbed flow parameters, on the boundary DC - to the quantities obtained behind an oblique shock wave attached to the leading edge, on the wing surface O_1D - the solid surface condition, on the plane of symmetry O_1B - $W \equiv 0$. For the initial flow parameters the values behind an oblique shodk wave attached to the leading edge of the V-shaped wing were taken. The calculation was terminated when

$$\max_{\ell,m} \{(f^{n+1}_{\ell,m}-f^n_{\ell,m})/f^n_{\ell,m}\} \leq \varepsilon$$

for every component of the vector f (in the calculations done $\varepsilon \approx 10^{-2} \div 10^{-3}$).

3. Results. To check the efficiency of the method a comparison of the results of calculations with experimental data [4] was made. In Fig. 3 the calculated pressure distribution $C_p = (P - P_\infty)/(1.2)\rho_\infty V_\infty^2$ along the wing surface is shown (solid lines), together with experimental data for $\psi = 29°30'$, $M_\infty = 3.95$, $Re = 6{,}8 \cdot 10^6$. There is good correlation between experimental and calculated data, excluding the region in front of the shock adjacent to the wall, where the experimental pressure is greater than the calculated pressure. Such a pressure overshoot was mentioned in [11] and may be explained [12] by the interaction of the shock with the boundary layer. In Fig. 4a,b the shapes of the isobars indicate flow regimes with Mach and regular reflection of the shock coming from the leading edge at the plane of symmetry. The isobars in Fig. 4b indicate quite a few reflections of the shock at the plane of symmetry and the wall. The dashed line shows the exact position of the plane shock wave attached to the leading edge.

Note that the pressure and the density behind the calculated plane shock differ from exact values by 3%. The width of the calculated shock is not more than 5 - 6 mesh intervals.

ANALYTICAL METHOD

1. Statement of the problem and basic equations. Let us consider a V-shaped wing in hypersonic flow (Fig. 1). In the system of coordinates chosen (Fig. 5) the equations of conical flow, transformed in terms of a new independent variable $\psi = \psi(\theta,\phi)$, (ψ = const is a stream surface in a conical flow [8]) are written in the form

$$\frac{w}{\cos\theta}\frac{\partial u}{\partial\phi} - v^2 - w^2 = 0$$

$$\frac{w}{\cos\theta}\frac{\partial v}{\partial\phi} + uv + w^2 \operatorname{tg}\theta = -\frac{1}{\rho\theta_\psi}\frac{\partial p}{\partial\psi}$$

$$\frac{k}{k-1}\frac{p}{\rho}\ \frac{u^2+v^2+w^2}{2} = C \tag{1.1}$$

$$\frac{\partial}{\partial\phi}\frac{p}{\rho k} = 0$$

$$\frac{\partial}{\partial\phi}\ln(\rho w\theta_\psi) + 2\frac{u}{w}\cos\theta = 0$$

$$w\theta_\psi = v\cos\theta$$

After applying the transformation $\theta = \varepsilon\bar{\theta}$, $v = \varepsilon\bar{v}$, $\rho = \varepsilon^{-1}\bar{\rho}$ where $\varepsilon = (k-1)/(k+1)$ is considered to be a small parameter, and appropriate evaluation of terms, cancelling the terms of order ε^{1+n} $(n > 0)$ in comparison with unity, the following boundary conditions at the attached shock wave 1 (Fig. 6) are

$$u^* = \cos\alpha[\cos\phi - \operatorname{tg}\alpha\sin\gamma\sin\phi - \varepsilon\theta^*\operatorname{tg}\alpha\cos\gamma]$$

$$\begin{aligned} v^* = & -\theta_\phi^*[\cos\alpha\sin\phi + \sin\alpha\sin\gamma\cos\phi] - (1+m_o)\sin\alpha\cos\gamma + \\ & + \varepsilon(1+m_o)\cos\alpha[\theta_\phi^*\sin\phi - \theta^*\cos\phi] \\ & + \varepsilon(1+m_o)\sin\alpha\cos\gamma[\theta^*\sin\phi + \theta_\phi^*\cos\phi] - \varepsilon\,\theta_\phi^{*2}\sin\alpha\cos\gamma \end{aligned}$$

$$w^* = -\cos\alpha[\sin\phi + \operatorname{tg}\alpha\sin\gamma\cos\phi + \varepsilon\operatorname{tg}\alpha\cos\gamma\,\theta_\phi^*]$$

$$p^* = \sin^2\alpha\cos^2\gamma + \varepsilon p_1^* \tag{1.2}$$

$$\begin{aligned} p_1^* = & \sin 2\alpha\cos\gamma[\theta^*\cos\phi - \theta_\phi^*\sin\phi] \\ & - \sin^2\alpha\sin 2\gamma[\theta^*\sin\phi + \theta_\phi^*\cos\phi] - \sin^2\alpha\cos^2\gamma - M_\infty^{-2} \end{aligned}$$

$$\rho^* = (1+m_o)^{-1}\ ,\quad m_o = \varepsilon^{-1}(1-\varepsilon)\csc^2\alpha\,\mathrm{sc}^2\gamma\, M_\infty^{-2}$$

Here $\gamma = 90° - \gamma_*^\circ$ where $\gamma°$ is an angle between the plane of symmetry and the wing surface. $\theta_{(\phi)}$ - describes the form of the shock wave.

Bearing in mind that the main term of the pressure P^* (1.2) is constant along the wing surface and cancelling the terms of order higher than ε in the system of equations (1.1) it follows that

$$\frac{\partial u}{\partial \phi} = w \quad , \quad w\frac{\partial v}{\partial \phi} + uv + w^2\theta = -\frac{1}{\rho\theta_\psi}\frac{\partial P_2}{\partial \psi}$$

$$u^2 + w^2 = \Delta^2(\psi) \quad , \quad \frac{\partial}{\partial \phi}\ell n(\rho w \theta_\psi) + 2\frac{u}{w} = 0 \quad , \quad \frac{\partial}{\partial \phi}\frac{P}{\rho^k} = 0 \tag{1.3}$$

$$w\theta_\phi = v \quad , \quad P = P^*(\phi,\varepsilon) + \varepsilon P_2(\phi,\psi,\varepsilon)$$

The solid boundary condition on the wing is

$$v = 0$$

The system of equations (1.3) and the boundary conditions (1.2) differ from the exact ones only by terms of higher order [8], whereas series expansion in a small parameter would yield a system of equations which differ from the exact one by terms of the first order.

2. The solution of the boundary problem. The first and the third equations of the system (1.3) can be immediately integrated, and using the boundary conditions (1.2) the components of the velocity u and w are written in the form

$$u = \Delta\cos\tau \quad , \quad w = \Delta\sin\tau$$

$$\Delta^2(\phi') = \cos^2\alpha[1 + \mathrm{tg}^2\alpha\sin^2\gamma + 2\varepsilon\,\mathrm{tg}\,\alpha\cos\gamma\,(\theta_\phi^{*'}\sin\phi' - \theta^{+'}\cos\phi') + \varepsilon\,\mathrm{tg}^2\alpha\sin 2\gamma(\theta_\phi^{*'}\cos\phi' + \theta^{*'}\sin\phi')]$$

$$\tau = -[\phi + f(\phi')] \quad , \quad f(\phi') = a_1 + \varepsilon b_1(\phi') \quad , \tag{2.1}$$

$$a_1 = \mathrm{arc}\sin\frac{\mathrm{tg}\,\alpha\sin\gamma}{\surd(1 + \mathrm{tg}^2\alpha\sin^2\gamma)}$$

$$b_1(\phi') = \frac{\mathrm{tg}\,\alpha\cos\gamma}{\surd 1 + \mathrm{tg}^2\alpha\sin^2\gamma)}[(\theta_\phi^{*'}\cos\phi' + \theta^{*'}\sin\phi') - \sin\gamma\,\mathrm{tg}\,\alpha(\theta_\phi^{*'}\sin\phi' - \theta^{*'}\cos\phi')]$$

The symbol (') indicates that corresponding values are taken at the point of intersection of the shock wave with a streamline $(\phi' = \phi)$. The formulae (2.1) are evidently true for every streamline going through the leading shock wave 1. The solution for uniform flow behind a plane shock wave attached to the leading edge is given by

$$\Delta_1{}^2 = K^2(1-2a_2L) \quad , \quad f_1(\phi') \equiv \sigma = a_1 + Ma_2 \quad , \quad K^2 = \cos^2\alpha(1+\mathrm{tg}^2\alpha\sin^2\gamma)$$

$$P_1{}^* = \sin^2\alpha\cos^2\gamma(1+2a_2N-\varepsilon) - \varepsilon M_\infty^{-2} \quad , \quad \rho^* = (1+m_o)^{-1}$$

$$\theta_1 = (1+m_o)\sin\alpha\cos\gamma\,R_1\sin(\phi+\sigma)\sin(\beta-\phi')/\Delta_1\sin(\beta+\sigma)\sin(\phi'+\sigma) \,,$$

$$R_1 = 1 + a_2N$$

$$v_1 = -u_1\theta_1 = -(1+m_o)\sin\alpha\cos\gamma\,R_1\sin 2(\phi+\sigma)\sin(\beta-\phi')/2\sin(\beta+\sigma)\sin(\phi'+\sigma)$$

$$a_2 = \delta[1+\delta(N+L)/K\sin(\beta+a_1)]/K\sin(\beta+a_1+M\delta/\sin(\beta+a_1)) \quad ,$$

$$N = \sin\beta + \mathrm{tg}\,\alpha\sin\gamma\cos\beta$$

$$L = tg^2\,\alpha\,\cos^2\gamma/(1 + tg^2\,\alpha\,\sin^2\gamma)N\ ,$$

$$M = (tg\,\alpha\,\sin\gamma\,\sin\beta - \cos\beta)L/N\ ,\quad \delta = \varepsilon\cos\alpha(1+m_o)$$

We construct a closed form solution of the problem providing there is Mach reflection of a plane shock wave from the plane of symmetry (Fig. 6). Inserting the internal shock $F(\phi)$, the boundary conditions for region 2 (Fig. 6) on the discontinuity F are written in the form

$$u_2^* = u_1\ ,\quad v_2^* = -\varepsilon^{-1}\Delta_1\{\sin(\phi+\sigma)F_\phi(1-\varepsilon)(1-M_{1n}^{-2}) +$$

$$\cos(\phi+\sigma)F[1-(1-\varepsilon)(1-M_{1n}^{-2}) + F_\phi^{\,2}]\}/(1 + F^2 + F_\phi^{\,2})$$

$$w_2^* = -\Delta_1\{\sin(\phi+\sigma)[1+F_\phi^{\,2}[1-(1-\varepsilon)(1-M_{1n}^{-2})]] +$$

$$+ FF_\phi\cos(\phi+\sigma)(1-\varepsilon)(1-M_{1n}^{-2})\}/(1+F^2+F_\phi^{\,2}) \qquad (2.2)$$

$$P_2^* = P_1^*[(1+\varepsilon)M_{1n}^2-\varepsilon]\ ,\quad \rho_2^* = (1+m_o)^{-1}[1-(1-\varepsilon)(1-M_{1n}^{-2})]^{-1}$$

$$M_{1n}^2 = \Delta_1^{\,2}[F_\phi\sin(\phi+\sigma) - F\cos(\phi+\sigma)]^2/a_1^{*2}(1+F^2+F_\phi^{\,2})$$

M_{1n} - Mach number of the flow normal to the shock in the region $1, a_1^*$ - sound velocity in region 1.

The variables Δ and τ which determine u and w through formulae (2.1) in region 2 are written in the form

$$\Delta_2^{\,2} = \Delta_1^{\,2}\{1+[\sin(\phi_1+\sigma)F_\phi' - \cos(\phi_1+\sigma)F']^2(1-\varepsilon)(1-M_{1n}^{-2}(\phi_1))\ \times$$

$$\times\ [(1-\varepsilon)(1-M_{1n}^{-2}(\phi_1))-2]/(1+F'^2+F_\phi'^2)\}$$

$$\tau_2 = -[\phi+\sigma-(1-\varepsilon)(1-M_{1n}^{-2}(\phi_1))[\sin(\phi_1+\sigma)F_\phi' - \cos(\phi_1+\sigma)F']\ \times$$

$$\times\ \{F'\sin\phi_1 + F_\phi'\cos\phi_1 + [\sin(\phi_1+\sigma)F_\phi' - \cos(\phi_1+\sigma)F']\ \times$$

$$[1 - \tfrac{1}{2}(1-\varepsilon)(1-M_{1n}^{-2}(\phi_1))]\}/\cos\sigma(1+F'^2+F_\phi'^2)]$$

Here $\phi_1 = \phi$ on the shock $F(\phi)$. An investigation of the triple point A (Fig. 6) yields the result that for $F_\phi(\phi_2) < 0$ the shock F is a strong one. We consider such a compressed layer flow structure when $F_\phi(\phi_2) > 0$. In this case the following interactions of the incident shock F with the wall are possible: 1) Mach reflection (or in a particular case a regular one), 2) no reflection $F_\phi(\phi_4) = \infty$ (Fig. 7). Further we consider only the second case. There are many known cases [3] when reflected shocks do not appear.*

Note further that we expect the parameter ϕ_1, which corresponds to the streamlines intersecting the plane shock wave and the shock F and going through to region 2, to be within the limits $\phi_4 \le \phi_1 \le \phi_2$. Such a condition makes it easy

*The proposed approach allows us to consider such flow structure where $F_\phi < 0$. But an investigation shows that a flow with a strong shock F in the point ϕ_2 cannot be realized.

to integrate the system (1.3) because now the implicit connection between ϕ_1 and ϕ' through the function describing the shape of the shock F $(\phi_2 \leq \phi' \leq \beta)$ is eliminated.

The relations (2.1) allow us to find an integral of the fourth equation of the system (1.3).

$$\frac{\rho\theta_\psi}{w} = \frac{\rho^*\theta_\psi^*}{w^*}$$

The symbol (*) denotes that corresponding values are calculated on the leading shock wave in region 3 or on the shock F. Integrating this equation with respect to ψ in the interval $\phi_4 \leq \phi \leq \phi_2$ from the shock F along the line ϕ = const yields the following expression for streamlines in region 2 $(\theta_1 = \varepsilon^{-1}F)$

$$\theta_2 = \theta_1 + \int_\phi^{\phi_1} \frac{\rho_2' w_2}{\rho_2 w_2} \theta'_{2\psi} \psi_{\phi_1} d\phi_1$$

Taking into consideration

$$\theta'_{2\psi}\psi_{\phi_1} = \frac{d\theta_2'}{d\phi_1} - \left(\frac{\partial\theta_2}{\partial\phi}\right)' \quad , \quad \frac{d\theta_2'}{d\phi_1} \equiv \theta'_{1\phi}$$

we find

$$\theta'_{2\psi}\psi_{\phi_1} = -\Delta_1[1-(1-\varepsilon)(1-M_{1n}^{-2}(\phi_1))][\sin(\phi_1+\sigma)\theta'_{1\phi} - \cos(\phi_1+\sigma)\theta'_1]/w_2' \tag{2.3}$$

From the system of equations (1.3) it follows that the expression for the pressure in region 2 is

$$P_2 = P_3^*(\phi) + \varepsilon P_{32}(\phi_1\phi_2) + \varepsilon P_{22}(\phi_1\phi_1) \tag{2.4}$$

Here the first index denotes the number of the region, the second one indicates an additional term for the pressure as found from the second equation of tye system (1.3).

Using the fourth equation of the system (1.3) and relations (2.2) and (2.4), we find

$$\frac{\rho_2'}{\rho_2} = [(1+\varepsilon)M_{1n}^2(\phi_1)-\varepsilon][1 + \frac{\varepsilon}{\sin^2\alpha\cos^2\gamma}(P_{11}^*-P_{31}^*(\phi)-P_{32}(\phi,\phi_2)-P_{22}(\phi,\phi_1))] \tag{2.5}$$

Substitution of (2.3) and (2.5) into the expression for θ_2 yields

$$\theta_2(\phi_1\phi_1) = \theta_1(\phi) - \int_\phi^{\phi_1} \frac{w_2}{w_2'^2} F_1 R_2 d\phi_1$$

$$F_1 = \Delta_1[1+\varepsilon(M_{1n}^2(\phi_1)-1)+\varepsilon(1-M_{1n}^{-2}(\phi_1))][\sin(\phi_1+\sigma)\theta'_{1\phi} - \cos(\phi_1+\sigma)\theta'_1] \tag{2.6}$$

$$R_2 = 1 + \frac{\varepsilon}{\sin^2\alpha\cos^2\gamma}[P_{11}^*-P_{31}^*(\phi)-P_{32}(\phi_1\phi_2)-P_{22}(\phi_1\phi_1)]$$

For streamlines in region 3

$$\theta_3(\phi,\phi') = \theta_2(\phi_1\phi_2) + \int_{\phi_2}^{\phi'} \frac{\rho_3{}'w_3}{\rho_3 w_3{}'}\theta'_{34}\,\psi_{\phi'}\,d\phi'$$

Taking into consideration

$$\frac{\rho_{3'}}{\rho_3} = 1 + \frac{\varepsilon}{\sin^2\alpha\cos^2\gamma}(P^{*'}_{31} - P^{*}_{31} - P_{32})$$

$$\theta'_{3\psi}\,\psi_{\phi'} = (1+m_o)\sin\alpha\cos\gamma\{1-\varepsilon[\frac{\text{ctg}\,\alpha}{\cos\gamma}[\theta^{*'}_{\phi}\sin\phi' - \theta^{*'}\cos\phi') +$$

$$\text{tg}\,\gamma(\theta^{*'}\sin\phi' + \theta^{*'}_{\phi}\cos\phi')]\}/w_3{}'$$

we get

$$\theta_3(\phi_1\phi') = \theta_2(\phi_1\phi_2) + (1+m_o)\sin\alpha\cos\gamma\int_{\phi_2}^{\phi'}\frac{w_3}{w'^2_3}R_3\,d\phi'$$

$$R_3 = 1-\varepsilon[\frac{\text{ctg}\,\alpha}{\cos\gamma}(\theta^{*'}_{\phi}\sin\phi' - \theta^{*'}\cos\phi') + \text{tg}\,\gamma(\theta^{*'}\sin\phi' + \theta^{*'}_{\phi}\cos\phi') -$$

$$- \frac{1}{\sin^2\alpha\cos^2\gamma}(P^{*'}_{31} - P^{*}_{31} - P_{32})] \tag{2.7}$$

Integrating the second equation of system (1.3) from the leading shock wave in region 3 where $P_{32} = 0$ along the line $\phi = \text{const}$ and retaining the principal terms, we get

$$P_{32} = \sin\alpha\cos\gamma\int_{\phi'}^{\phi}\frac{w_3^3}{w'^2_3}(\theta_{3\phi\phi} + \theta_3)\,d\phi' \tag{2.8}$$

Integrating the same equation in region 2 from the streamline $\phi_1 = \phi_2$ where $P_{22} = 0$, we get

$$P_{22} = -\int_{\phi_1}^{\phi_2}\frac{w_2^3}{w'^2_2}(\theta_{2\phi\phi} + \theta_2)\,F_2\,d\phi_1$$

$$F_2 = \Delta_1(1+m_o)^{-1}[\sin(\phi_1+\sigma)\theta'_{1\phi} - \cos(\phi_1+\sigma)\theta'_1] \tag{2.9}$$

Differentiating (2.7) and (2.6) twice and substituting the results into (2.8) and (2.9) we get expressions for P_{32} and P_{22} to the right and to the left of the point ϕ_4. In these expressions the main terms [8] are retained. Substituting P_{32} and P_{22} into (2.6), (2.7) and taking $\phi' = \phi$ in (2.7) we get the following integro-differential equation for the shock wave shape

$$z+\delta\{(1+m_o)^{-1}(1+2a_2N)L_1+2[(z\cos\phi - z'\sin\phi)- \text{tg}\,\alpha\sin\gamma\,(z\sin\phi +$$

$$+ z'\cos\phi)]L_2-L_3\} - 2\varepsilon\delta\{[(\sin\phi + \text{tg}\,\alpha\sin\gamma\cos\phi)(z'''+z') +$$

$$+ (\cos\phi - \text{tg}\,\alpha\sin\gamma\sin\phi)(z''+z)]L_4-2(\sin\phi + \text{tg}\,\alpha\sin\gamma\cos\phi)\times$$

$$\times(z''+z)L_5\} = \eta z_1 + \eta\varepsilon(z''_1+z_1)\Phi(\phi)L_6 \tag{2.10}$$

$$z = \varepsilon\,\theta^*\frac{\text{ctg}\,\alpha}{\cos\gamma}\quad,\quad z_1 = \varepsilon\,\theta_1\frac{\text{ctg}\,\alpha}{\cos\gamma}$$

$$L_1 = \int_{\zeta}^{\phi_2} \frac{w_2}{w_2'^2} F_{11}\, d\phi_1 \quad ,$$

$$L_2 = \int_{\phi_2}^{\phi} \frac{w_3}{w_3'^2}\, d\phi' - (1+m_o)^{-1} \int_{\zeta}^{\phi_2} \frac{w_2}{w_2'^2} F_{11}\, d\phi_1 \quad ,$$

$$L_3 = \int_{\phi_2}^{\phi} \frac{w_3}{w_3'^2} \{1+3[(z(\phi')\cos\phi' - z'(\phi')\sin\phi') - \mathrm{tg}\,\alpha \sin\gamma(z(\phi')\sin\phi' +$$

$$+ z'(\phi')\cos\phi')]\}\, d\phi'$$

$$L_4 = \int_{\zeta}^{\phi_2} \frac{w_2}{w_2'^2} F_{11}\, d\phi_1 \int_{\phi_2}^{\phi} [\frac{w_3^3}{w_3'^2} \int_{\phi_2}^{\phi'} \frac{w_3}{w_3'^2}\, d\phi']\, d\phi' - (1+m_o)^{-1}(\int_{\zeta}^{\phi_2} \frac{w_2}{w_2'^2} F_{11}\, d\phi_1)^2 \times$$

$$\times \int_{\phi_2}^{\phi} \frac{w_3^3}{w_3'^2}\, d\phi' + (1+m_o)^{-1} \int_{\zeta}^{\phi_2} [\frac{w_2}{w_2'^2} F_{11} \int_{\phi_1}^{\phi_2} [\frac{w_2^3}{w_2'^2} F_{21} \int_{\zeta}^{\phi_1} \frac{w_2}{w_2'^2} F_{11}\, d\phi_1] d\phi_1] \times$$

$$\times\, d\phi_1 - (1+m_o) \int_{\phi_2}^{\phi} [\frac{w_3^3}{w_3'^2} \int_{\phi'}^{\phi} [\frac{w_3^3}{w_3'^2} \int_{\phi_2}^{\phi'} \frac{w_3}{w_3'^2}\, d\phi'] d\phi'] d\phi' +$$

$$+ \int_{\zeta}^{\phi_2} \frac{w_2}{w_2'^2} F_{11}\, d\phi_1 \int_{\phi_2}^{\phi} [\frac{w_3}{w_3'^2} \int_{\phi'}^{\phi} \frac{w_3^3}{w_3'^2}\, d\phi'] d\phi' \quad ,$$

$$L_5 = \int_{\zeta}^{\phi_2} \frac{w_2}{w_2'^2} F_{11}\, d\phi_1 \int_{\phi_2}^{\phi} [\frac{w_3^3}{w_3'^2} \int_{\phi_2}^{\phi'} \frac{u_3}{w_3'^2}\, d\phi'] d\phi' - (1+m_o)^{-1} \int_{\zeta}^{\phi_2} \frac{w_2}{w_2'^2} F_{11}\, d\phi_1 \times$$

$$\times \int_{\zeta}^{\phi_2} \frac{u_2}{w_2'^2} F_{11}\, d\phi_1 \int_{\phi_2}^{\phi} \frac{w_3^3}{w_3'^2}\, d\phi' + (1+m_o)^{-1} \int_{\zeta}^{\phi_2} [\frac{w_2}{w_2'^2} F_{11} \int_{\phi_1}^{\phi_2} [\frac{w_3^3}{w_2'^2} F_{21} \times$$

$$\times \int_{\zeta}^{\phi_1} \frac{u_2}{w_2'^2} F_{11}\, d\phi_1] d\phi_1] d\phi_1 + \int_{\zeta}^{\phi_2} \frac{u_2}{w_2'^2} F_{11}\, d\phi_1 \int_{\phi_2}^{\phi} [\frac{w_3}{w_3'^2} \int_{\phi'}^{\phi} \frac{w_3^3}{w_3'^2}\, d\phi'] d\phi' -$$

$$- (1+m_o) \int_{\phi_2}^{\phi} [\frac{w_3}{w_3'^2} \int_{\phi'}^{\phi} [\frac{w_3^3}{w_3'^2} \int_{\phi_2}^{\phi'} \frac{u_3}{w_3'^2}\, d\phi'] d\phi'] d\phi' \quad ,$$

$$L_6 = \int_{\phi}^{\phi_2} \frac{w_2}{w_2'^2} F_{11}\, d\phi_1 \int_{\phi_2}^{\phi} \frac{w_3^3}{w_3'^2}\, d\phi' - \int_{\phi}^{\phi_2} [\frac{w_2}{w_2'^2} F_{11} \int_{\phi_1}^{\phi_2} \frac{w_2^3}{w_2'^2} F_{21}\, d\phi_1] d\phi_1 -$$

$$- (1+m_o) \int_{\phi_2}^{\phi} [\frac{w_3}{w_3'^2} \int_{\phi'}^{\phi} \frac{w_3^3}{w_3'^2}\, d\phi'] d\phi' \quad ,$$

$$F_{11} = \Delta_1^{-1} \frac{\mathrm{ctg}\,\alpha}{\cos\gamma} F_1 \quad , \qquad F_{21} = \Delta_1^{-1} \frac{\mathrm{ctg}\,\alpha}{\cos\gamma} F_2 \quad ,$$

$$\Phi(\phi) = 1 + \frac{\Delta_1[1-(1-\varepsilon)(1-M_{1n}^{-2})][\sin(\phi+\sigma)F_{\phi} - \cos(\phi+\sigma)F]}{\Delta_2(\phi)\sin[\phi+\sigma+f_2(\phi)](F_{\phi\phi}+F)} \{\mathrm{ctg}[\phi+\sigma+f_2(\phi)] f_{2\phi}(\phi) +$$

$$+ \Delta_2^{-1}\Delta_{2\phi}(\phi) + [1-(1-\varepsilon)(1-M_{1n}^{-2})]^{-1}(1-\varepsilon)(1-M_{1n}^{-2})_\phi + [(1+\varepsilon)M_{1n}^2-\varepsilon]^{-1} \times$$
$$\times [(1+\varepsilon)M_{1n}^2-\varepsilon]_\phi - [\sin(\phi+\sigma)F_\phi-\cos(\phi+\sigma)F]^{-1}\sin(\phi+\sigma)(F_{\phi\phi}+F)\}$$

Here $\eta = 1$, $\zeta = \phi$ in the interval $\phi_4 \leq \phi \leq \phi_2$, $\eta = 0$, $\zeta = \phi_4$ in the interval $0 \leq \phi \leq \phi_4$.

Because of the method used for integration of the system (1.3) Eq. (2.10) contains a compatibility condition along the slip line ϕ_2 coming from the triple point A (Fig. 7). The second necessary condition in the interval $[\phi_4,\phi_2]$ is that of the pressure P_2^*, calculated on transition through the shock F from region 1 to region 2, the pressure $P_2(\phi,\phi_1)|_{\phi_1=\phi}$ from (2.4), should be equal

$$P_3^*(\phi) + \varepsilon P_{32}(\phi_1\phi_2) + \varepsilon P_{22}(\phi,\phi) = P_1^*[(1+\varepsilon)M_{1n}^2 - \varepsilon] \qquad (2.11)$$

In this the integro-differential equation for the shock wave shape in the interval $[\phi_3,\phi_4]$ is obtained, namely, Eq. (2.10) at $\eta = 0$ and $\zeta = \phi_4$; also we find a system of two equations for the leading shock wave and the internal shock F in the interval $[\phi_4,\phi_2]$, these are Eq. (2.10) at $\eta = 1$, $\zeta = \phi$ and Eq. (2.11)

An investigation of P_{32} and P_{22} shows that for continuity of the pressure along the line $\phi = \phi_4$ we must have

$$(Z_1'' + Z_1)\,\Phi(\phi)\Big|_{\phi=\phi_4} = 0$$

An analysis of the expression $\Phi(\phi)$ (2.10) allows us to find the main term of the function $\Phi(\phi)$: $\Phi(\phi) \approx (1-M_{1n}^{-2})$ which is not zero. Thus it follows that

$$(Z_1'' + Z_1)\Big|_{\phi=\phi_4} = 0$$

Having a differential operator of the third order for the function Z in Eq. (2.10) in the interval $[\phi_3,\phi_4]$ and of the third and the second order for functions Z and Z_1, respectively, in the system of equations (2.10) and (2.11) in the interval $[\phi_4,\phi_2]$ we have to apply eight restrictions to the unknown functions. Two unknown boundaries ϕ_3 and ϕ_2 may be found using conditions of intersection of the shock waves with the plane of symmetry 11: $Z(\phi_3) = \mathrm{ctg}\,\alpha\,\csc\gamma\,\sin\phi_3$, and with the plane shock wave $Z_2 = a_2\sin(\beta-\phi)$ attached to the leading edge

$$Z(\phi_2) \equiv Z_1(\phi_2) = a_2 \sin(\beta-\phi_2)$$

Two conditions are known for the function Z_1 in the point ϕ_4 and the symmetry condition for the function Z in the point ϕ_3

$$Z'(\phi_3) = -\,\mathrm{tg}\,\gamma\,\mathrm{ctg}\,\alpha\,\mathrm{sc}\,\gamma + O(\varepsilon\gamma^2)$$

Thus there are five arbitrary conditions which may be used to connect $Z^{(n)}(\phi_4)$ for $n = 0,1,\ldots,4$. This means that on the line $\phi = \phi_4$ the functions P and θ as well as their derivatives are continuous.

3. Approximate solution and calculation results. To find the approximate solution we shall use some simplifications. In the integrands the functions of ϕ_1 are considered to be constant, in the functions W_i only the main terms are used believing the shock F to be of small intensity. After integration of the functionals and some transformations, the integro-differential equation for the function Z in the interval $[\phi_3,\phi_4]$ is reduced to an Euler's equation, the general solution of which is known. The system of equations (2.10) and (2.11) for the functions Z and Z_1 in the interval $[\phi_4,\phi_2]$ after similar transformations, is reduced to an essentially nonlinear system of differential equations, no simple solution of which is known. So in the interval $[\phi_4,\phi_2]$ the functions Z and Z_1 were expanded in

Taylor's series at the point ϕ_2. Four terms of the series were used. This allows us at the point ϕ_4 to link up the functions Z and their first and second derivatives only.

In Fig. 8a,b,c the curves of pressure distribution (solid lines) behind the shock wave are represented for $M_\infty = \infty$, $\beta = 45°$, $\gamma = 10°$, 15° and 20°. The curves 1,2,3 correspond to the angles of attack $\alpha = 10°, 15°$ and 20°.

In Fig. 8b a comparison of the pressure distribution calculated using the above theory with the pressure distribution (dashed line) resulting from numerical computation described in the first part of the paper is represented. There is good agreement between the shock positions. Some discrepancy in the pressure distributions may be explained by taking into consideration the somewhat smaller values of the pressure and density obtained by numerical computation behind the plane shock. Note that the shock *F* in the point ϕ_2 is a weak one and the velocity behind *F* is subsonic. On the shock wave in the interval $[\phi_3, \phi_2]$ there is a point of inflection which moves to the left as the angle γ increases and then disappears. After this the curvature of the shock wave is positive. In Fig. 9 a general picture of the flow past a V-shaped wind is represented for $M_\infty = \infty$, $\beta = 45°$, $\alpha = 20°$; $\gamma = 10°$, 15° and 20° in the cases 1, 2 and 3, respectively.

REFERENCES

1. Gonor, A. L. & Shvetz, A. I. "Investigation of pressure distribution on certain star shaped bodies for Mach number M = 4," Prik. Mekh. i Tekh. Fiz. 6, 1965.
2. Kücheman, D. "Hypersonic aircraft and their aerodynamic problems," Progress in Aeronautical Science, Pergamon Press, Vol. 5, 1965.
3. Gonor, A. L. & Shvetz, A. I. "Investigation of shock systems for flow past star shaped bodies," Izv. Akad Nauk SSSR, Mechanics of Liquids and Gases 3, 1966.
4. Gonor, A. L. & Shvetz, A. I. "Flow past V shaped wings in a supersonic stream for Mach number M = 3.9," Izv. Akad Nauk SSSR, Mechanics of Liquids and Gases 6, 1967.
5. Maikapar, G. I. "On the wave drag of non-axisymmetric bodies at supersonic speeds," Prik. Mat. i Mekh. 23, 2, 1959.
6. Gonor, A. L. "Exact solutions of the problem of flow past certain three dimensional bodies in a supersonic gas stream," Prik. Mat. i Mekh. 28, 5, 1964.
7. Gonor, A. L. "Certain three dimensional flows with Mach-type shock wave interactions," Izv. Akad Nauk SSSR, Mechanics of Liquids and Gases 6, 1966.
8. Gonor, A. L. "Flow past delta wings in a hypersonic stream," Prik. Mat. i Mekh. 34, 3, 1970.
9. Gonor, A. L. & Ostapenko, N. A. "Hypersonic flow past a delta wing of finite thickness," Izv. Akad Nauk SSSR, Mechanics of Liquids and Gases 3, 1970.
10. Rusanov, V. V. "Calculation of the interaction of shock waves with an obstacle," Zh. Vych. Mat. i Mat. Fiz. 2, 1961.
11. Zaitsev, Yu. I. & Keldysh, V. V. "Particular cases of flow near supersonic edges and lines of intersection of shock waves," Lecture Notes TsAGI 1, 1, 1970.
12. Panov, Yu. A. "Interaction of an incident three dimensional shock wave with a turbulent boundary layer," Izv. Akad Nauk SSSR, Mechanics of Liquids and Gases 3, 1968.

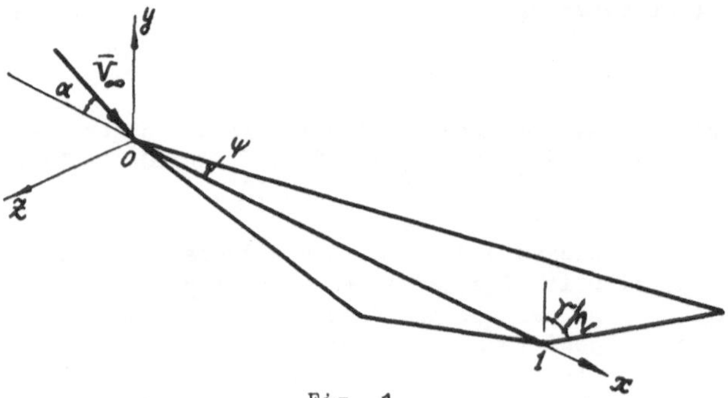

Fig. 1

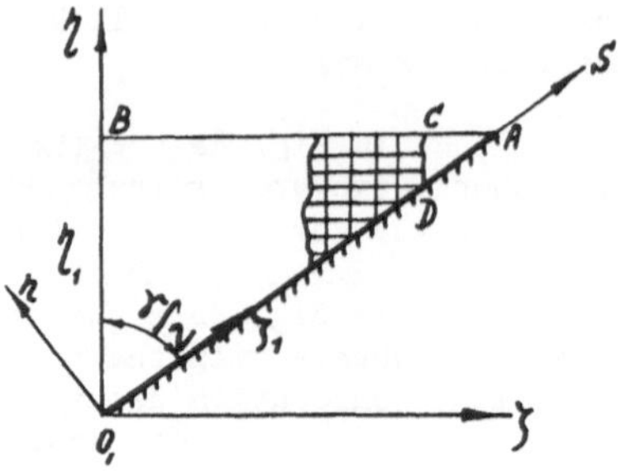

Fig. 2

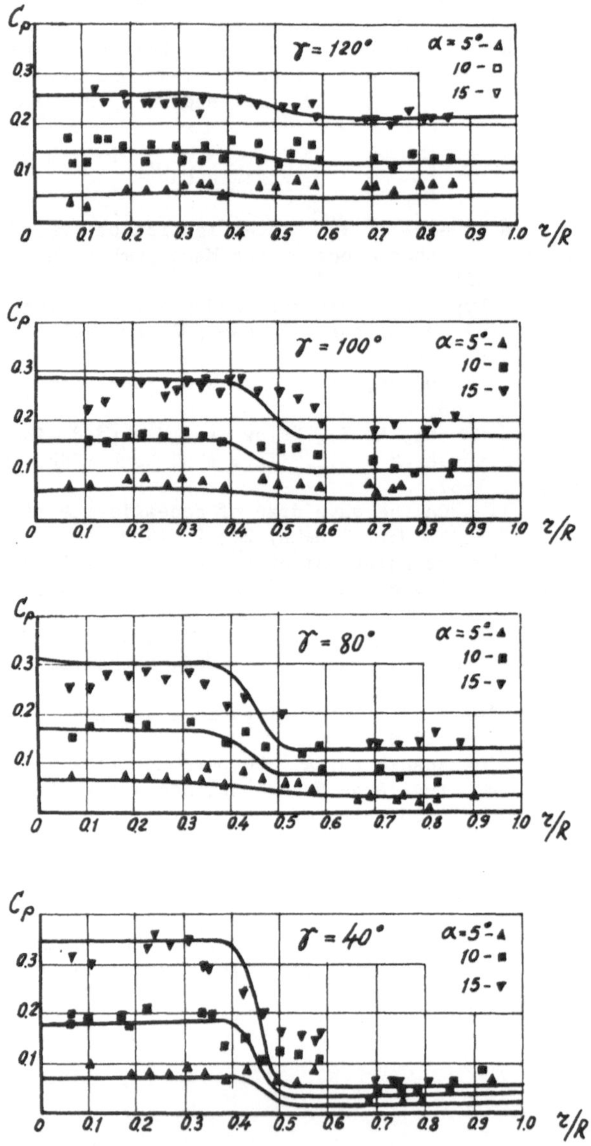

Fig. 3

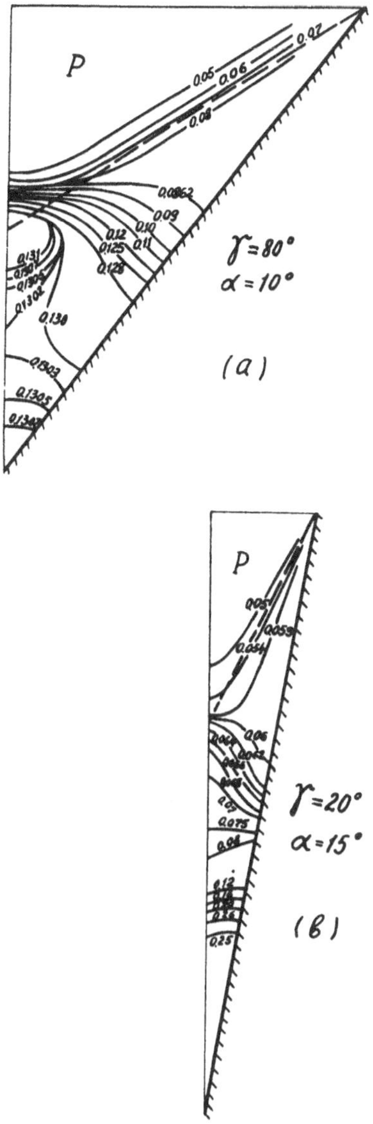

Fig. 4

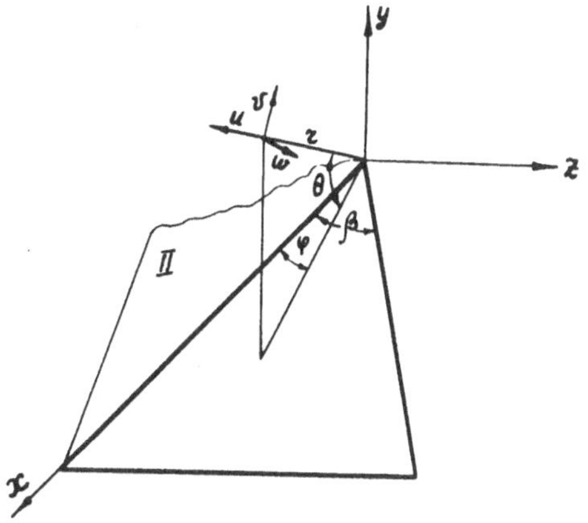

Fig. 5

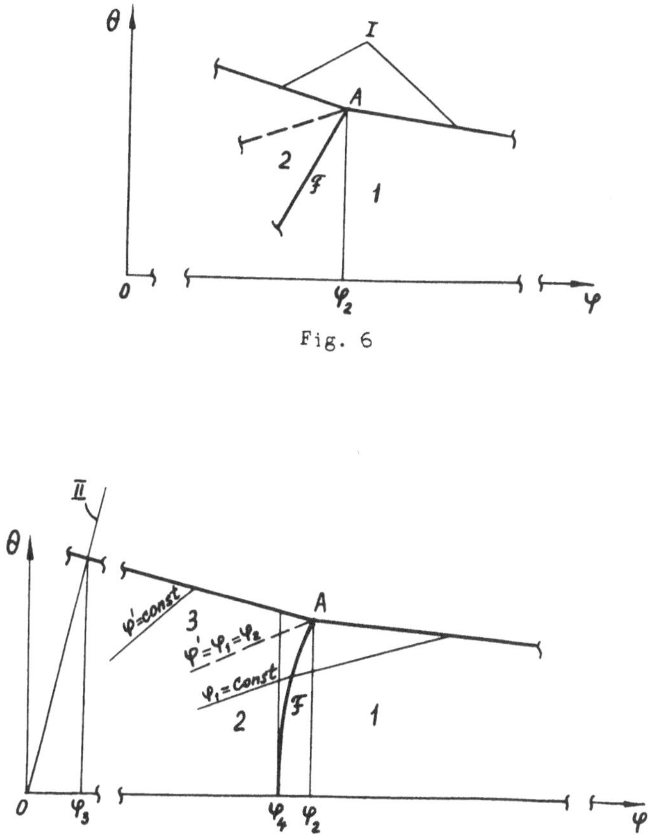

Fig. 6

Fig. 7

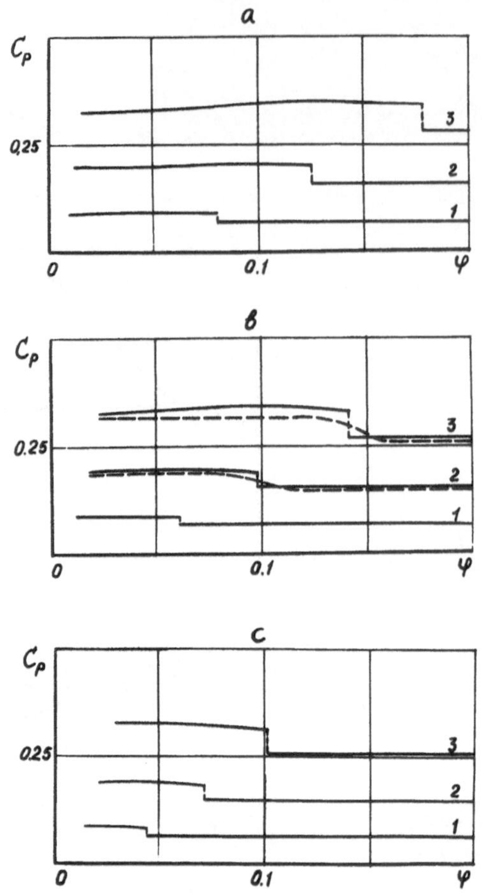

Fig. 8

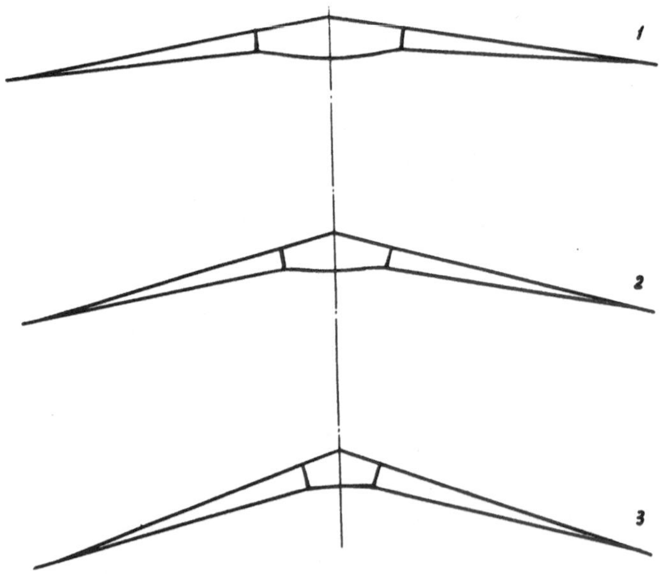

Fig. 9

Session VI

Navier-Stokes Equations. Fully Viscous Flows

K. Stewartson, Chairman

CALCUL DES ECOULEMENTS D'UN FLUIDE VISQUEUX INCOMPRESSIBLE

M. FORTIN*, R. PEYRET**, R. TEMAM*

1. INTRODUCTION

Nous proposons ici trois méthodes de résolution numérique des équations régissant les écoulements bidimensionnels d'un fluide visqueux incompressible. Dans ces méthodes les inconnues sont le vecteur vitesse et la pression. Le premier schéma qui sera décrit se rattache aux méthodes de perturbation de l'équation de la divergence exposées en [1], [2], [3], cette perturbation étant introduite ici d'une façon nouvelle. Cette méthode, telle que nous l'envisageons ici, n'est utilisable que pour le calcul d'écoulement stationnaire obtenu comme limite, lorsque le temps tend vers l'infini, d'une solution instationnaire. Par contre, les deux autres schémas qui seront présentés considèrent les équations d'évolution exactes et peuvent être employés au calcul d'un écoulement non stationnaire ; ces deux schémas sont basés sur la méthode de projection [4], [5], [6].

Le choix de la discrétisation a été conduit par le triple souci suivant : facilité de mise en oeuvre, précision et rapidité de la convergence vers l'état stationnaire. La question des grands nombres de Reynolds n'a pas été particulièrement considérée dans cette étude, cependant on a pu, du moins par une des méthodes proposées, atteindre des nombres de Reynolds modérément élevés.

Ces trois schémas ont été successivement utilisés pour le calcul de l'écoulement stationnaire dans une cavité carrée. Ce problème, qui a fait l'objet de nombreux travaux, est intéressant pour la comparaison des différentes méthodes.

Nous nous bornerons dans cette Note à décrire succintement les schémas et à donner quelques résultats, renvoyant à [7] pour une version détaillée de notre travail.

2. EQUATIONS

Les équations de Navier-Stokes régissant les écoulements non stationnaires d'un fluide visqueux incompressible dans un domaine borné Ω, de frontière Γ, sont :

$$\frac{\partial \vec{U}}{\partial t} - \frac{1}{\mathcal{R}} \Delta \vec{U} + \vec{U}.\nabla \vec{U} + \nabla p = \vec{f}, \tag{1a}$$

$$\nabla . \vec{U} = 0 \; ; \tag{1b}$$

$$\vec{U}|_{\Gamma} = \vec{\alpha}, \qquad \vec{U}|_{t=0} = \vec{U}_0 . \tag{1c}$$

Toutes les quantités sont adimensionnelles : $\vec{U}$ est le vecteur vitesse, p est la pression ; $\vec{f}$, force extérieure, est donnée, les conditions aux limites $\vec{\alpha}$ et initiale $\vec{U}_0$ sont des vecteurs donnés ; enfin, $\mathcal{R}$ est le nombre de Reynolds.

* Université de Paris-Orsay, 91-Orsay et IRIA, Domaine de Voluceau, 78-Rocquencourt.

** Département de Mécanique, Faculté des Sciences de Paris et ONERA, 92 - Chatillon-sous-Bagneux.

3. METHODE DE PERTURBATION DES EQUATIONS OU DE MINIMAX (SCHEMA I)

Considérons le problème stationnaire correspondant à (1). Certaines analogies entre ce problème et des problèmes d'optimisation sous contrainte (l'équation $\nabla.\vec{U}=0$ peut être considérée comme une contrainte) nous ont conduit à associer à ce problème stationnaire un problème d'évolution sans signification physique réelle et qu'on obtient comme système d'Arrow-Hurwicz (cf. [8]) d'un problème d'optimisation convenable (problème de "minimax"). On a pu ainsi introduire le système suivant :

$$-\frac{1}{\mathcal{R}}\Delta\vec{U} + \vec{U}.\nabla\vec{U} + \nabla p = \vec{f}, \tag{2a}$$

$$\frac{\partial p}{\partial t} + c^2\,\nabla.\vec{U} = 0\ ; \tag{2b}$$

$$\vec{U}|_\Gamma = \vec{\alpha}\ , \qquad p|_{t=0} = p_0\ . \tag{2c}$$

En formulant le problème sous une forme voisine on obtient, comme système d'Arrow-Hurwicz associé, le système suivant :

$$\frac{\partial\vec{U}}{\partial t} - \frac{1}{\mathcal{R}}\Delta\vec{U} + \vec{U}.\nabla\vec{U} + \nabla p = \vec{f}\ , \tag{3a}$$

$$\frac{\partial p}{\partial t} + c^2\,\nabla.\vec{U} = 0\ ; \tag{3b}$$

$$\vec{U}|_\Gamma = \vec{\alpha}\ , \qquad \vec{U}|_{t=0} = \vec{U}_0\ , \qquad p|_{t=0} = p_0\ . \tag{3c}$$

Dans ces problèmes les conditions initiales $\vec{U}_0$ et p_0 sont arbitraires et c^2 est une constante positive à choisir au mieux.

Aucun des deux systèmes (2) ou (3) ne décrit le régime transitoire du fluide mais diverses considérations nous permettent de penser que pour $t \to \infty$, $\vec{U}(t)$ et $p(t)$ convergent vers la solution des équations de Navier-Stokes stationnaires. Des résultats de ce type sont prouvés en [7] .

Les discrétisations de (2) et (3) conduisent à des schémas numériques équivalents sous des conditions que nous allons préciser. La vitesse $\vec{U}$ de composantes u,v est définie aux points x=ih,y=jh (i et j sont des entiers, h est le pas d'espace) et la pression p aux points $x=(i+\frac{1}{2})h$, $y=(j+\frac{1}{2})h$. L'équation (2b) (ou(3b)) est discrétisée de la façon suivante :

$$p^{n+1}_{i+1/2,\,j+1/2} = p^{n}_{i+1/2,\,j+1/2} - c^2 k\,(\nabla_h.\vec{U})^n_{i+1/2,\,j+1/2}\ , \tag{4}$$

où k est le pas de temps, t=nk et où

$$(\nabla_h.\vec{U})_{i+1/2,\,j+1/2} = \frac{1}{2h}(u_{i+1,j} - u_{i,j} + u_{i+1,j+1} - u_{i,j+1}) + \frac{1}{2h}(v_{i,j+1} - v_{i,j} + v_{i+1,j+1} - v_{i+1,j})\ . \tag{5}$$

L'équation (2a) est discrétisée de façon implicite et sa résolution est effectuée selon le processus itératif suivant (l'indice s se réfère à la $s^{\text{ième}}$ itérée) :

$$\vec{U}_{i,j}^{n+1,s+1} = \vec{U}_{i,j}^{n+1,s} + \frac{k}{\mathcal{R}h^2}\left(\vec{U}_{i+1,j}^{n+1,s} + \vec{U}_{i-1,j}^{n+1,s+1} + \vec{U}_{i,j+1}^{n+1,s} + \vec{U}_{i,j-1}^{n+1,s+1} - 4\,\vec{U}_{i,j}^{n+1,s}\right)$$

$$(6) \qquad - \frac{k}{2h}\left[u_{i,j}^{n+1,s}\left(\vec{U}_{i+1,j}^{n+1,s} - \vec{U}_{i-1,j}^{n+1,s+1}\right) + v_{i,j}^{n+1,s}\left(\vec{U}_{i,j+1}^{n+1,s} - \vec{U}_{i,j-1}^{n+1,s+1}\right)\right]$$

$$- k\,(\nabla_h p)_{i,j}^{n+1} + k\,\vec{f}_{i,j}^{n+1} ,$$

avec $\vec{U}_{i,j}^{n+1,0} = \vec{U}_{i,j}^{n}$.

Il n'est pas nécessaire de résoudre complètement (6) à chaque pas de temps et, en effectuant une seule itération (s=0) par cycle de temps, le schéma aux différences (6) donne une approximation consistante de l'équation (3a), équation qu'on aurait discrétisée de façon semi-implicite de telle sorte que pour calculer $\vec{U}_{i,j}^{n+1}$ on se serve des $\vec{U}_{i-1,j}^{n+1}$ et $\vec{U}_{i,j-1}^{n+1}$ déjà calculés.

Des études de stabilité [7] conduisent aux critères :

$$(7) \qquad k \leqslant \frac{\mathcal{R}h^2}{4} , \qquad \frac{k}{h}\left(|\vec{U}^n| + \sqrt{|\vec{U}^n|^2 + 4c^2}\right) < 1 .$$

4. METHODES DE PROJECTION (SCHEMAS II ET III).

4.1 - Schéma II.

Cette méthode résoud les équations non stationnaires (1). Elle s'apparente, dans une certaine mesure aux méthodes exposées en [9] et [10] . C'est un schéma de pas fractionnaires défini de la façon suivante :

$$(8a) \qquad \frac{1}{k}\left(\vec{U}^{n+1/2} - \vec{U}^{n}\right) - \frac{1}{\mathcal{R}}\left(\Delta\vec{U}\right)^n + \left(\vec{U}.\nabla\vec{U}\right)^n = \vec{f}^n ,$$

$$(8b) \qquad \frac{1}{k}\left(\vec{U}^{n+1} - \vec{U}^{n+1/2}\right) + (\nabla p)^{n+1} = 0 ,$$

$$(8b') \qquad (\nabla.\vec{U})^{n+1} = 0 ;$$

avec les conditions

$$(8c) \qquad \vec{U}^{n+1}\big|_{\Gamma} = \vec{\alpha}^{n+1} , \qquad \vec{U}^0 = \vec{U}_0 .$$

La discrétisation spatiale est identique à celle décrite au paragraphe précédent et ne sera pas explicitée ici (cf. [7]) ; il en est de même des conditions de stabilité.

On sait [4], en ce qui concerne l'étape (8b)-(8b'), que

$$(9) \qquad \vec{U}^{n+1} = P(\vec{U}^{n+1/2}) ,$$

où P est un opérateur de projection orthogonale dans l'espace fonctionnel $(L^2(\Omega))^2$.

Décrivons brièvement le déroulement du calcul : l'équation (8a) détermine $\vec{U}^{n+1/2}$, puis, en prenant la divergence de (8b) et compte-tenu de (8b'), on définit p^{n+1} comme solution du problème de Neumann:

$$(10a) \qquad (\Delta p)^{n+1} = \frac{1}{k}\,(\nabla.\vec{U})^{n+1/2} ,$$

$$(10b) \qquad \left(\frac{\partial p}{\partial \nu}\right)^{n+1}\Bigg|_{\Gamma} = \frac{1}{k}\left(\vec{U}^{n+1/2} - \vec{U}^{n+1}\right).\vec{\nu} \qquad \text{sur } \Gamma ,$$

où (10b) a été obtenue en projetant (8b) selon la normale $\vec{\nu}$ à Γ . Enfin, $\vec{U}^{n+1}$ est déterminé par (8b).

La présence de $\vec{U}^{n+1/2}$ nécessite d'imposer des conditions aux limites pour cette quantité en plus de celles relatives à $\vec{U}^{n+1}$. On écrira sur Γ :

$$(11) \qquad \vec{U}^{n+1}.\vec{\tau} = \vec{\alpha}^{n+1}.\vec{\tau} \;, \qquad \vec{U}^{n+1/2}.\vec{\tau} = \vec{\alpha}^{n+1}.\vec{\tau} + k\left(\frac{\partial p}{\partial \tau}\right)^{n} ,$$

pour les composantes tangentielles ($\vec{\tau}$ est le vecteur unitaire tangent à Γ) et

$$(12) \qquad \vec{U}^{n+1}.\vec{\nu} = \vec{U}^{n+1/2}.\vec{\nu} = \vec{\alpha}^{n+1}.\vec{\nu} ,$$

pour les composantes normales. Compte-tenu de (12), (10b) s'écrit :

$$(13) \qquad \left(\frac{\partial p}{\partial \nu}\right)^{n+1}\Big|_{\Gamma} = 0 .$$

Cette valeur de la dérivée normale de p n'a évidemment aucune signification physique mais l'on peut vérifier [7] que les valeurs de p en tout point intérieur (rappelons que p n'est pas calculé sur Γ) sont indépendantes de $\vec{U}^{n+1/2}.\vec{\nu}$ au bord Γ , car cette quantité apparaît à la fois dans l'un et l'autre des seconds membres de (10). La valeur de p sur Γ comme celle de sa dérivée normale pourraient être obtenues par extrapolation.

4.2 - Schéma III.

Il s'agit du schéma implicite de pas fractionnaires déjà présenté en [4] et [11] avec des améliorations sensibles relatives au traitement des conditions aux limites (cf. [6]). Nous renvoyons à ces références pour la description de cette méthode.

5. RESULTATS NUMERIQUES.

Les trois schémas ont été utilisés pour le calcul de l'écoulement stationnaire dans une cavité carrée dont l'un des côtés est en mouvement (le côté du carré est pris comme longueur de référence et la vitesse constante du côté en mouvement comme vitesse de référence).

On a pu comparer les résultats pour des nombres de Reynolds allant jusqu'à 400 (les méthodes II et III devenant très coûteuses en temps de machine pour des valeurs supérieures). L'état stationnaire était supposé atteint pour des résidus variant, selon les cas, entre 10^{-4} et 10^{-6}, le temps de calcul (IBM 360-91) variant, pour h=1/40, entre 10 s. et 90s.

Les résultats donnés par les méthodes I et II sont identiques à moins de 5‰ près, ce qui s'explique par la similitude des discrétisations. Par contre, en ce qui concerne la méthode III, bien que la discrétisation soit analogue, l'écart entre ses résultats et ceux des méthodes précédentes peut atteindre 5 % ; cette différence, traduisant une perte de précision, est dûe, en partie, à l'erreur de troncature introduite par la décomposition des opérateurs.

Seuls quelques résultats sont présentés dans cette Note, on en trouvera un ensemble plus complet dans la référence [7] .

Les figures 1,2 et 3 représentent respectivement les lignes de courant, les lignes iso-tourbillon et les isobares pour $\mathcal{R}$ =10 (schémas I, II et III) et pour $\mathcal{R}$ =1000 (schéma I). Plus précisément, les valeurs maximales ψ_{MAX} de la fonction de courant, calculées dans le cas $\mathcal{R}$ =10 par chacune des méthodes, sont données dans le tableau :

	Schéma I	Schéma II	Schéma III
ψ_{MAX}	0,0982	0,0986	0,0971

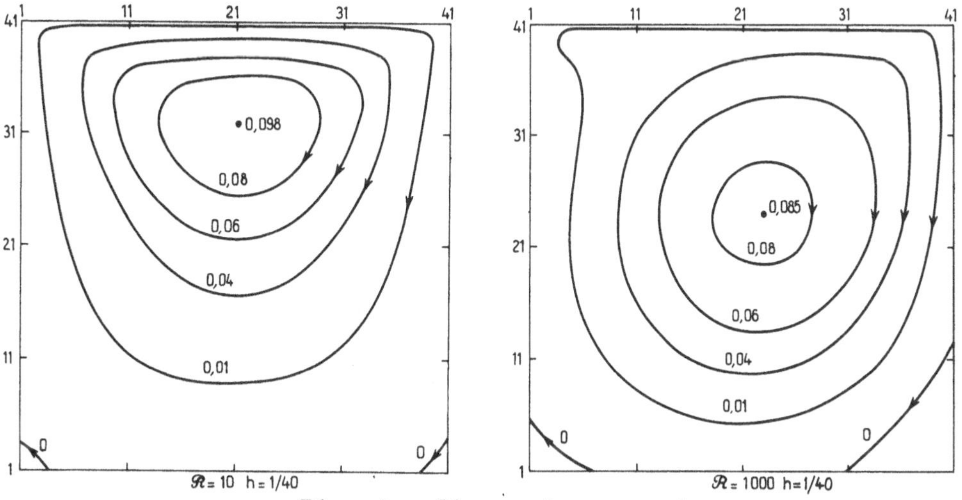

Fig. 1 - Lignes de courant

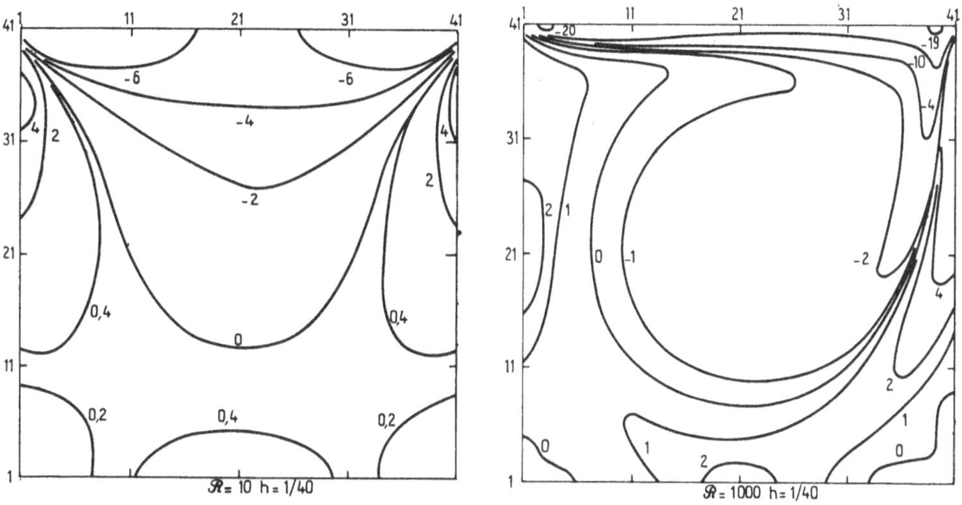

Fig. 2 - Lignes iso-tourbillon

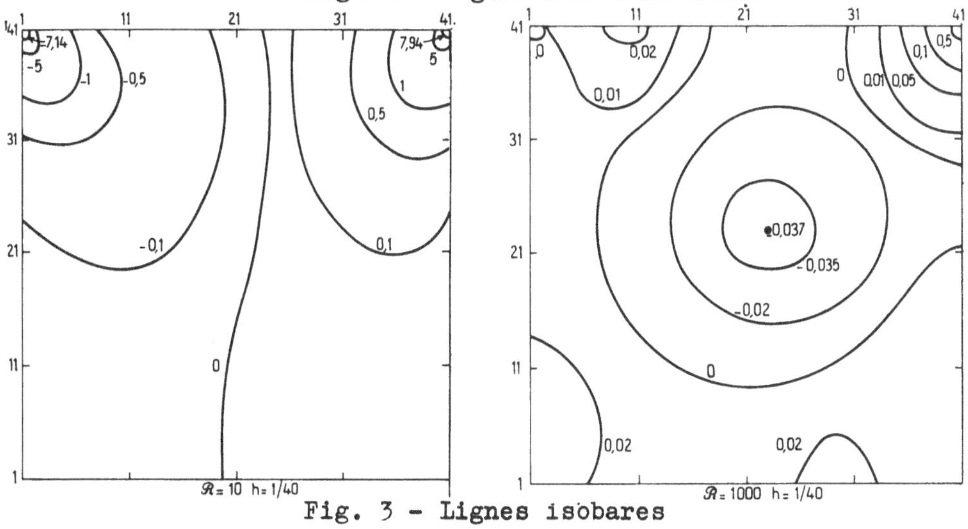

Fig. 3 - Lignes isobares

Des profils du module de vitesse sont représentés sur la figure 4.

On notera pour conclure, en ce qui concerne le calcul d'écoulements stationnaires, que :

-- Le schéma I, dont la mise en oeuvre est très facile, est le plus efficace aux "grands" nombres de Reynolds. La convergence vers l'état stationnaire est rapide ; la rapidité de cette convergence et la façon dont elle s'effectue dépendent du choix de c^2 (cf. [7]).

-- Le schéma II peut être rendu très rapide, à condition de ne pas effectuer, à chaque pas de temps, la résolution exacte du problème de Neumann pour p, c'est-à-dire en limitant le nombre d'itérations dans la résolution de (10) discrétisé.

-- Le schéma III présente une erreur de troncature proportionnelle à k et qui ne disparaît pas lorsque l'état stationnaire est atteint. De ce fait, l'avantage de la stabilité inconditionnelle de ce schéma (grands pas de temps) se trouve limité par des questions de précision. On a pu améliorer cette précision en divisant le pas de temps après stationnarisation et en calculant un nouvel état stationnaire, mais ce procédé est coûteux en temps.

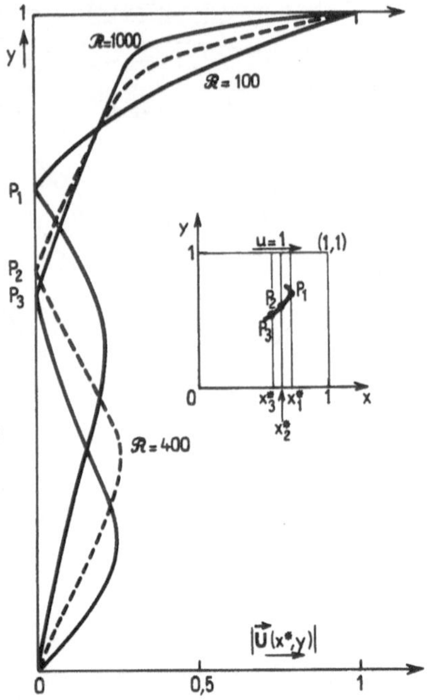

Fig. 4
Profils du module de vitesse

Signalons pour terminer que l'écoulement dans une cavité rectangulaire ainsi que celui dû à l'injection d'un fluide dans une conduite cylindrique avec élargissement (extension au cas à symétrie de révolution du schéma I) ont été calculés et sont présentés en [7] .

Enfin, la méthode II est actuellement utilisée pour le calcul de l'écoulement de culot [12] .

BIBLIOGRAPHIE

[1] Yanenko, N.N., Méthodes à pas fractionnaires, Armand Colin (1968)

[2] Chorin, A.J., J. Comput. Phys. 2, 12-26 (1967)

[3] Temam, R., Archiv. Rat. Mech. Analys. 32, 135-153 (1969)

[4] Temam, R., Archiv. Rat. Mech. Analys. 32, 377-385 (1969)

[5] Chorin, A.J., Math. Comput. 22, 745-762 (1968)

[6] Fortin, M., Thèse 3ème cycle, Faculté des Sciences d'Orsay (1970)

[7] Fortin, M., Peyret, R., Temam, R., à paraître

[8] Arrow, Hurwicz, Uzawa, Studies in non linear programming, Stanford Univ. Press (1958)

[9] Harlow, F.H., Welch, J.E., Phys. Fluids, 8, 2182-2189 (1965)

[10] Kuznetsov, B.G., Fluid Dynam. Trans. 4, 85-89 (1969)

[11] Fortin, M., Temam, R., 1ère Conf. Intern. Méthodes Numer. Dynam. Fluides, Novosibirsk (1969)

[12] Ladevèze, J., à paraître

A NUMERICAL METHOD FOR CALCULATING THE INITIAL FLOW PAST A CYLINDER IN A VISCOUS FLUID

S.C.R. Dennis and A.N. Staniforth

(Department of Applied Mathematics, University of Western Ontario, London, Ontario, Canada)

INTRODUCTION

The problem considered in this paper is that of finding the initial flow of a viscous, incompressible, fluid relative to a cylinder which suddenly starts to move in a direction at right angles to its axis with constant velocity. It is known that this problem can be treated by boundary-layer theory. The initial flow was given as the first two terms of a series in powers of the time from the start of the motion by Blasius (1908) and the results were extended by Goldstein and Rosenhead (1936). The first term is valid for all values of the Reynolds number R, but subsequent terms hold only in the limit $R \to \infty$. Recently Wang (1967) has attempted to extend the theory to lower Reynolds numbers, but little further theoretical progress has been made. The general theory does, however, indicate the initial flow structure.

One approach to the problem is to use numerical methods. This was considered by Payne (1958) for a circular cylinder. Most of the subsequent work has been on the circular cylinder. Recent papers by Son and Hanratty (1969), Thoman and Szewczyk (1969) have reviewed the literature on this problem. The basic procedure, which is valid for all Reynolds numbers, is to express the Navier-Stokes equations in terms of the stream function and vorticity and integrate them by a step-by-step procedure. For impulsively started cylinders, one of the difficulties is the specification of the initial conditions. Boundary-layer theory shows that the vorticity is initially infinite and confined to an infinitesimally thin region surrounding the cylinder surface. For this reason the direct application of finite-difference approximations to the equations does not give the initial flow correctly. In practice a finite initial vorticity distribution on the cylinder is obtained. Ingham (1968) has pointed out that very small time steps must be used in the integration to obtain good approximations to the initial flow.

In the present paper a method of numerical integration is proposed which utilizes the known boundary-layer structure of the initial flow. Boundary-layer coordinates are introduced into the equations for the vorticity and the stream function. The initial singularity in the vorticity is removed by a transformation. The final equations can then be integrated numerically for any Reynolds number, with time steps chosen to be virtually independent of the Reynolds number. The method is suitable for high R, and the case $R = \infty$ can be solved. It is valid, in theory, for general asymmetrical flow past a cylinder of any shape which can be mapped on to a straight line by a suitable conformal transformation.

An illustration is given for a symmetrical flow only. The initial motion past an impulsively started circular cylinder for $R = 100, 500, 10^3, 10^4$ and ∞ is calculated, assuming the flow to remain symmetrical. Here, R is the Reynolds number based on the diameter. The calculations are continued well beyond the time at which the flow separates, and the development of surface properties with time are in almost exact agreement with previous calculations by Collins and Dennis (to be published) on this problem. Details of some asymmetrical flows will be published later.

BASIC EQUATIONS

The problem is formulated for a cylinder of general shape which starts to move suddenly with velocity U at an angle α to the x axis as shown in Fig. 1(a). The axes of x and y are fixed in the cylinder, and motion relative to the cylinder is considered. A prior step is to make a conformal transformation of the form

$$\xi + i\eta = F(x + iy) \tag{1}$$

which maps the region outside the cylinder on to the semi-infinite strip shown in

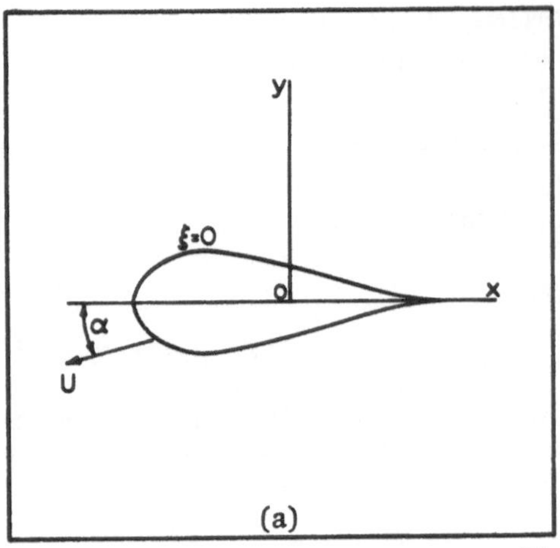

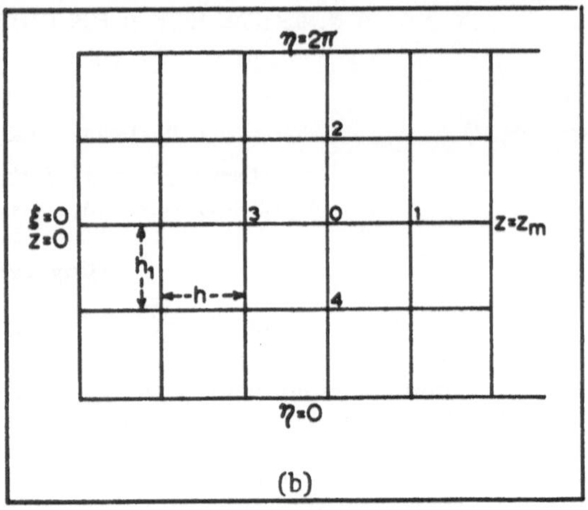

Fig. 1

Fig. 1(b). The object is to create a region which can be fitted exactly by a rectangular grid. A number of such transformations can be written down explicitly, e.g. for the cases of a circular cylinder, elliptic cylinder or finite flat plate, and Joukowski aerofoil.

It is assumed that lengths have been made dimensionless with respect to a representative dimension d and velocities with respect to U. The dependent variables are the dimensionless stream function ψ and the negative dimensionless vorticity ζ, defined by $\zeta = \partial u/\partial y - \partial v/\partial x$, where (u,v), given by $u = \partial\psi/\partial y$, $v = -\partial\psi/\partial x$, are dimensionless Cartesian velocity components. After the transformation (1), they satisfy the equations

$$\frac{\partial^2\psi}{\partial\xi^2} + \frac{\partial^2\psi}{\partial\eta^2} = \frac{\zeta}{H^2} \quad , \tag{2}$$

$$\frac{\partial\zeta}{\partial t} = H^2\left\{\frac{2}{R}\left(\frac{\partial^2\zeta}{\partial\xi^2} + \frac{\partial^2\zeta}{\partial\eta^2}\right) + \frac{\partial\psi}{\partial\xi}\frac{\partial\zeta}{\partial\eta} - \frac{\partial\psi}{\partial\eta}\frac{\partial\zeta}{\partial\xi}\right\} \quad , \tag{3}$$

where $$H^2 = (\partial\xi/\partial x)^2 + (\partial\xi/\partial y)^2 = (\partial\eta/\partial x)^2 + (\partial\eta/\partial y)^2 .$$

The Reynolds number R is defined by $R = 2Ud/\nu$, where ν is the kinematical viscosity. Finally, the time t in (3) is U/d times the actual time.

The functions ψ and ζ must be periodic functions of η of period 2π if the whole domain outside the cylinder is mapped on to the strip in Fig. 1(b) and thus

$$\psi(\xi,\eta,t) = \psi(\xi,\eta + 2\pi,t) \quad , \quad \zeta(\xi,\eta,t) = \zeta(\xi,\eta + 2\pi,t) . \tag{4}$$

The boundary conditions on the cylinder surface are

$$\psi = \partial\psi/\partial\xi = 0 \text{ when } \xi = 0 \quad , \quad t \geq 0 . \tag{5}$$

At large distances from the cylinder the stream relative to the x axis makes a constant angle α with it, so

$$u \to \cos\alpha \quad , \quad v \to \sin\alpha \quad , \quad \text{as } x^2 + y^2 \to \infty . \tag{6}$$

For the type of transformation (1) envisaged we find

$$x \sim Ke^{\xi}\cos\eta \quad , \quad y \sim Ke^{\xi}\sin\eta \quad , \quad \text{as } \xi \to \infty$$

and it can be shown that the conditions (6) give rise to the conditions

$$e^{-\xi}\partial\psi/\partial\xi \to K\sin(\eta - \alpha), \; e^{-\xi}\partial\psi/\partial\eta \to K\cos(\eta - \alpha), \quad \text{as } \xi \to \infty . \tag{7}$$

The quantity K is a constant which depends upon the particular details of the transformation (1). As a consequence of (7) it follows that

$$\zeta \to 0 \quad \text{as} \quad \xi \to \infty . \tag{8}$$

Finally, in any numerical scheme of integrating (3) it is necessary to state an initial condition for ζ, say

$$\zeta(\xi,\eta,0) = G(\xi,\eta) . \tag{9}$$

This condition is implied in the differential equations and boundary conditions already stated, but it is the precise specification of (9) and the subsequent use of finite-difference equations to approximate (2) and (3) which leads us to seek an alternative formulation for the solution for small and moderate times.

BOUNDARY-LAYER TRANSFORMATIONS

The nature of the boundary-layer solution at $t = 0$ suggests the transformations

$$\xi = kz \quad , \quad \psi = k\Psi \quad , \quad \zeta = \omega/k, \tag{10}$$

where $k = 2(2t/R)^{\frac{1}{2}}$. Equations (2) and (3) become

$$\frac{\partial^2\Psi}{\partial z^2} + k^2 \frac{\partial^2\Psi}{\partial\eta^2} = \frac{\omega}{H^2} , \tag{11}$$

$$H^2 \frac{\partial^2\omega}{\partial z^2} + 2z \frac{\partial\omega}{\partial z} + 2\omega = 4t \frac{\partial\omega}{\partial t} - \frac{8tH^2}{R} \frac{\partial^2\omega}{\partial\eta^2} - 4tH^2 \left(\frac{\partial\Psi}{\partial z}\frac{\partial\omega}{\partial\eta} - \frac{\partial\Psi}{\partial\eta}\frac{\partial\omega}{\partial z}\right) . \tag{12}$$

The advantage of this procedure is that equations (11) and (12) can be solved very easily by numerical methods near $t = 0$, and an initial expression $\omega = \Omega(z,\eta)$ for ω when $t = 0$ is readily obtained. This is found by solving (12) with the right side put equal to zero and with the quantity H put equal to its value $H_0(\eta)$ when $\xi = 0$, since by the first of (10), $\xi = 0$ when $t = 0$ for all z. It can be shown that the required solution which satisfies all the necessary conditions is

$$\Omega(z,\eta) = (4K/\pi^{\frac{1}{2}})H_0 e^{-z^2/H_0^2} \sin(\eta - \alpha) \tag{13}$$

and from this solution, which replaces (9) in effect, the step-by-step procedure can be started. Equation (13) is consistent with boundary-layer theory and a corresponding initial solution for Ψ, obtained by solving (11) with $k = 0$, may be found. This is not required, however, to start the integration procedure.

The boundary conditions (4), (5) and (8) are essentially unaltered when expressed in terms of the new variables and need not be re-stated. The conditions (7) become

$$e^{-kz}\, \partial\Psi/\partial z \to K \sin(\eta - \alpha),\ ke^{-kz}\partial\Psi/\partial\eta \to K \cos(\eta - \alpha),\ \text{as } z \to \infty. \tag{14}$$

A numerical method can be constructed which incorporates (14) into the solution procedure. It automatically gives the right initial condition that the external flow is potential flow.

NUMERICAL PROCEDURES

Equations (11) and (12) are solved within the finite rectangle shown in Fig. 1(b). The outer boundary is $z = z_m$, on which ω is taken to be zero. Equation (12) may be written as

$$t\partial\omega/\partial t = Q(z,\eta,t) . \tag{15}$$

For a given time step from t to $t + \Delta t$, a method similar to the Crank-Nicholson implicit procedure is used to approximate (15). We integrate both sides of (15) from t to $t + \Delta t$, using integration by parts on the left side, and then replace integrals by trapezoidal sums. This gives the result

$$(2t + \Delta t)\omega(z,\eta,t + \Delta t) - \Delta tQ(z,\eta,t + \Delta t) = (2t + \Delta t)\omega(z,\eta,t) + \Delta tQ(z,\eta,t). \tag{16}$$

The space derivatives of ω in Q are approximated by central differences. In the notation of Fig. 1(b) this gives

$$\begin{aligned} 2h(\partial\omega/\partial z)_0 &= \omega_1 - \omega_3 , \quad 2h_1(\partial\omega/\partial\eta)_0 = \omega_2 - \omega_4 , \\ h^2(\partial^2\omega/\partial z^2)_0 &= \omega_1 + \omega_3 - 2\omega_0 , \quad h_1^2(\partial^2\omega/\partial\eta^2)_0 = \omega_2 + \omega_4 - 2\omega_0 . \end{aligned} \tag{17}$$

The method of calculating the derivatives of Ψ will be mentioned shortly.

Equation (16) defines a matrix problem to determine $\omega(z,\eta,t + \Delta t)$. Since the space derivatives of $\Psi(z,\eta,t + \Delta t)$ occur in this problem, it must be solved jointly with some approximate analogue of equation (11) to determine $\Psi(z,\eta,t+\Delta t)$. It is not possible to describe fully the method which has been used, except that it is a generalization of a method proposed by Dennis and Chang (1969) for symmetrical flows. In this method the stream function was assumed to be represented by a Fourier sine series in the variable η. The generalization to the present case is the assumption

$$\Psi(z,\eta,t) = \sum_{n=1}^{\infty} \{f_n(z,t)\sin n\eta + g_n(z,t)\cos n\eta\} + \frac{1}{2} g_0(z,t) \ . \qquad (18)$$

If this is substituted in (11), we obtain the equations

$$\partial^2 f_n/\partial z^2 - n^2k^2 f_n = r_n(z,t) \qquad (n = 1,2,3, \ldots) \ ,$$
$$\partial^2 g_n/\partial z^2 - n^2k^2 g_n = s_n(z,t) \qquad (n = 0,1,2, \ldots) \ , \qquad (19)$$

where

$$r_n = \frac{1}{\pi}\int_0^{2\pi} (\omega/H^2)\sin n\eta \, d\eta \ , \quad s_n = \frac{1}{\pi}\int_0^{2\pi} (\omega/H^2)\cos n\eta \, d\eta \ . \qquad (20)$$

The solutions of equations (19) must satisfy

$$f_n = \partial f_n/\partial z = 0 \ , \quad g_n = \partial g_n/\partial z = 0 \ , \quad \text{when } z = 0. \qquad (21)$$

Also, it may be deduced from the conditions (14) and the differential equations (19) that

$$\int_0^{\infty} e^{-nkz} r_n dz = 2K\delta_n \cos\alpha \qquad (n = 1,2,3,\ldots) \ ,$$
$$\int_0^{\infty} e^{-nkz} s_n dz = -2K\delta_n \sin\alpha \qquad (n = 0,1,2,\ldots) \ , \qquad (22)$$

where

$$\delta_1 = 1 \ , \quad \delta_n = 0 \quad (n \neq 1) \ .$$

The method of solution of (19) and the use of the conditions (22) to calculate a boundary condition for the vorticity on $z = 0$ follows closely the procedure described by Dennis and Chang. The quantities $r_n(z,t)$ and $s_n(z,t)$ are calculated from the solution for ω for $z \neq 0$ and then $r_n(0,t)$, $s_n(0,t)$ are obtained by satisfying (22), expressed as numerical quadrature formulae. Then $\omega(0,\eta,t)$ is calculated from

$$\omega(0,\eta,t) = H^2 \Big[\sum_{n=1}^{\infty} \{r_n(0,t)\sin n\eta + s_n(0,t)\cos n\eta\} + \frac{1}{2} s_o(0,t) \Big] . \qquad (23)$$

From the solutions of (19) the space derivatives of Ψ are calculated. In any practical integration, the series (18) and (23) are approximated by a finite number n_o of terms, and this is a parameter of the solution. One advantage of this method of solution is that when $t = 0$, the boundary-layer solution gives rise to an expression for Ψ which can often be expressed very simply in the form (18). For example, for a circular cylinder

$$\Psi(z,\eta,0) = f_1(z,0)\sin\eta + g_1(z,0)\cos\eta, \qquad (24)$$

where f_1 and g_1 can be obtained exactly from boundary-layer theory. In this case the series (18) only builds up as t increases, and only a few terms are necessary for small t. Fuller details of this generalized method will be published elsewhere.

SYMMETRICAL FLOW PAST A CIRCULAR CYLINDER

The method has been tested for a circular cylinder, taking $\alpha = 0$ and assuming the flow to be symmetrical about the x axis. Only the region $0 \leq \eta \leq \pi$ need be considered, with

$$\Psi = \omega = 0 \quad \text{when} \quad \eta = 0, \pi \ .$$

The details of the transformation (1) are

$$\xi + i\eta = \log(x + iy) \ , \quad H = e^{-\xi},$$

which leads to the value $K = 1$. The representative dimension d is the radius of the cylinder a.

Calculations were carried out for $R = 100, 500, 10^3, 10^4$ and ∞. There are sev-

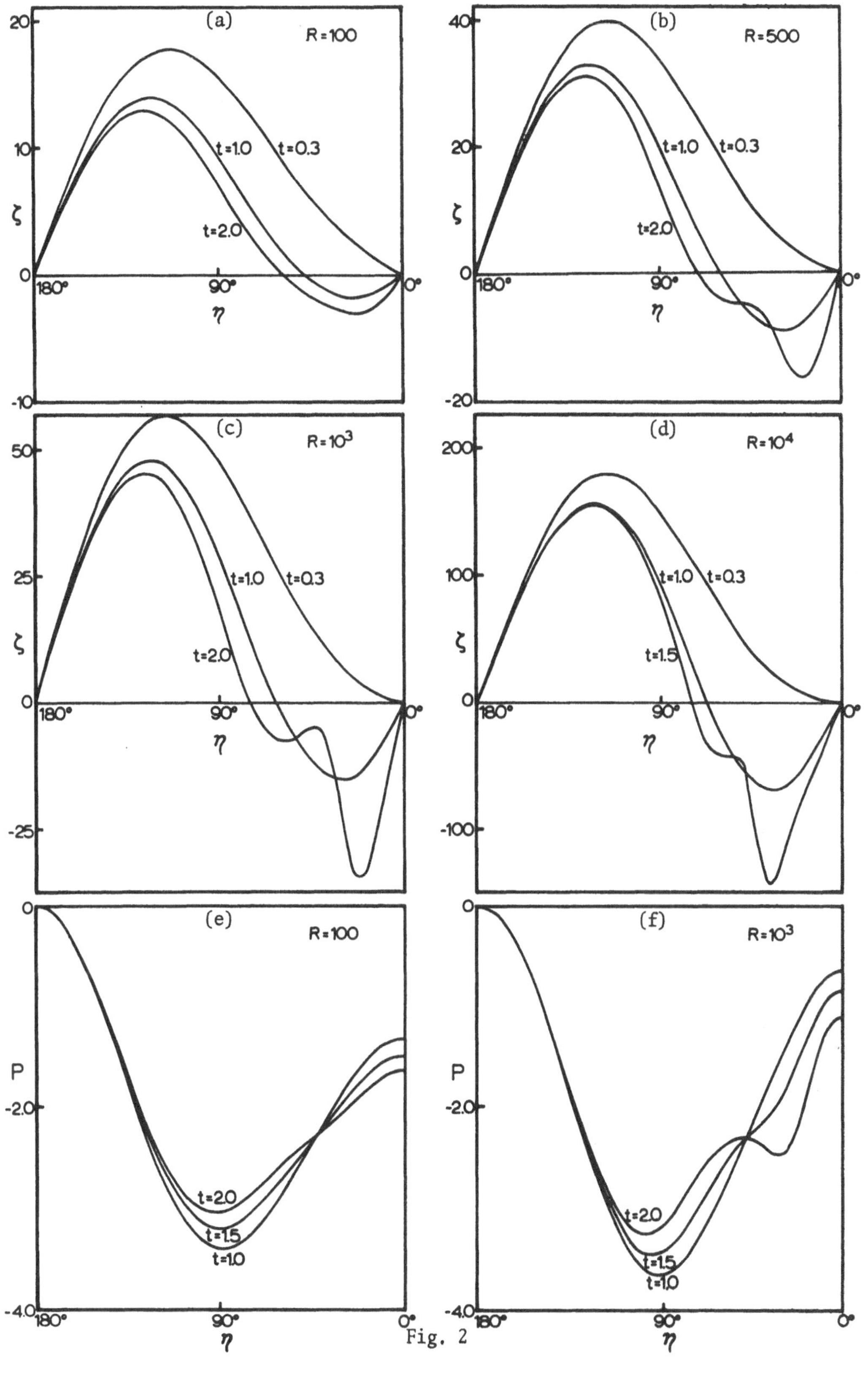
(a)
R=100
t=1.0
t=0.3
t=2.0
ζ
η
(b)
R=500
t=1.0
t=0.3
t=2.0
(c)
R=10³
t=1.0
t=0.3
t=2.0
(d)
R=10⁴
t=1.0
t=0.3
t=1.5
(e)
R=100
t=2.0
t=1.5
t=1.0
P
(f)
R=10³
t=2.0
t=1.5
t=1.0

Fig. 2

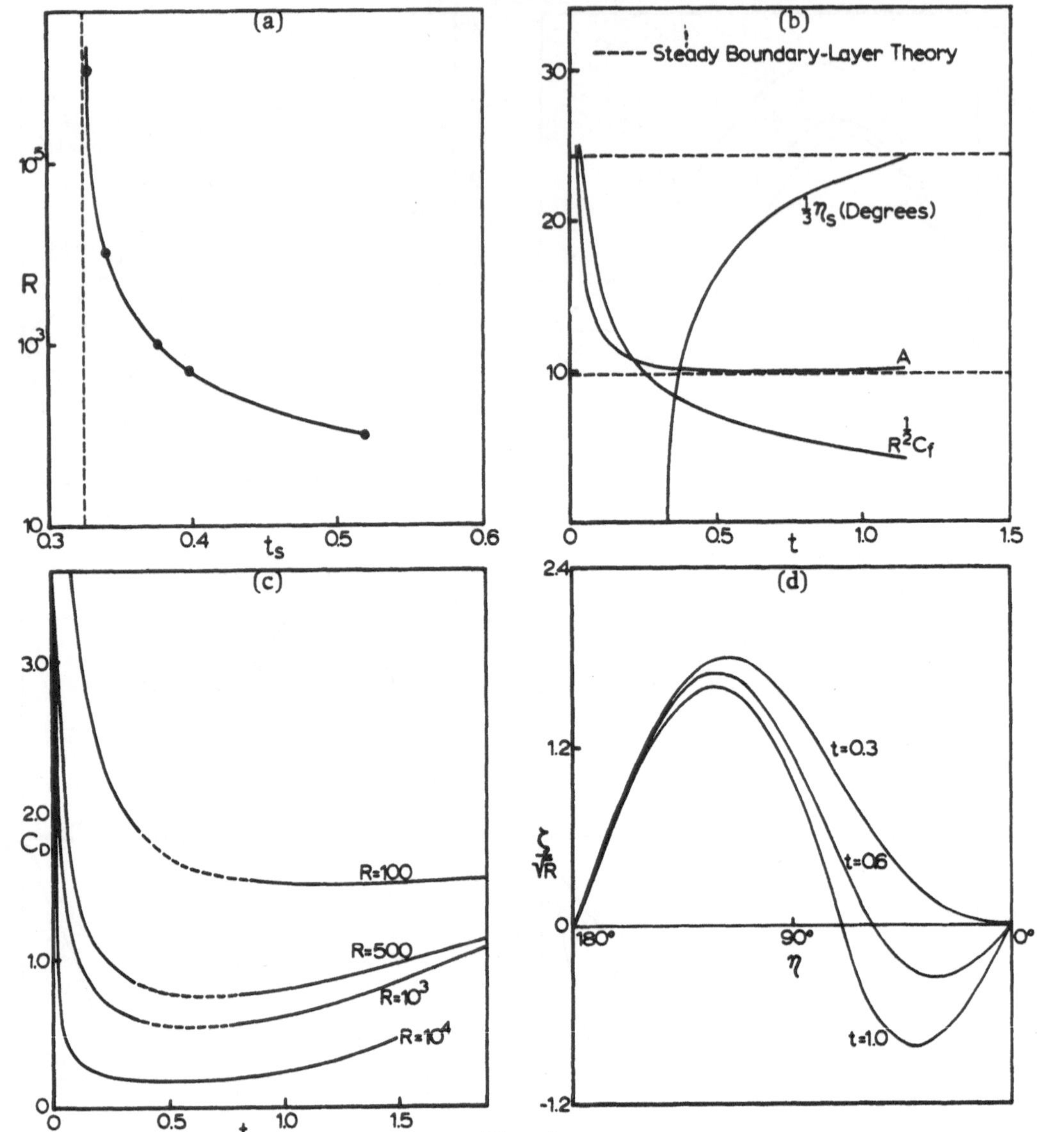

Fig. 3

eral parameters which can be varied, but the main results described in this section were calculated using $h = 0.05$, $h_1 = \pi/30$, $\Delta t = 0.025$, $n_0 = 30$ and $z_m = 5$. It may be seen from the initial solution (13) that $z_m = 5$ is satisfactory, at least for early times. Many checks on the solutions were applied by varying the parameters and the indication is that the solutions are reliable as far in time as they were continued. The solutions for $R = 100, 500$ and 10^3 were continued to $t = 2$, that for $R = 10^4$ to $t = 1.5$, and that for $R = \infty$ only to $t = 1.1$.

The main results are presented in Figs. 2 and 3. Parts (a) to (d) of Fig. 2 show the development of surface vorticity with time for the four finite values of R. It is initially infinite but rapidly decreases until eventually separation starts at the rear stagnation point. At some time thereafter, the vorticity starts to fluctuate in the separated region. This starts at earlier times as R increases. It is hardly noticeable at $t = 2$ for $R = 100$ and well advanced at $t = 1.5$ for $R = 10^4$. This type of behaviour has been noted by Son and Hanratty and by Thoman and Szewczyk. The present method breaks down soon after the more violent fluctuations set in. The iterations fail to converge. The pressure coefficient on the cylinder is calculated from

$$P(\eta) = \frac{p - p_\pi}{\frac{1}{2}\rho U^2} = \frac{1}{2t}\int_\eta^\pi \left(\frac{\partial\omega}{\partial z}\right)_{z=0} d\eta ,$$

where p is the pressure and p_π that at $\eta = \pi$. This coefficient is shown in parts (e) and (f) of Fig. 2 for $R = 100$ and 10^3, respectively.

The time of separation t_s is shown as a function of R in Fig. 3(a). A calculation at $R = 10^6$ is included. At $R = \infty$ there is good agreement with the value $t_s = 0.32$ calculated by Goldstein and Rosenhead. Fig. 3(c) gives the variation of total drag coefficient $C_D = D/(\rho U^2 a)$, where D is the drag and ρ the density. This was calculated from the formula

$$C_D = \frac{d}{dt}\left\{\int_0^\infty \int_0^\pi 2\omega e^{3kz} \sin\eta \, d\eta \, dz\right\}$$

obtained from the result given by Philips (1956). At lower values of R, C_D was found to fluctuate slightly with time shortly after separation has occurred and the dotted curves in Fig. 3(c) are mean curves in this region.

Some properties of the $R = \infty$ solution are given in Figs. 3(b) and 3(d). The friction drag coefficient C_f in Fig. 3(b) is calculated from

$$R^{\frac{1}{2}}C_f = \left(\frac{2}{t}\right)^{\frac{1}{2}} \int_0^\pi \omega(0,\eta,t)\sin\eta \, d\eta \quad .$$

The quantity η_s is the angle of separation. The quantity A is the magnitude of $R^{\frac{1}{2}}\partial c_f/\partial\eta$ at $\eta = \pi$, where c_f is the local coefficient of skin friction on the cylinder given by $c_f = 4\zeta(0,\eta,t)/R$. This is known to have the value $A = 9.86$ from steady boundary-layer theory with external potential flow. In Fig. 3(d) the variation of $R^{\frac{1}{2}}c_f/4$, which is equal to $\zeta/R^{\frac{1}{2}}$, is shown over the cylinder surface for the same case $R = \infty$. It was found to be impracticable to proceed beyond $t = 1.1$ in the case $R = \infty$. However, a very accurate solution procedure by Collins and Dennis (to be published) for the case of a circular cylinder has succeeded in integrating the equations to higher times in this case. In this solution there is no evidence of any fluctuation of surface vorticity such as is apparent in the cases of finite R. It is possible, but as yet by no means certain, that $R = \infty$ is a singular case. It is certain that the fluctuations are increased by the presence of the R^{-1} term in equation (12).

This work was supported by a grant from the National Research Council of Canada.

REFERENCES

Blasius, H. Zeit. Math. Phys., 56, 20-37 (1908).

Dennis, S.C.R. and Chang, G.-Z. Mathematics Research Center, University of Wisconsin Technical Summary Report No. 859, 1-89 (1969).

Goldstein, S. and Rosenhead, L. Proc. Camb. Phil. Soc., 32, 392-401 (1936).

Ingham, D.B. J. Fluid Mech., 31, 815-818 (1968).

Payne, R.B. J. Fluid Mech., 4, 81-86 (1958).

Philips, O.M. J. Fluid Mech., 1, 607-624 (1956).

Son, J.S. and Hanratty, T.J. J. Fluid Mech., 35, 369-386 (1969).

Thoman, D.C. and Szewczyk, A.A. Phys. Fluids Suppl. II, 76-86 (1969).

Wang, C.-Y. J. Math. Phys., 46, 195-202 (1967).

Wang, C.-Y. J. Appl. Mech., 34, 823-828 (1967).

AN ARBITRARY LAGRANGIAN-EULERIAN COMPUTING TECHNIQUE

C. W. Hirt*
Los Alamos Scientific Laboratory
University of California
Los Alamos, New Mexico

INTRODUCTION

The technique to be described, called ALE, is a combined Lagrangian and Eulerian computing method for the transient dynamics of an incompressible fluid. Because of the Lagrangian aspects of this technique it is applicable to flows with free surfaces or having material interfaces, but it also maintains Eulerian aspects to overcome undesirable grid distortions often associated with Lagrangian methods. This technique is referred to as an arbitrary Lagrangian or Eulerian computing method because there are three options for moving vertices: (1) they can flow with the fluid for Lagrangian computing, (2) they can remain fixed for Eulerian computing, or (3) they can move in an arbitrarily prescribed way to give a continuous rezoning capability.

The goal of this technique is to supply a tool for the study of complex flow problems where pure Eulerian or Lagrangian techniques cannot be used. For example, in the study of blood flowing through flexible arteries, the flexible walls of the artery are best represented by Lagrangian coordinate lines. However, the entire flow cannot be covered with a Lagrangian finite difference mesh for it would fold up as the blood streamed by. The ALE method could treat this problem by zoning the arterial walls with Lagrangian lines and then allowing Eulerian flow along the axis of the artery.

By combining the best features of Eulerian and Lagrangian methods, this new method is applicable to problems with many kinds of moving boundaries. In addition to boundaries that are driven by the fluid motions, as in the above example, the method is applicable to problems with prescribed boundary motions, such as might be used in the study of flow about a swimming fish or flow through a vibrating rubber hose. Also, since the ALE computing grid can always be rezoned to its original location, it can be used for purely Eulerian calculations where boundaries can then be treated as rigid or as input and output walls.

Each cycle of the ALE method consists of two phases. In phase I the Lagrangian equations of motion are solved in the same way as in the LINC method, Hirt, Cook, and Butler (1970), except that each new cell volume is made equal to its previous value, rather than its initial value. This is necessary because the volumes of cells may be changing in phase II, where a rezone calculation is performed. This second phase allows for the mesh to move relative to the fluid. The purpose of using two phases is to know the Lagrangian motion before a choice is made for rezoning. This is the most general case. In some instances, however, the rezoning can be determined in advance and it is unnecessary to perform the phase I calculations. For example, in a pure Eulerian calculation all vertices retain their initial position. Conversely, if a pure Lagrangian calculation is desired then the phase II steps are unnecessary since no rezoning is required.

---

*This work was performed under the auspices of the United States Atomic Energy Commission.

This paper will concentrate on the principle of the ALE technique without reference to specific finite difference approximations. In practice, various alternative approximations are possible. Those used to obtain the enclosed examples are described by Hirt and Amsden (1970).

FINITE DIFFERENCE MESH

To obtain numerical solutions the fluid is divided into a set of quadrilateral zones, with vertices denoted by indices (i,j), as shown in Fig. 1. Associated with each vertex are its coordinates (x_i^j, y_i^j), velocity (u_i^j, v_i^j), and mass M_i^j. Each cell has a density, ρ, volume, V, and pressure p, associated with its center. Cell centered quantities are labelled with indices $(i+\frac{1}{2}, j+\frac{1}{2})$. At the end of phase I calculations all quantities can be updated to the values they would have if a pure Lagrangian calculation is desired. Quantities at the end of phase I are denoted by the superscript L, e.g., $(u^L)_i^j$ denotes the horizontal velocity component of vertex (i,j) at the end of phase I.

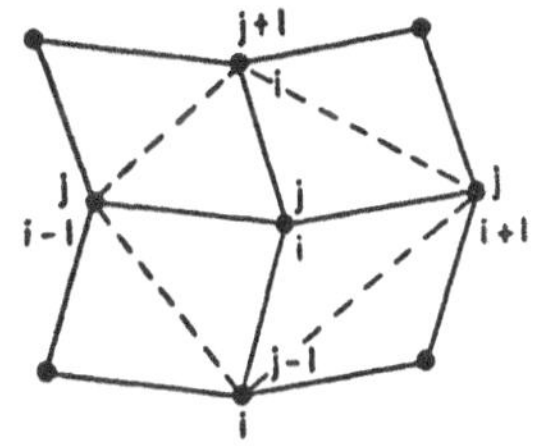

Fig. 1. Typical mesh configuration, with integration path indicated by dashed line.

THE BASIC EQUATIONS

The fluid dynamics is governed by a set of conservation laws that are conveniently written in integral form as

$$\frac{\partial}{\partial t}\int_V \rho dV - \int_S \rho\vec{U}_R\cdot\hat{n}dS = 0$$

$$\frac{\partial}{\partial t}\int_V \rho\vec{u}dV - \int_S \rho\vec{u}\vec{U}_R\cdot\hat{n}dS - \int_S \overline{\Pi}\cdot\hat{n}dS = 0 \quad . \tag{1}$$

The first equation expresses the conservation of mass and the second the conservation of momentum. Integrations are over a volume V or its surface S, which has the outward normal vector $\hat{n}$, and is moving relative to the fluid with velocity U_R.

The first term in the momentum equation is the time rate of change of momentum contained in V. The last term is the rate of change of momentum through the action of surface forces described by a stress tensor $\overline{\pi}$. The second term is the momentum change from a convective flux through the surface when the surface moves with velocity $\vec{U}_R$ with respect to the fluid.

DISCUSSION OF SOLUTION PROCEDURE

A. Phase I

Phase I is governed by the Lagrangian equations obtained from (1) by setting U_R to zero,

$$V^L = V^n$$

$$\frac{\partial}{\partial t}\int_V \rho\vec{u}dV = \int_S (-p\hat{n} + \overline{\Pi}\cdot\hat{n})\,dS + \int_V \vec{A}dV \quad . \tag{2}$$

The stress tensor has been split into three parts. The first part consists of pressure forces, the second, still denoted by $\overline{\Pi}$, contains forces conveniently defined at cell centers, and the last part, denoted by $\vec{A}$, contains forces conveniently defined at vertices. Π contains, for example, viscous and elastic forces, while $\vec{A}$ contains gravity or surface tension forces. Cell centered forces are those best approximated by surface integrals, and vertex centered forces are those best approximated by volume integrals.

To obtain a difference approximation for (2) at vertex (i,j), take as integration volume the area outlined by dashed lines in Figure 1. The time derivative on the left side of (2) is then differenced as

$$\frac{M_{in}^{L}\,\vec{u}^{L} - M_{in}^{n}\,\vec{u}^{n}}{\delta t} , \tag{3}$$

where M_{in} is the mass within the integration area. It is assumed that M_{in} is equal to twice the vertex mass, and the vertex mass is defined as one fourth the sum of neighboring cell masses. This assumption is not exact when the cells deviate from parallelograms. However, it is correct to first order in the mesh size, and it has been successfully used in the LINC method.

The calculation of $\vec{u}^{L}$ is carried out in two steps. A temporary updated velocity is calculated that does not contain pressure gradient forces,

$$\langle\vec{u}^{L}\rangle = \vec{u}^{n} + \frac{\delta t}{2M}\left\{\iint_S \Pi\cdot\hat{n}dS + \int_V \vec{A}dV\right\} \tag{4}$$

then after pressures are determined the final Lagrangian velocity is given by

$$\vec{u}^{L} = \langle\vec{u}^{L}\rangle - \frac{\delta t}{2M}\left\{\int_S p\hat{n}dS\right\} . \tag{5}$$

Any reasonable and conservative finite difference approximation can be used for the terms in curley brackets. The basic ALE technique is independent of this choice. Thus, for conciseness, we do not give explicit expressions for these terms. The interested reader will find complete details in the paper by Hirt and Amsden (1970).

B. The Pressure Equations

Pressures needed in (5) are obtained from the condition that the fluid remain incompressible. This is accomplished in the following way. The volume of a quadrilateral cell at the end of phase I is

$$V^{L} = 1/2 \sum_{k=1}^{4} x_{k}^{L}\,(y_{k+1}^{L} - y_{k-1}^{L}) , \tag{6}$$

where the index k labels the cell vertices counterclockwise. The vertex coordinates at the end of the Lagrangian phase are

$$x^{L} = x^{n} + \delta t\,\left(\frac{u^{n} + u^{L}}{2}\right)$$
$$y^{L} = y^{n} + \delta t\,\left(\frac{v^{n} + v^{L}}{2}\right) . \tag{7}$$

Inserting (7) into (6), together with the definition of $\vec{u}^{L}$ from (5), leads to an expression for the cell volume in terms of the unknown pressures. When we require each new cell volume to equal its value at the beginning of the cycle, this results in a set of coupled algebraic equations for the pressures. Although these equations are not linear, to a good approximation the nonlinear pressure terms can be omitted since they are of higher order in δt.

The pressure equation is easily solved by a Gauss-Seidel point over-relaxation technique. It would probably be faster to use an alternating direction method, although the computer programming required would be more complex. These equations are not easily solved by direct methods because of their variable coefficients.

Once the pressures have been determined, Lagrangian velocities are calculated using (5). If only a Lagrangian calculation is wanted then vertices are moved according to (7). Otherwise, rezone velocities must be specified and phase II calculations performed.

C. The Rezone Velocity

Having determined the Lagrangian velocities it is a simple matter to get rezone velocities. An example will illustrate the process. Suppose the vertices are to remain on specified vertical lines. They may move up and down along these lines, but not in a normal direction. To accomplish this all vertices are first moved to the positions they would occupy if they followed the fluid according to (7). The intersections of the new horizontal grid lines with vertical lines at the specified positions can be easily found. Rezone velocities are then calculated to move vertices to these intersection points.

Once rezone velocities have been assigned, vertices are moved according to the prescription,

$$\begin{aligned} x^{n+1} &= x^n + \delta t \left(\frac{u^n + u^L}{2} + U_R\right) \\ y^{n+1} &= y^n + \delta t \left(\frac{v^n + v^L}{2} + V_R\right) \quad . \end{aligned} \tag{9}$$

From these values new cell volumes and masses can be computed.

D. Phase II

It now remains to describe the phase II calculations, which account for the transfer of mass and momentum between cells during rezoning. These calculations are quite similar to those in phase I. A temporary velocity field is first calculated that accounts for the convective fluxes,

$$\langle \vec{u}^{n+1} \rangle = \left(\frac{M^n}{M^{n+1}}\right) \langle \vec{u}^L \rangle + \frac{\delta t}{2M^{n+1}} \left\{ \iint_S \rho \vec{u} U_R \cdot \hat{n} dS \right\} \quad . \tag{10}$$

Specific difference approximations for the convective terms are omitted. The first term on the right side of (10) contains the ratio of masses in successive time steps. This factor accounts for a change of mass, and hence momentum, incurred during rezoning. Notice also that this term involves the temporary velocity used in phase I. Thus, equation (10) is really an approximation for (1) without the pressure acceleration term.

Final velocities for the cycle are obtained by combining the temporary velocities (10) with the pressure acceleration,

$$\vec{u}^{n+1} = \langle \vec{u}^{n+1} \rangle - \frac{\delta t}{2M^{n+1}} \left\{ \iint_S p \hat{n} dS \right\} \quad . \tag{11}$$

The pressures needed in (11) are those that insure satisfaction of the incompressibility condition. That is, the final velocity field must possess a zero velocity divergence in every cell. The average divergence in a cell is approximated by the expression

$$\int_V \nabla \cdot \vec{u} dV \approx \frac{1}{2} \sum_{k=1}^{4} [u_k (y_{k+1} - y_{k-1}) + v_k (x_{k-1} - x_{k+1})] \tag{12}$$

where k labels the cell vertices counterclockwise.

Pressures are determined by substituting (11) into the right side of (12) and setting the resulting expression to zero. This guarantees that velocities determined from these pressures will satisfy the incompressibility condition.

Final velocities for the cycle are calculated from (11). This completes the updating of all quantities.

E. Boundary Conditions

Various types of boundaries are of interest if the ALE method is to be applicable to a wide variety of problems. In particular, it is desirable to have options for rigid free-slip or no-slip walls, free surfaces, input and output boundaries, and walls with specified motions. In all cases the imposed boundary conditions must be consistent with the finite difference approximations. Complete details for a variety of boundary conditions are contained in Hirt and Amsden (1970).

DISCUSSION OF EXAMPLES

A particularly simple example has been chosen to illustrate this new computing technique. A pressure pulse is used to set fluid in a rectangular tank into oscillation. This problem has been extensively studied analytically and numerically so it serves as a good test problem.

In the present example, the fluid initially occupies a region 12 units (cells) wide and 9 units (cells) high. A gravity acceleration of one unit acts downwards. At $t = 0$ an impulsive cosine pressure pulse is applied at the free surface, which sets the fluid in oscillatory motion. The motion is periodic, with period 12.8 units (64 time steps), but the surface crests are narrower and higher, while the troughs are broader and shallower, than predicted by simple linear theory. These are typical nonlinear effects.

To demonstrate the versatility of the ALE method, several calculations were made, each employing a different rezone option. In every case the computed results agree well with one another and with a LINC calculation adopted as a standard for comparison.

Figure (2a) shows the fluid cell configuration at $t = 3.0$ units, as determined from a purely Lagrangian calculation without rezoning. This time corresponds approximately to one quarter period when the fluid has reached its maximum amplitude. The calculation is essentially identical to that obtained using the LINC method.

Figure (2b) is the same calculation, except that each column of vertices was rezoned to lie beneath a surface vertex. This gives the accordion-like appearance to the mesh. The rezone velocities required for this calculation are easily obtained once the phase I calculations have been performed.

In Figure (2c) the vertices were rezoned to their initial horizontal positions by the method described in the text. Thus, the horizontal flow is Eulerian and the vertical flow is Lagrangian.

Again, in Figure (2d) vertices were rezoned to their initial horizontal positions, as in Fig. (2c), but they were also given a uniform vertical spacing in each column. This choice is probably the best one to use for this problem, as it maintains the most regular mesh shapes throughout the calculation.

To illustrate how ALE performs Eulerian calculations the bottom four rows of the mesh were continuously rezoned to their initial configuration. In this region, then, the flow is treated as Eulerian. Figure (2e) shows the resulting cell configuration at three quarters of a period. This is clearly not a good rezone method to use for the present problem, because the mesh has developed undesirable kinks at the boundary between the Eulerian and Lagrangian zones. However, these kinks can be eliminated by rezoning the upper rows in such a way that each vertex remains at the average position of its eight neighbors, Figure (2f). This type of rezoning is often useful in maintaining regular meshes.

ACKNOWLEDGMENT

Many helpful suggestions and all of the computer programming that made this study possible were contributed by A. A. Amsden and T. D. Butler. The author wishes to express his thanks to these colleagues.

REFERENCES

Hirt, C. W. and Amsden, A. A., to be published (1970).

Hirt, C. W., Cook, J. L., and Butler, T. D., Jour. Comp. Phys. 5, 103 (1970).

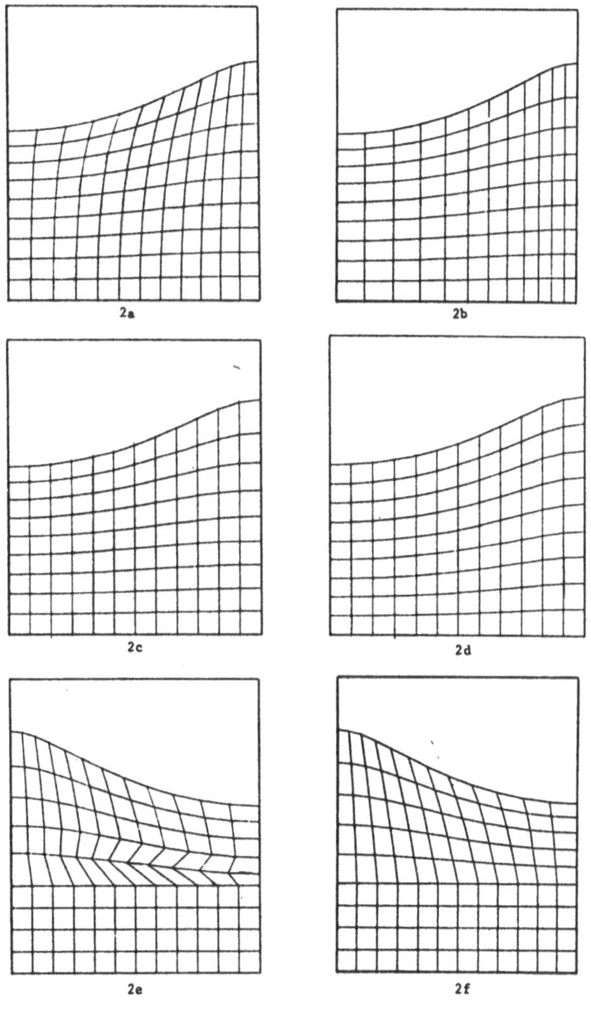

Figure 2. Six rezone options for a slosh problem.

COMPUTATION OF VISCOUS FLOW IN A CHANNEL BY THE METHOD OF SPLITTING

T. D. Taylor and E. Ndefo

Northrop Corporate Laboratories
Hawthorne, California

INTRODUCTION

Advances in the field of high speed computers have made it possible to consider numerical simulation of fluid flow systems by solving the exact conservation equations. As a result there is considerable interest in numerical techniques for integrating the Navier-Stokes equations. In this field, various numerical integration schemes have been attempted[1-3]. One of the most interesting is the method of splitting proposed by Yanenko.[4] In this technique, the N dimensional unsteady Navier-Stokes equations are reduced to a coupled set of N-one dimensional unsteady flow problems. As a consequence, implicit or high order explicit numerical schemes for 1-D unsteady equations can be used to perform the numerical integration. This paper summarizes the results of the study.

The problem chosen for application of the method of splitting was viscous incompressible flow in a two dimensional channel. The channel was first considered to have straight walls so that results could be checked with asympotic solutions. One wall was then adjusted to include a step so that separated flow would occur in the channel. The procedure was used to compute the flow for Reynolds numbers of 25 and 100 (based on maximum channel width). The computational steps taken to arrive at the results will now be discussed.

THE METHOD OF SPLITTING FOR INCOMPRESSIBLE FLOWS

In order to establish the method of splitting for incompressible flow, consider first the flow in the channel shown in Figure 1. This flow is described by the viscous flow equations which can be written in the dimensionless form

$$\frac{\partial u}{\partial x} + \frac{\partial v}{\partial y} = 0 \tag{1}$$

$$\frac{\partial u}{\partial t} + \frac{\partial u^2}{\partial x} + \frac{\partial uv}{\partial y} + \frac{\partial p}{\partial x} = \frac{1}{Re}\left\{\frac{\partial^2 u}{\partial x^2} + \frac{\partial^2 u}{\partial y^2}\right\} \tag{2}$$

$$\frac{\partial v}{\partial t} + \frac{\partial uv}{\partial x} + \frac{\partial v^2}{\partial y} + \frac{\partial p}{\partial y} = \frac{1}{Re}\left\{\frac{\partial^2 v}{\partial x^2} + \frac{\partial^2 v}{\partial y^2}\right\} \tag{3}$$

In these equations the variables have been made dimensionless by dividing velocities by the entrance velocity U_o, pressure by ρU_o^2, space dimensions by the channel width H and the time by H/U. Note also that ρ denotes the fluid density and Re the Reynolds Number $\frac{\rho U_o H}{\mu}$, where μ is the fluid viscosity.

These equations were integrated subject to the following initial and boundary conditions:

$$\text{at } t = 0,\ 0 \leq x \leq L/H,\ 0 \leq y \leq 1;\ u = 1,\ v = 0,\ p = p_o$$

$$\text{at } x = 0,\ 0 \leq y \leq 1;\ u = 1,\ v = 0,\ p = p_o \ \text{ for } t > 0$$

$$\text{at } x = L/H,\ 0 \leq y \leq 1;\ \frac{\partial u}{\partial x} = \frac{\partial v}{\partial x} = 0,\ \frac{\partial p}{\partial x} = -\beta(\text{const}) \text{ for } t > 0$$

Straight Channel

$$y = 0 \text{ or } 1,\ 0 \leq x \leq L/H;\ u = v = 0,\ \frac{\partial p}{\partial y} = \frac{1}{Re}\frac{\partial^2 v}{\partial y^2}$$

Step Channel

$$\text{at } y = h/H \text{ for } 0 \leq x \leq \ell/H,\ \text{at } y = 1 \text{ for } 0 \leq x \leq L/H$$

$$\text{and at } y = 0 \text{ for } \ell/H \leq x \leq \frac{L}{H};\ u = 0,\ v = 0,\ \frac{\partial p}{\partial y} = \frac{1}{Re}\frac{\partial^2 v}{\partial y^2}$$

$$\text{at } x = \frac{\ell}{H} \text{ for } 0 \leq y \leq h/H;\ u = 0,\ v = 0,\ \frac{\partial p}{\partial x} = \frac{1}{Re}\frac{\partial^2 u}{\partial x^2}$$

In these conditions, β is the prescribed pressure gradient, ℓ is the distance from the entrance to the step, L is the total length of the channel and h is the step height.

As written, equations (1) through (3) represent time dependent equations which can be solved explicitly for u and v by finite difference methods. The pressure p is implicit, however, since it does not appear in a time derivative. Yanenko suggested that for steady flows where the exact form of the transient is unimportant, another approach may be employed to determine p.

For this case Yanenko proposed that the continuity equation be modified to the form

$$\frac{\partial w}{\partial t} + \frac{\partial u}{\partial x} + \frac{\partial v}{\partial y} = 0 \tag{4}$$

$$\text{where } w = p + \frac{u^2 + v^2}{2}$$

Equations (2), (3) and (4) could then be integrated as initial value problems by employing the method of splitting to reduce the equations to the form

SET I

$$\frac{1}{2}\frac{\partial w_1}{\partial t} + \frac{\partial u}{\partial x} = 0$$

$$\frac{1}{2}\frac{\partial u}{\partial t} + \frac{\partial u^2}{\partial x} + \frac{\partial p}{\partial x} = \frac{1}{Re}\frac{\partial^2 u}{\partial x^2}$$

$$\frac{1}{2}\frac{\partial v}{\partial t} + \frac{\partial uv}{\partial x} = \frac{1}{Re}\frac{\partial^2 v}{\partial x^2}$$

SET II

$$\frac{1}{2}\frac{\partial w_2}{\partial t} + \frac{\partial v}{\partial y} = 0$$

$$\frac{1}{2}\frac{\partial u}{\partial t} + \frac{\partial uv}{\partial y} = \frac{1}{Re}\frac{\partial^2 u}{\partial y^2}$$

$$\frac{1}{2}\frac{\partial v}{\partial t} + \frac{\partial v^2}{\partial y} + \frac{\partial p}{\partial y} = \frac{1}{Re}\frac{\partial^2 v}{\partial y^2}$$

where $w_1 = p + \frac{u^2}{2}$ and $w_2 = p + \frac{v^2}{2}$

These sets form the basis for a numerical procedure. The principal is to integrate Set I, from a prescribed set of initial conditions, for a half time step along lines of constant y. The answers are then used as initial conditions for integrating Set II along lines of constant x for another half time step. The resulting answers are then taken to be the numerical solution of equations (2), (3) and (4) for a full time step.

Two additional steps must be taken before the integration can be accomplished. The first is splitting the boundary conditions and the second is selection of a differencing scheme. Splitting the boundary conditions is straightforward except at corners of the channel boundaries. To illustrate this, consider splitting the boundary conditions for the step channel. The results are:

SET I

at $x = 0$; $u = 1$, $v = 0$, $p = p_o$ for all y

at $x = L/H$; $\frac{\partial u}{\partial x} = \frac{\partial v}{\partial x} = 0$ and $\frac{\partial p}{\partial x} = -\beta$ for all y

at $x = \ell/H$; $u = 0$, $v = 0$ and $\frac{\partial p}{\partial x} = \frac{1}{Re}\frac{\partial^2 u}{\partial x^2}$ for $0 \leq y \leq \frac{h}{H}$

SET II

at $y = \frac{h}{H}$ for $0 \leq x \leq \frac{\ell}{H}$ and $y = 0$ for $\frac{\ell}{H} \leq x \leq \frac{L}{H}$; $u = 0$, $v = 0$ and $\frac{\partial p}{\partial y} = \frac{1}{Re}\frac{\partial^2 v}{\partial y^2}$

at $y = 1$ for $0 \leq x \leq \frac{L}{H}$; $u = 0$, $v = 0$ and $\frac{\partial p}{\partial y} = \frac{1}{Re}\frac{\partial^2 v}{\partial y^2}$

Upon examining these conditions it is apparent that at the corners of the boundaries the split conditions are not consistent. This can be overcome by using averaged conditions at the corners. For example, at the channel entrance the condition

$$p + \frac{\partial p}{\partial y} = p_o + \frac{1}{Re}\frac{\partial^2 v}{\partial y^2}$$

can be employed while at the step corners the condition

$$\frac{\partial p}{\partial x} + \frac{\partial p}{\partial y} = \frac{1}{Re}\left\{\frac{\partial^2 u}{\partial x^2} + \frac{\partial^2 v}{\partial y^2}\right\}$$ can be introduced.

At the end of the grid introducing the condition

$$\frac{\partial p}{\partial x} + \frac{\partial p}{\partial y} = -\beta + \frac{1}{Re}\frac{\partial^2 v}{\partial y^2}$$ eliminates the difficulty.

The selection of the numerical differencing procedure for integrating the split equations depends on the equation employed to determine pressure. In the test problems, various numerical methods were attempted in conjunction with different approaches to determine the pressure. These included variations of the scheme proposed by Yanenko in which w was taken to equal p or $p + \frac{u^2 + v^2}{2}$ with the damping term $\epsilon(t)\left\{\frac{\partial^2 p}{\partial x^2} + \frac{\partial^2 p}{\partial y^2}\right\}$ added to the continuity equation. All of these approaches failed to yield a steady state solution after 14,000 time steps even when implicit differencing was employed. The authors determined, however, that the best procedure for determining the pressure was to combine equations (1), (2) and (3) to yield

$$\frac{\partial^2 p}{\partial x^2} + \frac{\partial^2 p}{\partial y^2} = 2\left[\frac{\partial u}{\partial x}\frac{\partial v}{\partial y} - \frac{\partial v}{\partial x}\frac{\partial u}{\partial y}\right] \tag{5}$$

This equation is implicit, but does not have a time derivative for p. As a result, p cannot be determined by solving an initial value problem. This equation is solved at each half time step in conjunction with the split momentum equations for u and v. For simplicity, an explicit numerical scheme was used on the u and v equations with a three point formula for the second derivatives and central differencing of the convective terms. An implicit scheme could have been employed to increase accuracy since the split equations are one dimensional.

The Poisson type pressure equation (5) was differenced by a standard three point formula for the second derivatives and central differences for the first derivatives. The difference equations were solved both by Jacobi iteration and by over-relaxation. It was found that over-relaxation reduced the computation time by about a factor of three from that required by the simple Jacobi iteration procedure.

CALCULATION RESULTS

The splitting scheme which has been described was utilized to compute the flow within a straight channel for Reynolds numbers of 25 and a step channel for 25 and 100. The downstream pressure gradient β was chosen to have the value
$\beta = \frac{-12}{Re}\left(\frac{H-h}{H}\right)$.

For the calculations a step size of $\Delta x = 2\Delta y = 0.1$ was used for the straight channel while a variable step size ranging from $\Delta x = \Delta y = 0.01$ to $\Delta x = 0.1$ and $\Delta y = 0.09$ was used for the step channel. The arrangement of the grid for the step is shown in figure 1.

The time step for all the calculations was chosen to be $\Delta t = 0.001$. The calculations were performed on a CDC 6600 computer. The steady results for the u and v velocity profiles in the straight channel are shown in figures 2, 3 and 4. These results required 2,000 time steps to reach the steady state which amounted to five minutes of computer time when over-relaxation was used. A study of the results by employing conservation of mass to check for errors revealed the following facts regarding the accuracy of the numerical computation:

1) In computing the flow in the first (entrance) cell of the grid where the boundary conditions are discontinuous, the error in velocity resulted in a mass loss of 8%. By reducing the cell size a factor of ten, the mass loss was reduced to 5%. It appears, however, that in order to accommodate the singular boundary conditions with greater accuracy it is necessary to utilize a special numerical or analytical scheme.

2) The total error in the velocity profiles downstream of the first cell produced a 2% mass loss. This loss is what one would anticipate from an error estimate.

3) The maximum centerline velocity was 1.31. This is approximately 10% below that predicted by theory. This reduction is consistent with the 10% overall mass loss in the computation.

The results obtained for flow past a backward facing step in a channel are shown in figures 5 through 11. In figures 5a and 5b the u velocity profiles are shown for Reynolds numbers of 25 and 100. When these results were tested for conservation of mass they showed a 5% mass loss in the entrance cell and a total mass loss of 2% for the rest of the computation. Figure 6 shows the normal velocity profiles in the channel for a Reynolds number of 100 and figure 7 shows the pressure distribution on the lower wall. In figures 8a and 8b the streamline patterns are shown for the two different Reynolds numbers. These were obtained by solving the stream function equation by relaxation. A more detailed map of the streamlines near the base for a Reynolds number of 100 is shown in figure 9. Note that separation occurs at about 2/3 the step height instead of at the top. Figures 10 and 11 show selected results for the local behavior of velocity and pressure in the vicinity of the step. Note in figure 11 that the pressure on the face of the step first decreases sharply, then begins to increase until the adverse pressure gradient is sufficient to induce separation. The slight dip in pressure just downstream of the step indicates that the reverse flow near the base encounters adverse pressure gradient before it stagnates. Attempts were made to compare the step results with experiment, but detailed experimental data for flow in a channel with a step could not be found.

CONCLUSIONS

The method of splitting has been applied to calculate incompressible channel flow with and without separation. The results indicate that the method can be applied economically and without difficulty to compute small Reynolds number flows. The upper limit on the Reynolds number is unknown at this time. The method does, however, produce a loss of mass in regions of extreme gradients such as occur in the entrance regions of channels or on leading edges of plates. It is possible that this may be improved by changing to an implicit difference for integrating the split equations.

The study revealed that replacing the continuity equation with the Poission type pressure equation results in a trouble free computational scheme. This is with the provision that a rapid method such as over-relaxation be used to solve the pressure equation.

ACKNOWLEDGMENT

This work was supported by the Office of Naval Research under Contract N00014-70-C-0034.

REFERENCES

1. J. S. Allen and S. I. Cheng, Numerical solutions of the compressible Navier-Stokes equations for the laminar near wake, Phys. of Fluids, 13, (1970), 37.

2. P. J. Roache and T. J. Mueller, "Numerical solutions of compressible and incompressible laminar separated flows," AIAA Paper No. 68-741, presented at Los Angeles, California, June 1968.

3. F. N. Frenkiel and K. Stewartson, editors, "High-speed computing in fluid dynamics," (Proceedings of a Symposium), Phy. of Fluids Supplement, 1969.

4. N. N. Yanenko, "Method of fractional steps for solution of multidimensional problems of mathematical physics," Academy of Sciences SSR, Siberian Division, Novosibirsk, 1967.

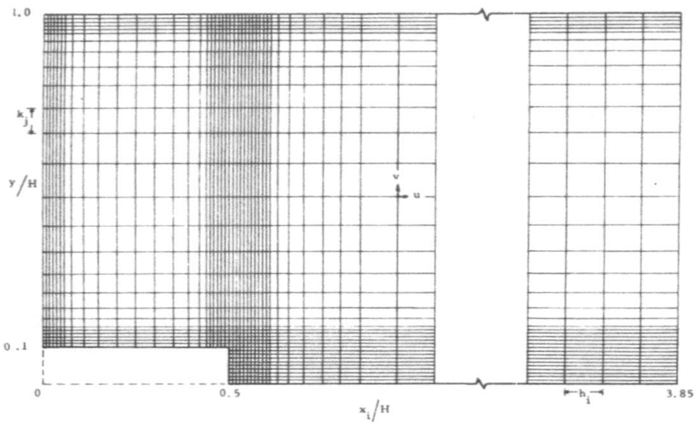

Fig. 1 Channel geometry and grid for step channel

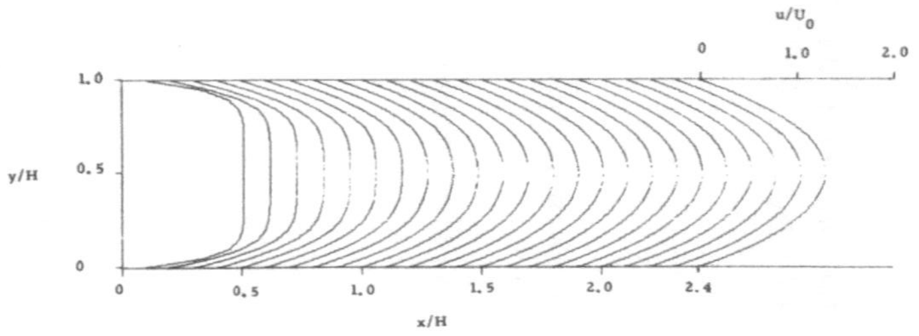

Fig. 2 u velocity profiles (Re = 25) for straight channel

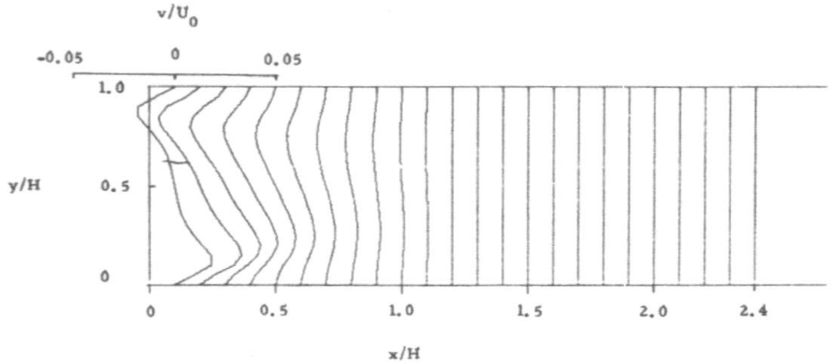

Fig. 3 v velocity profiles (Re = 25) for straight channel

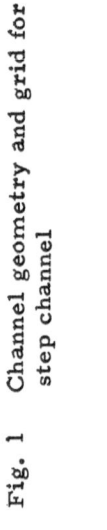

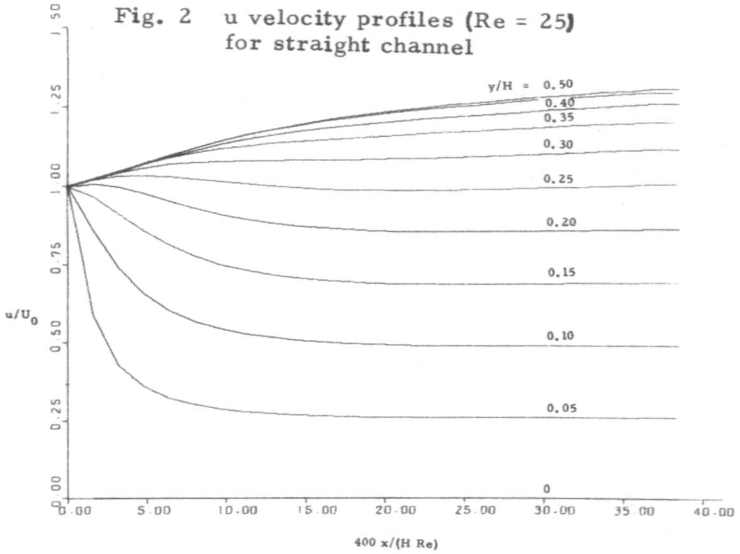

Fig. 4 Variation of u velocity along channel

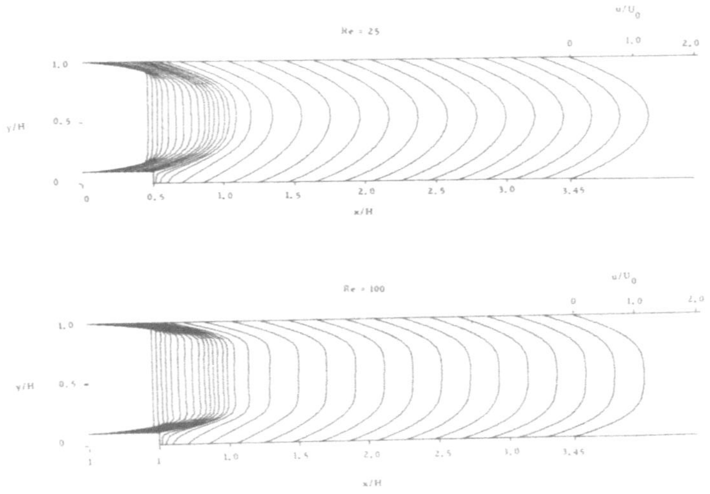

Fig. 5 u velocity profiles for channel with step

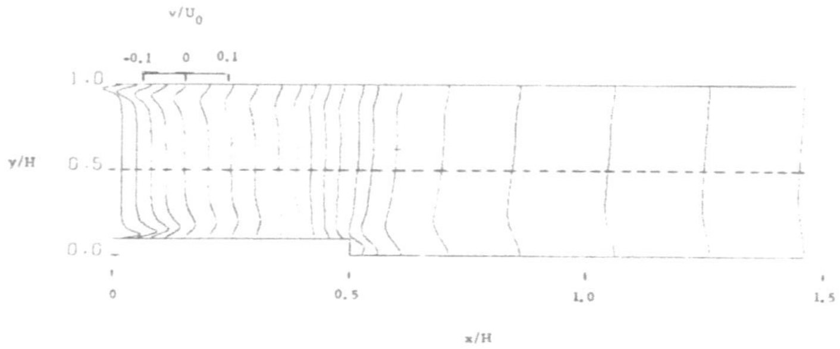

Fig. 6 v velocity profiles for channel with step (Re = 100)

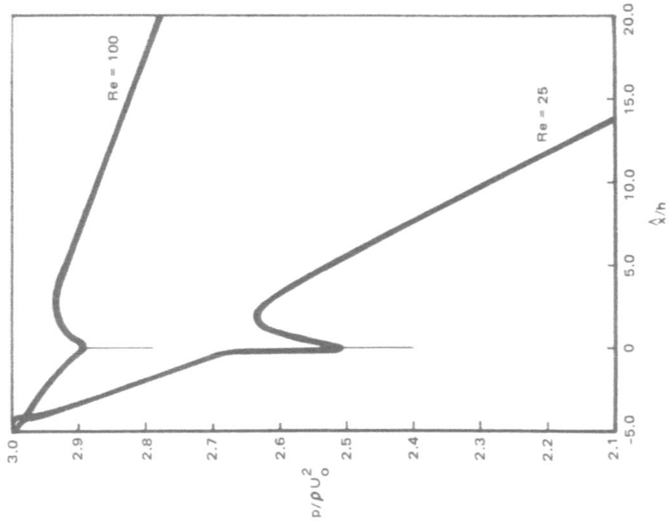

Fig. 7 Pressure distribution on lower wall

Re=100

Re = 25

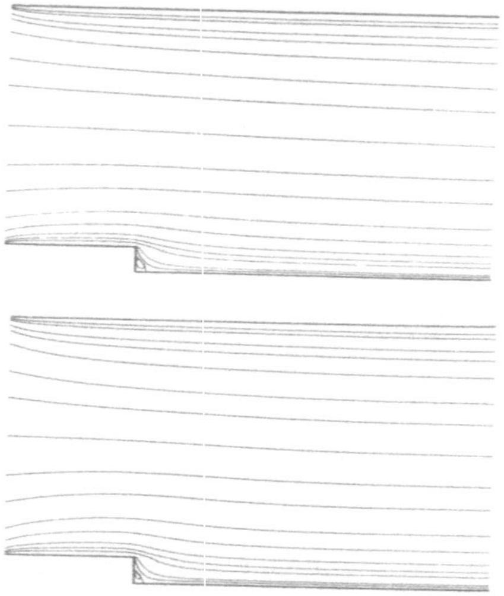

Fig. 8 Streamlines for channel with step

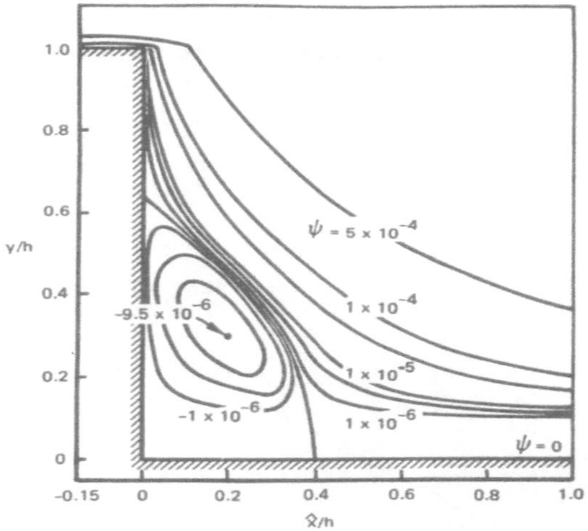

Fig. 9 Details of streamlines near the base (Re = 100)

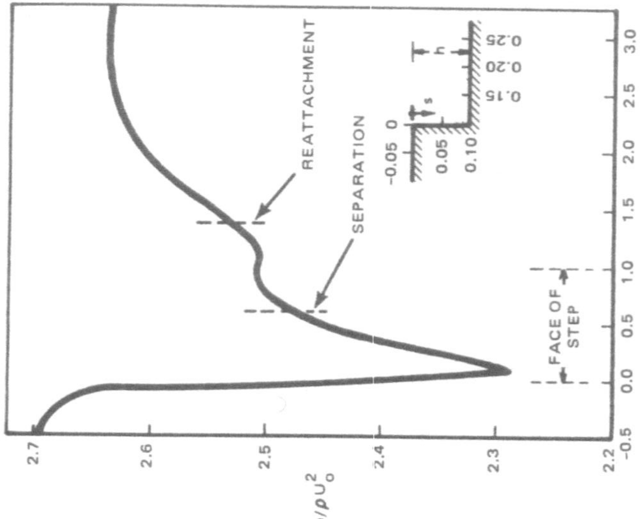

Fig. 10 Surface pressure distribution near the step (Re = 25)

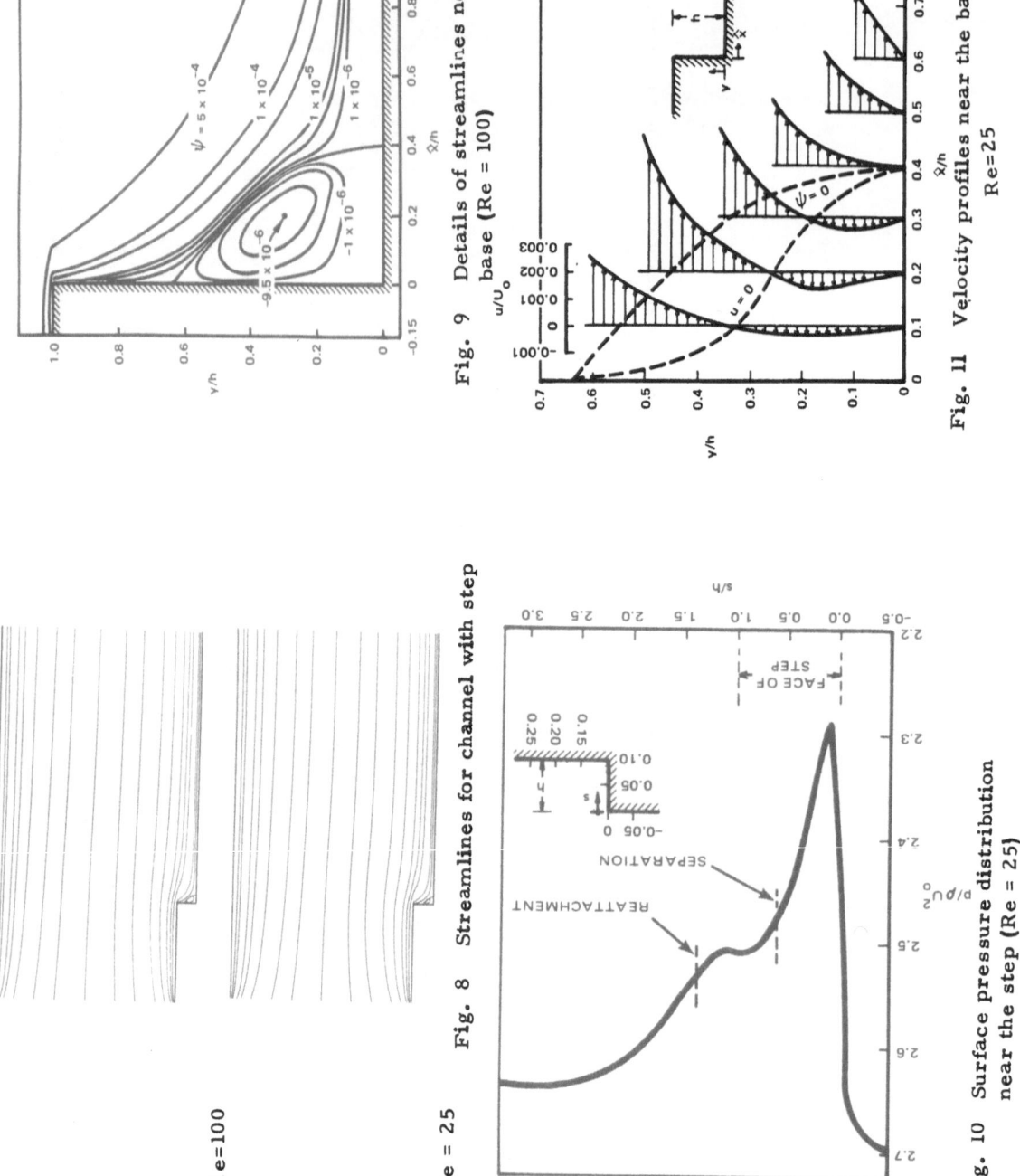

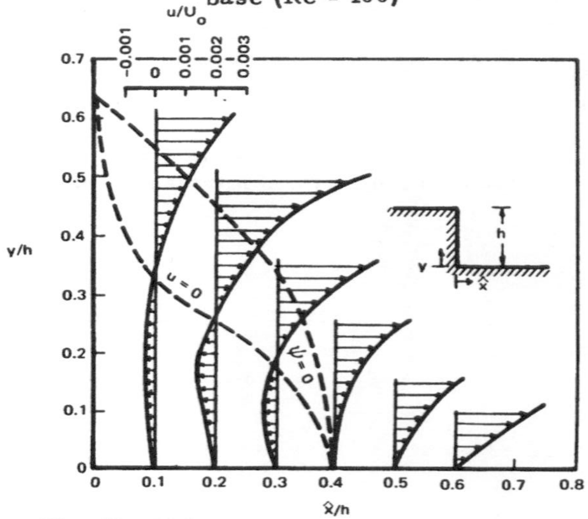

Fig. 11 Velocity profiles near the base Re=25

STUDY OF VORTEX FLOWS AT HIGH SWIRL
BY AN INTEGRAL METHOD USING EXPONENTIALS

Hartmut H. Bossel
Mechanical Engineering Department
University of California, Santa Barbara, Calif. 93106

ABSTRACT Rotationally symmetric quasi-cylindrical viscous incompressible vortex flows have been computed with different initial profiles and for a wide variety of swirl parameters and external axial velocity and circulation gradients. The computational method uses exponentials in the approximating functions for axial velocity and circulation profiles and for weighting functions. Several singular values of the swirl parameter S are identified, the most important being S_o and S_I which divide stable from stagnating, and supercritical from subcritical vortex flow, respectively.

INTRODUCTION

Very few exact solutions exist for viscous vortex flows [1]. They describe only a small portion of vortex flows observeable in nature and in experiments. More general flows must be dealt with numerically. Some previous approaches are described in [2, 3, 4, 5].

The object of the present research was exploration of possible solutions under a wide range of parameters, in particular swirl parameter $S = (dw/dr)_{ax}\, r_c/u_{ax}$ (see Fig. 1), initial profiles, and external axial velocity and circulation gradients. Since a large number of cases had to be computed, an economical method of computation was required. An N-parameter integral method was chosen. The method uses concepts previously applied successfully to boundary layer computation [6]. The partial differential equations are reduced to a set of ordinary differential equations for the parameters, eliminating stability problems and iterations. Qualitative answers can be obtained at minimal computing cost by one or two parameter solutions. Higher accuracy can be gained by increasing the number of parameters without any change in the program.

COMPUTATIONAL METHOD

Let coordinates and velocities in the vortex be as in Fig. 1. Introduce nondimensional (upper case) variables which are related to the dimensional (lower case) variables by

$$X = x/r_c$$

$$Y = R^2/2 = Re(r^2/2r_c^{\,2})$$

$$U = u/u_\infty$$

$$V = \sqrt{Re}\,(v/u_\infty)$$

$$W = w/u_\infty$$

where $Re = u_\infty r_c/\nu$ with u_∞ the free-stream velocity and r_c a representative core radius. Introduce $H = VR$ and $K = WR$ (circulation).

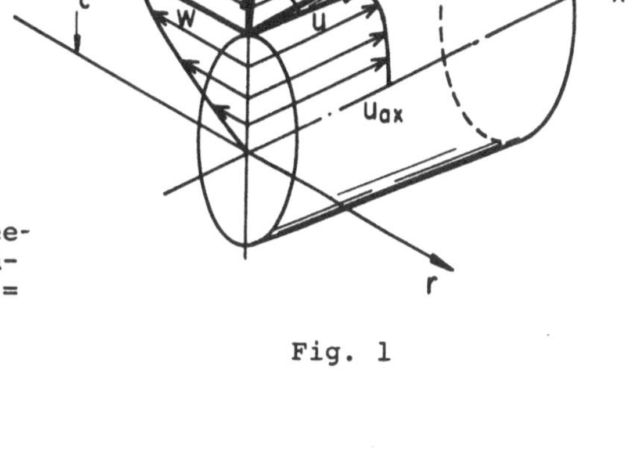

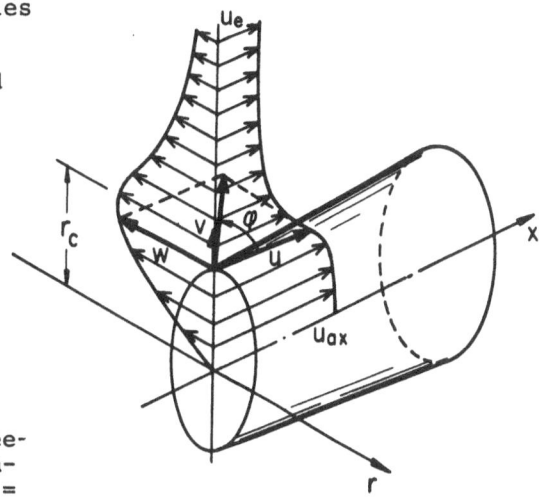

Fig. 1

The equations of quasicylindrical axisymmetric viscous incompressible vortex flow can be written as two equations for U and K [6]:

$$\frac{\partial}{\partial X}(UK) + \frac{\partial}{\partial Y}\left(HK - 2Y\frac{\partial K}{\partial Y} + 2K\right) = 0$$

$$\frac{\partial}{\partial X}\left(2Y^2U\frac{\partial U}{\partial Y} + \frac{K^2}{4}\right) + Y^2\frac{\partial^2}{\partial Y^2}\left(HU - 2Y\frac{\partial U}{\partial Y}\right) = 0 \qquad (1)$$

where

$$H = -\int_0^Y \frac{\partial U}{\partial X}\,dY$$

The equations are multiplied by weighting functions $g_k(Y)$ and $f_k(Y)$ satisfying $g_k(0)$, $f_k(0)$ = finite, and $g_k(\infty)$, $f_k(\infty) = 0$, and are then integrated in the Y-direction:

$$\frac{d}{dX}\int_0^\infty g_k UK dY - \int_0^\infty g_k' HK dY - \int_0^\infty (2g_k''Y + 4g_k')K dY = 0 \qquad (2)$$

$$\frac{d}{dX}\int_0^\infty (f_k'Y^2 + 2f_kY)U^2 dY - \frac{d}{dx}\int_0^\infty f_k\frac{K^2}{4}dY - \int_0^\infty (f_K''Y^2 + 4Yf_k' + 2f_k)HU dY$$

$$- \int_0^\infty (2f_k'''Y^3 + 14f_k''Y^2 + 20f_k'Y + 4f_k)U dY = 0$$

The integration is completed with the following choices for weighting functions and axial velocity and circulation approximations:

$$g_k(Y) = e^{-\sigma_k Y} \qquad k = 1, 2, \ldots, N$$

$$f_k(Y) = e^{-\sigma_k Y} \qquad k = 1, 2, \ldots, N+1$$

$$U(X,Y) = (1 - e^{-\alpha Y})[U_e(X) + \sum_{n=1}^{N} a_n(X)e^{-n\alpha Y}] + U_{ax}(X)e^{-\alpha Y} \qquad (3)$$

$$K(X,Y) = W(X,Y)R = (1 - e^{-\alpha Y})[K_e(X) + \sum_{n=1}^{N} b_n(X)e^{-n\alpha Y}]$$

The external velocity and circulation $U_e(X)$ and $K_e(X)$ are prescribed; the $a_n(X)$, $b_n(X)$ and the velocity on the axis $U_{ax}(X)$ are free parameters. Major reasons for these choices [6] are the exponential character of the flows, ease of analytical integration, and the fact that the approximations satisfy Weierstrass's approximation theorem after a coordinate transformation of the semi-infinite region $0 \leq Y < \infty$ into $1 \leq \eta < 0$, with $\eta = \exp(-\alpha Y)$.

Introduction of the approximating expressions and the weighting functions into equations (2) results in the following set of ordinary differential equations for U_{ax} and the parameters $a_n(X)$ and $b_n(X)$:

$$\sum_{n=1}^{N} \dot{a}_n c_{a,k} + \sum_{n=1}^{N} \dot{b}_n c_{b,k} + \dot{U}_{ax} c_{ax,k} = -\dot{U}_e c_{u,k} - \dot{K}_e c_{K,k} - c_k$$

$$k = 1, 2, \ldots, N + 1 \qquad (4)$$

$$\sum_{n=1}^{N} \dot{a}_n d_{a,k} + \sum_{n=1}^{N} \dot{b}_n d_{b,k} + \dot{U}_{ax} d_{ax,k} = -\dot{U}_e d_{u,k} - \dot{K}_e d_{K,k} - d_k$$

$$k = 1, 2, \ldots, N$$

The coefficients c and d in these equations are sums of products of present values of a_n, b_n, U_{ax}, U_e, K_e, and constant numbers computed only once at the beginning of the computations. The system (4) of (2N + 1) first order ordinary differential equations is first solved (by Gaussian elimination) for the derivative vector (da_n/dX, db_n/dX, dU_{ax}/dX). A standard method of integration then produces the a_n, b_n, and U_{ax} (the Runge-Kutta method has been used in the present work). Axial velocity U(X,Y), circulation K(X,Y) and swirl velocity W(X,Y) follow from relations (3).

Initial profile parameters a_n, b_n, and U_{ax} for the initial axial velocity and circulation profiles (3) are required to start the calculation. In the case of arbitrary profiles, these parameters are obtained as outlined in [6]. In the present study, simple exponential profiles have been used throughout, i.e.,

$$U_{initial}(Y) = U_e(1 - e^{-\alpha Y}) + U_{ax}e^{-\alpha Y}$$

and

$$K_{initial}(Y) = K_e(1 - e^{-\alpha Y}) \text{ or } K_e(1 - e^{-2\alpha Y})$$

Exponents $\alpha = 1$, and $\sigma_k = k$, $k = 1, 2, \ldots, N + 1$ were used in all of the calculations.

Results of the program have been compared with the computations of Hall [3] and with previous computations using polynomials in Y in the approximating functions [5]. There was good agreement in all of these cases. Critical runs have been repeated with different orders N of approximation. There appears to be convergence as N increases. Running times on the IBM 360/75 for typical cases are 5 to 15 seconds for N = 1 and 2, and 10 to 30 seconds for N = 3 and 4. All cases discussed here were run with N = 3, which appeared to be an efficient compromise between the conflicting demands of high accuracy and low computing time.

REPRESENTATIVE RESULTS

Three types of vortex flows are of particular interest: flow with initially uniform axial velocity profile, flow with a velocity excess at and near the axis over the freestream axial velocity ("leading edge vortex") and flow with velocity deficit at and near the axis ("trailing vortex"). All three were studied extensively with the present method. The results did not differ in any fundamental aspects, and the following discussion will therefore be limited to vortex flows having initially uniform axial velocity. Initial profiles were

$$U(Y) = 1 = \text{const}$$

$$K(Y) = W(Y)R = K_e(1 - e^{-Y})$$

The initial swirl parameter defined above follows as $S_i = 0.792\ K_e$ for this case.

The development of vortex flow at constant external axial velocity U_e and circulation K_e as a function of initial swirl parameter S_i is of basic interest. In computing such flows it is quickly found that many fail sooner or later, and that flows of different character develop for different choices of swirl parameter. Fig. 2 presents distances to failure as a function of swirl parameter. Several singular swirl values (S_I, S_{II}, S_{III}) can be identified, where no solutions can be obtained. Flows in region Ia are asymptotically stable (in the sense that no failure of the computation occurs), while no asymptotically stable flows were found in regions Ib, II, III, and IV (with the possible exception of $S_i \to \infty$). The existence of the singular points and the character of the solutions were confirmed by additional calculations with N = 2 and 4.

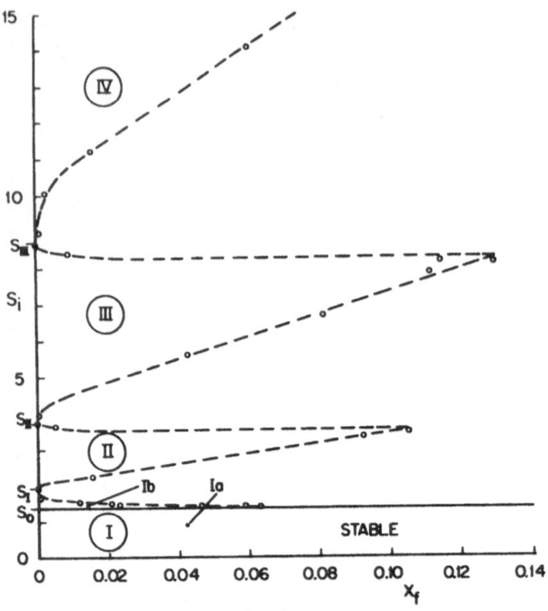

Fig. 2

Flows of different type show distinctively different behavior (Fig. 3). For zero external axial velocity and circulation gradients the velocity on the axis is always retarded for type I and III flows, and accelerated for type II and IV flows. In addition, a multilayer structure develops in the axial velocity profile as S_i is increased. Approximate streamline patterns for the different types of flow are also shown in Fig. 3. The swirl velocity profiles show nothing like the drastic reorganization of the axial velocity profiles.

Some understanding of the somewhat perplexing behavior of the solutions can be gained by considering inviscid results. Viscosity establishes rigid rotation on and near the axis of viscous vortices. It has been shown in [7] that in the region near the axis the Navier-Stokes equations are properly approximated by the (linear) equation of inviscid rotating flow. Consideration of the solutions to this equation shows that the solution changes from supercritical type (no

Fig. 3

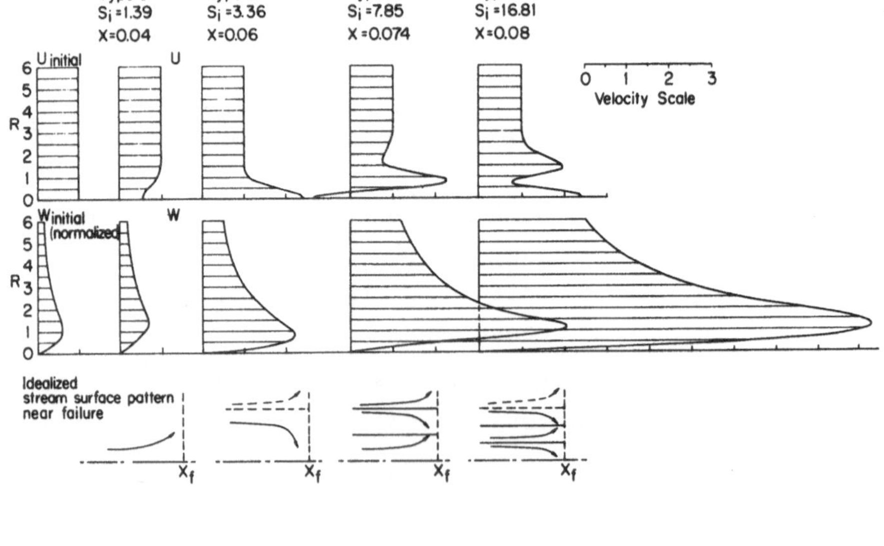

standing waves possible) to subcritical type (possibility of standing waves) when $S = j_{11}/2 = 3.8317/2 = 1.9159$ where j_{11} is the first zero of the Bessel function of order one. At this point the first wave (of infinite wave length) may appear. Since the regions of validity of the equation of inviscid rotating flow and viscous quasi-cylindrical vortex flow overlap at and near the axis, compatible results should be obtained from both. The viscous solution also changes character at $S_I = 1.91$.

By considering inviscid flow in rigid rotation, one can show [8] that this flow will tend to stagnate on the axis when $S \geq \sqrt{2}$, while flow with $S < \sqrt{2}$ is asymptotically stable. The same result is found from the present viscous computations: vortex flows with initially uniform axial velocity stagnate when $S > S_0 \approx 1.39$. This phenomenon is referred to as vortex breakdown; it is distinct from the vortex jump (also often called vortex breakdown [9]).

The physical significance of points S_{II} and S_{III} and corresponding solutions of type II, III, and IV is not clear at this point. Multi-layer vortex solutions are possible [10], and such vortices are observed in nature and in experiments. In the present computations such subcritical solutions have been maintained for considerable distance by application of external axial velocity and circulation gradients. However, the effect of downstream disturbances on the subcritical solution is not accounted for in using the parabolic model (1). The subcritical solutions will therefore not be discussed further here, and the remaining remarks will be concerned with supercritical flows of type I (i.e., $S_i < S_I \approx 1.91$).

The most important indicator of vortex behavior is a plot of velocity on the axis U_{ax} *vs.* distance X. Fig. 4 presents results for different choices of initial swirl ratio S_i. A swirl value of $S_0 = 1.39$ (theory: $\sqrt{2}$) separates the stable flows (Type Ia: $S < S_0$) from those that abruptly stagnate (Type Ib: $S_0 < S < S_I$).

Type Ia flow quickly adjusts to an almost constant value of axial velocity and thereafter behaves in a wake-like manner. The swirl still counteracts the normal spreading of the wake and the attendant diminishing of the axial velocity deficit; but eventually the axial velocity deficit again decreases, and the axial and swirl velocity profiles decay together.

Flows of Type Ib initially behave much like those of Type Ia, except that their rate of decrease of U_{ax} is steeper. Eventually, the swirl effects overwhelm the restoring tendencies of the wake, the drop of velocity on the axis steepens rapidly, and the computation fails. As a swirl value of $S = 1.91$ (theory: 1.9159) is approached, the computation fails increasingly sooner. No solutions can be obtained near this point.

The behavior of vortex flows under external velocity and circulation gradients is of considerable interest in cases where vortex flows are to be stabilized or destabilized. Computations by methods such as the present one can indicate what velocity and circulation gradients should be applied and where.

Fig. 5 presents the results of application of various positive and negative external axial velocity gradients to a Type Ib vortex flow of fixed initial swirl and with initially uniform axial velocity. This particular flow breaks down under zero external velocity gradient. The experimental observations (stabilizing effect of positive external axial velocity gradient) are

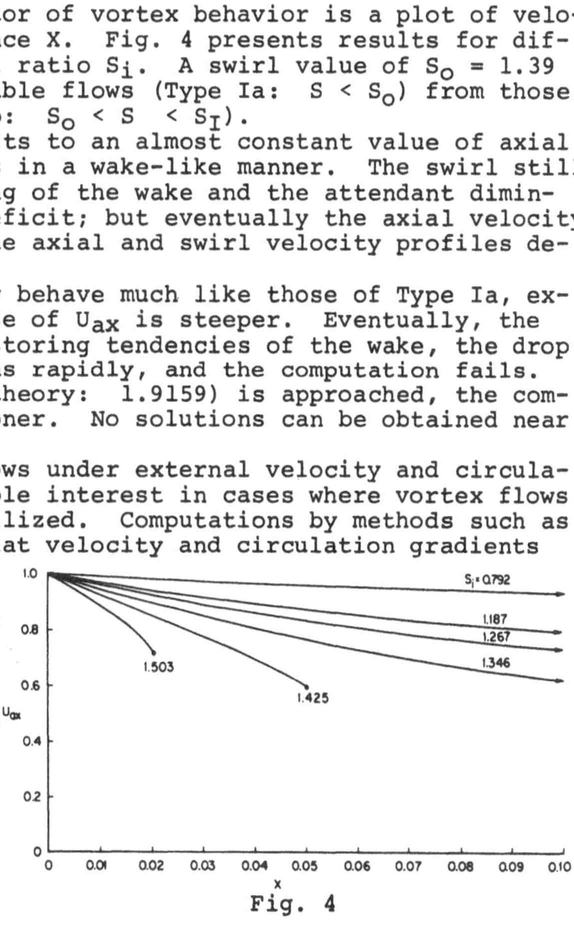

Fig. 4

confirmed, and the figure also illustrates the significant effect which even very small external velocity gradients can have on the velocity on the axis. Negative velocity gradients lower the critical swirl value and cause Type Ib flow to break down sooner. Note that a positive velocity gradient can stabilize a flow and convert it from Type Ib to Ia, even though the initial profile alone would indicate instability in a zero velocity gradient.

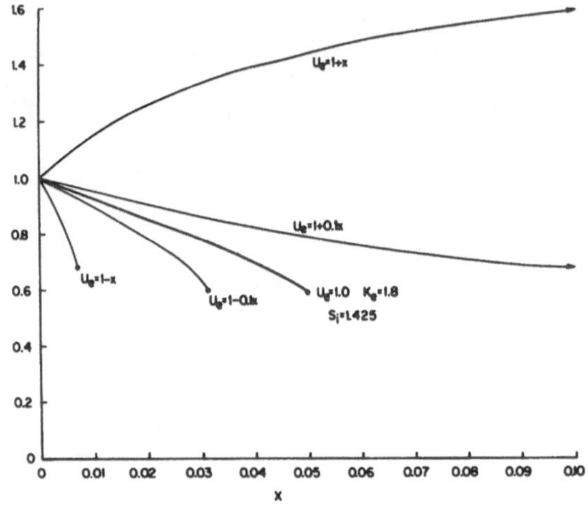

Fig. 5

Fig. 6 presents results of application of different circulation gradients to a Type Ia vortex. The effect of a moderate positive circulation gradient is qualitatively the same as for a positive velocity gradient. However, the effect reverses in Type Ib flow. Here a positive external circulation gradient results in earlier failure. Negative external circulation gradients can therefore be used to stabilize otherwise unstable Type Ib flows.

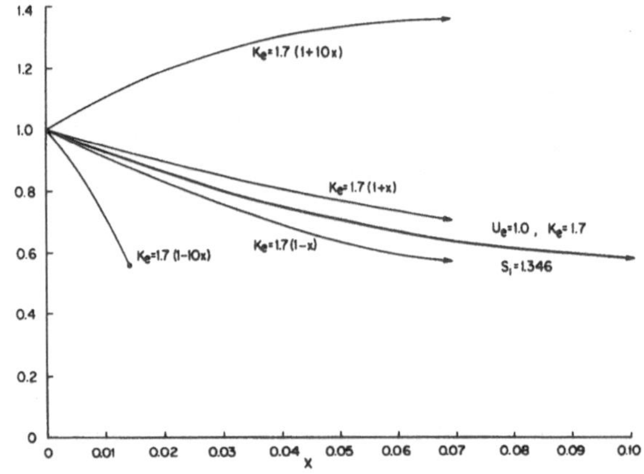

Fig. 6

CONCLUSIONS

Use of the described integral method in the computation of a very large number of vortex flows has been efficient and troublefree. With N = 3 (i.e., seven free parameters), the computations have been found to be quite accurate. The computations confirmed the inviscid results for a swirl value S_0 separating vortex flows which decay in a wake-like manner from those that stagnate on the axis; and a swirl value S_I separating supercritical from subcritical vortex flows. Other singular swirl values have been found which separate flows of different type and structure. The computations show strong effects of external axial velocity and circulation gradients. These effects can be used for stabilization or destabilization of vortex flows. They are often of opposite sign for different types of vortex flow.

REFERENCES

(1) Hall, M. G., in Progress in Aeronautical Sciences, D. Küchemann, Ed. (Pergamon Press, Ltd., Oxford, 1966), Vol. 7, p. 53.
(2) Gartshore, I. S., Nat. Res. Council, Canada, Rept. LR-378 (1963).
(3) Hall, M. G., RAE(Farnborough) Tech. Rept. 65106 (1965).
(4) Mager, A., AIAA Paper 70-51 (1970).
(5) Bossel, H. H., Univ. Calif. Berkeley, Eng. Rept. AS-67-14 (1967).
(6) Bossel, H. H., J. Computational Physics 5, 359 (1970).
(7) Bossel, H. H., Physics of Fluids 12, 498 (1969).
(8) Bossel, H. H., AIAA J. 6, 1192 (1968).
(9) Benjamin, T. B., J. Fluid Mech. 28, 65 (1967).
(10) Donaldson, C. duP., and Sullivan, R. D., Proc. Heat Transfer and Fluid Mechanics Institute 16 (1960).

RECENT EXTENSIONS TO THE MARKER-AND-CELL METHOD
FOR INCOMPRESSIBLE FLUID FLOWS

B. D. Nichols*
Los Alamos Scientific Laboratory
University of California
Los Alamos, New Mexico

INTRODUCTION

The Marker-and-Cell (MAC) method is a computing technique that calculates the transient dynamics of viscous, incompressible fluid flow. It utilizes a finite-difference technique to solve the continuity equation and the time dependent Navier-Stokes equations. Free-surface problems are calculable by this method, the position of the fluid being delineated by marker particles. These particles flow with the fluid through an Eulerian network of calculational cells. A detailed description of this method together with numerous examples is given by Welch, et al. (1965).

Although the MAC method has been highly successful in calculating free-surface fluid flows, as is evidenced by its widespread use, there are several directions in which improvements can be made. The purpose of this paper is to describe some of these extensions. These include the addition of the azimuthal component of velocity, the use of an improved differencing technique and, most notably, the improvement of the free-surface treatment.

DESCRIPTION OF THE EXTENSIONS

Azimuthal Velocity

The inclusion of the azimuthal component of velocity is straightforward. Axial symmetry is assumed. Only one additional equation, the azimuthal direction momentum equation, is required, and only one term representing a centripetal acceleration must be added to the radial direction momentum equation. The complete Navier-Stokes equations in cylindrical coordinates are given by Schlichting (1966). Although the addition of the azimuthal component of velocity is simple, it allows many interesting rotating fluid studies to be done. For example, we have begun a study of the formation of Taylor cells in a column of fluid contained between two concentric cylinders and having a free top surface.

ZIP Type Differencing

A special differencing technique, referred to as ZIP type, has been employed for the convection terms in the momentum equations. A description of this technique and its advantages has been discussed by Hirt (1968). He shows the principal advantage is the elimination of important negative diffusion-like truncation errors in Cartesian coordinates. In cylindrical coordinates, however, due to the radius factor, there are additional negative diffusion-like truncation error terms that are not eliminated. In either coordinate system, however, this differencing technique is also useful because it ensures conservation of momentum at rigid boundaries, whereas the difference method used in the original MAC method does not.

---

*This work was performed under the auspices of the United States Atomic Energy Commission.

Free-Surface Treatment

The most outstanding improvement in the MAC method has been in treatment of free-surface boundary conditions. The improvements are two-fold: (a) inclusion of the correct free-surface stress conditions and (b) use of a surface pressure interpolation scheme.

A. The work of Hirt and Shannon (1968) on free-surface stress conditions describes the inclusion of the correct normal stress conditions. For low Reynolds number flows, they demonstrated the improved accuracy that could be obtained with the inclusion of the correct normal stress condition. However, their results indicated that tangential stress conditions should also be satisfied. Both the normal and tangential stress conditions have now been added and a recalculation of the example used by Hirt and Shannon (1968) clearly demonstrates the improved accuracy to be obtained with the full stress conditions. This problem together with several other examples illustrating the improved free-surface treatments will be published in a paper by Hirt and Nichols (1970).

B. The pressure in cells containing the fluid surface can be variously determined. Originally the pressure in surface cells was set to zero. Later the inclusion of the correct normal stress condition gave rise to a non-zero surface pressure, but this was again specified at the cell center. Clearly any scheme that does not take into account the location of the free surface within the cell will be correct only when the surface is located at the center of the cell. To account for variable interpolation distances Chan, et al. (1969) developed a technique of delineating the fluid surface and appropriately adjusting the Poisson equation for pressure. The success of this method in calculating accurate wave motion was demonstrated for the case of a solitary wave reflecting from a rigid wall.

We have developed a somewhat different method that is easily implemented into existing MAC method computer programs. This scheme uses a pressure at the center of the surface cell that is a linear extrapolation from an adjacent full cell and a pressure of zero at the fluid surface. This addition enables the MAC method to calculate large amplitude surface motion while maintaining a very smooth and stable appearance. The technique also permits a more accurate treatment of low amplitude free surface motions.

To calculate the pressure in a surface cell the location of the surface within this cell must be known. A special set of consecutively numbered surface marker particles is used for this purpose. These surface particles are labeled separately from the interior marker particles but are moved with the local fluid velocity, in the same way as interior particles.

The location of the surface within the surface cell dictates the neighboring full cell with which the pressure interpolation is to be done. If there are no full neighbors, the pressure is set to zero. When more than one full cell is adjacent to a surface cell we must decide with which to interpolate. This is accomplished by drawing for each neighboring full cell a line segment from the center of this cell through the center of and to the far side of the surface cell, as illustrated in Fig. 1. We next determine the point of intersection of this line and the fluid surface by numbering the surface particles consecutively and choosing those two bracketing the line. The point of intersection of this line and the line segment connecting the two surface particles is easily determined. The distance from the center of the full cell to this point of intersection is denoted by d. If the surface does not cross the line connecting cell centers d is given its maximum value, $(3/2)\delta y$. If it crosses the line segment in the full cell, then d is set to $\delta y/2$. The reason for bounding d by these values is to prevent the development of numerical instabilities in the solution of the pressure equation. Finally, the full cell corresponding to the smallest value of $|\delta y-d|$ is chosen as the interpolation cell. This is the cell for which the intersection point is nearest the center of the surface cell.

The corresponding distance d is also used to calculate the interpolation number,

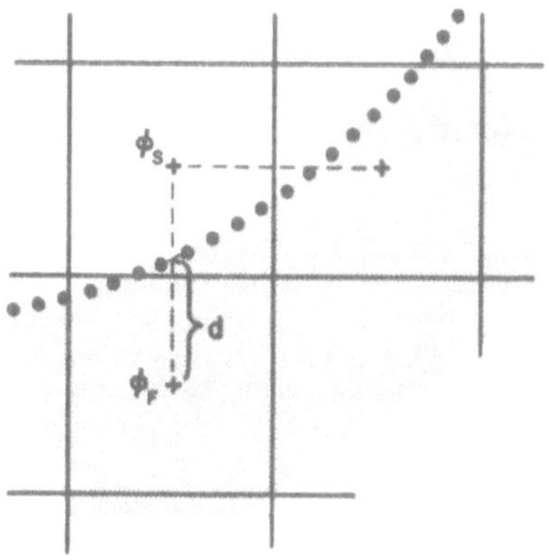

Fig. 1. A typical interpolation cell selection.

$\eta = \delta y/d$, associated with the surface cell. The interpolation cell and the interpolation number for every surface cell in the calculational mesh is determined only once each calculational cycle. Surface cell pressures are recalculated, however, after each iteration of the complete pressure field. Currently, a Gauss-Seidel method is used with successive-over-relaxation. The linear interpolation equation used to determine each surface cell pressure, ϕ_s, is

$$\phi_S = (1 - \eta)\ \phi_F \quad ,$$

where ϕ_F is the current value of the full cell pressure chosen for the interpolation.

Non-zero surface pressures can also be incorporated. For example, a pressure ϕ_a applied directly to the surface or the pressure ϕ_{ns} from the effects of the normal stress condition may be employed. If these are used the interpolation equation becomes

$$\phi_S = \eta\ (\phi_a + \phi_{ns}) + (1 - \eta)\ \phi_F \quad .$$

When the normal stress condition is used a new stability condition relating to the viscosity must be utilized. As reported by Welch, et al. (1965) the stability condition

$$2\nu\delta t < \frac{\delta x^2 \delta y^2}{\delta x^2 + \delta y^2}$$

was necessary to insure stability in the original MAC method. When the interpolation pressure technique is used with the normal stress condition, however, it is required for stability that

$$4\nu\delta t < \frac{\delta x^2 \delta y^2}{\delta x^2 + \delta y^2} \quad .$$

During the course of a calculation surface particles can be moved into a full cell, that is, into a cell whose neighboring cells all contain fluid. This situation can lead to calculational difficulty at later times and so it is necessary to delete these particles. Surface particles can also be too far apart to adequately define the surface. In this case it is necessary to insert particles. When either deletion or addition of particles occurs, a renumbering of particles ensures that they remain consecutively numbered.

DISCUSSION OF RESULTS

A good example showing the strong influence of the surface pressure interpolation technique is the problem of fluid sloshing in a rectangular tank. The sloshing can be initiated by a cosine pressure pulse applied at the surface. Figure 2 contrasts the surface smoothness, after 1/4 period of oscillation, between an original MAC calculation and a calculation using the new surface treatment. The irregularity of the surface in (a) is a result of the surface cell pressures being defined as zero at the center of each surface cell. This introduces fictitious fluid accelerations near the surface. The effect of these spurious velocities on the surface smoothness is clearly evident in (a). When the correct surface cell pressure is determined by the pressure interpolation technique a smooth surface is calculated as illustrated in (b).

The above problem involves a single, well defined surface configuration. This new pressure interpolation scheme, however, can also be used in situations where surfaces are colliding together; for example, when a liquid drop impacts on a standing pool of fluid or when a breaking wave collapses. The results of a splashing drop calculation are shown in Fig. 3. The initial plot of marker particles shows the drop at the moment of impact with the pool of fluid. The drop is 0.23 cm in diameter and has an initial downward velocity of 320 cm/sec. The depth of the pool is 0.19 cm. The kinematic viscosity is 1.0 cm^2/sec. The calculational mesh consists of 30 radial by 24 axial cells. These cells have square cross sections with side dimensions of 0.02 cm. Axial symmetry is assumed in this problem and only the right half of the problem is calculated. The left half is plotted as a mirror image.

The region where the drop makes contact with the standing fluid contains several full cells because of the coarseness of the computing mesh. In this region surface particles must be deleted prior to calculation of the pressure. After the deletion it is necessary to add some additional surface particles. These are placed on a straight line connecting the two surface particles on each side of the gap left by the deletion. As mentioned previously, the test to determine whether to delete or add particles is made each calculational cycle. This is done immediately after all particles have been moved and cells are reflagged as, for example, surface or full cells.

The breaking wave illustrated in Fig. 4 is generated by assigning an initial horizontal velocity of - 1.0 to the entire fluid configuration so that it runs into the rigid left-hand wall. This causes the fluid to run up the left wall, as shown in the plot at t = 0.9. When the gravitational force (g_z = - 1.0) overcomes the upward motion of the fluid it falls downward and to the right. The fluid column is in a position to collapse at t = 3.3. As the fluid surface changes each calculational cycle it is redefined by deletion of surface particles in full cells and addition of particles wherever needed. Finally, the fluid has completely collapsed in the plot at t = 3.81.

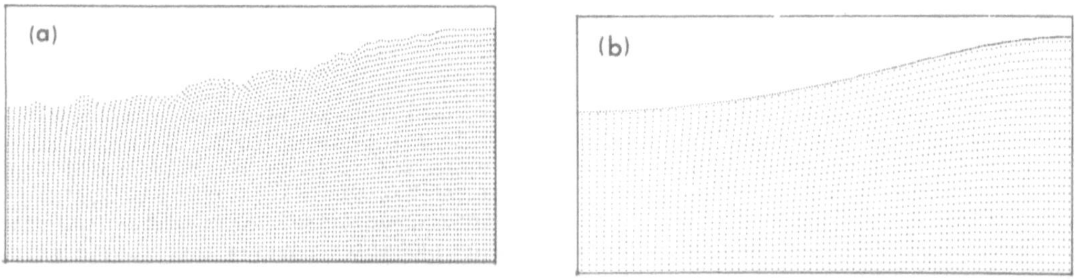

Fig. 2. Marker particle configurations for the sloshing wave problem at 1/4 period, t = 6.0, comparing (a) an original MAC calculation with (b) a pressure interpolation calculation.

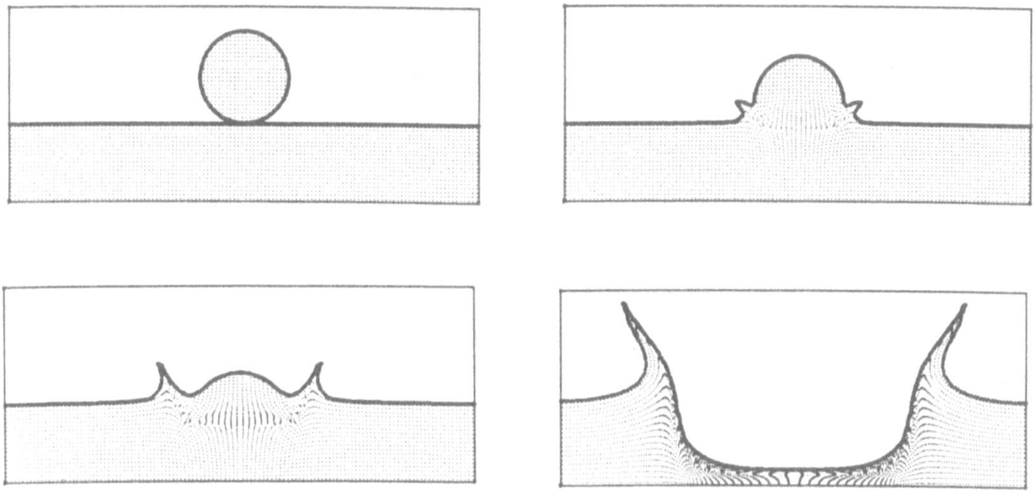

Fig. 3. Cross section of a drop splashing in a shallow pool. Times, in seconds, reading from left to right, are 0.0, 0.0002, 0.0005, and 0.0023.

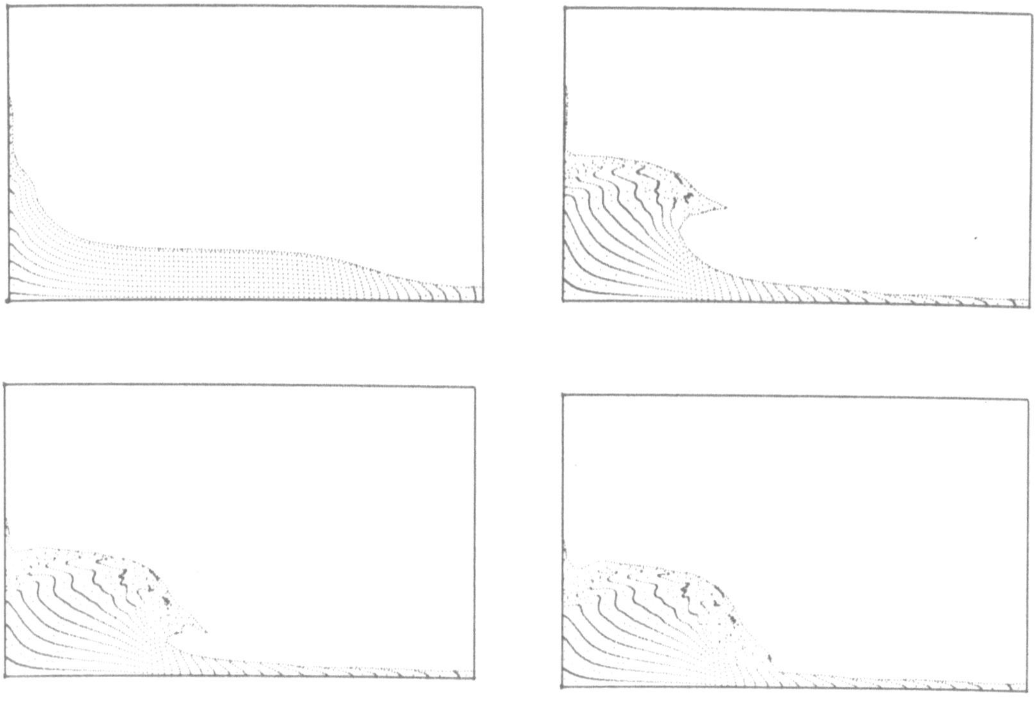

Fig. 4. Breaking wave at times 0.9, 3.3, 3.72, and 3.81, reading from left to right.

ACKNOWLEDGMENT

The author is indebted to C. W. Hirt and W. E. Pracht for their assistance and many valuable suggestions.

REFERENCES

Chan, R. K.-C., Street, R. L. and Strelkoff, T., "Computer Studies of Finite-Amplitude Water Waves," Stanford University Technical Report No. 104 (1969).

Hirt, C. W. J. Comp. Phys. 2, 339-355 (1968).

Hirt, C. W. and Nichols, B. D., to be published (1970).

Hirt, C. W. and Shannon, J. P. J. Comp. Phys. 2, 403-411 (1968).

Schlichting, "Boundary-Layer Theory," 6th ed., 63-64, McGraw-Hill Book Company, New York (1966).

Welch, J. E., Harlow, F. H., Shannon, J. P. and Daly, B. J., "The MAC Method," Los Alamos Scientific Laboratory Report, LA-3425 (1965).

A NUMERICAL STUDY OF TIME-DEPENDENT ROTATING FLOW IN A CYLINDRICAL CONTAINER AT LOW AND MODERATE REYNOLDS NUMBERS

W. R. Briley
United Aircraft Research Laboratories
East Hartford, Connecticut

and

H. A. Walls
Department of Mechanical Engineering
The University of Texas at Austin

INTRODUCTION

Numerical solutions to the Navier-Stokes equations are presented here for two problems involving the time-dependent laminar motion of a viscous incompressible fluid in a closed cylindrical container. The container (see Fig. 1) is a right circular cylinder closed by disks on both ends and is completely filled with fluid. The problems considered are: (1) a spin-down problem in which the fluid is brought to rest from an initial state of rigid-body rotation about the container axis by impulsively stopping the container, and (2) the opposite problem of spin-up, in which the fluid is brought from rest to rigid-body rotation by impulsively spinning the container about is axis. These problems are related to a problem studied by Greenspan and Howard (Ref. 4), who developed a linear analysis which describes the spin-up (or spin-down) of fluid in an arbitrary axisymmetric container whose constant initial angular velocity is impulsively changed by a small amount, assuming a large Reynolds number. Their analysis shows that spin-up under these conditions involves three phases: the formation of boundary layers on the endwalls, the spin-up of interior fluid by convective processes associated with the secondary flow, and, finally, the viscous decay of small inertial oscillations. Using boundary layer techniques, Greenspan and Weinbaum (Ref. 5) have extended this analysis to include first order nonlinear terms which account for larger changes in angular velocity. They concluded that nonlinear effects are of minor importance for changes of up to 50 percent of the initial angular velocity. An approximate analysis for the spin-up from rest of fluid in a cylindrical container has been given by Wedemeyer (Ref. 13) for cases with high Reynolds number.

The problems of spin-up from rest and spin-down to rest are fundamentally different in that, although the former remains laminar and axisymmetric over a wide range of Reynolds numbers, the latter motion is inherently unstable and becomes turbulent at relatively low Reynolds numbers. Due to its complexity, the problem of spin-down to rest has received very little attention; however, it is known that spin-down is prone to instability in the form of Taylor vortices near the cylindrical sidewall. Moreover, the difference in stability between spin-up from rest and spin-down to rest may in part be due to the differing character of the endwall boundary layers. During spin-down, the flow near the center of the endwalls a short time after the container is stopped is related to a steady flow problem in which a fluid rotates above a disk at rest. A solution of the Navier-Stokes equations for that problem was first obtained by Bodewadt (Ref. 2) on the basis of similarity assumptions which are applicable to an infinite disk. In his solution, the velocity profiles possess spatial oscillations in the axial direction which decay with increasing distance from the disk. The question of whether these oscillations exist in practice has been widely debated (see, for example, Refs. 6, 9, 10, and 11) because of the singular manner in which the boundary layer originates at infinity in Bodewadt's solution and the possibility of boundary layer separation and eruption at the axis of rotation. The flow near the center of the endwalls during spin-up is related to the reverse of Bodewadt's problem, that of an infinite disk rotating in a fluid which is otherwise at rest. A solution for this steady flow problem has been obtained by von Karman (Ref. 12) and others, and in contrast to Bodewadt's solution, the solution of von Karman possesses nonoscillatory velocity profiles. The purpose of this study was to investigate the character of spin-down and spin-up at low to moderate Reynolds number and to determine how the differing nature of the von Karman and Bodewadt flows carries over to their time-dependent counterparts near the end-walls during spin-down and spin-up. The numerical solutions show that oscillatory velocity profiles similar to those of Bodewadt do occur during spin-down at moderate Reynolds numbers and that they interact with the basic secondary flow so as to produce weak recirculation cells in the interior of the container. The oscillations in tangential velocity temporally preceed those in the radial and axial velocities. Nonoscillatory velocity profiles are present during spin-down at lower Reynolds numbers and during spin-up. At sufficiently high Reynolds number, Taylor vortex instability occurs near the cylindrical sidewall in the solutions for spin-down.

METHOD OF SOLUTION

The Navier-Stokes equations governing the axisymmetric flow of a viscous, incompressible fluid with constant properties were solved in terms of the θ-component of vorticity, the Stokes axisymmetric stream function, and the tangential component of velocity using a computational scheme based on a method proposed by Pearson (Ref. 8). Basically, the procedure used here is to solve the vorticity and tangential velocity equations over a time step using a modification of the alternating-direction-implicit (ADI) method of Peaceman and Rachford (Ref. 7) for parabolic equations. The stream function equation was solved by the ADI method of Peaceman and Rachford (Ref. 7) for elliptic equations. The latter ADI method was chosen in preference to the more popular method of successive over-relaxation by points because the ADI method results in a substantial reduction of the computing time for solving the stream function equation. Moreover, this relative savings in computer time increases as more grid points are added to the computational region (Birkhoff, et al. (Ref. 1)). The details of the numerical computations are available elsewhere (Briley, (Ref. 3)).

NUMERICAL RESULTS

In the discussion which follows, u , v , and w denote the r, θ , z-components of velocity in an inertial, cylindrical coordinate system whose axis coincides with that of the container (see Fig. 1). The solutions have been made nondimensional by making the following substitutions (a prime denotes a dimensional variable):

$$r' = Rr,\ z' = Hz,\ t' = R^2 t/\nu,\ u' = R\Omega u,\ v' = R\Omega v,\ w' = R\Omega w,\ \psi' = R^2 H\Omega\psi$$

where R is the container radius, H is one half the container height, Ω is the initial or final angular velocity, and ν is the kinematic viscosity. The foregoing scaling of variables results in two dimensionless parameters, the rotational Reynolds number, Re , and the geometry ratio, a, which are defined by

$$Re = R^2\Omega/\nu, \quad \alpha = R/H$$

Numerical solutions were calculated for geometry ratios, α, of 2.0 and 0.5, with corresponding grid sizes of 26 x 51 and 26 x 101 in the r, z-directions. Values of 10^{-3} to 10^{-4} were used for Δt.

To give an indication of the rate at which spin-up and spin-down occur at low to moderate Reynolds number, the rotational volume flowrate, Q, defined by

$$Q = \int_0^1\int_0^2 v \, dz \, dr$$

was computed from the numerical solutions and is plotted against time for several cases in Figs. 2 and 3. Numerical solutions for various values of Re and α are presented in Figs. 4 through 11 in the form of contour plots of the tangential velocity and stream function at selected values of time. The solutions for spin-down in a container with geometry ratio, α, of 2.0 are considered first. Figure 4 gives contour plots of the flow at $t = 0.01$ for Re = 66. Since the numerical solutions are symmetrical about the plane midway between the container ends, only the lower half of the flow cross section is shown. The flow in Fig. 4 is typical of that found at low Reynolds numbers. The secondary flow is weak and has little influence on the distribution of tangential velocity. The flow is dominated by diffusion, and there are no spatial oscillations in the velocity profiles near the center of the endwalls. In Fig. 5, the Reynolds number has been increased to 215 and the flow is shown at $t = 0.02$, shortly after the secondary flow reaches its maximum strength. Convective effects are more important at this Reynolds number, although the flow is still highly viscous. It can be seen from the contours of tangential velocity in Fig. 5 that the effect of the secondary flow is to convect higher angular momentum fluid from the interior of the container toward the cylindrical sidewall, where its angular momentum is reduced by viscosity. The fluid is then returned to the interior of the container through a thick boundary layer on the endwall, where its angular momentum is further reduced. In Fig. 6, the Reynolds number is 1000 and, at $t = 0.005$, nonlinear effects are definitely in evidence. Over the inner half of the endwall disk, the tangential velocity profiles possess spatial oscillations similar to those found in Bodewadt's (Ref. 4) steady solution for an infinite disk. Moreover, these oscillations interact with the basic secondary flow so as to produce a weak recirculation cell in the interior of the container. The recirculation cell and tangential velocity oscillations have disappeared by $t = 0.03$, when the velocities and thus the instantaneous Reynolds number have decreased.

Numerical solutions were also computed for spin-down in a container with a geometry ratio, α, of 0.5. Nonlinear effects similar to those above were again found at moderate Reynolds number. In Fig. 7, the Reynolds number is 405 and, at $t = 0.01$, both the tangential velocity and the stream function contours have oscillations similar to those found by Bodewadt. It is worth noting that the spatial oscillations in the tangential velocity temporally preceed those in the stream function; at $t = 0.005$, the tangential velocity contours possess the oscillations whereas the stream function contours do not. In Fig. 7 at $t = 0.01$, the fluid near the axis at the midplane ($r < 0.6$, $z = 1.0$) has just begun to spin-down from its initial state of rigid body rotation. At $t = 0.025$ in Fig. 8, the tangential velocity oscillations have reached the interior of the container and a weak recirculation cell has formed and covers much of the upper half of the cross section. At this time, the fluid at the midplane has spun-down to about two-thirds of its original angular velocity. The recirculation cell disappeared by $t = 0.035$, and thereafter the oscillatory behavior gradually decayed.

A numerical solution was computed for spin-down at $Re = 1167$ with $\alpha = 0.5$. Although the accuracy of this solution is questionable due to the mesh spacing, which is somewhat coarse for this Reynolds number, the solution at $t = 0.016$ is nevertheless included as Fig. 9 because it illustrates the occurrence of Taylor vortex instability along the cylindrical sidewall. The Taylor vortices appear in addition to the oscillatory velocity profiles discussed previously, and they persist until late in the spin-down process, although their strength is of course diminished. It is not known whether the physical flow is axisymmetric at this Reynolds number.

Representative contour plots for two spin-up solutions are shown in Figs. 10 and 11, and these are discussed only briefly. The flow at $t = 0.02$ for the solution with $Re = 215$ and $\alpha = 2.0$ is shown in Fig. 10. The secondary flow in this instance is similar in character to that during spin-down in Fig. 5 except for the reversal in flow direction. In both cases, the flow is highly viscous. In Fig. 11, the solution with $Re = 405$ and $\alpha = 0.5$ is shown at $t = 0.02$, when the secondary flow is fully developed. Contrary to the solutions for spin-down in Figs. 7 and 8, the secondary flow near the center of the endwall does not possess spatial oscillations and there are no recirculation cells. In the interior of the container, the tangential velocity is nearly independent of z. The flow is qualitatively similar to that assumed by Wedemeyer (Ref. 3) and is known to occur experimentally at higher Reynolds number.

COMPARISON WITH EXPERIMENT

In addition to examining the numerical solutions for validity by refining the mesh spacing and time increment, a limited number of experimental measurements of tangential velocity decay profiles were made for comparison with the numerical results. These profiles were measured with a calibrated constant-temperature hot-wire anemometer using oil as the fluid in an open container. The free surface of the oil corresponds to the midplane, $z = 1$, in the present notation. The twelve inch diameter container was carefully machined from aluminum pipe and fitted with an aluminum bottom plate. The container was mounted on a turntable assembly which was driven by a small electric motor and equipped with a brake. Free surface effects were small for the conditions of the experiments and were neglected.

The comparison between numerical and experimental results is presented in Figs. 12 and 13 for three sets of experimental conditions. The agreement is very good and is within the bounds of experimental error. This confirms the ability of the numerical method to follow with reasonable accuracy the time-dependent behavior of this fluid motion for the Reynolds numbers and mesh spacings used in Figs. 12 and 13. The single bump in the computed decay profile for $Re = 405$ in Fig. 13 was present in each experimental run, although at a slightly earlier time. The precise location of the bump is somewhat sensitive to Reynolds number, and a slight experimental variation in Reynolds number ($\pm$ 2 percent) is largely responsible for the scatter in the experimental data near the bump. The bump is associated with the oscillatory velocity profiles and recirculation cell which were discussed previously in connection with Figs. 6 through 8, and the experimental presence of the bump provides evidence that these features of the numerical solutions are physically real. Although the numerical solution for Re = 1000 with $\alpha = 2.0$ was not compared with experiment, it is believed to possess quantitative accuracy comparable to that of the solution for Re = 405 and $\alpha = 0.5$. The solution for Re = 1167 with $\alpha = 0.5$ must be viewed with caution, as mentioned earlier, even though the Reynolds number is comparable to that of the solution for Re = 1000 and $\alpha = 2.0$, because the physical mesh spacing used for Re = 1167, $\alpha = 0.5$ is coarser while the fluid motion is more vigorous.

ACKNOWLEDGMENT

The authors wish to acknowledge support by the University of Texas Computation Center and the support of W.R.B. by a National Science Foundation Traineeship.

REFERENCES

1. Birkhoff, G., Varga, R. S., and Young, D., Advances in Computers (Academic Press, New York, 1962), Vol. 3, p. 189.

2. Bodewadt, U. T., Z. angew. Math. Mech. 20, 241 (1940).

3. Briley, W. R., Dissertation, University of Texas at Austin, (1968).

4. Greenspan, H. P., and Howard, L. N., J. Fluid Mech. 17, 385 (1963).

5. Greenspan, H. P., and Weinbaum, S., J. Math. Phys. 44, 66 (1965).

6. Mager, A., Theory of Laminar Flows (Princeton Univ. Press, Princeton, N. J., 1964), Vol. IV, p. 286.

7. Peaceman, D. W., and Rachford, H. H., J. Soc. Indus. Appl. Math. 3, 28 (1955).

8. Pearson, C. E., J. Fluid Mech. 21, 611 (1965).

9. Rott, N., and Lewellen, W. S., Progress in Aeronautical Sciences (Pergamon Press, London, 1966), Vol. 7, p. 111.

10. Schwiderski, E. W., and Lugt, H. J., Phys. Fluids 7, 867 (1964).

11. Stewartson, K., Advances in Applied Mechanics (Academic Press, New York, 1960), Vol. VI, p. 1.

12. Von Karman, T., Z. angew. Math. Mech. 1, 233 (1921).

13. Wedemeyer, E. H., J. Fluid Mech. 20, 383 (1964).

FIG. 1

CYLINDRICAL CONTAINER AND COORDINATE SYSTEM (r' AND z' ARE DIMENSIONAL COORDINATES)

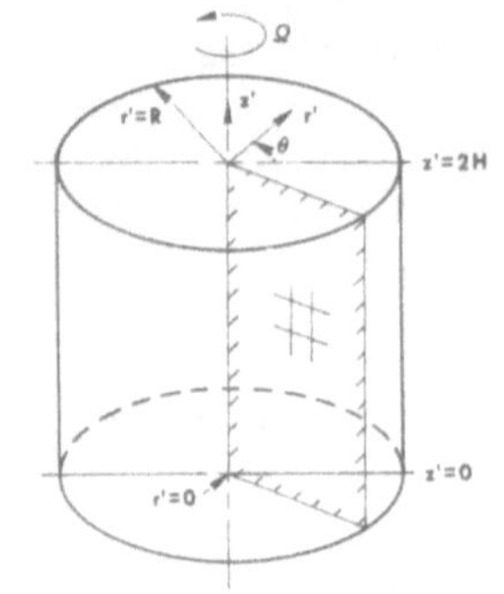

FIG. 2

DIMENSIONLESS ROTATIONAL VOLUME FLOWRATE Q VS. DIMENSIONLESS TIME t FOR SPIN-UP AND SPIN-DOWN WITH GEOMETRY RATIO α = 2.0

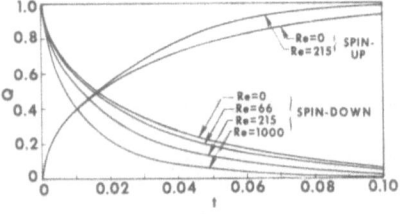

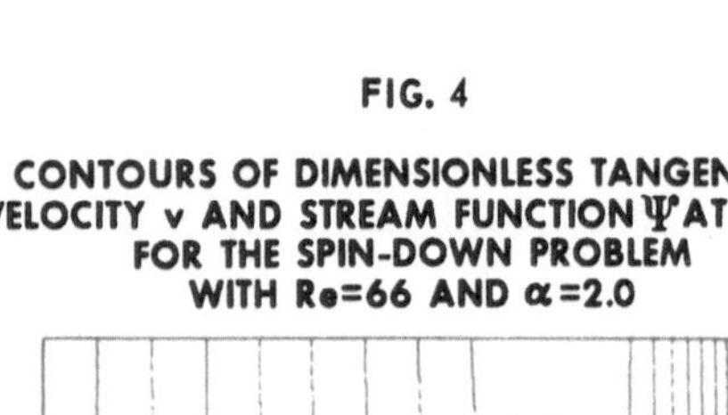

FIG. 3

DIMENSIONLESS ROTATIONAL VOLUME FLOWRATE Q VS DIMENSIONLESS TIME t FOR SPIN-UP AND SPIN-DOWN WITH GEOMETRY RATIO α=0.5

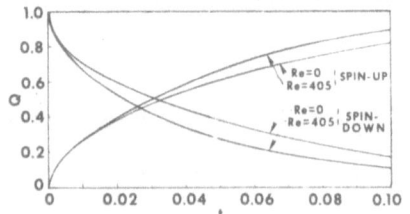

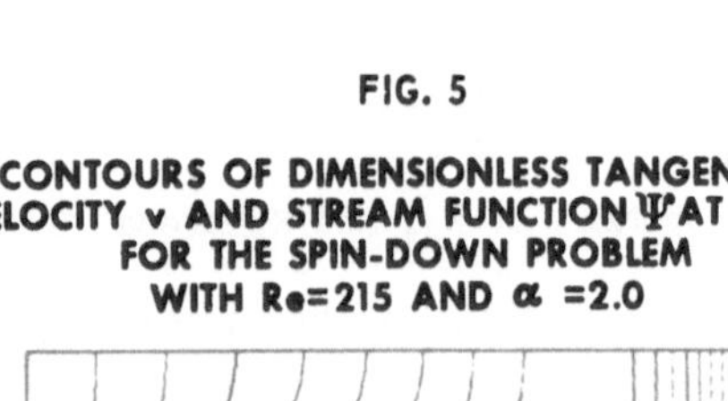

FIG. 4

CONTOURS OF DIMENSIONLESS TANGENTIAL VELOCITY v AND STREAM FUNCTION Ψ AT t=0.01 FOR THE SPIN-DOWN PROBLEM WITH Re=66 AND α=2.0

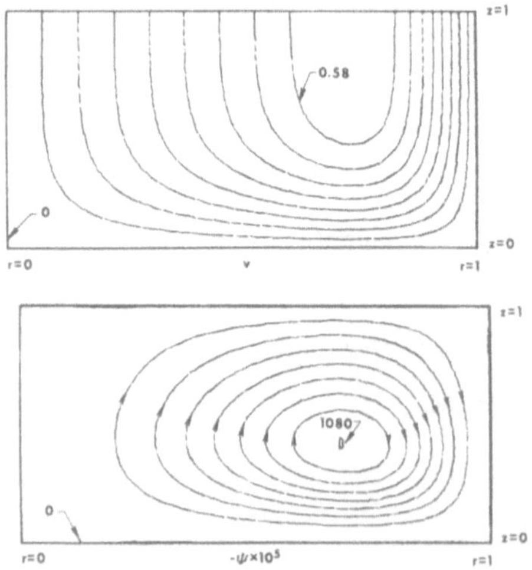

FIG. 5

CONTOURS OF DIMENSIONLESS TANGENTIAL VELOCITY v AND STREAM FUNCTION Ψ AT t=0.02 FOR THE SPIN-DOWN PROBLEM WITH Re=215 AND α =2.0

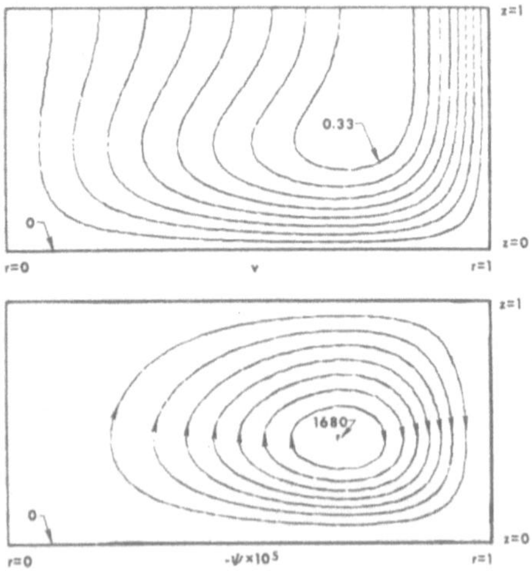

FIG. 6

CONTOURS OF DIMENSIONLESS TANGENTIAL VELOCITY v AND STREAM FUNCTION Ψ AT t=0.005 FOR THE SPIN-DOWN PROBLEM WITH Re=1000 AND α=2.0

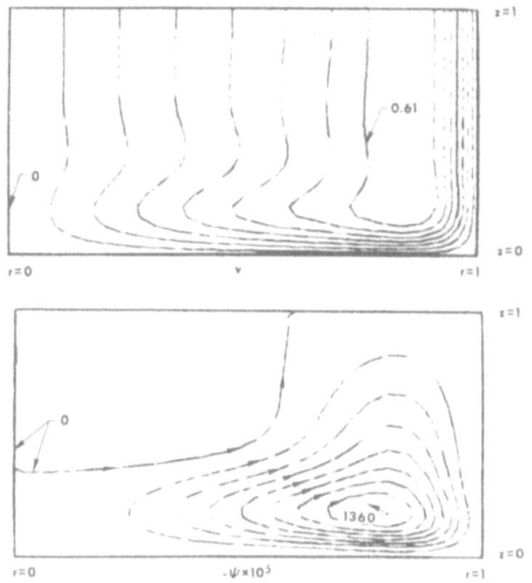

FIG. 7

CONTOURS OF DIMENSIONLESS TANGENTIAL VELOCITY v AND STREAM FUNCTION Ψ AT t =0.01 FOR THE SPIN-DOWN PROBLEM WITH Re=405 AND α=0.5

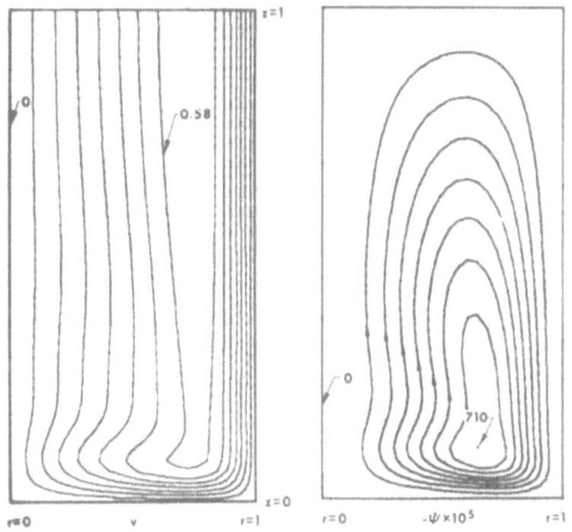

FIG. 8

CONTOURS OF DIMENSIONLESS TANGENTIAL VELOCITY v AND STREAM FUNCTION Ψ AT t=0.025 FOR THE SPIN-DOWN PROBLEM WITH Re=405 AND α=0.5

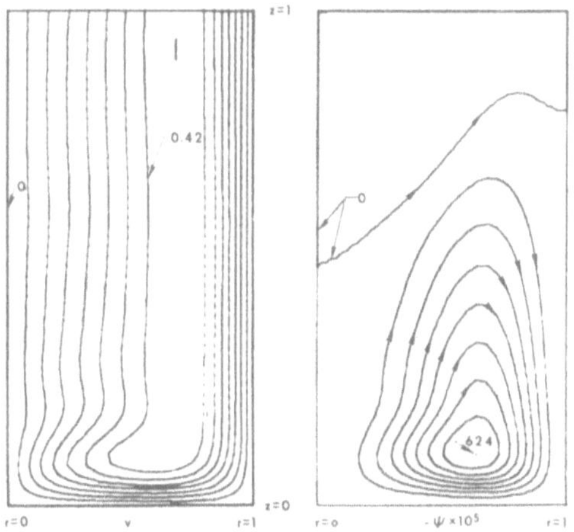

FIG. 9

CONTOURS OF DIMENSIONLESS TANGENTIAL VELOCITY v AND STREAM FUNCTION Ψ AT t =0.016 FOR THE SPIN-DOWN PROBLEM WITH Re=1167 AND α=0.5

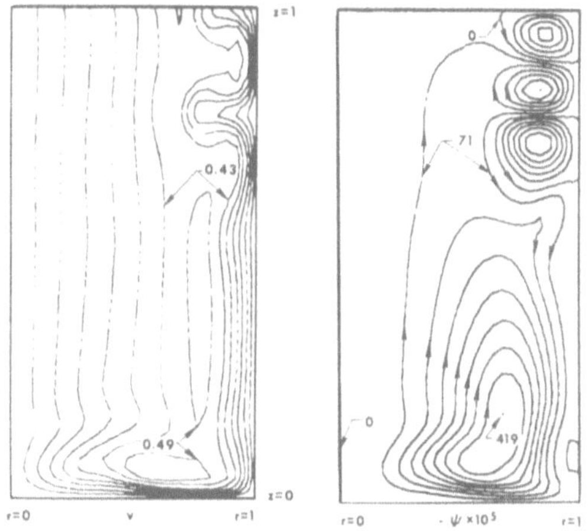

FIG. 10

CONTOURS OF DIMENSIONLESS TANGENTIAL VELOCITY v AND STREAM FUNCTION Ψ AT t=0.02 FOR THE SPIN-UP PROBLEM WITH Re=215 AND α=2.0

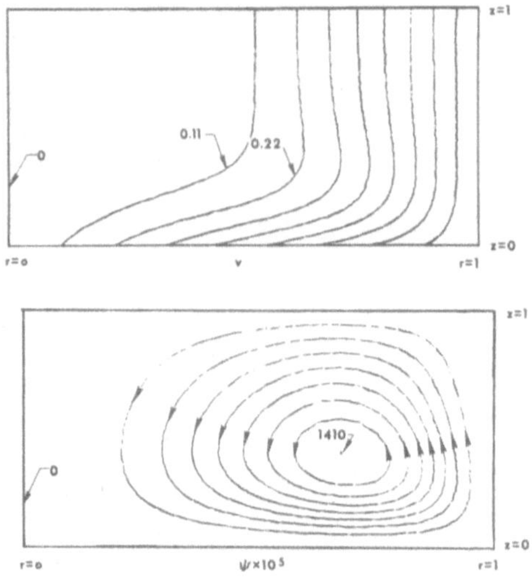

FIG. 11

CONTOURS OF DIMENSIONLESS TANGENTIAL VELOCITY v AND STREAM FUNCTION Ψ AT t=0.02 FOR THE SPIN-UP PROBLEM WITH Re=405 AND α=0.5

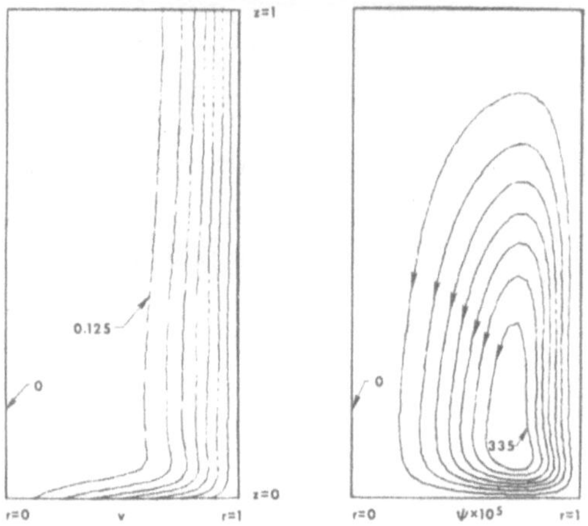

FIG. 12

DIMENSIONLESS TANGENTIAL VELOCITY v AT r=0.6, z=0.67 VS DIMENSIONLESS TIME t FOR SPIN-DOWN WITH α=0.5

EXPERIMENTAL MEASUREMENTS: ○Re=66, □Re=215

CURVES ARE COMPUTED SOLUTIONS.

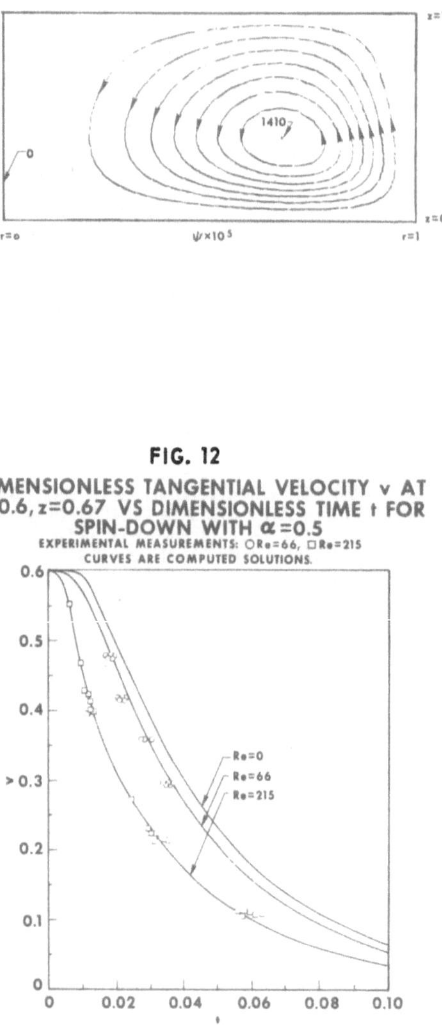

FIG. 13

DIMENSIONLESS TANGENTIAL VELOCITY v AT r=0.6, z=0.92 VS DIMENSIONLESS TIME t FOR SPIN-DOWN WITH α=2.0

EXPERIMENTAL MEASUREMENTS: △Re=405. CURVES ARE COMPUTED SOLUTIONS.

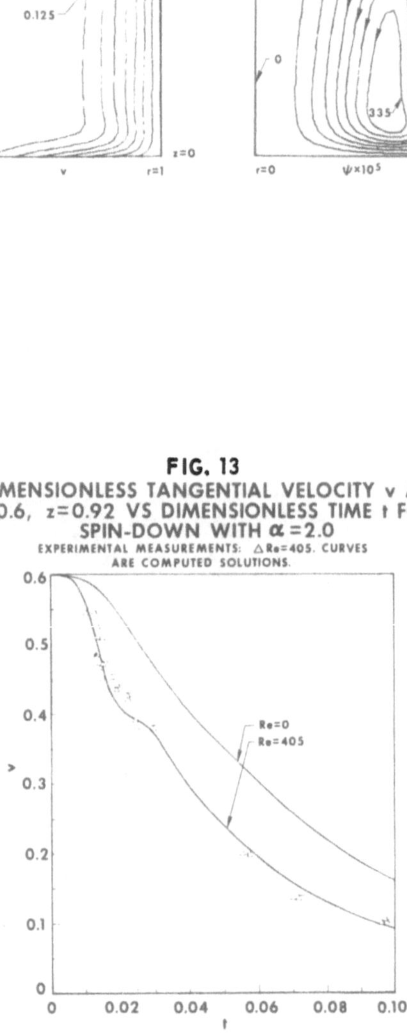

METHODES NUMERIQUES POUR L'ECOULEMENT STATIONNAIRE D'UN FLUIDE RIGIDE VISCO-PLASTIQUE INCOMPRESSIBLE.

R.GLOWINSKI [1]

RESUME :

Dans cet article, on étudie numériquement l'écoulement stationnaire d'un fluide rigide visco-plastique incompressible (fluide de BINGHAM) par des méthodes basées sur la notion de dualité en calcul des variations et plus généralement dans les inéquations variationnelles ; les méthodes utilisées, rigoureuses dans le cas de l'écoulement laminaire dans une conduite cylindrique, sont encore largement heuristiques dans le cas, plus complexe, d'un écoulement stationnaire bidimensionnel. On indiquera les résultats numériques obtenus dans l'étude de ces deux cas.

I- GENERALITES.

La définition d'un fluide de BINGHAM est donnée dans W.PRAGER [1] et rappelée dans DUVAUT-LIONS [1], Ch.6, où l'étude de l'écoulement d'un tel fluide, dans un domaine Ω de frontière Γ, est ramené à la recherche d'un champ de vitesse u, solution de l'inéquation variationnelle :

$$(1.1)\quad \begin{cases} (\frac{\partial u}{\partial t}, v-u) + a(u(t), v-u(t)) + b(u(t), u(t), v) + gj(v) - gj(u(t)) \\ \qquad \geq (f(t), v-u(t)) \\ \forall\, v \text{ tel que } \operatorname{Div} v = 0 , \quad v|_\Gamma = 0 \end{cases}$$

et vérifiant

$$(1.2)\quad \begin{cases} \operatorname{Div} u = 0 \\ u|_\Gamma = 0 \\ u(x,0) = u_0(x) \end{cases}$$

avec

$$(1.3)\quad \begin{cases} D_{ij} = \frac{1}{2}(\frac{\partial u_i}{\partial x_j} + \frac{\partial u_j}{\partial x_i}) \\ a(u,v) = 2\mu \int_\Omega D_{ij}\, u\, D_{ij}\, v \; dx \\ D_{II}(v) = \frac{1}{2}(D_{ij}\, v)^2 \\ j(v) = 2\int_\Omega \sqrt{D_{II}(v)}\; dx \\ b(u, v, w) = \int_\Omega u_i \frac{\partial v_j}{\partial x_i} w_j \; dx \\ (f,g) = \int_\Omega f_i\, g_i \; dx \end{cases}$$

les scalaires g et μ étant respectivement le seuil de plasticité et la viscosité du fluide de BINGHAM.

Remarque 1.1 : Dans le cas stationnaire, on a $\frac{\partial u}{\partial t} = 0$ dans (1.1). Dans DUVAUT-LIONS [1], on précise le cadre fonctionnel dans lequel on formule (1.1) et (1.2) et on démontre l'existence de solutions.

Remarque 1.2 : Si $g = 0$, on retrouve la formulation faible des équations de NAVIER-STOKES.

[1] IRIA, 78 - ROCQUENCOURT, France

/II/- ECOULEMENT STATIONNAIRE, LAMINAIRE DANS UNE CONDUITE CYLINDRIQUE.

Toutes les considérations de ce paragraphe sont détaillées dans J.CEA - R.GLOWINSKI [1].

II.1 POSITION DU PROBLEME

Soit Ω la section de la conduite et $\Gamma = \partial\Omega$, prenant Oz parallèle aux génératrices du cylindre, on a $u_1 = u_2 = 0$ et on pose $u_3 = u$, les relations (1.1) et (1.2) se réduisent alors à :

(2.1) $\mu\, a(u,v-u) + gj(v) - gj(u) \geq (c,v-u) \quad \forall v \in H_0^1(\Omega)$

avec

(2.2) $H_0^1(\Omega) = \left[v \mid v \in L^2(\Omega),\ \frac{\partial v}{\partial x_i} \in L^2(\Omega)\ ,\ v|_\Gamma = 0 \right]$

(2.3) $$\begin{cases} a(u,v) = \int_\Omega \operatorname{grad} u \cdot \operatorname{grad} v \; dx \\ j(v) = \int_\Omega |\operatorname{grad} v| \; dx \\ (f,v) = \int_\Omega fv \; dx \end{cases}$$

et où c est la chute linéique de pression, remplacée dans la suite, pour plus de généralité, par $f \in L^\infty(\Omega)$.

On montre, sans grande difficulté, que (2.1) est équivalent à la minimisation sur $H_0^1(\Omega)$ de la fonctionnelle

(2.4) $$J_0(v) = \mu\, a(v,v) + 2g\, j(v) - 2(c,v)$$

on considèrera donc le problème un peu plus général

(2.5) $$\begin{cases} \displaystyle\min_{v \in H_0^1(\Omega)} J(v) \\ J(v) = \mu\, a(v,v) + 2g\, j(v) - 2(f,v) \end{cases}$$

II.2 RESOLUBILITE DU PROBLEME CONTINU

La fonctionnelle J (de même que J_0) vérifiant :

i) J strictement convexe sur $H_0^1(\Omega)$

ii) J continue sur $H_0^1(\Omega)$

iii) $\displaystyle\lim_{\|v\|_{H_0^1} \to +\infty} J(v) = +\infty$

admet une solution optimale unique, c'est un résultat classique de la théorie de l'optimisation en dimension infinie pour lequel on renvoie, par exemple, à CEA [1], LIONS [1] .
Les propriétés de la solution sont étudiées dans MOSOLOV-MIASNIKOV [1], dans le cas f = c et DUVAUT-LIONS [1] dans le cas général.

II.3 UN RESULTAT DE DUALITE

Soit Λ le convexe de $(L^2(\Omega))^2$ défini par :

(2.6) $$\Lambda = \left\{ q \mid q = (q_1, q_2) \in (L^2(\Omega))^2,\ q_1^2 + q_2^2 \leq 1 \ \text{p.p.} \right\}$$

et $\mathcal{L} : H_0^1(\Omega) \times \Lambda \longrightarrow R$ définie par :

(2.7) $$\mathcal{L}(v,q) = \mu\, a(v,v) - 2(f,v) + 2g \int_\Omega q \cdot \operatorname{grad} v \; dx$$

on a le :

Théorème 2.1 : Si u est solution de (2.5) il existe $p \in \Lambda$ tel que :

$$(2.8) \quad \begin{cases} \mathcal{L}(u,q) \leq \mathcal{L}(u,p) \leq \mathcal{L}(v,p) \\ \forall\ q \in \Lambda,\ v \in H_0^1(\Omega) \\ p.\ \mathrm{grad}\ u = |\mathrm{grad}\ u| \end{cases}$$

il existe des démonstrations de ce théorème dûes à LIONS, TEMAM, CEA-GLOWINSKI, utilisant respectivement : le théorème de HAHN-BANACH, des résultats de T.R.ROCKAFELLAR, le théorème de KY FAN-SION.

Corollaire 2.1 : Tout couple (u,p) répondant au Théorème 2.1 est caractérisé par

$$(2.9) \quad \begin{cases} -\mu \Delta u - g\ \mathrm{div}\ p = f \\ u|_\Gamma = 0 \\ p.\ \mathrm{grad}\ u = |\mathrm{grad}\ u| \end{cases}$$

II.4 UNE VARIANTE DE L'ALGORITHME D'UZAWA

Le théorème 2.1 montre que le couple (u,p) est un point selle de la fonctionnelle $\mathcal{L}$ d'où l'idée naturelle d'appliquer à $\mathcal{L}$ l'algorithme d'UZAWA [1] qui, dans ce cas, prend la forme :

$$(2.10) \quad \begin{cases} p^0 \text{ donné} \\ -\mu \Delta u^{n+1} - g\ \mathrm{div}\ p^n = f \\ u^{n+1}|_\Gamma = 0 \\ p^{n+1} = P_\Lambda (p^n + \rho\ \mathrm{grad}\ u^{n+1}) \qquad (\rho > 0) \end{cases}$$

et où P_Λ est l'opérateur de projection sur Λ défini par :

$$(2.11) \quad (P_\Lambda(q))(x) \begin{cases} = q(x) & \text{si } |q(x)| \leq 1 \\ \dfrac{q(x)}{|q(x)|} & \text{si } |q(x)| > 1 \end{cases}$$

avec $|q(x)| = \sqrt{q_1^2(x) + q_2^2(x)}$

Dans CEA-GLOWINSKI [1], on démontre le

Théorème 2.2 : Si la constante ρ vérifie

$$(2.12) \qquad 0 < \rho < 2\frac{\mu}{g}$$

l'algorithme (2.10) est convergent et u^n tend vers la solution du problème (2.5).

Remarque 2.1 : On utilisera évidemment (2.10) sous une forme discrétisée par rapport aux variables d'espace.

II.5 MISE EN OEUVRE NUMERIQUE

Pour la discrétisation de (2.10) par des techniques de différences finies et les résultats de convergence correspondants, on renvoie à J.CEA - R.GLOWINSKI [1], disons simplement que si la solution approchée est définie sur un réseau de pas h, soit R_h, il est commode de définir p sur un réseau Q_h de même pas, mais décalé de $\frac{h}{2}$ par rapport à R_h (v. figure 2.1)

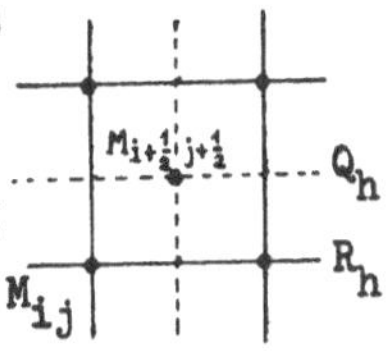

Figure 2.1

dans ces conditions, et avec les indexations habituelles, le terme div p est approché en M_{ij} par :

$$(2.13)\quad p_{ij} = \frac{1}{2h}\Big[(p^1_{i+\frac12\, j+\frac12} - p^1_{i-\frac12\, j+\frac12}) + (p^1_{i+\frac12\, j-\frac12} - p^1_{i-\frac12\, j-\frac12}) + (p^2_{i+\frac12\, j+\frac12} - p^2_{i+\frac12\, j-\frac12}) + (p^2_{i-\frac12\, j+\frac12} - p^2_{i-\frac12\, j-\frac12})\Big]$$

et le calcul de u^{n+1} à partir de p^n se fait en résolvant un problème de Dirichlet approché dont le second membre est calculé à l'aide de (2.13).

En ce qui concerne le calcul de p^{n+1} à partir de p^n et u^{n+1} on utilise :

$$(2.14)\quad \begin{cases} p^{n+\frac12}_{i+\frac12\, j+\frac12} = p^n_{i+\frac12\, j+\frac12} + \rho G_{i+\frac12\, j+\frac12} u^{n+1} \\[2ex] p^{n+1}_{i+\frac12\, j+\frac12} = \dfrac{p^{n+\frac12}_{i+\frac12\, j+\frac12}}{\sup(1, |p^{n+\frac12}_{i+\frac12\, j+\frac12}|)} \end{cases}$$

avec

$$(2.15)\quad \begin{cases} (G_{i+\frac12\, j+\frac12}u)_1 = \dfrac{1}{2h}\Big[(u_{i+1\, j+1} - u_{ij+1}) + (u_{i+1\, j} - u_{ij})\Big] \\[2ex] (G_{i+\frac12\, j+\frac12}u)_2 = \dfrac{1}{2h}\Big[(u_{i+1\, j+1} - u_{i+1\, j}) + (u_{ij+1} - u_{ij})\Big] \end{cases}$$

II.6 EXEMPLES TRAITES

Tous les exemples qui suivent ont été traités en utilisant l'analogue discret de l'algorithme (2.10) (les problèmes de Dirichlet étant résolus par surrelaxation avec paramètre optimal) le test d'arrêt utilisé étant

$$\sum_{ij} |u^{n+1}_{ij} - u^n_{ij}| \leqslant 10^{-4}$$

Exemple 1 : $\Omega =]0,1[\times]0,1[$. $\mu = 1$, $f = 10$.

Le paramètre ρ ayant sa valeur optimale, il y a convergence en 12 itérations lorsque $g = 1$, soit un temps de calcul de l'ordre de 1 s. sur IBM 360/91 pour $h = \frac{1}{20}$, soit environ 400 points de discrétisation.

Pour $\rho = 2$ (valeur limite donnée par (2.12) dans le cas continu) il y a encore convergence (136 itérations), mais divergence pour $\rho = 2.1$ (v. figure 2.6).

On peut démontrer que, dans le cas continu, $J(u) = -\|u\|^2_{\overset{\circ}{H}{}^1}$ d'où l'intérêt de la courbe de la figure 2.7 qui montre que $\|u\|_{\overset{\circ}{H}{}^1} \longrightarrow 0$ comme $|g - g_c|^{3/2}$

g_c étant la valeur critique de g , au-delà de laquelle $u = 0$.

Les figures 2.2 et 2.3 représentent les zones fluide et rigide (la zone rigide étant celle qui correspond à grad $u = 0$) pour $g = 1$ et $g = 1.8$.

Exemple 2 : Ω est le demi-disque de rayon $\frac{1}{2}$.

Pour $\mu = 1$, $f = 10$, $g = 0.75$, l'allure des zones fluide et rigide est indiquée figure 2.5.

III - ECOULEMENT STATIONNAIRE BI-DIMENSIONNEL.

Ce qui suit ne représente qu'une première approche, essentiellement expérimentale du problème et un grand nombre de points théoriques restent à justifier. Là encore, on va utiliser de façon essentielle la notion de dualité.

III.1 UN PREMIER RESULTAT DE DUALITE

Proposition 3.1 : Dans le cas des équations de Navier-Stokes, la pression p (en fait son opposé) apparaît comme le multiplicateur de Lagrange lié à la "contrainte" div $u = 0$.

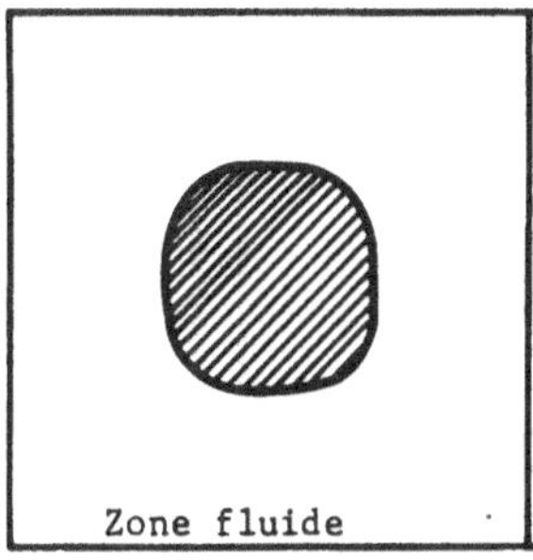

Fig. 2.2

$\mu = 1$, $f = 10$, $g = 1$

$u_{Max} = 0.291$

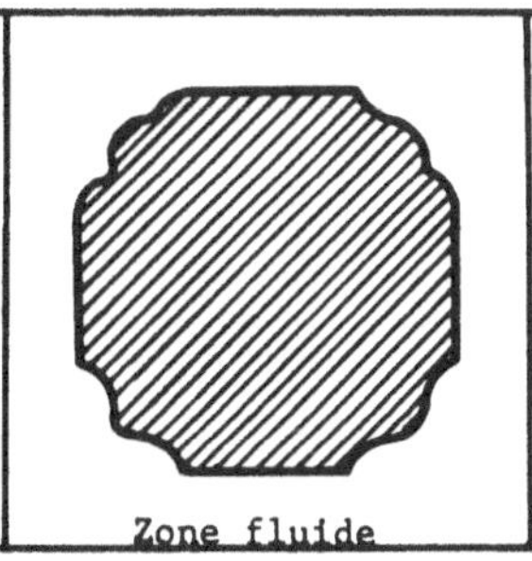

Fig. 2.3

$\mu = 1$, $f = 10$, $g = 1.8$

$u_{Max} = 0.080$

Conduite de section carrée

La zone rigide est hachurée

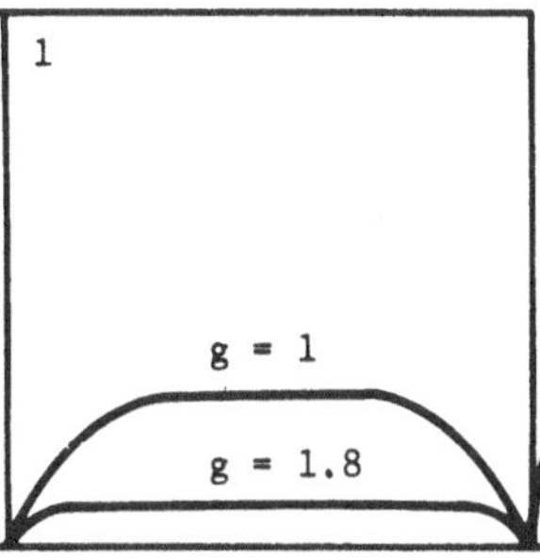

Fig. 2.4

$\mu = 1$, $f = 10$

Représentation de

$u(x, \frac{1}{2})$

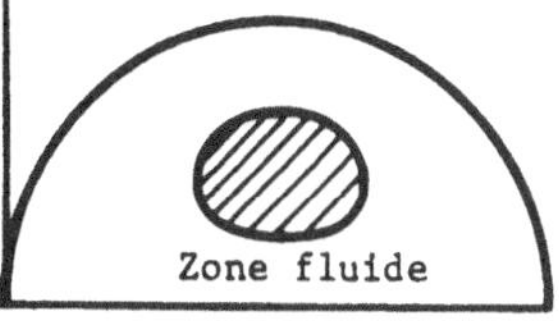

Fig. 2.5

$\mu = 1$, $f = 10$, $g = 0.75$

conduite semi-circulaire

La zone rigide

est hachurée

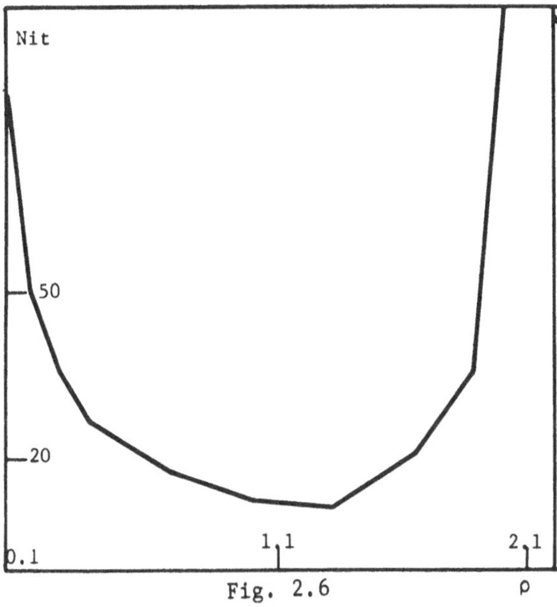

Fig. 2.6

Nombre d'itérations en fonction de ρ

$\Omega =]0,1[\times]0,1[$

$\mu = 1$, $F = 10$, $g = 1$

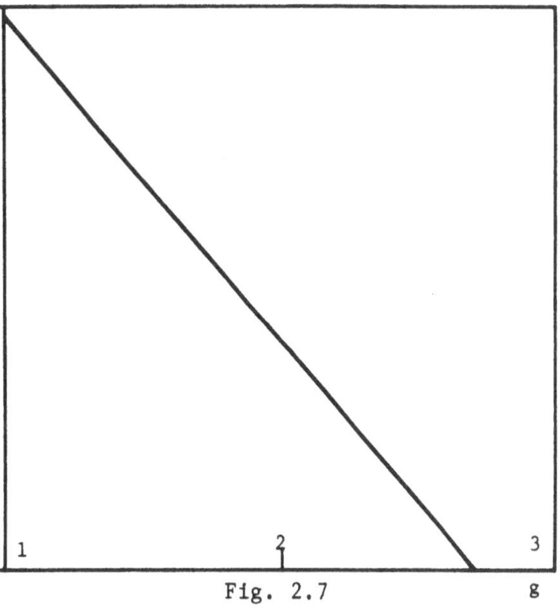

Fig. 2.7

Variation de $J(u)$ en fonction de g

$\Omega =]0,1[\times]0,1[$

$\mu = 1$, $f = 10$

(On a représenté $- (J(u))^{1/3}$)

Ce point de vue est examiné dans FORTIN - PEYRET - TEMAM [1] et l'algorithme d'UZAWA appliqué aux équations de Stokes (i.e. sans termes non linéaires) stationnaires, soit :

$$(3.1)\qquad \begin{cases} -\mu\Delta u^{n+1} + \operatorname{grad} p^n = f \\ u^{n+1}|_\Gamma = 0 \\ p^{n+1} = p^n - \rho \operatorname{div} u^{n+1} \end{cases}$$

est convergent pour ρ suffisamment petit et plus précisément $0 < \rho < \mu$, la situation étant plus compliquée lorsque l'on réintroduit les termes non-linéaires (v. à ce sujet FORTIN - PEYRET - TEMAM [1])

III.2 UN RESULTAT DE DUALITE POUR LES FLUIDES DE BINGHAM

Dans le cas $n = 2$ où le problème (1.1),(1.2) admet, sur $(0,T)$, une solution unique, si u_0 et f vérifient des hypothèses convenables (v. DUVAUT-LIONS [1] pour la démonstration de ce résultat), J.L.LIONS a démontré le résultat de dualité suivant qui constitue une généralisation du corollaire 2.1 :

Théorème 3.1 : Si u est la solution de (1.1) et (1.2) dans le cas $n = 2$, il existe des fonctions m_{ij} telles que :

$$(3.2)\qquad \begin{cases} m_{ij} \text{ mesurable de } (0,T) \longrightarrow L^\infty(\Omega) \text{ faible } * \\ m_{ij} = m_{ji} \\ (m_{ij}(t)\, m_{ij}(t))^{\frac{1}{2}} \leqslant \frac{g}{\sqrt{2}} \text{ p.p.} \end{cases}$$

$$(3.3)\qquad m_{ij}\, D_{ij}(u) + \frac{g}{\sqrt{2}}\,(D_{ij}(u)\, D_{ij}(u))^{\frac{1}{2}} = 0 \text{ p.p.}$$

$$(3.4)\qquad \frac{\partial u_i}{\partial t} - \mu\Delta u_i + u_j \frac{\partial u_i}{\partial x_j} + 2g\frac{\partial}{\partial x_j}(m_{ij}) = f_i - \frac{\partial p}{\partial x_i}$$

Réciproquement si on se donne u, p et des m_{ij} vérifiant (3.2),(3.3),(3.4) alors u est solution du problème (1.1),(1.2).

III.3 UNE GENERALISATION DE L'ALGORITHME D'UZAWA DANS LE CAS STATIONNAIRE

Dans le cas stationnaire, on va écrire (3.2),(3.3),(3.4) sous une forme plus commode en explicitant les D_{ij} d'où l'existence de q_1, q_2, q_3 tels que :

$$(3.5)\qquad \begin{cases} -\mu\Delta u + u\frac{\partial u}{\partial x} + v\frac{\partial u}{\partial y} - g\sqrt{2}\,\frac{\partial q_1}{\partial x} - g\,\frac{\partial q_3}{\partial y} = f_1 - \frac{\partial p}{\partial x} \\ -\mu\Delta v + u\frac{\partial v}{\partial x} + v\frac{\partial v}{\partial y} - g\,\frac{\partial q_3}{\partial x} - g\sqrt{2}\,\frac{\partial q_2}{\partial y} = f_2 - \frac{\partial p}{\partial y} \\ (u,v)|_\Gamma = 0 \\ q_1\frac{\partial u}{\partial x} + q_2\frac{\partial v}{\partial y} + \frac{q_3}{\sqrt{2}}\left(\frac{\partial u}{\partial y} + \frac{\partial v}{\partial x}\right) = \sqrt{\left(\frac{\partial u}{\partial x}\right)^2 + \left(\frac{\partial v}{\partial y}\right)^2 + \frac{1}{2}\left(\frac{\partial u}{\partial y} + \frac{\partial v}{\partial x}\right)^2} \\ q_1^2 + q_2^2 + q_3^2 \leqslant 1 \end{cases}$$

les relations (3.5) généralisent les relations (2.9) et l'algorithme d'UZAWA généralisé donne :

$$(3.6)\quad \begin{cases} p^o,\ q_1^o \text{ donnés} \\ -\mu\Delta u^{n+1} + u^{n+1}\dfrac{\partial u^{n+1}}{\partial x} + v^{n+1}\dfrac{\partial u^{n+1}}{\partial y} - g\sqrt{2}\dfrac{\partial}{\partial x}q_1^n - g\dfrac{\partial q_3^n}{\partial y} = f_1 - \dfrac{\partial p^n}{\partial x} \\ -\mu\Delta v^{n+1} + u^{n+1}\dfrac{\partial v^{n+1}}{\partial x} + v^{n+1}\dfrac{\partial v^{n+1}}{\partial y} - g\dfrac{\partial q_3^n}{\partial x} - g\sqrt{2}\dfrac{\partial q_2^n}{\partial y} = f_2 - \dfrac{\partial p^n}{\partial y} \\ p^{n+1} = p^n - \tau_p\left(\dfrac{\partial u^{n+1}}{\partial x} + \dfrac{\partial v^{n+1}}{\partial y}\right) \\ q_1^{n+\frac12} = q_1^n + \tau_q\dfrac{\partial u^{n+1}}{\partial x} \\ q_2^{n+\frac12} = q_2^n + \tau_q\dfrac{\partial v^{n+1}}{\partial y} \\ q_3^{n+\frac12} = q_3^n + \dfrac{\tau_q}{\sqrt{2}}\left(\dfrac{\partial u^{n+1}}{\partial y} + \dfrac{\partial v^{n+1}}{\partial x}\right) \\ q_i^{n+1} = \dfrac{q_i^{n+\frac12}}{\sup(1,\ |q^{n+\frac12}|)} \end{cases}$$

où $|q| = \sqrt{q_1^2 + q_2^2 + q_3^2}$

la convergence de (3.6) pour τ_p, τ_q convenablement choisis est encore à étudier.

III.4 DISCRETISATION

On a, là encore, avantage à définir p, q_1, q_2, q_3, sur le réseau Q_h décalé de R_h de $\frac{h}{2}$ (v.II.5), dans ces conditions, en M_{ij}, $\frac{\partial p}{\partial x}$ sera approchée par :

$$(3.7)\quad \left(\frac{\partial p}{\partial x}\right)_{ij} \simeq \frac{1}{2h}\left[(p_{i+\frac12\ j+\frac12} - p_{i-\frac12\ j+\frac12}) + (p_{i+\frac12\ j-\frac12} - p_{i-\frac12\ j-\frac12})\right]$$

et des expressions analogues pour les autres dérivées premières de p et q_i.

En $M_{i+\frac12\ j+\frac12}$, $\frac{\partial u}{\partial x}$ sera approchée par :

$$(3.8)\quad \left(\frac{\partial u}{\partial x}\right)_{i+\frac12\ j+\frac12} \simeq \frac{1}{2h}\left[(u_{i+1\ j+1} - u_{i\ j+1}) + (u_{i+1\,j} - u_{ij})\right]$$

et des expressions analogues pour les autres dérivées premières de u et v.

En ce qui concerne l'approximation en M_{ij} des termes non linéaires du type $u\frac{\partial u}{\partial x}$ on a utilisé :

$$(3.9)\quad \left(u\frac{\partial u}{\partial x}\right)_{ij} \simeq u_{ij}\frac{(u_{i+1\,j} - u_{i-1\,j})}{2h}$$

et expressions analogues pour les autres termes non linéaires, l'utilisation de schéma décentré avant ou arrière, suivant le signe de u_{ij}, v_{ij} n'améliorant pas la vitesse de convergence de l'algorithme (v. la remarque correspondante de FORTIN - PEYRET - TEMAM[1] dans le cas de NAVIER-STOKES) .

III.5 MISE EN OEUVRE DU DISCRETISE DE (3.6)

Le problème " primal " i.e. le calcul de u^{n+1}, v^{n+1} à partir de p^n , q_i^n se fait en résolvant un problème non linéaire, en fait, il n'y a pas de perte sur la vitesse de convergence si on se contente d'un balayage de type surrelaxation analogue à celui utilisé dans FORTIN - PEYRET - TEMAM [1], 3.2.3 , pour les équations de NAVIER-STOKES et c'est ce qui a été fait dans ce travail. On posera

(3.10) $$\tau_o = \frac{\omega h^2}{4\mu}$$

Les itérations "duales" i.e. le passage de u^{n+1}, v^{n+1} à partir de p^{n+1}, q_i^{n+1} sont du type gradient projeté et sont une simple transposition du cas continu donné par (3.6).

III.6 EXEMPLES TRAITES

Exemple 1 :

$\Omega =]0,1[\times]0,1[$, $f = 0$, avec les conditions aux limites

(3.11) $$\begin{cases} u(x,1) = u(x,0) = 1 \\ v(x,1) = v(x,0) = 0 \\ u(0,y) = u(1,y) = 0 \\ v(0,y) = v(1,y) = 0 \end{cases}$$

on a pris $h = \frac{1}{40}$, soient environ 1600 points de discrétisation pour u et v. Les calculs ont été faits pour $\mu = 0.1$ et $\mu = 1$ en faisant varier g .

Cas $\mu = 0.1$: $\omega = 1.8$, $\tau_o = 2.96 \times 10^{-3}$, $\tau_p = \tau_o$.

Pour $g = 0$ (cas de Navier-Stokes), en 1200 itérations, soit environ 90 s. d'IBM 360/91 on a $\sup(|u_{ij}^{n+1} - u_{ij}^n| , |v_{ij}^{n+1} - v_{ij}^n|) \leq 10^{-6}$, les résultats étant pratiquement stables dès la 500ème itération ; l'écoulement correspondant est représenté figure 3.1.

Pour $g = 0.5$, la valeur optimale de τ_q est approximativement $12\tau_o$ et l'on a $\sup(|u_{ij}^{n+1} - u_{ij}^n| , |v_{ij}^{n+1} - v_{ij}^n|) \simeq 4.10^{-5}$ à la 1700e itération ; des zones rigides apparaissent et la vitesse de l'écoulement dans la zone fluide est considérablement ralentie par rapport au cas $g = 0$. (v. figure 3.2)

Pour $g = 5$, la valeur optimale de τ_q est approximativement $0.9\,\tau_o$ et l'on a $\sup(|u_{ij}^{n+1} - u_{ij}^n| , |v_{ij}^{n+1} - v_{ij}^n|) \simeq 7.5 \times 10^{-5}$ à la 2000e itération ; la vitesse décroît très rapidement à partir des deux parois $y = 0$ et $y = 1$, le milieu étant rigidifié sauf au voisinage immédiat des parois $y = 0$ et $y = 1$.

On a représenté figure 3.3 la courbe donnant τ_q optimal en fonction de g , fonction bien représentée par $(\tau_q)_{opt} = \tau_o \times 5.4/g^{9/8}$.

Cas $\mu = 1$:

Dans le cas où τ_o, τ_p, τ_q correspondent à des valeurs optimales expérimentales, et toutes choses égales par ailleurs (g en particulier) la vitesse de convergence semble moins rapide que pour $\mu = 1$.

Exemple 2 : On a pris pour Ω la couronne circulaire $\frac{1}{2} < r < 1$, $f = 0$, les conditions aux limites étant

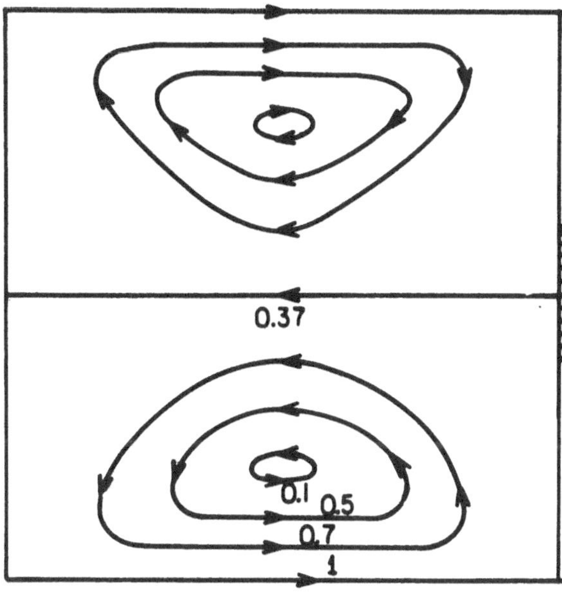

Fig. 3.1

$\mu = 0.1$ $g = 0.$

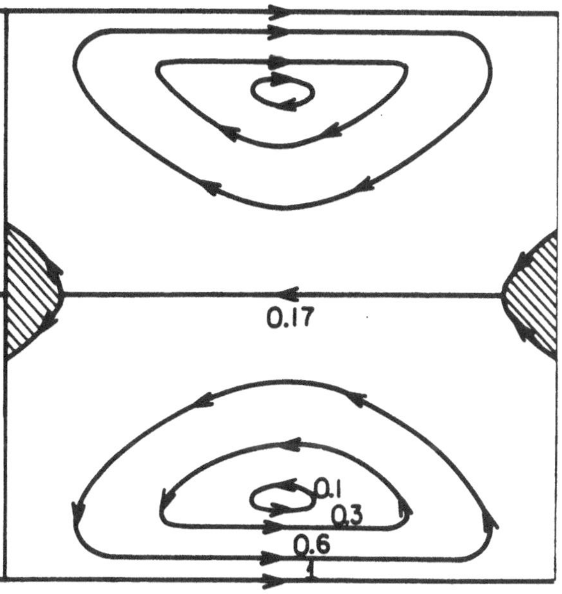

Fig. 3.2

$\mu = 0.1$ $g = 0.5.$

la zone rigide est hachurée.

Fig. 3.3

$\mu = 0.1$

$\tau_p = \tau_o$

représentation de τ_q optimal en fonction de $g (\tau_q = K \tau_o)$

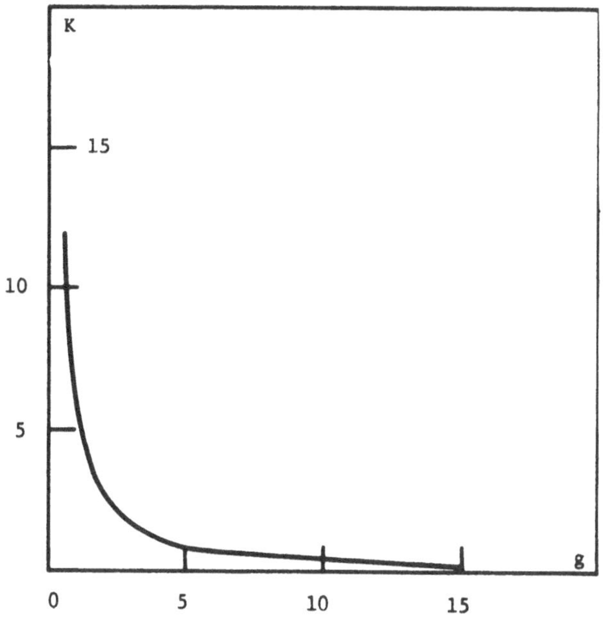

$$
(3.12) \qquad \left\{ \begin{aligned} u_\theta(1,\theta) &= u_r(1,\theta) = 0 \\ u_\theta(\tfrac{1}{2},\theta) &= 1 \\ u_r(\tfrac{1}{2},\theta) &= 0 \end{aligned} \right.
$$

ce qui conduit à un écoulement révolutif calculable analytiquement (en utilisant les résultats de dualité par exemple) ou numériquement par des méthodes analogues à celles considérées en II .

Le problème a été traité en cartésien, sans utiliser la propriété de symétrie et avec $h = \frac{1}{40}$ les résultats coïncident bien avec ceux obtenus analytiquement, sauf au voisinage de $r = \frac{1}{2}$ ce qui est dû au fait que l'on approche cette frontière intérieure par un polygone, il semble donc que l'on doit pouvoir améliorer ces résultats en discrétisant au voisinage de la frontière par des méthodes du type BRAMBLE-HUBBARD.

Remerciements : Je tiens à remercier MM. BEGIS et GOURSAT, Chercheurs à l'IRIA, pour la part active qu'ils ont pris à la programmation des résultats figurant dans ce travail et M. FORTIN pour les nombreuses discussions que nous avons pu avoir à propos des équations de NAVIER-STOKES.

BIBLIOGRAPHIE

J. CEA [1] : Théorie de l'optimisation - A paraître chez DUNOD.

J. CEA - R. GLOWINSKI [1] : Méthodes Numériques pour l'écoulement laminaire d'un fluide rigide visco-plastique incompressible. Article à paraître en 1971.

G. DUVAUT - J.L.LIONS [1] : Livre à paraître chez DUNOD.

FORTIN - PEYRET - TEMAM [1] : Conférence à ce Congrès. (à paraître)

J.L. LIONS [1] : Contrôle optimal de systèmes gouvernés par des équations aux dérivées partielles - DUNOD - 1968.

MOSSOLOV - MIASNIKOV [1] : Zones stationnaires de l'écoulement d'un fluide viscoplastique dans un cylindre - Maths Appl. et Mécanique, Tome 30, Moscou 1966 (en russe).

W. PRAGER [1] : Introduction to mechanics of continua - 1961 .

H. UZAWA [1] : Itérative methods for concave programming in "Studies in Linear and non linear programming" K.J.ARROW, L. HURWICZ, H. UZAWA - Stanford. Univ. Press. 1958.

NUMERICAL ANALYSIS AND MODELING OF SLIP FLOWS AT VERY HIGH MACH NUMBERS

Barry Bernard Novack and Hsien Kei Cheng
University of Southern California
Los Angeles, California 90007

I. INTRODUCTION

Nonlinear fluid mechanics problems of practical importance are often characterized by a variety of flow regimes and the many distinct regions attending them. Asymptotic theories or heuristic models may be constructed for these problems, but their validity is not easy to prove. In a number of instances, numerical solutions to the fuller equations (used as an experiment) are found very helpful in settling questions concerning the validity of the models. The application of numerical methods to these problems will demand, on the other hand, a level of accuracy higher than required for the standard problem. The analysis of viscous compressible flows around slender bodies in the hypersonic slip flow regime to be discussed below represents one numerical study of this kind.

Our numerical study is based on finite difference methods applied to a set of composite equations of the parabolic type, which are reduced from the Navier-Stokes equations for a compressible flow at a high Mach number and a high Reynolds number, allowing slip and temperature jump. The general framework, with an application to the flat plate problem in the rarefaction range $1/10 < \bar{V} < 1$, has been reported previously (Cheng and coworkers 1969, 1970; Cheng 1970). The present paper discusses the use of an alternate difference scheme and new results obtained for flows over concave surfaces, slender and very slender cones, as well as the flow behind a trailing edge. The study clarifies several questions arising from recent results obtained by Rubin and coworkers, 1969.

The accuracy requirement in the present problem has rendered the use of the unsteady approach to the Navier-Stokes equations undesirable because of practical limitations in the computer capacity and economy*. The use of the reduced equation system permits an order of magnitude saving in computing time, and therefore requires no apology. As brought out previously (Cheng, 1970), the present approach is limited at far downstream by the computer capacity and grid size requirement, and in the upstream, by the small deflection (thin layer) assumption and the breakdown of the continuum description. Therefore, comparisons with analyses based on other relevant models at both ends are essential in assessing the usefulness of our results. On the other hand, the same comparisons also shed light on the limitation and applicability of methods based on the other (kinetic as well as boundary layer) models, as demonstrated in Cheng's (1970) study.

2. REDUCTION TO INITIAL VALUE PROBLEM

The simplified equations, reduced from the Navier-Stokes equation under a small flow deflection angle θ, and subject to an error of order θ^2, are (Cheng, 1970):

Continuity: $\frac{\partial}{\partial x}(\rho u) + \nabla\cdot\rho\vec{q} = 0$

x-momentum: $\rho(u\frac{\partial}{\partial x} + \vec{q}\cdot\nabla)u = -\frac{\partial p}{\partial x} + \nabla\cdot\mu\nabla u$

Transverse momenta: Navier-Stokes, omitting $\frac{\partial}{\partial x}\mu\frac{\partial}{\partial x}\vec{q}$

Energy: $$\rho(u\frac{\partial}{\partial x} + \vec{q}\cdot\nabla)H = \nabla\cdot\frac{\mu}{Pr}\nabla\left[H - \frac{1}{2}(1-Pr)(u^2+q^2)\right] + \nabla\cdot\mu(\vec{q}\cdot\nabla)\vec{q} + \nabla\cdot(\mu' - \tfrac{2}{3}\mu)(\nabla\cdot\vec{q})\vec{q} \tag{2.1}$$

where x and u denote the distance and velocity components along an axis parallel to the mainstream, $\vec{q}$ and ∇ are the velocity vector and the gradient operator in the transverse plane, and ρ, p, H and Pr are the density, thermodynamic pressure, total enthalpy and Prandtl number, respectively.

---

*Results based on an unsteady approach have been obtained earlier by Butler (1967) for a semi-infinite flat plate in hypersonic flow. The level of accuracy attained is judged, however, to be marginal.

The reduction and justification of this composite system, as well as the influence of initial conditions and importance of the retained terms, has been discussed previously (Cheng and coworkers, 1969, 1970; Cheng 1970). A caloric perfect gas and a viscosity-temperature relation $\mu \propto T^{\omega}$ are assumed for simplicity.

The composite system, together with the thermal and caloric equations of state, can be solved as an initial value problem having only one inner boundary at the body surface.

In the results which follow, a nonpermeable slip boundary condition, based on a local application of the classical two-dimensional formula, is used. The boundary is not restricted to straight surfaces. Let F(x, y, z) = 0 be the equation of the surface of a body or wall; the boundary conditions are:

$$\left(u\frac{\partial}{\partial x} + \vec{q}\cdot\nabla\right)F = 0\,, \quad T - T_w = \lambda_j\,\hat{n}\cdot\nabla T$$
$$u = \lambda_s\,\vec{n}\cdot\nabla u\,, \quad \vec{q}\cdot\vec{t} = \lambda_s\,\vec{n}\cdot\nabla(\vec{q}\cdot\vec{t}) \tag{2.2}$$

where $\vec{t} = (\vec{n}\times\vec{i})$ is the tangential unit vector in the transverse plane, $\vec{n}$ is the unit outward surface normal, and $\vec{i}$ is the unit vector along the x-axis. The velocity and temperature slip coefficients, λ_s and λ_j, are of the same order as the mean free path $\lambda \sim \mu/\rho\sqrt{\gamma RT}$. These slip conditions require correction when the transverse dimension of the Knudsen layer δ_k is comparable to the body radius r_b (such as that found in the problem of a needle). However, a recent study of Ellinwood (1970) based on the BGK equation shows that the corrections are numerically small.

Previous studies applied to flat plate cases (Cheng, 1970) indicate the model is adequate for the rarefaction parameter $\bar{V} = \bar{\chi}/M_\infty^2 = M_\infty\sqrt{C_*/Re_x}$ less than 0.4.

3. NUMERICAL PROCEDURE BASED ON CRANK-NICHOLSON SCHEME

Numerical programs based on two difference schemes (Dufort-Frankel and Richtmyer No. 9; see Richtmyer and Morton, 1967) have been developed previously for the above system (Cheng and coworkers 1969, 1970). The procedure and its applications to be discussed below for the plane and axisymmetric flows are based on the Crank-Nicholson scheme (illustrated below) which permits a smaller truncation error and less computing time. The basic program has been developed for (2.1) in its non-divergence form, although a program with the Crank-Nicholson scheme has also been made to work on a divergence form of (2.1) earlier. For the relatively stringent accuracy requirement in the present applications, there appears no advantage in going into a divergence formulation.

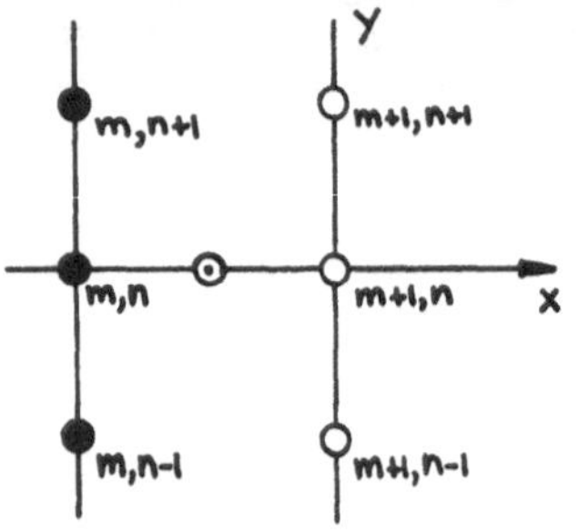

$$\frac{\partial S}{\partial x} = \frac{S_{m+1,n} - S_{m,n}}{\Delta x}$$
$$\frac{\partial S}{\partial y} = \frac{1}{2}\left[\frac{S_{m+1,n+1} - S_{m+1,n-1}}{2\Delta y} + \frac{S_{m,n+1} - S_{m,n-1}}{2\Delta y}\right]$$
$$\frac{\partial^2 S}{\partial y^2} = \frac{1}{2}\left[\frac{S_{m+1,n+1} - 2S_{m+1,n} + S_{m+1,n-1}}{(\Delta y)^2} + \frac{S_{m,n+1} - 2S_{m,n} + S_{m,n-1}}{(\Delta y)^2}\right]$$

In the present procedure, each of the equations, properly nondimensionalized, is integrated independently of the others, with all dependent variables, except one, forward extrapolated or taken at their calculated values. Uniform step and grid sizes are used in determining u, v, ρ and H in the order indicated; p is obtained from the equation of state. The pressure gradient in the main flow direction, $\partial p/\partial x$, is replaced by values extrapolated from the preceeding step, as in Cheng, 1970. Stepwise iteration cycles, consisting of a predictor-corrector operation, have been necessary at the larger step sizes to obtain an adequate continuous description of the pressure distribution near the shock.

Using a Gaussian reduction, modified to operate only on non-zero terms, the resulting equations in tridiagonal form are solved with additional equations supplied by the boundary conditions at the undisturbed region and at the body surface.

Boundary Logic

When the grid point does not coincide with the surface, the wall boundary condition is satisfied by a Taylor expansion to order $\Delta(y/\delta)^2$ to the first grid point above the surface.

Step and Grid Sizes; Computation Time

To ascertain the degree of accuracy of the present scheme, the flow over a flat plate at $M_\infty = 24.5$, $Re_L = 10^4$ is chosen, which allows direct comparison with results obtained in previous studies reported by Cheng (1970), based on the Dufort-Frankel and Richtmyer No. 9 schemes. Figure 1 shows a typical profile of thermodynamic pressure with grid density of the present results indicated by filled circles.

On the IBM 360-65, the program (with double precision) takes less than $2\frac{1}{2}$ minutes to reach $x/L = 1$ for $\Delta(x/L) = 0.005$ and $\Delta(y/\delta) = 0.01$ in most cases.

It has been observed that the program yields stable solutions for all cases considered until the surface slip velocity falls below 0.12, without further modifications in the program. The breakdown in the solutions for a flat plate occurs at $x/L = 2.3$, whereas this difficulty is delayed for the case of a 0.5° needle until $x/L = 15$, due to the large surface slip.

4. DISCUSSION OF RESULTS

To make direct comparison to previous results possible, all cases to be discussed are computed for the same flow conditions: Mach number = 24.5, Reynolds number/inch = 10,000, specific heat ratio = 1.4, Prandtl number = .75, and a wall to stagnation temperature ratio = .15. A linear viscosity law is assumed, with uniform conditions at the leading edge x = 0 and a diffuse reflection and full thermal accommodation. Both two-dimensional and axisymmetric bodies are studied.

In passing, we note that the essential features which differentiate the present program from the corresponding programs in the standard boundary layer analyses (see for example, Blottner (1970)) lies in the use of the transverse momentum equation and in the treatment of the tangent pressure gradient term $\partial p/\partial x$. The latter term is treated as being very small in the works of Rudman and Rubin (1968) and Rubin and coworkers (1969). As noted by Cheng and coworkers (1969, 1970), however, $\partial p/\partial x$ is an essential term at least at large $\bar{V}$. We report that, using the present program, suppression of $\partial p/\partial x$ from the equation reduces by 14% the surface pressure and by 20% the ordinate of the maximum pressures at $x/L = .75$ corresponding to $\bar{V} = .304$.

A simple application of the present program is the study of flow behind a flat plate with finite chord, which represents an idealized problem of wake development downstream of a trailing edge in hypersonic slip flow. A full solution to this flow must, of course, allow for the upstream influence of the trailing edge (T.E.) as well as a valid flow description for the immediate vicinity of the T.E. itself, which the present program is not equipped to do. The extensive literature on the local boundary-layer interaction (Neyland and Sychev, 1966; Riley and Stewartson, 1969; Messiter, 1970; Reeves and Lees, 1965; Ohrenberger and Baum, 1970; Garvine and Weinbaum, 1970) appears to indicate, however, that the extent of upstream influence is weak. The solution obtained for the present example with the T.E. located at $x/L = 1$ is described in Figure 2 for the pressure and tangential velocity profiles at various downstream stations. The results show that with the sudden relief of the wall shear, a "local expansion region" develops after the T.E., in which fluid particles experience a rapid pressure reduction and accelerate to higher velocities. Whereas the region influenced by the T.E. spreads outward, the outer flow region does not appear to be immediately affected by the T.E. To be sure, the latter behavior is not a result of the difference approximation since the implicit scheme used involves simultaneous determination of all points at the same station. Examination of the transverse velocity profiles (not shown) indicates the existence of a sizable enclosed area near the T.E. where the transverse velocity (v), and hence the streamline slope, is negative. The streamlines crossing the contour of v = o will therefore converge and diverge from the symmetry plane, signifying a "neck region"! We may also note that the shock and the outer flow region appear to be little perturbed by the T.E. at a downstream distance several times the chord. This is presumably a consequence of the relatively high momentum carried by the outer flow, which is characteristic of all hypersonic viscous flows past slender bodies.

In Figure 3 we present certain results of our study on flows over concave surfaces in the hypersonic slip flow regime, including a ramp at small incline angle. The surface values of the thermodynamic pressure on the concave surface $y/L = \tau(x/L)^2$ are presented in full line for $\tau = 0.04$ and $\tau = 0.0875$. For comparison, the flat plate result is included in dash line. The effect of introducing a smooth ramp as such, is seen to be quite large, since the ordinate of the two surfaces are rather small compared to the transverse dimension of the disturbed flow region, which is typically of the order $\delta/L \sim 0.2$. The influence of the concave surface is large enough for $\tau = 0.0875$ that $\partial p/\partial x$ becomes positive everywhere on the surface. Omission of $\partial p/\partial x$ from the momentum equation would yield substantially different solutions for this case. Also shown in Figure 3 is the pressure on a 5° ramp (or flap). Here, we invoke that separation does not occur and that, as in the previous discussion, the upstream influence of the corner is confined to a limit region which is small compared to the size of the flow region analyzed. Our assumptions are not inconsistent with current theoretical and experimental results on the viscous interactions (for example, see Stollery, 1970; Holden, 1970). The pressure curve for the 5° ramp (in dotted line) has a rapid rise after the corner, followed by a plateau (we have not obtained a solution for region further downstream to determine how the inviscid pressure plateau is reached). An oscillatory trend is noticeable in the approach to this plateau; an additional solution with a more refined step size (shown in dots) retains a weaker oscillatory pattern.

In Figures 4 and 5 we present certain results of a numerical study of axisymmetric flows over (unyawed) slender and very slender cones. The study is of interest in providing an assessment of results based on the theory of a needle (Stewartson, 1964; Ellinwood and Mirels, 1968; also see Cheng, 1969), and on a local similarity method (Mirels and Ellinwood, 1968), which do not take into account the slip effects. Figure 4 shows in full lines the thermodynamic pressure on cones as a function of $\bar{V}/\sin^2\theta_c$ for half angles $\theta_c = 10^o$, 5° and 0.5^o. The variable $\bar{V}/\sin^2\theta_c$ is essentially the local value of $(\delta/\theta_c)^2$. Included for comparison are the self similar solutions from Stewartson's (1964) theory of strong viscous interaction for a needle (modified by Ellinwood and Mirels for power law viscosity and Prandtl number effects), and Mirels and Ellinwood's results based on a local similarity method. It must be pointed out that good agreement in pressure of the present solutions with results based on the local similarity method is possible only when the surface velocity is small, since the latter does not account for slip effects, which is larger for an axisymmetric slender body than for a thin airfoil (see Cheng 1969). As compared to the needle solution, the difference is expected to be even larger because the needle solution has, in addition, an error of the order $1/\ln(\bar{V}/\theta_c^2)$ which cannot be treated as a small quantity unless $(\bar{V}/\theta_c^2)$ belongs to an order as large as 10^{10}.

Previously Rubin and coworkers (1969) have reported excellent agreement between their solution for a 5° cone and the non-slip, strong-interaction needle limit. However, their results reproduced in Figure 4 (in dash), appears to be close to the present solution, but clearly far from the needle solution. Their fortuitous conclusion could have resulted from uncorrected errors in Stewartson's original paper (see Solomon, 1967).

Figure 5 gives surface heat transfer rate and skin friction as a function of $(1 + 3.46\ T_w/T_o)\bar{V}/\sin^2\theta_c$ which permits a direct comparison with Mirels and Ellinwood's (1968) result. Unlike the pressure, skin friction and heat transfer rates (including work done by skin friction) from our solution appear to approach reasonably closely the non-slip, Mirels and Ellinwood's curves at downstream, with the exception of the skin friction for the 0.5^o cone. This apparent agreement is not too surprising since agreement of the present heat transfer and skin friction results with strong interaction limits have been found and explained for the flat plate (Cheng, 1970). We observe in passing, that the present solution appears to give a somewhat higher asymptotic heat transfer rate and skin friction than Mirels and Ellinwood's values. Whether the discrepancy results from the empiricism in the local-similarity method remains to be examined.

ACKNOWLEDGMENTS

The authors would like to acknowledge the helpful suggestions and assistance offered by Professor Richard Kaplan, Director, Systems Simulation Laboratory,

University of Southern California. This paper was under the sponsorship of the United States Air Force Office of Scientific Research, through AFOSR Research Grant AF-AFOSR-697-67.

REFERENCES

Blottner, F. G., 1970, AIAA J., 8, 193-205.

Butler, T. D., 1967, Phys. Fluids, 10, 1205-1215.

Cheng, H. K., 1969, USC Dept. of Aerospace Engineering USCAE Report 108; also see article by same author in Modern Developments in Gas Dynamics, Chapter 4, Plenum Press, 1969.

Cheng, H. K., 1970, to be published in the Proceedings of the Session of Numerical Methods in Gasdynamics of the Second International Colloquim on Gasdynamics of Explosions and Reactive Systems, Novosibirsk, August 14-23, 1969.

Cheng, H. K., Chen, S. Y., Mobley, R. and Huber, C., 1969, Rarefied Gas Dynamics, 6th Symposium, Vol. I, Academic Press, 451-464.

Cheng, H. K., Chen, S. Y., Mobley, R. and Huber, C., 1970, RAND Corp. Report RM-6193-PR.

Ellinwood, J. W., 1970, to be published in the Proceedings of the 7th International Symposium on Rarefied Gas Dynamics, Pisa, June 24-July 3, 1970.

Ellinwood, J. W. and Mirels, H., 1968, J. Fl. Mech., 34, 687-703.

Garvine, R. W. and Weinbaum, S., 1970, Proceedings of the 1969 Symposium of Viscous Interaction Phenomena in Supersonic and Hypersonic Flow, U. of Dayton, 427-462.

Holden, M. S., 1970, Proceedings of the 1969 Symposium of Viscous Interaction in Supersonic and Hypersonic Flow, U. of Dayton, 213-270.

Messiter, A. F., 1970, SIAM J. Appl. Math., 18, 241-257.

Mirels, H. and Ellinwood, J. W., 1968, AIAA J., 6, 2061-2070.

Neyland, V. Y. and Sychev, V. V., 1966, Melen. Zhidk. i Gaz., 4, 43-49.

Ohrenberger, J. T. and Baum, E., 1970, AIAA Preprint, 70-792.

Reeves, B. L. and Lees, L., 1965, AIAA J. 3, 2061-2075.

Richtmyer, R. D. and Morton, K. W., 1967, Difference Methods for Initial-Value Problems, 2nd ed., Interscience, New York, London and Sydney.

Riley, N. and Stewartson, K., 1969, J. Fl. Mech., 39, 193-207.

Rubin, S. G., Lin, T. C., Pierucci, M. and Rudman, S., 1969, AIAA J., 7, 1744-1751.

Rudman, S. and Rubin, S. G., 1968, AIAA J., 6, 1883-1890.

Solomon, J. M., 1967, U. S. Navy Ordnance Lab., White Oak Maryland, NOLTR66-225.

Stewartson, K., 1964, Phys. Fluids, 7, 667-675.

Stollery, J. L., 1970, to be published in J. Fl. Mech.; also see AIAA Preprint 70-782.

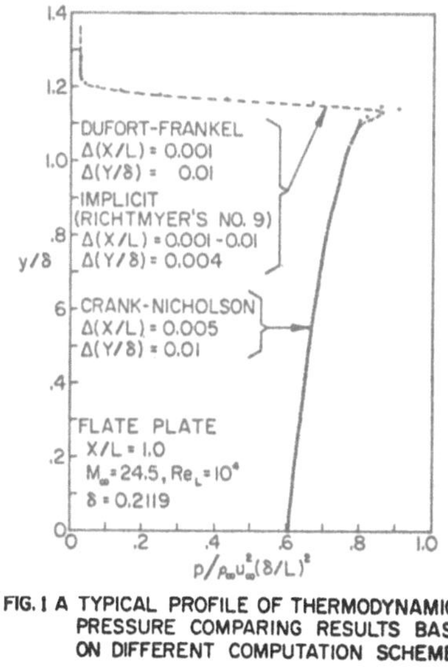

FIG. 1 A TYPICAL PROFILE OF THERMODYNAMIC PRESSURE COMPARING RESULTS BASED ON DIFFERENT COMPUTATION SCHEMES.

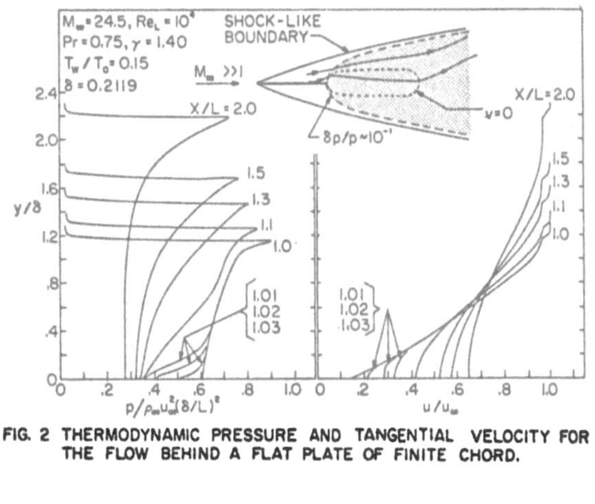

FIG. 2 THERMODYNAMIC PRESSURE AND TANGENTIAL VELOCITY FOR THE FLOW BEHIND A FLAT PLATE OF FINITE CHORD.

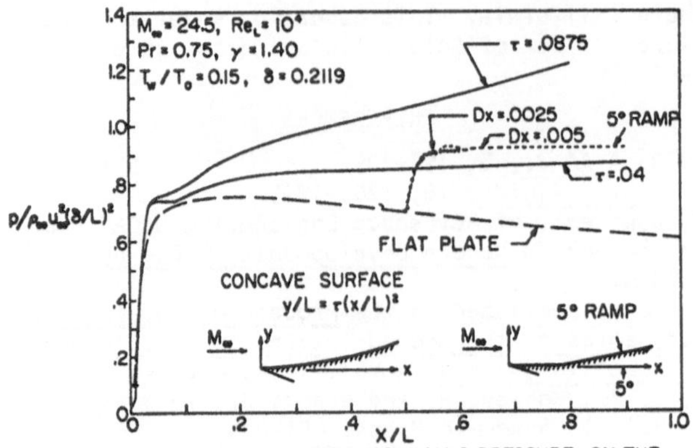

FIG. 3 SURFACE VALUES OF THERMODYNAMIC PRESSURE ON THE CONCAVE SURFACE $y/L = \tau(x/L)^2$ AND A 5° RAMP.

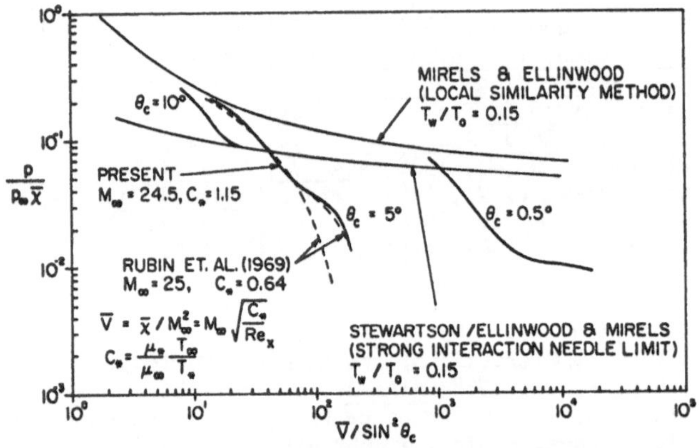

FIG. 4 THERMODYNAMIC PRESSURE DISTRIBUTIONS ON SLENDER CONES.

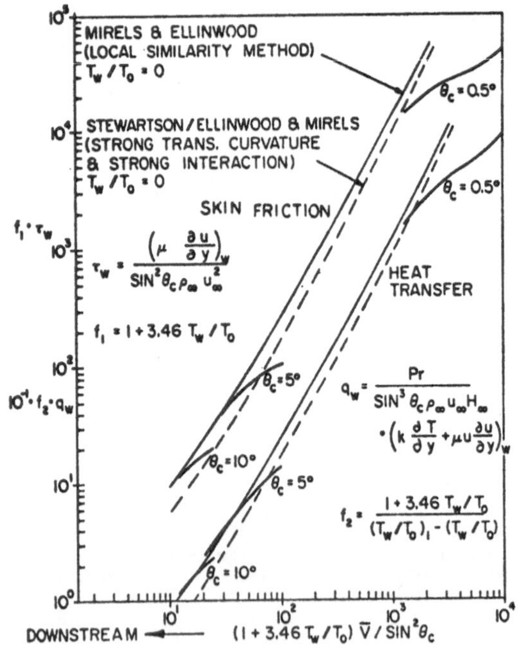

FIG. 5 LOCAL SKIN FRICTION AND HEAT TRANSFER ON SLENDER CONES.

Session VII

Incompressible Flow Problems

K. Roesner, Chairman

SOME NUMERICAL SOLUTIONS OF UNSTEADY FREE SURFACE WAVE PROBLEMS USING THE LAGRANGIAN DESCRIPTION OF THE FLOW

Christopher Brennen
California Institute of Technology

INTRODUCTION

Until very recently numerical solutions of unsteady, free surface flows invariably employed the Eulerian description of the motions. Perhaps the most widely used of these has been the marker and cell (MAC) technique developed by Fromm and Harlow (1963) and further refined by many others. In such a formulation the most difficult problem arises in attempting to reconcile the initially unknown shape and position of the free surface with a finite difference scheme and the necessity of determining derivatives at that surface (in a similar fashion few solutions exist with curved or irregular boundaries). But this difficulty can be surmounted by solving in a parametric plane in which the position and shape of the free surface are known in advance; such mappings have been successfully employed in steady flows (eg. Brennen (1969)). Whilst there are other possibilities (see John (1953), Brennen and Whitney (1970)) the Lagrangian description in its general form involves just such a parametric plane. The present paper describes briefly a numerical method for the solution of the Lagrangian equations of motion for the inviscid, planar flow of a homogeneous or inhomogeneous fluid, taking full advantage of the flexibility of choice of the Lagrangian coordinates (a,b). More details and other results can be found in Brennen and Whitney (1970). Very recently Hirt, Cook and Butler (1970) published details of a method which solves the Eulerian equations of motion in a fashion similar to the MAC technique but uses a Lagrangian tagging space.

BASIC EQUATIONS

The general inviscid dynamical equations of planar motion in Lagrangian form are (Lamb (1932)):

$$(X_{tt} - F)\begin{Bmatrix} X_a \\ X_b \end{Bmatrix} + (Y_{tt} - G)\begin{Bmatrix} Y_a \\ Y_b \end{Bmatrix} + \frac{1}{\rho}\begin{Bmatrix} P_a \\ P_b \end{Bmatrix} = 0 \qquad (1)$$

where X, Y are the cartesian coordinates of a fluid particle at time t, F, G are the components of extraneous force acting on it, P is pressure, ρ the density and (a,b), the Lagrangian coordinates, are any two quantities which serve to identify the particle and vary continuously from one particle to the next. Subscripts a,b,t denote differentiation. The equation of continuity is simply

$$\rho\ \partial(X,Y)/\partial(a,b) = \text{independent of time, } t \qquad (2)$$

If F, G have a potential and ρ, if not uniform, is a function only of P then from (1)

$$\frac{\partial}{\partial t}(U_a X_b - U_b X_a + V_a Y_b - Y_b Y_a) = -\frac{\partial \Gamma}{\partial t} = 0 \qquad (3)$$

where U, V are the velocities X_t, Y_t. It is easily shown that Γ is the vorticity multiplied by the Jacobian $(X_a Y_b - X_b Y_a)$ and can be calculated from the initial conditions. By introducing the vectors $Z = X + iY$, $W = U - iV$ the equations of motion, (3), and continuity ((2) differentiated w.r.t. t) conveniently combine to

$$Z_a W_b - Z_b W_a = \Gamma(a,b) \qquad (4)$$

In the case of an inhomogeneous fluid with no density diffusion (i.e., $\rho = \rho(a,b)$) equation (3) must be modified and the resulting equivalent of (4) is:

$$Z_a W_b - Z_b W_a = \Big[\Gamma(a,b)\Big]_{t=t_o} - \frac{1}{\rho}\int_{t_o}^{t} (X_{tt} - F)(\rho_b X_a - \rho_a X_b) + (Y_{tt} - G)(\rho_b Y_a - \rho_a Y_b)\, dt \qquad (5)$$

The integral represents the vorticity generated by the density gradients.

NUMERICAL METHOD FOR SOLUTION

The following method was designed to numerically solve the equations (4) or (5) of the preceding section. It was an implicit scheme with central differencing over a series of stations in time, t, distinguished by the integer superscript p. The values Z^{p+1} are determined by solving for the velocities, $Z_t = \overline{W}$, at midway stations, $p+\frac{1}{2}$, and then employing the numerical approximation:

$$Z^{p+1} = Z^{p} + \tau \overline{W}^{p+\frac{1}{2}} \quad (\text{error order } \tau^3 Z_{ttt}) \tag{6}$$

where τ is the time interval dividing stations p+1 and p.

The method is necessarily restricted to a finite body of fluid, S; this might, however, be part of a larger or infinite mass provided an "outer" approximate solution of sufficient accuracy was available to provide matching conditions at the interface. Then S need not be fixed in time. In a great many cases it is possible and convenient to choose S to be rectangular in the (a,b) plane. This rectangle (ABCD, figure 1) is divided into a set of elemental rectangles or 'cells' the motion

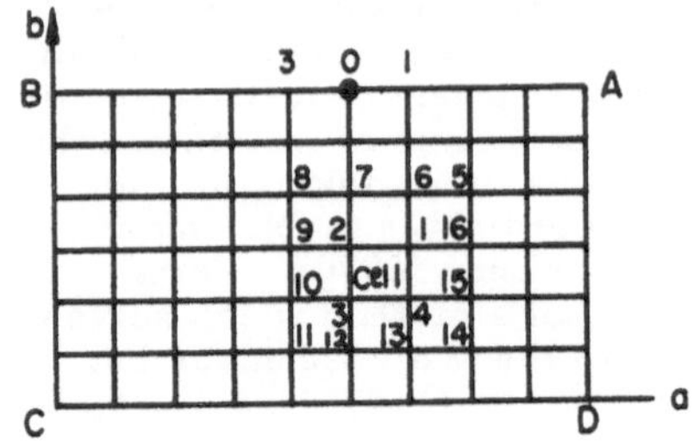

FIGURE 1

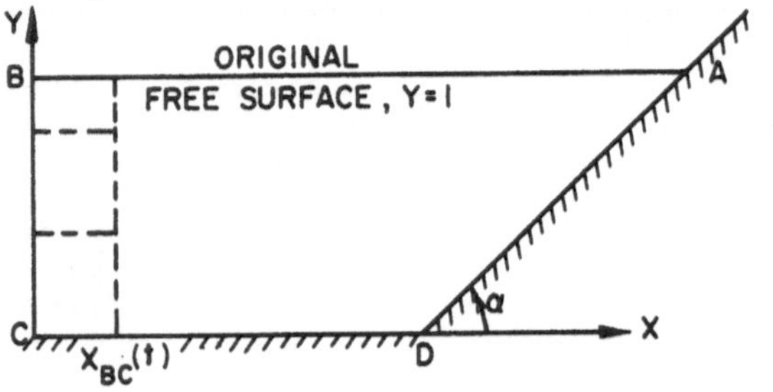

FIGURE 2

of each of which is to be followed by determining the Z values at all the nodes.

Equation (4) or (5) is discretized by integrating over the area of a general cell in the (a,b) plane using Taylor expansions about the center of that cell. This produces the first and second order terms of the Cell Equation, (7). The cell circulation Γ_c is calculated from the initial conditions at $t = t_o$. Subscripts refer to values at particular nodes surrounding the general cell as shown in figure 1.

$$(Z_2 - Z_4)(W_1 - W_3) - (Z_1 - Z_3)(W_2 - W_4) - 2\Gamma_c \qquad \text{First Order}$$

$$+\frac{1}{12}\{(W_{16}+W_9 - W_1 - W_2)(Z_1 - Z_2) - (W_{15}+W_{10} - W_3 - W_4)(Z_4 - Z_3) \qquad \text{Second Order Term}$$

$$+ (W_7 + W_{12} - W_2 - W_3)(Z_2 - Z_3) - (W_6 + W_{13} - W_1 - W_4)(Z_1 - Z_4)\} \qquad \text{if required}$$

$$+ i\tau\{(U_1 - U_3)(V_2 - V_4) - (U_2 - U_4)(V_1 - V_3)\} + 2i(A^p - A^o)/\tau \qquad \text{Continuity Corrections}$$

$$+ \theta^{p+\frac{1}{2}} \qquad \text{Inhomogeneous term (see later)}$$

$$= 0 = R_I + iR_c = R, \text{ The Cell Residual} \tag{7}$$

Since the values referred to are Z^p and $\{W,U,V\}^{p+\frac{1}{2}}$ the first of the continuity corrections is required to allow for this fact (see Brennen and Whitney (1970)). The second prevents accumulation of error over many time steps, A being the area of the cell. Then the equations (7), one for each cell, are to be solved for $W^{p+\frac{1}{2}}$, Z^p being known. Boundary conditions most often take the form of a relation connecting $U^{p+\frac{1}{2}}$ and $V^{p+\frac{1}{2}}$. Solid boundaries will be prescribed in the form $F(X,Y,t) = 0$ which leads to the relation $F(X^p+\tau U^{p+\frac{1}{2}}, Y^p+\tau V^{p+\frac{1}{2}}, t) = 0$. Dynamic free surface conditions are simply given through the equations of motion, (1). Thus if the line AB, figure 1 is a free surface, equation (1) leads to the following first order numerical constant pressure condition at a node such as 0, figure 1:

$$(X_1-X_3)^p(U_o^{p+\frac{1}{2}} - U_o^{p-\frac{1}{2}}) + (Y_1-Y_3)^p(V_o^{p+\frac{1}{2}} - V_o^{p-\frac{1}{2}} + \tau g) = 0 \tag{8}$$

where the only extraneous force is that due to gravity, g, in the negative Y direction. Again (8) connects $U_o^{p+\frac{1}{2}}$ to $V_o^{p+\frac{1}{2}}$, all other quantities being known.

Solution was effected by the iterative method of successive relaxation of the cells according to

$$\Delta W_1 = -\Delta W_3 = \omega i R\overline{(Z_1 - Z_3)}/8A \; ; \; \Delta W_2 = -\Delta W_4 = \omega i R\overline{(Z_2 - Z_4)}/8A \tag{9}$$

ω being an overrelaxation factor. These incremental velocity changes have a simple physical interpretation. They contain two components, one of pure stretching and one of pure rotation of the cell which respectively dissipate the continuity and circulation components of the cell residual, R. After one sweep over all cells, the boundary conditions were imposed and the process repeated to convergence.

If the fluid is inhomogeneous then further advantage can be taken of the flexibility in the choice of (a,b) by choosing $Z^o(a,b)$ so that ρ is some simple analytic function of a and b. For example, if ρ is constant on the free surface, AB (figure 1) and along the bed, CD, an appropriate choice of ρ may be $\rho = \rho_{CD}(1+\delta b)$. Then integration over the cell area yields the following expression for $\Theta^{p+\frac{1}{2}}$ in equation (7) corresponding to the integral term in equation (5):

$$\theta_{1234}^{p+\frac{1}{2}} = \theta_{1234}^{p-\frac{1}{2}} - \frac{1}{4}\ln(1-\mu)\ \mathrm{Real}\left[\sum_{N=1}^{4}\left\{(Z_1 - Z_2 - Z_3 + Z_4)^p (W_N^{p+\frac{1}{2}} - W_N^{p-\frac{1}{2}} - i\tau g)\right\}\right]$$
$$+\left\{1 + \left(\frac{1}{\mu} - \frac{1}{2}\right)\ln(1-\mu)\right\}\left[\mathrm{Real}\left\{\sum_{N=1}^{2}(Z_{2N-1} - Z_{2N})^p (W_{2N-1}^{p+\frac{1}{2}} + W_{2N}^{p+\frac{1}{2}} - W_{2N-1}^{p-\frac{1}{2}} - W_{2N}^{p-\frac{1}{2}} - 2i\tau g)\right\}\right]$$

where $\mu = \delta\Delta b/(1+\delta b_{34})$, b_{34} being the value of b on the side 34 of the cell and Δb the b difference across each and every cell.

More detail, including error and stability analyses are contained in Brennen and Whitney (1970).

SAMPLE SOLUTIONS

The feasibility and potential of the method have been tested in a variety of examples of free surface flow. In two simple cases of wave generation. one by vertical wall movement (wavemaker) and one by bed movement (tsunami model) the numerical results agreed satisfactorily with Lagrangian linearized solutions at small amplitudes and showed the divergences expected from non-linear effects as the wave height increased (Brennen and Whitney (1970)). Solutions involving the interactions of waves thus generated with various boundary geometries such as a beach or a shelf have also been obtained. Only two examples can be presented in the limited space available here. In both cases the results are non-dimensionalized using the original water depth, h, as typical length and $(h/g)^{\frac{1}{2}}$ as typical time.

In the first example fluid is originally at rest in the container ABCD, figure 2. The side BC then moves inward according to $X_{BC} = M\sin^2(\pi t/2T)$ in the interval $0 < t < T$ thereafter remaining at $X_{BC} = M$. This creates a wave which travels across the container and reacts with the beach. The positions of the free surface at a selected number of time stations are shown in figures 3 and 4; in the former $M = 0.30$, $T = 6\tau$, $\tau = 0.571$, $\alpha = 27^o$, in the latter the values are 0.6, 8τ, 0.571 and 18° respectively. The reaction with the beach is similar in both cases. Prior to maximum run-up the motions are fairly smooth. However the downwash and its associated fluid motions rapidly become rather violent. Positions $t/\tau = 21, 22$ of figure 3 and $t/\tau = 23, 25$ of figure 4 suggest that this causes 'downwash wave breaking'. By the last times shown the cells have become very distorted and the mesh points excessively widely spaced to allow further progress.

In figure 5, the fluid is originally at rest in a container, half of which is shown as ABCD. In this position it has a vertical, linear density gradient, $\rho = \rho_o(1+\delta Y)$, δ being negative. Symmetric with the center line, BC, a portion of the bed then begins to oscillate sinusoidally in time as shown in figure 6, the shape of the bed disturbance also being sinusoidal. With the same excitor frequency ($\omega = 0.125$) solutions were obtained for various δ with a view to observing

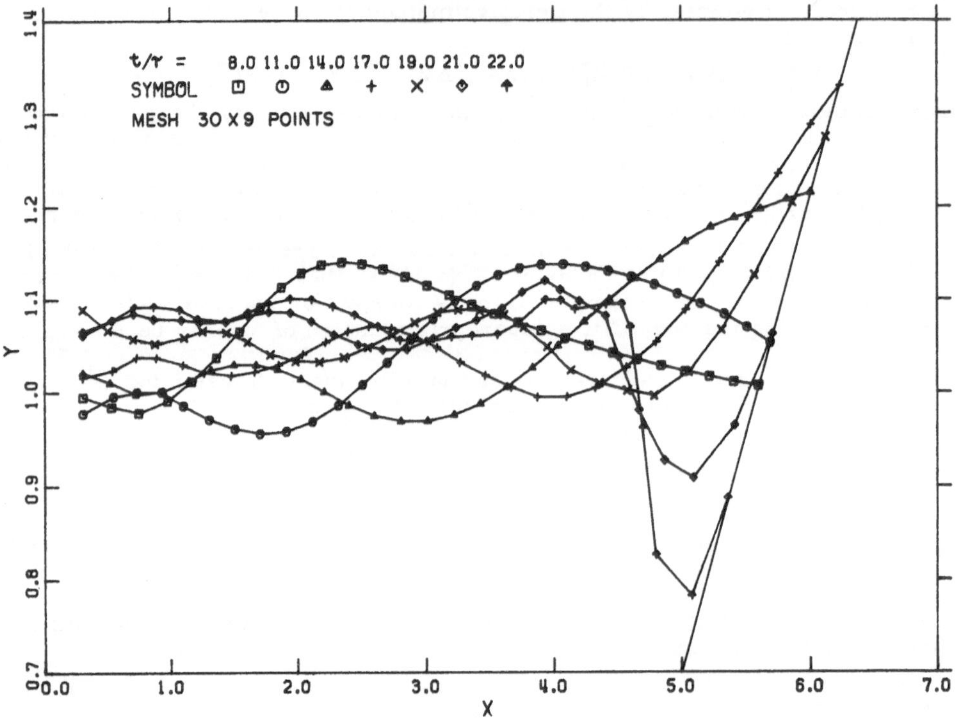

FIGURE 3

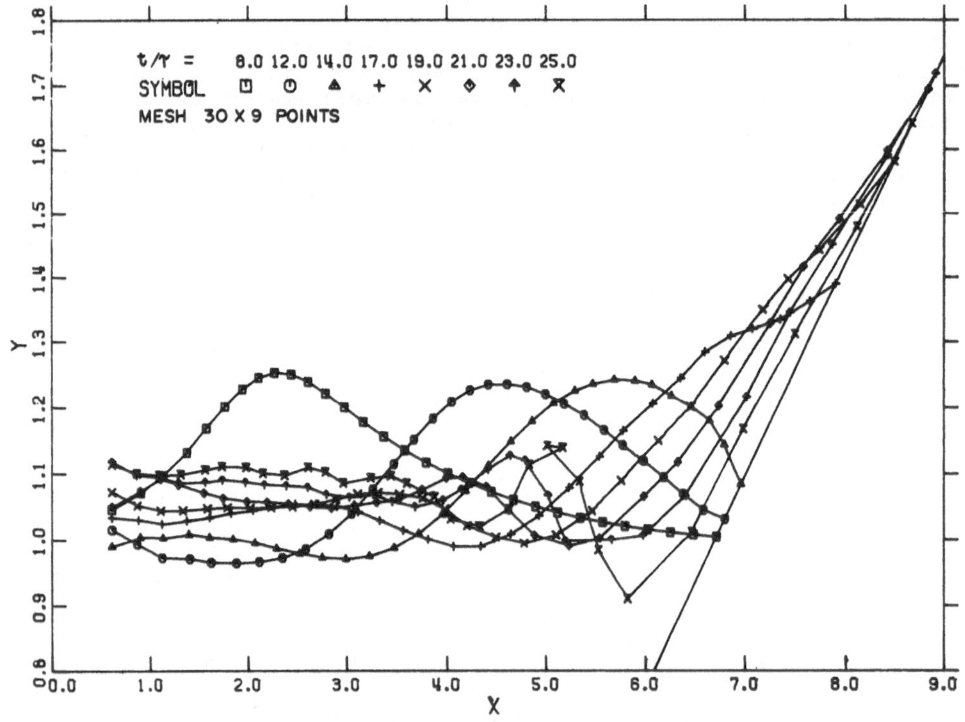

FIGURE 4

internal waves when the bed Väisälä frequency $N_o = (-g\delta)^{\frac{1}{2}}$ exceeded the excitor frequency. In figures 7 and 8, the configuration of three originally horizontal lines (at Y = 0.667, 0.883 and 1.00) at the half cycle (⊡), 3/4 cycle (⊙), full cycle (▲) and 1 1/4 cycle (+) time stations are shown for the cases $N_o/\omega = 0.8$ and 1.2. The profiles in the former case differ only slightly from the homogeneous

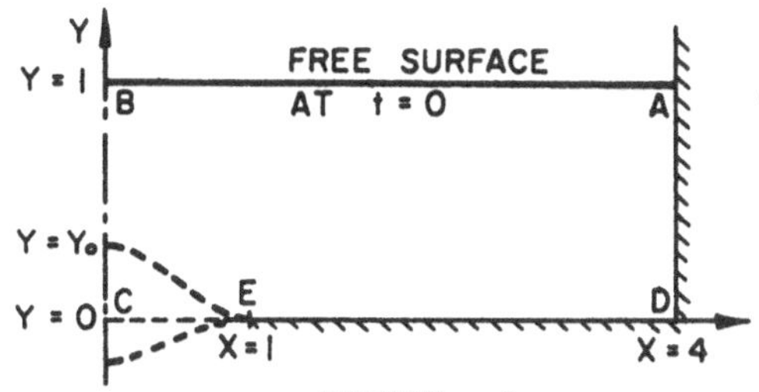

FIGURE 5

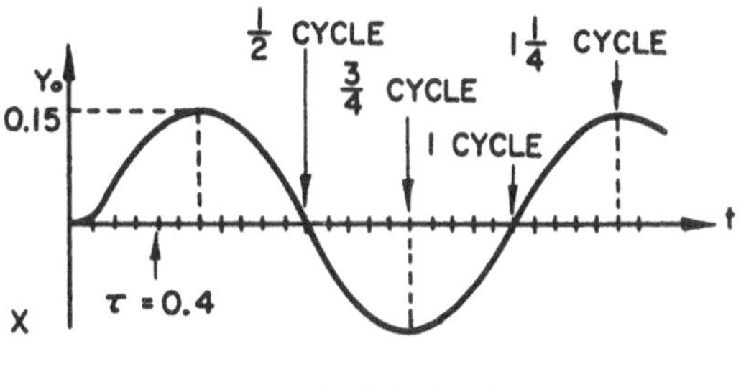

FIGURE 6

results with $N_o/\omega = 0$. Clearly the latter case is very different. Internal wave troughs can be observed in the 3/4 cycle profiles and peaks in the 1 1/4 cycle profiles. These peaks and troughs lie close to the line FF, drawn through the origin at an angle of $\tan^{-1}[(N_o/\omega)^2-1]^{\frac{1}{2}}$ to the horizontal (GG is drawn through the end of the excitor, E). This is the slope of the characteristic predicted by linearized theory (Wu (1966)) for a point disturbance. Positions of the cells after one cycle are shown in figure 9 with the lines of zero vorticity (.........) and the lines FF, GG superimposed. Also included is a line, HH, drawn so that its slope is everywhere $[(N/\omega)^2-1]^{\frac{1}{2}}$, N being the Väisälä frequency at each particular vertical

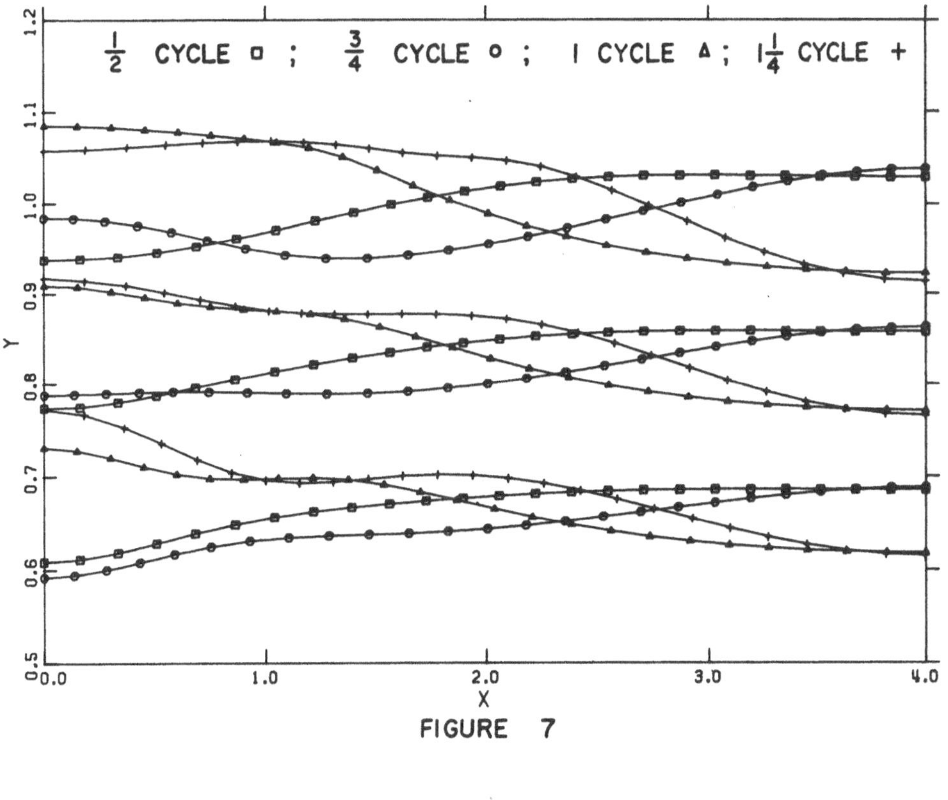

FIGURE 7

elevation ($N^2 = -g\delta/(1 + \delta(Y)_{t=0})$).

Other types of examples which have been only briefly investigated thus far are: the matching with a semi-infinite region in which some analytic solution is used; the inclusion of surface tension; extension to three dimensions. It is hoped to present such results in the near future.

This work was partially sponsored by the National Science Foundation under grant GK 2370 and by the Office of Naval Research. The author deeply appreciates the considerate help given by Professor T. Y. Wu and Dr. A. K. Whitney.

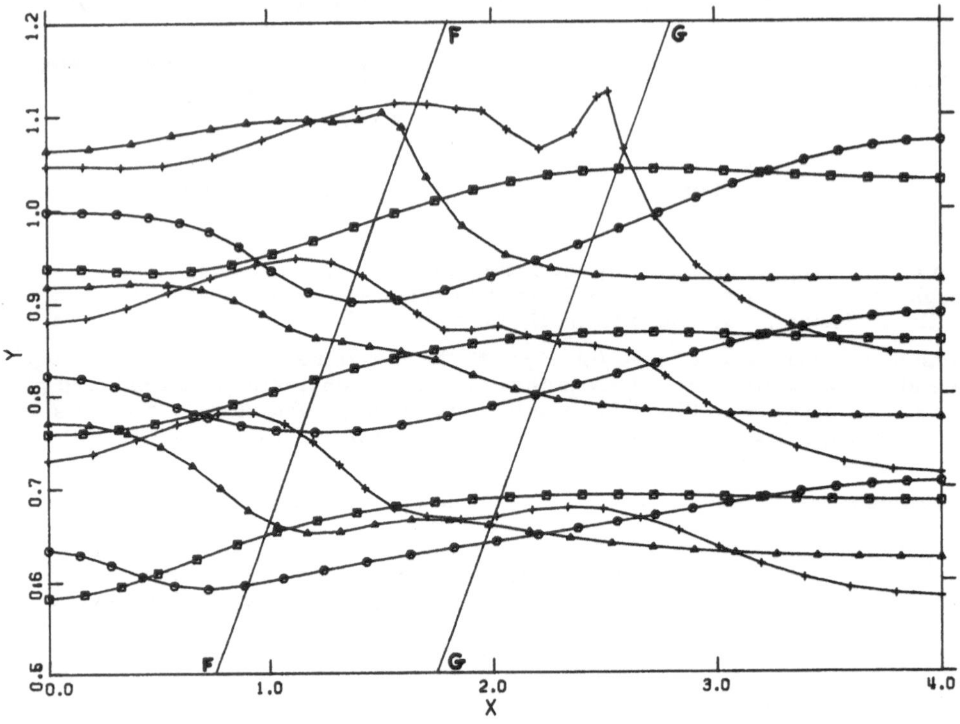

FIGURE 8

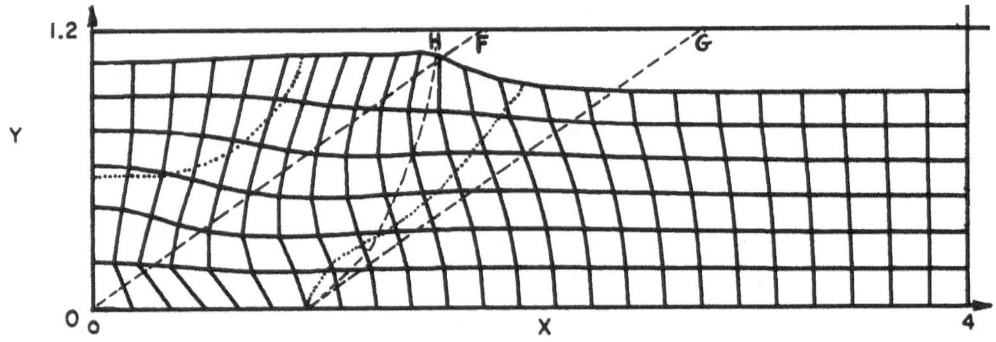

FIGURE 9

REFERENCES

Brennen, C. J. Fluid Mech., 37, 4 (1969).
Brennen, C. and Whitney, A. K. 8th O.N.R. Symposium on Naval Hydrodynamics, Aug. 1970.
Fromm, J. E. and Harlow, F. H. Physics of Fluids, 6 (1963).
Hirt, C. W., Cook, J. L. and Butler, T. D. J. Computational Physics, 5 (1970).
John, F. Comm. Pure and Appl. Math, 6 (1953).
Wu, T. Y. 6th O.N.R. Symposium on Naval Hydrodynamics, Sept. 1966.

NUMERICAL SOLUTION OF THE NON-LINEAR PROBLEMS OF UNSTEADY FLOWS IN OPEN CHANNELS

Prof. O. F. Vasiliev, Dr. Sc. (Eng.)
Institute of Hydrodynamics, Siberian Branch, U.S.S.R. Academy of Sciences
Novosibirsk

In this paper a survey is made of certain computational methods developed at the Institute of Hydrodynamics (Novosibirsk) for solving the following problems in open channel hydraulics:

1. Propagation of waves following rupture of a dam,
2. Propagation of waves of translation along a dry bed.

1. BASIC EQUATIONS AND GENERAL INFORMATION ON METHODS OF SOLUTION

The current state of computational mathematics and technology permit the successful solution of problems of unsteady flow in open channels within the shallow-water approximation in one space dimension, resulting in a hyperbolic system of quasi-linear equations [1].

For the study of flows with discontinuities (shock waves) the basic equations of unsteady one-dimensional flow in an open channel are given in so-called divergence form, which adequately reflects the physical conservation laws [2,3]. The first of these equations is the equation of mass flux, the second the equation of momentum flux:

$$\frac{\partial \omega}{\partial t} + \frac{\partial Q}{\partial x} = 0 \quad , \tag{1.1}$$

$$\frac{\partial Q}{\partial t} + \frac{\partial}{\partial x}\left(P - \frac{Q^2}{\omega}\right) = g\,\omega\left(i - \frac{1}{K^2}|Q|\,Q\right) + R_x \quad ; \tag{1.2}$$

in which

$$\omega = \int_0^h b(x,\xi)d\xi \quad , \quad P = g\int_0^h (h-\xi)b(x,\xi)d\xi \quad ,$$

$$R_x = g\int_0^h (h-\xi)\,\frac{\partial b(x,\xi)}{\partial x}\,d\xi \quad . \tag{1.3}$$

Here x is the coordinate of the cross section, t the time, $Q(x,t)$ the volumetric discharge, $\omega(x,h)$ the cross sectional area of the flow, $h = z(x,t) - z_o(x)$ the depth of flow, $z(x,t)$ the ordinate of the free surface of the flow measured from some horizontal datum, $z_o(x)$ the ordinate of the channel bottom, $b(x,\xi)$ the width of the cross section of the flow, $i(x) = - z_o'(x)$ the inclination of the bottom, $K(x,h)$ the conveyance, characterizing the hydraulic resistance of the channel, and g is the gravitational acceleration.

In the absence of discontinuities in the solution functions Eqs. (1.1) and (1.2) can be written in characteristic form. Taking the fundamental dependent variables as $z(x,t)$ and $Q(x,t)$, we can write:

$$\left[\frac{\partial Q}{\partial t} + (v \pm c)\frac{\partial Q}{\partial x}\right] + B(-v \pm c)\left[\frac{\partial z}{\partial t} + (v \pm c)\frac{\partial z}{\partial x}\right] =$$
$$= \left[Bi + \left(\frac{\partial \omega}{\partial x}\right)_h\right]v^2 - \frac{g\omega}{K^2}|Q|\,Q \quad , \quad \left[v = \frac{Q}{\omega} \quad , \quad c = \sqrt{\frac{g\omega}{B}}\right] . \tag{1.4}$$

in which v is the average velocity of flow, c is the celerity of propagation of small disturbances, $B = b(x,h)$ is the width of the free surface.

Analytic solutions to these equations can be found only in particular cases, for example, under the assumptions of a prismatic channel in the absence of resistance and with certain initial conditions satisfied. The solution of more general problems can be found only by the application of numerical methods. There exist several numerical methods:

1. The method of characteristics. This method of solution of unsteady open channel flow problems has been used extensively. The method allows the isolation of discontinuities in the solution for the front of a bore and thus makes possible solutions of high accuracy. However, as is well known, the method of characteristics is not convenient for programming.

2. Method of continuous calculation. In 1954 Peter Lax published an explicit difference scheme for the calculation of gas flows with discontinuities, constructed on a foundation of conservation laws. Noteworthy also were the explicit schemes of continuous calculation of Lax and Wendroff and V. F. Kuropatenko. Implicit continuous calculation schemes were proposed by S. K. Godunov and N. N. Yanenko and others. In contrast to the method of characteristics, the method of continuous calculation consists of quite standard operations and is relatively simple to program; its application, however, leads to smoothing of the discontinuities. In hydraulics problems, particularly, this method "smears out" the shock rather seriously.

3. Method of isolation of discontinuities. There is a well established method for isolating the discontinuity in a grid with constant step size in x and variable step in t, the latter being determined in the course of calculation. This method utilizes the advantages of isolating the discontinuity but suffers from the fact that the time step depends on the speed of propagation of the wave front. In 1961 at the IV-th All-Union Mathematical Conference, S. K. Godunov and K. A. Semendyaev proposed the concept of isolating discontinuities in conjunction with a movable grid. This method combines the virtue of the method of characteristics, which permits isolation of the discontinuity, with the standardization of operations by the method of continuous calculation, in which the time step is chosen independently of the step in x (within stability requirements). In this method the discontinuity is viewed as a moving boundary subject to associated discontinuity conditions. The computational grid in the x,t plane is constructed in conjunction with the movement of the boundaries. At the node points of the grid a difference scheme is written satisfying the conservation laws. This method permits the isolation of significant discontinuities while smoothing the insignificant ones (that is, treating them by a continuous calculation).

Application of both the method of continuous calculation and the method of isolation of the discontinuities to problems of propagation of bores in a real channel was associated with substantial difficulties. Primarily these difficulties stemmed from the non-prismatic character of the channel, reflecting changes in channel width and bottom inclination with distance along its trace.

Lying at the heart of the problem is the fact that in some cases the cross section of the channel can change in such a manner that the derivative of channel width $b(x,h)$ with respect to x suffers discontinuities at certain channel sections. The bottom line $z_0(x)$ can also be broken, leading to discontinuities in its derivative with respect to x. This is especially characteristic of artificial canals but can also take place in the course of applying certain simple means of approximating geometries of natural channels. These discontinuities in the derivatives of b and z_0 with respect to x lead to the generation of discontinuities in the derivatives of the solution functions of the equations. Even when the fundamental discontinuities of the solution functions are isolated, these circumstances, as well as the possibility of the occurrence of other less significant discontinuities in the solution, call for the application of a difference scheme of the "predictor-corrector" type, satisfying the basic conservation laws.

Serving as the foundation for such a scheme was the "predictor-corrector" scheme developed for calculations in Gas Dynamics [4]. The application of such a scheme to the calculation of bores in prismatic canals was free of difficulties and was carried out in references [2] and [3]. For non-prismatic channels, however, extensive additional work was required in constructing a difference scheme that would properly account for their geometrical peculiarities, referred to previously. The manner in which this question was resolved (by M. T. Gladyshev, V. G. Sudobicher and the writer) is detailed below.

2. PROPAGATION OF THE WAVE DUE TO THE RUPTURE OF A DAM

Let us examine one of the more typical problems of the theory of unsteady open channel flow--the problem of propagation of a flood wave out of a reservoir with a ruptured dam.

As is well known, in examining the dam break problem, the movement of the depression-wave front or weak discontinuity (1), and the bore front or strong discontinuity (2), can be clearly represented graphically in the form of the lines $\ell_1(t)$ and $\ell_2(t)$ in the x,t plane (Fig. 1). The first of these curves coincides with the backward characteristic. The trajectories of the discontinuities are determined by the equations

$$\frac{d\ell_1}{dt} = D_1 \quad , \quad \frac{d\ell_2}{dt} = D_2 \tag{2.1}$$

The speed of propagation of the weak discontinuity is

$$D_1 = v^- - c^- \qquad (c = \sqrt{\frac{g\omega}{B}}) \tag{2.2}$$

The associated discontinuity conditions at the strong discontinuity are:

$$(P^- - P^+)(\frac{1}{\omega^-} - \frac{1}{\omega^+}) + (v^- - v^+)^2 = 0 \tag{2.3}$$

$$D_2 = v^+ + \sqrt{\frac{\omega^-}{\omega^+}\,\frac{P^- - P^+}{\omega^- - \omega^+}} \quad . \tag{2.4}$$

Here, and in what follows, the superscript "-" refers to a quantity on the left side of the discontinuity, the "+" superscript refers to a right hand value.

If the problem is restricted to determination of the so-called generalized solution of the problem on the basis of the divergence equations (1) and (1.2) in the region ABCD (Fig. 2), without isolation of the discontinuities, the method of continuous calculation can be used. In the event that at both upstream and downstream bounding sections of the given reach of river $x = x_1$ and $x = x_2$, the flow is sub-critical $(v < c)$, then at each of these sections one of the following boundary conditions must be given:

$$z(x_1,t) = f_1(t) \quad , \tag{2.5}$$

$$Q(x_1,t) = f_2(t) \quad , \tag{2.6}$$

$$Q = \phi(z) \tag{2.7}$$

The last boundary condition is usually given in the closing section $x = x_2$. Initial conditions in the internal $x_1 < x < x_2$ are given in the form

$$z(x,0) = F_1(x) \quad , \quad Q(x,0) = F_2(x) \tag{2.8}$$

The first of these functions is characterized by a discontinuity at the point $x = 0$ (discontinuity in levels at the dam site).

In solving practical problems the initial flow can usually be assumed steady, and then in the regions of undisturbed flow, that is, below the curves $\ell_1(t)$ and $\ell_2(t)$ (Fig. 2),

$$z = z(x) \quad , \quad Q = \text{const.} \tag{2.9}$$

Consequently, in utilizing the method of isolation of discontinuities the region in which the solution is sought can be bounded from below by the curves $\ell_1(t)$ and $\ell_2(t)$, which represent, respectively, the left and right boundaries.

Let us note that in expressions (2.3) and (2.4) the quantities with the superscript "+" are known in view of (2.9) and expression (2.3) assumes a form analogous to relationship (2.7). Consequently, in application of both the continuous method of calculation and the method of isolation of the discontinuities the right boundary will be assumed to have a boundary condition written in the form (2.7).

Practical computational experience shows that in many cases sufficiently accurate results can be obtained by isolating only the strongest discontinuity. In this case the left boundary remains fixed and the solution is sought in the region AOCD.

In the case when the flood wave out of the reservoir is fed by flow through an opening in the body of the dam (for example, following partial rupture of the dam) further simplification of the problem is possible. In this case flow through the opening (especially in the initial stages of reservoir emptying) often occurs in such a manner that the water surface elevation in front of the dam does not affect the discharge through the opening. As long as the discharge is thus characterized, the flow upstream of the dam is independent of that on the downstream side and the calculation can be carried out in two stages. First the movement of water upstream of the dam can be calculated, giving the head discharge relationship for the opening. The variation with time of discharge through the opening $Q(0,t)$ so found can then be used as a left hand boundary condition of the type (2.6) at the dam section in order to compute the wave motion below the dam.

The hydraulic characteristics of the channel over the given reach of stream (or canal) are usually given at some finite number of sections with x coordinates x_m $(m = 1,2,\ldots,M)$. At each of these sections the bottom elevation is $z_o(x)$, the geometry of the cross section $b = b(x,h)$, and the conveyance $K = K(x,h)$, characterizing the hydraulic resistance of the channel. The determination of intermediate values of these channel characteristics can then be made by interpolation, by linear interpolation in the simplest case.

3. THE DIFFERENCE SCHEME

The foundation for the construction of a difference scheme consisted of the "predictor-corrector" scheme on a movable grid applied by S. K. Godunov et al.[4] to calculations in Gas Dynamics. The calculations for one time step are performed in this scheme in two stages. In the first stage the unknowns are found on an intermediate time layer, utilizing a non-divergent implicit scheme, and then in the second stage values are re-computed in accordance with an explicit divergent scheme. The computations at the intermediate time can be carried out in any variables utilizing a non-divergence form of the equations, for example, the characteristic form (1.4). The final computation is performed for the variables ω and Q, for which the original equations (1.1) and (1.2) written in divergence form are solved.

1. Construction of a Moving Grid. The moving grid is constructed as follows (Fig. 3a). Let the boundary locations $\ell_1{}^k$, $\ell_2{}^k$ at the time $t^k = k\tau$ be known and the grid functions $Q_n{}^k$ and $\omega_n{}^k$ at the points $x_n{}^k = \ell_1 + n\Delta^k$ be determined, where the step $\Delta^k = (\ell_1 - \ell_2)/N$, N is the number of intervals, $n = 0,1,\ldots,N$. The new boundaries are found on the intermediate time layer in accordance with formulas (3.1)

$$\ell_1^{k+\frac{1}{2}} = \ell_1^{k} + \varkappa\tau D_1^{k} \quad , \quad \ell_2^{k+\frac{1}{2}} = \ell_2^{k} + \varkappa\tau D_2 \quad (\varkappa \leq 1) \; . \tag{3.1}$$

Following computation of the step $\Delta^{k+\frac{1}{2}} = (\ell_2^{k+\frac{1}{2}} - \ell_1^{k+\frac{1}{2}})/N$ and determination of the coordinates of node points on the intermediate layer $x_n^{k+\frac{1}{2}} = \ell_1^{k+\frac{1}{2}} + n\Delta^{k+\frac{1}{2}}$, the grid functions $z_n^{k+\frac{1}{2}}$, $Q_n^{k+\frac{1}{2}}$ are found by the difference relationships described below.

Subsequently, utilizing the computed values of the solution functions on the intermediate layer, the boundary positions are re-computed

$$\ell_1^{k+1} = \ell_1^{k} + \tau D_1^{k+\frac{1}{2}} \quad , \quad \ell_2^{k+1} = \ell_2^{k} + \tau D_2^{k+\frac{1}{2}} \tag{3.2}$$

Following this, the coordinates of the node points on the main time layer $k+1$ are found from formulas analogous to those just mentioned, and values of the solution functions are determined at these points.

In order that the distance step Δ should not exceed some limiting value δ, the number of node points can be increased as required on particular time layers. The positions of all of the points on the given time layer are changed accordingly (Fig. 3b). The values of the solution functions at the new points can be found by linear interpolation between the old nodes.

2. Difference Relationships at Interior Node Points. Let us begin by writing relationships for the intermediate time. Since the grid is not rectangular but oblique, derivatives with respect to x and t are approximated as follows

$$\frac{\partial u}{\partial x} \approx \frac{u_{n+1}^{k+\frac{1}{2}} - u_{n-1}^{k+\frac{1}{2}}}{2\Delta^{k+\frac{1}{2}}} \; ,$$

$$\frac{\partial u}{\partial t} \approx \frac{u_n^{k+\frac{1}{2}} - u_n^{k}}{\varkappa\tau} - \frac{x_n^{k+\frac{1}{2}} - x_n^{k}}{\varkappa\tau} \; \frac{u_{n+1}^{k+\frac{1}{2}} - u_{n-1}^{k+\frac{1}{2}}}{2\Delta^{k+\frac{1}{2}}} \tag{3.3}$$

In accordance with Reference [6], in order to improve the stability of the difference scheme the resistance term is taken on the intermediate time layer, that is, at the instant $t^{k+\frac{1}{2}}$. Then the difference relationships approximating the characteristic equations (1.4) on the intermediate layer can be written:

$$\begin{aligned} &\frac{Q_n^{k+\frac{1}{2}} - Q_n^{k}}{\varkappa\tau} + a_n^k \frac{Q_{n+1}^{k+\frac{1}{2}} - Q_{n-1}^{k+\frac{1}{2}}}{2\Delta^{k+\frac{1}{2}}} \\ &- B_n^k(v_n^k \mp c_n^k)\left[\frac{z_n^{k+\frac{1}{2}} - z_n^{k}}{\varkappa\tau} + a_n^k \frac{z_{n+1}^{k+\frac{1}{2}} - z_{n-1}^{k+\frac{1}{2}}}{2\Delta^{k+\frac{1}{2}}} \right] = \\ &= \left[B_n^k \frac{z_o(x_{n-1}^k) - z_o(x_{n+1}^k)}{2\Delta^k} + \frac{\omega(x_{n+1}^k, h_n^k) - \omega(x_{n-1}^k, h_n^k)}{2\Delta^k}\right](v_n^k)^2 - \\ &- \left(\frac{g\omega}{K^2}\right)_n^k |Q_n^k| \left[Q_n^k + 2(Q_n^{k+\frac{1}{2}} - Q_n^k) - 2\left(\frac{Q}{K}\frac{\partial K}{\partial h}\right)_n^k (z_n^{k+\frac{1}{2}} - z_n^k)\right] \; , \end{aligned} \tag{3.4}$$

in which

$$a_n^k = v_n^k \pm c_n^k - (x_n^{k+\frac{1}{2}} - x_n^k)/\varkappa\tau \quad , \quad n = 1,2,\ldots,N-1$$

Values for the solution functions obtained at the intermediate layer are averaged in accordance with the formulas

$$f_{n+\frac{1}{2}} = [3(f_{n+1} + f_n) + f_{n+1} + f_{n-1}]/8 \quad , \quad (n = 2,3,\ldots,N-2)$$

$$f_{n+\frac{1}{2}} = (f_n + f_{n+1})/2 \quad , \quad (n = 1,\ N - 1) \tag{3.5}$$

(f represents any one of the functions being averaged).

This is followed by the operation of re-computation in accordance with difference relationships corresponding to the basic divergence equations (1.1) and (1.2):

$$\omega_n^{k+1} = \frac{1}{\Delta^{k+1}} [\omega_n^k \Delta^k - \tau(Q_{n+\frac{1}{2}}^{k+\frac{1}{2}} - Q_{n-\frac{1}{2}}^{k+\frac{1}{2}})$$

$$+ \omega_{n+\frac{1}{2}}^{k+\frac{1}{2}} (x_{n+\frac{1}{2}}^{k+1} - x_{n+\frac{1}{2}}^{k}) - \omega_{n-\frac{1}{2}}^{k+\frac{1}{2}} (x_{n-\frac{1}{2}}^{k+1} - x_{n-\frac{1}{2}}^{k})] \quad , \tag{3.6}$$

$$Q_n^{k+1} = \frac{1}{\Delta^{k+1}} \{Q_n^k \Delta^k - \tau[P(x_n^{k+\frac{1}{2}}, h_{n+\frac{1}{2}}^{k+\frac{1}{2}})$$

$$- P(x_n^{k+\frac{1}{2}}, h_{n-\frac{1}{2}}^{k+\frac{1}{2}}) + (\frac{Q^2}{\omega})_{n+\frac{1}{2}}^{k+\frac{1}{2}} - (\frac{Q}{\omega})_{n-\frac{1}{2}}^{k+\frac{1}{2}}]$$

$$+ Q_{n+\frac{1}{2}}^{k+\frac{1}{2}} (x_{n+\frac{1}{2}}^{k+1} - x_{n+\frac{1}{2}}^{k}) - Q_{n-\frac{1}{2}}^{k+\frac{1}{2}} (x_{n-\frac{1}{2}}^{k+1} - x_{n-\frac{1}{2}}^{k})$$

$$- \Omega_n^{k+\frac{1}{2}} [\frac{z_o(x_{n+\frac{1}{2}}^{k+\frac{1}{2}}) - z_o(x_{n-\frac{1}{2}}^{k+\frac{1}{2}})}{\Delta^{k+\frac{1}{2}}} + (\frac{|Q|\,Q}{K^2})_n^{k+\frac{1}{2}}] \frac{\tau}{2} (\Delta^k + \Delta^{k+1})\} \quad , \tag{3.7}$$

in which

$$\Omega_n^{k+\frac{1}{2}} = g\omega_n^{k+\frac{1}{2}} \quad , \quad \text{if} \quad \eta = \frac{|h_{n+\frac{1}{2}}^{k+\frac{1}{2}} - h_{n-\frac{1}{2}}^{k+\frac{1}{2}}|}{h_n^{k+\frac{1}{2}}} < \varepsilon$$

and

$$\Omega_n^{k+\frac{1}{2}} = [P(x_n^{k+\frac{1}{2}}, h_{n+\frac{1}{2}}^{k+\frac{1}{2}})$$

$$- P(x_n^{k+\frac{1}{2}}, h_{n-\frac{1}{2}}^{k+\frac{1}{2}})](h_{n+\frac{1}{2}}^{k+\frac{1}{2}} - h_{n-\frac{1}{2}}^{k+\frac{1}{2}})^{-1} \quad , \quad \text{if} \quad \eta \geq \varepsilon$$

Here ε is a small quantity. In the above the following relationship derived from the hydrostatic pressure distribution in the cross section was used:

$$\frac{\partial P}{\partial h} = g\omega \quad .$$

3. Relationships on the Boundaries of the Region. In addition to the boundary conditions described in Section 2 the characteristic equations (1.4) are also used at the boundaries of the region in which a solution is sought. The choice of the required characteristic equation is made as usual in accordance with the direction of approach of the characteristics to the boundary region.

The principle of construction of difference relationships approximating

(1.4) at the boundary points is the same as for interior points on the intermediate layer. Only the calculation of derivatives with respect to x differs. In the difference relationships for both the intermediate and main time steps derivatives with respect to x are taken at the time $K + \frac{1}{2}$ and are computed from two points (one on the boundary, the other, its interior neighbor).

Because all of the difference equations for the intermediate time must be linear with respect to z and Q, the non-linear relationships of the form (2.7) utilized at the right boundary are first linearized by expansion. The linearization is performed on the basis of known boundary values at the earlier time. In this fashion the boundary condition (2.7) is represented in the form

$$Q_N^{k+\frac{1}{2}} = \phi(z_N^k) + \left(\frac{d\phi}{dz}\right)_N^k (z_N^{k+\frac{1}{2}} - z_N^k) \quad . \tag{3.8}$$

On the main layer the non-linear relationships (2.7) are utilized without simplification and are solved iteratively by Newton's method.

4. Additional Remarks. The difference relationships for the intermediate time layer form a closed system of linear algebraic equations, the matrix of which has a tri-diagonal structure. This system is solved by the matrix double sweep method. The formulas for re-computation on the main layer are explicit and consequently calculations are performed for each point separately.

For stability of the scheme the parameter $\mathcal{H}$ must lie within the range of $\frac{1}{2} < \mathcal{H} \leq 1$.

It should also be noted that the aforementioned difference relationships were formulated for the most general case, in which the grid is movable in the x direction. For the particular case of stationary outer boundaries and fixed grid (this occurs in applications of the usual method of continuous calculations), the difference relationships are special cases of those above with appropriate simplifications.

EXAMPLES OF CALCULATION

Example 1. Let us consider the propagation of a bore in a non-prismatic channel following a sudden discharge through an unsubmerged orifice in the body of the dam. As was indicated at the end of Section 2, separate calculations of the wave motion above and below the dam are possible in this case. The first stage calculation yielding the flow upstream from the dam and the discharge through the opening showed that during the initial period of outflow (lasting about one hour) a virtually constant magnitude of discharge $Q(0,t) = 5.4 \times 10^5$ cubic meters per second obtains at the dam section. The initial conditions of steady flow in the channel comprised a discharge $Q = 1800\ m^3/sec$ and a depth h approximately equal to 4 meters (the precision of these data is of no particular significance in the given case). Hydraulic characteristics of the channel in the given reach of some 35 KM in length were given at six cross sections in the form of curves $b = b(x,h)$, $K = K(x,h)$, along with the corresponding bottom elevations $z_0(x)$. At intermediate stations these hydraulic channel characteristics were determined by linear interpolation. Figure 4 shows the variation in channel width b with various depths of flow $(h = 10, 20, 30, 40, 50, 60)$ at various stations along the reach.

In this fashion boundary conditions of the type (2.6) were established at the left hand boundary of the unsteady flow field downstream from the dam. It should be noted furthermore that in applying the method of bore isolation initial conditions in the vicinity of the point $x = 0$ were taken from the approximate analytic solution for the depths and discharges at an instant of time t_0 close to $t = 0$ (in practice such a calculation with neglect of resistance of bottom slope and non-prismatic channel was performed for $t_0 = 15$ sec). Any errors which might arise as a result of such an approximation of initial conditions quickly vanish with increasing t.

This calculation, with isolation of the discontinuity, was carried out with $\mathcal{H} = 1$. The time step τ varied from 0.25 seconds during the initial stages of wave movement to 60 seconds at large values of time. The distance step Δ changed correspondingly from 20 to 1500 meters (calculation began with a number of intervals $N = 20$). Figure 4 shows computed water surface profiles at various times. It is apparent that the height of the bore rapidly decreases (primarily as the result of resistance). In 30 minutes the wave covers a distance of 33 KM. For a comparison a continuous computation of the same problem (without isolation of the discontinuity) is shown on the same figure by open circles for the time $t = 30$ minutes. More or less significant discrepancies between the two solutions are evident only at the wave front. The continuous computation was carried out for the constant $\tau = 90$ sec, $\Delta = 1500$ m, $\mathcal{H} = 1$.

Example 2. Let us now examine the propagation of a bore into a reservoir located downstream from the dam. The variation of breadth with length for several depths h (h = 10, 20, 30 m) is shown in Fig. 5, while the dashed line represents the initial free surface width. For practical purposes it can be assumed that initially the water in the reservoir is at rest with a horizontal free surface: $Q(x,0) = 0$, $z(x,0) = 43.0$ m (while the depth varies from 3 to 30 meters). The boundary conditions at the upstream boundary (at the axis of the dam) were the same as in the previous example. In Fig. 5 are shown longitudinal water surface profiles obtained both by the method of discontinuity isolation (solid line) and by the method of continuous computation (circles) for successive instants of time $t = 20$ min., 1 hour, 2 hours. The wave traverses the entire length of the reservoir, some 100 KM, in the time of 2 hours 10 minutes.

The calculation with isolation of the discontinuity was carried out for $\mathcal{H} = 1$. The time step τ varied from 10 to 40 sec and the distance step Δ from 10 to 2000 m (the number of intervals N was constant at 50). For the continuous computations $\mathcal{H} = 1$, $\tau = 60$ sec, $\Delta = 1400$ m $(N = 70)$.

4. PROPAGATION OF A WAVE OF TRANSLATION OVER A DRY BED

The movable grid method leads to an extremely effective solution of a dam-break wave propagating in a dry channel, that is, when initially there is no water in the valley downstream from the dam. In this case the propagating transitory wave front below the dam comprises the right hand boundary of the domain of solution $\ell_2(t)$ in the x,t plane. Thus the very essence of the given problem requires the determination of position of one of the boundaries of the given region in the course of calculation and the superiority of the moving grid method in this case is clearly obvious. In order to perform the calculation, boundary conditions must be formulated on the right boundary $\ell_2(t)$, that is, at the front of the wave propagating in the dry channel. And this, as is known, is not such a simple matter in a problem with resistance, the latter affecting considerably the advance of the wave front.

There have been a few attempts at solving this question within the framework of shallow water theory. Noteworthy are the researches of G. B. Whitham (1955) and R. F. Dressler (1952), in which approximate analytic solutions accounting for resistance at the head of the wave were proposed. Whitham's scheme formed the basis of a numerical solution by the method of characteristics developed in France by Fauré and Nahas (1961). In contrast to these approaches the Italian hydraulicist Montouri [6,7] assumed that the classical hydraulics equations for unsteady flow in an open channel with the customary quadratic resistance term (that is, equations of the type (1.1), (1.2) or (1.4)), are valid in the entire region of flow including the neighborhood of the wave front moving along the dry bed. Utilizing the method of characteristics, Montouri obtained a boundary condition at the wave front simply by equating the velocity of propagation of the wave and the flow velocity at a section of some small depth near the wave front. The experimental corroboration of computed results was entirely satisfactory.

A detailed mathematical analysis of this problem in a formulation similar to that of Montouri was carried out by S. M. Shugrin and V. G. Sudobicher [8]. Following a determination of the behavior of the solution to the governing hydraulic equations with resistance in the vicinity of the wave front, these authors formulated the boundary conditions at the wave front as follows:

$$\frac{d\ell_2}{dt} = D_2 = v(\ell_2,t) \quad , \quad h(\ell_2,t) = 0 \qquad (4.1)$$

It is evident that as x approaches ℓ_2 the resistance term in the governing equations grows without limit and the point $x = \ell_2$ is a singularity. It proved possible, in the course of numerical solution of the problem, to circumvent this singularity easily by choosing the coefficients and right hand sides in the difference scheme equations not at the boundary point $x = \ell_2$ but in the neighboring point $x = \ell_2 - \Delta$. A comparison of the numerical solution with an exact analytic solution of a wave traveling with constant speed and form is shown in Fig. 5. This shows the good convergence of the difference solution to the exact one. The method described here was used to perform a series of practical calculations confirming its considerable effectiveness. V. G. Sudobicher applied this method to the run-up of waves on an inclined beach in the approximation of non-linear shallow water theory with resistance taken into account.

In conclusion I express my deep gratitude to M. T. Gladishev, V. G. Sudobicher and E. N. Shokina of the Department of Applied Hydrodynamics of the Institute of Hydrodynamics, who were of great help in preparing this paper.

REFERENCES

1. Stoker, J. J. Water Waves. Interscience Publishers, 1957.

2. Vasiliev, O. F., Gladyshev, M. T., Pritvits, N. A., Sudobicher, V. G. "Numerical methods for the calculation of shock wave propagation in open channels," XI Congress of the IAHR, Leningrad, 1965, V. 3, Paper 44.

3. Vasiliev, O. F. and M. T. Gladyshev. "On the calculation of steep fronted waves in open channels," Izv. Akad Nauk SSSR, Mechanics of Liquids and Gases (1966), No. 6.

4. Alaliken, G. B., S. K. Godunov, I. L. Kireev and L. A. Pliner. "The solution of one dimensional gas dynamics problems in moving grids," Moscow NAUKA (1970).

5. Vasiliev, O. R., T. A. Temnoeva, S. M. Shugrin. "Numerical method for the calculation of unsteady flows in open channels," Izv. Akad Nauk SSSR. Mechanics (1965), 2.

6. Montouri, C. "L'onda di un canale vuoto," Universita di Napoli, Facolta di Ingegneria Istituti Idraulici (1964), 189.

7. Montouri, C. "Introduction d'un debit constaut dans un canal vide," XI Congress of the IAHR, Leningrad (1965), V. #, Paper 55.

8. Sudobicher, V. G. and Shugrin, S. M. "Flow of water in dry channels," Izv. Akad Nauk (Siberian Branch) Technical Sciences (1968) 3, 13.

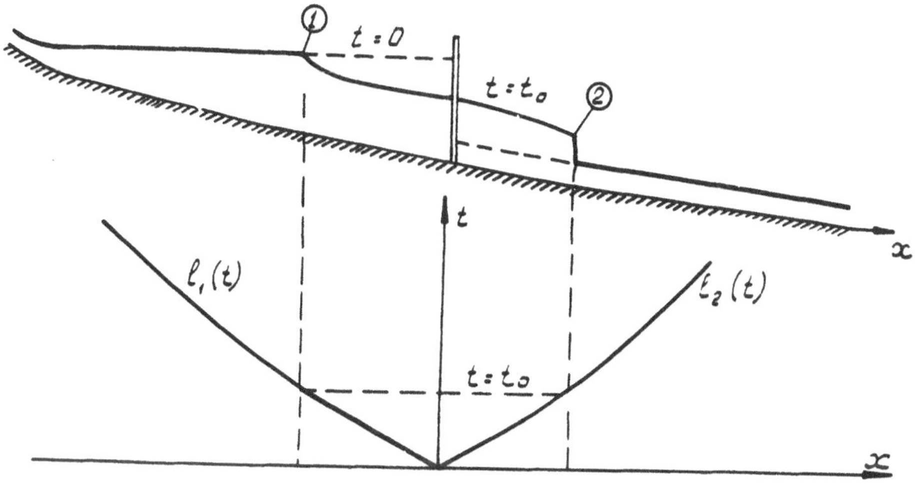

Fig. 1. Wave movement following rupture of a dam

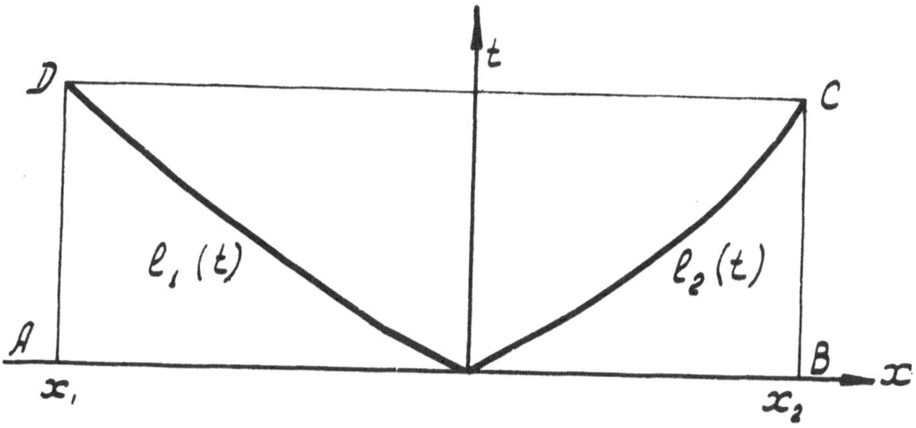

Fig. 2. Domain in which the solution is sought

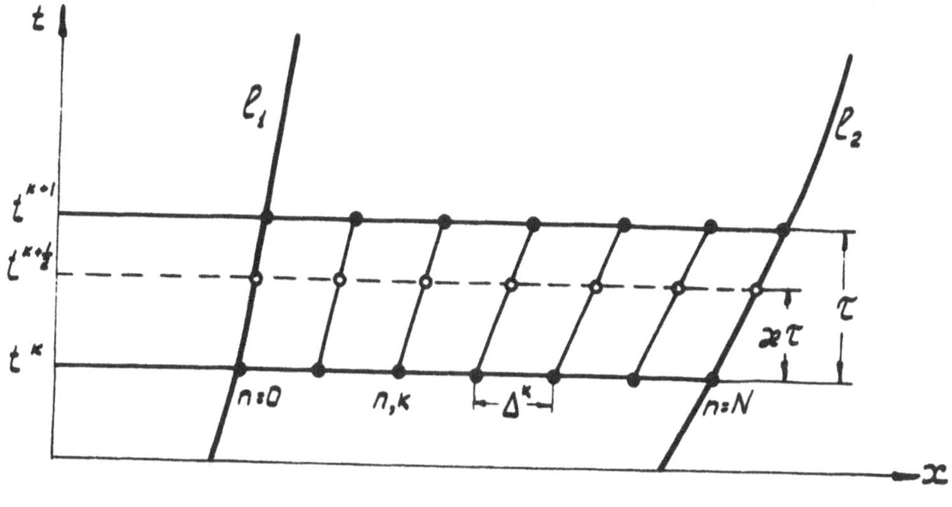

Fig. 3a. The moving grid

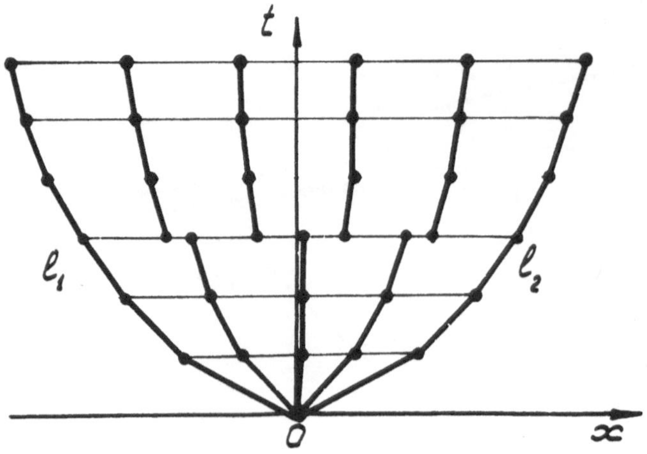

Fig. 3b. The moving grid

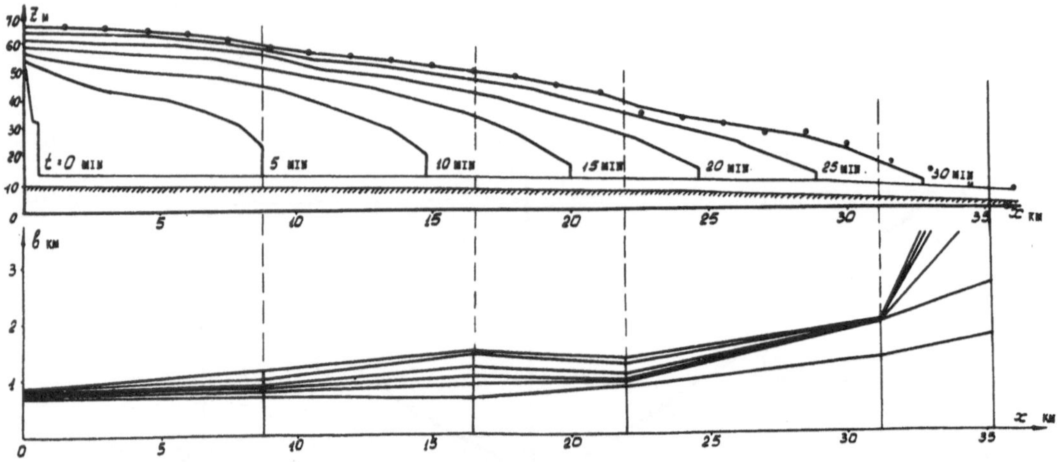

Fig. 4. Propagation of a bore in a non-prismatic channel

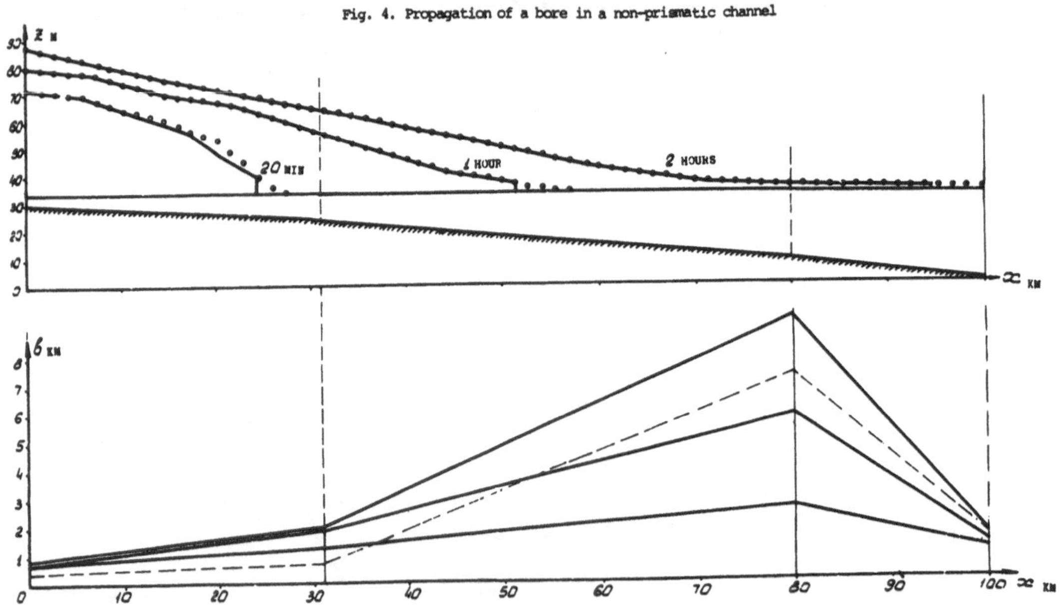

Fig. 5. Propagation of a bore in a reservoir

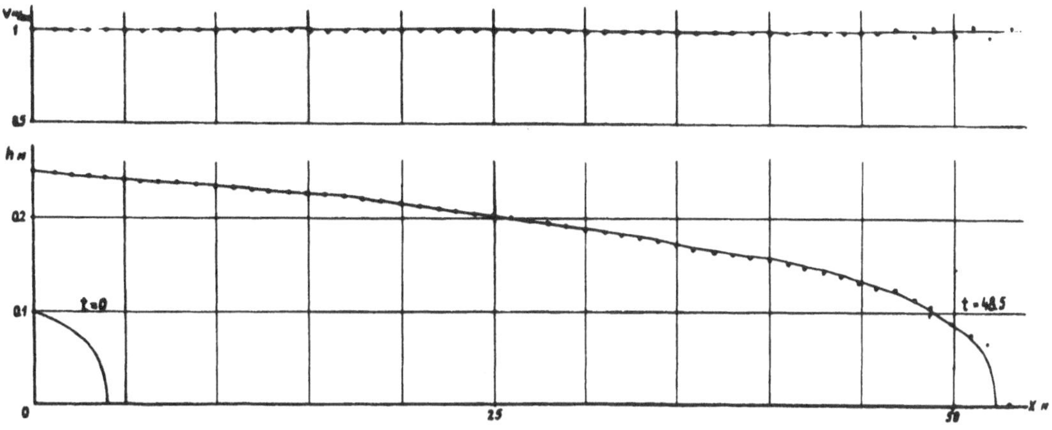

Fig. 6. A surge propagating on a dry bed

INCOMPRESSIBLE CALCULATIONS OF UNDERWATER EXPLOSION PHENOMENA

John W. Pritchett
Information Research Associates, Inc.
Berkeley, California

When an intense explosion occurs underwater, energy is first transferred from the point of burst to the immediately adjacent water mass by radiation. This heated fluid then expands and forms a strong shock wave which propagates away from the explosion point, carrying with it about half the explosion energy and leaving behind it a cavity which contains steam at extremely high temperature and pressure. If the explosion is deep enough, this cavity, or "bubble," will grow in size, the internal pressure and temperature dropping rapidly, to a maximum diameter determined by the explosion energy and the burst depth, and then collapse to a minimum size, re-expand, and continue to oscillate with decreasing amplitude and period as indicated in Figure 1. The reversal of the motion at the bubble minimum is so abrupt as to appear to be discontinuous on a time scale appropriate for the expansion-contraction cycle as a whole. During by far the greatest part of the expansion-contraction cycle, the bubble internal pressure is much less than the ambient hydrostatic pressure. Except for brief time intervals near bubble minima, after emission of the shock the bubble interface velocity is well below acoustic speed, and hence the water motion may be adequately treated as incompressible. Furthermore, the motion is dominated by the inertia of the water outside the bubble, rather than that of the gas inside. Consequently, the experimentally determined radius-time relation for a single cycle is fairly well represented by that of the expansion and contraction of an empty spherical cavity in an incompressible inviscid fluid under constant hydrostatic pressure (Cole, 1948). If the calculation is made slightly more complicated by including a homogeneously distributed ideal gas within the bubble which undergoes adiabatic expansion and compression, the radius-time relation that results will match experimental measurements within the bounds of experimental error. It may be shown, for the first cycle, that the maximum bubble radius and the oscillation period are given by

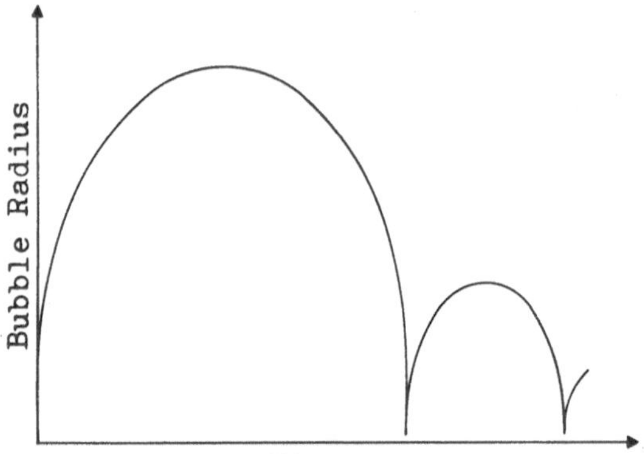

FIGURE 1
EXPLOSION BUBBLE PULSATION

$$R_{max} = J(E/P)^{1/3} \qquad T = K\sqrt{\rho}\,E^{1/3}/P^{5/6}$$

where R_{max} = maximum bubble radius for the first cycle
T = period of oscillation for the first cycle
P = hydrostatic pressure at burst point
ρ = density of water
E = energy available for first cycle bubble motion; about half the total explosion energy.

Thus, both the maximum radius and period increase with increasing explosion size and decrease with increasing depth of burst. The above dimensionless coefficients (J and K) vary only slightly with the explosion intensity and with the gas description (that is, γ, the adiabatic exponent) and may therefore be treated as quasi-constants.

For underwater explosions at hydrostatic pressures of order 1-100 atmospheres, $J \approx 0.58$ and $K \approx 1.12$.

At the bubble minimum, the assumptions of water incompressibility and adiabatic gas behavior (particularly the latter) become rather poor. Due to the intense recompression a pressure pulse is radiated which carries away a few percent of the bubble energy. More important, near the minimum, Taylor instability occurs at the interface, which both generates intense turbulence near the bubble (of which the energy is derived from bubble energy) and causes jets of water and spray to penetrate the bubble, cooling it and causing condensation of the steam atmosphere. Experimental results have therefore shown that the energy available for the second bubble cycle is only a small fraction of that of the first.

Classically, the effect of gravity (bubble buoyancy) has been treated by assuming that the bubble remains spherical, but translates upward as a whole. The upward momentum generated up to a time t (neglecting the density of the gas within the bubble) is given by:

$$p = \frac{4}{3} \pi\rho g \int_0^t [R(\tau)]^3 d\tau \qquad (g = \text{acceleration of gravity})$$

and therefore the system acquires momentum most rapidly near bubble maxima. At maxima, however, the upward momentum is distributed in a large volume of water, whereas at minima the momentum is concentrated in a relatively small region; thus, most of the upward motion occurs near minima. The momentum acquired during the first cycle is just proportional to $\rho g R_{max}^3 T$, or, in normalized form, gT^2/R_{max}, a reciprocal Froude number. Therefore, for a given explosion geometry (defined as the ratio of the burst depth to the first maximum bubble radius) the relative effect of bubble buoyancy will be more pronounced for larger explosions than for smaller ones.

It has long been realized that if the gravity effect is strong the assumption of spherical bubble form is poor. Under these circumstances, after the first maximum, the bubble bottom will collapse back toward the explosion point more rapidly than does the top, forming a jet of water which collides with the upper bubble interface. Thus, at the minimum, the bubble is toroidal, rather than spherical, in shape. This vertical central column may, if the gravity effect is strong enough, persist throughout the remainder of the bubble motion. These phenomena have been observed experimentally at laboratory scale in test chambers in which the air pressure is reduced or the effective acceleration of gravity is increased (or both) to obtain large gravity effects using very small explosions (Pritchett, 1966).

In order to compute the water flow around the bubble resulting from large explosions, it is thus clearly necessary to make fully two-dimensional calculations. For this purpose, finite-difference hydrodynamic codes have been created specifically to solve problems of this general character. The latest version of the scheme, described in Pritchett (1970), is quite elaborate, including temperature, salinity, and other scalar transport and diffusion. Fluid density (a function of temperature and salinity) may vary slightly in the Boussinesq approximation. The effects of turbulence are simulated using a separately-developed heuristic model of the "turbulent energy-scale of turbulence" type (Gawain and Pritchett, 1970). This code is currently being used to study the explosion debris transport from extremely deep underwater explosions. The calculations presented here, however, were performed using a prototype version of the code which solves only the unsteady incompressible axisymmetric Navier-Stokes equations in pressure-velocity (rather than stream function-vorticity) form. The numerical method is an explicit forward-time finite difference scheme. The momentum equation uses a nine-point second-order space difference representation of the advection terms in conservative form. For high Reynolds number calculations (such as the present one) a stabilizer viscosity is used which is automatically computed at each time step to be just sufficiently large to insure computational stability. This code is described in Pritchett (1967). Variable space intervals are permitted, and the time step varies as the calculation proceeds so as to satisfy stability requirements. Free-surface boundary conditions are treated using the MAC ("Marker-and-Cell") technique first developed by Harlow et.al.(1966) in which the fluid is represented by a large number of massless marker particles which move with the flow through the Eulerian grid and thereby specify the position of free surfaces. Two free surfaces are permitted, representing the air-water and the bubble-water interfaces. The air pressure is simply an input constant, and the bubble pressure is an essentially arbitrary function of bubble volume and/or time. In order to ascertain the extent to which such a method can produce accurate results for large underwater explosions, this prototype code was used to simulate the flow after the WIGWAM test.

The nuclear explosion WIGWAM, fired on 14 May 1955, consisted of the detonation of a 30 kiloton device 610 meters below the sea surface in deep Pacific Ocean waters. Underwater pressure-vs.-time measurements were made at several locations: in addition to the primary shockwave, other signals were recorded. Among them were pulses identifiable as the pressure waves propagated by bubble recompressions. These signals were emitted from the bubble at about 2.88, 5.5, and 7.3 seconds after the burst. These recompression pulses, unlike the primary shock, were weak, broad, and diffuse, so that the times of occurrence are rather uncertain; the above numbers may however be considered accurate within ± 0.1 second.

Visible early-time above-surface effects may be divided into three distinct phases (see Figure 2). The primary shock wave reached the water surface at about 0.4 seconds after the detonation. The reflection of the shock from the sea surface imparted upward velocity to the water; the surface layer subsequently broke up into fine spray droplets, forming the primary spray dome, which continued to rise (see Fig. 3). At about 3.1 seconds, the shock generated by the first bubble recompression reached the surface and interacted with the tenuous primary spray dome, accelerating it upward and forming the secondary spray dome (see Figure 4). No further spray domes were formed, as the subsequent pressure pulses were relatively weak. At about 13 seconds, a large dense mass of irregular plumes rose above the secondary spray dome (at that time 270 meters high) and reached a maximum height of 440 meters at 20 seconds before collapsing back toward the surface (see Figure 5). These plumes later proved to be highly contaminated with radioactive debris, which suggests that they were associated with the mass water motion caused by the upward migration of the explosion bubble.

Although no direct experimental information exists concerning the bubble motion at the WIGWAM test other than the indirect evidence mentioned above, considerable experimental work has been performed using both the laboratory-scale techniques mentioned earlier and relatively small high-explosive field tests. This work has led to the development of semi-empirical models for bubble be-

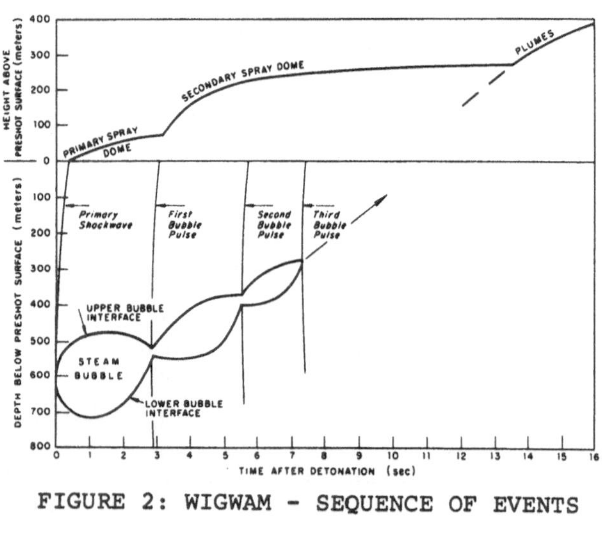

FIGURE 2: WIGWAM - SEQUENCE OF EVENTS

FIGURE 3: PRIMARY SPRAY DOME AT 2 SEC.

FIGURE 4: SECONDARY SPRAY DOME AT 6 SEC.

FIGURE 5: PLUMES AT 14 SEC.

havior from underwater explosions (see e.g. Snay, 1964). Although such models provide no information concerning the actual flow pattern to be expected, they predict with a fair degree of accuracy the depths and times of bubble minima as the bubble rises towards the surface. Thus, in this sense, they may be considered fairly reliable for prediction of WIGWAM bubble motion. Snay hypothesizes that the steam bubble experienced three pulsations, that the bubble then condensed completely at a depth of about 275 meters, and that the upward momentum thereafter resided in a rising ring vortex which proceeded to the surface and caused the spectacular plume eruption some seconds later.

For the numerical calculation, a computing mesh of size 24 cells (radial) by 102 cells (vertical) was used. Near the axis, the mesh spacing was 10 meters, but beyond a radius of 120 meters, and also well below the burst point and above the water surface, the mesh spacing increased rapidly, so that the outer boundary of the mesh was 540 meters from the axis. Total water depth was 1135 meters, burst depth was 610 meters, and total vertical mesh extent was 1640 meters. Initially, the fluid was at rest, but one empty cell (representing the initial bubble) contained a pressure of $2 \times 10^9 nt/m^2$ (19740 atmospheres).

The bubble expansion was taken as adiabatic, that is, PV^γ = constant, where γ for steam was taken as 4/3 (V = bubble volume; P = bubble pressure). Energy loss at minima was accounted for by changing the above constant so that the internal energy of the bubble atmosphere was never allowed to exceed a prescribed function of time. If it did, the constant was permanently reduced to accommodate the change. This loss function was taken from the experimental work of Snay. Air pressure was constant, and equal to one atmosphere ($1.013 \times 10^5 nt/m^2$). The integration was carried out through 766 time steps, which required a total of 5.5 hours of Control Data 6600 computer time.

One significant difference between the actual WIGWAM conditions and the numerical calculation was, of course, the presence of the boundaries. WIGWAM was fired in the open ocean, in very deep water, whereas the numerical analogue occurred in a large cylindrical "tank." At the "tank walls," the boundaries were considered "free-slip " but impenetrable. This tank was made quite large relative to the "size of the event" (that is, the steam bubble size) but would still be expected to exert some influence upon the results. Fortunately, considerable experimental data has been accumulated on the effect of tank boundaries on bubble behavior; the principal effect of both tank walls and a tank bottom seems to be a lengthening of the oscillation period of the bubble. The data may be represented by the empirical equation (Snay, 1964):

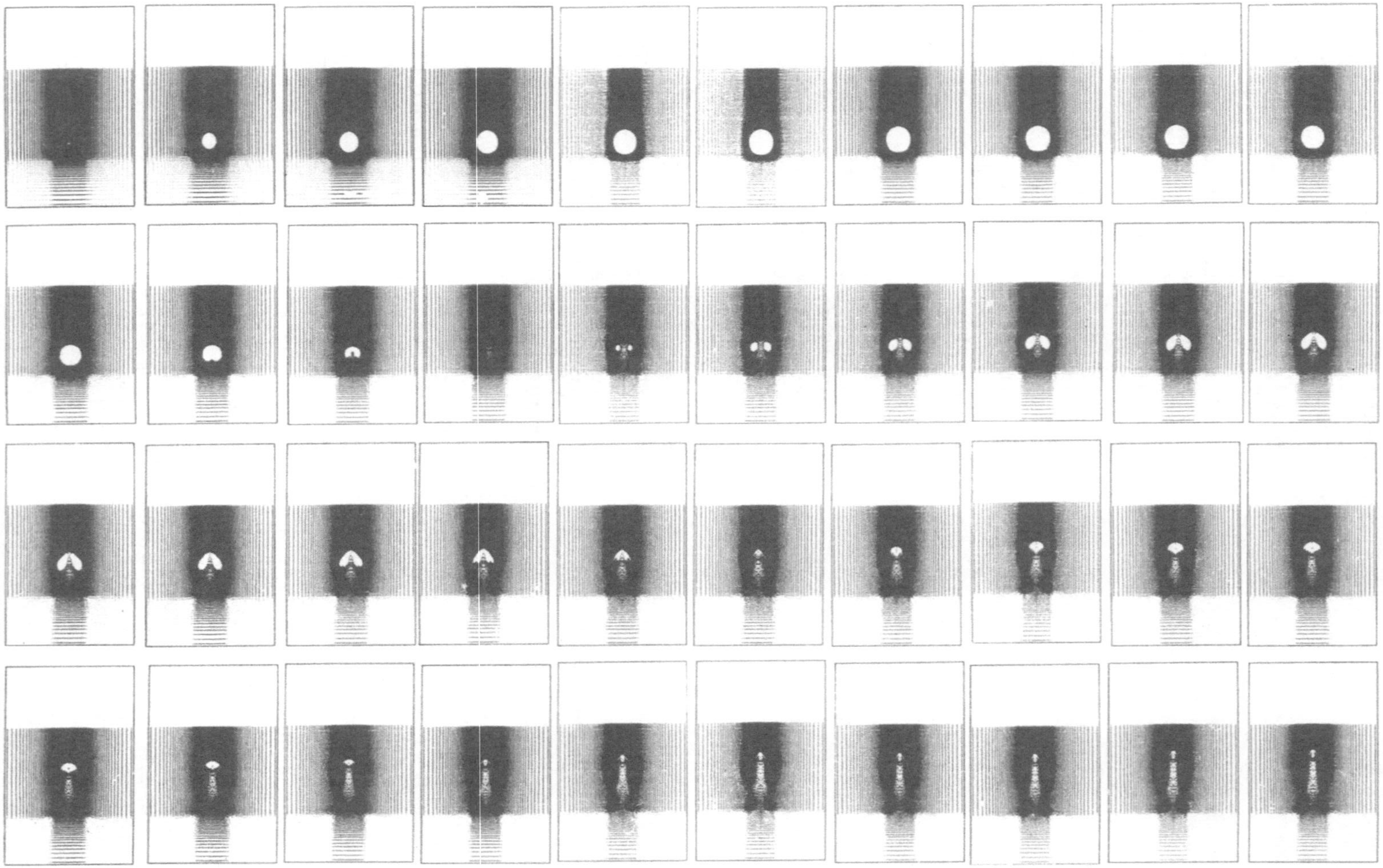

FIGURE 6: CALCULATED WIGWAM BUBBLE MOTION (0.2244 Seconds Between Frames)

$$T_T = T_F\ (1 + 0.216\eta + 0.783\eta^2)\ (1 + 0.15\xi)$$

where

T_T = first-cycle oscillation period in tank

T_F = first-cycle oscillation period in free water

η = ratio of maximum bubble radius to tank radius

ξ = ratio of maximum bubble radius to the distance to the bottom

If the ratio of the two periods (T_T/T_F) is used as a first-order correction factor to the time scale, we find that

$$t_{corrected} = t_{calculated}/1.1143$$

that is, a ten percent reduction. Hereafter, all calculated times given will be "corrected time."

Selected plots of marker particle positions are shown in Figure 6. The early expansion is essentially radial, but after the first maximum the bottom collapse begins, and the ring-or spherical-vortex-like character remains throughout the motion. After the final collapse, an energetic ring vortex remained and was translating upward rapidly when the problem was terminated (at about 8.7 seconds real time). The computed depths of the upper and lower bubble interfaces as functions of time are shown in Figure 7. Also shown are points corresponding to the times of bubble minima inferred from observed pressure-time records, and the depths of the tops of these minima as predicted by the semi-empirical relationships discussed earlier. It thus seems clear that the disagreements between calculations and observations are well within experimental uncertainty. A synopsis of quantitative comparisons between observed data and semi-empirical predictions on the one hand, and the numerical results on the other, is contained in Table I.

It is therefore believed that a fairly successful calculation has been performed of the water flow following the WIGWAM explosion. The extent of agreement with experiment was somewhat surprising to the author, considering the relatively primitive methods used. The more elaborate code (discussed earlier) will therefore presumably produce quite accurate results. It is concluded that this general approach can be useful in the study of mass-motion effects of large underwater explosions.

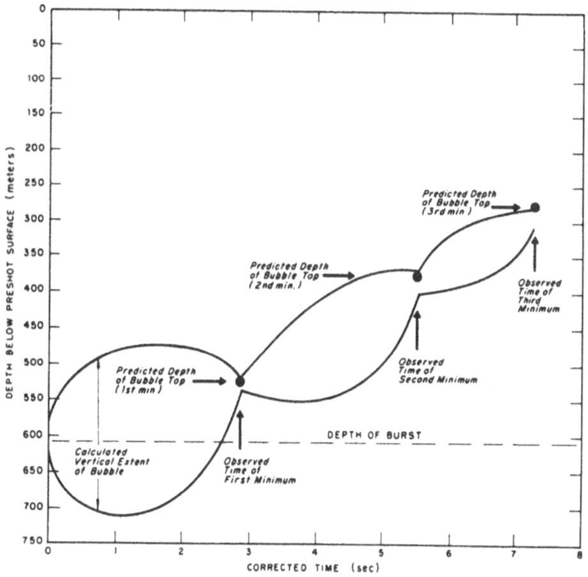

FIGURE 7: CALCULATED BUBBLE MIGRATION

ACKNOWLEDGEMENTS

This work was sponsored by the Defense Atomic Support Agency, Department of Defense, under Work Unit WU-NA-007, and by the Advanced Research Projects Agency, Department of Defense, under Order Numbers 961 and 1593.

TABLE I

Quantity	Estimated by Semi-Empirical Theory	Observed From Pressure-Time Measurements	Computed
1st Max Bubble Radius (m.)	115	---	113*
1st Min Depth (m.)	527	---	525
1st Min Time (sec.)	---	2.88	2.89
2nd Max Bubble Radius	89	---	95*
2nd Min Depth (m.)	377	---	375
2nd Min Time (sec.)	---	5.50	5.54
3rd Max Bubble Radius (m.)	52	---	58*
3rd Min Depth (m.)	275	---	280
3rd Min Time (sec.)	---	7.30	7.25

*Radius of sphere with same volume as bubble.

REFERENCES

Cole, R. H., Underwater Explosions, Princeton Univ. Press, Princeton, N. J. 1948.

Gawain, T. H. and J. W. Pritchett, "A Unified Heuristic Model of Fluid Turbulence," J. Comp. Phys., 5, 383, 1970.

Harlow, F. H., J. P. Shannon, B. J. Daly, J. E. Welch, "The MAC Method," Los Alamos Scientific Laboratory Report No. LA3425, 1966.

Pritchett, J. W., "Explosion Product Redistribution Mechanisms for Scaled Migrating Underwater Explosion Bubbles," Naval Radiological Defense Laboratory Report No. USNRDL-TR-1044, 1966.

Pritchett, J. W. "MACYL-A Two-Dimensional Cylindrical Coordinate Incompressible Hydrodynamic Code," Naval Radiological Defense Laboratory Report No. USNRDL-LR-67-97, 1967.

Pritchett, J. W., "The MACYL6 Hydrodynamic Code: A Numerical Method for Calculating Incompressible Axisymmetric Time-Dependent Free-Surface Fluid Flows at High Reynolds Number," Information Research Associates, Inc. Report No. IRA-TR-1-70, 1970.

Snay, H. G., "Model Tests and Scaling," Naval Ordnance Laboratory Report No. NOLTR-63-257, 1964.

THE DIGITAL SIMULATION OF WATER WAVES -- AN EVALUATION OF SUMMAC

Robert K.-C. Chan
Robert L. Street

Department of Civil Engineering
Stanford University, California

Jacob E. Fromm

IBM Research Division
San Jose, California

INTRODUCTION

At present several numerical methods are available for computing plane waves in shallow water and the nonlinear terms are included to some extent (See Street et al. [1970] for a detailed summary). However, the models are approximate and their applicability is limited to long waves of small, but finite, amplitude-to-depth ratio. An exact model should be employed if one wishes to study plane waves of considerable amplitude.

Chan and Street [1970a] proposed a computing technique for analyzing two-dimensional finite-amplitude water waves under transient conditions. The method, called the Stanford-University-Modified MAC (SUMMAC), is a modified version of the Marker-And-Cell method (MAC) developed by Welch, et al. [1966]. The essence of the modifications consists of a rigorous application of the pressure condition at the free surface and of an extrapolation of velocity components from the fluid interior so that inaccuracy in shifting the free surface is kept at a minimum. Thus, Chan and Street [1970a] outlined the basic features of SUMMAC and established its viability as an engineering research tool. The present work summarizes the earlier study, presents some essential new features, and through direct numerical tests, establishes realms of usage and bounds on expectations.

The differences, in approach and derivations, between the SUMMAC and the original MAC methods are examined. An analysis of the merits and limits of each method is also given. The currently implemented SUMMAC, while yielding accurate accounts of the free-surface dynamics, is limited to waves that are non-breaking and non-turbulent.

THE SUMMAC METHOD

The fluid is regarded as incompressible and the effect of viscosity on the macroscopic behavior of flow is considered to be negligible for water waves. To set up a computing network, the entire flow field is covered with a rectangular mesh of cells, each of dimensions δx and δy . The field-variable values are directly associated with these cells (Fig. 1). From Euler's equation of motion we can derive

$$u^{n+1}_{i+\frac{1}{2}j} = u^{*}_{i+\frac{1}{2}j} + \delta t \cdot g_x + \frac{\delta t}{\delta x}(p_{ij} - p_{i+1j}) , \tag{1}$$

$$v^{n+1}_{ij+\frac{1}{2}} = v^{*}_{ij+\frac{1}{2}} + \delta t \cdot g_y + \frac{\delta t}{\delta y}(p_{ij} - p_{ij+1}) , \tag{2}$$

and the conservation of mass gives

$$D^{n+1}_{ij} \equiv \frac{u^{n+1}_{i+\frac{1}{2}j} - u^{n+1}_{i-\frac{1}{2}j}}{\delta x} + \frac{v^{n+1}_{ij+\frac{1}{2}} - v^{n+1}_{ij-\frac{1}{2}}}{\delta y} = 0 . \tag{3}$$

Here x and y are the Cartesian coordinates; u and v are the velocity components; p is the pressure; g_x and g_y are the components of gravity acceleration; D is the velocity divergence and δt is the time increment. All variables are dimensionless. Variables with the superscript n+1 are associated with the n+1th time step while those without a superscript refer to the nth step. The superscript * indicates a convective term to be described below.

Substitution of Eqs. (1) and (2) into Eq. (3) gives the equation for pressure

$$p_{ij} = \frac{1}{Z}\left(\frac{p_{i+1j} + p_{i-1j}}{\delta x^2} + \frac{p_{ij+1} + p_{ij-1}}{\delta y^2} + R_{ij}\right), \tag{4}$$

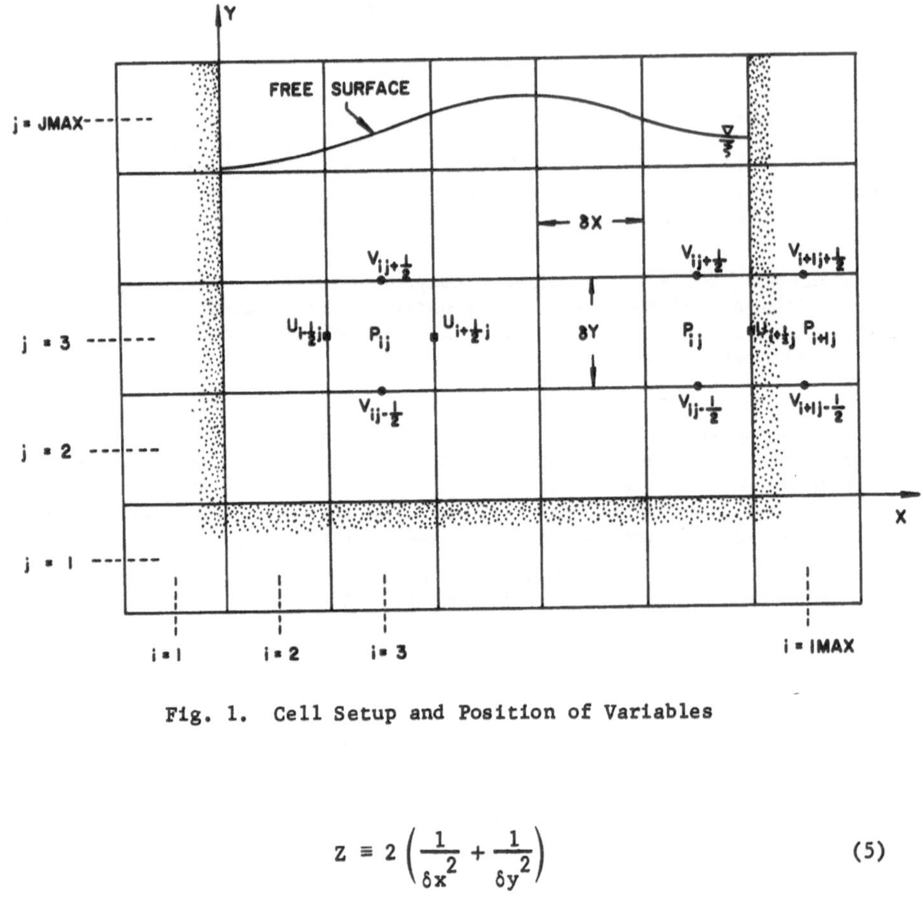

Fig. 1. Cell Setup and Position of Variables

where

$$Z \equiv 2\left(\frac{1}{\delta x^2} + \frac{1}{\delta y^2}\right) \tag{5}$$

and

$$R_{ij} \equiv -\frac{1}{\delta t}\left[\frac{\overset{*}{u}_{i+\frac{1}{2}j} - \overset{*}{u}_{i-\frac{1}{2}j}}{\delta x} + \frac{\overset{*}{v}_{ij+\frac{1}{2}} - \overset{*}{v}_{ij-\frac{1}{2}}}{\delta y}\right]. \tag{6}$$

It has been shown by Chan and Street [1970a] that the formula

$$p_{ij} = \frac{\eta_1\eta_2\eta_3\eta_4}{2(\eta_2\eta_4 + \eta_1\eta_3)}\left[\frac{\eta_3 p_1 + \eta_1 p_3}{\eta_1\eta_3\left(\frac{\eta_1 + \eta_3}{2}\right)} + \frac{\eta_4 p_2 + \eta_2 p_4}{\eta_2\eta_4\left(\frac{\eta_2 + \eta_4}{2}\right)} + R_{ij}\right] \tag{7}$$

must be used for cells near the free surface (Fig. 2).

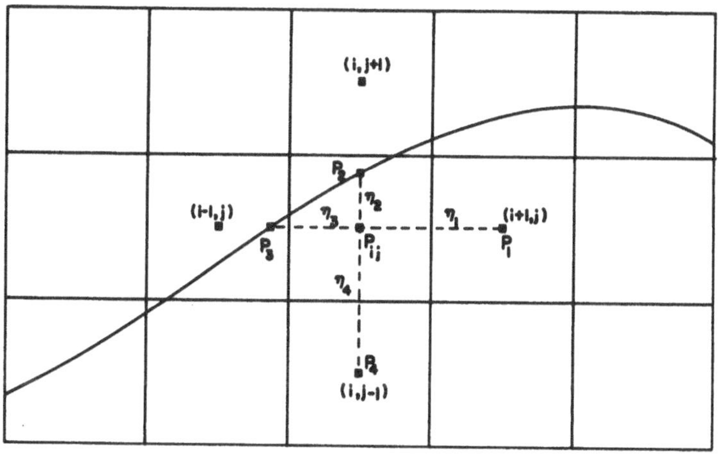

Fig. 2. Irregular Star for P Calculations

To compute the convective contribution term $u^*_{i+\frac{1}{2}j}$ we use Fromm's [1968] second-order upstream difference scheme. If $u > 0$ and $v > 0$ at the mesh point $(i+\frac{1}{2},j)$, then

$$u^*_{i+\frac{1}{2}j} = \tilde{u}_{i+\frac{1}{2}j-1} + \frac{\beta - 1}{2}(\tilde{u}_{i+\frac{1}{2}j-2} - \tilde{u}_{i+\frac{1}{2}j}) + \frac{(\beta - 1)^2}{2}(\tilde{u}_{i+\frac{1}{2}j-2} - 2\tilde{u}_{i+\frac{1}{2}j-1} + \tilde{u}_{i+\frac{1}{2}j}) \quad , \tag{8}$$

where

$$\tilde{u}_{i+\frac{1}{2}j} = u_{i-\frac{1}{2}j} + \frac{\alpha - 1}{2}(u_{i-\frac{3}{2}j} - u_{i+\frac{1}{2}j}) + \frac{(\alpha - 1)^2}{2}(u_{i-\frac{3}{2}j} - 2u_{i-\frac{1}{2}j} + u_{i+\frac{1}{2}j}) \quad , \tag{9}$$

$$\alpha \equiv \frac{u_{i+\frac{1}{2}j}\,\delta t}{\delta x} \quad ; \quad \beta \equiv \frac{v_{i+\frac{1}{2}j}\,\delta t}{\delta y} \quad . \tag{10}$$

Similar derivations can be made for $v^*_{ij+\frac{1}{2}}$ and other flow directions. Fromm's scheme helps reduce numerical dispersion in confined flow calculations and is stable according to a linear analysis for $|\alpha| \leq 1$ and $|\beta| \leq 1$.

In the original MAC, each free surface marker particle is advanced by

$$x_k^{n+1} = x_k^n + u_k^{n+1}\,\delta t \ , \tag{11}$$

$$y_k^{n+1} = y_k^n + v_k^{n+1}\,\delta t \ , \tag{12}$$

where x_k and y_k refer to the position of the kth particle and the particle velocities are interpolated from the velocity fields at the n+lth time step. These equations were derived from the Lagrangian point of view. Round-off errors are quite severe if this approach is used to simulate oscillatory waves. An alternative would be the Eulerian viewpoint, in which we divide the flow region by a number of vertical lines with equal spacing Δ. Let η be the height of each line measured from the reference level $y = 0$ to the free surface. Then from the kinematic condition

$$\frac{\partial \eta}{\partial t} = v - u \frac{\partial \eta}{\partial x} , \tag{13}$$

we can derive the "forward implicit" scheme

$$\frac{\eta_k^{n+1} - \eta_k^n}{\delta t} = v_k^{n+1} - u_k^{n+1} \left(\frac{\eta_{k+1}^{n+1} - \eta_{k-1}^{n+1}}{2\Delta} \right) \tag{14}$$

where u_k^{n+1} and v_k^{n+1} are the velocities at the upper end of the kth vertical line. The linear algegra problem represented by Eq. (14) involves a tri-diagonal matrix which can be readily solved by Gauss elimination.

NUMERICAL TESTS

We selected the problem of a solitary wave propagating in a channel of constant depth to directly test the new features of the SUMMAC method. It is well-known that this nonlinear wave should propagate without any change in its shape and height as long as its crest is far from the end walls of the channel. Thus, by computing the wave motion for 800 time steps and comparing the wave shape with the exact one, we can readily determine the relative merit of the scheme being tested.

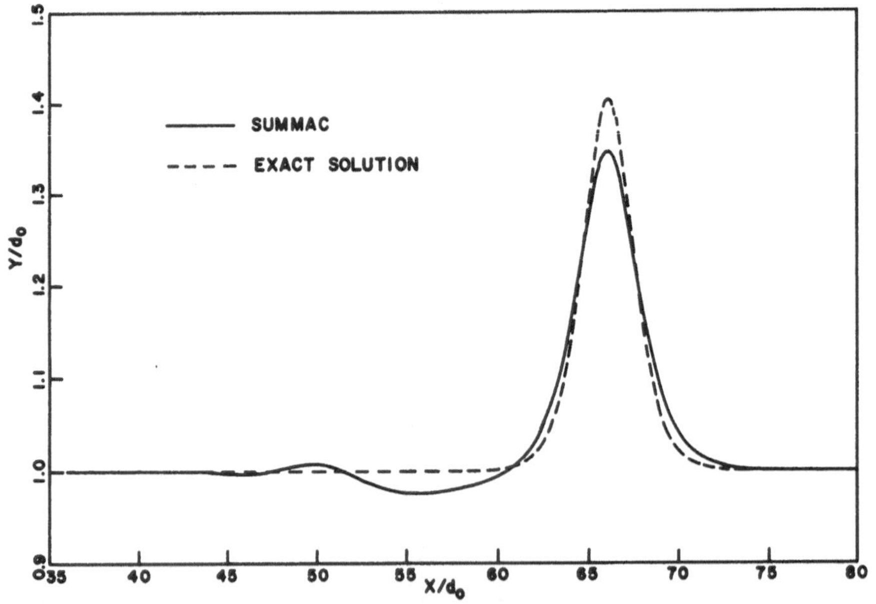

Fig. 3. Numerical Test Employing MAC Scheme

In Fig. 3 the result of using the original MAC formulas for computing u , v and shifting the free surface is compared with the exact solution. Although the difference equations for computing u and v possess linear instability [Chan and Street, 1970b], the use of the advanced-time velocities u_k^{n+1} , v_k^{n+1} in Eqs. (11) and (12) caused damping of the wave amplitude. Damping by this mechanism is seen to be excessive. Use of Eqs. (8), (9) and (14) gave a wave height that was only slightly damped; the overall wave profile is closer to the exact solution (Fig. 4). However, numerical dispersion, which is manifested by the small trailing waves, was found in both schemes.

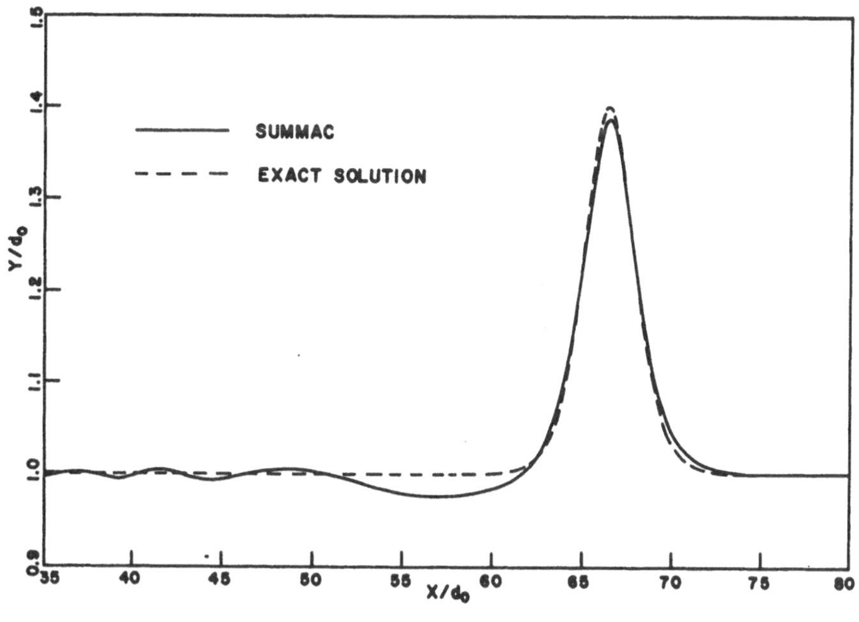

Fig. 4. Numerical Test with Eqs. 8, 9 and 14

CONCLUSION

While it is possible to employ the SUMMAC technique to attack a wide variety of water wave problems, some limitations must be noted. First, as a result of achieving a high degree of accuracy in applying the free surface pressure condition by using irregular stars [Eq. (7)], waves after breaking cannot be simulated. Second, as noted above, non-physical numerical dispersion must be separated from meaningful physical manifestations in the computed results. The second-order upstream difference approximation for convection terms does not help much to reduce numerical dispersion in the free surface problem considered here. Third, only non-turbulent flows are considered in our model. Although laminar viscous damping has little effect on large scale wave motions, energy dissipation due to turbulence can be significant. However, recent studies by Gawain and Pritchett [1970] show that it is feasible to model turbulence in the MAC framework.

REFERENCES

Chan, R. K.-C., and Street, R. L., Jour. Computational Physics, 6, 1 (1970a).

Chan, R. K.-C., and Street, R. L., Dept. of Civil Engrg. Tech. Rept. No. 135, Stanford University, Calif. (1970b).

Fromm, J. E., IBM Research Rept. RJ531, IBM Research Lab., San Jose, Calif. (1968).

Gawain, T. H., and Pritchett, J. W., Jour. Computational Physics, 6, 1 (1970).

Street, R. L., Chan, R. K.-C., and Fromm, J. E., 8th Symposium on Naval Hydrodynamics (1970).

Welch, J. E., Harlow, F. H., Shannon, J. P., and Daly, B. J., Los Alamos Scientific Lab. Rept. LA-3425 (1966).

LINC METHOD EXTENSIONS

Thomas Daniel Butler*
Los Alamos Scientific Laboratory
University of California
Los Alamos, New Mexico

A. INTRODUCTION

LINC is a numerical technique for transient incompressible flows that have free surfaces and material interfaces. In contrast to Eulerian methods in which the fluid moves through a fixed computing mesh, LINC is based on Lagrangian coordinates and the computing cells move with the fluid. It is best suited for flows that have free boundaries and do not undergo large distortions.

The basic method has been detailed in an earlier publication, Hirt, Cook, and Butler (1970). The purposes of this paper are to briefly summarize the calculational procedure and to point out refinements and extensions that have been developed since the first paper. These include the development of a mechanism to retard non-physical motions of the Lagrangian mesh as well as capabilities for calculating flows with surface tension, flows that involve several fluids, and flows that include elastic-plastic effects. Each of these extensions is illustrated with example problems that serve to proof-test the method.

B. DESCRIPTION OF THE METHOD

A brief description of the method is given here in order to familiarize the reader with its basic concepts. The equations of motion are the momentum equations, written in tensor notation where a double subscript indicates the usual summation convention,

$$\frac{Du_i}{Dt} = -\frac{1}{\rho}\frac{\partial p}{\partial x_i} + \frac{\partial}{\partial x_j}\Pi_{ij} + A_i \quad , \qquad (1)$$

and the Lagrangian coordinate equation

$$\frac{dr_i}{dt} = u_i \quad . \qquad (2)$$

Here, D/Dt is the time derivative along the path of a given fluid element; u is the velocity; ρ, the density; p, the pressure; Π_{ij}, the stress tensor deviator; A, acceleration resulting from body forces; and r_i is the position vector. These equations are approximated by finite difference equations related to the computing mesh. The mesh cells are quadrilateral Lagrangian elements embedded in the fluid and moving with it.

Positions, velocities, and body accelerations are defined at the vertices of the quadrilaterals with pressures and components of the stress tensor deviator defined at cell centers. The difference equations are derived by integrating both sides of Eq. (1) over a small volume and transforming appropriate volume integrals to surface integrals, or to line integrals in the case of two dimensional Cartesian coordinates. The integration path for a given vertex is formed by the diagonals of the four adjacent cells. During a time step, momentum fluxes into the volume enclosed by the diagonals are summed for the vertex. The pressure for a given cell is derived by

*This work was performed under the auspices of the United States Atomic Energy Commission.

constraining the volume of that cell to remain constant when the vertices are moved to new positions. This constraint leads to a Poisson-like equation to solve for the pressure field. The method of solution is a Gauss-Seidel iterative technique.

The process of solution through one time step is as follows: (1) All accelerations other than those arising from pressure forces are computed for each vertex. (2) These accelerations are used to compute source terms for the Poisson equation for the pressures. (3) The pressure field is determined. (4) The new velocities are computed and cell vertices are moved. (5) The time is advanced. This process is repeated each time step.

C. MESH REGULARIZING MECHANISM

In certain applications of the LINC method, it has been observed that non-physical motions of the cell vertices occur. These motions, found in many Lagrangian techniques, are induced by short wavelength disturbances that lead to an hourglass-shaped distortion of the computing cells. In the usual LINC procedure, there is no resistance to these motions and once the mode is excited, cell distortion continues to grow and ultimately renders the calculational results meaningless. A mechanism to retard the distortion has been developed. It is derived from a simple model of a Hooke's Law force with damping that tends to keep a vertex in the center of mass of its eight adjacent neighbors.

These additional forces are included in the calculation as another term in A_i of Eq. (1). The regularizing acceleration on a given vertex in tensor notation is

$$A_i' = k^2 \; (r_i' - r_i) + 2k \; (u_i' - u_i) \quad ,$$

in which k is a parameter, r' and u' are the centers of mass and velocity of the eight neighboring vertices, respectively. The value of k depends on the particular problem. Its choice is guided by the desire to keep the mesh regular without significantly altering the overall dynamics.

For many LINC applications, the hourglass distortions are not present. One case where they do appear is in the following example of a single inviscid fluid oscillating in a tank. The fluid occupies the bottom of a rectangular tank and is set into motion by a cosine wave pressure pulse delivered to the free surface at initial time. The left, bottom, and right boundaries of the computing mesh are rigid free-slip walls.

The pressure pulse produces a standing wave in the tank. The computed amplitudes and period of oscillation agree with the predictions of Tadjbakhsh and Keller (1960) who obtained a third order expansion solution of the non-linear equations.

The effectiveness of the mechanism on retarding the distortion is seen in Fig. 1. Taken successively, the frames show the initial mesh configuration, the configuration at t = 21.0 with no restoring force, the mesh at t = 21.0 with the force included, and the mesh at t = 59.0, approximately three periods of oscillation later. At t = 21.0, the fluid has completed 1.75 oscillations. By this time in the top right frame, the mesh becomes extremely distorted near the bottom boundary when there is no restoring force. This is not the case in the lower two frames where the mesh retains its integrity without significantly altering the dynamics.

D. SURFACE TENSION

The primary advantage of a Lagrangian technique is the ability to follow precisely the dynamics of free surfaces and material interfaces. Detailed knowledge of surface configurations permits the effects of surface tension to be included in the calculations in a straightforward manner. One method of accomplishing this is to treat the surface tension effects as an applied pressure at the free surface or interface. The applied pressure is of the form

$$P_{st} = -\frac{T}{R} \tag{3}$$

where T is the surface tension coefficient and R is the surface radius of curvature. Equation (3) is computed for each surface segment and is included as an additional pressure applied along that segment.

To evaluate this procedure in LINC, the following study has been made. When a liquid in an inverted tank is perturbed, the liquid falls from the tank under the influence of gravity. The effect of surface tension on the dynamics is to decrease the rate at which the fluid falls, and, with a sufficient amount of surface tension, the perturbed surface is stabilized and the liquid oscillates in the tank. It is desired to determine the critical value of T necessary to stabilize the surface and compare the LINC result with the linear theory prediction.

With the exception of the modifying effects at the free surface, the boundary conditions are the same as in the previous application. The direction of gravity is reversed, and the mesh is reduced in size to 8 cells in the horizontal and vertical directions. In order to permit the inviscid liquid to start from rest at a small but finite amplitude, the dynamics are initiated by a cosine wave pressure pulse on the surface, and the fluid is brought to rest when the amplitude of the surface displacement is 0.1. At this time ($t = 6.0$), the value of T is changed from its initial value, 1.7, to study the effects of various values of T on the subsequent growth of the disturbance.

The results of a sequence of problems are summarized in Fig. (2). This figure contains plots of the position of the top surface vertex at $x = L$ for various values of T ranging from 1.60 to 1.80. It is seen that the perturbation continues to grow for $T = 1.60$ and grows less rapidly for $T = 1.64$. For $T = 1.65$, the surface begins an oscillation of low frequency. The $T = 1.80$ case shows a nearly complete oscillation of the surface. These graphs indicate that the approximate critical value of T necessary to stabilize the surface is $T = 1.65$. This agrees to within 2 per cent of the value of 1.62 predicted by the linear theory.

E. MULTI-FLUID PROBLEMS

The LINC method is well suited for multi-fluid problems because material interfaces are defined by Lagrangian cell boundaries. The method outlined by Hirt, Cook, and Butler (1970) for choosing the integration contours for vertices along the interface has proven successful in test problems. Briefly, the paths for these vertices are composed of two half contours, one in each material, lying along the cell diagonals and traversing the interface. The direction of traversing the interface for one material is in the opposite sense to the direction of the traverse for the other, and, since the stress is continuous at the boundary, these two contributions to the momentum flux cancel. The resulting path is the usual one defined by the diagonals of the four adjacent cells of a given vertex.

Comparisons between the LINC method and the two-fluid Marker-and-Cell (MAC) technique of Daly (1967) have been made for the case of two inviscid, superposed fluids with a density ratio of 2:1. The interface is in an unstable configuration with the direction of gravity from the heavier to lighter fluid. Details of the problem are given on pages 299-301 of Daly's paper. The interface undergoes a cosine wave disturbance at initial time and the heavier fluid falls into the lighter one. There is no surface tension acting at the interface.

In Fig. 3 are seen plots of the mesh configuration at three times ($t = 0$, 0.050, and 0.112) during the LINC calculation and the corresponding configuration of the MAC calculation at $t = 0.112$. The horizontal interface deforms into a growing amplitude cosine wave and forms the falling spike and rising bubble that are characteristic of the non-linear flow. With increasing amplitude, the Lagrangian cells become more distorted. Finally, the distortion becomes so severe that the computing cells in the region of the spike become inverted causing the LINC calculation to be terminated. Close examination of the amplitude as a function of time for the two methods shows very good agreement for the entire time of the calculation, even when the cell distortions are most severe. A comparison of the interface configuration at t = .112 also reveals very good agreement between the two methods.

Elastic-Plastic Flows

Strength of materials properties for deformable media can be incorporated in the model with relative ease because of its Lagrangian basis. Wilkins (1969) has included the effects of elasticity and plasticity in a Lagrangian compressible flow technique. Hirt and Shannon (1970) are adapting this approach for the LINC method. The approach is a finite difference incremental strain theory. This is convenient because changes in the strain of a given fluid element over a time step result in incremented changes in the stress. Thus, the total stress for an element at a given time reflects its prior deformations. Plastic flow results when the elastic limit for an element is exceeded (i.e., when the total stress is greater than the yield stress, Y_o, of the material).

In LINC, strength of materials effects are included by adding an additional term to the stress tensor Π_{ij} of Eq. (1). The three components, defined for each cell, are derived from the cell deformations. When the total deformation stress exceeds Y_o, the stress components are adjusted such that the total stress is the yield value, and the calculation proceeds in the usual fashion.

Examples of the effects of elasticity and plasticity are seen in Fig. 4. The problem is that of the dynamics of a plate resulting from the initial velocity profile seen in the first frame of this figure. The plate is assumed to be isotropic; work hardening and viscous effects are neglected in these calculations. The succeeding frames show the mesh configuration at the same time for three different values of Y_o: ∞, 0.02, 0. The displacements of the mesh are magnified by a factor of 10 in these frames. For the no yield case, the plate is set into oscillation by the loading. The frequency and amplitude of oscillation for the standing wave are in agreement with the predictions of linear theory for this case. The $Y_o = 0.02$ case shows a double humped configuration of the top surface. This appearance is a result of the plate yielding first at the corners of the mesh while middle region is still flowing elastically. Eventually, the entire plate yields and the spikes continue to grow. The final frame shows the result of a flow with zero yield strength, that is, an ideal fluid.

ACKNOWLEDGMENTS

The author wishes to thank J. L. Cook and J. P. Shannon who did the computer programming and especially C. W. Hirt for his many valuable suggestions.

F. REFERENCES

Daly, B. J., Phys. Fluids 10, 297-307 (1967).

Hirt, C. W., Cook, J. L., and Butler, T. D., J. Comp. Phys. 5, 103-124 (1970).

Hirt, C. W., and Shannon J. P., (Manuscript in preparation).

Tadjbakhsh, I. and Keller, J. B., J. Fluid Mech. 8, 442-451 (1960).

Wilkens, M. L., University of California, Livermore Report UCRL-7322, Rev. I (1969).

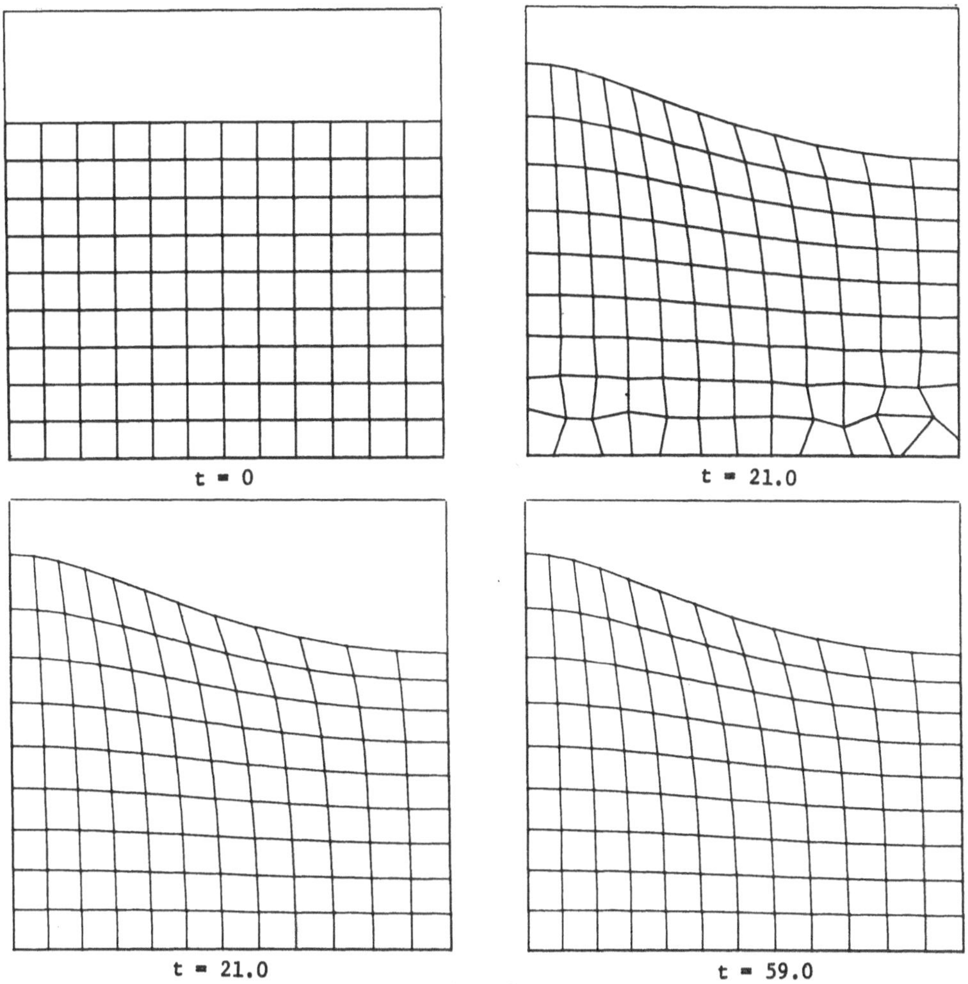
t = 0
t = 21.0
t = 21.0
t = 59.0

Fig. 1

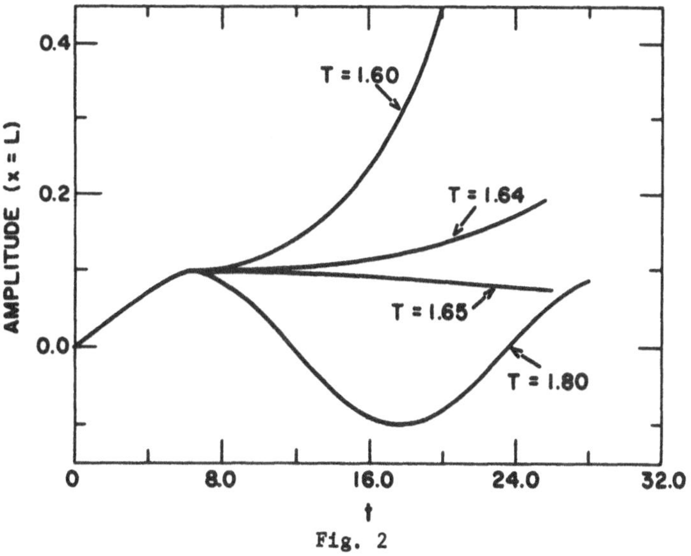
0.4
0.2
0.0
AMPLITUDE (x = L)
T = 1.60
T = 1.64
T = 1.65
T = 1.80
0
8.0
16.0
24.0
32.0
t

Fig. 2

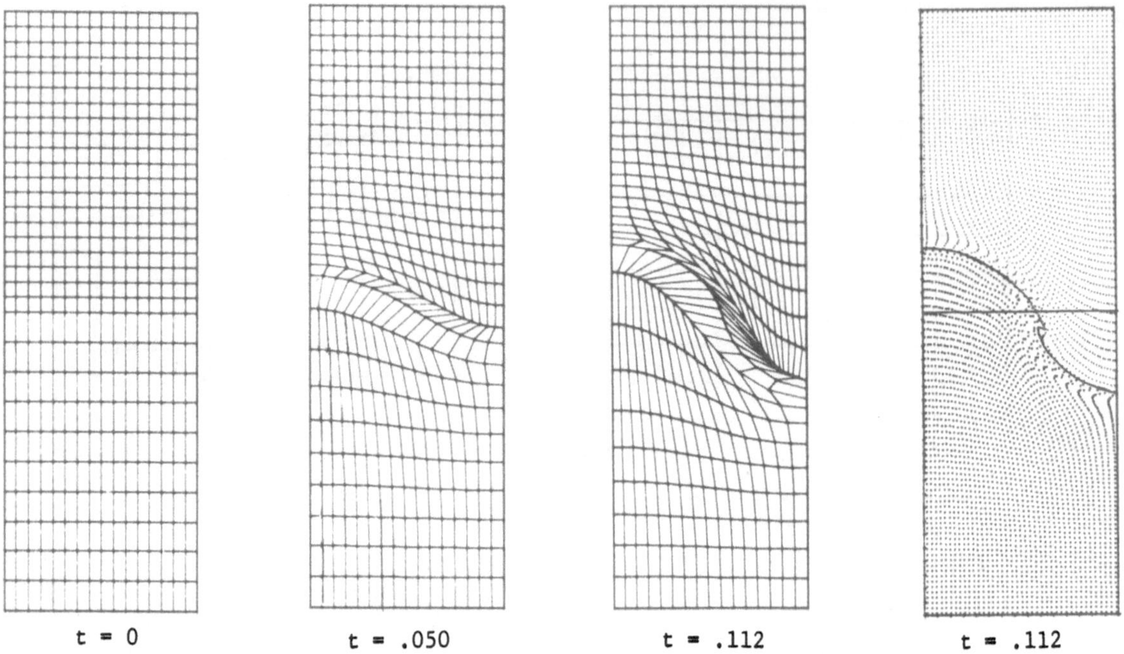

Fig. 3

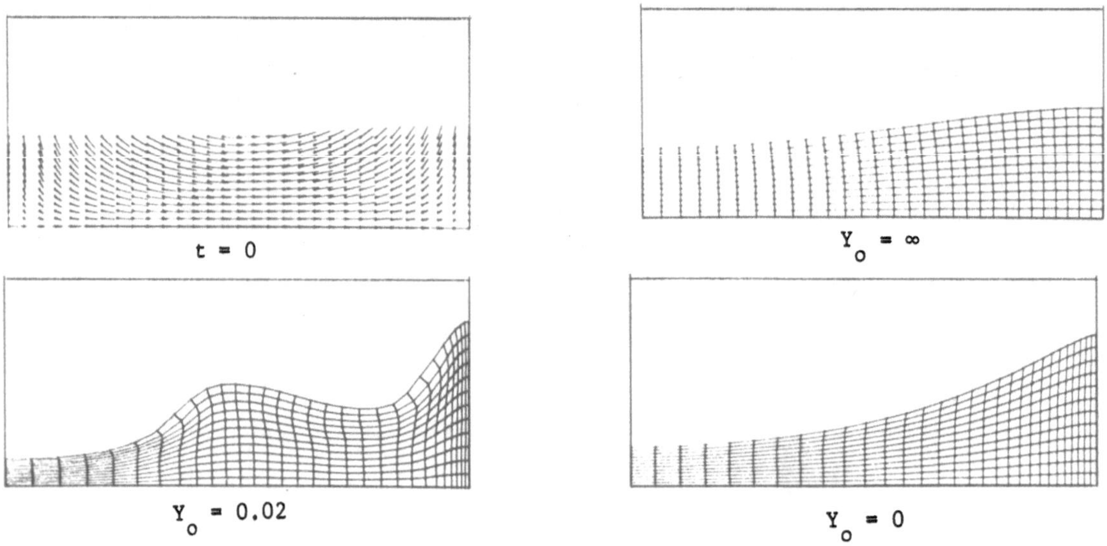

Fig. 4

AN EXACT NUMERICAL SOLUTION OF THE SOLITARY WAVE

Theodor Strelkoff

(University of California at Davis, Davis, California, U.S.A)

The solitary wave, first studied in the field by J. Scott Russell, is an elevation of the water surface moving with essentially constant form at a celerity dependent upon the ratio a of its height to the depth of water upon which it propagates. The fore and aft portions of its profile are symmetrical, and gradually approach the undisturbed water-surface elevation at large distances from the crest. While the solitary wave can be observed upon a running stream, the essential features are retained when considering propagation in standing water. It is under these latter circumstances that the fluid motion in the wave is most nearly irrotational and most accurately subject to analysis by potential-flow techniques.

Earlier solutions have stemmed from approximations of either a physical nature, such as assumption of small wave height, small slopes, etc., or of a convenient mathematical nature, such as the postulate that a solitary wave is approximated by the movement of a doublet a short distance below the surface of a ponded liquid. In the present analysis, the flow is assumed two-dimensional and irrotational, and the channel bottom is assumed horizontal in order to satisfy the condition that the profile propagate without change in form. Within this context, the problem is exactly formulated, i.e., described exactly by an integro-differential equation in the wave profile. Numerical values for the profile can be obtained to any desired degree of precision by choosing appropriate quadrature and differentiation formulas and by choosing a sufficiently small step length.

FORMULATION OF THE INTEGRO-DIFFERENTIAL EQUATION

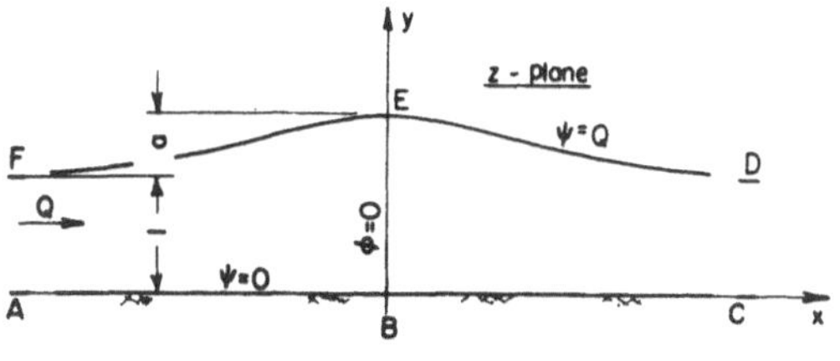

Fig. 1

The inherently unsteady wave motion described above is made steady by adopting a coordinate system centered on the crest of the wave and moving with its celerity (See Fig. 1). All lengths and velocities pertinent to the problem are expressed in nondimensional form by referring these, respectively, to the undisturbed depth h and to $(gh)^{\frac{1}{2}}$, in which g is the ratio of weight to mass. Furthermore, the problem is formulated in the complex-potential plane $w = \phi + i\psi$ [expressed in nondimensional form through division by $h(gh)^{\frac{1}{2}}$] so that the flow field appears as an infinite strip of constant width Q = C [Q is the nondimensional discharge across any section of the (steady) wave, and C is the corresponding velocity at infinity]. The origin of the w-plane is placed on the channel bottom directly below the crest of the wave (Fig. 2). The dimensionless pressure head p in any point in the flow at elevation $y(\phi,\psi)$ above the bottom is given by the Bernoulli equation

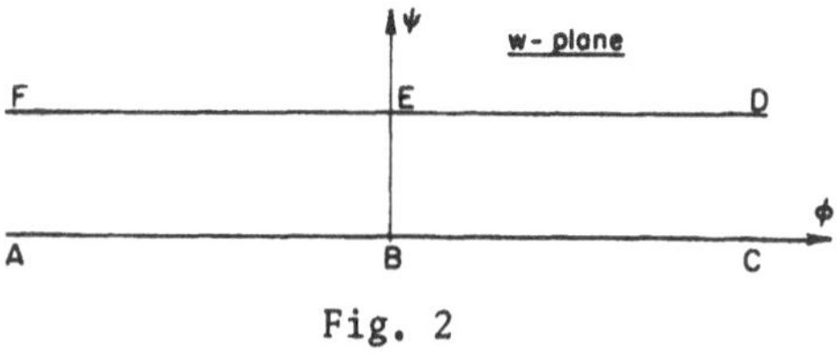

Fig. 2

$$p + y + \frac{1/2}{\left(\frac{\partial y}{\partial \phi}\right)^2 + \left(\frac{\partial y}{\partial \psi}\right)^2} = 1 + \frac{Q^2}{2} \qquad (1)$$

On the free surface p = 0, and $y(\phi,Q) \equiv 1+\eta(\phi)$, where η is defined as free-surface elevation above the undisturbed water level, so that

$$\eta(\phi) + \frac{1/2}{(\eta')^2 + \left[\frac{\partial y}{\partial \psi}(\phi,Q)\right]^2} = \frac{Q^2}{2} \qquad (2)$$

Now $\partial y/\partial\psi$ can be found in terms of an integral of η, weighted by an appropriate kernel function, by making use of the Schwarz (Hilbert) transformation, which relates the value of a complex function anywhere in the half plane to an integral of its imaginary part, appropriately weighted, over the real axis. The interior of the infinite-strip flow field of the w-plane can be mapped into the upper half of an auxiliary t-plane, $t = \xi + i\nu$, by the transformation

$$t = e^{\pi w/Q} \tag{3}$$

The free surface $\psi=Q$ corresponds to the left semi-axis of reals, and the channel bottom $\psi=0$ conforms to the right semi-axis (Fig. 3). The origin of the t-plane corresponds to the flow at $\phi=-\infty$.

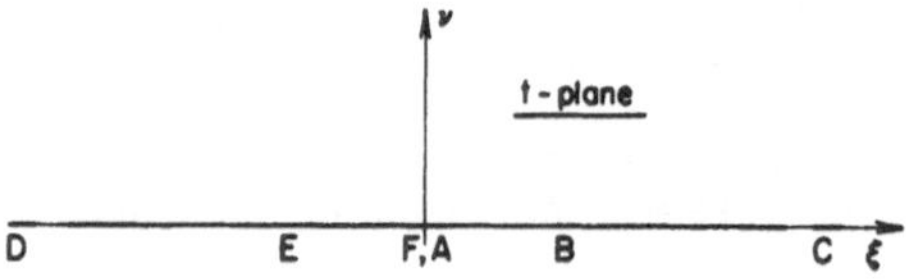

Fig. 3

Then, with $z=x+iy$, the Schwarz-Hilbert transformation gives in the upper half plane

$$z(t) = \frac{1}{\pi}\int_{-\infty}^{\infty} \frac{y(\chi,0)\,d\chi}{\chi - t} + \text{const.} \tag{4}$$

The singularity that is encountered in the integrand function at $\chi=\xi$ when t is real is easily removed from Eq. 4 to yield:

$$z(t) = \frac{1}{\pi}\int_{-\infty}^{\infty} \frac{y(\chi,0)-y(\xi,0)}{\chi - t}\,d\chi + i\,y(\xi,0) + \text{const} \tag{5}$$

in which the second term on the right-hand side is obtained for arbitrary t in the upper half plane. The real part of dz/dt, for points on the real axis, is found from Eq. 5 to be

$$\frac{\partial y}{\partial \nu} = \frac{1}{\pi}\int_{-\infty}^{\infty} \frac{y(\chi,0)-y(\xi,0)}{(\chi-\xi)^2}\,d\chi$$

In the case of the solitary wave, $y(\chi,0) = 0$ for $\chi > 0$; furthermore, $\partial y/\partial\nu$ is sought only on the free surface $\xi < 0$. Then the normal derivative can be written

$$\frac{\partial y}{\partial \nu} = \frac{1}{\pi}\int_{-\infty}^{0} \frac{y(\chi,0)-y(\xi,0)\,d\chi}{(\chi-\xi)^2} + \frac{y(\xi,0)}{\pi\xi}$$

For points on the free surface

$$\frac{\partial y}{\partial \psi} = \frac{\partial y}{\partial \nu}\,\frac{\pi\xi}{Q}\;;\quad \xi = -e^{\frac{\pi\phi}{Q}}\;;\quad \chi = -e^{\frac{\pi\tau}{Q}}\;;\quad d\chi = \frac{\pi\chi}{Q}\,d\tau;$$

where τ is the variable of integration in the w-plane, so that

$$\frac{\partial y}{\partial \psi} = \frac{y(\phi,Q)}{Q} - \frac{\pi}{2Q^2}\int_{-\infty}^{\infty} \frac{y(\tau,Q)-y(\phi,Q)}{\cosh\frac{\pi(\tau-\phi)}{Q} - 1}\,d\tau \tag{6}$$

The interval of integration is split at $\tau=\phi$ and new variables introduced: $\sigma=\phi-\tau$ $(\tau<\phi)$; $\sigma=\tau-\phi$ $(\tau>\phi)$, to yield the final result

$$\frac{\partial y}{\partial \psi}(\phi,Q) = \frac{1+\eta(\phi)}{Q} - \frac{\pi}{Q^2}\int_0^{\infty} \frac{\eta(\phi+\sigma)+\eta(\phi-\sigma)-2\eta(\phi)}{e^{\frac{\pi\sigma}{Q}}+e^{\frac{-\pi\sigma}{Q}}-2}\,d\sigma = \frac{1+\eta(\phi)}{Q} + \int_0^{\infty} g(\phi,\sigma)\,d\sigma \tag{7}$$

Substitution of (7) into (2) yields, in fact, the non-linear integro-differential equation in $\eta(\phi)$ that defines the wave profile for any given value of the parameter Q.

NUMERICAL SOLUTION

The integrand function in (7) is smooth and quite well behaved for wave heights lower than the maximum (See Fig. 4), and can be approximated to any desired degree of precision by means of a quadrature formula of any order. The method of solution--essentially that of Fredholm for his integral equations of the first kind--requires, however, that the ordinates in the formula be equally spaced. The exponential in the denominator of the integrand permits truncation of the integral at some large value of σ without loss of accuracy.

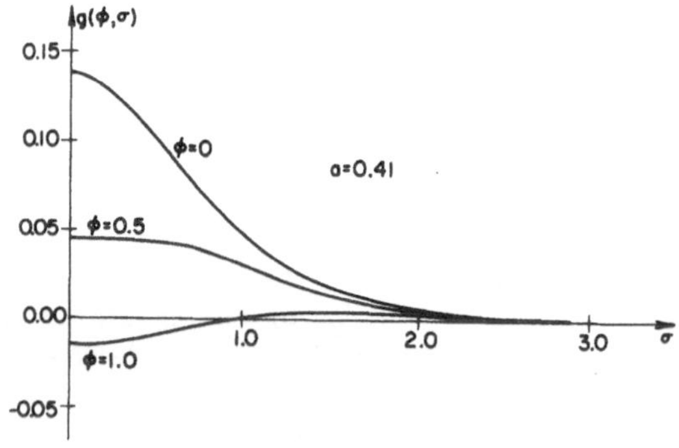

Fig. 4

Thus, to any desired precision,

$$\int_0^\infty g(\phi,\sigma)d\sigma = \delta\sigma \sum_{j=1}^{M} g(\phi,\sigma_j)w_j \tag{8}$$

in which

$$\sigma_j = (j-1)\delta\sigma$$

The quantity w_j is a weighting factor, and $\delta\sigma$ and M are chosen small enough and large enough, respectively, to maintain the desired precision of results. The writer used Simpson's rule to perform the quadrature, so that

$$w_j = \begin{cases} 1/3 \;....\; j = 1, M \\ 4/3 \;....\; j = 2, 4, 6, ..., M-1 \\ 2/3 \;....\; j = 3, 5, 7, ..., M-2 \end{cases} \tag{9}$$

In keeping with the second-order Simpson's rule, the derivative was approximated by the centered finite-difference formula

$$\eta'(\phi_k) = \frac{\eta_{k+1}-\eta_{k-1}}{2 \cdot \delta\phi} \tag{10}$$

in which $\eta_k \equiv \eta(\phi_k)$ and

$$\phi_k = (k-1)\;\delta\phi \tag{11}$$

with any desired accuracy being achieved by choosing $\delta\phi$ small enough. However, it is essential to the method that

$$\delta\phi = \delta\sigma = \Delta \tag{12}$$

so that Eq. 2 can be written

$$\eta_k + \frac{1/2}{\left[\frac{\eta_{k+1}-\eta_{k-1}}{2\Delta}\right]^2 + \left[\frac{1+\eta_k}{Q} - \sum_{j=1}^{M} \frac{\pi w_j \Delta}{2Q^2} \frac{\eta_{k+j-1}+\eta_{k-j+1}-\eta_k}{\cosh\left[\frac{\pi}{Q}(j-1)\Delta\right] - 1}\right]^2} = \frac{Q^2}{2} \tag{13}$$

Equation 13, representing the free-surface zero-pressure condition at ϕ_k, is an algebraic relation amongst 2M-1 consecutive values of η, centered on η_k. The same equation written for ϕ_{k+1} contains the same unknowns as the first, plus one more. A set of N such equations written for N consecutive values of ϕ, uniformly spaced over a region of interest containing the bulk of the wave profile, comprises a highly nonlinear system that contains, moreover, more unknowns than equations (See Fig. 5). The number of unknowns is, of course, reduced by taking advantage of the

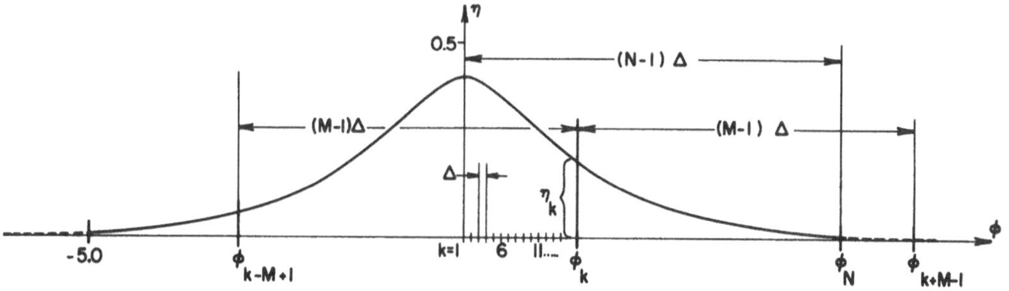

Fig. 5

symmetry of the profile, i.e., $\eta_{2-k}=\eta_k$. Furthermore, again because of the exponential in the denominator of the integrand function, those unknown η corresponding to values of ϕ lying outside the region of interest--in Eq. 13, k+j>N--are of great importance in determining only a small portion of the profile, lying near the outer boundary of the region, i.e., for large $k \leq N$, provided only that $(N-1)\Delta$ is chosen great enough, so that the boundary is well down the flank of the wave. Thus, those η values (k+j>N) can be approximated (dotted curve, Fig. 5) by some relation like the Boussinesq hyperbolic-secant-squared approximate solution, or simply set to zero, or equal to the boundary value η_N. The result of this operation, which does not affect the accuracy of any portion of interest in the wave, is to reduce the number of unknowns to equal the number of equations and make the system, in principle, solvable. Equation 13 can thus be interpreted as the k-th equation of a system of N nonlinear algebraic equations in the η_k (k = 1, 2,...N). While it is possible to solve this system by means of standard computer-library routines, a special method of successive approximation was devised which proved to be rapidly convergent and extremely effective.

The n-th approximation to the wave profile η_k^n (k = 1, 2,...,N) results in a distribution of pressure on the nominal free-stream boundary

$$p_k^n = \frac{Q^2}{2} - \eta_k^n - H_{v_k}^{\;n} \tag{14}$$

in which the dimensionless velocity head $H_{v_k}^{\;n}$ is given by the second term of the left-hand side of Eq. 13 with the n-th approximation to the η-function substituted. An improvement $c(\phi)$ in the distribution of η

$$\eta_k^{n+1} = \eta_k^n + c_k^n \tag{15}$$

will change the pressure at each point approximately by the amount

$$\delta p_k^n = - c_k^n + 4(H_{v_k}^{\;n})^2 \left\{ y_{\phi_k}^{\;n} \frac{c_{k+1}^n - c_{k-1}^n}{2\Delta} + y_{\psi_k}^{\;n} \left[\frac{1+c_k}{Q} - \sum_{j=1}^{M} \frac{\Delta\pi w_j}{2Q^2} \, \frac{c_{k+j-1}^n + c_{k-j+1}^n - 2c_k^n}{\cosh\left[\frac{\pi}{Q}(j-1)\Delta\right] - 1} \right] \right\} \tag{16}$$

as is seen by taking the first differential of Eq. 14. With δp_k^n set to $- p_k^n$

$$\delta p_k^n = - p_k^n \tag{17}$$

Eq. 16 is seen to be a representative member of a system of N linear algebraic equations in the c_k^n. Solution of this linear system of N equations by standard computer-library subroutines yields the correction fuction to be added to any given approximation of the wave profile to improve it. By monitoring p_k^n, it can be determined when to stop the iterations. In practice, 3-4 iterations was found to be sufficient to reduce the p_k to < 10^{-6} (See Fig. 6, which shows for a typical case the dimensionless pressure distributions on the nominal surface streamline in 3 successive approximations).

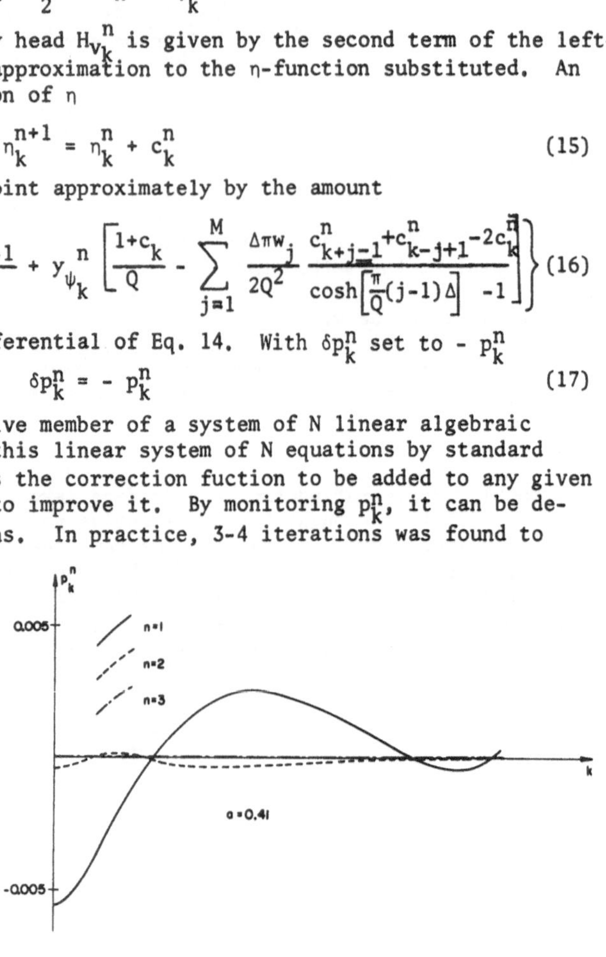

Fig. 6

RESULTS

This method was used to obtain wave profiles in the range, 0.1 < a < 0.75. A typical profile is shown in Fig. 5. For the lower values of relative wave height a, the first approximation of η was derived from the Boussinesq approximate solution to the solitary wave; the commensurate value of celerity was used for the parameter Q in the numerical solution. At the larger values of a, the difference between the first (Boussinesq) approximation of the wave profile and the correct solution was so great, that the successive approximations did not converge at all. Consequently, in this case, first approximations were obtained by using a wave profile known to be correct for some particular value of the discharge parameter, and then incrementing the latter by some amount. The greater the height of the wave, the smaller is the increment permissible, for the successive approximations to converge to a new profile. At the highest values of a determined by this technique the permissible increment in Q was less than 0.005. In this way, it is possible to creep up towards the solution of the cusped wave of maximum possible height (bearing in mind that at the higher values of a, Δ must be substantially decreased to maintain good accuracy, because of the great curvature at the crest), approaching the limiting case ever more slowly.

THE LIMITING CASE OF MAXIMUM POSSIBLE WAVE HEIGHT

Two special problems arise in this case. The integrand function g, in Eq. 7 instead of being smooth now has a cusp, at which point, furthermore, the slope becomes infinite. This problem is solved by seeking out the behavior of g in the immediate vicinity of the cusp and choosing an appropriate quadrature formula for one step Δ on each side of the cusp, retaining the previous formula (Simpson's rule) for the remainder of the interval of integration. In practice, this means that Eqs. 9 are replaced by an alternate set of conditions, in which, now the weighting factor w depends upon both j and k.

The second problem stems from the fact that Q is no longer a free parameter, the choice of which determines $\eta(0)$ and the remainder of the profile. In the wave of maximum height, the velocity head vanishes at the crest, so that, simply,

$$a_{max} = \eta(0) = \frac{Q^2}{2} \tag{18}$$

At the same time, there is but one, unknown value of Q, which corresponds to the wave of maximum height. This value is found as closely as desired by a method, detailed below, which depends upon knowledge of the limiting behavior of η as $\phi \to 0$.

As was shown by Stokes on the basis of Bernoulli's equation with zero pressure on the free surface and the known potential for flow in a corner, the crest of the wave of maximum possible height comprises a cusp, which in the immediate vicinity of $\phi = 0$ has the form of two straight lines, intersecting at an angle of 120°. The profile, then, as (positive) $\phi \to 0$, approaches the limiting function

$$\eta = \frac{Q^2}{2} - \frac{1}{2}\left(\frac{3}{2}\right)^{2/3} \phi^{2/3} \quad (0 \leq \phi << 1) \tag{19}$$

Indeed, with s the distance from the crest along the free surface (See Fig. 7) and V the magnitude of the velocity vector on the free surface,

$$\frac{d\eta}{d\phi} = \frac{d\eta}{ds}\frac{ds}{d\phi} = \frac{1}{V}\frac{d\eta}{ds} = \frac{\sin 30°}{V} \tag{20}$$

Furthermore, from Fig. 7, it is apparent that

$$\frac{\frac{V^2}{2}}{s} = \sin 30° \tag{21}$$

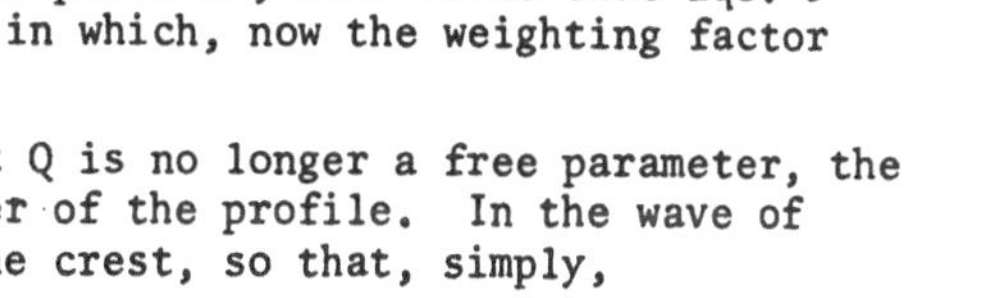

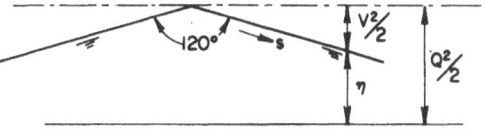

Fig. 7

and therefore,

$$\frac{d\eta}{d\phi} = \frac{1}{2}\frac{1}{\sqrt{s}}$$

But

$$\phi = \int_0^s V\,ds = \int_0^s s^{1/2}\,ds = \frac{2}{3}s^{3/2}$$

and so

$$\frac{d\eta}{d\phi} = \frac{1}{2}\left(\frac{2}{3}\right)^{1/3}\frac{1}{\phi^{1/3}} \tag{22}$$

from which Eq. 19 follows by integration.

From Eqs. 19 and 22, it is apparent that the integrand function g in Eq. 7 possesses a cusp at which the slope of the function becomes infinite (See Fig. 8), g itself remaining continuous. With Eq. 19 in hand, it is easy to determine appropriate values for the weighting function w_{kj} for values of j in the vicinity of k. Thus Eqs. 9 are replaced by the following:

$$w_{kj} = \frac{w_l + w_r}{2} \left\{ \begin{array}{l} w_l = \begin{cases} 0.8 & (j=k) \\ 1.2 & (j=k+1) \\ 0.0 & (j=1) \\ 1.0 & (j \neq \text{any of above}) \end{cases} \\ w_r = \begin{cases} 0.8 & (j=k) \\ 1.2 & (j=k-1) \\ 0.0 & (j=M) \\ 1.0 & (j \neq \text{any of above}) \end{cases} \end{array} \right\} \quad (23)$$

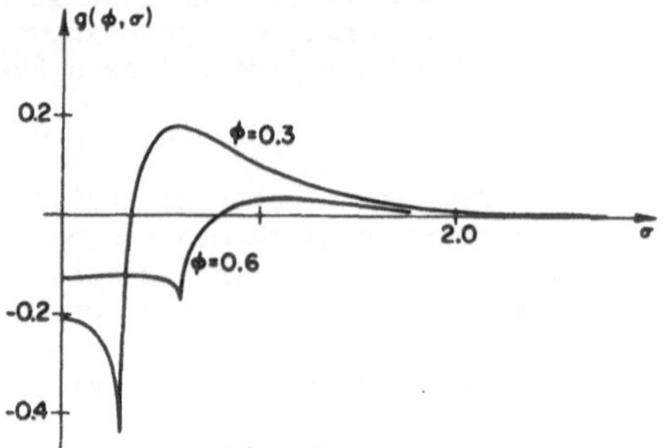

Fig. 8

in which the trapezoidal rule is used for the smooth portions of the integrand function.

Now, successive approximations for the cusped profile can be obtained just as before, omitting from consideration, however, the point $\eta(0)$ which is determined directly by the choice of Q (Eq. 18). The proper choice of Q, now, is made by a simple trial and error procedure that points very rapidly towards the correct value. With any given Q, let the discrete values of η at $\phi = \Delta, 2\Delta, 3\Delta, \ldots$ as given by Eq. 19 be denoted $\eta_k^{\circ}(Q)$; let the values of η obtained in solution of the algebraic equations of the type of Eq. 13 be denoted $\eta_k^{*}(Q)$; and form the difference of these two values

$$\varepsilon(k,Q) = \eta_k^{*} - \eta_k^{\circ} \quad (24)$$

The behavior of ε near the crest points to the correct value of Q. As can be deduced from Fig. 9 and the recollection that the 120° angle at the cusp determined the weighting factors in Eq. 23, only the correct value of Q will yield in solution a self-consistant wave profile, i.e., one that cusps at the proper 120° angle. The value of Q obtained in this way was 1.304 and the corresponding maximum relative wave height was

$$a_{max} = 1.85$$

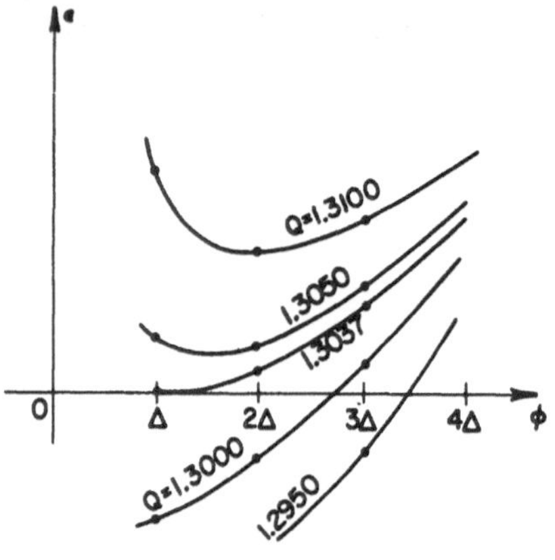

Fig. 9

****

An approach similar to that leading to Eqs. 13 and 16 was taken, independently of the writer, in 1969, by J. Byatt-Smith (1969), who presented computed profiles for the solitary wave of less than maximum height.

Reference:

1. Byatt-Smith, J.G.B., Proc. Roy. Soc. Lond. A. 315, 405-418 (1970).

NUMERICAL CALCULATION OF FLUID FLOWS AT ARBITRARY MACH NUMBER

Francis H. Harlow, Anthony A. Amsden and Cyril W. Hirt*
Los Alamos Scientific Laboratory
University of California
Los Alamos, New Mexico

INTRODUCTION

For the numerical solution of time-dependent, multidimensional problems in fluid dynamics, there are various approximation techniques [see, for example, Harlow (1969)], all of which are restricted in their scope of applicability. Some, for example, are useful for high-speed (supersonic or compressible) flows, while others can be applied to low speed (far subsonic or incompressible) flows. Our purpose here is to describe an Implicit Continuous-fluid Eulerian (ICE) method that is applicable to flows in which all fluid speeds can occur, either simultaneously or at different stages in the evolution of the problem.

A typical example of the utility of such a numerical solution technique is the problem of high-speed impact, in which a supersonic collision (of a large meteorite into a lake, for example) degenerates into subsonic flow, and the nature of the final incompressible waves are to be investigated. Likewise, material dynamics near an explosion will change from supersonic to far subsonic during many kinds of interesting circumstances. As an opposite type of example, the far-subsonic collapse of material may create a fast jet that interacts with other parts of the flow in a highly compressible fashion. Also, the non-isotropic Alfvén waves in a plasma can produce the effect of far subsonic flow in one direction with supersonic flow in another, so that here, too, is a circumstance in which a modified ICE technique can be used for numerical solutions.

The ICE methodology is described here in association with an Eulerian mesh of computational cells, which form the basis for the finite-difference approximation to the full, time-dependent, non-linear equations of fluid dynamics. The ideas can also be adapted to a Lagrangian mesh, or to any of a variety of combined Eulerian-Lagrangian coordinate systems. For a concise illustration of the principles, we restrict the discussion to a discussion without equations. Harlow, Amsden and Hirt (1970) give a more general presentation of the equations, appropriate for confined or free surface motions in cylindrical coordinates with axial symmetry, together with listings and flow sheets for the computer program.

The underlying idea is that the changes in the fluid configuration, which take place in a sequence of brief time intervals or cycles, must be calculated from an implicit formulation, in which the unknown new density and velocity values occur not only in the time derivative terms but also in certain crucial terms throughout the rest of the equations. As a result, the solution is somewhat more complicated than in the simple explicit techniques for high-speed flow. For all Mach numbers, it is necessary to solve a finite difference Poisson equation, which sometimes will be amenable to direct solution, but in many cases will require an iterative procedure.

Thus, although the calculations may be somewhat more time-consuming than with the more restricted methods, the ICE technique allows for the investigation of prob-

---

*This work was performed under the auspices of the United States Atomic Energy Commission.

lems not previously tractable. Beyond its capability for encompassing the entire range of Mach numbers, the ICE method offers no advantages over such techniques as MAC [Harlow and Welch (1965)] or SMAC [Amsden and Harlow (1970)] for incompressible flows, or FLIC [Gentry, Martin and Daly (1966)] for supersonic flows. Indeed, in the limit of infinite sound speed, the ICE technique becomes essentially identical to MAC, while for supersonic flow, it is an implicit, staggered-mesh variant of FLIC; but in neither of those extremes does ICE effectively compete with the limiting method, being generally slower and more extravagant in the use of computer memory.

FEATURES OF THE PRESENT VERSION

Harlow and Amsden (1968) described an earlier version of the ICE method, from which the present form differs in a number of respects:

1. An arbitrary equation of state can be included. This requires separation of the term proportional to the square of the sound speed, in which the density variation is treated implicitly.

2. The dependent variable in the Poisson equation is more meaningful. For this we choose the separated pressure function described above.

3. The finite-difference Poisson equation has been reduced from a nine-point to a five-point form, and the coefficients are considerably simplified.

4. The complete viscous stress tensor is incorporated, rather than a simple artificial viscosity.

5. The crucial terms contain a time-centering option, by which some undesirable low-order truncation error terms can be removed.

6. Free-surface problems, as well as confined flows, can be calculated.

7. The program can be used for studies of both plane and cylindrically symmetric configurations.

CALCULATIONAL EXAMPLES

To show the range of capabilities of the ICE method, we illustrate calculations for both low-speed and high-speed impact of a spherical projectile. The former has been favorably compared with calculations for a completely incompressible fluid performed with the SMAC method [Amsden and Harlow (1970)]. The latter likewise compares well with PIC method calculations [Amsden (1966)].

The initial marker particle configuration for each calculation is shown in Fig. 1, where a spherical projectile of radius 0.115 cm and downward velocity of 320 cm/sec is impinging upon a cylindrical target of thickness 0.4 cm. Both the projectile and the target have 1 gm/cc density. (In the case of low-speed flow, this corresponds to a splashing drop.) The left wall is the axis of symmetry and the other three walls of the computing region are rigid but allow free slippage along them.

Figures 2 and 3 are sequences taken from the early stages of the collision process, the former being the low speed and the latter the high speed. The times correspond exactly between the two figures: 0.00026, 0.00052, 0.00076, 0.00103, 0.00152, and 0.00252 seconds after the initial impact of Fig. 1. The one and only difference in the two calculations is the square of the input zero-temperature sound speed: $a^2 = 10^7$ cm^2/sec^2 for the low speed problem and $a^2 = 1$ cm^2/sec^2 for the high speed.

Note that by time 0.00052 seconds in Fig. 2, the fluid level on the right has already begun to rise noticeably, as it must in order to conserve the volume of an incompressible fluid. In the high-speed limit, however, the fluid level on the right remains undisturbed beyond the time of the last frame, illustrating the high compressibility of the fluid. The distortion of the uniformly aligned particles in the target material is a useful indicator of the progress of the shock through it.

The later stages of these calculations are discussed by Harlow, Amsden and Hirt (1970).

REFERENCES

Amsden, A. A., "The Particle-in-Cell Method for Calculation of the Dynamics of Compressible Fluids," Los Alamos Scientific Laboratory Report No. LA-3466 (1966).

Amsden, A. A. and Harlow, F. H., "The SMAC Method," Los Alamos Scientific Laboratory Report No. LA-4370 (1970).

Gentry, R. A., Martin, R. E., and Daly, B. J. Journal of Computational Physics 1, 87 (1966).

Harlow, F. H. and Amsden, A. A. Journal of Computational Physics 3, 80 (1968).

Harlow, F. H., Amsden, A. A., and Hirt, C. W. To be published.

Harlow, F. H., "Numerical Methods for Fluid Dynamics, An Annotated Bibliography," Los Alamos Scientific Laboratory Report No. LA-4281 (1969).

Harlow, F. H. and Welch, J. E. Physics of Fluids 8, 2182 (1965); see also Harlow (1969), pp. 8, 9.

Fig. 1. Initial configuration for Figures 2 and 3.

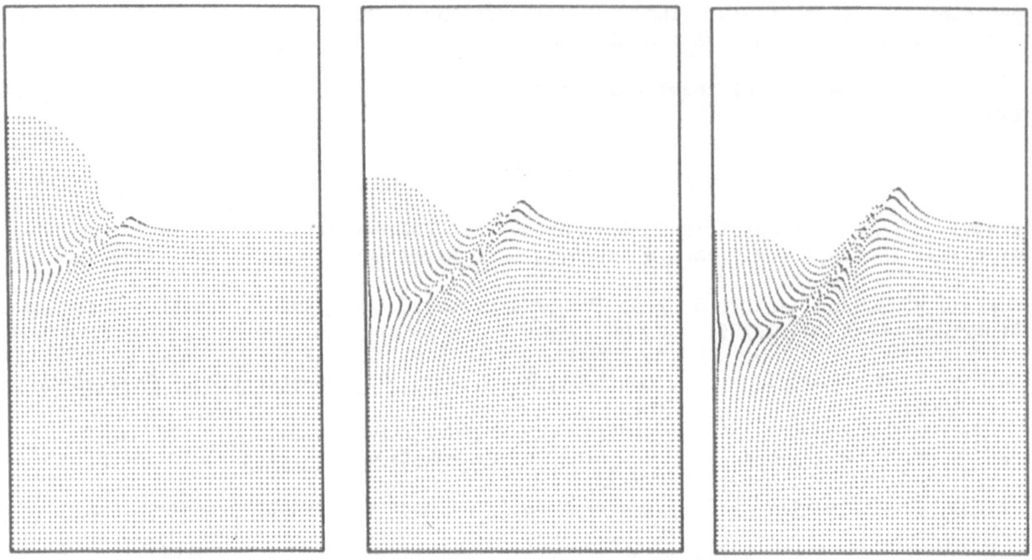

Fig. 2

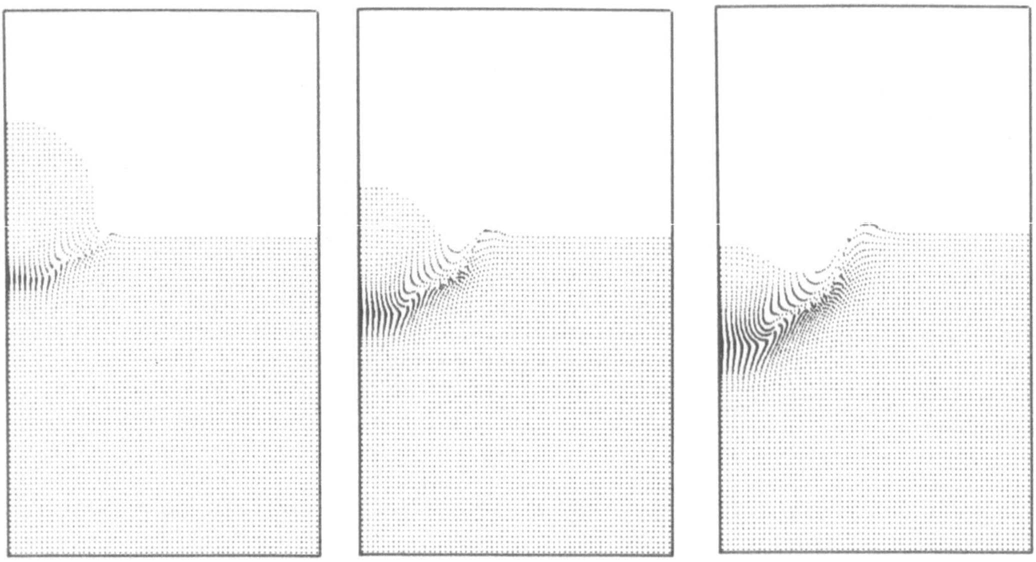

Fig. 3

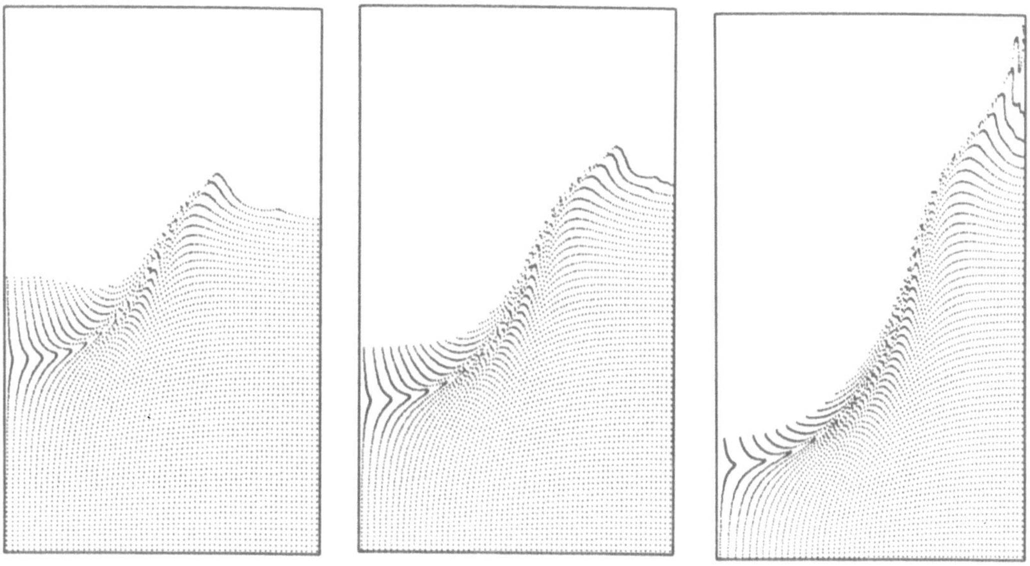

Fig. 2 Contd.

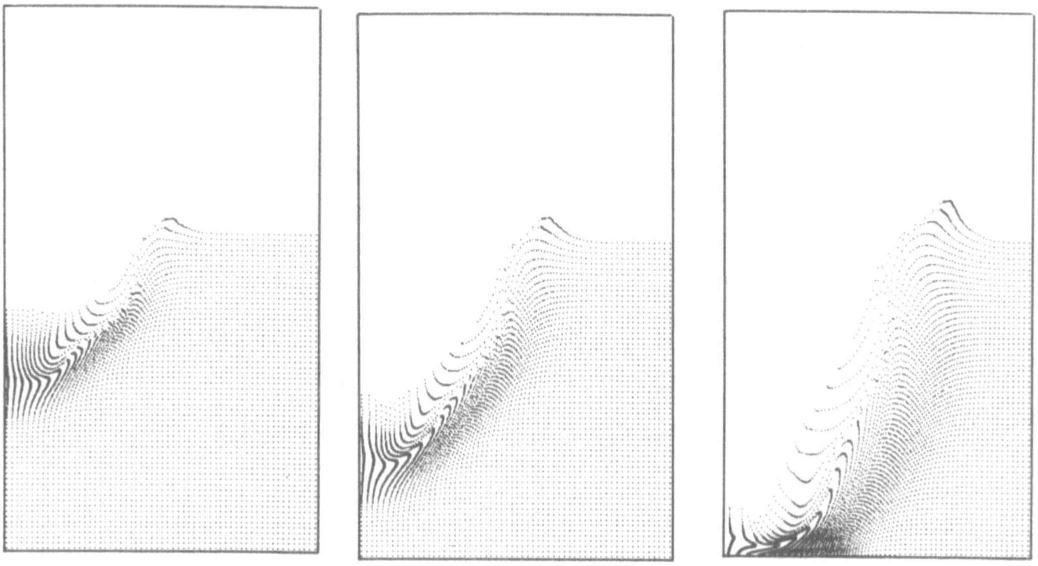

Fig. 3 Cond.

IMPLICIT SOLUTION OF CREEPING FLOWS, WITH APPLICATION TO CONTINENTAL DRIFT

William E. Pracht*
Los Alamos Scientific Laboratory
University of California
Los Alamos, New Mexico

INTRODUCTION

The numerical solution of low Reynolds number incompressible flow problems is complicated by numerical stability restrictions related to the dominance of the viscous terms in the flow equations. Several numerical methods, for example those by Harlow and Welch (1965), Hirt, et al. (1970), and Fromm and Harlow (1963), have been developed to calculate the time dependent flow of incompressible fluids; but as a result of these restrictions they are not readily applicable to problems for which the flow Reynolds number ($\equiv LU/\nu$) < 1. Here L and U represent a typical dimension and velocity of the flow, and ν is the kinematic viscosity.

This paper describes a numerical technique, called MACRL, that applies to such problems. It is an extension of the Marker-and-Cell (MAC) method of Harlow and Welch (1965), which is a finite-difference technique for solving the full time dependent Navier-Stokes equations. Like MAC, MACRL uses an Eulerian formulation for these equations, together with a set of marker particles that move with the local fluid velocity.

The basic feature of MACRL that distinguishes it from other incompressible flow methods is the implicit form of the finite-difference equations, together with several newly developed numerical procedures required for solution of the implicit system of equations. This is in contrast to the MAC difference equations, where advanced time velocities are expressed entirely in terms of old time quantities. That system is called explicit, and is stable provided $\nu\, \delta t/\delta x^2 < 1/4$, where δt is the finite-difference time step and δx is the cell size. The consequence of this for problems in which the viscous forces are large compared to the inertial forces is that if δx is chosen small for accuracy, δt must be so small that a prohibitively large number of time cycles is required to complete a problem.

To overcome this, MACRL is based on a system of equations in which the velocities that occur in the dominant viscous terms are expressed implicitly, i.e., as a function of the field variables at the advanced cycle time. As a result the equations are unconditionally stable. Solution for the new velocities requires a somewhat more complicated technique, but this form offers advantages in time step size which overcome the disadvantage of solution complexity.

OUTLINE OF PROCEDURE

The MACRL method is based on a finite-difference approximation to the full Navier Stokes equations together with the incompressibility condition. We use the following equations:

*This work was performed under the auspices of the United States Atomic Energy Commission.

$$\frac{\partial u}{\partial t} + \frac{\partial u^2}{\partial x} + \frac{\partial uv}{\partial y} = - \frac{\partial \phi}{\partial x} + g_x - 2 \frac{\partial}{\partial x} [\nu \frac{\partial v}{\partial y}] + \frac{\partial}{\partial y} [\nu (\frac{\partial u}{\partial y} + \frac{\partial v}{\partial x})] \tag{1}$$

$$\frac{\partial v}{\partial t} + \frac{\partial uv}{\partial x} + \frac{\partial v^2}{\partial y} = - \frac{\partial \phi}{\partial y} + g_y + \frac{\partial}{\partial x} [\nu (\frac{\partial v}{\partial x} + \frac{\partial u}{\partial y})] - 2 \frac{\partial}{\partial y} [\nu \frac{\partial u}{\partial x}] \tag{2}$$

$$\frac{\partial u}{\partial x} + \frac{\partial v}{\partial y} = 0 \tag{3}$$

where g_x and g_y are the x and y components of a body acceleration, and ϕ the ratio of pressure to constant density. In discussing the derivation of the MACRL finite-difference equations, it is appropriate to begin with a simplified version in which the coefficient of viscosity is constant. This version of MACRL is described in detail by Pracht (1970) so that only a brief outline is presented here. This will serve as a basis for a discussion of the more complicated case to follow.

From the finite-difference analog of Eqs. (1) and (2) we obtain

$$\frac{D_{i,j}^{n+1} - D_{i,j}^{n}}{\delta t} = \frac{1}{\delta x^2} [2 (\phi)_{i,j}^{n+1} - (\phi)_{i+1,j}^{n+1} - (\phi)_{i-1,j}^{n+1}]$$

$$+ \frac{1}{\delta y^2} [2 (\phi)_{i,j}^{n+1} - (\phi)_{i,j+1}^{n+1} - (\phi)_{i,j-1}^{n+1}] - (Q)_{i,j}^{n} \quad , \tag{4}$$

where

$$Q_{i,j}^{n} \equiv \frac{1}{\delta x^2} [(u^2)_{i+1,j}^{n} - 2 (u^2)_{i,j}^{n} + (u^2)_{i-1,j}^{n}]$$

$$+ \frac{1}{\delta y^2} [(v^2)_{i,j+1}^{n} - 2 (v^2)_{i,j}^{n} + (v^2)_{i,j-1}^{n}]$$

$$+ \frac{2}{\delta x \delta y} [(uv)_{i+\frac{1}{2},j+\frac{1}{2}}^{n} - (uv)_{i-\frac{1}{2},j+\frac{1}{2}}^{n}$$

$$- (uv)_{i+\frac{1}{2},j-\frac{1}{2}}^{n} + (uv)_{i-\frac{1}{2},j-\frac{1}{2}}^{n}]$$

and

$$D_{i,j} \equiv \frac{u_{i+\frac{1}{2},j} - u_{i-\frac{1}{2},j}}{\delta x} + \frac{v_{i,j+\frac{1}{2}} - v_{i,j-\frac{1}{2}}}{\delta y} \quad .$$

The subscripts refer to finite-difference cell locations, $x = i\delta x$, $y = j\delta y$, and the superscript n counts time cycles, $t = n\delta t$. Cell centers are designated by integer values of i and j, and cell edges by half integer values (e.g., $i+\frac{1}{2},j$ designates the center of the right-hand side of cell i,j).

Equation (3) implies that we should require $D_{i,j}^{n+1} = 0$, so that Eq. (4) is a Poisson equation for pressures which may require an iterative process for solution. In order to avoid accumulation of errors resulting from terminating the iteration process with a finite convergence criterion, the term with $D_{i,j}^{n}$ is retained, in a corrective procedure described by Hirt and Harlow (1967). Substitution of these pressures into the momentum equations then gives the advanced time velocities. Since in the finite-difference approximation of these equations, (n+1)-time velocities are expressed in terms of (n+1)-time quantities, i.e., velocities in the viscous terms, an implicit technique is also required for solution. For this we use an iteration process analogous to that which may be required for the pressure.

In summary, each cycle is divided into several steps:

1) A field of pressures and velocities, together with a set of marker particle coordinates, are recorded either from initial conditions or the preceding cycle.

2) Equation (4) is solved by a Gauss-Seidel or Liebman relaxation procedure for (n+1)-time pressures.

3) These pressures are inserted into the finite-difference momentum equations derived from Eqs. (1) and (2), to solve for the velocities. In some cases, described below, steps 2 and 3 must be iterated together.

4) New marker particle coordinates are found according to a local average of velocities nearest each particle.

5) Output in a variety of forms can be recorded if desired.

This, then in principle, completes the advancement of the fluid configuration for one cycle of calculation. There are, however, several circumstances in which additional features are required. These include the very low Reynolds number problems, flows adjacent to no-slip walls, and examples with variable viscosity. These extensions are first discussed for the relatively simple case of free-slip boundary conditions.

Free-Slip Walls

The absence of (n+1)-time velocity terms in Eq. (4) implies that it can be solved by itself, with the results then inserted into the finite-difference momentum equations for (n+1)-time velocity solution. Indeed, this is the case when free-slip boundary conditions are used, so that the boundary conditions on pressure likewise contain no (n+1)-time velocity terms. Another implication can be illustrated by expressing Eqs. (1) and (2) in the form of a vorticity transport equation. This equation is independent of ϕ so that any scalar field could be inserted into the momentum equations with a resulting velocity field that carries the correct vorticity. The significance of this fact becomes apparent when one is applying the MACRL technique to problems with very low Reynolds numbers.

For problems in which Re << 1, the time derivative and convective terms can be omitted from the momentum equations. As a result the pressure equation becomes independent of the condition in Eq. (3), so that an additional procedure each calculational cycle is necessary to insure continued volume conservation. The reason for this can be seen by considering the equations,

$$\frac{\partial \phi}{\partial x} = \nu \left(\frac{\partial^2 u}{\partial y^2} - \frac{\partial^2 v}{\partial x \partial y}\right) + g_x \tag{5}$$

$$\frac{\partial \phi}{\partial y} = \nu \left(\frac{\partial^2 v}{\partial x^2} - \frac{\partial^2 u}{\partial x \partial y}\right) + g_y \tag{6}$$

where only the dominant terms of the momentum equations have been retained.

An expression for ϕ is obtained as before but the result in this case is

$$0 = \frac{1}{\delta x^2}\left[2(\phi)^{n+1}_{i,j} - (\phi)^{n+1}_{i+1,j} - (\phi)^{n+1}_{i-1,j}\right] + \frac{1}{\delta y^2}\left[2(\phi)^{n+1}_{i,j} - (\phi)^{n+1}_{i,j+1} - (\phi)^{n+1}_{i,j-1}\right] \; . \tag{7}$$

This is now simply Laplace's equation for ϕ, from which the time rate of change of D term is absent. Because the incompressibility condition is not incorporated in the derivation, the resulting velocities, would not, in general, conserve volume. As we have seen, however, for problems with free-slip walls the solution would yield a velocity field with the correct vorticity.

This suggests a procedure in which step 3 is divided into two phases. During the first phase, tentative velocities are calculated using pressures found by iterating Eq. (4), with $D_{i,j}^{n+1} = 0$, to convergence. The velocity field contains the correct vorticity but is not conservative of volume. The second phase of step 3 corrects these tentative velocities in such a way as to preserve the vorticity. To accomplish this, we define a potential, ψ, such that

$$u_{i+\frac{1}{2},j}^{n+1} - \tilde{u}_{i+\frac{1}{2},j}^{n+1} = \frac{\psi_{i,j} - \psi_{i+1,j}}{\delta x} \tag{8}$$

and

$$v_{i,j+\frac{1}{2}}^{n+1} - \tilde{v}_{i,j+\frac{1}{2}}^{n+1} = \frac{\psi_{i,j} - \psi_{i,j+1}}{\delta y} \quad , \tag{9}$$

where $\tilde{u}$ and $\tilde{v}$ are the velocities following phase 1 of step 3, and u and v are the final correct velocities. The expression for ψ derived from Eqs. (8) and (9) is

$$D_{i,j}^{n+1} - \tilde{D}_{i,j}^{n+1} = \frac{1}{\delta x^2} [2\,\psi_{i,j} - \psi_{i+1,j} - \psi_{i-1,j}] + \frac{1}{\delta y^2} [2\psi_{i,j} - \psi_{i,j+1} - \psi_{i,j-1}] \quad . \tag{10}$$

Solution of Eq. (10) with $D_{i,j}^{n+1}$ set to zero gives a set of ψ which when inserted back into Eq. (8) and (9) produces the final correct velocities.

<u>No-Slip Walls</u>

For no-slip wall boundary conditions the situation is somewhat different. The reason for this is that the boundary conditions on pressure are in this case expressed in terms of derivatives of (n+1)-time velocities. This means that a simultaneous iteration of the pressure equation and momentum equations is required. In addition, for problems in which Re << 1, the tentative ϕ, u, v calculation must be simultaneously iterated with the potential function solution for ψ.

SAMPLE CALCULATIONS

The properties of the MACRL method can best be illustrated by a discussion of the results of several calculational examples. The first example, shown in Fig. (1), illustrates the applicability of MACRL to the low Reynolds number flow of a fluid having free surfaces. This sequence represents the slow viscous slumping motion that would occur if a rectangular block of tar or pitch were placed on a non-slip surface and left undisturbed. The thin empty layer toward the bottom edges of the block is a result of the course resolution used in the calculation. This layer is less than a computational cell in depth and is therefore treated as though it contained fluid despite the fact that the marker particles remain about half a cell width from the bottom boundary.

A conservation check of this free surface calculation reveals that the incompressibility condition is everywhere satisfied, i.e., that $D_{i,j}$ remains negligibly small for each cell and that the total volume remains constant throughout the run to within the accuracy that can be measured from the particle configuration plots.

The next two examples are from preliminary stages in the development of a numerical model for studying some of the features of mantle flow associated with the new global tectonics. There exists considerable geological evidence in favor of the hypothesis of large scale convective cells within the earth's upper mantle, but relatively little work has been done on developing a complete theoretical model. The work of Turcotte and Oxburgh (1970) in modeling the cellular convection problem provides considerable insight into the phenomenon, but was restricted by the requirement that the viscosity be constant.

In order to model this convection problem several additional features must be included. They are heat transport, bouyancy effects, and variable viscosity. Heat transport is included in the calculation by coupling the heat flow equation with the momentum equations. The heat equation used includes both conduction and convection terms, together with frictional dissipation and radioactivity source terms. The bouyancy effects resulting from variations in temperature are accounted for by a Boussinesq approximation.

The additional solution each calculational cycle of the heat transport and viscosity equations is accomplished in a straightforward manner. Whereas the expression for ϕ given by Eq. (4) in the constant viscosity case contained no (n+1)-time velocity terms, for variable viscosity a variety of these terms appear. As a result, even for free-slip boundaries, it is necessary to simultaneously solve the pressure equation and the momentum equations.

Figure (2) shows the steady state velocity vector configuration of a thermal convection problem in which the viscosity is constant. In this case, the flow Reynolds number is of the order of 10^{-24}, and the Rayleigh number is about $5x10^3$.

Figure (3) represents a more realistic model of mantle convection in which the viscosity depends on temperature and pressure according to a function suggested by Gordon (1965). At this stage of development, the viscosity profile was held fixed in time. The effects of this type of variation are illustrated in Fig. (3), which shows the vortices developing from an initial hot spot in the lower left-hand corner. Confinement of the large velocities to the upper half of the computing region (representing 1000 km of depth) is a result of a viscosity minimum in Gordon's profile at a depth of about 500 km.

Following calculations will allow for full variation of viscosity with temperature and pressure and for the effects of phase transitions.

ACKNOWLEDGMENT

The author gratefully acknowledges the many valuable contributions to this work made by Dr. F. H. Harlow.

REFERENCES

Fromm, J. E. and Harlow, F. H., Phys. Fluids 6, 975 (1963).

Gordon, R. B., J. Geophysical Research 70, 2413 (1965).

Harlow, F. H., and Welch, J. E., Phys. Fluids 8, 2182 (1965).

Hirt, C. W., Cook, J. L. and Butler, T. D., Journal of Computational Physics 5, 103 (1970).

Hirt, C. W. and Harlow, F. H., Journal Computational Physics 2, 114 (1967).

Pracht, W. E., Journal Computational Physics, accepted for publication (1970).

Turcotte, D. L. and Oxburgh, E. R., J. Fluid Mech. 28, 29 (1967).

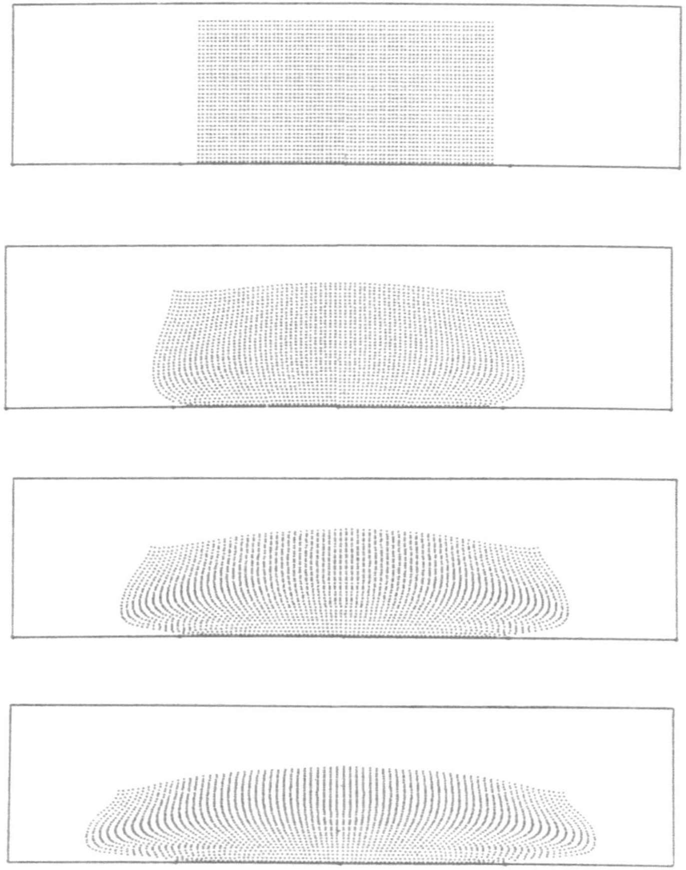

Fig. 1. Calculation of a sagging block of tar or pitch.

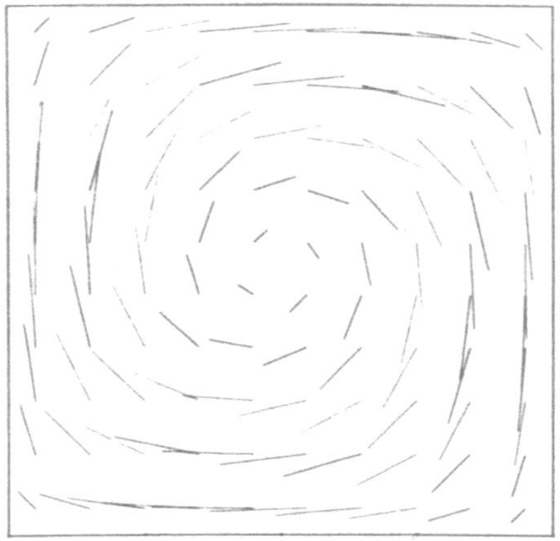

Fig. 2. Symmetric vortex in the simplified, constant viscosity model of the earth's mantle.

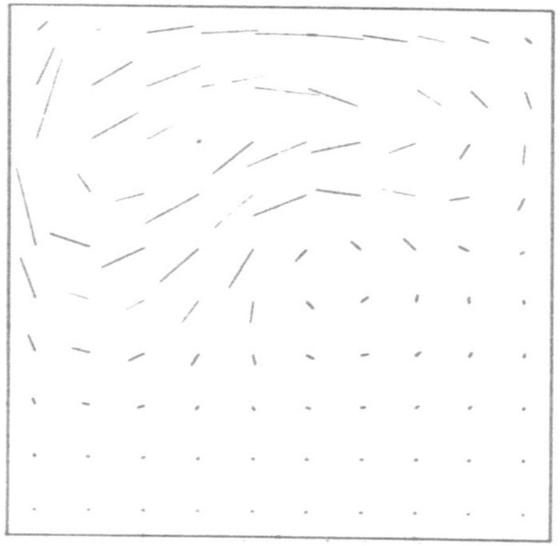

Fig. 3. Development of vortices in the earth's mantle with the viscosity formula of Gordon (1965).

THE NUMERICAL SOLUTION OF UNSTEADY FREE SURFACE FLOWS BY CONFORMAL MAPPING

Arthur K. Whitney
California Institute of Technology

INTRODUCTION

Exact analytical study of the motion of an inviscid fluid with a free surface is difficult for two reasons: (1) the position of the free surface is not known a priori, it must be found as part of the solution; and (2) the kinematic and dynamic boundary conditions at the free surface are nonlinear. Although much progress has been made through the use of various approximate analytical techniques, such as nonlinear shallow-water wave theory and higher-order infinitesimal wave theory, the continuing evolution of high-speed digital computers has encouraged the use of numerical techniques for this difficult problem.

At present, numerical methods for the solution of the exact equations describing the irrotational motion of an inviscid fluid are limited to two spatial dimensions. These methods are well represented by the work of Chan and Street (1970), Hirt, et al. (1970), and Brennen and Whitney (1970). Chan and Street's method (SUMMAC), which is based on the Marker-and-Cell (MAC) technique developed by Welch, et al. (1966), employs an Eulerian description of the fluid motion. Hirt, et al., work basically with Eulerian equations, although the positions of the fluid particles are followed in a Lagrangian manner. Brennen and Whitney solve the full Lagrangian equations of motion. All three methods have one thing in common: the governing field and boundary equations are differenced and solved on a two-dimensional grid of points.

This paper introduces a new formulation for the problem of the unsteady planar motion of an inviscid, incompressible fluid with a free surface. The region which the fluid occupies at any time is mapped conformally to a known, fixed parametric space. If the parametric space has a simple geometry and if the velocities and boundaries of the fluid are known at a particular time, the information needed to step forward in time (namely, the time-derivatives of the boundary velocity and position) may be found exactly by functional analysis. These solutions depend only on boundary information, thus reducing the number of spatial dimensions to one.

Since the details of this technique have not previously been published, the major portion of the paper is devoted to mathematical formulation. An example of the use of this method for numerical calculations is given at the end of the paper.

MATHEMATICAL FORMULATION

We consider the irrotational motion of a fluid in a basin, such as that in Fig. 1(a). Gravity is directed in the negative y direction. We denote the region

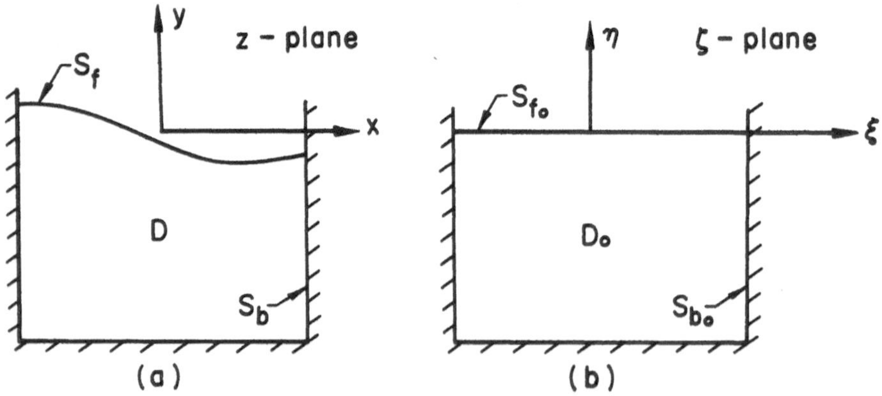

FIG. 1

which the fluid occupies by D, the wetted solid boundary by S_b, and the free surface by S_f.

In D, the fluid velocities satisfy the continuity and irrotationality conditions, thus the complex velocity, $w \equiv u - iv$, is an analytic function of $z = x + iy$, i.e.,

$$w = w(Z,t). \tag{1}$$

Here, u and v are the x and y velocities and t is the time.

On either an S_b or S_f boundary the fluid velocity normal to the surface must equal the rate of translation of the surface normal to itself. If the surface is represented by $F(x,y,t) = 0$, this may be written as $q_n = -F_t/\sqrt{F_x^2 + F_y^2}$.

If the fluid pressure is p and the atmospheric pressure is $p_o(x,t)$, then the dynamic condition on S_f is simply

$$p = p_o \quad \text{on} \quad S_f \tag{2}$$

neglecting surface tension. It is convenient to write Eq. (2) in terms of the gradient of the pressure. To do this, we imagine the fluid motion to be frozen at a particular time. Since Eq. (2) holds at all points on S_f, it follows that

$$\vec{s} \cdot \nabla p = \vec{s} \nabla p_o, \tag{3}$$

where $\vec{s}$ is a unit vector tangent to S_f and $\nabla = (\partial/\partial x, \partial/\partial y)$. We now eliminate ∇p in Eq. (3) by using the equations of motion. The final form of the dynamic boundary condition becomes

$$\vec{s} \cdot (\vec{q}_t + \vec{q} \cdot \nabla \vec{q} + g\vec{k}) = -\frac{1}{\rho} \vec{s} \cdot \nabla p_o, \tag{4}$$

in which $\vec{q} = (u,v)$, $\vec{k}$ is a unit vector in the y direction, g is the gravitational acceleration, and ρ is the density of the fluid.

Now, suppose that D is mapped to a fixed region D_o in (ξ,η)-space (see Fig. 1(b)) by

$$Z = Z(\zeta,t), \tag{5}$$

where $\zeta = \xi + i\eta$. This transformation necessarily contains t as a parameter since the region D changes with time.

As a result of Eq. (5), the complex velocity depends on ζ and t. This functional dependence is denoted by capital letters, i.e.,

$$W(\zeta,t) = U(\xi,\eta,t) - iV(\xi,\eta,t) \equiv w(Z(\zeta,t),t).$$

Next, consider the transformation of the kinematic and dynamic boundary conditions. To save space, we assume that S_b and S_f map to line segments parallel to the ξ-axis. (Results for boundaries which map to other orientations are similar and will be presented when needed.) With this assumption the unit normal and tangent vectors have the following representation:

$$\vec{n} = (-y_\xi, x_\xi)/|Z_\zeta| \;, \quad \vec{s} = (x_\xi, y_\xi)/|Z_\zeta| \;. \tag{6}$$

On S_{b_o}, there are two kinematic conditions which must be satisfied--one on the transform, and one on the velocities. These can be written as

$$\begin{aligned}(-x_t y_\xi + y_t x_\xi)/|Z_\zeta| &= (-Uy_\xi + Vx_\xi)/|Z_\zeta| \\ &= -F_t/\sqrt{F_x^2 + F_y^2} \quad \text{on} \quad S_{b_o}\end{aligned} \tag{8}$$

A simple geometrical interpretation of these equations follows from Fig. 2 and Eqs. (6).

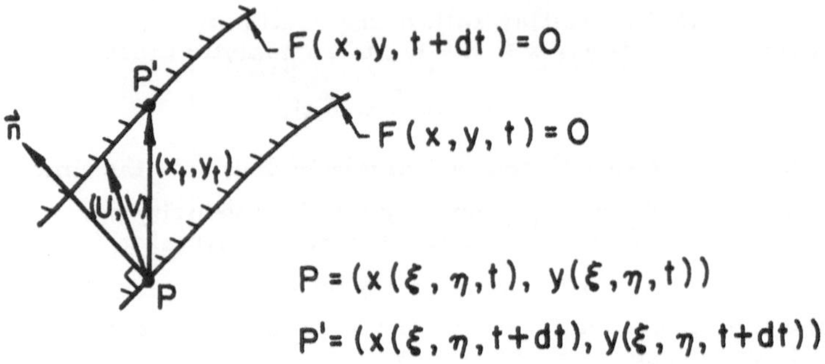

FIG. 2

To understand Eqs. (8), it is important to distinguish between the movement of a fluid element, which is along (U, V), and the movement of a boundary point, which is along (x_t, y_t). Eqs. (8) simply state that the normal components of the velocities of the fluid element and boundary point are equal to each other <u>and</u> to the rate of translation of the surface normal to itself.

On S_{f_o}, F is not prescribed, so there is only one kinematic condition, namely

$$(U - x_t)y_\xi - (V - y_t)x_\xi = 0 \quad \text{on } S_{f_o} . \tag{9}$$

The dynamic boundary condition on S_{f_o} follows by a straightforward transformation of Eq. (4) and becomes

$$U_t x_\xi + (V_t + g)y_\xi + (U - x_t)U_\xi + (V - y_t)V_\xi = \frac{1}{\rho} p_{ox} x_\xi \quad \text{on } S_{f_o} \tag{10}$$

Eqs. (9) and (10) are written conveniently as

$$\operatorname{Im} G = \operatorname{Im}(\overline{W}/Z_\zeta) \tag{11}$$

$$\operatorname{Re} H = -\frac{1}{2}|W|_\xi^2 - \frac{1}{\rho} p_{ox} x_\xi , \tag{12}$$

on S_{f_o}

where

$$G(\zeta) \equiv Z_t/Z_\zeta \quad \text{and} \quad H(\zeta) \equiv (W_t - ig)Z_\zeta - W_\zeta Z_t . \tag{13}$$

The boundary conditions for G and H on S_{b_o} are found from Eqs. (8)

$$\begin{aligned} \operatorname{Im} G &= -F_t/(|Z_\zeta|\sqrt{F_x^2 + F_y^2}) \\ \operatorname{Im} H &= (Vx_t - Uy_t)_\xi + (|Z_\zeta|F_t/\sqrt{F_x^2 + F_y^2})_t \end{aligned} \tag{14}$$

For stationary boundaries, $F_t = 0$, (U, V) and (x_t, y_t) are both parallel to $\vec{s}$, so that Eqs. (14) simplify to

$$\left.\begin{aligned} \operatorname{Im} G &= 0 \\ \operatorname{Im} H &= 0 \end{aligned}\right\} \text{for stationary } S_b \ (S_{b_o} \parallel \xi\text{-axis}) \tag{15}$$

If a fixed boundary S_b maps to a line segment parallel to the η-axis, then instead of Eqs. (15),

$$\left.\begin{array}{l}\operatorname{Re} G = 0 \\ \operatorname{Re} H = 0\end{array}\right\} \text{ for stationary } S_b \ (S_{b_o} \parallel \eta\text{-axis}) \tag{16}$$

A practical use of the above formulation is perhaps best illustrated by the following example.

AN EXAMPLE

Consider the motion of an infinitely deep fluid between vertical barriers. The region D is bounded by a free surface with mean level, $y = 0$, and by the two lines, $x = 0$ and π. D_o is chosen to be $\eta \leq 0$, $0 \leq \xi \leq \pi$. We consider the case $p_o \equiv 0$ and set $g = 1$.

Since Eqs. (16) hold for $x = 0, \pi$, the functions G and H may be continued to the entire lower half ζ-plane by the reflection

$$G(-\overline{\zeta},t) = -\overline{G(\zeta,t)}, \quad H(-\overline{\zeta},t) = -\overline{H(\zeta,t)}$$

and the periodic continuation

$$G(\zeta + 2\pi n,t) = G(\zeta,t) \ , \quad H(\zeta + 2\pi n,t) = H(\zeta,t), \ n = \pm 1, \pm 2, \ldots$$

If we consider the right-hand sides of Eqs. (11) and (12) to be known, the solutions for G and H are found to be (e.g., see Woods (1961))

$$G(\zeta) = -\frac{1}{2\pi}\int_{-\pi}^{\pi} \operatorname{Im}(\overline{W}/Z_\xi) \cot\frac{1}{2}(\lambda - \zeta)\, d\lambda \tag{17}$$

$$H(\zeta) = -\frac{1}{2\pi i}\int_{-\pi}^{\pi} \frac{1}{2}|W|_\xi^2 \cot\frac{1}{2}(\lambda - \zeta)\, d\lambda - i \tag{18}$$

Z_t and W_t may be found from Eqs. (13), (17), (18). Evaluating these expressions on $\zeta = \xi - i0$, we have

$$Z_t = Z_\xi\left\{i\operatorname{Im}(\overline{W}/Z_\xi) - \frac{1}{2\pi}\fint_{-\pi}^{\pi} \operatorname{Im}(\overline{W}/Z_\xi) \cot\frac{1}{2}(\lambda - \xi)\, d\lambda\right\} \tag{19}$$

$$W_t = i\left(1 - \frac{1}{Z_\xi}\right) + \frac{Z_t}{Z_\xi}W_\xi + \frac{1}{Z_\xi}\left\{\frac{1}{2}|W|_\xi^2 + \frac{i}{\pi}\fint_{-\pi}^{\pi}\frac{1}{2}|W|_\xi^2 \cot\frac{1}{2}(\lambda - \xi)\, d\lambda\right\} \tag{20}$$

where $\fint$ denotes Cauchy principle value. The procedure for numerical solution now follows: The integrals in Eqs. (19) and (20) are replaced by N-point quadratures involving evaluation of the integrand at ξ_k, $k = 1,\ldots,N$. Derivatives with respect to ξ are approximated by differences. The resulting expressions for Z_t and W_t are evaluated at the same N points, ξ_k, thus leading to the system of 2N differential equations for $Z_k(t) \equiv Z(\xi_k,t)$ and $W_k(t)$, which can be solved by standard techniques.

In Fig. 3, waveforms are plotted for the problem in which the initial shape of the free surface is given, initial velocities are zero, and the subsequent motion is sought. In this case, the initial wave shape is a second-order standing wave solution and the numerical calculations are compared to this solution for $t > 0$.

For these calculations, the ξ_k are equally spaced and $N = 25$. Simpson's rule, modified for the singular behavior of the integrand, was used for the quadrature. The differential equations were solved by the fourth-order Milne predictor-corrector scheme, with starting values calculated by the fourth-order Runga-Kutta method. These numerical solutions compare favorably with the second-order theory.

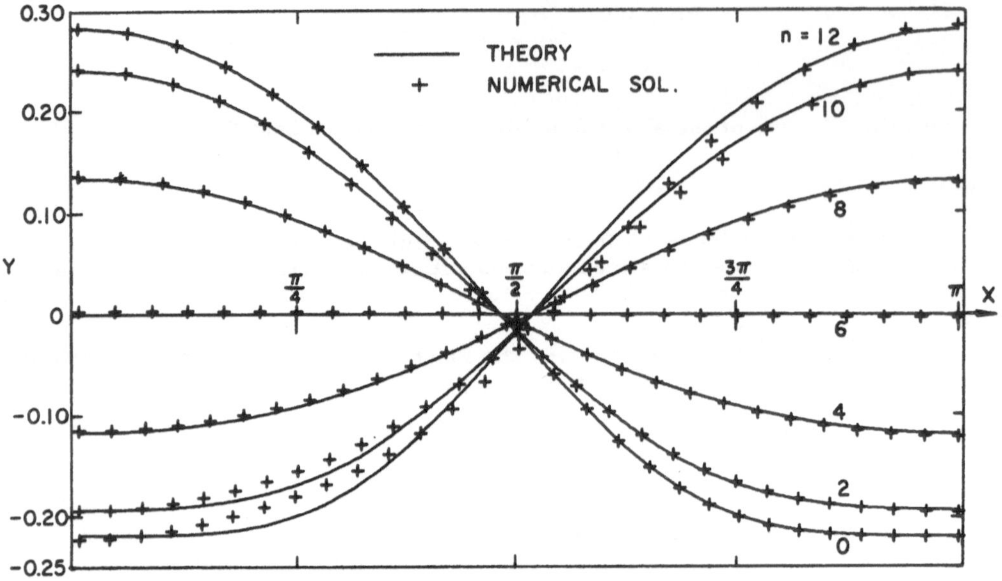

Fig. 3 A comparison of numerical and theoretical solutions at times $t = n(2\pi/24)$.

CONCLUSION

The practical numerical application of the method described in this paper is in its formative stages, so it is too early to make a final judgement as to its real use. The technique of time-dependent conformal mapping does seem to hold promise, however, since the number of spatial dimensions reduces to only one for a wide class of domains which may even be infinite in extent.

REFERENCES

Brennen, C., and Whitney, A. K. (1970) Unsteady, Free Surface Flows; Solutions Employing the Lagrangian Description of the Motion. 8th Symposium on Naval Hydrodynamics, Pasadena.

Chan, R. K. C., and Street, R. L. (1970) SUMMAC--A Numerical Model for Water Waves. Stanford C.E. Dept. T.R. No. 135, Stanford, California.

Hirt, C. W., Cook, J. L., and Butler, T. D. (1970) A Lagrangian Method for Calculating the Dynamics of an Incompressible Fluid with Free Surface. J. Compt. Physics, 5, 1.

Welch, J. E., Harlow, F. H., Shannon, J. D. and Daly, B. J. (1966) The MAC Method--A Computer Technique for Solving Viscous, Incompressible, Transient Fluid-Flow Problems Involving Free Surfaces. Los Alamos Sci. Lab. Rep. LA-3 25.

Woods, L. C. (1961) The Theory of Subsonic Plane Flow. Cambridge University Press.

This work was partially sponsored by the National Science Foundation under grant GK 2370 and by the Office of Naval Research under contract N00014-67-A-0094-0012.

Beschaffenheit der Manuskripte

Die Manuskripte werden photomechanisch vervielfältigt; sie müssen daher in sauberer Schreibmaschinenschrift mit ausreichend großer Type geschrieben sein. Handschriftliche Formeln bitte nur mit schwarzer Tusche eintragen. Notwendige Korrekturen sind bei dem bereits geschriebenen Text entweder durch Überkleben des alten Textes vorzunehmen oder aber müssen die zu korrigierenden Stellen mit weißem Korrekturlack abgedeckt werden. Die reproduktionsfähigen Abbildungen (in Originalgröße) sollen in den Text eingeklebt werden. Falls das Manuskript oder Teile desselben neu geschrieben werden müssen, ist der Verlag bereit, dem Autor bei Erscheinen seines Bandes einen angemessenen Betrag zu zahlen. Die Autoren erhalten 50 Freiexemplare.

Zur Erreichung eines möglichst optimalen Reproduktionsergebnisses ist es erwünscht, daß bei der vorgesehenen Verkleinerung der Manuskripte der Text auf einer Seite in der Breite möglichst 18 cm und in der Höhe 26,5 cm nicht überschreitet. Entsprechende Satzspiegelvordrucke werden vom Verlag gern auf Anforderung zur Verfügung gestellt.

Manuskripte, in englischer, deutscher oder französischer Sprache abgefaßt, sind einzureichen bei: Springer-Verlag, 6900 Heidelberg, Postfach 1780.

Cette série a pour but de donner des informations rapides, de niveau élevé, sur des développements récents en physique, aussi bien dans la recherche que dans l'enseignement supérieur. On prévoit de publier.

1. des versions préliminaires de travaux originaux et de monographies
2. des cours spéciaux portant sur un domaine nouveau ou sur des aspects nouveaux de domaines classiques
3. des rapports de séminaires
4. des conférences faites lors de congrès ou de colloques

En outre il est prévu de publier dans cette série, si la demande le justifie, des rapports de séminaires et des cours multicopiés ailleurs mais déjà épuisés.

Dans l'intérêt d'une diffusion rapide, les contributions auront souvent un caractère provisoire; le cas échéant, les démonstrations ne seront données que dans les grandes lignes. Les travaux présentés pourront également paraître ailleurs. Une réserve suffisante d'exemplaires sera toujours disponible. En permettant aux personnes intéressées d'être informées plus rapidement, les éditeurs Springer espèrent, par cette série de «prépublications», rendre d'appréciables services aux instituts de physique. Les annonces dans les revues spécialisées, les inscriptions aux catalogues et les copyrights rendront plus facile aux bibliothèques la tâche de réunir une documentation complète.

Présentation des manuscrits

Les manuscrits, étant reproduits par procédé photomécanique, doivent être soigneusement dactylographiés type assez grand. Il est recommandé d'écrire à l'encre de Chine noire les formules non dactylographiées. Les corrections nécessaires doivent être effectuées soit par collage du nouveau texte sur l'ancien soit en recouvrant les endroits à corriger par du vernis correcteur blanc. Les illustrations; en dimension originale, préparées pour reproduction sont à insérer dans le texte. S'il s'avère nécessaire d'écrire de nouveau le manuscrit, soit complètement, soit en partie, la maison d'édition se déclare prête à verser à l'auteur, lors de la parution du volume, le montant des frais correspondants. Les auteurs recoivent 50 exemplaires gratuits.

Pour obtenir une reproduction optimale il est désirable que le texte dactylographié sur une page ne dépasse pas 26,5 cm en hauteur et 18 cm en largeur. Sur demande la maison d'edition met à la disposition des auteurs du papier spécialement preparé.

Les manuscrits en anglais, allemand ou français peuvent être adressés à Springer-Verlag, 6900 Heidelberg, Postfach 1780.